Fundamentals of Laser Spectroscopy for Atoms and Diatomic Molecules

For the engineer or scientist using spectroscopic laser diagnostics to investigate gas-phase media or plasmas, this book is an excellent resource for gaining a deeper understanding of the physics of radiative transitions. While a background in quantum mechanics is beneficial to the reader, the book presents a comprehensive review of the relevant aspects of the field, extensively covering atomic and molecular structure alongside radiative transitions.

The author employs effective Hamiltonians and Hund's case (a) basis wavefunctions to develop the energy level structure of diatomic molecules. These techniques also form the basis for treating radiative transitions in diatomic molecules. Recent advancements in quantum chemistry, enabling readers to calculate absolute single-photon and Raman transition strengths, are also presented.

Illustrated with detailed example calculations of molecular structure and transition rates, this self-contained reference for spectroscopic data analysis will appeal to professionals in mechanical, aerospace, and chemical engineering, and in applied physics and chemistry.

Robert P. Lucht is the Bailey Distinguished Professor at Purdue University. His research projects range from the physics of emerging laser techniques to applications of laser diagnostics in practical combustion devices such as gas turbines. Professor Lucht is a fellow of Optica, AIAA, ASME, and the Combustion Institute. He also received the AIAA Aerodynamic Measurement Technology Award.

"This is an excellent and essential book for anyone seeking a detailed understanding of the fundamental physics pertaining to absorption, emission, and Raman spectroscopy of diatomic molecules. It is a must-have for laser spectroscopists studying the gas phase."

Christopher Goldenstein, Purdue University

"This book presents a thorough foundation for both linear and nonlinear spectroscopy of atoms and diatomic molecules, including examples relevant to combustion science and high-temperature gas physics. The author develops both classical and semiclassical quantum models that describe atomic and molecular spectral properties and interactions with electric fields and lasers. The detailed derivations provide deep insight into the underlying mechanisms. At each step, the author makes the connection between derived expressions and relevant applications, reflecting his long leadership in the development and application of advanced spectroscopic methods. This book serves as a welcome benchmark for the community."

Richard B. Miles, Texas A&M University

"Here is an invaluable resource for graduate students and researchers wishing to study the fundamental theory and finer details of gas-phase laser spectroscopy. It describes atomic and molecular structure, the physics behind matter and laser radiation interaction, and the theory of specific spectroscopic techniques, all in one comprehensive, in-depth text."

Chloe E. Dedic, University of Virginia

Fundamentals of Laser Spectroscopy for Atoms and Diatomic Molecules

ROBERT P. LUCHT
Purdue University

CAMBRIDGE
UNIVERSITY PRESS

CAMBRIDGE
UNIVERSITY PRESS

Shaftesbury Road, Cambridge CB2 8EA, United Kingdom

One Liberty Plaza, 20th Floor, New York, NY 10006, USA

477 Williamstown Road, Port Melbourne, VIC 3207, Australia

314–321, 3rd Floor, Plot 3, Splendor Forum, Jasola District Centre, New Delhi – 110025, India

103 Penang Road, #05-06/07, Visioncrest Commercial, Singapore 238467

Cambridge University Press is part of Cambridge University Press & Assessment, a department of the University of Cambridge.

We share the University's mission to contribute to society through the pursuit of education, learning and research at the highest international levels of excellence.

www.cambridge.org
Information on this title: www.cambridge.org/9781108837927

DOI: 10.1017/9781108936514

When citing this work, please include a reference to the DOI 10.1017/9781108936514

First published 2025

A catalogue record for this publication is available from the British Library.

Library of Congress Cataloging-in-Publication Data
Names: Lucht, Robert P., 1954- author.
Title: Fundamentals of laser spectroscopy for atoms and diatomic molecules / Robert P. Lucht, Purdue University, Indiana.
Description: Cambridge, United Kingdom ; New York, NY : Cambridge University Press, 2024. | Includes bibliographical references and index.
Identifiers: LCCN 2024011580 (print) | LCCN 2024011581 (ebook) | ISBN 9781108837927 (hardback) | ISBN 9781108936514 (epub)
Subjects: LCSH: Laser spectroscopy. | Atomic spectroscopy. | Molecular spectroscopy.
Classification: LCC QC454.L3 L83 2024 (print) | LCC QC454.L3 (ebook) | DDC 539.7/44–dc23/eng/20240409
LC record available at https://lccn.loc.gov/2024011580
LC ebook record available at https://lccn.loc.gov/2024011581

ISBN 978-1-108-83792-7 Hardback

Dedicated to my late parents Alta and George Lucht

Contents

	Preface	*page* ix
1	**Introduction: Classical Theory of Electric Dipole Radiative Interactions**	1
2	**Atomic Structure and the Quantum Mechanics of Angular Momentum**	16
3	**Structure of Diatomic Molecules**	56
4	**Quantum Mechanical Analysis of the Interaction of Laser Radiation with Electric Dipole Resonances**	120
5	**Quantum Mechanical Analysis of Single-Photon Electric Dipole Resonances for Diatomic Molecules**	160
6	**Absorption and Emission Spectroscopy**	236
7	**Raman Spectroscopy**	259
8	**Coherent Anti-Stokes Raman Scattering (CARS) Spectroscopy**	306
	Appendix 1 Spherical Harmonics and Radial Wavefunctions for One-Electron Atoms	360
	Appendix 2 Clebsch–Gordan Coefficients, Dipole Moments, and Spontaneous Emission Coefficients for the 2p–1s *Transition in Atomic Hydrogen*	362
	Appendix 3 Properties and Values for Selected 3j *Symbols*	365
	Appendix 4 Properties and Values for Selected 6j *Symbols*	369
	Appendix 5 Allowed LS *Coupling Terms for Equivalent* d^2 Electrons	371
	Appendix 6 Derivation of the Higher-Order Density Matrix Elements for Doublet and Triplet Electronic Levels	376

Appendix 7 Einstein Coefficients for Spontaneous Emission for the $X^2\Pi$–$A^2\Sigma^+$ (0,0) Bands of OH and NO and the $X^3\Sigma^-$–$A^3\Pi$ (0,0) Band of NH 384
Appendix 8 Effect of Hyperfine Splitting on Radiative Transition Rates 393
Appendix 9 Voigt Function Values 399
References 405
Index 415

Preface

Spectroscopic laser diagnostics have been applied to the study of flames, plasmas, and fluid dynamics for over 50 years, and there have been many books that summarize the application of laser diagnostics in these systems. In books such as *Laser Diagnostics for Combustion Temperature and Species* by Alan C. Eckbreth, different laser techniques are discussed, including some discussion of the physics of interaction of the laser with atomic and molecular resonances.

The focus of this book is the physics of radiative transitions for gas-phase spectroscopic diagnostics. The idea for this book grew out of my own experience in trying to educate myself in the physics of radiative transitions. As an undergraduate at Purdue University, I wrote a report entitled "Fundamentals of Absorption Spectroscopy for Selected Diatomic Flame Radicals," with Richard Peterson and my late advisor Normand Laurendeau as coauthors. This was my first exposure to the literature of spectroscopic transitions and was a tremendous learning experience for me. It also turned out that it was useful for others in the combustion diagnostics community, which was a very young and expanding community at that time.

The book begins in Chapter 1 with a discussion of the energy and interactions of a classical electric dipole; this includes a discussion of the interaction of a classical dipole with a monochromatic laser field – to introduce the concepts of absorption and stimulated emission – and the radiation emitted by an oscillating dipole – to introduce the concept of spontaneous emission. Chapter 2 contains an introduction to atomic structure, and essential quantum mechanical concepts are introduced in a discussion of the structure of atomic hydrogen. Most significantly, the quantum treatment of angular momentum is discussed, and *3j* symbols and the Wigner-Eckart theorem are introduced. The structure of diatomic molecules is discussed in Chapter 3. The treatment of energy levels in diatomic molecules is based on the effective Hamiltonian method featuring the use of Hund's case (a) basis wavefunctions. Electric dipole resonance transitions in atoms are discussed in Chapter 4, and the Wigner–Eckart theorem is applied. Electric dipole resonance transitions in diatomic molecules are covered in Chapter 5. The treatment of electronic transitions in diatomic molecules again features the use of Hund's case (a) basis wavefunctions. Tables of rotational line strengths are developed in a different kind of format incorporating the weighting coefficients for the different Hund's case (a) basis wavefunctions. Herman–Wallis effects for electronic transitions are discussed in detail and incorporate Rydberg–Klein–Rees (RKR) calculations of vibration–rotation wavefunctions and quantum chemistry calculations of the dependence of the electronic transition moment on internuclear separation. The physics of absorption and emission processes is discussed in Chapter 6. Detailed example problems are included at the conclusions of Chapters 6, 7, and 8. The focus of Chapter 7 is Raman spectroscopy. Placzek polarizability theory is developed using irreducible spherical tensors. Following a similar approach to Chapter 5, Herman–Wallis effects for Raman transitions are discussed in detail and incorporate RKR calculations of vibration–rotation wavefunctions and quantum chemistry calculations of the dependence of the Raman polarizability on internuclear separation. Finally, in Chapter 8 the third-order nonlinear coherent anti-Stokes Raman scattering (CARS) susceptibility is derived in detail. The physics of both nanosecond and femtosecond CARS is discussed.

I would like to acknowledge the support of my wife Martha and my two daughters Kimberly and Heather during the rather lengthy process of writing this book. I would like to thank my graduate advisors Donald Sweeney and the late Normand Laurendeau for introducing me to the fascinating field of laser spectroscopy. My colleagues Drs. Roger Farrow and Larry Rahn at Sandia National Laboratories introduced me to the absorbing world of nonlinear optics in general and CARS in particular. I would also like to thank my numerous graduate students who have made my academic career such a pleasure. I would like to thank in particular my PhD students Robert Hancock, Ken Bertagnolli, Mark Woodmansee, Steve Green, Fred Schauer, Terrence Meyer, Joel Kuehner, Tom Reichardt, Sean Kearney, Robert Foglesong, Sukesh Roy, Rodolfo Barron-Jimenez, Ning Chai, Waruna Kulatilaka, Aman Satija, Mathew Thariyan, Warren Lamont, Daniel Richardson, Carson Slabaugh, Gurneesh Jatana, Alfredo Tuesta, Clare Dennis, Levi Thomas, Neil Rodrigues, Mingming Gu, Will Senior, Ziqiao Chang, and Ben Murdock. These students worked with me on the theory, development, and application of advanced laser spectroscopic techniques, such as CARS spectroscopy, laser-induced fluorescence (LIF) spectroscopy, and polarization spectroscopy; this research forms the motivation for the material presented in this book. I would especially like to thank Ziqiao Chang and Will Senior for proofreading the final version of the manuscript.

1 Introduction

Classical Theory of Electric Dipole Radiative Interactions

1.1 Introduction

The classical theory of the interaction of light with the electron clouds of atoms and molecules will be discussed in this chapter. The discussion will begin with the interaction of a steady electric field with a collection of point charges, leading to the development of terms describing the electric dipole and quadrupole moments. The classical Lorentz model is then introduced to describe the interaction of an oscillating electric field with the electron cloud of an atom, and the concepts of absorption and emission are introduced. The propagation of a light wave through a medium with electric dipoles is then discussed. Finally, the classical theory of radiation from an oscillating dipole is discussed.

1.2 Interaction of a Collection of Charges with a Steady Electric Field

Before beginning a discussion of the interaction of atoms and molecules with the oscillating electric field associated with a light wave, we will first consider the interaction of a collection of point charges with a steady electric field. The result of this analysis will be the separation of the contributions of the electric dipole and quadrupole moments to the energy of a system of point charges in a steady electric field. The point charges in an atom or molecule are associated with the electrons $(-e)$ and the nuclei $(+Ze)$ (C), where Z is the number of protons in each nucleus. Now consider a point charge q located at position r (m) in a steady electric field $\boldsymbol{E}(\boldsymbol{r})$ (V/m or J/C-m), where

$$\boldsymbol{E}(\boldsymbol{r}) = -\nabla\Phi(\boldsymbol{r}) \tag{1.1}$$

and $\Phi(\boldsymbol{r})$ (J/C) is the scalar potential. The potential energy of the point charge q(C) is given by

$$V = q\Phi(\boldsymbol{r}). \tag{1.2}$$

Following Struve (1988), the scalar potential can be expanded in a Taylor series about $\boldsymbol{r} = 0$:

$$\begin{aligned}\Phi(\boldsymbol{r}) = \Phi(\mathbf{0}) &+ x\left[\frac{\partial\Phi}{\partial x}(\mathbf{0})\right] + y\left[\frac{\partial\Phi}{\partial y}(\mathbf{0})\right] + z\left[\frac{\partial\Phi}{\partial z}(\mathbf{0})\right] \\ &+ \frac{1}{2}\left\{x^2\left[\frac{\partial^2\Phi}{\partial x^2}(\mathbf{0})\right] + y^2\left[\frac{\partial^2\Phi}{\partial y^2}(\mathbf{0})\right] + z^2\left[\frac{\partial^2\Phi}{\partial z^2}(\mathbf{0})\right]\right. \\ &\left.+2xy\left[\frac{\partial^2\Phi}{\partial x\partial y}(\mathbf{0})\right] + 2yz\left[\frac{\partial^2\Phi}{\partial y\partial z}(\mathbf{0})\right] + 2zx\left[\frac{\partial^2\Phi}{\partial z\partial x}(\mathbf{0})\right]\right\} + \cdots.\end{aligned} \tag{1.3}$$

Equation (1.3) can be written in more compact form as

$$\Phi(\boldsymbol{r}) = \Phi(\mathbf{0}) + \boldsymbol{r}\bullet[\nabla\Phi(\mathbf{0})] + \frac{1}{2}\sum_i\sum_j x_i x_j \frac{\partial}{\partial x_i}\left[\frac{\partial\Phi}{\partial x_j}(\mathbf{0})\right] + \cdots. \tag{1.4}$$

The electric field is calculated from the gradient of the potential,

$$\boldsymbol{E}(\mathbf{0}) = -\nabla\Phi(\mathbf{0}) \tag{1.5}$$

and

$$E_i(\mathbf{0}) = -\frac{\partial\Phi}{\partial x_i}(\mathbf{0}). \tag{1.6}$$

Substituting Eqs. (1.5) and (1.6) into (1.4), we obtain

$$\Phi(\boldsymbol{r}) = \Phi(\mathbf{0}) - \boldsymbol{r}\bullet\boldsymbol{E}(\mathbf{0}) - \frac{1}{2}\sum_i\sum_j x_i x_j \frac{\partial E_j}{\partial x_i}(\mathbf{0}) + \cdots. \tag{1.7}$$

For a point charge q, the interaction energy is thus given by

$$V = q\Phi(\mathbf{0}) - q\boldsymbol{r}\bullet\boldsymbol{E}(\mathbf{0}) - \frac{q}{2}\sum_i\sum_j x_i x_j \frac{\partial E_j}{\partial x_i}(\mathbf{0}) + \cdots. \tag{1.8}$$

For a collection of N point charges q_n with associated position vectors $\boldsymbol{r}_n$, the energy due to the interaction with the potential $\Phi(\boldsymbol{r})$ is given by

$$V = \Phi(\mathbf{0})\sum_{n=1}^{N} q_n - \left(\sum_{n=1}^{N} q_n\boldsymbol{r}_n\right)\bullet\boldsymbol{E}(\mathbf{0}) - \sum_{n=1}^{N}\frac{q_n}{2}\sum_i\sum_j x_{ni}x_{nj}\frac{\partial E_j}{\partial x_{ni}}(\mathbf{0}) + \cdots, \tag{1.9}$$

where

$$\boldsymbol{r}_n = x_n\hat{\boldsymbol{x}} + y_n\hat{\boldsymbol{y}} + z_n\hat{\boldsymbol{z}} = x_{n1}\hat{\boldsymbol{x}}_1 + x_{n2}\hat{\boldsymbol{x}}_2 + x_{n3}\hat{\boldsymbol{x}}_3. \tag{1.10}$$

The electric dipole moment is given by

$$\boldsymbol{\mu} = \sum_{n=1}^{N} q_n\boldsymbol{r}_n. \tag{1.11}$$

The components of the electric quadrupole moment tensor for a given charge n can be written as (Struve, 1988)

$$Q_{ij}^{(n)} = \frac{q_n}{2}\left(3x_{ni}x_{nj} - r_n^2\delta_{ij}\right). \tag{1.12}$$

Substituting Eqs. (1.11) and (1.12) into Eq. (1.9), we obtain

$$V = \Phi(\mathbf{0})\sum_{n=1}^{N} q_n - \boldsymbol{\mu}\cdot\boldsymbol{E}(\mathbf{0}) - \frac{1}{6}\sum_{n=1}^{N}\sum_{i}\sum_{j} Q_{ij}^{(n)}\frac{\partial E_j}{\partial x_{ni}}(\mathbf{0}) + \cdots. \tag{1.13}$$

The first term in Eq. (1.13) is the product of the sum of the charges and the scalar potential at $\boldsymbol{r} = \mathbf{0}$, the second term is the dot product of the dipole moment of the charge distribution with the electric field, and the third term describes the interaction of the quadrupole moment of the electric field with the gradients of the electric field.

1.3 The Lorentz Classical Electron Oscillator Model

For an atom or molecule interacting with an electric field oscillating at the very high frequencies associated with visible radiation, the electric dipole moment term is dominant compared to the quadrupole term, except for transitions where electric dipole transitions are forbidden. We begin our discussion of the electric dipole interaction of laser radiation with atoms and molecules by considering the classical electron oscillator model originally developed by Lorentz (2011). The Lorentz classical electron oscillator (CEO) model was developed before the advent of quantum mechanics and our modern picture of the atom consisting of negatively charged electrons orbiting a massive and positively charged nucleus. The CEO model is schematically illustrated in Figure 1.1. In the absence of an external electric field, the center of the electron charge cloud coincides with the positively charged nucleus of the atom, and the electric dipole moment $\boldsymbol{\mu}$ (C-m) of the atom is zero. When a high-frequency oscillating electric field is applied, it is assumed that the electron cloud can respond to the applied force $-eE_x(t)$ but that the position of the massive nucleus is

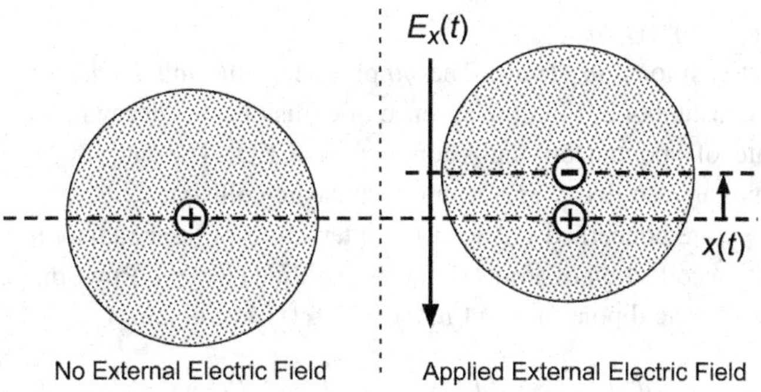

Figure 1.1 Schematic illustration of the CEO model.

essentially unchanged by the interaction. It is further assumed that the restoring force between the electron cloud and the nucleus is linearly proportional to the displacement of the electron cloud from its equilibrium position. The response of the electron cloud to the oscillating electric field is given by

$$m_e \frac{d^2x(t)}{dt^2} = -Kx(t) - eE_x(t), \tag{1.14}$$

where m_e (kg) is the electron mass, x(m) the electron displacement, K (N/m) the magnitude of the restoring force, e (C) the magnitude of the electron charge, and $E_x(t)$ (J/C-m) the electric field amplitude in the x-direction. Equation (1.14) can be rewritten as

$$\frac{d^2x(t)}{dt^2} + \omega_a^2 x(t) = -\left(\frac{e}{m_e}\right) E_x(t). \tag{1.15}$$

The term $\omega_a = \sqrt{K/m_e}$ (s^{-1}) is identified as the resonant frequency of the CEO. The motion of the electron cloud is damped by processes such as spontaneous emission, and we therefore introduce a damping constant Γ (s^{-1}):

$$\frac{d^2x(t)}{dt^2} + \Gamma\frac{dx(t)}{dt} + \omega_a^2 x(t) = -\left(\frac{e}{m_e}\right) E_x(t). \tag{1.16}$$

Before discussing the solution of Eq. (1.16) in the presence of an applied electric field, we consider the solution of the equation given a nonzero initial displacement $x(0)$ and zero applied electric field. The solution to Eq. (1.16) under these conditions is given by

$$x(t) = x(0) \exp\left[-\frac{\Gamma t}{2}\right] [\exp(-i\omega_{a1}t) + \exp(+i\omega_{a1}t)]. \tag{1.17}$$

The electron amplitude given by Eq. (1.17) contains an oscillating term with frequency ω_{a1} given by

$$\omega_{a1} = \sqrt{\omega_a^2 - \left(\frac{\gamma}{2}\right)^2}. \tag{1.18}$$

For optical transitions, $\omega_{a1} \cong \omega_a$. The amplitude of the initial displacement decays with a time constant of $2/\Gamma$. In the absence of collisions, the decay rate Γ is associated with the rate of spontaneous emission for the CEO. The amplitude of the initial displacement can also decay as a result of inelastic collisions.

Usually we are concerned not with the interaction of laser radiation with a single atom but with a collection of atoms in a small volume element. Rewriting Eq. (1.16) in terms of the electric dipole moment $\mu_x(t) = -ex(t)$, we obtain

$$\frac{d^2\mu_x(t)}{dt^2} + \Gamma\frac{d\mu_x(t)}{dt} + \omega_a^2 \mu_x(t) = \left(\frac{e^2}{m_e}\right) E_x(t). \tag{1.19}$$

The macroscopic polarization per unit volume $p_x(t)$ (C-m/m^3) is given by summing over all the atoms in a small volume element and then dividing by the volume,

$$p_x(t) = \frac{1}{\forall}\sum_{i=1}^{N} \mu_{xi}(t). \tag{1.20}$$

The decay rate of the macroscopic polarization is in general greater than the decay rate of the dipole moments for the individual molecules because of elastic pure dephasing collisions. Consider a group of three atoms with electron clouds oscillating in phase at time zero. The amplitude of the macroscopic polarization will be a maximum at time zero. At time t_1, one of the atoms undergoes an elastic pure dephasing collision. As a result of the collision, it is assumed that the phase of the electron cloud oscillation is randomized. On average, the individual dipole moment of the atom that undergoes the collision no longer contributes to the macroscopic polarization, and the amplitude of the macroscopic polarization decreases. This is illustrated in Figure 1.2, where the results of pure dephasing collisions at times t_2 and t_3 are also shown.

Consider N_0 atoms with dipole moments oscillating in phase at time t_0. At time $t - t_0$, $N_0 - \tilde{N}(t)$ of these atoms will have undergone pure dephasing collisions, where $\tilde{N}(t)$ is given by

$$\tilde{N}(t) = N_0 \exp\left(-\frac{t - t_0}{T_2}\right) = N_0\left[-Q_{pd}(t - t_0)\right]. \tag{1.21}$$

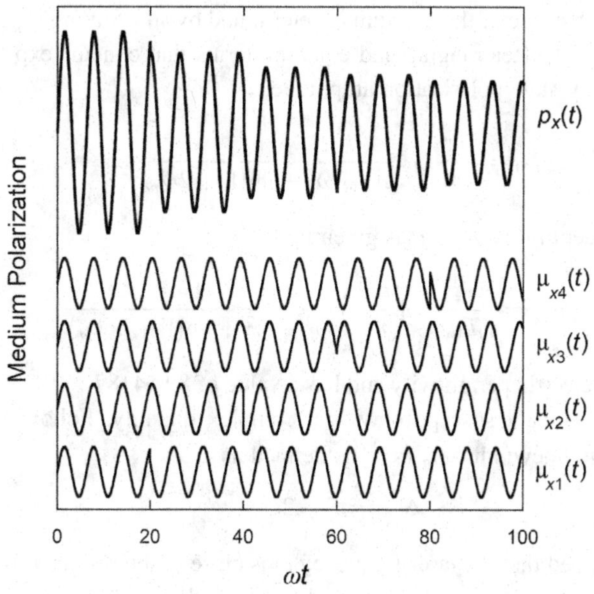

Figure 1.2 Effect of dephasing collisions on the macroscopic dipole polarization of the medium.

The parameters T_2 (s) and Q_{pd} (s^{-1}) are the characteristic time and rate coefficient, respectively, for pure dephasing collisions. After time t, the macroscopic polarization of the medium will be given by

$$\begin{aligned} p_x(t) &= \tilde{n}(t)\mu_x(t) \\ &= n_0\mu_x(t_0)\exp\left[-\left(\frac{\gamma}{2}+Q_{pd}\right)(t-t_0)+i\omega_a(t-t_0)+i\varphi_0\right]+\text{c.c.}, \end{aligned} \tag{1.22}$$

where the number densities $\tilde{n}(t)$ and n_0 (m^{-3}) are given by $\tilde{N}(t)/\forall$ and $N_0/\forall$, respectively, and the abbreviation c.c. denotes the complex conjugate of the preceding term. The initial phase of the oscillators at time t_0 is φ_0. Comparing Eqs. (1.17) and (1.22), it is apparent that the macroscopic polarization decays with a rate coefficient of $\frac{\Gamma}{2}+Q_{pd}$ as compared to a rate coefficient of $\frac{\Gamma}{2}$ for a single atomic dipole.

Incorporating the effects of the pure dephasing collisions, we rewrite Eq. (1.19) as

$$\frac{d^2p_x(t)}{dt^2}+(\Gamma+2Q_{pd})\frac{dp_x(t)}{dt}+\omega_a^2p_x(t)=\left(\frac{ne^2}{m_e}\right)E_x(t). \tag{1.23}$$

Consider the response of the medium to an oscillating electric field given by

$$E_x(t)=\frac{1}{2}\left[E_{0x}\exp(+i\omega t)+E_{0x}^*\exp(-i\omega t)\right]. \tag{1.24}$$

Assume that the applied electric field at angular frequency ω induces a polarization at the same frequency,

$$p_x(t)=\frac{1}{2}\left[P_{0x}\exp(+i\omega t)+P_{0x}^*\exp(-i\omega t)\right]. \tag{1.25}$$

The steady-state response of the medium is determined by substituting Eqs. (1.24) and (1.25) into Eq. (1.23). Rearranging and equating terms that contain exp $(+i\omega t)$, we solve for the steady-state polarization amplitude,

$$P_{0x}=\frac{ne^2}{m_e}E_{0x}\frac{1}{\omega_a^2-\omega^2+i\omega(\Gamma+2Q_{pd})}. \tag{1.26}$$

The resonant susceptibility $\chi_{res}(\omega)$ is given by

$$\chi_{res}(\omega)=\frac{P_{0x}}{\varepsilon_0E_{0x}}=\left(\frac{ne^2}{m_e}\right)\frac{1}{\omega_a^2-\omega^2+i\omega(\Gamma+2Q_{pd})}, \tag{1.27}$$

where ε_0 is the dielectric permittivity and has a value of 8.854187×10^{-12} $\mathrm{C^2/J\,m}$ for free space. The resonant susceptibility is a complex quantity. Following Siegman (1986), the atomic linewidth $\Delta\omega_a$ (s^{-1}) is defined as

$$\Delta\omega_a=\Gamma+2Q_{pd}. \tag{1.28}$$

Further, it is assumed that the laser frequency ω is close to the resonant frequency ω_a and the following approximation is assumed to be valid:

$$\omega_a^2-\omega^2=(\omega_a+\omega)(\omega_a-\omega)\cong2\omega_a(\omega_a-\omega)\quad\text{for }\omega_a\cong\omega. \tag{1.29}$$

Using Eqs. (1.28) and (1.29), we rewrite Eq. (1.27) as

$$\chi_{res}(\omega) = \frac{ne^2}{\varepsilon_0 m_e}\left[\frac{1}{2\omega_a(\omega_a - \omega) + i\omega_a\Delta\omega_a}\right] = -i\frac{ne^2}{\varepsilon_0 m_e \omega_a \Delta\omega_a}\frac{1}{1 + i\Delta x}, \tag{1.30}$$

where the normalized detuning Δx is given by

$$\Delta x = 2\frac{\omega - \omega_a}{\Delta\omega_a}. \tag{1.31}$$

The real and imaginary components of the susceptibility are given by

$$\chi_{res}(\omega) = \chi'(\omega) + i\chi''(\omega) = -\chi_0''\left[\frac{\Delta x}{1 + (\Delta x)^2} + i\frac{1}{1 + (\Delta x)^2}\right], \tag{1.32}$$

where

$$\chi_0'' = \frac{ne^2}{\varepsilon_0 m_e \omega_a \Delta\omega_a}. \tag{1.33}$$

The real or dispersive component of the susceptibility is given by

$$\chi'(\omega) = -\chi_0''\frac{\Delta x}{1 + (\Delta x)^2} \tag{1.34}$$

and the imaginary or absorbing component of the susceptibility is given by

$$\chi''(\omega) = -\chi_0''\frac{1}{1 + (\Delta x)^2}. \tag{1.35}$$

The normalized line shapes for the real and imaginary components of the resonant susceptibility are plotted in Figure 1.3. The real part of the susceptibility oscillates in phase with the applied electric field, and the imaginary part of the susceptibility oscillates 90° out of phase with the applied electric field. The real part of the susceptibility is much greater than the imaginary part of the susceptibility when the laser field is far from resonance ($\Delta x \gg 1$). Because the real part of the susceptibility oscillates in phase with the applied field, the integral of $\boldsymbol{F}\cdot\mathbf{v} = -e\boldsymbol{E}\cdot\mathbf{v}$ over a complete cycle of the electric field is zero, and there is no energy exchange with the applied field. For a laser field in exact resonance ($\Delta x = 0$), the response is purely imaginary, and the integral of $-e\boldsymbol{E}\cdot\mathbf{v}$ over a complete cycle of the applied field is nonzero.

As discussed in detail by Siegman (1986), at this point some quantum mechanical results can be incorporated into this purely classical picture of the interaction of the electron cloud with the applied field. The resonant frequency ω_a in Eq. (1.33) can be written as

$$\omega_a = \frac{(\varepsilon_e - \varepsilon_g)}{\hbar}, \tag{1.36}$$

where ε_e and ε_g are the energies (J) of the upper (excited) and lower (ground) quantum states for an allowed radiative transition. Furthermore, the number density n in Eq. (1.33) is replaced by the number density difference $(n_g - n_e)$. We can rewrite Eq. (1.33) as

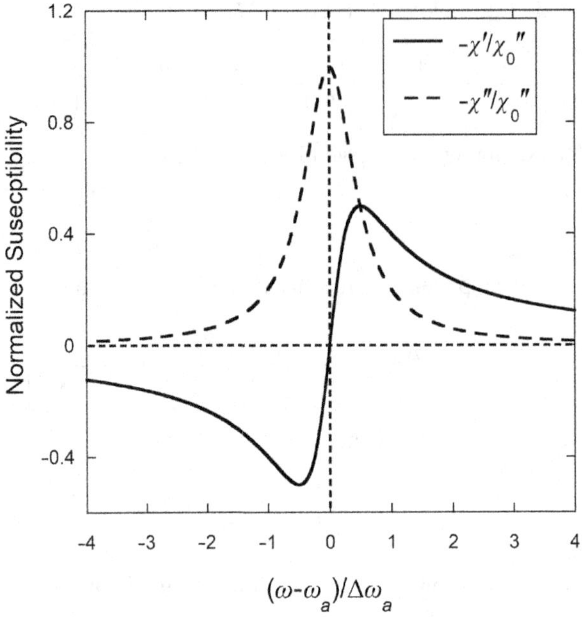

Figure 1.3 Normalized real and imaginary components of the complex resonant susceptibility.

$$\chi_0'' = \frac{(n_g - n_e)e^2}{\varepsilon_0 m_e \omega_a \Delta\omega_a}. \tag{1.37}$$

Whether the electron cloud gains or loses energy will depend on the relative phase of the applied field and the electron motion. The electron cloud gains energy by absorption from the field when the population of the lower energy level of the transition is greater than the population of the upper energy level of the transition; it loses energy by stimulated emission when the population of the upper energy level of the transition is greater than the population of the lower energy level of the transition.

To illustrate the energy exchange between the field and the oscillating electron cloud, consider the case of exact resonance, $\omega = \omega_a, \Delta x = 0$. If the incident plane wave electric field is given by

$$\boldsymbol{E}(t) = E_{0x}\cos(\omega t)\hat{\boldsymbol{x}}, \tag{1.38}$$

where E_{0x} is a real and constant, then from Eqs. (1.27) and (1.32) we obtain

$$P_{0x} = -i\varepsilon_0\,\chi_0''\,E_{0x}. \tag{1.39}$$

From Eq. (1.25), the macroscopic polarization is given by

$$\begin{aligned} p_x(t) &= \frac{1}{2}\left[-i\varepsilon_0\,\chi_0''\,E_{0x}\exp(+i\omega t) + i\varepsilon_0\,\chi_0''\,E_{0x}\exp(-i\omega t)\right] \\ &= \varepsilon_0\,\chi_0''\,E_{0x}\sin(\omega t). \end{aligned} \tag{1.40}$$

We can define a displacement of the electron cloud for each atom in the medium as

$$x_e(t) = -\frac{p_x(t)}{(n_g + n_e)e} = -\frac{\varepsilon_0 \chi_0'' E_{0x} \sin(\omega t)}{(n_g + n_e)e}. \tag{1.41}$$

The velocity of the electron cloud is given by

$$\mathrm{v}_e(t) = \dot{x}_e(t) = -\frac{\varepsilon_0 \chi_0'' E_{0x}}{(n_g + n_e)e} \omega \cos(\omega t). \tag{1.42}$$

Substituting for χ_0'' in Eqs. (1.41) and (1.42) using Eq. (1.37), we obtain

$$x_e(t) = -\frac{eE_{0x}(n_g - n_e)}{m_e \omega \Delta\omega_a (n_g + n_e)} \sin(\omega t) \tag{1.43}$$

and

$$\mathrm{v}_e(t) = \dot{x}_e(t) = -\frac{eE_{0x}(n_g - n_e)}{m_e \Delta\omega_a (n_g + n_e)} \cos(\omega t). \tag{1.44}$$

Normalized values of the electric field, the position and velocity of the electron cloud, and the force acting on the electron cloud will now be plotted for the case of a "normal" population distribution $(n_g > n_e)$ and a population inversion $(n_g < n_e)$. The normalized input electric field is given by $E_x(t)/E_{0x}$. The normalized force on the electron cloud is given by

$$\tilde{F}_x(t) = \frac{F_x(t)}{eE_{0x}} = -\frac{eE_x(t)}{eE_{0x}} = -\cos(\omega t). \tag{1.45}$$

The normalized electron cloud displacement is given by

$$\tilde{x}_e(t) = \frac{1}{2} \frac{x_e(t) m_e \omega \Delta\omega_a (n_g + n_e)}{eE_{0x}|n_g - n_e|} = -\frac{(n_g - n_e)}{2|n_g - n_e|} \sin(\omega t). \tag{1.46}$$

The factor of $\frac{1}{2}$ is introduced in Eq. (1.46) to make it easier to see the difference between the electric field and the electron cloud displacement. The normalized electron cloud velocity is given by

$$\tilde{\mathrm{v}}_e(t) = \frac{1}{2} \frac{\mathrm{v}_e(t) m_e \Delta\omega_a (n_g + n_e)}{eE_{0x}|n_g - n_e|} = -\frac{(n_g - n_e)}{2|n_g - n_e|} \cos(\omega t). \tag{1.47}$$

Again, the factor of $\frac{1}{2}$ is introduced to make it easier to see the difference between the normalized force on the electron cloud and the normalized electron velocity.

The normalized position of the electron cloud is plotted as a function of the normalized electric field in Figure 1.4. The "normal" population distribution case is shown in Figure 1.4a, and the population inversion case is shown in Figure 1.4b. Note that in both cases the electron position is oscillating 90° out of phase with the driving electric field.

The normalized velocity of the electron cloud is plotted as a function of the normalized force on the electron in Figure 1.5. The "normal" population distribution

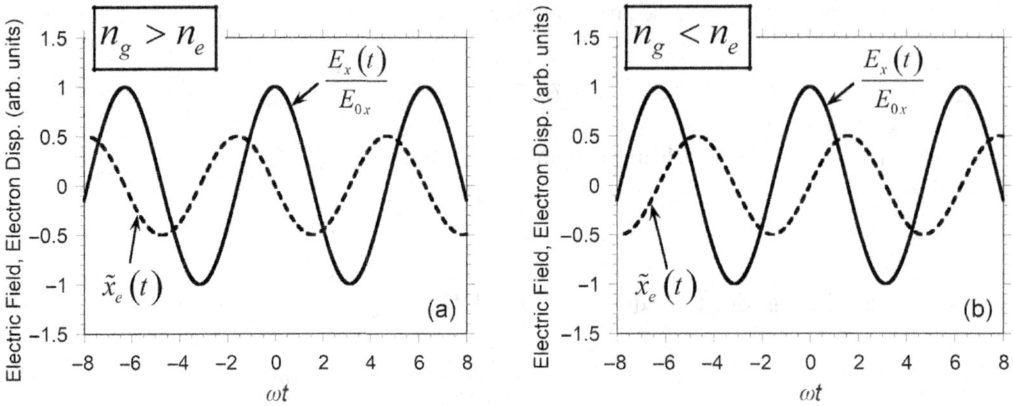

Figure 1.4 Normalized electric field and electron cloud displacement for (a) a "normal" population distribution and (b) a population inversion.

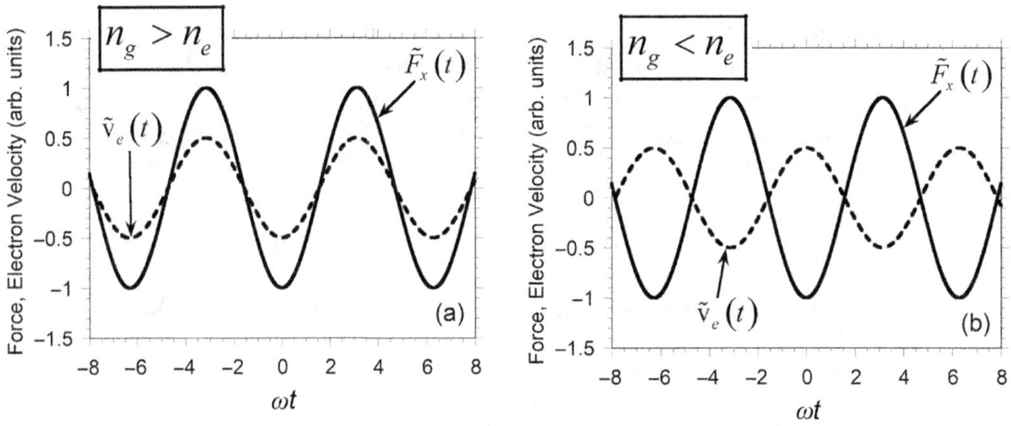

Figure 1.5 Normalized force on the electron and electron velocity for (a) a "normal" population distribution and (b) a population inversion.

case is shown in Figure 1.5a, and the population inversion case is shown in Figure 1.5b. For the "normal" population distribution case, the velocity of the electron cloud is in phase with the force imposed by the driving field. Thus, the electron cloud continually gains energy due to its interaction with the driving field, which results in a decrease of energy for the driving electric field, or in other words through stimulated absorption of the input plane wave electromagnetic field. Conversely, for the population inversion case, the velocity of the electron cloud is 180° out of phase with the driving field. The electron cloud continually loses energy, resulting in an increase of energy for the input electromagnetic field. The medium is said to exhibit "gain" in the case of a population inversion due to the phenomenon of stimulated emission.

There are further refinements to the CEO model that would be necessary to develop a model of the resonance that is rigorously correct. Rather than incorporate these quantum mechanical details at this point, we will discuss the quantum structure of atoms and molecules in Chapters 2 and 3 and the quantum mechanical theory of resonance interactions in Chapter 4.

1.4 Propagation of an Applied Laser Field through an Absorbing or Emitting Medium

In Section 1.2 expressions for the macroscopic polarization of a medium were developed. In this section the effect of the medium's polarization on the propagation of a plane wave electromagnetic field is discussed. The equation for a plane wave propagating in free space is given by (Siegman, 1986)

$$\nabla^2 \boldsymbol{E} - \mu_0 \varepsilon_0 \frac{\partial^2 \boldsymbol{E}}{\partial t^2} = \frac{\partial^2 \boldsymbol{E}}{\partial z^2} - \mu_0 \varepsilon_0 \frac{\partial^2 \boldsymbol{E}}{\partial t^2} = 0, \tag{1.48}$$

where the magnetic permeability μ_0 has a value of $4\pi \times 10^{-7}$ J s^2/C^2 m for free space. The speed of light in free space is given by

$$c = c_0 = \frac{1}{\sqrt{\mu_0 \varepsilon_0}}. \tag{1.49}$$

For a plane wave polarized in the x-direction and propagating in the z-direction, the solution to the wave equation is given by

$$\boldsymbol{E}(z,t) = \hat{\boldsymbol{x}} \left\{ \frac{1}{2} E_0 \exp[+i(kz - \omega t)] + \frac{1}{2} E_0^* \exp[-i(kz - \omega t)] \right\}, \tag{1.50}$$

where the propagation constant k (m^{-1}) is given by

$$k = \frac{2\pi}{\lambda}, \tag{1.51}$$

where λ(m) is the wavelength of the plane wave. The angular frequency ω is given by

$$\omega = \frac{2\pi c}{\lambda} = 2\pi \nu, \tag{1.52}$$

where ν (Hz) is the optical frequency of the plane wave.

For propagation in a dielectric medium containing atoms or molecules with electric dipole resonance transitions, the wave equation becomes

$$\frac{\partial^2 \boldsymbol{E}}{\partial z^2} - \mu_0 \varepsilon_0 \frac{\partial^2 \boldsymbol{E}}{\partial t^2} = \mu_0 \frac{\partial^2 \boldsymbol{P}}{\partial t^2}, \tag{1.53}$$

where

$$\boldsymbol{P} = \chi_{res}(\omega) \varepsilon_0 \boldsymbol{E}. \tag{1.54}$$

We will assume that a plane wave defined by Eq. (1.50) enters the dielectric medium at $z = 0$, i.e.,

$$E(0,t) = \hat{x}\left[\frac{1}{2}E_0 \exp(-i\omega t) + \text{c.c.}\right]. \tag{1.55}$$

The solution for the wave equation in the dielectric medium is

$$E(z,t) = \frac{1}{2}\hat{x}E_0 \exp\{+i[(k + \Delta k_{res})z - \omega t]\} \exp(+\alpha_{res}z) + \text{c.c.}, \tag{1.56}$$

where

$$\Delta k_{res} = \left(\frac{k}{2}\right)\chi'(\omega) \tag{1.57}$$

and

$$\alpha_{res} = \left(\frac{k}{2}\right)\chi''(\omega). \tag{1.58}$$

The effect of the medium's polarization is thus to induce both a phase shift and an amplitude change in the propagating EM wave. The phase shift is due to the real part of the resonant susceptibility, and the amplitude change is due to the imaginary part of the susceptibility.

1.5 Emission of Electromagnetic Radiation by the Classical Electron Oscillator

In this section the electromagnetic field radiated by the classical electron oscillator is discussed. For an arbitrary distribution of charge and current, the scalar and vector potentials, $\Phi(\boldsymbol{r},t)$ and $\boldsymbol{A}(\boldsymbol{r},t)$, respectively, are given by (Becker 1964; Marion and Heald 1980)

$$\Phi(\boldsymbol{r},t) = \frac{1}{4\pi\varepsilon_0}\iiint \frac{\rho(\boldsymbol{r}',\tau)}{|\boldsymbol{r}-\boldsymbol{r}'|}\,dx'dy'dz' = \frac{1}{4\pi\varepsilon_0}\int_{\mathcal{V}} \frac{\rho(\boldsymbol{r}',\tau)}{|\boldsymbol{r}-\boldsymbol{r}'|}\,d\mathcal{V}', \tag{1.59}$$

$$\boldsymbol{A}(\boldsymbol{r},t) = \frac{\mu_0}{4\pi}\int_{\mathcal{V}} \frac{\boldsymbol{J}(\boldsymbol{r}',\tau)}{|\boldsymbol{r}-\boldsymbol{r}'|}\,d\mathcal{V}', \tag{1.60}$$

where ρ is the charge density (C/m^3), $\boldsymbol{J}$ is the current density (A/m^2), and the variable $\tau(\text{s})$ is the retarded time,

$$\tau = t - \frac{|\boldsymbol{r}-\boldsymbol{r}'|}{c}. \tag{1.61}$$

For the calculation of the radiation field from the oscillating dipole, we assume that the dipole is confined to a volume element at the origin with a characteristic dimension d that is very small compared to our region of interest $(d \ll r = |\boldsymbol{r}|)$. The coordinate

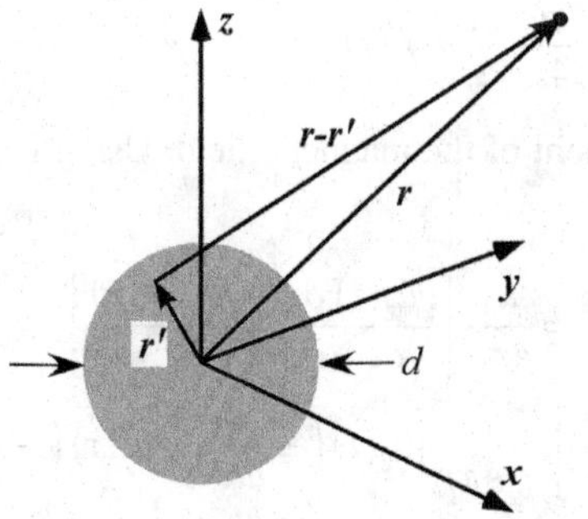

Figure 1.6 Geometry for the calculation of the electromagnetic radiation emitted by a distribution of current confined within a region with a characteristic dimension d.

system for the calculations is shown in Figure 1.6. The dipole is located the origin of the coordinate system so that

$$r = |\boldsymbol{r}| = \sqrt{x^2 + y^2 + z^2}. \tag{1.62}$$

With this assumption we can rewrite Eq. (1.60) as

$$\boldsymbol{A}(\boldsymbol{r}, t) = \frac{\mu_0}{4\pi r} \int_{\forall} \boldsymbol{J}(\boldsymbol{r}', \tau) d\forall'. \tag{1.63}$$

The current density is the product of charge density and velocity,

$$\boldsymbol{J}(\boldsymbol{r}', \tau) = \rho(\boldsymbol{r}', \tau)\mathbf{v}(\boldsymbol{r}', \tau). \tag{1.64}$$

For a collection of discrete charges within the volume element, the integral in Eq. (1.63) reduces to

$$\boldsymbol{A}(\boldsymbol{r}, t) = \frac{\mu_0}{4\pi r} \int_{\forall} \rho(\boldsymbol{r}', \tau)\mathbf{v}(\boldsymbol{r}', \tau) d\forall' = \frac{\mu_0}{4\pi r} \sum_i q_i \mathbf{v}_i(\boldsymbol{r}', \tau). \tag{1.65}$$

Specializing to the case where an electron (cloud) is executing harmonic motion about an atomic nucleus, we obtain

$$\boldsymbol{A}(\boldsymbol{r}, t) = \frac{\mu_0}{4\pi r} [-e\mathbf{v}_e(\boldsymbol{r}', \tau)] = \frac{\mu_0}{4\pi r} \left[-e \frac{d\boldsymbol{r}_e(\boldsymbol{r}', \tau)}{d\tau} \right] = \frac{\mu_0}{4\pi r} \frac{d\boldsymbol{\mu}(\boldsymbol{r}', \tau)}{d\tau}. \tag{1.66}$$

The magnetic field of the emitted radiation is given by

$$\begin{aligned} \boldsymbol{B}(\boldsymbol{r}, t) &= \nabla \times \boldsymbol{A}(\boldsymbol{r}, t) = \frac{\mu_0}{4\pi} \nabla \times \left[\frac{1}{r} \frac{d\boldsymbol{\mu}(\tau)}{d\tau} \right] \\ &= \frac{\mu_0}{4\pi} \nabla \left(\frac{1}{r} \right) \times \frac{d\boldsymbol{\mu}(\tau)}{d\tau} + \frac{\mu_0}{4\pi r} \nabla \times \left(\frac{d\boldsymbol{\mu}(\tau)}{d\tau} \right) \\ &= -\frac{\mu_0}{4\pi r^3} \boldsymbol{r} \times \dot{\boldsymbol{\mu}}(\tau) + \frac{\mu_0}{4\pi r} \nabla \times \dot{\boldsymbol{\mu}}(\tau). \end{aligned} \tag{1.67}$$

In the far field, the first term on the right-hand side is negligible compared to the second term. Eliminating the first term, we obtain

$$\boldsymbol{B}(\boldsymbol{r},t) = \frac{\mu_0}{4\pi r}\nabla \times \dot{\boldsymbol{\mu}}(\tau). \tag{1.68}$$

At this point we solve for the x-component of the magnetic field. Using Eq. (1.62) we obtain

$$\begin{aligned} B_x(\boldsymbol{r},t) &= \frac{\mu_0}{4\pi r}\left[\frac{\partial \dot{\mu}_z(\tau)}{\partial y} - \frac{\partial \dot{\mu}_y(\tau)}{\partial z}\right] = \frac{\mu_0}{4\pi r}\left[\frac{\partial \dot{\mu}_z(\tau)}{\partial \tau}\frac{\partial \tau}{\partial y} - \frac{\partial \dot{\mu}_y(\tau)}{\partial \tau}\frac{\partial \tau}{\partial z}\right] \\ &= \frac{\mu_0}{4\pi r}\left[\ddot{\mu}_z(\tau)\frac{\partial \tau}{\partial y} - \ddot{\mu}_y(\tau)\frac{\partial \tau}{\partial z}\right] = \frac{\mu_0}{4\pi r}\left[\ddot{\mu}_z(\tau)\left(-\frac{y}{cr}\right) - \ddot{\mu}_y(\tau)\left(-\frac{z}{cr}\right)\right] \\ &= \frac{\mu_0}{4\pi c r^2}\left[\ddot{\mu}_y(\tau)z - \ddot{\mu}_z(\tau)y\right]. \end{aligned} \tag{1.69}$$

A similar analysis for the y- and z-components of the magnetic field results in

$$\begin{aligned} B_y(\boldsymbol{r},t) &= \frac{\mu_0}{4\pi c r^2}\left[\ddot{\mu}_z(\tau)x - \ddot{\mu}_x(\tau)z\right] \\ B_z(\boldsymbol{r},t) &= \frac{\mu_0}{4\pi c r^2}\left[\ddot{\mu}_x(\tau)y - \ddot{\mu}_y(\tau)x\right]. \end{aligned} \tag{1.70}$$

Generalizing, we obtain

$$\boldsymbol{B}(\boldsymbol{r},t) = \frac{\mu_0}{4\pi c r^2}\ddot{\boldsymbol{\mu}}(\tau) \times \boldsymbol{r}. \tag{1.71}$$

The electric field due to the dipole is given by

$$\boldsymbol{E}(\boldsymbol{r},t) = \boldsymbol{B}(\boldsymbol{r},t) \times \frac{\boldsymbol{r}}{r} = \frac{\mu_0}{4\pi c r^3}(\ddot{\boldsymbol{\mu}}(\tau) \times \boldsymbol{r}) \times \boldsymbol{r}. \tag{1.72}$$

Now consider an electron cloud executing harmonic motion along the z-axis. The polarization and the second derivative of the polarization are given by

$$\boldsymbol{\mu}(\tau) = a\cos(\omega\tau)\hat{z} \quad \ddot{\boldsymbol{\mu}}(\tau) = -\omega^2 a\cos(\omega\tau)\hat{z}, \tag{1.73}$$

where $a(\mathrm{m})$ is the amplitude of the harmonic motion of the electron cloud (the symbol μ_0 would have been more appropriate for the amplitude but would be too easy to confuse with the magnetic permeability). The magnetic and electric fields due to the harmonic motion of the dipole are given by

$$\boldsymbol{B}(\boldsymbol{r},t) = \frac{\mu_0}{4\pi c r^3}\left[-\omega^2 a\cos(\omega\tau)\hat{z} \times \boldsymbol{r}\right] = -\frac{\mu_0\omega^2 a\cos(\omega\tau)}{4\pi c r}\sin\theta\hat{\boldsymbol{\varphi}}, \tag{1.74}$$

$$\boldsymbol{E}(\boldsymbol{r},t) = \frac{\mu_0}{4\pi c r^3}\left[-\omega^2 a\cos(\omega\tau)\hat{z} \times \boldsymbol{r}\right] \times \boldsymbol{r} = -\frac{\mu_0\omega^2 a\cos(\omega\tau)}{4\pi c r}\sin\theta\hat{\boldsymbol{\theta}}, \tag{1.75}$$

The Poynting vector for the dipole is given by

$$\boldsymbol{S}(\boldsymbol{r},t) = \boldsymbol{E}(\boldsymbol{r},t) \times \boldsymbol{H}(\boldsymbol{r},t) = \frac{\mu_0^2\omega^4 a^2\cos^2(\omega\tau)}{16\pi^2 c^2 r^2}\sin^2\theta. \tag{1.76}$$

We calculate the total power W radiated by the dipole oscillator by integrating over all solid angles and averaging over a single cycle of the oscillator. The result is

$$W = \frac{d\varepsilon_d}{dt} = -\frac{\omega^4 a^2}{12\pi\varepsilon_0 c^3}. \tag{1.77}$$

The energy associated with the harmonic oscillation of the electron cloud is given by

$$\varepsilon_d = \frac{1}{2}Ka^2 = \frac{1}{2}m_e\omega^2 a^2. \tag{1.78}$$

Substituting Eq. (1.78) into Eq. (1.77) and rearranging, we obtain

$$W = \frac{d\varepsilon_d}{dt} = -\frac{e^2\omega^2\varepsilon_d}{6\pi m_e\varepsilon_0 c^3}. \tag{1.79}$$

If an oscillator has an initial energy ε_{d0} in the absence of an applied field, the time dependence of the dipole energy is given by

$$\varepsilon_d = \varepsilon_{d0}\exp\left(-\gamma_{rad,ceo}t\right), \tag{1.80}$$

where

$$\gamma_{rad,ceo} = \frac{e^2\omega^2}{6\pi m_e\varepsilon_0 c^3}. \tag{1.81}$$

The radiative decay rate $\gamma_{rad,ceo}$ (s^{-1}) is proportional to the square of the angular frequency of the electric dipole operator. For electric dipole transitions in atoms and molecules, spontaneous emission is responsible for the decay of the induced electric dipole, and the spontaneous emission rate coefficient Γ_{spe} (s^{-1}) is analogous to the radiative decay rate. The radiative decay rate of the classical electron oscillator is an upper limit on the decay rate Γ_{spe} (s^{-1}) for an actual atomic or molecular electric dipole transition.

2 Atomic Structure and the Quantum Mechanics of Angular Momentum

2.1 Introduction

Although the material discussed in Chapter 2 is similar to material contained in most introductory textbooks on quantum mechanics, the discussion of atomic structure here serves several purposes. Angular momentum and angular momentum coupling algebra are introduced, and these quantities will be central to later discussions of radiative transitions. Much of the nomenclature for radiative transitions is introduced.

The chapter begins with the introduction of the two-particle Schrödinger wave equation (SWE) and the solution of this equation for the hydrogen atom. The orbital angular momentum of the electron results from the SWE solution. However, the electron spin angular momentum is not predicted by the SWE solution. The electron spin angular momentum arises naturally in the solution of the relativistic Dirac equation. The solutions of the Dirac equation for the hydrogen atom are discussed briefly, but a full discussion of these solutions is beyond the scope of this book. The Pauli spinors are introduced, and the SWE wavefunctions are modified to account for the spin of the electron.

The structure of multielectron atoms is then discussed. The transition from the analysis of a one-electron atom to a multielectron atom is discussed in detail by Weissbluth (1978) and Cowan (1981). The Hamiltonians for multielectron atoms for both low-Z and high-Z atoms are discussed. In this chapter the discussion will be focused on low-Z atoms for which Russell–Saunders or *LS* coupling is appropriate. The total, orbital, and spin angular momenta of the electrons for multielectron atoms are shown to be very similar to the total, orbital, and spin angular momenta for hydrogen atoms, described by the same set of quantum numbers. Alternate coupling schemes including *jj* coupling and pair coupling are briefly discussed.

The analysis of the angular momentum quantum states for multielectron atoms is the subject of the latter part of the chapter. This analysis is extremely useful for the determination of selection rules and relative transition strengths between multielectron quantum states, and for understanding energy level splitting based on the angular momentum quantum numbers of the levels. The chapter continues with a discussion of Euler's angles and the quantum mechanical rotation operator. Angular momentum coupling algebra, the Clebsch–Gordan coefficients, and *3j* symbols are then introduced. The Wigner–Eckart theorem is discussed, and the use of irreducible spherical tensors for the evaluation of quantum mechanical matrix elements is discussed in detail. The chapter concludes with a quantitative analysis of spin–orbit splitting in the calcium atom.

2.2 The Postulates of Quantum Mechanics: Wavefunctions and Operators

In quantum mechanics, information about the state of a physical system is wholly contained in the system wavefunction $\Psi(\boldsymbol{r}, t)$. The wavefunction is in general a complex quantity. The product of the wavefunction and its complex conjugate is the probability density for the system, and the wavefunction is normalized such that

$$\iiint_{all\ space} \Psi^*(\boldsymbol{r}, t)\Psi(\boldsymbol{r}, t)d\forall = 1. \tag{2.1}$$

Properties of the system are determined by operating on the wavefunction, and operators are associated with the dynamical variables of the system such as position ($\boldsymbol{r}$), momentum ($\boldsymbol{p}$), energy (ε), and angular momentum ($\boldsymbol{L}$). The operators associated with these dynamical variables are listed in Table 2.1. The average value or the expectation value for a dynamical variable is given by

$$\langle O \rangle = \iiint_{all\ space} \Psi^*(\boldsymbol{r}, t)\left[\hat{O}\Psi(\boldsymbol{r}, t)\right]d\forall = 1. \tag{2.2}$$

The caret above the variable is used to indicate an operator. The operator $\hat{O}$ acts on the wavefunction $\Psi(\boldsymbol{r}, t)$ to the right of the operator.

Dirac (1958) introduced *bra* and *ket* notation as a useful shorthand notation, and the Dirac notation is used extensively in this book. The wavefunction $\Psi(\boldsymbol{r}, t)$ is represented by the *ket*

$$\Psi(\boldsymbol{r}, t) = |\Psi\rangle. \tag{2.3}$$

Table 2.1 Quantum mechanical operators

Dynamical variable	Operator	
$\boldsymbol{r}$	$\hat{\boldsymbol{r}} = \boldsymbol{r}$	
$\boldsymbol{p}$	$\hat{\boldsymbol{p}} = -i\hbar\nabla$	$\hat{p}_x = -i\hbar\frac{\partial}{\partial x}$
$p^2 = \boldsymbol{p} \cdot \boldsymbol{p}$	$\hat{p}^2 = -\hbar^2\nabla^2$	
ε	$\hat{\varepsilon} = i\hbar\frac{\partial}{\partial t}$	
$\boldsymbol{L} = \boldsymbol{r} \times \boldsymbol{p}$	$\hat{\boldsymbol{L}} = \hat{\boldsymbol{r}} \times \hat{\boldsymbol{p}}$	$\hat{L}_x = y\hat{p}_z - z\hat{p}_y$

The Dirac inner product is denoted by

$$\langle \Psi | \Psi \rangle = \iiint_{all\ space} \Psi^*(\boldsymbol{r},t)\Psi(\boldsymbol{r},t)d\forall = 1. \tag{2.4}$$

Equation (2.4) serves to define the *bra* in Dirac notation. The expectation value is written as

$$\langle O \rangle = \langle \Psi | \hat{O} | \Psi \rangle = \langle \Psi | \hat{O}\Psi \rangle. \tag{2.5}$$

The inner product $\langle \Psi_b | \Psi_a \rangle$ of two different wavefunctions is not necessarily equal to unity. If the inner product is zero, the functions are said to be orthogonal. If $\langle \Psi_b | \Psi_a \rangle = 0$ and $\langle \Psi_a | \Psi_a \rangle = \langle \Psi_b | \Psi_b \rangle = 1$, the functions are said to be orthonormal. In Section 3.3 we will discuss the representation of a general wavefunction of a quantum mechanical system as a superposition of orthonormal basis wavefunctions or eigenstates of the system. Any real physical quantity in quantum mechanics has an associated Hermitian operator. For a Hermitian operator,

$$O_{ba} = \langle \Psi_b | \hat{O}\Psi_a \rangle = \langle \hat{O}\Psi_b | \Psi_a \rangle = \langle \Psi_a | \hat{O}\Psi_b \rangle^*. \tag{2.6}$$

The Hermitian conjugate $\hat{O}^\dagger$ of an operator is defined by

$$\left\langle \Psi_b | \hat{O}^\dagger \Psi_a \right\rangle = \langle \hat{O}\Psi_b | \Psi_a \rangle = \langle \Psi_a | \hat{O}\Psi_b \rangle^*. \tag{2.7}$$

All real physical quantities in quantum mechanics are represented by Hermitian operators. The expectation value for a physical quantity associated with a Hermitian operator is a real quantity, as shown below:

$$\langle O \rangle = \langle \Psi | \hat{O}\Psi \rangle = \langle \hat{O}\Psi | \Psi \rangle = \langle \Psi | \hat{O}\Psi \rangle^*. \tag{2.8}$$

The Hermitian conjugate of a sum of operators is given by the sum of the Hermitian conjugates,

$$\left(\hat{O}_1 + \hat{O}_2 + \cdots \hat{O}_N\right)^\dagger = \hat{O}_1^\dagger + \hat{O}_2^\dagger + \cdots \hat{O}_N^\dagger. \tag{2.9}$$

The Hermitian conjugate of the product of operators is given by product of the Hermitian conjugates with the order of operations reversed,

$$\left(\hat{O}_1\hat{O}_2 \ldots \hat{O}_N\right)^\dagger = \hat{O}_N^\dagger \ldots \hat{O}_2^\dagger\hat{O}_1^\dagger. \tag{2.10}$$

The commutator of two operators is defined by

$$\left[\hat{O}_1, \hat{O}_2\right] = \hat{O}_1\hat{O}_2 - \hat{O}_2\hat{O}_1. \tag{2.11}$$

Two operators commute if the value of the commutator for these operators is zero. The product of two Hermitian operators is Hermitian if and only if the two operators commute,

$$\left(\hat{O}_1\hat{O}_2\right)^\dagger = \hat{O}_2\hat{O}_1. \tag{2.12}$$

2.3 The Time-Independent, Two-Particle Schrödinger Wave Equation

Consider the two-particle system shown in Figure 2.1. The quantum mechanical Hamiltonian for the system is given by

$$H = \frac{\hat{\boldsymbol{p}}_1^2}{2m_1} + \frac{\hat{\boldsymbol{p}}_2^2}{2m_2} + V_{12}(r,t), \tag{2.13}$$

where the momentum operator $\hat{\boldsymbol{p}}_i$ operates on the coordinates of particle i, and the potential function $V_{12}(r,t)$ is a function of the distance r between the particles (external potentials and interactions are neglected). In Cartesian coordinates the Schrödinger wave equation can then be written as

$$-\frac{\hbar^2}{2m_1}\nabla_1^2\Psi(\boldsymbol{r},t) - \frac{\hbar^2}{2m_2}\nabla_2^2\Psi(\boldsymbol{r},t) + V_{12}(r,t)\Psi(\boldsymbol{r},t) = \varepsilon_{op}\Psi(\boldsymbol{r},t) = i\hbar\frac{\partial\Psi(\boldsymbol{r},t)}{\partial t}, \tag{2.14}$$

where the energy operator $\varepsilon_{op} = i\hbar\frac{\partial}{\partial t}$. We assume that the wavefunction $\Psi(\boldsymbol{r},t)$ can be written as the product of a function that depends only on the spatial coordinates and a function the depends only on time,

$$\Psi(\boldsymbol{r},t) = \psi(\boldsymbol{r})T(t), \tag{2.15}$$

and that the potential function depends only on the spatial coordinates and is not explicitly dependent on time. Substituting Eq. (2.15) into Eq. (2.14), performing the partial differentiations, dividing by $\Psi(\boldsymbol{r},t)$, and rearranging, we obtain

$$-\frac{\hbar^2}{2m_1\psi(\boldsymbol{r})}\nabla_1^2\psi(\boldsymbol{r}) - \frac{\hbar^2}{2m_2\psi(\boldsymbol{r})}\nabla_2^2\psi(\boldsymbol{r}) + V_{12}(r) = \frac{i\hbar}{T(t)}\frac{\partial T(t)}{\partial t}. \tag{2.16}$$

The left-hand side (LHS) of Eq. (2.16) depends only on the spatial coordinates and the right-hand side (RHS) depends only on time. Therefore, this equation can only be satisfied if both are equal to a separation constant α. Setting the RHS of Eq. (2.16) equal to the separation constant α and solving for $T(t)$, we obtain

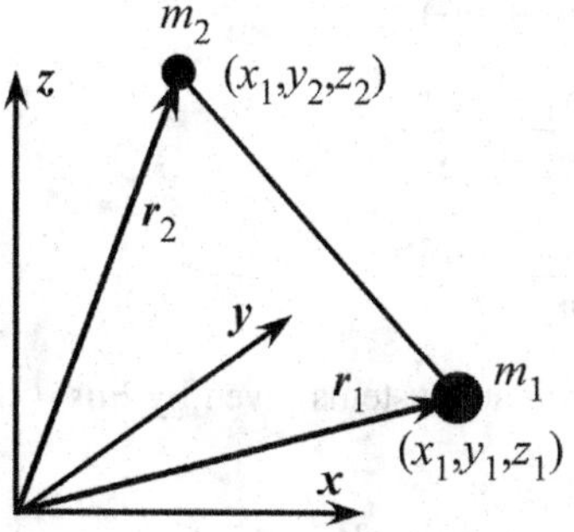

Figure 2.1 Schematic diagram of the two-particle system.

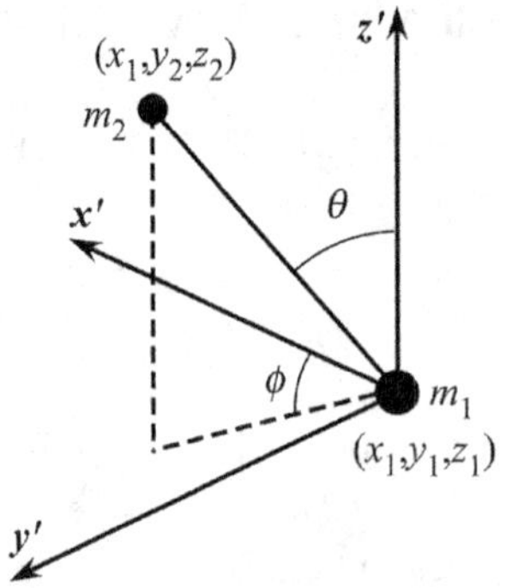

Figure 2.2 Two-particle spherical coordinate system.

$$T(t) = \exp\left(-\frac{i\alpha t}{\hbar}\right). \tag{2.17}$$

The LHS of the equation can be written in Cartesian coordinates as

$$-\frac{\hbar^2}{2m_1}\left[\frac{\partial^2\psi(\mathbf{r})}{\partial x_1^2}+\frac{\partial^2\psi(\mathbf{r})}{\partial y_1^2}+\frac{\partial^2\psi(\mathbf{r})}{\partial z_1^2}\right]-\frac{\hbar^2}{2m_2}\left[\frac{\partial^2\psi(\mathbf{r})}{\partial x_2^2}+\frac{\partial^2\psi(\mathbf{r})}{\partial y_2^2}+\frac{\partial^2\psi(\mathbf{r})}{\partial z_2^2}\right] \\ +V_{12}(r)\psi(\mathbf{r}) = \alpha\psi(\mathbf{r}). \tag{2.18}$$

We now attach a spherical coordinate system to particle 1 as shown in Figure 2.2. We obtain the following relations for the particle coordinates:

$$x_2 = x_1 + x = x_1 + r\sin\theta\cos\phi. \tag{2.19}$$

$$y_2 = y_1 + y = y_1 + r\sin\theta\sin\phi. \tag{2.20}$$

$$z_2 = z_1 + z = z_1 + r\cos\theta. \tag{2.21}$$

The center of mass coordinates for the two-particle system are given by

$$X = \frac{m_1x_1 + m_2x_2}{(m_1+m_2)}. \tag{2.22}$$

$$Y = \frac{m_1y_1 + m_2y_2}{(m_1+m_2)}. \tag{2.23}$$

$$Z = \frac{m_1z_1 + m_2z_2}{(m_1+m_2)}. \tag{2.24}$$

Using the relations between the two coordinate systems given by Eqs. (2.19)–(2.24), we can show that

$$-\frac{\hbar^2}{2m_1}\frac{\partial^2\psi}{\partial x_1^2}-\frac{\hbar^2}{2m_2}\frac{\partial^2\psi}{\partial x_2^2} = -\frac{\hbar^2}{2M}\frac{\partial^2\psi}{\partial X^2}-\frac{\hbar^2}{2\mu}\frac{\partial^2\psi}{\partial x^2}, \tag{2.25}$$

$$-\frac{\hbar^2}{2m_1}\frac{\partial^2\psi}{\partial y_1^2}-\frac{\hbar^2}{2m_2}\frac{\partial^2\psi}{\partial y_2^2}=-\frac{\hbar^2}{2M}\frac{\partial^2\psi}{\partial Y^2}-\frac{\hbar^2}{2\mu}\frac{\partial^2\psi}{\partial y^2}, \tag{2.26}$$

$$-\frac{\hbar^2}{2m_1}\frac{\partial^2\psi}{\partial z_1^2}-\frac{\hbar^2}{2m_2}\frac{\partial^2\psi}{\partial z_2^2}=-\frac{\hbar^2}{2M}\frac{\partial^2\psi}{\partial Z^2}-\frac{\hbar^2}{2\mu}\frac{\partial^2\psi}{\partial z^2}, \tag{2.27}$$

where the total mass M and reduced mass μ for the two-particle system are given by

$$M = m_1 + m_2. \tag{2.28}$$

$$\mu = \frac{m_1 m_2}{m_1 + m_2}. \tag{2.29}$$

Substituting Eqs. (2.25)–(2.27) into Eq. (2.18), we obtain

$$-\frac{\hbar^2}{2M}\left(\frac{\partial^2\psi}{\partial X^2}+\frac{\partial^2\psi}{\partial Y^2}+\frac{\partial^2\psi}{\partial Z^2}\right)-\frac{\hbar^2}{2\mu}\left(\frac{\partial^2\psi}{\partial x^2}+\frac{\partial^2\psi}{\partial y^2}+\frac{\partial^2\psi}{\partial z^2}\right)+(V_{12}-\varepsilon)\psi=0. \tag{2.30}$$

The first term on the LHS of Eq. (2.30) depends only on the center-of-mass coordinates of the two-particle system, and the second term depends only on the internal coordinates. We write the wavefunction as the product of a wavefunction depending only on center-of-mass (external) coordinates and a wavefunction depending only on internal coordinates:

$$\psi(x_1,y_1,z_1,x_2,y_2,z_2)=\psi(X,Y,Z,x,y,z)=\psi_e(X,Y,Z)\psi_i(x,y,z). \tag{2.31}$$

The energy is written as the sum of an energy term depending only on center-of-mass coordinates and an energy term depending only on internal coordinates:

$$\varepsilon(x_1,y_1,z_1,x_2,y_2,z_2)=\varepsilon(X,Y,Z,x,y,z)=\varepsilon_e(X,Y,Z)+\varepsilon_i(x,y,z). \tag{2.32}$$

We further assume that the potential function V_{12} is a function only of internal coordinates,

$$V_{12}=V_{12}(x,y,z). \tag{2.33}$$

Substituting Eqs. (2.31)–(2.33) into Eq. (2.30), we obtain two separate equations,

$$-\frac{\hbar^2}{2M}\left(\frac{\partial^2\psi_e}{\partial X^2}+\frac{\partial^2\psi_e}{\partial Y^2}+\frac{\partial^2\psi_e}{\partial Z^2}\right)-\varepsilon_e\psi_e=-\frac{\hbar^2}{2M}\nabla_e^2\psi_e-\varepsilon_e\psi_e=0. \tag{2.34}$$

$$-\frac{\hbar^2}{2\mu}\left(\frac{\partial^2\psi_i}{\partial x^2}+\frac{\partial^2\psi_i}{\partial y^2}+\frac{\partial^2\psi_i}{\partial z^2}\right)+(V_{12}-\varepsilon_i)\psi_i=-\frac{\hbar^2}{2\mu}\nabla_i^2\psi_i+(V_{12}-\varepsilon_i)\psi_i=0. \tag{2.35}$$

The solution to Eq. (2.34) is essentially the particle-in-a-box solution. The solution is dependent on the boundary conditions and gives rise to the translational energy states associated with the bulk motion of the two-particle system. The solution of Eq. (2.35) is dependent on the definition of the potential function of the two-particle system. The hydrogen atom solution, discussed in Section 2.3, results when V_{12} is a central potential function associated with the attractive electric force acting between the proton and electron. For a diatomic molecule, the two particles are the nuclei of the diatomic molecule, and the attractive force between the two nuclei is provided by the electron cloud. The potential function is approximately a harmonic oscillator potential for small displacements of the nuclei from the equilibrium internuclear position. The structure of the diatomic molecule is discussed in Chapter 3.

2.4 The Hydrogen Atom

The quantum mechanical analysis of atomic hydrogen is incredibly important because it is one of the few problems for which analytical solutions of the Schrödinger and Dirac equations can be derived. The solution of the Schrödinger wave equation results in the definition of the quantum number associated with orbital angular momentum and the projection of the angular orbital momentum on the laboratory z-axis. The spin or intrinsic angular momentum of the electron results from the solution of the relativistic Dirac equation for the hydrogen atom and is incorporated in our analysis by multiplying the wavefunctions that result from the solution of the Schrödinger wave equation by Pauli spinors.

2.4.1 Solution of the Schrödinger Wave Equation

The attractive force between the electron and proton is described by a central potential function

$$V_{12}(r) = -\frac{e^2}{4\pi\varepsilon_0 r}. \tag{2.36}$$

A spherical coordinate system is used as the natural coordinate system for the central potential system. In spherical coordinates Eq. (2.35) becomes

$$\begin{aligned} -\frac{\hbar^2}{2\mu}\left\{\frac{1}{r^2}\frac{\partial}{\partial r}\left(r^2\frac{\partial}{\partial r}\right)+\frac{1}{r^2\sin\theta}\frac{\partial}{\partial\theta}\left(\sin\theta\frac{\partial}{\partial\theta}\right)+\frac{1}{r^2\sin^2\theta}\frac{\partial^2}{\partial^2\phi}\right\}\psi_i \\ +[V_{12}(r)-\varepsilon_i]\psi_i = 0. \end{aligned} \tag{2.37}$$

Equation (2.37) is again solved using separation of variables. The wavefunction $\psi_i(r,\theta,\phi)$ is written as

$$\psi_i(r,\theta,\phi) = R(r)Y(\theta,\phi). \tag{2.38}$$

Substituting into Eq. (2.38), we obtain

$$\begin{aligned} 0 = Y(\theta,\phi)\frac{1}{r^2}\frac{\partial}{\partial r}\left[r^2\frac{\partial R(r)}{\partial r}\right] + R(r)\frac{1}{r^2\sin\theta}\frac{\partial}{\partial\theta}\left[\sin\theta\frac{\partial Y(\theta,\phi)}{\partial\theta}\right] \\ +R(r)\frac{1}{r^2\sin^2\theta}\left[\frac{\partial^2 Y(\theta,\phi)}{\partial^2\phi}\right] + \frac{2\mu}{\hbar^2}[\varepsilon_i - V_{12}(r)]R(r)Y(\theta,\phi). \end{aligned} \tag{2.39}$$

Dividing through by $R(r)\,Y(\theta,\phi)$ and rearranging, we obtain

$$\frac{1}{R}\frac{d}{dr}\left[r^2\frac{dR}{dr}\right]+\frac{2\mu r^2}{\hbar^2}[\varepsilon_i - V_{12}(r)] = -\frac{1}{Y}\left\{\frac{1}{\sin\theta}\frac{\partial}{\partial\theta}\left[\sin\theta\frac{\partial Y}{\partial\theta}\right]+\frac{1}{r^2\sin^2\theta}\left[\frac{\partial^2 Y}{\partial^2\phi}\right]\right\}. \tag{2.40}$$

The LHS is a function of r only, and the RHS is a function of angular variables only. Setting both the RHS and LHS of Eq. (2.40) equal to a separation variable β and rearranging, we obtain the following two equations:

$$\frac{d}{dr}\left[r^2\frac{dR(r)}{dr}\right]+\left\{\frac{2\mu r^2}{\hbar^2}[\varepsilon_i - V_{12}(r)]-\beta\right\}R(r)=0. \tag{2.41}$$

$$\frac{1}{\sin\theta}\frac{\partial}{\partial\theta}\left[\sin\theta\frac{\partial Y(\theta,\phi)}{\partial\theta}\right]+\frac{1}{\sin^2\theta}\left[\frac{\partial^2 Y(\theta,\phi)}{\partial^2\phi}\right]+\beta Y(\theta,\phi)=0. \tag{2.42}$$

Equation (2.42) is once again solved by separation of variables. Dividing through by $Y(\theta,\phi)=\Theta(\theta)\Phi(\phi)$ and rearranging, we obtain

$$\frac{\sin\theta}{\Theta(\theta)}\frac{d}{d\theta}\left[\sin\theta\frac{d\Theta(\theta)}{d\theta}\right]+\beta\sin^2\theta=\frac{1}{\Phi(\phi)}\frac{d^2\Phi(\phi)}{d^2\phi}. \tag{2.43}$$

The LHS is a function only of θ, and the RHS is a function only of ϕ, and we set both sides equal to a separation variable η. Again, this results in two equations,

$$\frac{1}{\sin\theta}\frac{d}{d\theta}\left[\sin\theta\frac{d\Theta(\theta)}{d\theta}\right]+\left(\beta-\frac{\eta}{\sin^2\theta}\right)\Theta(\theta)=0. \tag{2.44}$$

$$\frac{d^2\Phi(\phi)}{d^2\phi}+\eta\Phi(\phi)=0. \tag{2.45}$$

The solutions to Eqs. (2.44) and (2.45) result in the definition of new quantum numbers associated with the orbital angular momentum of the electron. The solution to Eq. (2.45) is

$$\Phi(\phi)=C\exp(i\sqrt{\eta}\phi). \tag{2.46}$$

However, the requirement that $\Phi(\phi) = \Phi(\phi+2\pi)$ means that the factor $\sqrt{\eta}$ can take on only integer values; this leads to the definition of the quantum number m_l associated with the projection of the orbital angular momentum of the electron on the z-axis. We rewrite Eq. (2.46) as

$$\Phi(\phi)=C\exp(im_l\phi). \tag{2.47}$$

At this point we introduce the operator for orbital angular momentum,

$$\hat{\boldsymbol{L}}=\hat{\boldsymbol{r}}\times\hat{\boldsymbol{p}}=\boldsymbol{r}\times(-i\hbar\nabla). \tag{2.48}$$

In Cartesian coordinates the components of the orbital angular momentum operator are

$$\hat{L}_x = y\hat{p}_z - z\hat{p}_y = -i\hbar\left(y\frac{\partial}{\partial z}-z\frac{\partial}{\partial y}\right). \tag{2.49}$$

$$\hat{L}_y = z\hat{p}_x - x\hat{p}_z = -i\hbar\left(z\frac{\partial}{\partial x} - x\frac{\partial}{\partial z}\right). \tag{2.50}$$

$$\hat{L}_z = x\hat{p}_y - y\hat{p}_x = -i\hbar\left(x\frac{\partial}{\partial y} - y\frac{\partial}{\partial x}\right). \tag{2.51}$$

$$\hat{L}^2 = -\hbar^2\left[x^2\left(\frac{\partial^2}{\partial^2 y}\right) - 2xy\left(\frac{\partial^2}{\partial x\partial y}\right) + y^2\left(\frac{\partial^2}{\partial^2 x}\right) + \ldots\right]. \tag{2.52}$$

It can also be shown that the orbital angular momentum operator obeys the following equation:

$$\hat{\boldsymbol{L}} \times \hat{\boldsymbol{L}} = i\hat{\boldsymbol{L}}. \tag{2.53}$$

The orbital angular momentum operator can also be expressed in spherical coordinates. In spherical coordinates the z-components of the angular momentum are given by

$$\hat{L}_z = -i\hbar\frac{\partial}{\partial\phi}. \tag{2.54}$$

Operating on the ϕ-dependent component of the wavefunction $\Phi(\phi)$ gives the following:

$$L_{z,op}\Phi(\phi) = m_l\hbar\Phi(\phi). \tag{2.55}$$

The angular momentum about the z-axis is quantized and is an eigenvalue. The quantum state for the two-particle system is characterized by the quantum number m_l. The quantum number m_l is often termed a projection quantum number because it is the projection of the orbital angular momentum of the electron on the laboratory-fixed z-axis. Note that the result of the operation of the operator for the z-component of angular momentum on the wavefunction is a number times the wavefunction. This number is the eigenvalue of the z-component of angular momentum, and it is ***precisely*** defined for the wavefunction.

Now consider the quantum number l, which is found from the solution of the separated SWE for the function $\Theta(\theta)$. Separation of variables leads to the following equation:

$$\frac{1}{\sin\theta}\frac{d}{d\theta}\left(\sin\theta\frac{d\Theta}{d\theta}\right) + \left(\alpha - \frac{\beta}{\sin^2\theta}\right)\Theta = 0. \tag{2.56}$$

The solution to Eq. (2.56) is discussed in detail by Walecka (2013); the discussion of the boundary conditions and physics of the solution is especially illuminating. The solution to Eq. (2.56) is a Legendre polynomial which is a function both of the quantum number l, for orbital angular momentum, and of m_l, the quantum number that describes the projection of the angular momentum on the z-axis,

$$\begin{aligned} &\alpha = l(l+1) \\ &\Theta(\theta) = P_l^{|m_l|}(\cos\theta) \qquad l = |m_l|,\ |m_l|+1,\ |m_l|+2,\ \ldots, \end{aligned} \tag{2.57}$$

where

$$P_l^{|m|}(x) = \frac{1}{2^l l!}\left(1 - x^2\right)^{|m|/2} \frac{d^{|m|+l}(x^2 - 1)^l}{dx^{|m|+l}}. \tag{2.58}$$

Operating on the function $\Theta(\theta)$ with the operator $\hat{L}^2$ expressed in spherical coordinates gives the eigenvalues for the total angular momentum squared,

$$\hat{L}^2\Theta(\theta) = l(l+1)\hbar^2\Theta(\theta). \tag{2.59}$$

The total angular momentum $\boldsymbol{L}$ is quantized and has the magnitude

$$|\boldsymbol{L}| = \sqrt{l(l+1)}\hbar. \tag{2.60}$$

The normalized product of $\Theta(\theta)$ and $\Phi(\phi)$ is the spherical harmonic function $Y_{lm}(\theta,\phi)$, where

$$Y_{lm}(\theta,\phi) = \sqrt{(-1)^{m+|m|}}\sqrt{\frac{2l+1}{4\pi}}\sqrt{\frac{(l-|m|)!}{(l+|m|)!}}\, P_l^{|m|}(\cos\theta)\exp(im\phi), \tag{2.61}$$

where the associated Legendre function $P_l^{|m|}(\cos\theta)$ is given in terms of the ordinary Legendre polynomial $P_l(\cos\theta)$ by (Zare, 1988)

$$P_l^{|m|}(\cos\theta) = (\sin\theta)^{|m|}\left[\frac{d}{d(\cos\theta)}\right]^{|m|} P_l(\cos\theta). \tag{2.62}$$

$$P_l(\cos\theta) = \frac{1}{2^l l!}\left[\frac{d}{d(\cos\theta)}\right]^l \left(\cos^2\theta - 1\right)^l. \tag{2.63}$$

The spherical harmonics are of crucial importance in the study of atomic and molecular structure and the interaction of laser radiation with resonance transitions (Weissbluth, 1978). The spherical harmonics for $l \leq 3$ are listed in Appendix A1.

The spherical harmonics are orthonormal functions,

$$\int_{\phi=0}^{2\pi}\int_{\theta=0}^{\pi} Y_{lm}^*(\theta,\phi)Y_{l'm'}(\theta,\phi)\sin\theta\, d\theta\, d\phi = \delta_{ll'}\,\delta_{mm'}. \tag{2.64}$$

It can also be shown that

$$Y_{lm}^*(\theta,\phi) = (-1)^m Y_{l,-m}(\theta,\phi). \tag{2.65}$$

$$Y_{lm}(\pi-\theta,\pi-\phi) = (-1)^l Y_{l,m}(\theta,\phi). \tag{2.66}$$

This coordinate transformation $\theta \to \pi - \theta, \phi \to \pi - \phi$ in Eq. (2.66) corresponds to an inversion $(x,y,z) \to (-x,-y,-z)$. Functions that do not change sign under this inversion are said to be even parity functions, and functions that do change sign are odd parity functions. The parity of the spherical harmonic function $Y_{lm}(\theta,\phi)$ is determined by the orbital angular quantum number l.

The equation for the radial part of the wavefunction $R(r)$ is given by

$$\begin{aligned}\frac{1}{r^2}\frac{d}{dr}\left[r^2\frac{dR(r)}{dr}\right]-\frac{l(l+1)}{r^2}R(r)+\frac{2\mu}{\hbar^2}[\varepsilon-V_{12}(r)]R(r)\\ =\frac{1}{r^2}\frac{d}{dr}\left[r^2\frac{dR(r)}{dr}\right]-\frac{l(l+1)}{r^2}R(r)+\frac{2\mu}{\hbar^2}\left[\varepsilon+\frac{e^2}{4\pi\varepsilon_0 r}\right]R(r)=0.\end{aligned} \tag{2.67}$$

The solution to Eq. (2.67) is rather complicated, but an analytical solution does exist; the case of atomic hydrogen (or hydrogen-like atoms) is unique in that an analytical solution of the Schrödinger equation exists. The solution for Eq. (2.67), again discussed in great detail by Walecka (2013), is given by

$$R_{nl}(\rho)=e^{-\rho/2}\rho^{\ell}L_{n+l}^{2l+1}(\rho), \tag{2.68}$$

where the principal quantum number n has been introduced as an eigenvalue in the solution,

$$\rho=2\beta r \qquad \beta^2=-\frac{2\mu\varepsilon}{\hbar^2}>0, \tag{2.69}$$

and $L_{n+l}^{2l+1}(\rho)$ is a Laguerre polynomial. For a given value of the principal quantum number n, the values of the orbital angular momentum quantum number l are constrained such that

$$l=0,1,2,\ldots,n-1. \tag{2.70}$$

The total wavefunction for atomic hydrogen is given by

$$\begin{aligned}|nlm_l\rangle &= \psi_{nlm_l}(r,\theta,\phi)=R_{nl}(r)Y_{lm_l}(\theta,\phi)=R_{nl}(r)\Theta(\theta)\Phi(\phi)\\ &= R_{nl}(r)P_l^{|m_l|}(\cos\theta)\exp(im_l\phi).\end{aligned} \tag{2.71}$$

The radial functions $R_{nl}(r)$ for $n\leq 3$ are listed in Appendix A1. The radial functions are normalized such that

$$\int_0^{\infty}R_{n'l'}^{*}(r)R_{nl}(r)r^2dr=\delta_{nn'}\delta_{ll'}. \tag{2.72}$$

The normalization conditions represented by Eqs. (2.64) and (2.72) guarantee that the overall wavefunction is orthonormal,

$$\int_0^{\infty}\int_{\phi=0}^{2\pi}\int_{\theta=0}^{\pi}\psi_{n'l'm'}^{*}(r,\theta,\phi)\psi_{nlm}(r,\theta,\phi)r^2\sin\theta dr d\theta d\phi=\delta_{nn'}\delta_{ll'}\delta_{mm'}, \tag{2.73}$$

where the differential volume element in spherical coordinates is $r^2\sin\theta\, dr\, d\theta\, d\phi$.

The energy levels in the SWE solution are given by

$$\varepsilon=-\frac{e^4\mu}{32\pi^2\varepsilon_0^2\hbar^2}\frac{1}{n^2}=\varepsilon_n \qquad n=1,2,3,\ldots. \tag{2.74}$$

The energy of a particular quantum state $|n,l,m_l\rangle$ is dependent only on the principal quantum number n, and the degeneracy of each level is given by n^2.

2.4.2 Intrinsic Angular Momentum of the Electron

The SWE is a nonrelativistic equation. The Dirac wave equation includes relativistic effects, and its analysis and solutions are much more complicated than the SWE. The Dirac equation can be solved exactly for atomic hydrogen, or more correctly for the quantum states of the electron in atomic hydrogen. The intrinsic angular momentum or spin of the electron arises naturally in the solution of the Dirac equation. The spin of the electron is given by

$$|\boldsymbol{S}|^2 = S^2 = s(s+1)\hbar^2, \tag{2.75}$$

where $s = \frac{1}{2}$ for the electron. The projection quantum number m_s for the electron spin is given by

$$m_s = \pm\frac{1}{2}. \tag{2.76}$$

With the inclusion of intrinsic spin, the wavefunction of the electron can no longer be treated solely as a function of spatial coordinates. The wavefunction is instead represented as the product of a function of spatial coordinates (resulting from the solution of the Schrödinger wave equation) and a Pauli spinor (Weissbluth, 1978),

$$\psi(\boldsymbol{r}, m_s) = \phi(\boldsymbol{r})\xi(m_s) = R_{nl}(r)Y_{lm_l}(\theta,\varphi)\zeta(m_s), \tag{2.77}$$

where the spinor kets are defined as

$$\zeta\left(s=\frac{1}{2},\ m_s=+\frac{1}{2}\right) = |\alpha\rangle = \left|\frac{1}{2}\,\frac{1}{2}\right\rangle = \begin{pmatrix}1\\0\end{pmatrix}, \tag{2.78}$$

$$\zeta\left(s=\frac{1}{2},\ m_s=-\frac{1}{2}\right) = |\beta\rangle = \left|\frac{1}{2}\,-\frac{1}{2}\right\rangle = \begin{pmatrix}0\\1\end{pmatrix}, \tag{2.79}$$

and the spinor bras are defined as

$$\langle\alpha| = (1 \quad 0). \tag{2.80}$$

$$\langle\beta| = (0 \quad 1). \tag{2.81}$$

It is easy to show that

$$\langle\alpha|\beta\rangle = \langle\beta|\alpha\rangle = 0, \quad \langle\alpha|\alpha\rangle = \langle\beta|\beta\rangle = 1. \tag{2.82}$$

The wavefunction for atomic hydrogen can be expressed in terms of either the coupled or uncoupled representation. In the coupled representation, the orbital angular momentum $\boldsymbol{L}$ and spin angular momentum $\boldsymbol{S}$ of the electron combine to give the total electron angular momentum $\boldsymbol{J} = \boldsymbol{L} + \boldsymbol{S}$. The projection of the quantum number $\boldsymbol{J}$ on the z-axis is labeled m_J. The spin and orbital angular momentum of the electron couple to give the total angular momentum J, where

$$J = l \pm s = l \pm \frac{1}{2}. \tag{2.83}$$

The transformation between the uncoupled representation and the coupled representation is determined using Clebsch–Gordan coefficients,

$$|lsJm_J\rangle = \sum_{m_l m_s} |lm_l sm_s\rangle\langle lm_l sm_s|lsJm_J\rangle. \tag{2.84}$$

$$|lm_l sm_s\rangle = \sum_{Jm_J} |lsJm_J\rangle\langle lsJm_J|lm_l sm_s\rangle. \tag{2.85}$$

The terms $\langle lm_l sm_s|lsJm_J\rangle$ and $\langle lsJm_J|lm_l sm_s\rangle$ are the Clebsch–Gordan (C–G) coefficients; the C–G coefficients in Eqs. (2.84) and (2.85) are zero unless $m_J = m_l + m_s$. The C–G coefficients for angular momentum coupling in the $2p$ level of atomic hydrogen are tabulated in Appendix A2.

The C–G coefficients are real quantities and are related to the $3j$ symbols by the following formulae (Weissbluth, 1978):

$$\begin{pmatrix} j_1 & j_2 & j \\ m_1 & m_2 & m \end{pmatrix} = \frac{(-1)^{j_1-j_2-m}}{\sqrt{2j+1}} \langle j_1 m_1 j_2 m_2|j_1 j_2 j - m\rangle \tag{2.86}$$

and

$$\langle j_1 m_1 j_2 m_2|j_1 j_2 jm\rangle = (-1)^{j_1-j_2+m}\sqrt{2j+1}\begin{pmatrix} j_1 & j_2 & j \\ m_1 & m_2 & -m \end{pmatrix}. \tag{2.87}$$

Using Eq. (2.87), we can immediately write

$$\langle lm_l sm_s|lsJm_J\rangle = (-1)^{s-l+m_J}\sqrt{2J+1}\begin{pmatrix} l & s & J \\ m_l & m_s & -m_J \end{pmatrix}. \tag{2.88}$$

Fine structure splitting due to the spin–orbit interaction of the energy levels is also predicted by the solution of the Dirac equation. The energy of states with the same values for l and s but different values of J will differ in energy due to spin–orbit coupling. In the nonrelativistic limit of the Dirac equation for atomic hydrogen, the Hamiltonian for the spin–orbit coupling effect is given by (Weissbluth, 1978)

$$H_{SO} = -\frac{e\hbar^2}{2m_e^2c^2r}\frac{d\varphi}{dr}\mathbf{L}\bullet\mathbf{S} = \xi(r)\mathbf{L}\bullet\mathbf{S}, \tag{2.89}$$

where

$$\varphi = \frac{Ze}{4\pi\varepsilon_0 r} \tag{2.90}$$

and

$$\xi(r) = \frac{Ze\hbar^2}{8\pi\varepsilon_0 m_e^2c^2r^3}. \tag{2.91}$$

The energy of the spin–orbit interaction for a quantum state characterized by the quantum numbers n, l, s, J, and m_J is given by

$$\varepsilon_{SO} = \langle nlsJm_J|\xi(r)\hat{\boldsymbol{L}}\bullet\hat{\boldsymbol{S}}|nlsJm_J\rangle = \langle nl|\xi(r)|nl\rangle\langle lsJm_J|\hat{\boldsymbol{L}}\bullet\hat{\boldsymbol{S}}|lsJm_J\rangle, \tag{2.92}$$

where the operator $\hat{\boldsymbol{L}}\bullet\hat{\boldsymbol{S}}$ operates only on the angular variables. The total angular momentum is the sum of the orbital and spin angular momenta,

$$\hat{\boldsymbol{J}}^2 = (\hat{\boldsymbol{L}} + \hat{\boldsymbol{S}})^2 = \hat{\boldsymbol{L}}^2 + 2\hat{\boldsymbol{L}}\bullet\hat{\boldsymbol{S}} + \hat{\boldsymbol{S}}^2. \tag{2.93}$$

Therefore, we can write

$$\hat{\boldsymbol{L}}\bullet\hat{\boldsymbol{S}} = \frac{1}{2}\left(\hat{\boldsymbol{J}}^2 - \hat{\boldsymbol{L}}^2 - \hat{\boldsymbol{S}}^2\right). \tag{2.94}$$

Substituting Eq. (2.94) into Eq. (2.92), we obtain

$$\begin{aligned}\varepsilon_{SO} &= \langle nl|\xi(r)|nl\rangle\langle lsJm_J|\frac{1}{2}\left(\hat{\boldsymbol{J}}^2 - \hat{\boldsymbol{L}}^2 - \hat{\boldsymbol{S}}^2\right)|lsJm_J\rangle \\ &= \frac{\langle\xi(r)\rangle_{nl}}{2}[J(J+1) - l(l+1) - s(s+1)].\end{aligned} \tag{2.95}$$

The term $\langle\xi(r)\rangle_{nl}$ is given by (Weissbluth, 1978)

$$\langle\xi(r)\rangle_{nl} = \frac{Z^4e^2}{8\pi\,\varepsilon_0\,a_0^3\,m_e^2\,c^2\,n^3 l\,(l+1)\left(l+\frac{1}{2}\right)}. \tag{2.96}$$

For the case where $n = 3$, $l = 2$, and $s = 1/2$ for atomic hydrogen ($Z = 1$),

$$\frac{\langle\xi(r)\rangle_{32}\hbar^2}{2} = 1.434 \times 10^{-25}\ J. \tag{2.97}$$

The spin–orbit interaction energy for the different terms is thus given by

$$\begin{aligned}\varepsilon_{SO}\left(3^2D_{5/2}\right) &= \frac{\langle\xi(r)\rangle_{32}\hbar^2}{2}(2) = 2.868 \times 10^{-25}\ J. \\ \varepsilon_{SO}\left(3^2D_{3/2}\right) &= \frac{\langle\xi(r)\rangle_{32}\hbar^2}{2}(-3) = -4.302 \times 10^{-25}\ J.\end{aligned} \tag{2.98}$$

The calculated spin–orbit slitting of the $3^2D_{5/2}$ and $3^2D_{3/2}$ levels is 0.0361 cm^{-1}, in close agreement with experiment. For the same states in doubly ionized lithium (Li^{++}, Z = 3), the calculated and experimental spin–orbit splittings for these two levels are 2.924 cm^{-1} and 2.929 cm^{-1}, respectively. The magnitude of the spin–orbit splitting is 81 times as great in doubly ionized lithium as in atomic hydrogen due to the Z^4 dependence in Eq. (2.96).

Hyperfine splitting is also observed because the proton has an intrinsic spin that interacts with the electron orbital and spin angular momenta. Finally, the Lamb shift results from interaction of the electron with the vacuum fluctuations of the electromagnetic field; the quantized nature of the electromagnetic field must be taken into account for the calculation of the Lamb shift. This is the domain of quantum electrodynamics and is outside the scope of this book.

2.4.3 The Generalized Angular Momentum

The orbital angular momentum arises naturally in the SWE solution for atomic hydrogen. The orbital angular momentum operator relation given in Eq. (2.53) can be used to define a generalized angular momentum operator,

$$\hat{\boldsymbol{J}} \times \hat{\boldsymbol{J}} = i\hbar\hat{\boldsymbol{J}}. \tag{2.99}$$

The generalized angular momentum operator components have the matrix representations

$$\hat{J}_x = \frac{1}{2}\hbar\sigma_x = \frac{1}{2}\hbar\begin{pmatrix} 0 & 1 \\ 1 & 0 \end{pmatrix}. \tag{2.100}$$

$$\hat{J}_y = \frac{1}{2}\hbar\sigma_y = \frac{1}{2}\hbar\begin{pmatrix} 0 & -i \\ i & 0 \end{pmatrix}. \tag{2.101}$$

$$\hat{J}_z = \frac{1}{2}\hbar\sigma_z = \frac{1}{2}\hbar\begin{pmatrix} 1 & 0 \\ 0 & -1 \end{pmatrix}. \tag{2.102}$$

The raising $\hat{J}_+$ and lowering $\hat{J}_-$ operators are also defined:

$$\hat{J}_+ = \hat{J}_x + i\hat{J}_y = \hbar\begin{pmatrix} 0 & 1 \\ 0 & 0 \end{pmatrix}. \tag{2.103}$$

$$\hat{J}_- = \hat{J}_x - i\hat{J}_y = \hbar\begin{pmatrix} 0 & 0 \\ 1 & 0 \end{pmatrix}. \tag{2.104}$$

The square of the angular momentum operator is given by

$$\hat{\boldsymbol{J}}^2 = \hat{J}_x^2 + \hat{J}_y^2 + \hat{J}_z^2 = \frac{3}{4}\hbar^2\begin{pmatrix} 1 & 0 \\ 0 & 1 \end{pmatrix}. \tag{2.105}$$

The results of using the generalized angular momentum operators to analyze the spinors $|\alpha\rangle$ and $|\beta\rangle$ are as follows:

$$\hat{J}_z|\alpha\rangle = \frac{1}{2}\hbar\begin{pmatrix} 1 & 0 \\ 0 & -1 \end{pmatrix}\begin{pmatrix} 1 \\ 0 \end{pmatrix} = \frac{1}{2}\hbar\begin{pmatrix} 1 \\ 0 \end{pmatrix} = \frac{1}{2}\hbar|\alpha\rangle = m_s\hbar|\alpha\rangle. \tag{2.106}$$

$$\hat{J}_z|\beta\rangle = \frac{1}{2}\hbar\begin{pmatrix} 1 & 0 \\ 0 & -1 \end{pmatrix}\begin{pmatrix} 0 \\ 1 \end{pmatrix} = -\frac{1}{2}\hbar\begin{pmatrix} 0 \\ 1 \end{pmatrix} = -\frac{1}{2}\hbar|\beta\rangle = m_s\hbar|\beta\rangle. \tag{2.107}$$

$$\hat{\boldsymbol{J}}^2|\alpha\rangle = \frac{3}{4}\hbar^2\begin{pmatrix} 1 & 0 \\ 0 & 1 \end{pmatrix}\begin{pmatrix} 1 \\ 0 \end{pmatrix} = \frac{3}{4}\hbar^2\begin{pmatrix} 1 \\ 0 \end{pmatrix} = \frac{3}{4}\hbar^2|\alpha\rangle = s(s+1)\hbar^2|\alpha\rangle. \tag{2.108}$$

$$\hat{\boldsymbol{J}}^2|\beta\rangle = \frac{3}{4}\hbar^2\begin{pmatrix} 1 & 0 \\ 0 & 1 \end{pmatrix}\begin{pmatrix} 0 \\ 1 \end{pmatrix} = \frac{3}{4}\hbar^2\begin{pmatrix} 0 \\ 1 \end{pmatrix} = \frac{3}{4}\hbar^2|\beta\rangle = s(s+1)\hbar^2|\beta\rangle. \tag{2.109}$$

It can also be shown that

$$\hat{J}_+|\alpha\rangle = 0 \qquad \hat{J}_+|\beta\rangle = \hbar\,|\alpha\rangle. \tag{2.110}$$

$$\hat{J}_-|\alpha\rangle = \hbar|\beta\rangle \qquad \hat{J}_-|\beta\rangle = 0. \tag{2.111}$$

The operators $\hat{J}_+$, $\hat{J}_-$, and $\hat{J}_z$ can also be represented as differential operators,

$$\hat{J}_+ = \hbar|\alpha\rangle\frac{\partial}{\partial|\beta\rangle} \quad \hat{J}_- = \hbar|\beta\rangle\frac{\partial}{\partial|\alpha\rangle} \quad \hat{J}_z = \frac{1}{2}\hbar|\alpha\rangle\frac{\partial}{\partial|\alpha\rangle} - \frac{1}{2}\hbar|\beta\rangle\frac{\partial}{\partial|\beta\rangle}, \tag{2.112}$$

where

$$\frac{\partial|\alpha\rangle}{\partial|\alpha\rangle} = \frac{\partial|\beta\rangle}{\partial|\beta\rangle} = 1 \quad \frac{\partial|\beta\rangle}{\partial|\alpha\rangle} = \frac{\partial|\alpha\rangle}{\partial|\beta\rangle} = 0. \tag{2.113}$$

The total angular momentum operator can be written as

$$\begin{aligned} \hat{\boldsymbol{J}}^2 &= \hat{J}_x^2 + \hat{J}_y^2 + \hat{J}_z^2 = \left(\hat{J}_+ + \hat{J}_-\right)^2 + \left[i\left(\hat{J}_- - \hat{J}_+\right)^2\right] + \hat{J}_z^2 \\ &= \frac{1}{2}\hat{J}_+\hat{J}_- + \frac{1}{2}\hat{J}_-\hat{J}_+ + \hat{J}_z^2. \end{aligned} \tag{2.114}$$

It can be shown by substituting in Eq. (2.114) using Eq. (2.112) that the differential form of $\hat{\boldsymbol{J}}^2$ is given by

$$\hat{\boldsymbol{J}}^2 = \frac{1}{4}\hbar^2\left(|\alpha\rangle^2\frac{\partial^2}{\partial|\alpha\rangle^2} + 2|\alpha\rangle|\beta\rangle\frac{\partial^2}{\partial|\alpha\rangle\partial|\beta\rangle} + |\beta\rangle^2\frac{\partial^2}{\partial|\beta\rangle^2}\right) + \frac{3}{4}\hbar^2\left(|\alpha\rangle\frac{\partial}{\partial|\alpha\rangle} + |\beta\rangle\frac{\partial}{\partial|\beta\rangle}\right). \tag{2.115}$$

Zare (1988) notes that a generalized wavefunction $|JM\rangle$ can be constructed using $|\alpha\rangle$ and $|\beta\rangle$:

$$|JM\rangle = \frac{|\alpha\rangle^{J+M}|\beta\rangle^{J-M}}{\sqrt{(J+M)!(J-M)!}}. \tag{2.116}$$

The wavefunction $|JM\rangle$ is an eigenfunction of $\hat{\boldsymbol{J}}^2$ and $\hat{J}_z$,

$$\hat{\boldsymbol{J}}^2|JM\rangle = J(J+1)|JM\rangle. \tag{2.117}$$

$$\hat{J}_z|JM\rangle = M|JM\rangle. \tag{2.118}$$

Operating with the raising $\hat{J}_+$ and lowering $\hat{J}_-$ operators on $|JM\rangle$ gives the following results:

$$\hat{J}_+|JM\rangle = \sqrt{J(J+1) - M(M+1)}|JM+1\rangle. \tag{2.119}$$

$$\hat{J}_-|JM\rangle = \sqrt{J(J+1) - M(M-1)}|JM-1\rangle. \tag{2.120}$$

The derivation of Eq. (2.117) is as follows:

$$\begin{aligned}
\hat{j}^2\,|JM\rangle &= \frac{1}{4}\hbar^2\left(|\alpha\rangle^2\frac{\partial^2|JM\rangle}{\partial|\alpha\rangle^2}+2|\alpha\rangle|\beta\rangle\frac{\partial^2|JM\rangle}{\partial|\alpha\rangle\partial|\beta\rangle}+|\beta\rangle^2\frac{\partial^2|JM\rangle}{\partial|\beta\rangle^2}\right)\\
&+\frac{3}{4}\hbar^2\left(|\alpha\rangle\frac{\partial|JM\rangle}{\partial|\alpha\rangle}+|\beta\rangle\frac{\partial|JM\rangle}{\partial|\beta\rangle}\right)\\
&=\frac{1}{4}\hbar^2\left(|\alpha\rangle^2(J+M)(J+M-1)\frac{|\alpha\rangle^{J+M-2}|\beta\rangle^{J-M}}{\sqrt{(J+M)!(J-M)!}}\right.\\
&+2|\alpha\rangle|\beta\rangle(J+M)(J-M)\frac{|\alpha\rangle^{J+M-1}|\beta\rangle^{J-M-1}}{\sqrt{(J+M)!(J-M)!}}\\
&\left.+|\beta\rangle^2(J-M)(J-M-1)\frac{|\alpha\rangle^{J+M}|\beta\rangle^{J-M-2}}{\sqrt{(J+M)!(J-M)!}}\right)\\
&+\frac{3}{4}\hbar^2\left(|\alpha\rangle(J+M)\frac{|\alpha\rangle^{J+M-1}|\beta\rangle^{J-M}}{\sqrt{(J+M)!(J-M)!}}+|\beta\rangle(J-M)\frac{|\alpha\rangle^{J+M}|\beta\rangle^{J-M-1}}{\sqrt{(J+M)!(J-M)!}}\right)\\
&=\frac{1}{4}\hbar^2[(J+M)(J+M-1)+2(J+M)(J-M)+(J-M)(J-M-1)]|JM\rangle\\
&+\frac{3}{4}\hbar^2[(J+M)+(J-M)]|JM\rangle\\
&=\frac{1}{4}\hbar^2\left[J^2+2MJ+M^2-J-M+2J^2-2M^2+J^2-2MJ+M^2-J+M\right]|JM\rangle\\
&+\frac{3}{4}\hbar^2(2J)|JM\rangle\\
&=\frac{1}{4}\hbar^2\left[4J^2-2J\right]|JM\rangle+\frac{3}{4}\hbar^2(2J)|JM\rangle=\hbar^2J(J+1)|JM\rangle. \qquad (2.121)
\end{aligned}$$

Equations (2.118)–(2.120) can be proved using a similar analysis.

2.5 Multielectron Atoms

The discussion of the structure of multielectron atoms in this section is based on the Pauli exclusion principle and the results of approximate solutions of the SWE for multielectron atoms. The structure of multielectron atoms is discussed in great detail in many excellent textbooks (e.g., Cowan, 1981; Shore & Menzel, 1968; Slater, 1960a, 1960b; Sobelman 1992; Weissbluth, 1978).

2.5.1 Electronic Configurations for Multielectron Atoms

The Pauli exclusion principle is developed from a consideration of the symmetry properties of the wavefunction for fermions. The spin-$\frac{1}{2}$ electrons are fermions, and for a system such as a multielectron atom where these electrons interact, no two electrons may occupy the same quantum state. The electron structure of multielectron atoms is greatly complicated by the Pauli exclusion principle; if the Pauli exclusion principle did not apply, all electrons in the atom would tend to exist in the lowest possible energy state.

The results of approximate solutions of the SWE for multielectron atoms indicate that the same set of quantum numbers $(n,\ l,\ m_l,\ m_s)$ that apply to the hydrogen atom also have meaning for a multielectron atom. Any individual electron in the multielectron atom will experience a potential function that is approximately spherically symmetric. The attractive potential due to the nucleus will be spherically symmetric, of course, but the average repulsive potential due to the other electrons in the atom will also be approximately spherically symmetric. This is the central field approximation. The development of the Hamiltonian for multielectron atoms is discussed in detail by Weissbluth (1978). Briefly, for atoms with low Z, the Hamiltonian may be written as the sum of two terms, the Coulomb attraction terms between the individual electrons and the nucleus, and the Coulomb repulsion terms between pairs of electrons. The spin–orbit interaction term is regarded as a perturbation for low-Z atoms. For high-Z atoms, the spin–orbit interaction is dominant compared to the electron repulsion term, and the electron repulsion term is regarded as a perturbation.

For given values of n, the radial part of the wavefunction exhibits maxima at nearly the same value of r. Electrons that have the same value of n are referred to as a **shell**. For a hydrogen atom, the energy is dependent only on the quantum number n if effects such as the spin–orbit interaction are neglected. For a multielectron atom, the energy of each electron depends on both quantum numbers n and l because the potential $V_i(r)$ that each electron sees is only approximately a centrally symmetric Coulomb potential. The configuration $n_i l_i^{N_i} n_j l_j^{N_j} \ldots$ for a multielectron atom is used to designate the principal quantum numbers n and l and the number of electrons in each subshell. The shell (n) and subshell (l) designations for the state of the electrons are given by

Shell designation:	K	L	M	$N\ldots$
n:	1	2	3	4...

Subshell designation:	s	p	d	$f\ldots$
l:	0	1	2	3

There are $2(2l+1)$ states associated with each subshell.

To determine the configuration of electrons in the ground state of a multielectron atom, two rules must be followed. First, the electrons must occupy the available subshells of lowest energy; from experiment, the energy ordering is

$$1s, 2s, 2p, 3s, 3p, 4s, 3d, 4p, 5s, 4d, 5p, 6s, \ldots.$$

Second, the maximum number of electrons that may occupy any subshell is $2(2l+1)$, determined by the Pauli exclusion principle. The configurations of the ground states of several atoms are as follows:

H	$1s^1$	(superscript gives number of electrons in a subshell)
He	$1s^2$	
F	$1s^2 2s^2 2p^5$	
Ne	$1s^2 2s^2 2p^6$	
Na	$1s^2 2s^2 2p^6 3s$	

2.5.2 Approximate Wavefunctions for Multielectron Atoms

The theoretical development of the computation of wavefunctions for multielectron atoms is discussed in great detail in Cowan (1981). Here the basic results of that treatment will be summarized, with the goal of later providing insight into the selection rules and intensities of radiative transitions in multielectron atoms. The Hamiltonian for the multielectron atom is given by (Cowan, 1981)

$$\begin{aligned} H &= H_{kin} + H_{elec-nucl} + H_{elec-elec} + H_{SO} + \cdots \\ &= -\sum_i \nabla_i^2 - \sum_i \frac{2Z}{4\pi\varepsilon_0 r_i} + \sum_{i>j}\sum_j \frac{2}{4\pi\varepsilon_0 r_{ij}} + \sum_i \xi(r_i)(\boldsymbol{l}_i \bullet \boldsymbol{s}_i) + \cdots . \end{aligned} \tag{2.122}$$

The Hamiltonian in Eq. (2.122) includes terms for kinetic energy, Coulomb attraction between the nucleus and the ith electron, Coulomb repulsion between the ith and jth electrons, and the spin–orbit interaction.

The basic idea is that the wavefunction for individual electrons in a multielectron atom will have the same form as the wavefunction for an electron in a single-electron atom, except for the radial component. Therefore, we can write, for the ith with quantum numbers $(n_i,\ l_i,\ m_{li},\ m_{si})$ in the multielectron atom,

$$\varphi(\xi_i) = \phi(\boldsymbol{r}_i)\zeta(m_{si}) = R_{n_i l_i}(r_i) Y_{l_i m_{li}}(\theta_i, \varphi_i)\zeta(m_{si}), \tag{2.123}$$

where the symbol ξ_i in this case is taken to represent both the position $\boldsymbol{r}_i$ and the spin orientation m_{si} of the ith electron. The spherical harmonic function $Y_{l_i m_{li}}(\theta_i, \varphi_i)$ and the spinor $\zeta(m_{si})$ will have same form as for the one-electron atom, because it is assumed that the time-averaged field experienced by an individual electron will be spherically symmetric; this is the central field approximation. The radial function $R_{n_i l_i}(r_i)$ will be different because of the addition of the electron repulsion term in the multielectron Hamiltonian and can only be determined by numerical computation. The wavefunctions $\varphi(\xi_i)$ will be orthonormal,

$$\langle \varphi(\xi_i), \varphi(\xi_j) \rangle = \delta_{ij} = \left\langle \varphi_{nlm_l m_s}, \varphi_{n'l'm_l'm_s'} \right\rangle = \delta_{nn'}\delta_{ll'}\delta_{m_l m_l'}\delta_{m_s m_s'}. \tag{2.124}$$

The overall wavefunction for the multielectron atom can then be written as

$$\psi = \varphi_1(\xi_1)\varphi_2(\xi_2)\varphi_3(\xi_3)\varphi_4(\xi_4)\cdots. \tag{2.125}$$

However, the wavefunction for the multielectron atom must take into account the indistinguishability of electrons. Electrons are fermions, and the wavefunction must be antisymmetric with respect to the exchange of any two electrons. Therefore, for an atom with two electrons, the appropriate wavefunction would be

$$\Psi = \frac{1}{\sqrt{2}}[\varphi_1(\xi_1)\varphi_2(\xi_2) - \varphi_1(\xi_2)\varphi_2(\xi_1)], \tag{2.126}$$

where the factor of $1/\sqrt{2}$ is necessary for the to fulfill the normalization condition $\langle\Psi|\Psi\rangle = 1$. Upon the exchange of the coordinates for the two electrons, the wavefunction becomes

$$\Psi' = \frac{1}{\sqrt{2}}[\varphi_1(\xi_2)\varphi_2(\xi_1) - \varphi_1(\xi_1)\varphi_2(\xi_2)] = -\Psi, \tag{2.127}$$

so that the wavefunction is antisymmetric upon exchange of the electrons. In general, the wavefunction for a multielectron atom with N electrons can be written as a Slater determinant (Cowan, 1981),

$$\Psi = \frac{1}{\sqrt{N!}}\begin{vmatrix} \varphi_1(\xi_1) & \varphi_1(\xi_2) & \varphi_1(\xi_3) & \varphi_1(\xi_4) & \cdots \\ \varphi_2(\xi_1) & \varphi_2(\xi_2) & \varphi_2(\xi_3) & \varphi_2(\xi_4) & \cdots \\ \varphi_3(\xi_1) & \varphi_3(\xi_2) & \varphi_3(\xi_3) & \varphi_3(\xi_4) & \cdots \\ \varphi_4(\xi_1) & \varphi_4(\xi_2) & \varphi_4(\xi_3) & \varphi_4(\xi_4) & \cdots \\ \vdots & \vdots & \vdots & \vdots & \ddots \end{vmatrix}. \tag{2.128}$$

For a multielectron atom with $N = 3$, the wavefunction is given by

$$\begin{aligned}\Psi = \frac{1}{\sqrt{6}}[&\varphi_1(\xi_1)\varphi_2(\xi_2)\varphi_3(\xi_3) - \varphi_1(\xi_1)\varphi_2(\xi_3)\varphi_3(\xi_2) - \varphi_1(\xi_2)\varphi_2(\xi_1)\varphi_3(\xi_3) \\ &+ \varphi_1(\xi_2)\varphi_2(\xi_3)\varphi_3(\xi_1) + \varphi_1(\xi_3)\varphi_2(\xi_1)\varphi_3(\xi_2) - \varphi_1(\xi_3)\varphi_2(\xi_2)\varphi_3(\xi_1)].\end{aligned} \tag{2.129}$$

2.5.3 Angular Momentum Coupling and Spectroscopic Term Classification for Multielectron Atoms

LS Coupling

Angular momentum coupling is more complicated in multielectron atoms than in hydrogen atoms. In a multielectron atom each electron can be assigned a set of quantum numbers (n, l, m_l, m_s). However, for given values of n and l, different spectroscopic terms can be observed based on the different ways in which the orbital and spin angular momenta of the individual electrons couple.

Two limiting cases of angular momentum coupling observed in atoms are Russell–Saunders or *LS* coupling and *jj* coupling. For the light atoms that are typical of combustion media, the energy level structure (and, as will be discussed in Chapter 4, the radiative transitions) are described by *LS* coupling. In general, for the *LS* coupling case, the orbital angular momenta for each of the individual electrons add, resulting in a total orbital angular momentum $\boldsymbol{L}$, and the spin angular momenta for each of the individual electrons add, resulting in a total spin angular momentum $\boldsymbol{S}$. The total orbital and spin angular momenta then add to give the total angular momentum $\boldsymbol{J}$:

$$\boldsymbol{L} = \sum \boldsymbol{l}_i \quad \boldsymbol{S} = \sum \boldsymbol{s}_i \quad \boldsymbol{J} = \boldsymbol{L} + \boldsymbol{S}. \tag{2.130}$$

The quantum numbers J, L, and S (total, orbital, spin) are defined, and the magnitudes of the angular momenta are given by

$$|\boldsymbol{L}| = \sqrt{L(L+1)}\hbar \quad |\boldsymbol{S}| = \sqrt{S(S+1)}\hbar \quad |\boldsymbol{J}| = \sqrt{J(J+1)}\hbar. \tag{2.131}$$

Table 2.2. Analysis of allowed quantum states for a p^2 configuration of equivalent electrons (the electrons are in the same subshell)

m_{l1}	m_{s1}	m_{l2}	m_{s2}	M_L	M_S	$M = M_L + M_S$
−1	−1/2	−1	+1/2	−2	0	−2
		0	+1/2	−1	0	−1
		0	−1/2	−1	−1	−2
		+1	+1/2	0	0	0
		+1	−1/2	0	−1	−1
−1	+1/2	0	+1/2	−1	+1	0
		0	−1/2	−1	0	−1
		+1	+1/2	0	+1	+1
		+1	−1/2	0	0	0
0	−1/2	0	+1/2	0	0	0
		+1	+1/2	+1	0	+1
		+1	−1/2	+1	−1	0
		+2	+1/2	+2	0	+2
		+2	−1/2	+2	−1	+1
0	+1/2	+1	+1/2	+1	+1	+2
		+1	−1/2	+1	0	+1
+1	−1/2	+1	+1/2	+2	0	+2

The quantum number J can take on the following values:

$$J = (L+S),\quad (L+S-1),\quad \ldots \,|\,L-S\,|\,. \tag{2.132}$$

If $L < S$, there will be $2L + 1$ possible values of J. If $S < L$, there will be $2S + 1$ values of J. The number of values of J is referred to as the multiplicity. The electronic energy levels of atoms are referred to as **terms** and are designated by the term symbol

$$^{2S+1}L_J.$$

The symbol n is the shell designation. The value for the orbital angular momentum L is designated this time by capital letters –

Symbol:	S	P	D	F	$G\ldots$
L:	0	1	2	3	4...

as compared to the lowercase letters used to designate a particular electron configuration.

As discussed in detail by Cowan (1981), some terms that might exist in LS coupling are forbidden by the Pauli exclusion principle for equivalent electrons $n_i l_i^{N_i}$. Cowan analyzes the case where the valence electrons have an np^2 configuration, as is the case for the ground state of the neutral carbon atom. The procedure outlined by Cowan (1981) is illustrated here for an np^2 configuration in Table 2.2. The possible values of m_l are ± 1 and 0, and the possible values of m_s are $\pm\frac{1}{2}$.

Counting the number occurrences of different (M_L, M_S) pairs in Table 2.2, we can then construct the following matrix:

M_L \ M_S	−1	0	+1
−2		1	1
−1	1	2	1
0	1	3	1
+1	1	2	1
+2		1	1

This matrix in turn can be represented as the sum of the following submatrices:

M_L \ M_S	−1	0	+1
−2		1	
−1		1	
0		1	
+1		1	
+2		1	

M_L \ M_S	−1	0	+1
−2			
−1	1	1	1
0	1	1	1
+1	1	1	1
+2			

M_L \ M_S	−1	0	+1
−2			
−1			
0		1	
+1			
+2			

These submatrices, in order, correspond to ^1D, ^3P, and ^1S terms. The terms ^3D, ^1P, and ^3S, which might occur if the p electrons were in different subshells,

Table 2.3 Level structure for atomic nitrogen (Kramida et al., 2022)

Configuration	Term	J	Level energy (cm^{-1})
$2s^2 2p^3$	$^4S^{\circ}$	3/2	0.000
$2s^2 2p^3$	$^2D^{\circ}$	5/2	19,224.464
		3/2	19,233.177
$2s^2 2p^3$	$^2P^{\circ}$	1/2	28,838.920
		3/2	28,839.306
$2s^2 2p^2(^3P)3s$	^4P	1/2	83,284.070
		3/2	83,317.830
		5/2	83,364.620
$2s^2 2p^2(^3P)3s$	^2P	1/2	86,137.350
		3/2	86,220.510
$2s2p^4$	^4P	5/2	88,107.260
		3/2	88,151.170
		1/2	88,170.570
$2s^2 2p^2(^3P)3p$	$^2S^{\circ}$	1/2	93,581.550
$2s^2 2p^2(^3P)3p$	$^4D^{\circ}$	1/2	94,770.889
		3/2	94,793.490
		5/2	94,830.890
		7/2	94,881.820
$2s^2 2p^2(^3P)3p$	$^4P^{\circ}$	1/2	95,475.310
		3/2	95,493.690
		5/2	95,532.150
$2s^2 2p^2(^3P)3p$	$^4S^{\circ}$	3/2	96,750.840
$2s^2 2p^2(^3P)3p$	$^2D^{\circ}$	3/2	96,787.680
		5/2	96,864.050
$2s^2 2p^2(^3P)3p$	$^2P^{\circ}$	1/2	97,770.180
		3/2	97,805.840
$2s^2 2p^2(^1D)3s$	^2D	5/2	99,663.427
		3/2	99,663.912

are forbidden by the Pauli exclusion principle. A similar analysis is performed in Appendix A6 for the d^2 equivalent electron case. For the d^2 configuration, the allowed terms are the ^1G, ^3F, ^1D, ^3P, and ^1S terms. A complete listing of allowed terms for subshells with equivalent electrons is given in table 4-2 in Cowan (1981).

An excellent source for tables of atomic energy levels is the NIST Atomic Spectra Database (Kramida et al., 2022). The NIST tables contain information both for neutral atoms and for ionized atoms. An example of the type of information contained in these tables is shown in Table 2.3, which contains information on the 13 lowest-energy terms of atomic nitrogen. For atomic nitrogen, terms up to energies of 110,000 cm^{-1} are described well by *LS* coupling. Table 2.3 lists the electron configuration, the term classification, the total angular momentum quantum number *J* for each level, and the energy for each level. For electron configurations with two open subshells, the *LS* term configuration of the lower energy subshell is also indicated. The symbol $^{\circ}$ to the right of the term symbol indicates that the energy levels in the term have odd parity; if the symbol is absent, the levels have even parity. The parity of the levels is determined by the sum of the orbital angular momentum quantum numbers in the electron configuration $n_i l_i^{N_i} n_j l_j^{N_j} \ldots$. If the sum $N_i l_i + N_j l_j + \cdots$ is even, then the parity of each of the levels associated with that configuration is even; if the sum is odd, then the parity of each level is odd.

jj Coupling

For very heavy atoms, the spin–orbit interaction term in Eq. (2.122) becomes strong relative to the Coulombic repulsion term, and *jj* coupling may describethe energy level structure more accurately than *LS* coupling. For *jj* coupling, the orbital angular momentum $\boldsymbol{l}_i$ and the spin angular momentum $\boldsymbol{s}_i$ first couple to give an angular momentum $\boldsymbol{j}_i$ for each electron, and then the individual $\boldsymbol{j}_i$ couple to give a total angular momentum $\boldsymbol{J}$ for the electron configuration. For *jj* coupling, the set of quantum numbers that apply is $(n,\ l,\ j,\ m_j)$, and no two electrons in the atom can have the same set of these quantum numbers according to the Pauli principle. For a two-electron configuration, the *jj* coupling term symbol is given by $[j_1, j_2]_J$. For electron configurations with more than two electrons, the coupling can occur in arbitrary order (Cowan, 1981). For example, for a configuration with three electrons outside of closed subshells, the coupling order might be

$$\begin{aligned} &\boldsymbol{l}_i + \boldsymbol{s}_i = \boldsymbol{j}_i && i = 1,2,3 \\ &\boldsymbol{j}_1 + \boldsymbol{j}_2 = \boldsymbol{J}' && \boldsymbol{j}_3 + \boldsymbol{J}' = \boldsymbol{J} \\ \text{or } &\boldsymbol{j}_1 + \boldsymbol{j}_3 = \boldsymbol{J}' && \boldsymbol{j}_2 + \boldsymbol{J}' = \boldsymbol{J} \\ \text{or } &\boldsymbol{j}_2 + \boldsymbol{j}_3 = \boldsymbol{J}' && \boldsymbol{j}_1 + \boldsymbol{J}' = \boldsymbol{J}. \end{aligned} \tag{2.133}$$

As discussed by Cowan (1981), *jj* coupling is very important in atomic nuclei but applies for a few very rare cases for the electron structure for high-*Z* atoms.

Pair Coupling

Pair coupling is a type of coupling that is much more common than *jj* coupling for atoms. Pair coupling is discussed in detail by Cowan (1981) and Martin et al. (2002). One common type of pair coupling is *jK* coupling, in which two electrons exist outside of closed subshells. One inner electron is in general much more tightly bound than the outer electron, and so the most significant energy interaction is the spin–orbit interaction of the more tightly bound electron. The spin–orbit interaction for the outer electron is generally small as the spin–orbit interaction energy is approximately proportional to l^{-3} according to Eq. (2.96) for atomic hydrogen. The coupling scheme for the inner electron 1 and the outer electron 2 is as follows:

$$\begin{aligned} \boldsymbol{l}_1 + \boldsymbol{s}_1 &= \boldsymbol{j}_1. \\ \boldsymbol{j}_1 + \boldsymbol{l}_2 &= \boldsymbol{K}. \\ \boldsymbol{K} + \boldsymbol{s}_2 &= \boldsymbol{J}. \end{aligned} \tag{2.134}$$

According to Cowan (1981), *jK* pair coupling is characteristic of excited electron configurations for noble gases and for "carbon group elements (C, Si, Ge, Sn, Pb)."

Table 2.4 Structure of energy levels for the $1s^22s^22p5d$ and $1s^22s^22p5f$ electron configurations for atomic carbon (Kramida et al., 2022)

Configuration	Term	J	Level energy (cm^{-1})
$2s^22p5d$	$^1D^\circ$	2	86,185.233
$2s^22p5d$	$^3F^\circ$	2	86,317.69
		3	86,327.186
		4	86,369.385
$2s^22p5d$	$^3D^\circ$	1	86,362.54
		2	86,389.425
		3	86,397.83
$2s2p\left(^2P^\circ_{1/2}\right)5f$	$^2[5/2]$	3	86,412.030
		2	86,412.131
$2s2p\left(^2P^\circ_{1/2}\right)5f$	$^2[7/2]$	3	86,414.566
		4	86,414.766
$2s^22p5d$	$^1F^\circ$	3	86,449.208
$2s2p\left(^2P^\circ_{3/2}\right)5f$	$^2[7/2]$	3	86,469.544
		4	86,469.725
$2s2p\left(^2P^\circ_{3/2}\right)5f$	$^2[5/2]$	3	86,482.673
		2	86,482.794
$2s2p\left(^2P^\circ_{3/2}\right)5f$	$^2[9/2]$	5	86,488.024
		4	86,488.356
$2s^22p5d$	$^1P^\circ$	1	86,491.441
$2s2p\left(^2P^\circ_{3/2}\right)5f$	$^2[3/2]$	1	86,498.523
		2	86,498.753
$2s^22p5d$	$^3P^\circ$	0	86,506.713
		1	86,519.506
		2	86,523.19

Examination of the NIST tables for atomic carbon (C I) shows numerous examples of jK coupling for excited electron configurations. The energy levels for the excited electron configurations $1s^22s^22p5d$ and $1s^22s^22p5f$ are listed in Table 2.4, and the energy level structure is depicted schematically in Figure 2.3. The energy level structure for the electron configuration $1s^22s^22p5d$ is characteristic of LS coupling, while the energy level structure of $1s^22s^22p5f$ is characteristic of jK pair coupling. The difference in the energy level structure between the two configurations is quite striking.

2.6 The Rotation Operator and the Rotation Matrix

A further discussion of angular momentum and operators associated with angular momentum is warranted because of the importance of these operators in the

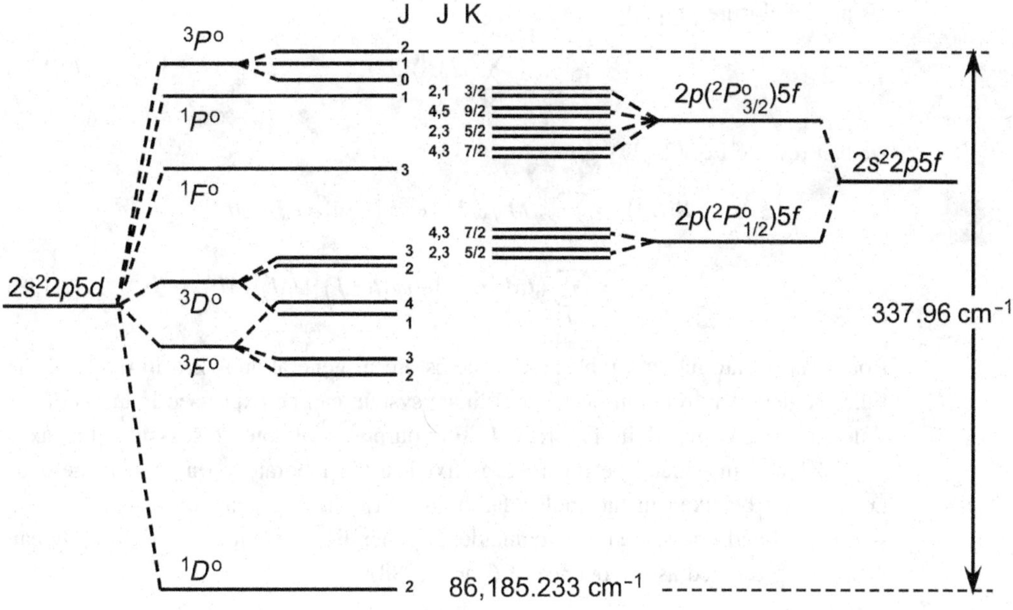

Figure 2.3 Energy level structure of the $1s^22s^22p5d$ and $1s^22s^22p5f$ electron configurations for atomic carbon. The energy level spacings shown in the figure are approximate. For example, the splitting of the energy levels with the same value of K but different values of J for the $1s^22s^22p5f$ electron configuration is greatly exaggerated for clarity.

calculation of the energy level structure in atoms and molecules and the interaction of laser radiation with resonance transitions. A full discussion is outside the scope of this work, but there are many excellent books on the subject of angular momentum in quantum mechanics (Edmonds, 1960; Rose, 1957; Zare, 1988).

2.6.1 The Rotation Operator and Euler's Angles

The rotation operator $\hat{R}$ is defined in terms of the transformation of a wavefunction as the coordinate system is rotated through an angle φ about a unit vector $\boldsymbol{n}$,

$$\psi' = \hat{R}(\boldsymbol{n}, \varphi)\psi. \tag{2.135}$$

Rose (1957) shows that the rotation operator is unitary and can be written in terms of the angular momentum operator

$$\hat{R}(\boldsymbol{n}, \varphi) = \exp\left(-i\varphi\boldsymbol{n} \cdot \hat{\boldsymbol{J}}\right) \tag{2.136}$$

and that in fact this can serve as a means of defining the angular momentum operator. The rotation matrix is introduced by considering the properties of a wavefunction $|JM\rangle$ under rotation,

$$\hat{R}|JM\rangle = \exp\left(-i\varphi\boldsymbol{n} \cdot \hat{\boldsymbol{J}}\right)|JM\rangle. \tag{2.137}$$

Using the closure property

$$1 = \sum_{M'} |JM'\rangle\langle JM'|, \tag{2.138}$$

we can rewrite Eq. (2.137) as

$$\begin{aligned}\hat{R}|JM\rangle &= \sum_{M'} |JM'\rangle\langle JM'| \exp\left[-i\varphi(\boldsymbol{n}\bullet\hat{\boldsymbol{J}})\right]|JM\rangle \\ &= \sum_{M'} \langle JM'| \exp\left[-i\varphi(\boldsymbol{n}\bullet\hat{\boldsymbol{J}})\right]|JM\rangle|JM'\rangle.\end{aligned} \tag{2.139}$$

Rotations for atomic and molecular systems are in general analyzed in terms of the Euler angles. Any rotation of the coordinate system can be expressed in terms of the Euler angles, depicted in Figure 2.4. For purposes of later discussion, the axes (X, Y, Z) will in general be regarded as fixed in the laboratory frame, and the axes (x, y, z) will be fixed in the molecular frame. Prior to the generalized rotation, the space-fixed and body-fixed axes coincide. A generalized rotation of a rigid body can then be represented as a sequence of (Zare, 1988)

(1) a rotation through an angle ϕ about the vertical axis Z
(2) a rotation through an angle θ about the line of nodes N
(3) a rotation through an angle χ about the new vertical axis z

where the rotations as shown are positive and counterclockwise. In Rose (1957) the angles $(\alpha\beta\gamma)$ are used in place of $(\phi\theta\chi)$. The rotation operator can thus be written as

$$\begin{aligned}\hat{R} &= \exp(-i\chi\hat{\boldsymbol{n}}_z\bullet\hat{\boldsymbol{J}})\exp(-i\theta\hat{\boldsymbol{n}}_N\bullet\hat{\boldsymbol{J}})\exp(-i\phi\hat{\boldsymbol{n}}_Z\bullet\hat{\boldsymbol{J}}) \\ &= \exp(-i\chi J_z)\exp(-i\theta J_N)\exp(-i\phi J_Z).\end{aligned} \tag{2.140}$$

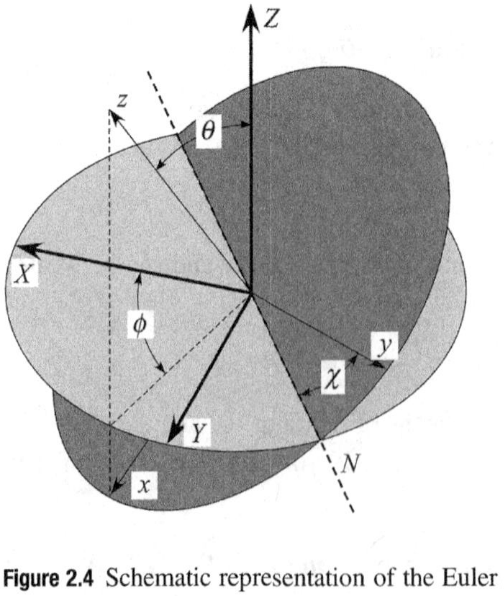

Figure 2.4 Schematic representation of the Euler angles.

It is shown in Zare (1988) and Rose (1957) that the rotation operator can also be expressed strictly in terms of rotations about the axes in the laboratory frame of reference:

$$\begin{aligned}\hat{R} &= \exp\left(-i\phi\hat{\boldsymbol{n}}_Z \bullet \hat{\boldsymbol{J}}\right)\exp\left(-i\theta\hat{\boldsymbol{n}}_N \bullet \hat{\boldsymbol{J}}\right)\exp\left(-i\chi\hat{\boldsymbol{n}}_Z \bullet \hat{\boldsymbol{J}}\right) \\ &= \exp\left(-i\phi J_Z\right)\exp\left(-i\theta J_Y\right)\exp\left(-i\chi J_Z\right).\end{aligned} \tag{2.141}$$

From Figure 2.4, it is apparent that a generalized rotation can also be represented by

(1) a rotation through an angle χ about the fixed-space vertical axis Z
(2) a rotation through an angle θ about the fixed-space axis Y
(3) a rotation through an angle ϕ about the fixed-space vertical axis Z

Now consider the effect of the rotation operator as expressed in Eq. (2.141) on the matrix elements introduced in Eq. (2.139):

$$\langle JM'|\hat{R}(\phi,\theta,\chi)|JM\rangle = \langle JM'|\exp(-i\phi J_Z)\exp(-i\theta J_Y)\exp(-i\chi J_Z)|JM\rangle. \tag{2.142}$$

We can evaluate this expression by expanding the exponential form of the elements of the rotation operator:

$$\begin{aligned}\exp\left(-i\chi J_Z\right)|JM\rangle &= \sum_{j=0}^{\infty}\frac{1}{j!}(-i\chi)^j(J_Z)^j|JM\rangle \\ &= \sum_{j=0}^{\infty}\frac{1}{j!}(-i\chi M)^j|JM\rangle \\ &= \exp\left(-i\chi M\right)|JM\rangle.\end{aligned} \tag{2.143}$$

Therefore Eq. (2.142) becomes

$$\langle JM'|\hat{R}(\phi,\theta,\chi)|JM\rangle = \exp(-i\chi M)\langle JM'|\exp(-i\phi J_Z)\exp(-i\theta J_Y)|JM\rangle. \tag{2.144}$$

The unitary rotation operator $\exp(-i\phi J_Z)$ is also a Hermitian operator. Using Eq. (2.6) we can rewrite Eq. (2.144) as

$$\begin{aligned}\langle JM'|\hat{R}(\phi,\theta,\chi)|JM\rangle &= \left\langle \exp\left(-i\theta J_Y\right)JM\right|\exp\left(i\phi J_Z\right)|JM'\rangle^*\exp\left(-i\chi M\right) \\ &= \exp\left(-i\phi M'\right)\left\langle \exp\left(-i\theta J_Y\right)JM|JM'\right\rangle^*\exp\left(-i\chi M\right) \\ &= \exp\left(-i\phi M'\right)\langle JM'|\exp\left(-i\theta J_Y\right)|JM\rangle\exp\left(-i\chi M\right).\end{aligned} \tag{2.145}$$

The rotation matrix operator is defined as

$$\begin{aligned}D^J_{M'M}(\phi,\theta,\chi) &= \exp\left(-i\phi M'\right)\langle JM'|\exp\left(-i\theta J_Y\right)|JM\rangle\exp\left(-i\chi M\right) \\ &= \exp\left(-i\phi M'\right)d^J_{M'M}(\theta)\exp\left(-i\chi M\right),\end{aligned} \tag{2.146}$$

where $d^J_{M'M}(\theta)$ is given by

$$\begin{aligned}d^J_{M'M}(\theta) &= [(J+M)!(J-M)!(J+M')!(J-M')!] \\ &\times\sum_{j=0}^{\infty}\left\{\frac{(-1)^j}{(J-M'-j)!(J+M-j)!(j+M'-M)!j!}\right. \\ &\times\left.\left[\cos\left(\frac{\theta}{2}\right)\right]^{2J+M-M'-2j}\left[-\sin\left(\frac{\theta}{2}\right)\right]^{M'-M+2j}\right\}.\end{aligned} \tag{2.147}$$

The θ-rotation operator $d^J_{M'M}(\theta)$ is a real quantity and is discussed in detail in Rose (1957), Edmonds (1960), and Zare (1988). From analysis of Eq. (2.147) and related expressions, it can be shown that

$$d^J_{M'M}(\theta) = d^J_{MM'}(-\theta) = (-1)^{M-M'} d^J_{MM'}(\theta). \tag{2.148}$$

It can also be shown that when the signs of M and M' are changed and the quantities are interchanged, then

$$d^J_{M'M}(\theta) = d^J_{-M,-M'}(\theta) = (-1)^{M'-M} d^J_{-M',-M}(\theta). \tag{2.149}$$

Using Eq. (2.149) we can also write

$$\begin{aligned} D^{J*}_{MM'}(\phi,\theta,\chi) &= \exp(i\phi M) d^J_{MM'}(\theta) \exp(i\chi M') \\ &= \exp(i\phi M)(-1)^{M-M'} d^J_{-M,-M'}(\theta) \exp(i\chi M') \\ &= (-1)^{M-M'} D^J_{-M,-M'}(\phi,\theta,\chi). \end{aligned} \tag{2.150}$$

Inverting Eq. (2.150) we find

$$D^J_{MM'}(\phi,\theta,\chi) = (-1)^{M'-M} D^{J*}_{-M,-M'}(\phi,\theta,\chi). \tag{2.151}$$

An alternative expression for $d^J_{M'M}(\theta)$ that will be useful for evaluating some important integrals of $D^J_{M'M}(\phi,\theta,\chi)$ is given by Edmonds (1960):

$$d^J_{M'M}(\theta) = \left[\frac{(J+M')!(J-M')!}{(J+M)!(J-M)!}\right]^{1/2} \left[\cos\left(\frac{\theta}{2}\right)\right]^{M'+M} \left[\sin\left(\frac{\theta}{2}\right)\right]^{M'-M} P^{(M'-M,M'+M)}_{J-M'}(\cos\theta). \tag{2.152}$$

The Jacobi polynomial $P^{(\alpha,\beta)}_n(x)$ is given by

$$P^{(\alpha,\beta)}_n(x) = \frac{(-1)^n}{2^n n!}(1-x)^{-\alpha}(1+x)^{-\beta} \frac{d^n\left[(1-x)^{\alpha+n}(1+x)^{\beta+n}\right]}{dx^n}. \tag{2.153}$$

The Jacobi polynomial is normalized such that

$$\begin{aligned} &\int_{-1}^{1} (1-x)^\alpha (1+x)^\beta P^{(\alpha,\beta)}_n(x) P^{(\alpha,\beta)}_m(x) dx \\ &= \frac{2^{\alpha+\beta+1}}{2n+\alpha+\beta+1} \frac{\Gamma(n+\alpha+1)\Gamma(n+\beta+1)}{\Gamma(n+1)\Gamma(n+\alpha+\beta+1)} \delta_{nm}. \end{aligned} \tag{2.154}$$

The Jacobi polynomial $P^{(M,M)}_{L-M}(x)$ can be written in terms of the associated Legendre function $P^{-M}_L(x)$:

$$P^{(M,M)}_{L-M}(x) = (-2)^M \frac{L!}{(L-M)!}(1-x^2)^{-M/2} P^{-M}_L(x). \tag{2.155}$$

Therefore, we can express the θ-rotation operator $d^L_{M0}(\theta)$ as

$$d^L_{M0}(\theta) = (-1)^M \left[\frac{(L+M)!}{(L-M)!}\right]^{1/2} P^{-M}_L(\cos\theta) = \left[\frac{(L-M)!}{(L+M)!}\right]^{1/2} P^M_L(\cos\theta). \tag{2.156}$$

Combining Eqs. (2.61), (2.146), and (2.156), we obtain the following relation between the rotation matrix element $D^L_{M0}(\phi,\theta,\chi)$ and the spherical harmonic $Y_{LM}(\phi,\theta)$:

$$D^L_{M0}(\phi,\theta,\chi) = (-1)^M\left[\frac{4\pi}{(2L+1)}\right]^{1/2} Y^*_{LM}(\phi,\theta). \tag{2.157}$$

The inverse of the rotation operator is given by the product of the rotation operators in inverse order,

$$\hat{R}^{-1} = \exp(+i\chi J_Z)\exp(+i\theta J_Y)\exp(+i\phi J_Z). \tag{2.158}$$

Because the rotation operator is a unitary Hermitian operator, we can write

$$\langle JM'|\hat{R}^{-1}(-\chi,-\theta,-\phi)|JM\rangle = \langle JM|\hat{R}(\phi,\theta,\chi)|JM'\rangle^*. \tag{2.159}$$

Thus, we can write

$$\begin{aligned} D^J_{M'M}(-\chi,\,-\theta,\,-\phi) &= \exp(i\phi M')\langle JM'|\exp(i\theta J_Y)|JM\rangle\exp(i\chi M) \\ &= \exp(i\phi M')d^J_{M'M}(-\theta)\exp(i\chi M). \end{aligned} \tag{2.160}$$

Using Eq. (2.148) we can write

$$D^J_{M'M}(-\chi,-\theta,-\phi) = \exp(i\chi M)d^J_{MM'}(\theta)\exp(i\phi M') = D^{J*}_{MM'}(\phi,\theta,\chi). \tag{2.161}$$

2.6.2 The Rotation Matrices for Coupled States

The coupling of two states in the uncoupled representation $|J_1M_1\rangle$ and $|J_2M_2\rangle$ to produce a resultant state $|J_3M_3\rangle$ in the coupled representation is given by

$$|J_3M_3\rangle = \sum_{M_1M_2}\langle J_1M_1,J_2M_2|J_3M_3\rangle|J_1M_1\rangle|J_2M_2\rangle, \tag{2.162}$$

where the Clebsch–Gordan coefficient $\langle J_1M_1,J_2M_2|J_3M_3\rangle$ has been introduced in Eq. (2.84). In Eq. (2.162), the Clebsch–Gordan coefficient is zero unless $M_1+M_2=M_3$. Similarly, we can write the states in the uncoupled representation as

$$|J_1M_1\rangle|J_2M_2\rangle = \sum_{J_3}\langle J_3M_3|J_1M_1,J_2M_2\rangle|J_3M_3\rangle, \tag{2.163}$$

where again $M_1+M_2=M_3$. Applying a rotation $\widehat{\boldsymbol{R}}$ to both sides of Eq. (2.163), we obtain

$$\begin{aligned} &\sum_{N_1}\sum_{N_2} D^{J_1}_{N_1M_1}(\omega)|J_1N_1\rangle D^{J_2}_{N_2M_2}(\omega)|J_2N_2\rangle \\ &\quad = \sum_{J_3}\sum_{N_3}\langle J_3M_3|J_1M_1,J_2M_2\rangle D^{J_3}_{N_3M_3}(\omega)|J_3N_3\rangle. \end{aligned} \tag{2.164}$$

Multiplying both sides of Eq. (2.164) by $\langle J_1N_1|\langle J_2N_2|$ and simplifying, we obtain

$$D^{J_1}_{N_1M_1}(\omega)D^{J_2}_{N_2M_2}(\omega) = \sum_{J_3}\langle J_1M_1,J_2M_2|J_3M_3\rangle\langle J_1N_1,J_2N_2|J_3N_3\rangle D^{J_3}_{N_3M_3}(\omega), \tag{2.165}$$

where $N_3 = N_1 + N_2$ and we have used the relation $\langle J_1M_1, J_2M_2|J_3M_3\rangle = \langle J_3M_3|J_1M_1, J_2M_2\rangle$.

2.6.3 Angular Integrals Involving Two or Three Rotation Matrix Elements

The main importance of Eq. (2.165) is that, as will be shown in this section, it allows us to evaluate integrals containing multiple rotation matrix elements or spherical harmonics, and these integrals are needed to evaluate single-photon, two-photon, and Raman transition line strengths. The differential angular element of interest is

$$d\omega = d\phi \sin\theta \, d\theta \, d\chi, \tag{2.166}$$

where the ranges of integration are from 0 to 2π for $d\phi$ and $d\chi$, and 0 to π for $d\theta$. Consider the integral over the entire ranges of integration for an integrand consisting of the product of two rotation matrices,

$$\int_0^{2\pi} d\phi \int_0^{\pi} \sin\theta \, d\theta \int_0^{2\pi} d\chi D^{J_1 *}_{N_1M_1}(\phi,\theta,\chi) D^{J_2}_{N_2M_2}(\phi,\theta,\chi) = \int d\omega D^{J_1 *}_{N_1M_1}(\omega) D^{J_2}_{N_2M_2}(\omega). \tag{2.167}$$

The complex conjugate of the first rotation matrix element can be eliminated using Eq. (2.150). Using Eq. (2.165) we can write

$$\begin{aligned}
&\int d\omega (-1)^{N_1-M_1} D^{J_1}_{-N_1,-M_1}(\omega) D^{J_2}_{N_2M_2}(\omega) \\
&\quad = \sum_{J_3} \langle J_1 - M_1, J_2M_2|J_3M_3\rangle \langle J_1 - N_1, J_2N_2|J_3N_3\rangle (-1)^{N_1-M_1} \int d\omega D^{J_3}_{N_3M_3}(\omega),
\end{aligned} \tag{2.168}$$

where $M_3 = -M_1 + M_2$ and $N_3 = -N_1 + N_2$. The integral on the right-hand side of Eq. (2.168) can be written

$$\begin{aligned}
&\int d\omega D^{J_3}_{N_3M_3}(\omega) \\
&\quad = \int_0^{2\pi} \exp[-i(N_2 - N_1)\phi] \, d\phi \int_0^{\pi} d^{J_3}_{N_3M_3} \sin\theta \, d\theta \int_0^{2\pi} \exp[-i(M_2 - M_1)\chi] \, d\chi \\
&\quad = 4\pi^2 \delta_{N_2N_1} \delta_{M_2M_1} \int_0^{\pi} d^{J_3}_{00} \sin\theta \, d\theta = 4\pi^2 \delta_{N_2N_1} \delta_{M_2M_1} \int_0^{\pi} D^{J_3}_{00} \sin\theta \, d\theta.
\end{aligned} \tag{2.169}$$

Substituting for $D^{J_3}_{00}$ using Eq. (2.157), we obtain

$$\begin{aligned}
\int_0^{\pi} D^{J_3}_{00} \sin\theta \, d\theta &= \left[\frac{4\pi}{(2J_3 + 1)}\right]^{1/2} \int_0^{\pi} Y^*_{J_30}(\theta) \sin\theta \, d\theta \\
&= \left[\frac{4\pi}{(2J_3 + 1)}\right]^{1/2} \int_0^{\pi} Y^*_{J_30}(\theta) \left[\sqrt{4\pi} Y_{00}(0)\right] \sin\theta \, d\theta \\
&= 4\pi \left[\frac{1}{(2J_3 + 1)}\right]^{1/2} \int_0^{\pi} Y^*_{J_30}(\theta) Y_{00}(\theta) \sin\theta \, d\theta = 2\delta_{J_30},
\end{aligned} \tag{2.170}$$

where in Eq. (2.170) we have made use of the relations $Y_{00}(\phi,\theta) = Y_{00}(\theta) = 1/\sqrt{4\pi}$ and

$$\delta_{LK} = \int_0^{2\pi} d\phi \int_0^{\pi} Y^*_{L0}(\phi,\theta) Y_{K0}(\phi,\theta) \sin\theta d\theta = 2\pi \int_0^{\pi} Y^*_{L0}(\theta) Y_{K0}(\theta) \sin\theta d\theta. \quad (2.171)$$

Combining Eqs. (2.167), (2.168), (2.169), and (2.170), we obtain

$$\int d\omega D^{J_1*}_{N_1M_1}(\omega) D^{J_2}_{N_2M_2}(\omega) = 8\pi^2 (-1)^{N_1-M_1} \langle J_1 - M_1, J_2M_1|00\rangle \langle J_1 - N_1, J_2N_1|00\rangle. \quad (2.172)$$

The Clebsch–Gordan coefficients can be evaluated by expressing them in terms of *3j* symbols and consulting Table A.1:

$$\begin{aligned} \langle J_1 - M_1, J_2M_1|00\rangle &= (-1)^{J_1-J_2} \begin{pmatrix} J_1 & J_2 & 0 \\ -M_1 & M_1 & 0 \end{pmatrix} = \frac{(-1)^{J_1-M_1}\delta_{J_1J_2}}{\sqrt{2J_1+1}}. \\ \langle J_1 - N_1, J_2N_1|00\rangle &= \frac{(-1)^{J_1-N_1}\delta_{J_1J_2}}{\sqrt{2J_1+1}}. \end{aligned} \quad (2.173)$$

Substituting for the Clebsch–Gordan coefficients using Eq. (2.173), we obtain

$$\int d\omega D^{J_1*}_{N_1M_1}(\omega) D^{J_2}_{N_2M_2}(\omega) = \frac{8\pi^2}{2J_1+1} \delta_{J_1J_2}\delta_{M_1M_2}\delta_{N_1N_2}. \quad (2.174)$$

Another important integral is the integral of the triple product of rotation matrix elements over the entire space spanned by the Euler angles,

$$\begin{aligned} &\int d\omega D^{J_3}_{N_3M_3}(\omega) D^{J_2}_{N_2M_2}(\omega) D^{J_1}_{N_1M_1}(\omega) \\ &= \sum_J \langle J_1M_1J_2M_2|JM_1+M_2\rangle \langle J_1N_1J_2N_2|JN_1+N_2\rangle \int d\omega D^{J_3}_{N_3M_3}(\omega) D^{J}_{N_1+N_2,M_1+M_2}(\omega), \end{aligned} \quad (2.175)$$

where Eq. (2.165) has been used to simplify the expression. The integral on the RHS of Eq. (2.175) can be evaluated using the results of Eq. (2.174) to yield

$$\int d\omega\, D^{J_3*}_{N_3M_3}(\phi,\theta,\chi) D^{J}_{N_1+N_2,M_1+M_2}(\phi,\theta,\chi) = \frac{8\pi^2}{2J_3+1} \delta_{M_1+M_2,M_3}\delta_{N_1+N_2,N_3}\delta_{JJ_3}. \quad (2.176)$$

Substituting Eq. (2.176) into Eq. (2.175) and simplifying, we obtain

$$\int d\omega\, D^{J_3*}_{N_3M_3}(\omega) D^{J_2}_{N_2M_2}(\omega) D^{J_1}_{N_1M_1}(\omega) = \frac{8\pi^2}{2J_3+1} \langle J_1M_1J_2M_2|J_3M_3\rangle \langle J_1N_1J_2N_2|J_3N_3\rangle. \quad (2.177)$$

Expressing the Clebsch–Gordan coefficients in Eq. (2.177) in terms of *3j* symbols, we obtain

$$\int d\omega\, D^{J_3*}_{N_3M_3}(\omega) D^{J_2}_{N_2M_2}(\omega) D^{J_1}_{N_1M_1}(\omega)$$
$$= 8\pi^2(-1)^{2J_1-2J_2+M_3+N_3} \begin{pmatrix} J_1 & J_2 & J_3 \\ M_1 & M_2 & -M_3 \end{pmatrix} \begin{pmatrix} J_1 & J_2 & J_3 \\ N_1 & N_2 & -N_3 \end{pmatrix}. \tag{2.178}$$

From Eq. (2.150) we can write

$$D^{J_3*}_{N_3M_3}(\omega) = (-1)^{N_3-M_3} D^{J_3}_{-N_3,-M_3}(\omega). \tag{2.179}$$

Substituting Eq. (2.179) into Eq. (2.178) and simplifying, we obtain

$$\begin{aligned}
&\int d\omega\, D^{J_3}_{-N_3,-M_3}(\omega) D^{J_2}_{N_2M_2}(\omega) D^{J_1}_{N_1M_1}(\omega) \\
&= 8\pi^2(-1)^{2J_1-2J_2+2M_3} \begin{pmatrix} J_1 & J_2 & J_3 \\ M_1 & M_2 & -M_3 \end{pmatrix} \begin{pmatrix} J_1 & J_2 & J_3 \\ N_1 & N_2 & -N_3 \end{pmatrix} \\
&= 8\pi^2(-1)^{2(J_1+M_1)-2(J_2+M_2)} \begin{pmatrix} J_1 & J_2 & J_3 \\ M_1 & M_2 & -M_3 \end{pmatrix} \begin{pmatrix} J_1 & J_2 & J_3 \\ N_1 & N_2 & -N_3 \end{pmatrix} \\
&= 8\pi^2 \begin{pmatrix} J_1 & J_2 & J_3 \\ M_1 & M_2 & -M_3 \end{pmatrix} \begin{pmatrix} J_1 & J_2 & J_3 \\ N_1 & N_2 & -N_3 \end{pmatrix}.
\end{aligned} \tag{2.180}$$

Then, substituting N_3 for $-N_3$ and M_3 for $-M_3$ gives

$$\int d\omega\, D^{J_3}_{N_3M_3}(\omega) D^{J_2}_{N_2M_2}(\omega) D^{J_1}_{N_1M_1}(\omega) = 8\pi^2 \begin{pmatrix} J_1 & J_2 & J_3 \\ M_1 & M_2 & M_3 \end{pmatrix} \begin{pmatrix} J_1 & J_2 & J_3 \\ N_1 & N_2 & N_3 \end{pmatrix}. \tag{2.181}$$

The integral of the triple product of spherical harmonic functions can be obtained in a straightforward manner using Eqs. (2.65) and (2.157):

$$\begin{aligned}
&\int d\omega D^{J_3}_{N_30}(\omega) D^{J_2}_{N_20}(\omega) D^{J_1}_{N_10}(\omega) \\
&= 2\pi(-1)^{N_1+N_2+N_3} \left[\frac{(4\pi)^3}{(2J_3+1)(2J_2+1)(2J_1+1)}\right]^{1/2} \int d\Omega Y^*_{J_3N_3}(\Omega) Y^*_{J_2N_2}(\Omega) Y^*_{J_1N_1}(\Omega) \\
&= 2\pi \left[\frac{(4\pi)^3}{(2J_3+1)(2J_2+1)(2J_1+1)}\right]^{1/2} \int d\Omega Y_{J_3N_3}(\Omega) Y_{J_2N_2}(\Omega) Y_{J_1N_1}(\Omega) \\
&= 8\pi^2 \begin{pmatrix} J_1 & J_2 & J_3 \\ 0 & 0 & 0 \end{pmatrix} \begin{pmatrix} J_1 & J_2 & J_3 \\ N_1 & N_2 & N_3 \end{pmatrix}.
\end{aligned} \tag{2.182}$$

Rearranging Eq. (2.182) and substituting L_1, L_2, L_3 for J_1, J_2, J_3 and M_1, M_2, M_3 for N_1, N_2, N_3, we obtain

$$\int d\Omega Y_{L_3M_3}(\Omega)Y_{L_2M_2}(\Omega)Y_{L_1M_1}(\Omega) = \int_0^{2\pi} d\phi \int_0^{\pi} \sin\theta d\theta Y_{L_3M_3}(\phi,\theta)Y_{L_2M_2}(\phi,\theta)Y_{L_1M_1}(\phi,\theta)$$
$$= \left[\frac{(2L_3+1)(2L_2+1)(2L_1+1)}{4\pi}\right]^{1/2}$$
$$\times \begin{pmatrix} L_1 & L_2 & L_3 \\ 0 & 0 & 0 \end{pmatrix} \begin{pmatrix} L_1 & L_2 & L_3 \\ M_1 & M_2 & M_3 \end{pmatrix}. \tag{2.183}$$

2.7 Irreducible Spherical Tensors

Irreducible spherical tensors are central to the discussion of radiative transitions. The irreducible spherical tensor has a rank k, and the number of tensor components is $2k + 1$. An important property of irreducible spherical tensors is that there is no mixing of tensors of different rank upon rotation of the coordinate system. For example, an irreducible spherical tensor of rank 0 is a scalar and is invariant under a coordinate system rotation.

2.7.1 Definition of Irreducible Spherical Tensor Operators

The irreducible spherical tensor operator $T_k(\hat{\boldsymbol{A}})$ of rank k is defined by the relation

$$R(\omega)T_{kp}(\hat{\boldsymbol{A}})R^{-1}(\omega) = \sum_{p'} T_{kp'}(\hat{\boldsymbol{A}}) D^k_{p'p}(\omega), \tag{2.184}$$

where $\hat{\boldsymbol{A}}$ is a quantum mechanical operator, and the irreducible spherical tensor has $k+1$ components $T_{kp}(\boldsymbol{A})$. An alternative definition of the irreducible spherical tensor operator is given by (Brown & Carrington, 2003)

$$\left[\hat{J}_Z, T_{kp}(\hat{\boldsymbol{A}})\right] = pT_{kp}(\hat{\boldsymbol{A}}). \tag{2.185}$$

$$\left[\hat{J}_\pm, T_{kp}(\hat{\boldsymbol{A}})\right] = \sqrt{k(k+1) - p(p\pm 1)}T_{kp\pm 1}(\hat{\boldsymbol{A}}). \tag{2.186}$$

The equivalence of these two definitions of the irreducible spherical tensor operators is shown in detail by Rose (1957). Irreducible spherical tensor operators $T_{00}(\hat{\boldsymbol{A}})$ of rank zero are invariant under rotation:

$$R(\omega)T_{00}(\hat{\boldsymbol{A}})R^{-1}(\omega) = T_{00}(\hat{\boldsymbol{A}}) D^0_{00}(\omega) = T_{00}(\hat{\boldsymbol{A}}). \tag{2.187}$$

As discussed by Zare (1988), these zero-rank tensor operators or scalar operators are of considerable significance because the components of the Hamiltonian operator are invariant under rotation, and thus the components of the Hamiltonian will all be scalar operators.

Examples of first-rank tensor operators include the position operator $\hat{\boldsymbol{r}}$, the dipole moment operator $\hat{\boldsymbol{\mu}}$, and the angular momentum operator $\hat{\boldsymbol{J}}$. The components of the irreducible spherical tensor operator for angular momentum are given by (Brown & Carrington, 2003)

$$T_{11}(\hat{\boldsymbol{J}}) = -\frac{1}{\sqrt{2}}\left(\hat{J}_X + i\hat{J}_Y\right) = -\frac{1}{\sqrt{2}}\hat{J}_+. \tag{2.188}$$

$$T_{10}(\hat{\boldsymbol{J}}) = \hat{J}_Z. \tag{2.189}$$

$$T_{1-1}(\hat{\boldsymbol{J}}) = \frac{1}{\sqrt{2}}\left(\hat{J}_X - i\hat{J}_Y\right) = \frac{1}{\sqrt{2}}\hat{J}_-. \tag{2.190}$$

The most important second-rank irreducible spherical tensor operator that will be discussed in this book is the Raman polarizability tensor, discussed in detail in Chapter 7.

2.7.2 The Wigner–Eckart Theorem

The Wigner–Eckart theorem is a very important theorem concerning the matrix elements of irreducible spherical tensor operators. It allows one to separate and generalize the geometric aspects of physical problems. The Wigner–Eckart theorem is given by

$$\langle \alpha JM | T_{kq}(\hat{\mathbf{A}}) | \alpha' J' M' \rangle = (-1)^{J-M} \begin{pmatrix} J & k & J' \\ -M & q & M \end{pmatrix} \langle \alpha J \| T_k(\hat{\mathbf{A}}) \| \alpha' J' \rangle, \tag{2.191}$$

where $\langle \alpha J \| T_k(\hat{\mathbf{A}}) \| \alpha' J' \rangle$ is a reduced operator matrix element that does not depend on the projection quantum numbers or on the orientation of the quantum mechanical system. All of the angular dependence in the quantum system is contained in the *3j* symbol and in the $(-1)^{J-M}$ phase factor. There are several different ways of deriving or proving the Wigner–Eckart theorem, including the following proof given by Brown and Carrington (2003). The matrix element is a scalar and is thus invariant under rotations. The effect of rotation on the spherical tensor matrix element can be calculated using the rotation operator matrix elements for the bra, operator, and ket of the spherical tensor matrix element:

$$\langle \alpha JM | T_{kq}(\hat{\mathbf{A}}) | \alpha' J' M' \rangle = \sum_N \sum_p \sum_{N'} D^{J*}_{NM}(\omega) D^k_{pq}(\omega) D^{J'}_{N'M'}(\omega) \langle \alpha JN | T_{kp}(\hat{\mathbf{A}}) | \alpha' J' N' \rangle. \tag{2.192}$$

The next step in the proof is to integrate both sides over all the Euler angles,

$$\int \langle \alpha JM | T_{kq}(\hat{\mathbf{A}}) | \alpha' J' M' \rangle d\omega = \int \sum_{N,p,N'} D^{J*}_{NM}(\omega) D^k_{pq}(\omega) D^{J'}_{N'M'}(\omega) \langle \alpha JN | T_{kp}(\hat{\mathbf{A}}) | \alpha' J' N' \rangle d\omega. \tag{2.193}$$

The integral on the left-hand side is easy to evaluate since $\langle \alpha JM | T_{kq}(\hat{\mathbf{A}}) | \alpha' J' M' \rangle$ is just a number. The integral on the right-hand side can be evaluated using Eq. (2.178), resulting in

$$8\pi^2 \langle \alpha JM | T_{kq}(\hat{\mathbf{A}}) | \alpha' J' M' \rangle$$
$$= 8\pi^2 \sum_{N,p,N'} (-1)^{2J'-2k+N+M} \begin{pmatrix} J' & k & J \\ M' & q & -M \end{pmatrix} \begin{pmatrix} J' & k & J \\ N' & p & -N \end{pmatrix} \langle \alpha JN | T_{kp}(\hat{\mathbf{A}}) | \alpha' J' N' \rangle. \tag{2.194}$$

Dividing both sides by $8\pi^2$ and performing an odd permutation of the columns in the $3j$ symbols, we obtain

$$\begin{aligned}
&\langle\alpha JM|T_{kq}(\hat{\mathbf{A}})|\alpha' J'M'\rangle \\
&= \sum_{N,p,N'} (-1)^{2J'-2k+N+M}(-1)^{2(J'+k+J)}\begin{pmatrix} J & k & J' \\ -M & q & M' \end{pmatrix}\begin{pmatrix} J & k & J' \\ -N & p & N' \end{pmatrix}\langle\alpha JN|T_{kp}(\hat{\mathbf{A}})|\alpha' J'N'\rangle \\
&= \sum_{N,p,N'} (-1)^{2J+N+M}\begin{pmatrix} J & k & J' \\ -M & q & M' \end{pmatrix}\begin{pmatrix} J & k & J' \\ -N & p & N' \end{pmatrix}\langle\alpha JN|T_{kp}(\hat{\mathbf{A}})|\alpha' J'N'\rangle \\
&= \sum_{N,p,N'} (-1)^{2J-N-M+2(N+M)}\begin{pmatrix} J & k & J' \\ -M & q & M' \end{pmatrix}\begin{pmatrix} J & k & J' \\ -N & p & N' \end{pmatrix}\langle\alpha JN|T_{kp}(\hat{\mathbf{A}})|\alpha' J'N'\rangle \\
&= (-1)^{J-M}\begin{pmatrix} J & k & J' \\ -M & q & M' \end{pmatrix}\left\{\sum_{N,p,N'} (-1)^{J-N}\begin{pmatrix} J & k & J' \\ -N & p & N' \end{pmatrix}\langle\alpha JN|T_{kp}(\hat{\mathbf{A}})|\alpha' J'N'\rangle\right\}.
\end{aligned} \tag{2.195}$$

The last term on the right-hand side in parentheses is independent of orientation because a summation over the projection quantum numbers N and N' and the spherical tensor components p is performed. The term in parentheses is the reduced density matrix element,

$$\langle\alpha J\|T_k(\hat{\mathbf{A}})\|\alpha' J'\rangle = \sum_{N,p,N'} (-1)^{J-N}\begin{pmatrix} J & k & J' \\ -N & p & N' \end{pmatrix}\langle\alpha JN|T_{kp}(\hat{\mathbf{A}})|\alpha' J'N'\rangle. \tag{2.196}$$

The value of the reduced operator matrix element is almost never determined from Eq. (2.196). Rather, it is usually determined from experiment. The final result is written as

$$\langle\alpha JM|T_{kq}(\hat{\mathbf{A}})|\alpha' J'M'\rangle = (-1)^{J-M}\begin{pmatrix} J & k & J' \\ -M & q & M' \end{pmatrix}\langle\alpha J\|T_k(\hat{\mathbf{A}})\|\alpha' J'\rangle. \tag{2.197}$$

2.7.3 Reduced Matrix Elements for Systems with Coupled Angular Momentum

Angular momentum coupling is very important in determining the structure of atoms and molecules. The case where an operator acts on one of the angular momentum components but not the other in a coupling scheme is frequently encountered. In atomic spectroscopy, for example, the electric dipole operator interacts with the electronic orbital angular momentum but not with the spin of the electrons or the nucleus. The reduced matrix element for an operator that acts only on the angular momentum $\boldsymbol{J_1}$ in a coupled scheme where $\boldsymbol{J_1}+\boldsymbol{J_2}=\boldsymbol{J}$ is given by

$$\begin{aligned}
\langle\alpha J_1J_2J\|T_k(\hat{\mathbf{A}}_1)\|\alpha' J_1'J_2'J'\rangle &= \delta_{J_2J_2'}(-1)^{J_1+J_2+J'+k}\sqrt{(2J'+1)(2J+1)} \\
&\times\begin{Bmatrix} J_1 & J & J_2 \\ J' & J_1' & k \end{Bmatrix}\langle\alpha J_1\|T_k(\hat{\mathbf{A}}_1)\|\alpha' J_1'\rangle.
\end{aligned} \tag{2.198}$$

The expressions for evaluation of the products of tensor operators are discussed in detail in a number of texts (Brown & Carrington, 2003; Edmonds, 1960; Judd, 1975; Weissbluth, 1978; Zare, 1988). The scalar product of irreducible tensor operators is defined by

$$T_k(\hat{\boldsymbol{A}}) \bullet T_k(\hat{\boldsymbol{B}}) = \sum_p (-1)^p T_{kp}(\hat{\boldsymbol{A}}) T_{k-p}(\hat{\boldsymbol{B}}). \tag{2.199}$$

The scalar product of two commuting spherical tensors that act on different angular momenta is given by

$$\begin{aligned} &\langle \alpha J_1 J_2 J M_J \| T_k(\hat{\boldsymbol{A}}_1) \bullet T_k(\hat{\boldsymbol{A}}_2) \| \alpha' J_1' J_2' J' M'_J \rangle \\ &= \delta_{JJ'} \delta_{M_J M_J'} (-1)^{J_1'+J_2+J} \begin{Bmatrix} J_1 & J_2 & J \\ J_2' & J_1' & k \end{Bmatrix} \sum_{\alpha''} \langle \alpha J_1 \| T_k(\hat{\boldsymbol{A}}_1) \| \alpha'' J_1' \rangle \langle \alpha'' J_2 \| T_k(\hat{\boldsymbol{A}}_2) \| \alpha' J_2' \rangle. \end{aligned} \tag{2.200}$$

2.7.4 Spherical Tensor Analysis of the Spin–Orbit Interaction

For calculation of the spin–orbit interaction for a multielectron atom we must evaluate the following matrix element (Weissbluth, 1978):

$$\varepsilon_{SO} = \left\langle \alpha LSJM_J \middle| \sum_i \xi(r_i) \hat{\boldsymbol{l}}_i \bullet \hat{\boldsymbol{s}}_i \middle| \alpha LSJM_J \right\rangle. \tag{2.201}$$

Because the term $\xi(r_i)$ depends only on the radial components, we can write

$$\varepsilon_{SO} = \zeta(\alpha LS) \langle \alpha LSJM_J | \hat{\boldsymbol{L}} \bullet \hat{\boldsymbol{S}} | \alpha LSJM_J \rangle. \tag{2.202}$$

Equating $T_k(\hat{\boldsymbol{A}}_1)$, $T_k(\hat{\boldsymbol{A}}_2)$, J_1, and J_2 with $\hat{\boldsymbol{L}}$, $\hat{\boldsymbol{S}}$, L, and S, respectively, in Eq. (2.200), we obtain

$$\begin{aligned} &\langle \alpha J_1 J_2 J M_J | T_k(\hat{\boldsymbol{A}}_1) \bullet T_k(\hat{\boldsymbol{A}}_2) | \alpha' J_1' J_2' J' M'_J \rangle = \langle \alpha LSJM_J | T_k(\hat{\boldsymbol{L}}) \bullet T_k(\hat{\boldsymbol{S}}) | \alpha LSJM_J \rangle \\ &= (-1)^{L+S+J} \begin{Bmatrix} L & S & J \\ S & L & 1 \end{Bmatrix} \sum_{\alpha''} \langle \alpha L \| T_k(\hat{\boldsymbol{L}}) \| \alpha'' L \rangle \langle \alpha'' S \| T_k(\hat{\boldsymbol{S}}) \| \alpha S \rangle. \end{aligned} \tag{2.203}$$

Evaluating the *6j* symbol using Eq. (A4.9) in Appendix A4, we obtain

$$\begin{Bmatrix} L & S & J \\ S & L & 1 \end{Bmatrix} = (-1)^{L+S+J+1} \frac{2[L(L+1) + S(S+1) - J(J+1)]}{[2L(2L+1)(2L+2)(2S)(2S+1)(2S+2)]^{1/2}}. \tag{2.204}$$

The reduced matrix elements can be determined by using the Wigner–Eckart theorem:

$$\langle \alpha LM_L | T_{1p}(\hat{\boldsymbol{L}}) | \alpha LM_L \rangle = (-1)^{L-M_L} \begin{pmatrix} L & 1 & L \\ -M_L & p & M_L \end{pmatrix} \langle \alpha L \| T_{1p}(\hat{\boldsymbol{L}}) \| \alpha L \rangle. \tag{2.205}$$

Evaluating the $p = 0$ component, we obtain

$$\begin{aligned}\langle \alpha L M_L | T_{10}(\hat{\mathbf{L}}) | \alpha'' L M_L \rangle &= \langle \alpha L M_L | \hat{L}_Z | \alpha'' L M_L \rangle \\ &= M_L \hbar \langle \alpha L M_L | \alpha'' L M_L \rangle = M_L \hbar \delta_{\alpha\alpha''}.\end{aligned} \tag{2.206}$$

The $3j$ symbol for $p = 0$ is given by

$$\begin{pmatrix} L & 1 & L \\ -M_L & 0 & M_L \end{pmatrix} = \frac{(-1)^{L-M_L} M_L}{[L(L+1)(2L+1)]^{1/2}}. \tag{2.207}$$

Substituting Eqs. (2.206) and (2.207) into Eq. (2.205) and rearranging, we obtain

$$\begin{aligned}\langle \alpha L \| T_1(\hat{\mathbf{L}}) \| \alpha L \rangle &= \frac{\langle \alpha L M_L | T_{10}(\hat{\mathbf{L}}) | \alpha L M_L \rangle}{(-1)^{L-M_L} \begin{pmatrix} L & 1 & L \\ -M_L & p & M_L \end{pmatrix}} = \frac{M_L \hbar}{(-1)^{L-M_L} \left\{ \frac{(-1)^{L-M_L} M_L}{[L(L+1)(2L+1)]^{1/2}} \right\}} \\ &= [L(L+1)(2L+1)]^{1/2} \hbar.\end{aligned} \tag{2.208}$$

Similarly,

$$\langle \alpha S \| T_1(\hat{\mathbf{S}}) \| \alpha S \rangle = [S(S+1)(2S+1)]^{1/2} \hbar. \tag{2.209}$$

Substituting Eqs. (2.204), (2.208), and (2.209) into Eq. (2.203), we obtain

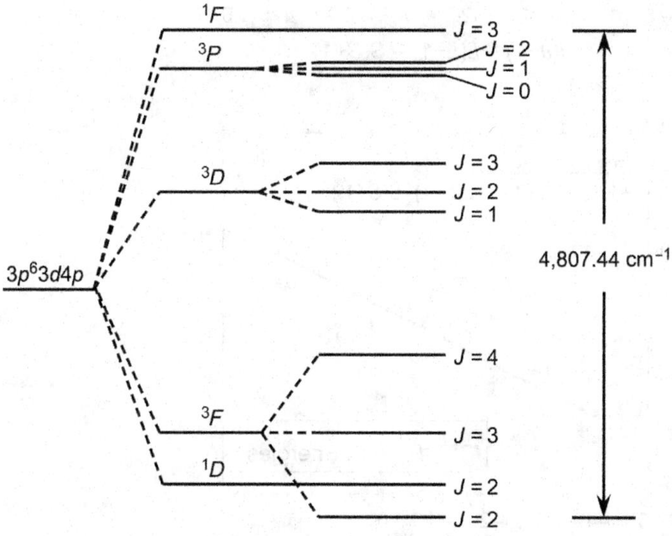

Figure 2.5 Energy level diagram for the multiplet terms arising from the $3p^6 3d4p$ electron configuration in the calcium atom. The energy splitting is approximately to scale.

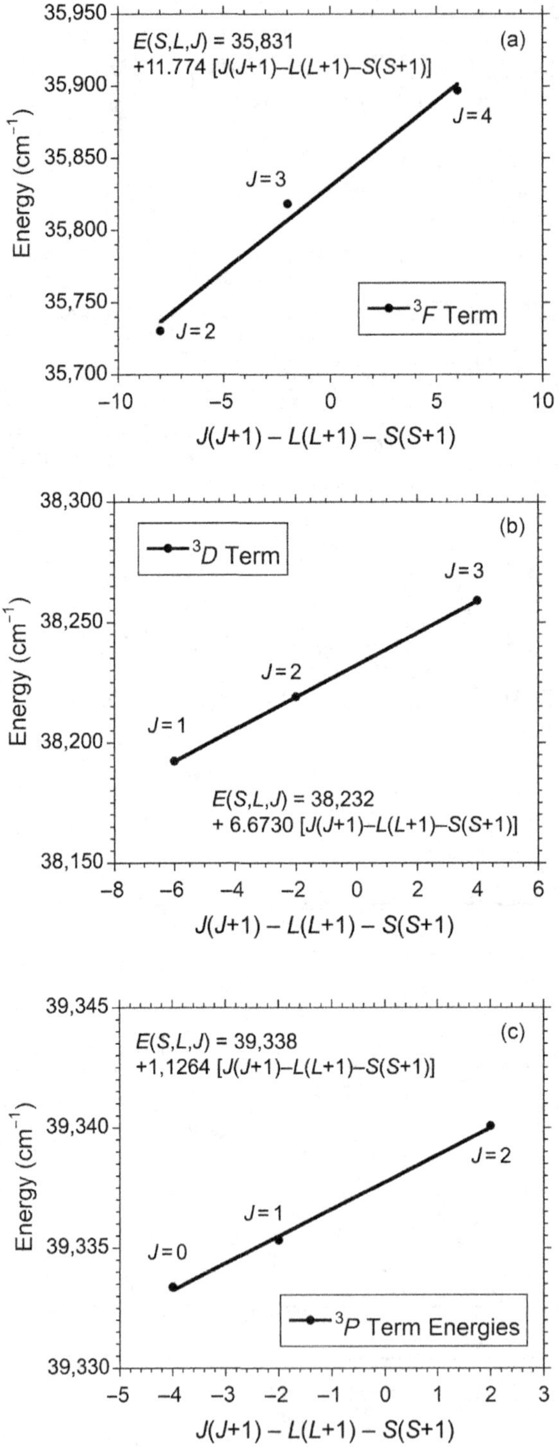

Figure 2.6 Energy level splittings in the calcium atom for the (a) ^3F, (b) ^3D, and (c) ^3P terms.

$$
\begin{aligned}
&\langle \alpha LSJM_J | T_k(\hat{L}) \bullet T_k(\hat{S}) | \alpha LSJM_J \rangle \\
&= (-1)^{L+S+J} \left\{ (-1)^{L+S+J+1} \frac{2[L(L+1)+S(S+1)-J(J+1)]}{[2L(2L+1)(2L+2)(2S)(2S+1)(2S+2)]^{1/2}} \right\} \\
&\quad \times \left\{ [L(L+1)(2L+1)]^{1/2} \hbar \right\} \left\{ [S(S+1)(2S+1)]^{1/2} \hbar \right\} \\
&= -\frac{1}{2}[L(L+1)+S(S+1)-J(J+1)]\hbar^2 \\
&= \frac{1}{2}[J(J+1)-L(L+1)-S(S+1)]\hbar^2.
\end{aligned}
\tag{2.210}
$$

The spin–orbit interaction energy is thus given by

$$
\varepsilon_{SO} = \frac{1}{2}\zeta(\alpha LS)\hbar^2[J(J+1)-L(L+1)-S(S+1)]. \tag{2.211}
$$

The energy level structure for five different terms arising from the $3p^6 3d4p$ electron configuration in the calcium atom are shown in Figure 2.5. A sixth term, the ^1P term, is also possible but does not appear in the NIST tables (Kramida et al., 2022). The energies of the different J levels for the terms are shown in Figure 2.6, along with a least squares fit of the level energies to Eq. (2.211). The fit to the level energies is very good in each case, although the values of $\zeta(\alpha LS)$ are very different for each of the three terms: $\zeta(\alpha 31) = 11.774$ cm^{-1}, $\zeta(\alpha 21) = 6.673$ cm^{-1}, and $\zeta(\alpha 11) = 1.1264$ cm^{-1}. The magnitude of the splitting for each term is dependent on the integral over the radial coordinate of the multielectron atom. The dependence of $\zeta(\alpha LS)$ on the structure of the multielectron atom is discussed in more detail by Weissbluth (1978).

3 Structure of Diatomic Molecules

3.1 Introduction

The structure of diatomic molecules is discussed in this chapter. We begin with a classical analysis of the vibrational, rotational, and translational modes of energy storage for a vibrating rotor. The Born–Oppenheimer approximation is then introduced. It is assumed that the electronic wavefunction is dependent only on the coordinates of these nuclei, and not on their momentum. The change in the electronic structure as a function of nuclear coordinates is represented as a potential well with a minimum energy at the equilibrium internuclear spacing. The formation of this potential well allows us to reformulate the two-particle SWE, and the nuclear rotational and vibrational energy modes and quantum numbers are a result of the solution of this two-particle SWE for the diatomic molecule.

The electronic structure of diatomic molecules is then discussed in detail. The coupling of the orbital and spin angular momentum of the electrons and the angular momentum associated with the nuclear rotation is discussed, with an emphasis on Hund's cases (a) and (b). The rotational wavefunctions for diatomic molecules in the limits of Hund's cases (a) and (b) and in the case intermediate between the two cases are then discussed in detail. For molecules that are of importance in combustion diagnostics, such as OH, CH, CN, and NO, the electronic levels are intermediate between Hund's cases (a) and (b).

These rotational wavefunctions for Hund's cases (a) and (b) are the wavefunctions of a symmetric rotor and can be expressed in terms of the rotational matrix elements introduced in Chapter 2. In this book we will use Hund's case (a) as the basis wavefunctions, and linear combinations of these wavefunctions will be used to represent wavefunctions for electronic levels intermediate between cases (a) and (b) (e.g., $^2\Pi$, $^2\Delta$, …, $^3\Pi$, $^3\Delta$, … and for electronic levels that are case (b) (e.g., $^2\Sigma$, $^3\Sigma$, …). The choice of case (a) wavefunctions as the basis set is typical although case (b) wavefunctions can also be used as a basis set. These wavefunctions will also be used in Chapter 4 to calculate rotational line strengths for radiative transitions.

3.2 Classical Analysis of Nuclear Motion – Energy of Translation, Rotation, and Vibration

Consider a diatomic molecule AB as two heavy balls, the nuclei of atoms A and B, connected by a stiff (but massless) spring, as depicted in Figure 3.1. The potential energy due to the spring can be modeled as a harmonic-oscillator potential,

$$V(r_{AB}) = \frac{1}{2}k(r_{AB} - r_e)^2. \tag{3.1}$$

The total mechanical energy of system is then given by

$$\varepsilon = \frac{1}{2}m_A\left(\dot{x}_A^2 + \dot{y}_A^2 + \dot{z}_A^2\right) + \frac{1}{2}m_B\left(\dot{x}_B^2 + \dot{y}_B^2 + \dot{z}_B^2\right) + V(r_{AB}). \tag{3.2}$$

The center of mass of the molecule is defined by the following equations:

$$\begin{aligned}
(m_A + m_B)X_{CM} &= m_A x_A + m_B x_B \quad \Rightarrow \quad x_B = [(m_A + m_B)X_{CM} - m_A x_A]/m_B. \\
(m_A + m_B)Y_{CM} &= m_A y_A + m_B y_B \quad \Rightarrow \quad y_B = [(m_A + m_B)Y_{CM} - m_A y_A]/m_B. \\
(m_A + m_B)Z_{CM} &= m_A z_A + m_B z_B \quad \Rightarrow \quad z_B = [(m_A + m_B)Z_{CM} - m_A z_A]/m_B.
\end{aligned} \tag{3.3}$$

We now introduce a spherical coordinate system with its origin at particle #1, as shown in Figure 3.1:

$$\begin{aligned}
x_B &= x_A + r_{AB}\sin\theta\cos\phi. \\
y_B &= y_A + r_{AB}\sin\theta\sin\phi. \\
z_B &= z_A + r_{AB}\cos\theta.
\end{aligned} \tag{3.4}$$

Equation (3.2) can then be manipulated to obtain separate translational, rotational, and vibrational energy terms. Combining Eqs. (3.3) and (3.4), we obtain:

$$\begin{aligned}
x_A &= X_{CM} - \frac{m_B}{m_A + m_B} r_{AB}\sin\theta\cos\phi. \\
y_A &= Y_{CM} - \frac{m_B}{m_A + m_B} r_{AB}\sin\theta\sin\phi. \\
z_A &= Z_{CM} - \frac{m_B}{m_A + m_B} r_{AB}\cos\theta.
\end{aligned} \tag{3.5}$$

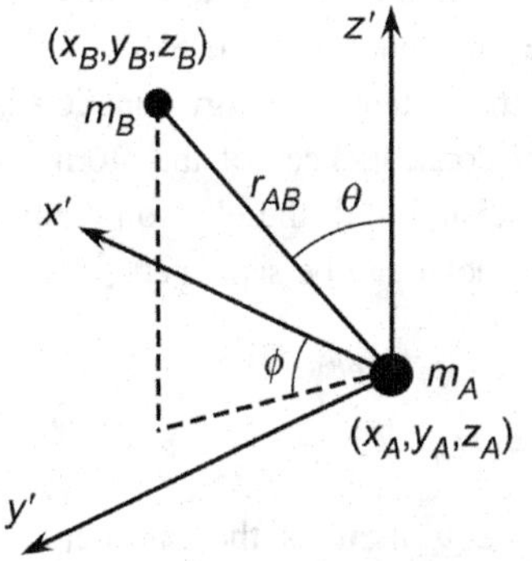

Figure 3.1 Coordinate system for analysis of the energy modes of diatomic molecule AB.

$$
\begin{aligned}
x_B &= X_{CM} + \frac{m_A}{m_A + m_B} r_{AB} \sin\theta \cos\phi. \\
y_B &= Y_{CM} + \frac{m_A}{m_A + m_B} r_{AB} \sin\theta \sin\phi. \\
z_B &= Z_{CM} + \frac{m_A}{m_A + m_B} r_{AB} \cos\theta.
\end{aligned}
\tag{3.6}
$$

Note that $X_{CM}, Y_{CM}, Z_{CM}, r_{AB}, \theta$, and ϕ are all time dependent. Taking the derivatives of the above expressions and performing some algebra results in the following relation:

$$
\varepsilon = \underbrace{\frac{1}{2} m \left(\dot{X}_{CM}^2 + \dot{Y}_{CM}^2 + \dot{Z}_{CM}^2\right)}_{translation} + \underbrace{\frac{1}{2}\mu r_{AB}^2 \left[\dot{\theta}^2 + \left(\sin^2\theta\right)\dot{\phi}^2\right]}_{rotation} + \underbrace{\left[\frac{1}{2}\mu \dot{r}_{AB}^2 + V(r_{AB})\right]}_{vibration}
\tag{3.7}
$$

where

$$
m = m_A + m_B \qquad \mu = \frac{m_A m_B}{m_A + m_B}.
\tag{3.8}
$$

The first term in Eq. (3.7) is the kinetic energy associated with the center of mass motion of the molecule, the second term is the kinetic energy associated with rotational motion about the center of mass, and the third term is the combined kinetic and potential energy associated with the vibrational motion of the nuclei about the equilibrium internuclear distance r_e.

3.3 The Born–Oppenheimer Approximation

A diatomic molecule is formed as two atoms, initially at large separation, are brought together to form a stable molecule. When the electron clouds of the atoms start to interact, the electron configuration will change so that the potential energy of the system decreases. In the Born–Oppenheimer approximation, it is assumed that the electrons move so rapidly compared to the massive nuclei that the electron configuration is able to adapt instantaneously to the changes in the internuclear separation. The electron configuration and energy levels can in principle be calculated from the SWE, although this is a much more complicated problem because the potential energy function is no longer spherically symmetric. The most important consequences of the Born–Oppenheimer approximation are (1) that the energies associated with the electron configuration and the vibrational and rotational motion of the nuclei can be separated,

$$
\varepsilon_{internal} = \varepsilon_{el} + \varepsilon_{vib} + \varepsilon_{rot},
\tag{3.9}
$$

$$
\hat{H}_{internal} = \hat{H}_{el} + \hat{H}_{vib} + \hat{H}_{rot},
\tag{3.10}
$$

and (2) that the total wavefunction can be written as the product of electronic, vibrational, and rotational wavefunctions,

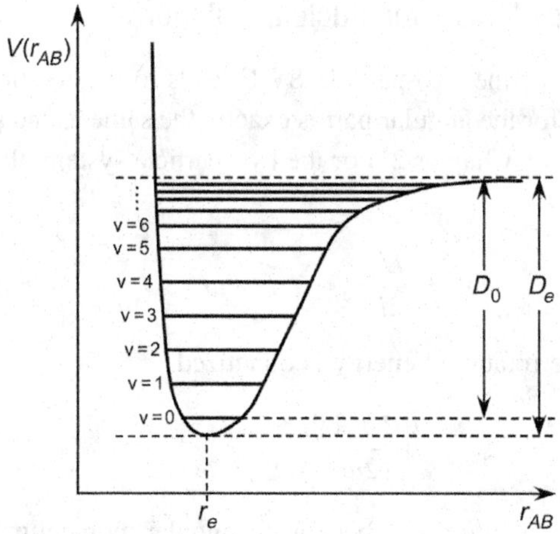

Figure 3.2 Potential function for a diatomic molecule AB.

$$\psi_{internal} = \psi_{el} \psi_{vib} \psi_{rot}. \tag{3.11}$$

The Born–Oppenheimer approximation is discussed in great detail by Brown and Carrington (2003).

Because it is assumed that the electron configuration can adapt rapidly to any change in internuclear separation r_{AB}, a set of electron energy levels ε_n^{el} can be assigned to each value of r_{AB}. These energy levels $\varepsilon_n^{el}(r_{AB})$ reach a minimum at the equilibrium internuclear separation distance $r_{AB} = r_e$ and form the potential function that allows the molecule to store energy associated with the vibrational motion of the nuclei. This potential function is given by

$$F_{A-B} = -\frac{dV(r_{AB})}{dr_{AB}} = -\frac{d\varepsilon_n^{el}(r_{AB})}{dr_{AB}} \tag{3.12}$$

and is depicted schematically in Figure 3.2. For large separations r_{AB}, the atoms do not interact and the molecule does not exist. As r_{AB} decreases, the nuclei experience an attractive force $F_{AB} < 0$. At $r_{AB} = r_e$, $F_{AB} = 0$. For $r_{AB} < r_e$, the nuclei experience a repulsive force $F_{AB} > 0$. The form of the potential is approximately harmonic near the potential minimum,

$$V(r) = \frac{1}{2}k(r - r_e)^2. \tag{3.13}$$

As shown in Section 3.4, the SWE analysis of the potential function results in a set of vibrational energy levels that are depicted in Figure 3.2. The energy difference between the potential minimum and the continuum is D_e, the electronic binding energy. The energy difference between the lowest vibrational energy level and the continuum is D_0, the dissociation energy. The difference $D_e - D_0$ is the vibrational zero-point energy.

3.3.1 Wavefunctions and Energy Levels for Nuclear Rotation

Solution of the angular part of the two-particle SWE leads to the rotational energy level structure. The solution for the angular part is exactly the same as the solution for the hydrogen atom discussed in Chapter 2. For the two-particle system, the rotational energy ε_{rot} is given by

$$\varepsilon_{rot} = \frac{1}{2} I\omega^2 = \frac{L^2}{2I} \qquad I = \mu r^2. \tag{3.14}$$

As shown in Section 2.4, the rotational energy is quantized,

$$\varepsilon_{rot} = \frac{[l(l+1)]\hbar^2}{2\mu r^2}. \tag{3.15}$$

For a diatomic molecule, the quantum number for the angular momentum associated with nuclear rotation is typically denoted by R rather than by l. The rotational energy levels for the diatomic molecule are thus given by

$$\varepsilon_{rot} = \frac{[R(R+1)]\hbar^2}{2\mu r^2}. \tag{3.16}$$

Now recall that for a particular value of R, there can be many values of m_R, that is, for a particular value of R, there are many different states corresponding to different projections of R on the z-axis, all of which have the same energy,

$$m_R = 0, \quad \pm 1, \quad \pm 2, \quad \ldots, \quad \pm R. \tag{3.17}$$

For each value of R there are $2R+1$ values of m_R,

$$g_R = 2R + 1. \tag{3.18}$$

In reality the notation m_R is almost never encountered in the literature. The nuclear rotation angular momentum $\boldsymbol{R}$ couples vectorially with the spin and orbital angular momenta of the electrons in the molecules to give the total angular momentum $\boldsymbol{J}$.

3.3.2 Vibrational Energy Levels and Wavefunctions for the Harmonic Oscillator Potential

With the solution of the angular part of the SWE, the radial part of the SWE can be reduced to

$$\left\{ \frac{1}{r^2} \frac{d}{dr} \left(r^2 \frac{dR(r)}{dr} \right) \right\} + \frac{2\mu}{\hbar^2} [\varepsilon_{vib} - V_i(r)] R(r) = 0. \tag{3.19}$$

Consider the harmonic oscillator potential function

$$V(r) = \varepsilon^{el}(r) = \frac{1}{2} k (r - r_e)^2. \tag{3.20}$$

This is equivalent to a Hooke's law force field with

$$F = -k(r - r_e) \qquad k = 4\pi^2 \mu \nu_{vib}^2. \tag{3.21}$$

Again, the solution to the SWE is rather complicated. The radial part of the wavefunction can be expressed in terms of Hermite polynomials:

$$x = r - r_e \qquad \xi = \sqrt{\frac{4\pi^2 \mu \nu_{vib}}{h}} x \qquad \psi_{\mathrm{v}}^{vib}(\xi) = u_{\mathrm{v}}(\xi) = H_{\mathrm{v}}(\xi) \exp(-\xi^2/2). \tag{3.22}$$

The important result is that the vibrational energy of the system is quantized,

$$\varepsilon_{vib} = \left(\mathrm{v} + \frac{1}{2}\right) h \sqrt{\frac{k}{2\pi^2 \mu}} = \left(\mathrm{v} + \frac{1}{2}\right) h \nu_{vib} \quad \mathrm{v} = 0,\ 1,\ 2,\ 3,\ \ldots, \tag{3.23}$$

and the degeneracy of each vibration level is equal to 1 (they are nondegenerate)

$$g_{\mathrm{v}} = 1. \tag{3.24}$$

3.3.3 Further Considerations of Vibrational and Rotational Energy Levels for Diatomic Molecules

For the idealized rigid-rotor, harmonic oscillator, the rotational and vibrational term energies are given by

$$F(J) = \frac{\varepsilon_{rot}}{hc} = B_e J(J+1) \qquad B_e = \frac{\hbar}{4\pi c \mu r_e^2}. \tag{3.25}$$

$$G(\mathrm{v}) = \frac{\varepsilon_{vib}}{hc} = \omega_e \left(\mathrm{v} + \frac{1}{2}\right) \qquad \omega_e = \frac{1}{2\pi c}\sqrt{\frac{k}{\mu}}. \tag{3.26}$$

However, these expressions are not accurate enough for actual diatomic molecules. The vibrational potential well for the diatomic molecule is described more accurately by the Morse potential (Morse, 1929),

$$V(r) = D_e\{1 - \exp[-\beta(r - r_e)]\}^2, \tag{3.27}$$

where D_e is the dissociation energy of the molecule and β is a constant with units of inverse length. The vibrational level term energies that result from a solution of the two-body SWE with the Morse potential are given by

$$G(\mathrm{v}) = \omega_e \left(\mathrm{v} + \frac{1}{2}\right) - \omega_e x_e \left(\mathrm{v} + \frac{1}{2}\right)^2, \tag{3.28}$$

where

$$\omega_e = \frac{\beta}{\pi c}\sqrt{\frac{D_e}{2\mu}} \qquad \omega_e x_e = \frac{hc\omega_e}{4D_e}. \tag{3.29}$$

From a perturbation analysis of the rotational energy levels, Morse (1929) also concluded that the rotational term energy can be more accurately described by

$$F_v(J) = B_v J(J+1) - D_v J^2(J+1)^2, \tag{3.30}$$

where the rotational constant B_v is given by (Pekeris, 1934)

$$B_v = B_e - \alpha_e\left(v + \frac{1}{2}\right) \qquad B_e = \frac{\hbar}{4\pi c\mu r_e^2} \qquad \alpha_e = 6\frac{B_e}{\omega_e}\left[\sqrt{B_e x_e \omega_e} - B_e\right], \tag{3.31}$$

and the centrifugal stretching constant D_v is given by (Morse, 1929; Pekeris, 1934)

$$D_v = \frac{4B_e^3}{\omega_e^2}. \tag{3.32}$$

This perturbation analysis will be discussed in much more detail in Section 3.4.

3.4 Electronic Energy Levels and the Coupling of Electronic and Nuclear Rotation Angular Momenta for Diatomic Molecules

The analysis of the energy level structure and spectroscopy of diatomic molecules is complicated by the coupling of the angular momentum $\boldsymbol{R}$ associated with nuclear rotation with the orbital angular momentum $\boldsymbol{L}$ and spin angular momentum $\boldsymbol{S}$ of the electrons. The coupling of the nuclear rotation angular momentum and the orbital and spin angular momenta of the electrons is characterized by Hund's coupling cases. In this section we will consider in detail Hund's cases (a) and (b), which are by far the most common of the coupling cases and apply for molecules such as NO, OH, CH, CN, and NH that are important for combustion diagnostics.

3.4.1 Rotational Wavefunctions for Symmetric Top Molecules

In general a rigid body will be characterized by three principal moments of inertia, I_{aa}, I_{bb}, and I_{cc}, referenced to rotations about the principal inertial axes x, y, and z, respectively, which are fixed to the molecule. The Hamiltonian for the rigid body rotation is given by (Brown & Carrington, 2003)

$$\hat{H}_R = A\hat{P}_x^2 + B\hat{P}_y^2 + C\hat{P}_z^2, \tag{3.33}$$

where the parameters A, B, and C are given by

$$A = \frac{\hbar^2}{2I_{aa}} \qquad B = \frac{\hbar^2}{2I_{bb}} \qquad C = \frac{\hbar^2}{2I_{cc}} \qquad A \geq B \geq C. \tag{3.34}$$

Rotations about the space-fixed axes X, Y, and Z commute with rotations about the molecule-fixed axes x, y, and z, so the quantum numbers J and M which are referenced to the space-fixed axes, are still good quantum numbers for the rigid body rotation. The wavefunction $\psi_{JM}(\phi, \theta, \chi)$ is assumed to be an eigenfunction of the Hamiltonian

H_R. Consider the effects of a rotation of the rigid body through angles $(\phi_2, \theta_2, \chi_2)$ on the wavefunction $\psi_{JM}(\phi_1, \theta_1, \chi_1)$. The effects of a rotation through angles $(\phi_2, \theta_2, \chi_2)$ is described in terms of the rotation matrices,

$$\hat{\boldsymbol{R}}_2 \psi_{JM}(\phi_1, \theta_1, \chi_1) = \psi'_{JM}(\phi_1, \theta_1, \chi_1) = \sum_K D^J_{KM}(\phi_2, \theta_2, \chi_2) \psi_{JK}(\phi_1, \theta_1, \chi_1). \quad (3.35)$$

The value of the transformed wavefunction in the new orientation for the coordinates $(\phi_1, \theta_1, \chi_1)$ is the same as the value of the original wavefunction at the coordinates $(\phi_3, \theta_3, \chi_3)$ in the original orientation that are carried into $(\phi_1, \theta_1, \chi_1)$ by the rotation $\hat{\boldsymbol{R}}_2$,

$$\psi'_{JM}(\phi_1, \theta_1, \chi_1) = \psi_{JM}(\phi_3, \theta_3, \chi_3) = \sum_K D^J_{KM}(\phi_2, \theta_2, \chi_2) \psi_{JK}(\phi_1, \theta_1, \chi_1). \quad (3.36)$$

Now consider the case where the body-fixed axes x, y, and z are initially coincident with the space-fixed axes X, Y, and Z; i.e., $\phi_1 = \theta_1 = \chi_1 = 0$. Now consider that rotation that carries the coordinates $\phi_1 = \theta_1 = \chi_1 = 0$ into $\phi_2 = -\phi$, $\theta_2 = -\theta$, and $\chi_2 = -\chi$. Then we can write

$$\psi_{JM}(\phi, \theta, \chi) = \sum_K D^J_{KM}(-\chi, -\theta, -\phi) \psi_{JK}(0,0,0) = \sum_K D^{J*}_{MK}(\phi, \theta, \chi) \psi_{JK}(0,0,0). \quad (3.37)$$

This is the general equation for the wavefunction of an asymmetric top molecule. For an oblate symmetric top molecule, $A = B$, and Eq. (3.33) can be written as

$$\hat{H}_R = B\hat{P}_x^2 + B\hat{P}_y^2 + B\hat{P}_z^2 - B\hat{P}_z^2 + C\hat{P}_z^2 = B\hat{\boldsymbol{P}}^2 - (B - C)\hat{P}_z^2. \quad (3.38)$$

For a prolate symmetric top molecule, $C = B$, and Eq. (3.33) can be written as

$$\hat{H}_R = B\hat{\boldsymbol{P}}^2 + (A - B)\hat{P}_z^2. \quad (3.39)$$

In both cases the total angular momentum and the projection of the angular momentum on the body-fixed z axis commute with $\hat{H}_R$, and they are thus quantized constants of motion. Therefore, we can write

$$\hat{P}_z \psi_{JM}(\phi, \theta, \chi) = -i\hbar \frac{\partial \psi_{JM}(\phi, \theta, \chi)}{\partial \chi} = \hbar K \psi_{JM}(\phi, \theta, \chi) \quad (3.40)$$

for a particular value of K between $-J$ and $+J$. Equation (3.40) implies that for a symmetric top molecule, the wavefunction $\psi_{JM}(\phi, \theta, \chi)$ is an eigenfunction of the angular momentum operator associated with projection on the internuclear axis. For the symmetric top molecule, the rotation matrix element $D^{J*}_{MK}(\phi, \theta, \chi)$ can thus be nonzero only for that one particular value of K, and the symmetric top wavefunction can be written as

$$\psi_{JKM}(\phi, \theta, \chi) = D^{J*}_{MK}(\phi, \theta, \chi) \psi_{JK}(0,0,0). \quad (3.41)$$

The wavefunction $\psi_{JM}(\phi, \theta, \chi)$ is normalized such that

$$\int \psi^*_{JM}(\phi,\theta,\chi)\psi_{JM}(\phi,\theta,\chi)d\omega = \int D^J_{MK}(\phi,\theta,\chi)\psi^*_{JK}(0,0,0)D^{J*}_{MK}(\phi,\theta,\chi)\psi_{JK}(0,0,0)d\omega$$
$$= \psi^*_{JK}(0,0,0)\psi_{JK}(0,0,0)\int D^J_{MK}(\phi,\theta,\chi)D^{J*}_{MK}(\phi,\theta,\chi)d\omega = 1. \tag{3.42}$$

Using Eq. (2.172), we obtain

$$1 = \psi^*_{JK}(0,0,0)\psi_{JK}(0,0,0)\frac{8\pi^2}{2J+1}. \tag{3.43}$$

The normalized symmetric top wavefunction is therefore given by

$$\psi_{JKM}(\phi,\theta,\chi) = \frac{1}{2\pi}\sqrt{\frac{2J+1}{2}}D^{J*}_{MK}(\phi,\theta,\chi). \tag{3.44}$$

The chief application for our development of the symmetric top wavefunction is that the wavefunction for Hund's case (a) will be a symmetric top wavefunction with quantum number K replaced by the quantum number associated with the projection of the sum of the orbital and spin angular momenta of the electron cloud on the internuclear axis.

3.4.2 Hund's Case (a)

Angular momentum coupling in Hund's case (a) is depicted schematically in Figure 3.3. For atomic systems the orbital angular momentum $\boldsymbol{L}$ is quantized because the electrons move in a spherically symmetric potential. For a diatomic molecule the potential field is axially symmetric, and only the projection of $\boldsymbol{L}$ on the internuclear axis is quantized,

$$L_{AB} = \Lambda\hbar, \tag{3.45}$$

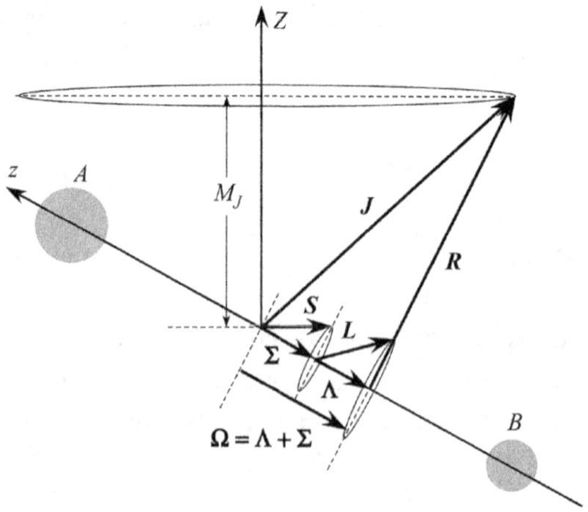

Figure 3.3 Schematic representation of Hund's case (a).

where Λ can be either positive or negative, indicating that the projection can be oriented in either direction along the internuclear axis. For example, for the NO molecule the projection can point toward the O atom or toward the N atom. The spectroscopic symbol for a molecular electronic level is based on the magnitude of the projection quantum number $|\Lambda|$,

$\lvert\Lambda\rvert$	0	1	2	3	4....
Symbol	Σ	Π	Δ	Φ	Γ....

Similarly, the projection of the spin S is quantized along the z-axis,

$$S_{AB} = \Sigma\hbar. \tag{3.46}$$

The quantum number Σ may assume any value from $-S$ to S, $\Sigma = 0, \pm 1, \ldots \pm S$. The electronic multiplet structure is a consequence of spin–orbit coupling effects. The energy of the electronic level is determined principally by the values of S and Λ, and secondarily by the value of Σ. The multiplicity of an electronic level characterized by S and Λ is thus 1 for $\Lambda = 0$ and $2S + 1$ for $|\Lambda| > 0$. In Hund's case (a), the projections $\boldsymbol{\Lambda}$ and $\boldsymbol{\Sigma}$ add vectorially to give the total electronic angular momentum projection vector $\boldsymbol{\Omega}$. The total electronic angular momentum projection vector $\boldsymbol{\Omega}$ couples with the nuclear rotation vector $\boldsymbol{R}$ to give the total angular momentum $\boldsymbol{J}$ (neglecting the coupling with the spin vector $\boldsymbol{I}$ of the nuclei). The projection M_J of $\boldsymbol{J}$ on the space-fixed axis Z is quantized with $2J + 1$ possible values ranging $-J$ to $+J$.

The basis function for a Hund's case (a) diatomic molecule can be written as

$$|\eta\Lambda^s; \mathrm{v}; S\Sigma; J\Omega M_J\rangle = |\eta\Lambda^s\rangle\,|\mathrm{v}\rangle\,|S\Sigma\rangle\,|J\Omega M_J\rangle, \tag{3.47}$$

where $|n\Lambda^s\rangle$ is the electronic orbital basis function, $|\mathrm{v}\rangle$ is the vibrational basis function, $|S\Sigma\rangle$ is the electronic spin basis function, and $|J\Omega M_J\rangle$ is the symmetric top rotational basis function. As noted by Hougen (1970), the wavefunction expression given in Eq. (3.47) is an approximation, and the validity of the approximation is based, among other factors, on the validity of the Born–Oppenheimer approximation. The symmetric top wavefunction is given by

$$|J\Omega M_J\rangle = \psi_{J\Omega M_J}(\phi,\theta,\chi) = \frac{1}{2\pi}\sqrt{\frac{2J+1}{2}}D^{J*}_{M_J\Omega}(\phi,\theta,\chi). \tag{3.48}$$

Explicit expressions for the wavefunctions $|n\Lambda^s\rangle$ and $|S\Sigma\rangle$ are not usually given, although Brown and Carrington (2003) discuss these wavefunctions for simple molecules like H_2^+ and H_2. The parity of the case (a) wavefunctions is discussed in detail by Brown and Carrington (2003), Zare (1988), and Hougen (1970) and in great detail by Brown et al. (1978). The parity of a quantum state is related to the space-fixed inversion operator E^* and is defined by

$$E^*\psi(X,Y,Z) = \psi(-X,-Y,-Z). \tag{3.49}$$

Quantum states have well defined parity such that

$$E^*\psi(X,Y,Z) = \psi(-X,-Y,-Z) = \pm\psi(X,Y,Z). \tag{3.50}$$

Brown and Carrington (2003) discuss the relations for inversion of coordinates in the molecule-fixed coordinate system for case (a) diatomic molecules in detail. The basic results of this analysis are

$$E^*|n\Lambda^s\rangle = (-1)^{\Lambda+s}|n-\Lambda^s\rangle. \tag{3.51}$$

$$E^*|\mathrm{v}\rangle = |\mathrm{v}\rangle. \tag{3.52}$$

$$E^*|S\Sigma\rangle = (-1)^{S-\Sigma}|S-\Sigma\rangle. \tag{3.53}$$

$$E^*|J\Omega M_J\rangle = (-1)^{J-\Omega}|J-\Omega M_J\rangle. \tag{3.54}$$

In Eq. (3.51), the parameter s is even for Σ^+ electronic levels, odd for Σ^- levels, and 0 for electronic levels with $|\Lambda| > 0$. Combining Eqs. (3.51)–(3.54), we obtain

$$E^*|n\Lambda^s;\mathrm{v};S\Sigma;J\Omega M_J\rangle = (-1)^{s+S-2\Sigma+J}|n-\Lambda^s;\mathrm{v};S-\Sigma;J-\Omega M_J\rangle. \tag{3.55}$$

In Eq. (3.55) the relation $\Omega = \Lambda + \Sigma$ has been used. Note that $\Sigma = S \pm n$, where n is an integer. Therefore $S - 2\Sigma = -S \pm 2n$, and Eq. (3.55) can be written as

$$E^*|\eta\Lambda^s;\mathrm{v};S\Sigma;J\Omega M_J\rangle = (-1)^{J-S+s}|\eta-\Lambda^s;\mathrm{v};S-\Sigma;J-\Omega M_J\rangle. \tag{3.56}$$

However, as is obvious from Eq. (3.56), the Hund's case (a) basis function is not an eigenfunction of the parity operator as required for quantum states. Instead, linear combinations of the basis functions given by Eq. (3.47) are used to form wavefunctions that are eigenfunctions of the parity operator,

$$|n\Lambda^s;\mathrm{v};JM_J;+\rangle = \frac{1}{\sqrt{2}}[|n\Lambda^s;\mathrm{v};S\Sigma;J\Omega M_J\rangle + (-1)^{\rho}|n-\Lambda^s;\mathrm{v};S-\Sigma;J-\Omega M_J\rangle] \tag{3.57}$$

$$|n\Lambda^s;\mathrm{v};JM_J;-\rangle = \frac{1}{\sqrt{2}}[|n\Lambda^s;\mathrm{v};S\Sigma;J\Omega M_J\rangle - (-1)^{\rho}|n-\Lambda^s;\mathrm{v};S-\Sigma;J-\Omega M_J\rangle], \tag{3.58}$$

where the positive and negative parity eigenfunctions are given by Eqs. (3.57) and (3.58), respectively, and $\rho = J - S + s$. An exception to this formulation of the basis functions is the case where $\Lambda = \Sigma = 0$. In this case the basis function is given by

$$|\eta 0;\mathrm{v};JM_J;\pm\rangle = \frac{1}{2}\left[|\eta 0;\mathrm{v};S0;J0M_J\rangle \pm (-1)^{J+s}|\eta 0;\mathrm{v};S0;J0M_J\rangle\right]. \tag{3.59}$$

The form of Eq. (3.59) is different because the case (a) wavefunctions on the RHS of the formula are not orthonormal.

The calculation of transition strengths from the wavefunctions for diatomic molecules will be discussed in Chapter 4. In preparation for that discussion, the calculation of Hamiltonian matrix elements will now be discussed. The following matrix element expressions will be quite useful in the evaluation of these matrix elements (Hougen, 1970; Kleiman, 1998; Zare, 1988; Zare et al., 1973):

$$\langle J\Omega M_J|\hat{J}^2|J\Omega M_J\rangle = J(J+1) \tag{3.60}$$

$$\langle J\Omega M_J|\hat{J}_Z|J\Omega M_J\rangle = M_J \tag{3.61}$$

$$\langle J\Omega M_J|\hat{J}_z|J\Omega M_J\rangle = \Omega \tag{3.62}$$

$$\langle J\Omega\mp 1 M_J|\hat{J}'_\pm|J\Omega M_J\rangle = [J(J+1) - \Omega(\Omega\mp 1)]^{1/2} \tag{3.63}$$

$$\langle S\Sigma|\hat{S}^2|S\Sigma\rangle = S(S+1) \tag{3.64}$$

$$\langle S\Sigma|\hat{S}_z|S\Sigma\rangle = \Sigma \tag{3.65}$$

$$\langle S\Sigma \pm 1|\hat{S}'_\pm|S\Sigma\rangle = [S(S+1) - \Sigma(\Sigma \pm 1)]^{1/2} \tag{3.66}$$

$$\langle n\Lambda|\hat{L}^2|n\Lambda\rangle = L(L+1) \tag{3.67}$$

$$\langle n\Lambda|\hat{L}_z|n\Lambda\rangle = \Lambda \tag{3.68}$$

$$\langle n\Lambda \pm 1|\widehat{L'}_\pm|n\Lambda\rangle = [L(L+1) - \Lambda(\Lambda \pm 1)]^{1/2}, \tag{3.69}$$

where $\hat{J}'_\pm = \hat{J}_x \pm i\hat{J}_y$, $\hat{S}'_\pm = \hat{S}_x \pm i\hat{S}_y$ and $\hat{L}'_\pm = \hat{L}_x \pm i\hat{L}_y$; these operator symbols are primed as a reminder that the components are evaluated in a reference system attached to the diatomic molecule. The operator $\hat{\boldsymbol{J}}$ is given by

$$\begin{aligned}\hat{\boldsymbol{J}} = \hat{J}_x\hat{\boldsymbol{e}}_x + \hat{J}_y\hat{\boldsymbol{e}}_y + J_z\hat{\boldsymbol{e}}_z &= \frac{(\hat{J}_x + i\hat{J}_y)}{\sqrt{2}}\frac{(\hat{\boldsymbol{e}}_x - i\hat{\boldsymbol{e}}_y)}{\sqrt{2}} + \frac{(\hat{J}_x - i\hat{J}_y)}{\sqrt{2}}\frac{(\hat{\boldsymbol{e}}_x + i\hat{\boldsymbol{e}}_y)}{\sqrt{2}} + \hat{J}_z\hat{\boldsymbol{e}}_z \\ &= \frac{\hat{J}'_+}{\sqrt{2}}\hat{\boldsymbol{e}}_+ + \frac{\hat{J}'_-}{\sqrt{2}}\hat{\boldsymbol{e}}_- + \hat{J}_z\hat{\boldsymbol{e}}_z.\end{aligned} \tag{3.70}$$

Note that $\hat{\boldsymbol{J}} = \hat{\boldsymbol{J}}^*$. Expressions analogous to Eq. (3.70) apply for $\hat{\boldsymbol{S}}$ and $\hat{\boldsymbol{L}}$ as well. The operators $\hat{J}'_-$, $\hat{S}'_+$, and $\hat{L}'_+$ are raising operators,

$$\hat{J}'_-|J\Omega M_J\rangle = [J(J+1) - \Omega(\Omega+1)]^{1/2}|J\Omega + 1 M_J\rangle \tag{3.71}$$

$$\hat{S}'_+|S\Sigma\rangle = [S(S+1) - \Sigma(\Sigma+1)]^{1/2}|S\Sigma + 1\rangle \tag{3.72}$$

$$\hat{L}'_+|n\Lambda\rangle = [L(L+1) - \Lambda(\Lambda+1)]^{1/2}|n\Lambda + 1\rangle, \tag{3.73}$$

and $\hat{J}'_+$, $\hat{S}'_-$, and $\hat{L}'_-$ are lowering operators,

$$\hat{J}'_+|J\Omega M_J\rangle = [J(J+1) - \Omega(\Omega-1)]^{1/2}|J\Omega - 1 M_J\rangle. \tag{3.74}$$

$$\hat{S}'_-|S\Sigma\rangle = [S(S+1) - \Sigma(\Sigma-1)]^{1/2}|S\Sigma - 1\rangle. \tag{3.75}$$

$$\hat{L}'_-|n\Lambda\rangle = [L(L+1) - \Lambda(\Lambda-1)]^{1/2}|n\Lambda - 1\rangle. \tag{3.76}$$

The anomalous behavior of the operator $\hat{J}'_\pm$ is discussed at length in Zare (1988), Hougen (1970), Brown and Carrington (2003), and in a recent review article by Parigger and Hornkohl (2010).

3.4.3 The Effective Hamiltonian Approach

The effective Hamiltonian approach for the calculation of the energies of rotational levels within a given vibrational level of a diatomic molecule is discussed in detail in chapter 7 of Brown and Carrington (2003). Briefly, the approach is to calculate matrix elements between Hund's case (a) $\langle \eta\Lambda'; \mathrm{v}; S\Sigma'; J\Omega' M_J | \hat{H}' | \eta\Lambda; \mathrm{v}; S\Sigma; J\Omega M_J\rangle$ wavefunctions (or sometimes Hund's case (b) basis wavefunctions) for the interactions $\hat{H}'$ that affect the energies of the rotational levels within a given vibrational level. The basis wavefunction for a given rotational level $|{}^{2S+1}X_\Omega; \Lambda; \mathrm{v}; JM_J; \pm\rangle$ is represented as the sum of the Hund's case (a) wavefunctions as shown above, because the Hund's case (a) wavefunctions $|\eta\Lambda; \mathrm{v}; S\Sigma; J\Omega M_J\rangle$ do not possess a definite parity. The effective Hamiltonian approach allows the determination of the basis wavefunction coefficients and the energies of each level based on the requirements that the rotational level wavefunctions are orthonormal. This aspect of the effective Hamiltonian approach will be explored in detail in Section 3.5 for singlet, doublet, and triplet electronic levels. In the remainder of this section, we will develop the expressions for the density matrix elements for the Hund's case (a) wavefunctions.

3.4.4 Matrix Elements for the Rotational Hamiltonian

The most important energy interaction is the energy associated with the rotation of the nuclei about the an axis perpendicular to the internuclear axis passing through the center of mass of the molecule. There are two different approaches for calculating this Hamiltonian, referred to as the R^2 formulation and the N^2 formulation (Brown et al., 1987). The R^2 formulation is more fundamental, but the N^2 formulation leads to simpler expressions, especially when centrifugal distortion effects are considered. The R^2 formulation gives rise to terms that are not dependent on the rotational quantum number J that are incorporated in the electronic and/or vibration term energies.

The N^2 form of the rotational Hamiltonian operator for the diatomic molecules is given by

$$\hat{H}_R = B(r)\hat{\boldsymbol{N}}^2 = B(r)\left[(\hat{\boldsymbol{J}} - \hat{\boldsymbol{S}})\right]^2 = B(r)\left[(\hat{\boldsymbol{J}} - \hat{\boldsymbol{S}})\right] \bullet \left[\left(\hat{\boldsymbol{J}}^* - \hat{\boldsymbol{S}}^*\right)\right]$$
$$= B(r)\left(\hat{\boldsymbol{J}}^2 - 2\hat{\boldsymbol{J}} \bullet \hat{\boldsymbol{S}} + \hat{\boldsymbol{S}}^2\right) = B(r)\left[\hat{\boldsymbol{J}}^2 - 2\hat{J}_z\hat{S}_z + \hat{\boldsymbol{S}}^2 - 2\left(\frac{\hat{J}'_+\hat{S}'_-}{2} + \frac{\hat{J}_-\hat{S}_+}{2}\right)\right]. \tag{3.77}$$

Collecting terms in Eq. (3.77), we obtain

$$\hat{H}_R = \hat{H}_{R1} + \hat{H}_{R2}. \tag{3.78}$$

$$\hat{H}_{R1} = B(r)\left(\hat{\boldsymbol{J}}^2 - 2\hat{J}_z\hat{S}_z + \boldsymbol{S}^2\right). \tag{3.79}$$

$$\hat{H}_{R2} = -B(r)\left(\hat{J}'_+\hat{S}'_- + \hat{J}'_-\hat{S}'_+\right). \tag{3.80}$$

Using Eqs. (3.60)–(3.76), we can evaluate the relevant matrix elements for the rotational Hamiltonian terms for an electronic level with a given value of Λ. All of the terms in Eq. (3.77) that do not include a lowering or raising operator are collected in the operator $\hat{H}_{R1}$. Consequently, the operator $\hat{H}_{R1}$ has a nonzero density matrix element only when the case (a) quantum states are identical, i.e., $\Omega' = \Omega$ and $\Sigma' = \Sigma$. The matrix element is given by

$$\begin{aligned}&\langle \eta\Lambda; \mathrm{v}; S\Sigma'; J\Omega' M_J|\hat{H}_{R1}|\eta\Lambda; \mathrm{v}; S\Sigma; J\Omega M_J\rangle \\ &\quad = B_\mathrm{v}[J(J+1) - 2\Omega\Sigma + S(S+1)]\delta_{\Sigma\Sigma'}\delta_{\Omega\Omega'},\end{aligned} \tag{3.81}$$

where

$$B_\mathrm{v} = \langle \mathrm{v}|B(r)|\mathrm{v}\rangle. \tag{3.82}$$

The operator $\hat{H}_{R2}$, on the other hand, couples case (a) quantum states where $\Lambda' = \Lambda$, $\Sigma' = \Sigma \pm 1$, and $\Omega' = \Omega \pm 1$. For $\Sigma' = \Sigma + 1, \Omega' = \Omega + 1$, we obtain

$$\begin{aligned}&\langle \eta\Lambda; \mathrm{v}; S\Sigma+1; J\Omega+1 M_J|\hat{H}_{R2}|\eta\Lambda; \mathrm{v}; S\Sigma; J\Omega M_J\rangle \\ &= -B_\mathrm{v}\langle \eta\Lambda'; \mathrm{v}; S\Sigma+1; J\Omega+1 M_J|\left(\hat{J}'_+\hat{S}'_- + \hat{J}'_-\hat{S}'_+\right)|\eta\Lambda; \mathrm{v}; S\Sigma; J\Omega M_J\rangle \\ &= -B_\mathrm{v}[J(J+1) - \Omega(\Omega+1)]^{1/2}[S(S+1) - \Sigma(\Sigma+1)]^{1/2}.\end{aligned} \tag{3.83}$$

For $\Sigma' = \Sigma - 1, \Omega' = \Omega - 1$, we obtain

$$\begin{aligned}&\langle \eta\Lambda; \mathrm{v}; S\Sigma-1; J\Omega-1 M_J|\hat{H}_{R2}|\eta\Lambda; \mathrm{v}; S\Sigma; J\Omega M_J\rangle \\ &\quad = -B_\mathrm{v}[J(J+1) - \Omega(\Omega-1)]^{1/2}[S(S+1) - \Sigma(\Sigma-1)]^{1/2}.\end{aligned} \tag{3.84}$$

The R^2 form of the rotational Hamiltonian operator for the diatomic molecules is given by

$$\begin{aligned}\hat{H}_R &= B(r)\hat{\boldsymbol{R}}^2 = B(r)\left[(\hat{\boldsymbol{J}} - \hat{\boldsymbol{S}}) - \hat{\boldsymbol{L}}\right]^2 = B(r)\left[(\hat{\boldsymbol{J}} - \hat{\boldsymbol{S}}) - \hat{\boldsymbol{L}}\right]\cdot\left[\left(\hat{\boldsymbol{J}}^* - \hat{\boldsymbol{S}}^*\right) - \hat{\boldsymbol{L}}^*\right] \\ &= B(r)\left[\left(\hat{\boldsymbol{J}}^2 - 2\hat{\boldsymbol{J}}\cdot\hat{\boldsymbol{S}} + \hat{\boldsymbol{S}}^2\right) - \hat{\boldsymbol{L}}\cdot(\hat{\boldsymbol{J}} - \hat{\boldsymbol{S}}) - (\hat{\boldsymbol{J}} - \hat{\boldsymbol{S}})\cdot\hat{\boldsymbol{L}} + \hat{\boldsymbol{L}}^2\right] \\ &= B(r)\left[\hat{\boldsymbol{J}}^2 - 2\hat{J}_z\hat{S}_z + \hat{\boldsymbol{S}}^2 - 2\left(\frac{\hat{J}'_+\hat{S}'_-}{2} + \frac{\hat{J}_-\hat{S}_+}{2}\right) - \hat{L}_z(\hat{J}_z - \hat{S}_z) - \frac{\hat{L}'_+\left(\hat{J}'_- + \hat{S}'_-\right)}{2} - \frac{\hat{L}'_-\left(\hat{J}'_+ + \hat{S}'_+\right)}{2}\right. \\ &\qquad \left. - (\hat{J}_z - \hat{S}_z)\hat{L}_z - \frac{\left(\hat{J}'_- + \hat{S}'_-\right)\hat{L}'_+}{2} - \frac{\hat{L}'_-\left(\hat{J}'_+ + \hat{S}'_+\right)\hat{L}'_-}{2} + \frac{\hat{L}'_+\hat{L}'_-}{2} + \frac{\hat{L}'_-\hat{L}'_+}{2} + \hat{L}_z^2\right].\end{aligned} \tag{3.85}$$

These operators commute, so that, for example, $\hat{L}'_+\left(\hat{J}'_- + \hat{S}'_-\right) = \left(\hat{J}'_- + \hat{S}'_-\right)\hat{L}'_+$; hence, the expression in Eq. (3.77) simplifies considerably. Collecting terms in Eq. (3.77), we obtain

$$\hat{H}_R = \hat{H}_{R1} + \hat{H}_{R2} + \hat{H}_{R3} + \hat{H}_{R4}. \tag{3.86}$$

$$\hat{H}_{R1} = B(r)\left[\hat{\boldsymbol{J}}^2 - 2\hat{J}_z\hat{S}_z + \boldsymbol{S}^2 - 2(\hat{J}_z - \hat{S}_z)\hat{L}_z + \hat{L}_z^2\right]. \tag{3.87}$$

$$\hat{H}_{R2} = -B(r)\left(\hat{J}'_+\hat{S}'_- + \hat{J}'_-\hat{S}'_+\right). \tag{3.88}$$

$$\hat{H}_{R3} = -B(r)\left[\left(\hat{J}'_+ - \hat{S}'_+\right)\hat{L}'_- + \left(\hat{J}'_- - \hat{S}'_-\right)\hat{L}'_+\right]. \tag{3.89}$$

$$\hat{H}_{R4} = \frac{1}{2}B(r)\left(\hat{L}'_+\hat{L}'_- + \hat{L}'_-\hat{L}'_+\right). \tag{3.90}$$

Using Eqs. (3.60)–(3.76), we can evaluate the relevant matrix elements for the rotational Hamiltonian terms for an electronic level with a given value of Λ. All of the terms in Eq. (3.77) that do not include a lowering or raising operator are collected in the operator $\hat{H}_{R1}$. Consequently, the operator $\hat{H}_{R1}$ has a nonzero density matrix element only when the case (a) quantum states are identical, i.e., $\Omega' = \Omega$ and $\Sigma' = \Sigma$. The matrix element is given by

$$\begin{aligned}
&\langle \eta\Lambda; \mathrm{v}; S\Sigma'; J\Omega' M_J|\hat{H}_{R1}|\eta\Lambda; \mathrm{v}; S\Sigma; J\Omega M_J\rangle \\
&\quad = B_\mathrm{v}\left[J(J+1) - 2\Omega\Sigma + S(S+1) - 2(\Omega-\Sigma)\Lambda + \Lambda^2\right]\delta_{\Sigma\Sigma'}\delta_{\Omega\Omega'} \\
&\quad = B_\mathrm{v}\left[J(J+1) - 2\Omega\Sigma + S(S+1) - 2(\Omega-\Sigma)(\Omega-\Sigma) + (\Omega-\Sigma)^2\right]\delta_{\Sigma\Sigma'}\delta_{\Omega\Omega'} \\
&\quad = B_\mathrm{v}\left[J(J+1) + S(S+1) - \Omega^2 - \Sigma^2\right]\delta_{\Sigma\Sigma'}\delta_{\Omega\Omega'},
\end{aligned} \tag{3.91}$$

where again

$$B_\mathrm{v} = \langle \mathrm{v}|B(r)|\mathrm{v}\rangle. \tag{3.92}$$

The operator $\hat{H}_{R2}$, on the other hand, couples case (a) quantum states where $\Lambda' = \Lambda$, $\Sigma' = \Sigma \pm 1$, and $\Omega' = \Omega \pm 1$. For $\Sigma' = \Sigma + 1, \Omega' = \Omega + 1$, we obtain

$$\begin{aligned}
&\langle \eta\Lambda; \mathrm{v}; S\Sigma+1; J\Omega+1 M_J|\hat{H}_{R2}|\eta\Lambda; \mathrm{v}; S\Sigma; J\Omega M_J\rangle \\
&\quad = -B_\mathrm{v}\langle \eta\Lambda'; \mathrm{v}; S\Sigma+1; J\Omega+1 M_J|\left(\hat{J}'_+\hat{S}'_- + \hat{J}'_-\hat{S}'_+\right)|\eta\Lambda; \mathrm{v}; S\Sigma; J\Omega M_J\rangle \\
&\quad = -B_\mathrm{v}[J(J+1) - \Omega(\Omega+1)]^{1/2}[S(S+1) - \Sigma(\Sigma+1)]^{1/2}.
\end{aligned} \tag{3.93}$$

For $\Sigma' = \Sigma - 1, \Omega' = \Omega - 1$, we obtain

$$\begin{aligned}
&\langle \eta\Lambda; \mathrm{v}; S\Sigma-1; J\Omega-1 M_J|\hat{H}_{R2}|\eta\Lambda; \mathrm{v}; S\Sigma; J\Omega M_J\rangle \\
&\quad = -B_\mathrm{v}[J(J+1) - \Omega(\Omega-1)]^{1/2}[S(S+1) - \Sigma(\Sigma-1)]^{1/2}.
\end{aligned} \tag{3.94}$$

The matrix elements for the operator $\hat{H}_{R3}$ are always zero. This is illustrated by evaluating the matrix element for just the first term $B(r)\hat{J}'_+\hat{L}'_-$ in $\hat{H}_{R3}$,

$$\begin{aligned}
&\langle \eta\Lambda; \mathrm{v}; S\Sigma'; J\Omega' M_J|B(r)\hat{J}'_+\hat{L}'_-|\eta\Lambda; \mathrm{v}; S\Sigma; J\Omega M_J\rangle \\
&\quad = B_\mathrm{v}[J(J+1) - \Omega(\Omega-1)]^{1/2}[L(L+1) - \Lambda(\Lambda-1)]^{1/2} \\
&\quad \times\langle \eta\Lambda; S\Sigma'; J\Omega' M_J|\eta\Lambda-1; S\Sigma; J\Omega+1 M_J\rangle = 0.
\end{aligned} \tag{3.95}$$

The matrix elements for the fourth term $\hat{H}_{R4}$ are given by

$$\begin{aligned}&\langle \eta\,\Lambda; \text{v}; S\Sigma'; J\Omega' M_J|\hat{H}_{R4}|\eta\,\Lambda; \text{v}; S\Sigma; J\Omega M_J\rangle\\&=\frac{1}{2}\langle \eta\,\Lambda; \text{v}; S\Sigma'; J\Omega' M_J|B(r)\left(\hat{L}'_+\hat{L}'_- + \hat{L}'_-L_+\right)|\eta\,\Lambda; \text{v}; S\Sigma; J\Omega M_J\rangle\\&=\frac{B_\text{v}}{2}\{[L(L+1)-\Lambda(\Lambda-1)]+[L(L+1)-\Lambda(\Lambda+1)]\}\delta_{\Sigma'\Sigma}\delta_{\Omega'\Omega}.\end{aligned} \tag{3.96}$$

The matrix elements do not depend on J, Σ, or Ω and thus represent a constant offset in the electronic term energy. This term is discussed in Hougen (1970) and is given the symbol $B\langle L_\perp^2\rangle$ in that monograph. This term is usually incorporated into the electronic term energy T_e.

3.4.5 Matrix Elements for the Spin–Orbit Interaction

The Hamiltonian operator for spin–rotation coupling is given by

$$\hat{H}_{SO} = A(r)\hat{\boldsymbol{L}}\bullet\hat{\boldsymbol{S}} = A(r)\left[\hat{L}_z\hat{S}_z + \frac{1}{2}\left(\hat{L}'_+\hat{S}'_- + \hat{L}'_-\hat{S}'_+\right)\right]. \tag{3.97}$$

The second term $\frac{1}{2}A(r)\left(\hat{L}'_+\hat{S}'_- + \hat{L}'_-\hat{S}'_+\right)$ makes no contribution to the spin–orbit interaction energy, as can be seen by evaluating the matrix element for $A(r)\hat{L}'_+\hat{S}'_-$,

$$\begin{aligned}&\langle \eta\,\Lambda; \text{v}; S\Sigma'; J\Omega' M_J|A(r)\hat{L}'_+\hat{S}'_-|\eta\,\Lambda; \text{v}; S\Sigma; J\Omega M_J\rangle\\&=A_\text{v}[L(L+1)-\Lambda(\Lambda+1)]^{1/2}[S(S+1)-\Sigma(\Sigma-1)]^{1/2}\\&\times\langle \eta\,\Lambda; \text{v}; S\Sigma'; J\Omega' M_J|\eta\,\Lambda+1; \text{v}; S\Sigma-1; J\Omega M_J\rangle = 0.\end{aligned} \tag{3.98}$$

The matrix element for the first term $A(r)\hat{L}_z\hat{S}_z$ is given by

$$\langle \eta\,\Lambda; \text{v}; S\Sigma'; J\Omega' M_J|A(r)\hat{L}_z\hat{S}_z|\eta\,\Lambda; \text{v}; S\Sigma; J\Omega M_J\rangle = (A_\text{v}\Lambda\Sigma)\delta_{\Sigma\Sigma'}\delta_{\Lambda\Lambda'}, \tag{3.99}$$

where

$$A_\text{v} = \langle \text{v}|A(r)|\text{v}\rangle. \tag{3.100}$$

3.4.6 Matrix Elements for Spin–Rotation Coupling

The Hamiltonian operator for spin–rotation coupling is given by

$$\begin{aligned}\hat{H}_{SR} &= \gamma(r)\hat{\boldsymbol{N}}\bullet\hat{\boldsymbol{S}} = \gamma(r)(\hat{\boldsymbol{J}}-\hat{\mathbf{S}})\bullet\hat{\boldsymbol{S}} = \gamma(r)\left(\hat{\boldsymbol{J}}\bullet\hat{\boldsymbol{S}} - \hat{\boldsymbol{S}}^2\right)\\&= \gamma(r)\left(\hat{J}_z\hat{S}_z - \hat{\boldsymbol{S}}^2\right) + \frac{\gamma(r)}{2}\left(\hat{J}'_+\hat{S}'_- + \hat{J}'_-\hat{S}'_+\right) = \hat{H}_{SR1} + \hat{H}_{SR2}.\end{aligned} \tag{3.101}$$

The spin–rotation operator $\hat{H}_{SR1}$ couples states with $\Sigma' = \Sigma$,

$$\begin{aligned}&\langle \eta\Lambda; S\Sigma'; \text{v}; J\Omega' M_J|\hat{H}_{SR1}|\eta\Lambda; S\Sigma; \text{v}; J\Omega M_J\rangle\\&= \langle \eta\Lambda; S\Sigma'; \text{v}; J\Omega' M_J|\gamma(r)\left(\hat{J}_z\hat{S}_z - \hat{\boldsymbol{S}}^2\right)|\eta\Lambda; S\Sigma; \text{v}; J\Omega M_J\rangle\\&= \gamma_\text{v}[\Omega\Sigma - S(S+1)]\delta_{\Sigma\Sigma'}\delta_{\Omega\Omega'},\end{aligned} \tag{3.102}$$

where

$$\gamma_\mathrm{v} = \langle \mathrm{v}|\gamma(r)|\mathrm{v}\rangle. \tag{3.103}$$

The spin–rotation operator $\hat{H}_{SR2}$ couples states with $\Sigma' = \Sigma \pm 1$,

$$\begin{aligned}
&\langle \eta\Lambda S\Sigma' \mathrm{v} J\Omega' M_J|\hat{H}_{SR2}|\eta\Lambda; S\Sigma; \mathrm{v}; J\Omega M_J\rangle \\
&= \langle \eta\Lambda; S\Sigma \pm 1; \mathrm{v}; J\Omega \pm 1 M_J|\frac{\gamma(r)}{2}\left(\hat{J}_+\hat{S}_- + \hat{J}_-\hat{S}_+\right)|\eta\Lambda; S\Sigma; \mathrm{v}; J\Omega M_J\rangle \\
&= \frac{\gamma_\mathrm{v}}{2}[J(J+1) - \Omega(\Omega \pm 1)]^{1/2}[S(S+1) - \Sigma(\Sigma \pm 1)]^{1/2}.
\end{aligned} \tag{3.104}$$

3.4.7 Matrix Elements for Spin–Spin Coupling

The Hamiltonian operator for spin–spin coupling is given by

$$\hat{H}_{SS} = \frac{2}{3}\lambda(r)\left(3\hat{S}_z^2 - \hat{S}^2\right). \tag{3.105}$$

The spin–spin operator $\hat{H}_{SS}$ couples states with $\Sigma' = \Sigma\Sigma' = \Sigma$,

$$\begin{aligned}
&\langle \eta\Lambda; S\Sigma'; \mathrm{v}; J\Omega' M_J|\frac{2}{3}\lambda(r)\left(3\hat{S}_z^2 - \hat{S}^2\right)|\eta\Lambda; S\Sigma; \mathrm{v}; J\Omega M_J\rangle \\
&= 2\lambda_\mathrm{v}\left[\Sigma^2 - \frac{S(S+1)}{3}\right]\delta_{\Sigma\Sigma'}\delta_{\Omega\Omega'},
\end{aligned} \tag{3.106}$$

where

$$\lambda_\mathrm{v} = \langle \mathrm{v}|\lambda(r)|\mathrm{v}\rangle. \tag{3.107}$$

Note from Eq. (3.106) that for doublet states, $S = |\Sigma| = \frac{1}{2}$, the spin–spin energy is zero.

3.4.8 Matrix Elements for the Rotational Centrifugal Distortion Hamiltonian

In the N^2 formulation, the Hamiltonian operator for the first-order correction for centrifugal distortion is given by

$$\begin{aligned}
\hat{H}_{CD} &= -D(r)\hat{N}^4 = -D(r)\left[\hat{F}_{CD1} + \hat{F}_{CD2}\right]^2 \\
&= -D(r)\left[\hat{F}_{CD1}^2 + \hat{F}_{CD2}^2 + \hat{F}_{CD1}\hat{F}_{CD2} + \hat{F}_{CD2}\hat{F}_{CD1}\right].
\end{aligned} \tag{3.108}$$

$$\hat{F}_{CD1} = \hat{H}_{R1}/B(r) = \hat{\boldsymbol{J}}^2 - 2\hat{J}_z\hat{S}_z + \boldsymbol{S}^2. \tag{3.109}$$

$$\hat{F}_{CD2} = \hat{H}_{R2}/B(r) = -\left(\hat{J}'_+\hat{S}'_- + \hat{J}'_-\hat{S}'_+\right). \tag{3.110}$$

We can now evaluate the case (a) matrix elements for each term on the right-hand side of Eq. (3.108). The first term, $-D(r)\hat{F}_{CD1}^2$, couples states with $\Sigma' = \Sigma, \Omega' = \Omega$. The diagonal matrix elements are given by

$$\begin{aligned}&\langle \eta\Lambda; \mathrm{v}; S\Sigma'; J\Omega' M_J| - D(r)\hat{F}^2_{CD1}|\eta\Lambda; \mathrm{v}; S\Sigma; J\Omega M_J\rangle \\ &= -D_\mathrm{v}[J(J+1) + S(S+1) - 2\Omega\Sigma]^2 \delta_{\Sigma\Sigma'}\delta_{\Omega\Omega'},\end{aligned} \tag{3.111}$$

where

$$D_\mathrm{v} = \langle \mathrm{v}|D(r)|\mathrm{v}\rangle. \tag{3.112}$$

Evaluation of the second term, $-D(r)\hat{F}^2_{CD2}$ in Eq. (3.108), results in both diagonal and off-diagonal matrix elements. The operator $\hat{F}^2_{CD2}$ is given by

$$\hat{F}^2_{CD2} = \left(-\hat{J}'_+\hat{S}'_- - \hat{J}'_-\hat{S}'_+\right)^2 = \hat{J}'_+\hat{S}'_-\hat{J}'_+\hat{S}'_- + \hat{J}'_-\hat{S}'_+\hat{J}'_-\hat{S}'_+ + \hat{J}'_+\hat{S}'_-\hat{J}'_-\hat{S}'_+ + \hat{J}'_-\hat{S}'_+\hat{J}'_+\hat{S}'_-. \tag{3.113}$$

The first term on the right-hand side of Eq. (3.113) couples states with $\Sigma' = \Sigma - 2, \Omega' = \Omega - 2$,

$$\begin{aligned}&\langle \eta\Lambda; \mathrm{v}; S\Sigma - 2; J\Omega - 2M_J| - D(r)\hat{J}'_+\hat{S}'_-\hat{J}'_+\hat{S}'_-|\eta\Lambda; \mathrm{v}; S\Sigma; J\Omega M_J\rangle \\ &= -D_\mathrm{v}[J(J+1) - \Omega(\Omega-1)]^{1/2}[J(J+1) - (\Omega-1)(\Omega-2)]^{1/2} \\ &\times[S(S+1) - \Sigma(\Sigma-1)]^{1/2}[S(S+1) - (\Sigma-1)(\Sigma-2)]^{1/2}.\end{aligned} \tag{3.114}$$

The second term on the right-hand side of Eq. (3.113), on the other hand, couples states with $\Sigma' = \Sigma + 2, \Omega' = \Omega + 2$,

$$\begin{aligned}&\langle \eta\Lambda; \mathrm{v}; S\Sigma + 2; J\Omega + 2M_J| - D(r)\hat{J}'_-\hat{S}'_+\hat{J}'_-\hat{S}'_+|\eta\Lambda; \mathrm{v}; S\Sigma; J\Omega M_J\rangle \\ &= -D_\mathrm{v}[J(J+1) - \Omega(\Omega+1)]^{1/2}[J(J+1) - (\Omega+1)(\Omega+2)]^{1/2} \\ &\times[S(S+1) - \Sigma(\Sigma+1)]^{1/2}[S(S+1) - (\Sigma+1)(\Sigma+2)]^{1/2}.\end{aligned} \tag{3.115}$$

The third and fourth terms on the right-hand side of Eq. (3.113) couple states with $\Sigma' = \Sigma, \Omega' = \Omega$,

$$\begin{aligned}&\langle \eta\Lambda; \mathrm{v}; S\Sigma; J\Omega M_J| - D(r)\left(\hat{J}'_+\hat{S}'_-\hat{J}'_-\hat{S}'_+ + \hat{J}'_-\hat{S}'_+\hat{J}'_+\hat{S}'_-\right)|\eta\Lambda; \mathrm{v}; S\Sigma; J\Omega M_J\rangle \\ &= -D_\mathrm{v}[J(J+1) - \Omega(\Omega+1)]^{1/2}[J(J+1) - (\Omega+1)\Omega]^{1/2} \\ &\times[S(S+1) - \Sigma(\Sigma+1)]^{1/2}[S(S+1) - (\Sigma+1)\Sigma]^{1/2} \\ &-D_\mathrm{v}[J(J+1) - \Omega(\Omega-1)]^{1/2}[J(J+1) - (\Omega-1)\Omega]^{1/2} \\ &\times[S(S+1) - \Sigma(\Sigma-1)]^{1/2}[S(S+1) - (\Sigma-1)\Sigma]^{1/2} \\ &= -D_\mathrm{v}\left\{\left[J(J+1) - \Omega^2 - \Omega\right]\left[S(S+1) - \Sigma^2 - \Sigma\right]\right. \\ &\left.+\left[J(J+1) - \Omega^2 + \Omega\right]\left[S(S+1) - \Sigma^2 + \Sigma\right]\right\} \\ &= -2D_\mathrm{v}\left\{\left[J(J+1) - \Omega^2\right]\left[S(S+1) - \Sigma^2\right] + \Omega\Sigma\right\}.\end{aligned} \tag{3.116}$$

The value of the term given in Eq. (3.116) in general will be negligible compared to the term given in Eq. (3.111) because it has a $J(J+1)$ dependence rather than a $J^2(J+1)^2$ dependence.

The third term on the right-hand side of Eq. (3.108), $-D(r)\hat{F}_{CD1}\hat{F}_{CD2}$, can be expressed as two separate terms $-D(r)\hat{F}_{CD1}\hat{J}'_+\hat{S}'_-$, which couples states with

$\Sigma' = \Sigma - 1, \Omega' = \Omega - 1$, and $D(r)\hat{F}_{CD1}\hat{J}'_{-}\hat{S}'_{+}$, which couples states with $\Sigma' = \Sigma + 1, \Omega' = \Omega + 1$:

$$\begin{aligned}
&\langle \eta\Lambda; \mathrm{v}; S\Sigma-1; J\Omega-1M_J|D(r)\hat{F}_{CD1}\hat{J}'_{+}\hat{S}'_{-}|\eta\Lambda; \mathrm{v}; S\Sigma; J\Omega M_J\rangle \\
&\quad = [J(J+1)-\Omega(\Omega-1)]^{1/2}[S(S+1)-\Sigma(\Sigma-1)]^{1/2} \\
&\qquad \times\langle \eta\Lambda; \mathrm{v}; S\Sigma-1; J\Omega-1M_J|D(r)\hat{F}_{CD1}|\eta\Lambda; \mathrm{v}; S\Sigma-1; J\Omega-1M_J\rangle \\
&\quad = D_{\mathrm{v}}[J(J+1)-\Omega(\Omega-1)]^{1/2}[S(S+1)-\Sigma(\Sigma-1)]^{1/2} \\
&\qquad \times[J(J+1)+S(S+1)-2(\Omega-1)(\Sigma-1)].
\end{aligned} \tag{3.117}$$

$$\begin{aligned}
&\langle \eta\Lambda; \mathrm{v}; S\Sigma+1; J\Omega+1M_J|D(r)\hat{F}_{CD1}\hat{J}'_{-}\hat{S}'_{+}|\eta\Lambda; \mathrm{v}; S\Sigma; J\Omega M_J\rangle \\
&\quad = [J(J+1)-\Omega(\Omega+1)]^{1/2}[S(S+1)-\Sigma(\Sigma+1)]^{1/2} \\
&\qquad \times\langle \eta\Lambda; \mathrm{v}; S\Sigma+1; J\Omega+1M_J|D(r)\hat{F}_{CD1}|\eta\Lambda; \mathrm{v}; S\Sigma+1; J\Omega+1M_J\rangle \\
&\quad = D_{\mathrm{v}}[J(J+1)-\Omega(\Omega+1)]^{1/2}[S(S+1)-\Sigma(\Sigma+1)]^{1/2} \\
&\qquad \times[J(J+1)+S(S+1)-2(\Omega+1)(\Sigma+1)].
\end{aligned} \tag{3.118}$$

The fourth term on the right-hand side of Eq. (3.108), $-D(r)\hat{F}_{CD2}\hat{F}_{CD1}$, can be expressed as two separate terms $-D(r)\hat{J}'_{+}\hat{S}'_{-}\hat{F}_{CD1}$, which couples states with $\Sigma' = \Sigma - 1, \Omega' = \Omega - 1$, and $D(r)\hat{J}'_{-}\hat{S}'_{+}\hat{F}_{CD1}$, which couples states with $\Sigma' = \Sigma + 1, \Omega' = \Omega + 1$:

$$\begin{aligned}
&\langle \eta\Lambda; \mathrm{v}; S\Sigma-1; J\Omega-1M_J|D(r)\hat{J}'_{+}\hat{S}'_{-}\hat{F}_{CD1}|\eta\Lambda; \mathrm{v}; S\Sigma; J\Omega M_J\rangle \\
&\quad = [J(J+1)+S(S+1)-2\Omega\Sigma] \\
&\qquad \times\langle \eta\Lambda; \mathrm{v}; S\Sigma-1; J\Omega-1M_J|D(r)\hat{J}'_{+}\hat{S}'_{-}|\eta\Lambda; \mathrm{v}; S\Sigma; J\Omega M_J\rangle \\
&\quad = D_{\mathrm{v}}[J(J+1)+S(S+1)-2\Omega\Sigma] \\
&\qquad \times[J(J+1)-\Omega(\Omega-1)]^{1/2}[S(S+1)-\Sigma(\Sigma-1)]^{1/2}.
\end{aligned} \tag{3.119}$$

$$\begin{aligned}
&\langle \eta\Lambda; \mathrm{v}; S\Sigma+1; J\Omega+1M_J|D(r)\hat{J}'_{-}\hat{S}'_{+}\hat{F}_{CD1}|\eta\Lambda; \mathrm{v}; S\Sigma; J\Omega M_J\rangle \\
&\quad = [J(J+1)+S(S+1)-2\Omega\Sigma] \\
&\qquad \times\langle \eta\Lambda; \mathrm{v}; S\Sigma+1; J\Omega+1M_J|D(r)\hat{J}'_{-}\hat{S}'_{+}|\eta\Lambda; \mathrm{v}; S\Sigma; J\Omega M_J\rangle \\
&\quad = D_{\mathrm{v}}[J(J+1)+S(S+1)-2\Omega\Sigma] \\
&\qquad \times[J(J+1)-\Omega(\Omega+1)]^{1/2}[S(S+1)-\Sigma(\Sigma+1)]^{1/2}.
\end{aligned} \tag{3.120}$$

Combining Eqs. (3.117) and (3.119), we obtain

$$\begin{aligned}
&\langle \eta\Lambda; \mathrm{v}; S\Sigma-1; J\Omega-1M_J|-D(r)\left(\hat{F}_{CD1}\hat{F}_{CD2}+\hat{F}_{CD2}\hat{F}_{CD1}\right)|\eta\Lambda; \mathrm{v}; S\Sigma; J\Omega M_J\rangle \\
&\quad = D_{\mathrm{v}}[J(J+1)-\Omega(\Omega-1)]^{1/2}[S(S+1)-\Sigma(\Sigma-1)]^{1/2} \\
&\qquad \times[2J(J+1)+2S(S+1)-2\Omega\Sigma-2(\Omega-1)(\Sigma-1)] \\
&\quad = 2D_{\mathrm{v}}[J(J+1)-\Omega(\Omega-1)]^{1/2}[S(S+1)-\Sigma(\Sigma-1)]^{1/2} \\
&\qquad \times[J(J+1)+S(S+1)-2\Omega\Sigma+(\Omega+\Sigma)-1].
\end{aligned} \tag{3.121}$$

Combining Eqs. (3.118) and (3.120), we obtain

$$\begin{aligned}
&\langle \eta\Lambda; \mathrm{v}; S\Sigma+1; J\Omega+1 M_J| - D(r)\left(\hat{F}_{CD1}\hat{F}_{CD2}+\hat{F}_{CD2}\hat{F}_{CD1}\right)|\eta\Lambda; \mathrm{v}; S\Sigma; J\Omega M_J\rangle \\
&\quad = D_{\mathrm{v}}[J(J+1)-\Omega(\Omega+1)]^{1/2}[S(S+1)-\Sigma(\Sigma+1)]^{1/2} \\
&\qquad \times[2J(J+1)+2S(S+1)-2\Omega\Sigma-2(\Omega+1)(\Sigma+1)] \\
&\quad = 2D_{\mathrm{v}}[J(J+1)-\Omega(\Omega+1)]^{1/2}[S(S+1)-\Sigma(\Sigma+1)]^{1/2} \\
&\qquad \times[J(J+1)+S(S+1)-2\Omega\Sigma-(\Omega+\Sigma)-1].
\end{aligned} \tag{3.122}$$

To summarize, for Hund's case (a) states with states with $\Sigma'=\Sigma, \Omega'=\Omega$, from Eqs. (3.111) and (3.116),

$$\begin{aligned}
&\langle \eta\Lambda; \mathrm{v}; S\Sigma; J\Omega M_J|\hat{H}_{CD}|\eta\Lambda; \mathrm{v}; S\Sigma; J\Omega M_J\rangle \\
&\quad = -D_{\mathrm{v}}\left\{[J(J+1)+S(S+1)-2\Omega\Sigma]^2+2\left[J(J+1)-\Omega^2\right]\left[S(S+1)-\Sigma^2\right]+2\Omega\Sigma\right\}.
\end{aligned} \tag{3.123}$$

For states with $\Lambda=\Lambda', \Sigma'=\Sigma\pm1, \Omega'=\Omega\pm1$, from Eqs. (3.121) and (3.122),

$$\begin{aligned}
&\langle \eta\Lambda; \mathrm{v}; S\Sigma\pm1; J\Omega\pm1 M_J|\hat{H}_{CD}|\eta\Lambda; \mathrm{v}; S\Sigma; J\Omega M_J\rangle \\
&\quad = 2D_{\mathrm{v}}[J(J+1)-\Omega(\Omega\pm1)]^{1/2}[S(S+1)-\Sigma(\Sigma\pm1)]^{1/2} \\
&\qquad \times[J(J+1)+S(S+1)-4\Omega\Sigma\mp2(\Omega+\Sigma)-2].
\end{aligned} \tag{3.124}$$

For states with $\Lambda=\Lambda', \Sigma'=\Sigma\pm2,\ \Omega'=\Omega\pm2$, from Eqs. (3.114) and (3.115),

$$\begin{aligned}
&\langle \eta\Lambda; \mathrm{v}; S\Sigma\pm2; J\Omega\pm2 M_J|\hat{H}_{CD}|\eta\Lambda; \mathrm{v}; S\Sigma; J\Omega M_J\rangle \\
&\quad = -D_{\mathrm{v}}[J(J+1)-\Omega(\Omega\pm1)]^{1/2}[J(J+1)-(\Omega\pm1)(\Omega\pm2)]^{1/2} \\
&\qquad \times[S(S+1)-\Sigma(\Sigma\pm1)]^{1/2}[S(S+1)-(\Sigma\pm1)(\Sigma\pm2)]^{1/2}.
\end{aligned} \tag{3.125}$$

In the R^2 formulation, for Hund's case (a) states with $\Sigma'=\Sigma, \Omega'=\Omega$,

$$\begin{aligned}
&\langle \eta\Lambda; \mathrm{v}; S\Sigma; J\Omega M_J|\hat{H}_{CD}|\eta\Lambda; \mathrm{v}; S\Sigma; J\Omega M_J\rangle \\
&\quad = -D_{\mathrm{v}}\left\{\left[J(J+1)+S(S+1)-\Omega^2-\Sigma^2\right]^2\right. \\
&\qquad \left.+2\left[J(J+1)-\Omega^2\right]\left[S(S+1)-\Sigma^2\right]+2\Omega\Sigma\right\}.
\end{aligned} \tag{3.126}$$

For states with $\Lambda=\Lambda', \Sigma'=\Sigma\pm1, \Omega'=\Omega\pm1$, from Zare et al. (1973),

$$\begin{aligned}
&\langle \eta\Lambda; \mathrm{v}; S\Sigma\pm1; J\Omega\pm1 M_J|\hat{H}_{CD}|\eta\Lambda; \mathrm{v}; S\Sigma; J\Omega M_J\rangle \\
&\quad = 2D_{\mathrm{v}}[J(J+1)-\Omega(\Omega\pm1)]^{1/2}[S(S+1)-\Sigma(\Sigma\pm1)]^{1/2} \\
&\qquad \times[J(J+1)+S(S+1)-\Omega(\Omega\pm1)-\Sigma(\Sigma\pm1)-1].
\end{aligned} \tag{3.127}$$

For states with $\Lambda=\Lambda', \Sigma'=\Sigma\pm2,\ \Omega'=\Omega\pm2$,

$$\begin{aligned}
&\langle \eta\Lambda; \mathrm{v}; S\Sigma\pm2; J\Omega\pm2 M_J|\hat{H}_{CD}|\eta\Lambda; \mathrm{v}; S\Sigma; J\Omega M_J\rangle \\
&\quad = -D_{\mathrm{v}}[J(J+1)-\Omega(\Omega\pm1)]^{1/2}[J(J+1)-(\Omega\pm1)(\Omega\pm2)]^{1/2} \\
&\qquad \times[S(S+1)-\Sigma(\Sigma\pm1)]^{1/2}[S(S+1)-(\Sigma\pm1)(\Sigma\pm2)]^{1/2}.
\end{aligned} \tag{3.128}$$

3.4.9 Matrix Elements for Centrifugal Distortion in the Spin–Orbit Interaction Hamiltonian

The Hamiltonian operator for centrifugal distortion effects in the spin–orbit interaction is given by (Brown et al., 1978)

$$\hat{H}_{SOD} = \frac{1}{2}A_D(r)\left[\hat{N}^2\hat{L}\bullet\hat{S} + \hat{L}\bullet\hat{S}\hat{N}^2\right]$$
$$= \frac{1}{2}A_D(r)\left\{\hat{N}^2\left[\hat{L}_z\hat{S}_z + \frac{1}{2}\left(\hat{L}'_+\hat{S}'_- + \hat{L}'_-\hat{S}'_+\right)\right] + \left[\hat{L}_z\hat{S}_z + \frac{1}{2}\left(\hat{L}'_+\hat{S}'_- + \hat{L}'_-\hat{S}'_+\right)\right]\hat{N}^2\right\}. \tag{3.129}$$

The term $\hat{N}^2 = (\hat{H}_{R1} + \hat{H}_{R2})/B(r) = \hat{F}_{CD1} + \hat{F}_{CD2}$. As shown above, the operator $\frac{1}{2}\left(\hat{L}'_+\hat{S}'_- + \hat{L}'_-\hat{S}'_+\right)$ will result in zero contribution because both terms in the operator change the value of Λ. The operator $\hat{F}_{CD1}$ will couple case (a) quantum states where $\Lambda' = \Lambda$, $\Sigma' = \Sigma$ and $\Omega' = \Omega$.

$$\begin{aligned}
&\langle \eta\Lambda; \mathrm{v}; S\Sigma; J\Omega M_J|\frac{1}{2}A_D(r)\left(\hat{F}_{CD1}\hat{L}_z\hat{S}_z + \hat{L}_z\hat{S}_z\hat{F}_{CD1}\right)|\eta\Lambda; \mathrm{v}; S\Sigma; J\Omega M_J\rangle \\
&= \Lambda\Sigma\langle \eta\Lambda; \mathrm{v}; S\Sigma; J\Omega M_J|\frac{1}{2}A_D(r)\hat{F}_{CD1}|\eta\Lambda; \mathrm{v}; S\Sigma; J\Omega M_J\rangle \\
&+\langle \eta\Lambda; \mathrm{v}; S\Sigma; J\Omega M_J|\frac{1}{2}A_D(r)\hat{L}_z\hat{S}_z|\eta\Lambda; \mathrm{v}; S\Sigma; J\Omega M_J\rangle \\
&\times[J(J+1) + S(S+1) - 2\Omega\Sigma] \\
&= A_{D\mathrm{v}}\Lambda\Sigma[J(J+1) + S(S+1) - 2\Omega\Sigma],
\end{aligned} \tag{3.130}$$

where

$$A_{D\mathrm{v}} = \langle \mathrm{v}|A_D(r)|\mathrm{v}\rangle \tag{3.131}$$

and the results of Eqs. (3.81), (3.98), and (3.99) have been used to evaluate the matrix element in Eq. (3.130). The term $\hat{F}_{CD2} = -\left(\hat{J}'_+\hat{S}'_- + \hat{J}'_-\hat{S}'_+\right)$ will couple case (a) quantum states where $\Lambda' = \Lambda$, $\Sigma' = \Sigma \pm 1$, and $\Omega' = \Omega \pm 1$,

$$\begin{aligned}
&\langle \eta\Lambda; \mathrm{v}; S\Sigma\pm1; J\Omega\pm1 M_J|\frac{1}{2}A_D(r)\left(\hat{F}_{CD2}\hat{L}_z\hat{S}_z + \hat{L}_z\hat{S}_z\hat{F}_{CD2}\right)|\eta\Lambda; \mathrm{v}; S\Sigma; J\Omega M_J\rangle \\
&= \langle \eta\Lambda; \mathrm{v}; S\Sigma\pm1; J\Omega\pm1 M_J|\frac{1}{2}A_D(r)\hat{F}_{CD2}|\eta\Lambda; \mathrm{v}; S\Sigma; J\Omega M_J\rangle\Lambda\Sigma \\
&-\langle \eta\Lambda; \mathrm{v}; S\Sigma\pm1; J\Omega\pm1 M_J|\frac{1}{2}A_D(r)\hat{L}_z\hat{S}_z|\eta\Lambda; \mathrm{v}; S\Sigma\pm1; J\Omega\pm1 M_J\rangle \\
&\times[J(J+1) - \Omega(\Omega\pm1)]^{1/2}[S(S+1) - \Sigma(\Sigma\pm1)]^{1/2} \\
&= -A_{D\mathrm{v}}\Lambda\left(\Sigma\pm\frac{1}{2}\right)[J(J+1) - \Omega(\Omega\pm1)]^{1/2}[S(S+1) - \Sigma(\Sigma\pm1)]^{1/2}.
\end{aligned} \tag{3.132}$$

3.4.10 Matrix Elements for the Hamiltonian for Λ-Doubling in $^{2S+1}\Pi$ Electronic Levels

The energy perturbation associated with the phenomenon of Λ-doubling is calculated using second-order perturbation theory. For Π electronic levels, levels with $\Lambda = \pm1$

are coupled with levels with $\Lambda = \mp 1$ by orbital angular momentum operators acting through intermediate $\Sigma(\Lambda = 0)$ levels (Brown et al., 1987). The effective Hamiltonian operator for Λ-doubling in Π electronic levels is given by (Brown & Carrington 2003; Brown & Merer, 1979)

$$\hat{H}_\Lambda = \frac{q_v}{2}\left(\hat{J}_+'^2 + \hat{J}_-'^2\right) - \left(q_v + \frac{p_v}{2}\right)\left(\hat{J}_+'\hat{S}_+' + \hat{J}_-'\hat{S}_-'\right) + \left(\frac{q_v}{2} + \frac{p_v}{2} + \frac{o_v}{2}\right)\left(\hat{S}_+'^2 + \hat{S}_-'^2\right), \tag{3.133}$$

with the understanding that the effective Hamiltonian will couple levels Λ and Λ' such that $\Lambda' - \Lambda = \pm 2$. The first term will couple case (a) quantum states where $\Sigma' = \Sigma$ and $\Omega' = \Omega \pm 2$,

$$\begin{aligned}\langle \eta \mp 1; v; S\Sigma; J\Omega \mp 2 M_J| &\frac{q_v}{2}\left(\hat{J}_+'^2 + \hat{J}_-'^2\right)|\eta \pm 1; v; S\Sigma; J\Omega M_J\rangle \\ &= \frac{q_v}{2}[J(J+1) - \Omega(\Omega \mp 1)]^{1/2}[J(J+1) - (\Omega \mp 1)(\Omega \mp 2)]^{1/2}.\end{aligned} \tag{3.134}$$

The second term will couple case (a) quantum states where $\Sigma' = \Sigma \pm 1$, and $\Omega' = \Omega \mp 1$,

$$\begin{aligned}\langle \eta \mp 1; v; S\Sigma \pm 1; J\Omega \mp 1 M_J| &-\left(q_v + \frac{p_v}{2}\right)\left(\hat{J}_+'\hat{S}_+' + \hat{J}_-'\hat{S}_-'\right)|\eta \pm 1; v; S\Sigma; J\Omega M_J\rangle \\ &= -\left(q_v + \frac{p_v}{2}\right)[J(J+1) - \Omega(\Omega \mp 1)]^{1/2}[S(S+1) - \Sigma(\Sigma \pm 1)]^{1/2}.\end{aligned} \tag{3.135}$$

The third term will couple case (a) quantum states where $\Sigma' = \Sigma \pm 2$, and $\Omega' = \Omega$,

$$\begin{aligned}\langle \eta \mp 1; v; S\Sigma \pm 2; J\Omega M_J| &\left(\frac{q_v}{2} + \frac{p_v}{2} + \frac{o_v}{2}\right)\left(\hat{S}_+'^2 + \hat{S}_-'^2\right)|\eta \pm 1; v; S\Sigma; J\Omega M_J\rangle \\ &= \left(\frac{q_v}{2} + \frac{p_v}{2} + \frac{o_v}{2}\right)[S(S+1) - \Sigma(\Sigma \pm 1)]^{1/2}[S(S+1) - (\Sigma \pm 1)(\Sigma \pm 2)]^{1/2}.\end{aligned} \tag{3.136}$$

The effective Hamiltonian operator for Λ-doubling in Δ electronic levels is given by Brown et al. (1987). The effects of Λ-doubling in Δ electronic levels are much less significant than for Π electronic levels because Λ-doubling is a fourth-order perturbation effect in Δ electronic levels; levels with $\Lambda = \pm 2$ are coupled with levels with $\Lambda = \mp 2$ by orbital angular momentum operators.

3.4.11 Hund's Case (b)

Angular momentum coupling in Hund's case (b) is depicted schematically in Figure 3.4. Just as for case (a), the projection of $\boldsymbol{L}$ on the internuclear axis is quantized,

$$L_{AB} = \Lambda\hbar. \tag{3.137}$$

Again, Λ can be either positive or negative, indicating that the projection can be oriented in either direction along the internuclear axis. Unlike case (a), however, the projection of the electronic spin $\boldsymbol{S}$ is not quantized along the z-axis; rather, the spin is

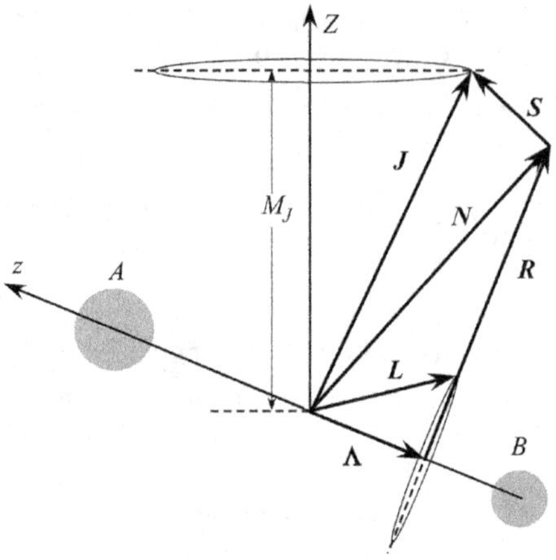

Figure 3.4 Schematic representation of Hund's case (b).

uncoupled from the internuclear axis. The electronic angular momentum projection $\boldsymbol{\Lambda}$ and the nuclear rotation $\boldsymbol{R}$ couple to give the resultant angular momentum vector $\boldsymbol{N}$. The angular momentum vector $\boldsymbol{N}$ then couples with the electronic spin angular momentum vector $\boldsymbol{S}$ to give the total angular momentum vector $\boldsymbol{J}$. The wavefunction $|\eta\Lambda; \mathrm{v}; N\Lambda SJM_J\rangle$ for a Hund's case (b) diatomic molecule can be written as a summation over case (a) wavefunctions (Brown & Carrington, 2003):

$$|\eta\Lambda; \mathrm{v}; N\Lambda SJM_J\rangle$$
$$= \sum_{\Sigma=-S}^{+S} (-1)^{J-S+\Lambda}\sqrt{2N+1}\begin{pmatrix} J & S & N \\ \Omega & -\Sigma & -\Lambda \end{pmatrix} |\eta\Lambda\rangle\,|\mathrm{v}\rangle\,|S\Sigma\rangle\,|J\Omega M_J\rangle. \tag{3.138}$$

3.5 Rotational Level Hamiltonians and Energies

The term symbol for the electronic levels of the diatomic molecule characterized by Hund's case (a) is given by

$$^{2S+1}\Lambda_{\Omega}. \tag{3.139}$$

The term $2S+1$ is termed the multiplicity of the electronic level. For molecules characterized by Hund's case (b) rather than case (a), the subscript Ω is omitted. The consideration of the energy level structure of diatomic molecules is discussed in detail in Mavrodineanu and Boiteux (1965). Here, the energy level structure will be discussed for particular electronic terms, following the treatments of Zare (1988) and Hougen (1970).

In the treatment that follows, the Hamiltonian $\hat{H}_0$ will include the electronic and vibrational Hamiltonians,

$$\hat{H}_0 = \hat{H}_{elec} + \hat{H}_{vib}, \tag{3.140}$$

and the Hamiltonian $\hat{H}'$ will include the rotation, spin–orbit, and spin–rotation Hamiltonians, and, in principle, terms accounting for centrifugal distortion of all of these energy terms,

$$\begin{aligned} \hat{H}' &= \hat{H}_R + \hat{H}_{SO} + \hat{H}_{SR} + \hat{H}_{SS} + \cdots \\ &= B(r)\hat{\boldsymbol{N}}^2 + A(r)\hat{\boldsymbol{L}} \bullet \hat{\boldsymbol{S}} + \gamma(r)\hat{\boldsymbol{N}} \bullet \hat{\boldsymbol{S}} + \frac{2}{3}\lambda(r)\left(3\hat{S}_z^2 - \hat{\boldsymbol{S}}^2\right) + \cdots, \end{aligned} \tag{3.141}$$

where $\hat{H} = \hat{H}_0 + \hat{H}'$.

3.5.1 Singlet Electronic Levels

$^1\Sigma$ Electronic Levels

For $^1\Sigma$ electronic levels, $\boldsymbol{L} = \boldsymbol{S} = 0$. From Eqs. (3.57) and (3.58), the wavefunction for the molecule is given by

$$\left|^1\Sigma^{\pm}; \Lambda^s; \text{v}; JM_J; +\right\rangle = \frac{1}{2}\left[|\eta 0; \text{v}; 00; J0M_J\rangle + (-1)^{J+s}|\eta 00; \text{v}; 00; J0M_J\rangle\right]. \tag{3.142}$$

$$\left|^1\Sigma^{\pm}; \Lambda^s; \text{v}; JM_J; -\right\rangle = \frac{1}{2}\left[|\eta 0; \text{v}; 00; J0M_J\rangle - (-1)^{J+s}|\eta 0; \text{v}; 00; J0M_J\rangle\right]. \tag{3.143}$$

Note that a factor of $\frac{1}{2}$ is used instead of $\frac{1}{\sqrt{2}}$ as in Eqs. (3.57) and (3.58). The parameter s is 0 for $^1\Sigma^+$ electronic levels and 1 for $^1\Sigma^-$ electronic levels. For $^1\Sigma^+$ electronic levels, therefore, levels with even values of J have positive parity and levels with odd values of J have negative parity; all $^1\Sigma^+$ rotational levels are e levels. The situation is reversed for $^1\Sigma^-$ electronic levels: levels with even and odd values of J have negative and positive parity, respectively, and all $^1\Sigma^-$ rotational levels are f levels. The rotational energy level structure for $^1\Sigma^+$ and $^1\Sigma^-$ electronic levels is shown in Figure 3.5.

The rotational energies of the levels can be determined using the Hund's case (a) matrix elements for the effective Hamiltonians discussed in Section 3.4.2. For rotational levels in $^1\Sigma$ electronic levels, only the Hamiltonians associated with nuclear rotational motion will have nonzero values. Applying the operator $\hat{H}_{R1}$ to the wavefunctions shown in Eqs. (3.142) and (3.143), we obtain

$$\begin{aligned} &\langle \eta 0; \text{v}; 00; J0M_J|\hat{H}_{R1}|\eta 0; \text{v}; 00; J0M_J\rangle \\ &= \langle \eta 0; \text{v}; 00; J0M_J|B(r)\hat{\boldsymbol{J}}^2|\eta 0; \text{v}; 00; J0M_J\rangle \\ &= \langle \text{v}|B(r)\hat{\boldsymbol{J}}^2|\text{v}\rangle\langle J0M_J|\hat{\boldsymbol{J}}^2|J0M_J\rangle = B_\text{v}J(J+1). \end{aligned} \tag{3.144}$$

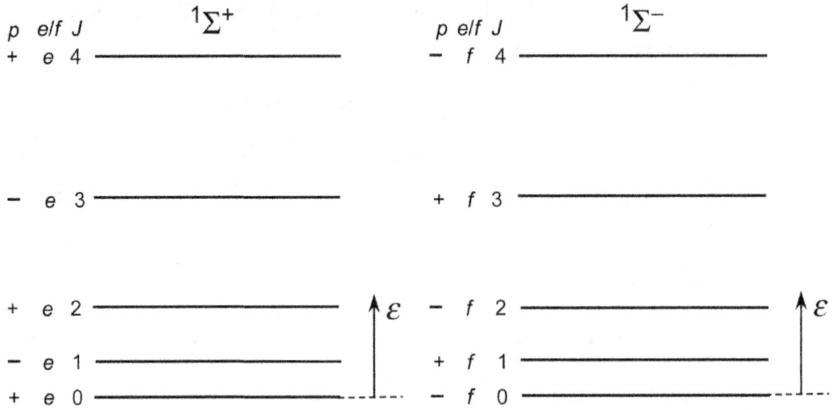

Figure 3.5 Rotational energy level structure in the vibrational bands of $^1\Sigma^+$ and $^1\Sigma^-$ electronic levels. For $^1\Sigma^+$ electronic levels, all levels are *e* levels; for $^1\Sigma^-$ electronic levels, all levels are *f* levels. The energy level spacing as depicted is only approximate.

The Hamiltonian associated with centrifugal distortion of the mechanical rotation also contributes to the energy of the rotational levels. From Eq. (3.123), we obtain

$$\langle \eta 0; \mathrm{v}; 00; J\, 0 M_J | \hat{H}_{CD} | \eta 0; \mathrm{v}; 00; J\, 0 M_J \rangle = -D_{\mathrm{v}} J^2 (J+1)^2. \tag{3.145}$$

The energy associated with $\hat{H}_0$ is the sum of the electronic term energy and the vibrational energy,

$$\varepsilon_0 = \langle \eta 0; \mathrm{v}; 00; J\, 0 M_J | \hat{H}_0 | \eta 0; \mathrm{v}; 00; J\, 0 M_J \rangle = T_e + G(\mathrm{v}). \tag{3.146}$$

The total energy for the rotational level is thus given

$$\varepsilon = \varepsilon_0 + \varepsilon' = T_e + G(\mathrm{v}) + B_{\mathrm{v}} J(J+1) - D_{\mathrm{v}} J^2 (J+1)^2. \tag{3.147}$$

$^1\Pi$ Electronic Levels

For $^1\Pi$ electronic levels, $S = 0$ and $|\Lambda| = 1$. The wavefunction for the molecule can be written as a linear combination of Hund's case (a) wavefunctions,

$$\left|^1\Pi; \mathrm{v}; J M_J; \pm\right\rangle = \frac{1}{\sqrt{2}} \left[|\eta\, 1; \mathrm{v}; 0\,0; J\, 1\, M_J\rangle \pm (-1)^J |\eta - 1; \mathrm{v}; 00; J - 1\, M_J\rangle \right], \tag{3.148}$$

where in this case $\rho = J - S + s = J$. For even J, the positive and negative parity levels are *e* and *f* levels, respectively. For odd J, the positive and negative parity levels are *f* and *e* levels, respectively. The rotational level structure is depicted schematically in Figure 3.6.

The energy of each of the wavefunctions $\left|^1\Pi; \mathrm{v}; J M_J; \pm\right\rangle$ can be found using the case (a) matrix elements for mechanical rotation, centrifugal distortion of mechanical rotation, and Λ-doubling. For mechanical rotation, the matrix elements are given by Eq. (3.81):

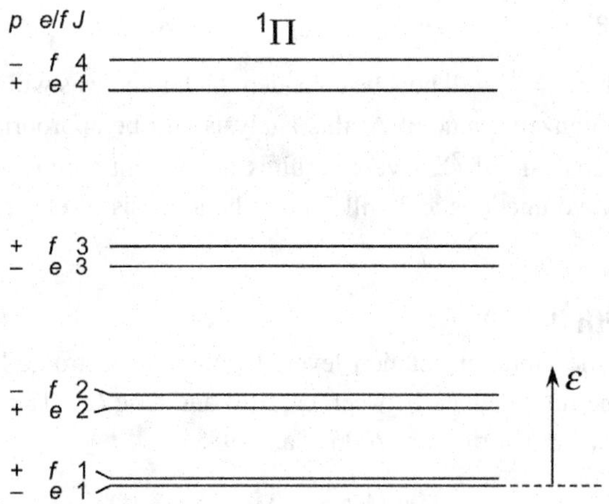

Figure 3.6 Rotational energy level structure in the vibrational bands of $^1\Pi$ electronic levels. The energy level spacing as depicted is only approximate. The Λ-doubling as depicted is characteristic of $q_v > 0$.

$$\begin{aligned}\left\langle {}^1\Pi; v; JM_J; \pm\middle|\hat{H}_{R1}\middle|{}^1\Pi; v; JM_J; \pm\right\rangle &= \frac{1}{2}\langle \eta 1; v; 00; J\,1M_J|\hat{H}_{R1}|\eta 1; v; 00; J1M_J\rangle \\ &+\frac{1}{2}(-1)^{2J}\langle \eta-1; v; 00; J-1M_J|\hat{H}_{R1}|\eta-1; v; 00; J-1M_J\rangle \\ &= B_v[J(J+1)+S(S+1)-2\Omega\Sigma] = B_vJ(J+1).\end{aligned} \tag{3.149}$$

Incorporating the effects of centrifugal stretching using Eq. (3.123), we obtain

$$\begin{aligned}\left\langle {}^1\Pi; v; JM_J; \pm\middle|\hat{H}_{R1}\middle|{}^1\Pi; v; JM_J; \pm\right\rangle &= \frac{1}{2}\langle \eta 1; v; 00; J\,1M_J|\hat{H}_{R1}|\eta 1; v; 00; J1M_J\rangle \\ &+\frac{1}{2}(-1)^{2J}\langle \eta-1; v; 00; J-1M_J|\hat{H}_{R1}|\eta-1; v; 00; J-1M_J\rangle \\ &= -D_v\left\{[J(J+1)+S(S+1)-2\Omega\Sigma]^2+2\left[J(J+1)-\Omega^2\right]\left[S(S+1)-\Sigma^2\right]+2\Omega\Sigma\right\} \\ &= -D_v[J(J+1)]^2.\end{aligned} \tag{3.150}$$

The energy associated with Λ-doubling can be found using Eq. (3.134):

$$\begin{aligned}&\left\langle {}^1\Pi; v; JM_J; \pm\middle|\hat{H}_{R1}\middle|{}^1\Pi; v; JM_J; \pm\right\rangle \\ &= \pm\frac{1}{2}(-1)^J\{\langle \eta-1; v; 00; J-1M_J|\hat{H}_{LD}|\eta 1; v; 00; J1M_J\rangle \\ &+\langle \eta 1; v; 00; J\,1M_J|\hat{H}_{R1}|\eta-1; v; 00; J-1M_J\rangle\} \\ &= \pm\frac{1}{2}(-1)^J\frac{q_v}{2}\left\{[J(J+1)-1(1-1)]^{1/2}[J(J+1)-(1-1)(1-2)]^{1/2}\right. \\ &\left.+[J(J+1)-(-1)(-1+1)]^{1/2}[J(J+1)-(-1+1)(-1+2)]^{1/2}\right\} \\ &= \pm(-1)^J\frac{q_v}{2}J(J+1).\end{aligned} \tag{3.151}$$

3.5.2 Doublet Electronic Levels

Doublet electronic levels $\left(S=\frac{1}{2}\right)$ will now be considered. Our analysis will begin with electronic levels with a nonzero value of Λ; this analysis will be appropriate for $^2\Pi$, $^2\Delta$, $^2\Phi$, …levels. The analysis of $^2\Sigma$ levels requires a different formulation of the Hund's case (a) basis wavefunctions and will follow the analysis of electronic levels with $\Lambda > 0$.

Electronic Levels with $\Lambda > 0$

The wavefunctions for the vibration-rotation levels for the ^2X electronic levels with $\Lambda > 0$ are assumed to be intermediate between case (a) and case (b). The wavefunctions are given by (Brown & Carrington, 2003; Zare,1988)

$$|F_1;\mathrm{v};JM;\pm\rangle = S_{\Lambda-1/2,N+1/2}\left|{}^2X_{\Lambda-1/2};\Lambda;\mathrm{v};JM;\pm\right\rangle + S_{\Lambda+1/2,N+1/2}\left|{}^2X_{\Lambda+1/2};\Lambda;\mathrm{v};JM;\pm\right\rangle, \tag{3.152}$$

$$|F_2;\mathrm{v};JM;\pm\rangle = S_{\Lambda-1/2,N-1/2}\left|{}^2X_{\Lambda-1/2};\Lambda;\mathrm{v};JM;\pm\right\rangle + S_{\Lambda+1/2,N-1/2}\left|{}^2X_{\Lambda+1/2};\Lambda;\mathrm{v};JM;\pm\right\rangle, \tag{3.153}$$

where

$$\begin{aligned}\left|{}^2X_{\Lambda-1/2};\Lambda;\mathrm{v};JM_J;\pm\right\rangle &= \frac{1}{\sqrt{2}}\Big[|\Lambda;S\,\Sigma_-;\mathrm{v};J\,\Omega_- M_J\rangle \\ &\quad \pm(-1)^{J-\frac{1}{2}+s}\left|-\Lambda;\frac{1}{2}-\Sigma_-;\mathrm{v};J-\Omega_- M_J\right\rangle\Big],\end{aligned} \tag{3.154}$$

$$\begin{aligned}\left|{}^2X_{\Lambda+1/2};\Lambda;\mathrm{v};JM_J;\pm\right\rangle &= \frac{1}{\sqrt{2}}\left[\left|\Lambda;\frac{1}{2}\Sigma_+;\mathrm{v};J\Omega_+M_J\right\rangle\right. \\ &\quad \left.\pm(-1)^{J-\frac{1}{2}+s}\left|-\Lambda;\frac{1}{2}-\Sigma_+;\mathrm{v};J-\Omega_+M_J\right\rangle\right],\end{aligned} \tag{3.155}$$

where $s = 1$ for $^2\Sigma^-$ electronic levels, $s = 0$ for $^2\Sigma^+$ electronic levels and for ^2X electronic levels with $\Lambda \geq 1$, $\Sigma_\pm = \pm\frac{1}{2}$, and $\Omega_\pm = \Lambda \pm \frac{1}{2}$. The wavefunctions must be written as the sums of two case (a) basis wavefunctions $|\Lambda;\mathrm{v};S\Sigma;J\Omega M_J\rangle$ and $|-\Lambda;\mathrm{v};S-\Sigma;J-\Omega M\rangle$ because the case (a) basis wavefunctions do not exhibit a definite parity. The wavefunctions given in Eqs. (3.152) and (3.153) are orthonormal, so that

$$1 = S^2_{\Lambda-1/2,N+1/2} + S^2_{\Lambda+1/2,N+1/2}. \tag{3.156}$$

$$1 = S^2_{\Lambda-1/2,N-1/2} + S^2_{\Lambda+1/2,N-1/2}. \tag{3.157}$$

Both the energies of the levels and the coupling constants $S_{\Lambda-1/2,N+1/2}$, $S_{\Lambda+1/2,N+1/2}$, $S_{\Lambda-1/2,N-1/2}$, and $S_{\Lambda+1/2,N-1/2}$ can be determined from an analysis of the Hamiltonian of the system and from the requirement that $|F_1;\mathrm{v};JM_J;+\rangle$ and $|F_2;\mathrm{v};JM_J;+\rangle$ are orthogonal,

$$\langle F_2;\mathrm{v};JM_J;+|F_1;\mathrm{v};JM_J;+\rangle = S_{\Lambda-1/2,N-1/2}S_{\Lambda-1/2,N+1/2} + S_{\Lambda+1/2,N-1/2}S_{\Lambda+1/2,N+1/2} = 0. \tag{3.158}$$

Combined with Eqs. (3.156) and (3.157), this results in the relations

$$S_{\Lambda+1/2,N-1/2} = S_{\Lambda-1/2,N+1/2}, \tag{3.159}$$

$$S_{\Lambda-1/2,N-1/2} = -S_{\Lambda+1/2,N+1/2}, \tag{3.160}$$

although we could have placed the negative sign in Eq. (3.159) instead of Eq. (3.160).

The next step in determining the coefficients and the energies of the F_1 and F_2 wavefunctions is to operate on the wavefunctions in Eqs. (3.152) and (3.153) by multiplying by the ${}^2X_{\Lambda-1/2}$ and ${}^2X_{\Lambda+1/2}$ basis wavefunctions. Dealing first with the F_1 wavefunctions, we can write

$$\hat{H}'|F_1;\mathrm{v};JM_J;\pm\rangle = \varepsilon_1|F_1;\mathrm{v};JM_J;\pm\rangle. \tag{3.161}$$

In Eq. (3.161), the Hamiltonian is given by

$$\hat{H}' = \hat{H}_R + \hat{H}_{SO} + \hat{H}_{SR} + \hat{H}_{CD} + \hat{H}_{SD}, \tag{3.162}$$

where the terms $\hat{H}_R, \hat{H}_{SO}, \hat{H}_{SR}, \hat{H}_{CD}$, and $\hat{H}_{SD}$ are associated with mechanical rotation of the nuclei, the spin–orbit interaction, the spin–rotation interaction, centrifugal distortion of the mechanical rotation, and centrifugal distortion of the spin–orbit interaction, respectively. The results of the operation of these terms on Hund's case (a) wavefunctions are discussed in detail by Zare et al. (1973), Brown and Carrington (2003), and Brown et al. (1978). Multiplying Eq. (3.161) by $\langle {}^2X_{\Lambda-1/2};\Lambda;\mathrm{v};JM_J;\pm|$ and simplifying, we obtain

$$\begin{aligned}\langle {}^2X_{\Lambda-1/2};\Lambda;\mathrm{v};JM_J;\pm|\hat{H}|F_1;\mathrm{v};JM;\pm\rangle &= S_{\Lambda-1/2,N+1/2}H_{11} + S_{\Lambda+1/2,N+1/2}H_{12}\\ = \varepsilon_1\langle {}^2X_{\Lambda-1/2};\Lambda;\mathrm{v};JM_J;\pm|F_1;\mathrm{v};JM;\pm\rangle &= \varepsilon_1 S_{\Lambda-1/2,N+1/2}.\end{aligned} \tag{3.163}$$

The density matrix elements H_{11} and H_{12} are evaluated using the results tabulated in Zare et al. (1973),

$$\begin{aligned}H_{11} &= \langle {}^2X_{\Lambda-1/2};\Lambda;\mathrm{v};JM;\pm|\hat{H}_R + \hat{H}_{SO} + \hat{H}_{SR} + \hat{H}_{CD} + \hat{H}_{SD}|{}^2X_{\Lambda-1/2};\Lambda;\mathrm{v};JM;\pm\rangle\\
&= \frac{1}{2}\Bigg[\left\langle \Lambda;\frac{1}{2}\Sigma_- = \frac{1}{2};\mathrm{v};J\Omega_- = \frac{1}{2}M\middle|\hat{H}\middle|\Lambda;\frac{1}{2}\Sigma_- = \frac{1}{2};\mathrm{v};J\Omega_- = \frac{1}{2}M\right\rangle\\
&\quad +(-1)^{2p}\left\langle -\Lambda;\frac{1}{2}\frac{1}{2};\mathrm{v};J-\frac{1}{2}M\middle|\hat{H}\middle|-\Lambda;\frac{1}{2}\frac{1}{2};\mathrm{v};J-\frac{1}{2}M\right\rangle\\
&\quad \pm(-1)^{p}\left\langle -\Lambda;\frac{1}{2}\frac{1}{2};\mathrm{v};J-\frac{1}{2}M\middle|\hat{H}\middle|\Lambda;\frac{1}{2}-\frac{1}{2};\mathrm{v};J\frac{1}{2}M\right\rangle\\
&\quad \pm(-1)^{p}\left\langle \Lambda;\frac{1}{2}-\frac{1}{2};\mathrm{v};J\frac{1}{2}M\middle|\hat{H}\middle|-\Lambda;\frac{1}{2}\frac{1}{2};\mathrm{v};J-\frac{1}{2}M\right\rangle\Bigg]\\
&= B_\mathrm{v}[J(J+1) + S(S+1) - 2\Omega_-\Sigma_-] + \frac{1}{2}A_\mathrm{v}[\Lambda\Sigma_- + (-\Lambda)(-\Sigma_-)]\\
&\quad +\frac{1}{2}\gamma_\mathrm{v}[\Sigma_-\Omega_- + (-\Sigma_-)(-\Omega_-) - 2S(S+1)]\\
&\quad -\frac{1}{2}D_\mathrm{v}\left\{2[J(J+1) + S(S+1) - 2\Omega_-\Sigma_-]^2 + 4\left[J(J+1) - \Omega_-^2\right]\left[S(S+1) - \Sigma_-^2\right] + 4\Omega_-\Sigma_-\right\}\\
&\quad +\frac{1}{2}A_{D\mathrm{v}}[\Lambda\Sigma_- + (-\Lambda)(-\Sigma_-)][J(J+1) + S(S+1) - 2\Omega_-\Sigma_-]\\
&= \left(B_\mathrm{v} - \frac{1}{2}A_{D\mathrm{v}}\Lambda\right)(z+\Lambda) - \frac{1}{2}A_\mathrm{v}\Lambda - \frac{1}{2}\gamma_\mathrm{v}(\Lambda+1) - D_\mathrm{v}\left[z^2 + (1+2\Lambda)z\right],\end{aligned} \tag{3.164}$$

where $\rho = J - S + s = J - \frac{1}{2}$, $\Omega_- = \Lambda - \frac{1}{2}$, and the constants B_v, A_v, D_v, γ_v, and A_{Dv} are the spectroscopic parameters for energies associated with mechanical rotation of the nuclei, spin–orbit interaction, centrifugal distortion, spin–rotation interaction, and centrifugal distortion of the spin–orbit interaction, respectively. The parameter z is given by

$$z = J(J+1) + \frac{1}{4} = \left(J + \frac{1}{2}\right)^2. \tag{3.165}$$

A similar analysis results in the following expression for H_{12}:

$$\begin{aligned}
H_{12} &= \left\langle {}^2X_{\Lambda-1/2}; \Lambda; v; JM_J; \pm \middle| \hat{H}_R + \hat{H}_{SO} + \hat{H}_{SR} + \hat{H}_{CD} + \hat{H}_{SD} \middle| {}^2X_{\Lambda+1/2}; \Lambda; v; JM_J; \pm \right\rangle \\
&= \frac{1}{2}\left[\left\langle \Lambda; v; \frac{1}{2}\Sigma_-; J\Omega_- M_J \middle| \hat{H}' \middle| \Lambda; v; \frac{1}{2}\Sigma_+; J\Omega_+ M_J \right\rangle \right. \\
&\quad \left. + \left\langle -\Lambda; v; \frac{1}{2} - \Sigma_-; J - \Omega_- M_J \middle| \hat{H}' \middle| -\Lambda; v; \frac{1}{2} - \Sigma_+; J - \Omega_+ M_J \right\rangle \right] \\
&= \left[-B_v - A_{Dv}\left(\Sigma_- + \frac{1}{2}\right)\right][J(J+1) - \Omega_-\Omega_+]^{1/2}[S(S+1) - \Sigma_-\Sigma_+]^{1/2} \\
&\quad + 2D_v[J(J+1) - \Omega_-\Omega_+]^{1/2}[S(S+1) - \Sigma_-\Sigma_+]^{1/2} \\
&\quad \times [J(J+1) + S(S+1) - 4\Omega_-\Sigma_- - 2(\Omega_- + \Sigma_-) - 2] \\
&\quad + \frac{1}{2}\gamma_v[J(J+1) - \Omega_-\Omega_+]^{1/2}[S(S+1) - \Sigma_-\Sigma_+]^{1/2} \\
&= \left(-B_v + \frac{1}{2}\gamma_v + 2D_v z\right)\sqrt{z - \Lambda^2}
\end{aligned} \tag{3.166}$$

and

$$\begin{aligned}
H_{21} &= \left\langle {}^2X_{\Lambda+1/2}; \Lambda; v; JM_J; \pm \middle| \hat{H}_R + \hat{H}_{SO} + \hat{H}_{SR} + \hat{H}_{CD} + \hat{H}_{SD} \middle| {}^2X_{\Lambda-1/2}; \Lambda; v; JM_J; \pm \right\rangle \\
&= \left(-B_v + \frac{1}{2}\gamma_v + 2D_v z\right)\left(z - \Lambda^2\right)^{1/2} = H_{12}.
\end{aligned} \tag{3.167}$$

Multiplying Eq. (3.161) by $\left\langle {}^2X_{\Lambda+1/2}; n\Lambda; v; JM_J; \pm \right|$ and simplifying, we obtain

$$\begin{aligned}
&\left\langle {}^2X_{\Lambda+1/2}; \Lambda; v; JM_J; \pm \middle| \hat{H}_R + \hat{H}_{SO} + \hat{H}_{SR} + \hat{H}_{CD} + \hat{H}_{SD} \middle| F_1; v; JM_J; \pm \right\rangle \\
&\quad = S_{\Lambda-1/2,N+1/2}H_{21} + S_{\Lambda+1/2,N+1/2}H_{22} \\
&\quad = \varepsilon_1 \left\langle {}^2X_{\Lambda+1/2}; \Lambda; v; JM_J; \pm \middle| F_1; v; JM_J; \pm \right\rangle = \varepsilon_1 S_{\Lambda+1/2,N+1/2}.
\end{aligned} \tag{3.168}$$

Performing an analysis similar to that performed for the calculation of H_{11} and H_{12}, we obtain

$$
\begin{aligned}
H_{22} &= \langle {}^2X_{\Lambda+1/2};\Lambda;\mathrm{v};JM_J;\pm|\hat{H}_R+\hat{H}_{SO}+\hat{H}_{SR}+\hat{H}_{CD}+\hat{H}_{SD}|{}^2X_{\Lambda+1/2};\Lambda;\mathrm{v};JM_J;\pm\rangle \\
&= \frac{1}{2}\left[\left\langle \Lambda;\frac{1}{2}\Sigma_+=\frac{1}{2};\mathrm{v};J\Omega_+=\frac{3}{2}M_J\middle|\hat{H}'\middle|\Lambda;\frac{1}{2}\Sigma_+=\frac{1}{2};\mathrm{v};J\Omega_+=\frac{3}{2}M_J\right\rangle\right. \\
&\quad \left.+(-1)^{2p}\left\langle -\Lambda;\frac{1}{2}-\frac{1}{2};\mathrm{v};J-\frac{3}{2}M_J\middle|\hat{H}'\middle|-\Lambda;\frac{1}{2}-\frac{1}{2};\mathrm{v};J-\frac{3}{2}M_J\right\rangle\right] \\
&= B_\mathrm{v}[J(J+1)+S(S+1)-2\Omega_+\Sigma_+]+\frac{1}{2}A_\mathrm{v}[\Lambda\Sigma_++(-\Lambda)(-\Sigma_+)] \\
&\quad +\frac{1}{2}\gamma_\mathrm{v}[\Sigma_+\Omega_++(-\Sigma_+)(-\Omega_+)-2S(S+1)] \\
&\quad -\frac{1}{2}D_\mathrm{v}\left\{2[J(J+1)+S(S+1)-2\Omega_+\Sigma_+]^2+4\left[J(J+1)-\Omega_+^2\right]\left[S(S+1)-\Sigma_+^2\right]+4\Omega_+\Sigma_+\right\} \\
&\quad +\frac{1}{2}A_{D\mathrm{v}}[\Lambda\Sigma_++(-\Lambda)(-\Sigma_+)][J(J+1)+S(S+1)-2\Omega_+\Sigma_+] \\
&= \left(B_\mathrm{v}+\frac{1}{2}A_{D\mathrm{v}}\Lambda\right)(z-\Lambda)+\frac{1}{2}A_\mathrm{v}\Lambda+\frac{1}{2}\gamma_\mathrm{v}(\Lambda-1)-D_\mathrm{v}\left[z^2+(1-2\Lambda)z\right].
\end{aligned}
\tag{3.169}
$$

Now turning our attention to the F_2 wavefunctions,

$$
\hat{H}'|F_2;\mathrm{v};JM_J;\pm\rangle = \varepsilon_2|F_2;\mathrm{v};JM_J;\pm\rangle. \tag{3.170}
$$

Multiplying Eq. (3.170) by $\langle {}^2X_{\Lambda-1/2};\Lambda;\mathrm{v};JM_J;\pm|$ and simplifying, we obtain

$$
\begin{aligned}
&\langle {}^2X_{\Lambda-1/2};\Lambda;\mathrm{v};JM_J;\pm|\hat{H}'|F_2;\mathrm{v};JM_J;\pm\rangle = S_{\Lambda-1/2,N-1/2}H_{33}+S_{\Lambda+1/2,N-1/2}H_{34} \\
&\quad = \varepsilon_2\langle {}^2X_{\Lambda-1/2};\Lambda;\mathrm{v};JM_J;\pm|F_2;\mathrm{v};JM_J;\pm\rangle = \varepsilon_2 S_{\Lambda-1/2,N-1/2}.
\end{aligned}
\tag{3.171}
$$

Multiplying Eq. (3.170) by $\langle {}^2X_{\Lambda+1/2};\Lambda;\mathrm{v};JM_J;\pm|$ and simplifying, we obtain

$$
\begin{aligned}
&\langle {}^2X_{\Lambda+1/2};\Lambda;\mathrm{v};JM_J;\pm|\hat{H}'|F_2;\mathrm{v};JM_J;\pm\rangle = S_{\Lambda-1/2,N-1/2}H_{43}+S_{\Lambda+1/2,N-1/2}H_{44} \\
&\quad = \varepsilon_2\langle {}^2X_{\Lambda+1/2};n\Lambda;\mathrm{v};JM_J;\pm|F_2;\mathrm{v};JM_J;\pm\rangle = \varepsilon_2 S_{\Lambda+1/2,N-1/2}.
\end{aligned}
\tag{3.172}
$$

Note that $H_{33}=H_{11}$, $H_{44}=H_{22}$, and $H_{34}=H_{43}=H_{12}=H_{21}$.

We can write this set of four equations in matrix form as

$$
\begin{bmatrix} H_{11}-\varepsilon_1 & H_{12} \\ H_{21} & H_{22}-\varepsilon_1 \end{bmatrix}\begin{bmatrix} S_{\Lambda-1/2,N+1/2} \\ S_{\Lambda+1/2,N+1/2} \end{bmatrix} = \begin{bmatrix} 0 \\ 0 \end{bmatrix} \tag{3.173}
$$

and

$$
\begin{bmatrix} H_{33}-\varepsilon_2 & H_{34} \\ H_{43} & H_{44}-\varepsilon_2 \end{bmatrix}\begin{bmatrix} S_{\Lambda-1/2,N-1/2} \\ S_{\Lambda+1/2,N-1/2} \end{bmatrix} = \begin{bmatrix} 0 \\ 0 \end{bmatrix}. \tag{3.174}
$$

As discussed in Zare (1988), the system of two equations will have a solution only if the determinant $(H_{11}-\varepsilon)(H_{22}-\varepsilon)-H_{12}H_{21}$ of the secular matrix $\begin{bmatrix} H_{11}-\varepsilon & H_{12} \\ H_{21} & H_{22}-\varepsilon \end{bmatrix}$ is zero,

$$\varepsilon^2 - \varepsilon(H_{11} + H_{22}) + H_{11} + H_{22} - H_{12}^2 = 0. \tag{3.175}$$

Solving Eq. (3.175) for ε we obtain

$$\varepsilon = \left(\frac{H_{11} + H_{22}}{2}\right) \pm \left[\left(\frac{H_{11} - H_{22}}{2}\right)^2 + H_{12}^2\right]^{1/2} = B_v z - \gamma_v - D_v\left(z^2 + z\right)$$
$$\pm\left[\Lambda\left[B_v - \frac{1}{2}(\gamma_v + A_v + A_{Dv}) - 2D_v\right]^2 + \left(-B_v + \frac{1}{2}\gamma_v + 2D_v z\right)^2 \left(z - \Lambda^2\right)\right]^{1/2}. \tag{3.176}$$

We obtain exactly the same result as expressed in Eq. (3.176) if we start with Eq. (3.174). The positive root corresponds to F_2 levels with $N = J + \frac{1}{2}$, and the negative root corresponds to F_1 levels with $N = J - \frac{1}{2}$. In the high-J limit, neglecting centrifugal distortion, the energy of the rotational level is given by

$$\varepsilon \cong B_v\left[J(J+1) \pm \sqrt{J(J+1)}\right]. \tag{3.177}$$

Taking the positive root and substituting $J = N - \frac{1}{2}$ into Eq. (3.177), as is appropriate for F_2 levels, we obtain

$$\varepsilon \cong B_v\left[\left(N - \frac{1}{2}\right)\left(N + \frac{1}{2}\right) + \sqrt{\left(N - \frac{1}{2}\right)\left(N + \frac{1}{2}\right)}\right] \cong \hbar^2 B_v\left(N^2 + N\right) = \hbar^2 B_v N(N+1). \tag{3.178}$$

Taking the negative root and substituting $J = N + \frac{1}{2}$ into Eq. (3.176), as is appropriate for F_1 levels, we obtain

$$\varepsilon \cong B_v\left[\left(N + \frac{1}{2}\right)\left(N + \frac{3}{2}\right) - \sqrt{\left(N + \frac{1}{2}\right)\left(N + \frac{3}{2}\right)}\right] \cong B_v\left(N^2 + 2N - N\right) = B_v N(N+1). \tag{3.179}$$

By associating the positive root with F_2 levels and the negative root with F_1 levels, we obtain the correct limiting forms for the rotational energy (Zare, 1988).

Returning to Eq. (3.173), we have two equations in two unknowns for the coefficients for the F_1 levels, subject to the normalization condition that $1 = S_{\Lambda-1/2,N+1/2}^2 + S_{\Lambda+1/2,N+1/2}^2$:

$$(H_{11} - \varepsilon_1)S_{\Lambda-1/2,N+1/2} + H_{12}S_{\Lambda+1/2,N+1/2} = 0. \tag{3.180}$$

$$H_{21}S_{\Lambda-1/2,N+1/2} + (H_{22} - \varepsilon_1)S_{\Lambda+1/2,N+1/2} = 0. \tag{3.181}$$

Solving Eq. (3.180) for $S_{\Lambda-1/2,N+1/2}$, we obtain

$$S_{\Lambda-1/2,N+1/2}^2 = \frac{H_{12}^2}{\left[\left(H_{11} - \varepsilon_{F_1}\right)^2 + H_{12}^2\right]} = \frac{H_{12}^2}{\left\{\left(\frac{H_{11}-H_{22}}{2}\right) + \left[\left(\frac{H_{11}-H_{22}}{2}\right)^2 + H_{12}^2\right]^{1/2}\right\}^2 + H_{12}^2}. \tag{3.182}$$

Substituting for H_{11}, H_{12}, and H_{22} using Eqs. (3.164), (3.166), and (3.169), respectively, and then rearranging, we obtain

$$S^2_{\Lambda-1/2,N+1/2} = \frac{1}{2}\left(1 - \frac{\beta}{X}\right), \tag{3.183}$$

where

$$\beta = \frac{1}{2}(H_{11} - H_{22}) = \Lambda\left[B_v - \frac{1}{2}A_v - \frac{1}{2}\gamma_v - \alpha\left(\frac{1}{2}A_{Dv} + 2D_v\right)\right], \tag{3.184}$$

$$X = \left[\left(\frac{H_{11} - H_{22}}{2}\right)^2 + H_{12}^2\right]^{1/2}$$
$$= \sqrt{\left(J + \frac{1}{2}\right)^2 \left[B_v - \frac{1}{2}\gamma_v - 2D_v\alpha\right]^2 + \Lambda^2(A_v + \alpha A_{Dv})\left[-B_v + \frac{1}{2}\gamma_v + 2D_v\alpha + \frac{1}{4}(A_v + \alpha A_{Dv})\right]}, \tag{3.185}$$

and

$$S_{\Lambda-1/2,N+1/2} = \sqrt{\frac{1}{2}\left(1 - \frac{\beta}{X}\right)}. \tag{3.186}$$

Choosing the positive square root, we obtain

$$\alpha = J(J+1) + \frac{1}{4} - \Lambda^2 = \left(J + \frac{1}{2}\right)^2 - \Lambda^2. \tag{3.187}$$

The coefficient $S_{\Lambda+1/2,N+1/2}$ is given by

$$S_{\Lambda+1/2,N+1/2} = +\sqrt{1 - S^2_{\Lambda-1/2,N+1/2}} = +\sqrt{\frac{1}{2}\left(1 + \frac{\beta}{X}\right)}. \tag{3.188}$$

A similar procedure for the F_2 levels gives the results

$$S_{\Lambda+1/2,N-1/2} = S_{\Lambda-1/2,N+1/2} = \sqrt{\frac{1}{2}\left(1 - \frac{\beta}{X}\right)}, \tag{3.189}$$

$$S_{\Lambda-1/2,N-1/2} = -S_{\Lambda+1/2,N+1/2} = -\sqrt{\frac{1}{2}\left(1 + \frac{\beta}{X}\right)}, \tag{3.190}$$

where the signs in Eqs. (3.189)–(3.190) have been chosen to be consistent with the results in Kleiman et al. (1998); note that there is a typographical error for the results given for the $^2\Pi$ levels in Zare (1988). It is straightforward to confirm that these coefficients satisfy the normalization condition as expressed in Eq. (3.158).

For electronic levels with $\Lambda \geq 1$, additional Λ-doubling terms can be added to Eqs. (3.184) and (3.185), although these are not expected to have a significant effect on the calculated values of the wavefunction coefficients $S_{\Lambda\pm1/2,N\pm1/2}$. An expression similar to Eq. (3.185) was developed by Bennett (1970), but that expression did not include the spin–orbit centrifugal distortion term A_{Dv}; the spin–orbit centrifugal distortion term and the spin–rotation coupling term γ_v are highly correlated, and usually only one of these is listed while the other is set to zero.

For electronic levels with $\Lambda \geq 1$, the rotational levels (J, N) are split into two levels with different parities and slightly different energies by Λ – doubling. For electronic levels with $\Lambda = 1$, the matrix element for the lambda-doubling operator between two states characterized by $\Omega = \Omega_- = \Lambda - 1/2 = 1/2$ is given by

$$\begin{aligned}
H_{11\Lambda\pm} &= \langle {}^2\Pi_{1/2}; \Lambda; v; JM_J; \pm | \hat{H}_\Lambda | {}^2\Pi_{1/2}; \Lambda; v; JM_J; \pm \rangle \\
&= \frac{1}{2}\Bigg[\left\langle 1; v; \frac{1}{2} - \frac{1}{2}; J\frac{1}{2}M_J \middle| \hat{H}_\Lambda \middle| 1; v; \frac{1}{2} - \frac{1}{2}; J\frac{1}{2}M_J \right\rangle \\
&\quad + (-1)^{2p}\left\langle -1; v; \frac{1}{2}\frac{1}{2}; J - \frac{1}{2}M_J \middle| \hat{H}_\Lambda \middle| -1; v; \frac{1}{2}\frac{1}{2}; J - \frac{1}{2}M_J \right\rangle \\
&\quad \pm (-1)^{p}\left\langle -1; v; \frac{1}{2}\frac{1}{2}; J - \frac{1}{2}M_J \middle| \hat{H}_\Lambda \middle| 1; v; \frac{1}{2} - \frac{1}{2}; J\frac{1}{2}M_J \right\rangle \\
&\quad \pm (-1)^{p}\left\langle 1; v; \frac{1}{2} - \frac{1}{2}; J\frac{1}{2}M_J \middle| \hat{H}_\Lambda \middle| -1; v; \frac{1}{2}\frac{1}{2}; J - \frac{1}{2}M_J \right\rangle\Bigg] \\
&= \mp(-1)^{p}\left(\frac{1}{2}q_v + \frac{1}{4}p_v\right)\left[J(J+1) - \left(\frac{1}{2}\right)\left(\frac{1}{2} - 1\right)\right]^{1/2}\left[\frac{1}{2}\left(\frac{1}{2}+1\right) - \left(-\frac{1}{2}\right)\left(\frac{1}{2}\right)\right]^{1/2} \\
&\quad \mp(-1)^{p}\left(\frac{1}{2}q_v + \frac{1}{4}p_v\right)\left[J(J+1) - \left(-\frac{1}{2}\right)\left(-\frac{1}{2} + 1\right)\right]^{1/2}\left[\frac{1}{2}\left(\frac{1}{2}+1\right) - \left(\frac{1}{2}\right)\left(-\frac{1}{2}\right)\right]^{1/2} \\
&= \mp(-1)^{J-\frac{1}{2}}\left(q_v + \frac{1}{2}p_v\right)\left(J + \frac{1}{2}\right),
\end{aligned} \tag{3.191}$$

where we have used Eq. (3.135) to evaluate the nonzero terms in Eq. (3.191). In terms of *e* and *f* levels, Eq. (3.191) can be written as

$$H_{11\Lambda e/f} = \mp\left(q_v + \frac{1}{2}p_v\right)\left(J + \frac{1}{2}\right), \tag{3.192}$$

where the minus and plus signs apply to *e* and *f* levels, respectively. The density matrix element for the lambda-doubling operator between two positive or negative parity states characterized by $\Omega = \Omega_+ = \Lambda + 1/2 = 3/2$ is zero, as shown below:

$$\begin{aligned}
H_{22\Lambda\pm} &= \langle {}^2X_{3/2}; \Lambda; v; JM_J; \pm | \hat{H}_\Lambda | {}^2X_{3/2}; \Lambda; v; JM_J; \pm \rangle \\
&= \frac{1}{2}\Bigg[\left\langle 1; v; \frac{1}{2}\frac{1}{2}; J\frac{3}{2}M_J \middle| \hat{H}_\Lambda \middle| \Lambda; v; \frac{1}{2}\frac{1}{2}; J\frac{3}{2}M_J \right\rangle \\
&\quad + (-1)^{2p}\left\langle -1; v; \frac{1}{2} - \frac{1}{2}; J - \frac{3}{2}M_J \middle| \hat{H}_\Lambda \middle| -1; v; \frac{1}{2} - \frac{1}{2}; J - \frac{3}{2}M_J \right\rangle \\
&\quad + (-1)^{p}\left\langle -1; v; \frac{1}{2} - \frac{1}{2}; J - \frac{3}{2}M_J \middle| \hat{H}_\Lambda \middle| 1; v; \frac{1}{2}\frac{1}{2}; J\frac{3}{2}M_J \right\rangle \\
&\quad + (-1)^{p}\left\langle 1; v; \frac{1}{2}\frac{1}{2}; J\frac{3}{2}M_J \middle| \hat{H}_\Lambda \middle| -1; v; \frac{1}{2} - \frac{1}{2}; J - \frac{3}{2}M_J \right\rangle\Bigg].
\end{aligned} \tag{3.193}$$

These terms are characterized by $\Sigma' = \Sigma,\ \Omega' = \Omega$ or by $\Sigma' = \Sigma \pm 1,\ \Omega' = \Omega \pm 3$, so $H_{22\Lambda\pm} = 0$ for $^2\Pi$ electronic levels. The off-diagonal density matrix element $H_{12\Lambda\pm}$ for the lambda-doubling operator is given by

$$\begin{aligned}
H_{12\Lambda\pm} &= \langle {}^2X_{1/2};\Lambda;\mathrm{v};JM_J;\pm|\hat{H}_\Lambda|{}^2X_{3/2};\Lambda;\mathrm{v};JM_J;\pm\rangle \\
&= \frac{1}{2}\left[\pm(-1)^p\left\langle -1;\mathrm{v};\frac{1}{2}\frac{1}{2};J-\frac{1}{2}M_J\right|\frac{q_\mathrm{v}}{2}\left(\hat{J}'_+\hat{J}'_+\right)\left|1;\mathrm{v};\frac{1}{2}\frac{1}{2};J\frac{3}{2}M_J\right\rangle\right. \\
&\quad \left.\pm(-1)^p\left\langle 1;\mathrm{v};\frac{1}{2}-\frac{1}{2};J\frac{1}{2}M_J\right|\frac{q_\mathrm{v}}{2}\left(\hat{J}'_-\hat{J}'_-\right)\left|-1;\mathrm{v};\frac{1}{2}-\frac{1}{2};J-\frac{3}{2}M_J\right\rangle\right] \\
&= \pm(-1)^p\frac{1}{4}q_\mathrm{v}\left[J(J+1)-\left(\frac{3}{2}\right)\left(\frac{1}{2}\right)\right]^{1/2}\left[J(J+1)-\left(\frac{1}{2}\right)\left(-\frac{1}{2}\right)\right]^{1/2} \\
&\quad \pm(-1)^p\frac{1}{4}q_\mathrm{v}\left[J(J+1)-\left(-\frac{3}{2}\right)\left(-\frac{1}{2}\right)\right]^{1/2}\left[J(J+1)-\left(-\frac{1}{2}\right)\left(+\frac{1}{2}\right)\right]^{1/2} \\
&= \pm(-1)^{J-\frac{1}{2}}\frac{1}{2}q_\mathrm{v}\left[\left(J+\frac{1}{2}\right)^2-1\right]^{1/2}\left(J+\frac{1}{2}\right),
\end{aligned} \tag{3.194}$$

where we have used Eq. (3.134) to evaluate the nonzero terms in Eq. (3.194). We can use a similar analysis to show that $H_{21\Lambda\pm} = H_{12\Lambda\pm}$. In terms of e and f levels, Eq. (3.194) can be written as

$$H_{12\Lambda e/f} = \pm\frac{1}{2}q_\mathrm{v}\left[\left(J+\frac{1}{2}\right)^2-1\right]^{1/2}\left(J+\frac{1}{2}\right), \tag{3.195}$$

where the minus and plus signs apply to e and f levels, respectively.

These terms given in Eqs. (3.191)–(3.194) can be added directly to Eqs. (3.164), (3.166), (3.167), and (3.169) for the determination of modified values of these terms. Similarly, the energies for the positive and negative parity Λ-doubled levels can be found by adding the terms given in Eqs. (3.191)–(3.194) into Eq. (3.176).

The matrix elements for higher-order corrections connected with centrifugal distortion for the rotational energy and centrifugal distortion corrections for spin–rotation coupling, spin–orbit interaction, and Λ-doubling are listed in Stark et al. (1994). The calculation of higher-order correction terms is simplified using the relation (Gottfried & Yan, 2003; p. 33)

$$\langle k|\hat{B}\hat{A}|k'\rangle = \sum_{k''}\langle k|\hat{B}|k''\rangle\langle k''|\hat{A}|k'\rangle. \tag{3.196}$$

This relation is now applied for the calculation of the first-order centrifugal distortion terms for spin–orbit interaction. In this case the operator is

$$\hat{H}_{SOD} = \frac{1}{2}A_D(r)\left[\hat{\boldsymbol{N}}^2\hat{\boldsymbol{L}}\bullet\hat{\boldsymbol{S}} + \hat{\boldsymbol{L}}\bullet\hat{\boldsymbol{S}}\hat{\boldsymbol{N}}^2\right]. \tag{3.197}$$

The matrix elements that are to be used in the calculation of H_{11SOD}, H_{12SOD}, and H_{22SOD} are given by

$$\frac{H_{11R}}{B_v} = F_{11R} = z + \Lambda \quad F_{22R} = z - \Lambda \quad F_{12R} = F_{21R} = -\left(z - \Lambda^2\right)^{1/2} \tag{3.198}$$

for the mechanical rotation energy operator and

$$\frac{H_{11SO}}{A_v} = F_{11SO} = -\frac{1}{2}\Lambda \quad F_{22SO} = \frac{1}{2}\Lambda \quad F_{12SO} = F_{21SO} = 0 \tag{3.199}$$

for the spin–orbit interaction operator. The density matrix terms H_{11SOD}, H_{12SOD}, and H_{22SOD} are given by

$$\begin{aligned} H_{11SOD} &= \frac{1}{2}A_{Dv}(F_{SO11}F_{R11} + F_{R11}F_{SO11} + F_{SO12}F_{R11} + F_{R12}F_{SO21}) = A_{Dv}F_{11SO}F_{11R} \\ &= -\frac{1}{2}A_{Dv}\Lambda(z + \Lambda) \end{aligned} \tag{3.200}$$

$$\begin{aligned} H_{22SOD} &= \frac{1}{2}A_{Dv}(F_{SO22}F_{R22} + F_{SO21}F_{R12} + F_{R22}F_{SO22} + F_{R21}F_{SO12}) = A_{Dv}F_{22SO}F_{22R} \\ &= \frac{1}{2}A_{Dv}\Lambda(z - \Lambda) \end{aligned} \tag{3.201}$$

and

$$\begin{aligned} H_{12SOD} &= \frac{1}{2}A_{Dv}(F_{12SO}F_{22R} + F_{11SO}F_{12R} + F_{12R}F_{22SO} + F_{11R}F_{12SO}) \\ &= -\frac{1}{2}A_{Dv}\left(z - \Lambda^2\right)^{1/2}\left(-\frac{1}{2}\Lambda + \frac{1}{2}\Lambda\right) = 0. \end{aligned} \tag{3.202}$$

The matrix elements for mechanical rotation, spin–rotation interaction, spin–orbit interaction, and for higher-order centrifugal distortion correction terms are listed in Table 3.1. The details of the derivation of the higher-order correction factors are listed in Appendix A5. As an example of the analysis outlined above, the energy level structure of the $X^2\Pi(\mathrm{v} = 0)$ levels of OH and NO was calculated. The spectroscopic constants, term energies, and basis state coefficients for the $X^2\Pi(\mathrm{v} = 0)$ vibrational levels of NO are listed in Tables 3.2 and 3.3, and these same quantities are listed for the $X^2\Pi(\mathrm{v} = 0)$ vibrational level of OH in Tables 3.4 and 3.5. The rotational energy level structures for $^2\Pi$ levels with $A_v > 0$ and $A_v < 0$ are depicted schematically in Figure 3.7.

$^2\Sigma$ Electronic Levels

If the electronic levels are $\Sigma^\pm$ levels, then the expressions in Eqs. (3.154) and (3.155) cannot be applied directly. For $\Sigma^\pm$ levels, for a given value of $N = J - \frac{1}{2}$, the F_1 state will have positive parity if $\rho = J - \frac{1}{2} + s$ is an even number and negative parity if ρ is an odd number. For F_1 states the coefficients $S_{\Lambda-1/2,N+1/2}$ and $S_{\Lambda+1/2,N+1/2}$ are given by

$$S_{\Lambda-1/2,N+1/2} = S_{\Lambda+1/2,N+1/2} = \frac{1 \pm (-1)^\rho}{4}, \tag{3.203}$$

where in Eqs. (3.203) the + and − signs apply for positive and negative parity states, respectively. For positive parity states, Eq. (3.152) reduces to

Table 3.1 Matrix elements for doublet electronic levels for mechanical rotation, spin–rotation interaction, spin–orbit interaction, and for higher order centrifugal distortion correction terms. The parameter $z = J(J+1) + \frac{1}{4} = (J + \frac{1}{2})^2$

Energy interaction	H_{11}	H_{22}	$H_{12} = H_{21}$
Rotational	$B_v(z+\Lambda)$	$B_v(z-\Lambda)$	$-B_v\sqrt{z-\Lambda^2}$
Rot CD1	$-D_v[z^2+(1+2\Lambda)z]$	$-D_v[z^2+(1-2\Lambda)z]$	$2D_v z\sqrt{z-\Lambda^2}$
Rot CD2	$H_v[z^3+(3+3\Lambda)z^2+\Lambda z]$	$H_v[z^3+(3-3\Lambda)z^2 \ -\Lambda z]$	$-H_v(z-\Lambda^2)^{1/2}\times[3z^2+z]$
Rot CD3	$-L_v[z^4+(6+3\Lambda)z^3+(1+4\Lambda)z^2]$	$-L_v[z^4+(6+3\Lambda)z^3+(1+3\Lambda-7\Lambda^2)z^2]$	$L_v(z-\Lambda^2)^{1/2}\times[4z^3+4z^2]$
Spin–rotation	$-\frac{1}{2}\gamma_v(\Lambda+1)$	$\frac{1}{2}\gamma_v(\Lambda-1)$	$\frac{1}{2}\gamma_v(z-\Lambda^2)^{1/2}$
SR CD1	$-\frac{1}{2}\gamma_{Dv}(2z+\Lambda z+\Lambda)$	$\frac{1}{2}\gamma_{Dv}(-2z+z\Lambda+\Lambda)$	$\frac{1}{2}\gamma_{Dv}(z-\Lambda^2)^{1/2}(z+1)$
SR CD2	$-\frac{1}{2}\gamma_{Hv}[z^2(3+\Lambda)+z(1+3\Lambda)]$	$-\frac{1}{2}\gamma_{Hv}[z^2(-1+\Lambda)+z(1+3\Lambda-2\Lambda^2)-2\Lambda^2]$	$\frac{1}{2}\gamma_{Hv}(z-\Lambda^2)^{1/2}\times[z^2+z(1-\Lambda)-\Lambda]$
Spin–orbit	$-\frac{1}{2}A_v\Lambda$	$\frac{1}{2}A_v\Lambda$	0
SO CD1	$-\frac{1}{2}A_{Dv}\Lambda(z+\Lambda)$	$\frac{1}{2}A_{Dv}\Lambda(z-\Lambda)$	0
SO CD2	$-\frac{1}{2}A_{Hv}\Lambda(z+\Lambda)^2$	$\frac{1}{2}A_{Hv}\Lambda(z-\Lambda)^2$	$\frac{1}{2}A_{Hv}(z-\Lambda^2)^{1/2}\Lambda^2$

Table 3.2 Spectroscopic constants (MHz) for the rotational levels of the $X^2\Pi(\mathrm{v}=0)$ vibrational level of NO. Spectroscopic constants extracted from Varberg et al. (1999). Constants can be converted to units of cm^{-1} by dividing by $c = 2.99792458 \times 10^4\ \mathrm{cm}/\mu\mathrm{s}$

T_v	0.00000	A_v	3,691,913.855	p_v	350.405443
B_v	50,848.13072	γ_v	−193.9879	$p_{D\mathrm{v}}$	3.78
D_v	0.16414119	$\gamma_{D\mathrm{v}}$	0.0015822	q_v	2.82100
$H_\mathrm{v} \times 10^8$	3.774	$A_{D\mathrm{v}}$	0.000	$q_{D\mathrm{v}}$	4.370

$$\begin{aligned}|F_1;\mathrm{v};JM;+\rangle &= \frac{1+(-1)^\rho}{4\sqrt{2}}\left[\left|0;\frac{1}{2}-\frac{1}{2};\mathrm{v};J-\frac{1}{2}M\right\rangle+(-1)^\rho\left|0;\frac{1}{2}\frac{1}{2};\mathrm{v};J\frac{1}{2}M\right\rangle\right]\\ &+\frac{1+(-1)^\rho}{4\sqrt{2}}\left[\left|0;\frac{1}{2}\frac{1}{2};\mathrm{v};J\frac{1}{2}M\right\rangle+(-1)^\rho\left|0;\frac{1}{2}-\frac{1}{2};\mathrm{v};J-\frac{1}{2}M\right\rangle\right]\\ &=\frac{1}{\sqrt{2}}\left|0;\frac{1}{2}\frac{1}{2};\mathrm{v};J\frac{1}{2}M\right\rangle+\frac{1}{\sqrt{2}}\left|0;\frac{1}{2}-\frac{1}{2};\mathrm{v};J-\frac{1}{2}M\right\rangle\end{aligned} \tag{3.204}$$

for even ρ, and the wavefunction is zero for odd ρ. For negative parity states, Eq. (3.152) reduces to the same result

$$\begin{aligned}|F_1;\mathrm{v};JM;-\rangle &= \frac{1-(-1)^\rho}{4\sqrt{2}}\left[\left|0;\frac{1}{2}-\frac{1}{2};\mathrm{v};J-\frac{1}{2}M_J\right\rangle-(-1)^\rho\left|0;\frac{1}{2}\frac{1}{2};\mathrm{v};J-\frac{1}{2}M_J\right\rangle\right]\\ &+\frac{1-(-1)^\rho}{4\sqrt{2}}\left[\left|0;\frac{1}{2}\frac{1}{2};\mathrm{v};J\frac{1}{2}M\right\rangle-(-1)^\rho\left|0;\frac{1}{2}-\frac{1}{2};\mathrm{v};J-\frac{1}{2}M\right\rangle\right]\\ &=\frac{1}{\sqrt{2}}\left|0;\frac{1}{2}\frac{1}{2};\mathrm{v};J\frac{1}{2}M\right\rangle+\frac{1}{\sqrt{2}}\left|0;\frac{1}{2}-\frac{1}{2};\mathrm{v};J-\frac{1}{2}M\right\rangle\end{aligned} \tag{3.205}$$

for odd ρ, and the wavefunction is zero for even ρ. For $^2\Sigma^+$ electronic levels, all F_1 levels are e levels; for $^2\Sigma^-$ electronic levels, all F_1 levels are f levels. The rotational structures of $^2\Sigma^+$ and $^2\Sigma^-$ electronic levels are depicted schematically in Figure 3.8.

The situation is different for F_2 states. For a given value of $N = J + \frac{1}{2}$, the F_2 state will have negative parity if $\rho = J - \frac{1}{2} + s$ is an even number and positive parity if ρ is an odd number. For F_2 states, the coefficients $S_{-1/2,N-1/2}$ and $S_{1/2,N-1/2}$ are given by

$$S_{\Lambda-1/2,N-1/2} = -\frac{1\mp(-1)^\rho}{4}, \tag{3.206}$$

$$S_{\Lambda+1/2,N-1/2} = \frac{1\mp(-1)^\rho}{4}, \tag{3.207}$$

where in Eqs. (3.206) and (3.207) the − and + signs apply for positive and negative parity states, respectively. Substituting Eqs. (3.206) and (3.207) into Eq. (3.153) for a positive parity state with odd ρ, we obtain

$$\begin{aligned}|F_2;\mathrm{v};JM;+\rangle &= -\frac{1-(-1)^\rho}{4\sqrt{2}}\left[\left|0;\frac{1}{2}-\frac{1}{2};\mathrm{v};J-\frac{1}{2}M\right\rangle+(-1)^\rho\left|0;\frac{1}{2}\frac{1}{2};\mathrm{v};J\frac{1}{2}M\right\rangle\right]\\ &\quad\frac{1-(-1)^\rho}{4\sqrt{2}}\left[\left|0;\frac{1}{2}\frac{1}{2};\mathrm{v};J\frac{1}{2}M\right\rangle+(-1)^\rho\left|0;\frac{1}{2}-\frac{1}{2};\mathrm{v};J-\frac{1}{2}M\right\rangle\right]\\ &=\frac{1}{\sqrt{2}}\left|0;\frac{1}{2}\frac{1}{2};\mathrm{v};J\frac{1}{2}M\right\rangle-\frac{1}{\sqrt{2}}\left|0;\frac{1}{2}-\frac{1}{2};\mathrm{v};J-\frac{1}{2}M\right\rangle.\end{aligned} \tag{3.208}$$

Table 3.3 Level energies (cm^{-1}) and Hund's case (a)-basis wavefunction coefficients for the $X^2\Pi(v = 0)$ vibrational level of NO. The Hund's case (a) coefficients $S_{\Omega,J}$ are listed for the *e* levels

J	F_{1e}	F_{1f}	F_{2e}	F_{2f}	$S_{1/2,N+1/2}$	$S_{3/2,N+1/2}$	$S_{1/2,N-1/2}$	$S_{3/2,N-1/2}$
0.5	0.000	0.012			1.0000	0.0000		
1.5	5.010	5.034	124.914	124.914	0.9997	0.0246	−0.0246	0.9997
2.5	13.364	13.400	133.514	133.515	0.9992	0.0400	−0.0400	0.9992
3.5	25.062	25.110	145.554	145.555	0.9985	0.0547	−0.0547	0.9985
4.5	40.104	40.163	161.032	161.033	0.9976	0.0690	−0.0690	0.9976
6.5	80.220	80.301	202.299	202.301	0.9953	0.0969	−0.0969	0.9953
7.5	105.293	105.386	228.085	228.087	0.9939	0.1105	−0.1105	0.9939
8.5	133.710	133.813	257.303	257.306	0.9923	0.1238	−0.1238	0.9923
9.5	165.470	165.584	289.951	289.956	0.9906	0.1370	−0.1370	0.9906
10.5	200.574	200.698	326.028	326.034	0.9887	0.1499	−0.1499	0.9887
11.5	239.020	239.155	365.530	365.539	0.9867	0.1626	−0.1626	0.9867
12.5	280.809	280.953	408.456	408.466	0.9846	0.175	−0.175	0.9846
13.5	325.940	326.093	454.801	454.814	0.9823	0.1872	−0.1872	0.9823
14.5	374.412	374.575	504.564	504.580	0.9800	0.1990	−0.1990	0.9800
15.5	426.225	426.396	557.742	557.760	0.9776	0.2106	−0.2106	0.9776
16.5	481.378	481.558	614.330	614.351	0.9751	0.2219	−0.2219	0.9751
17.5	539.869	540.058	674.325	674.351	0.9725	0.2330	−0.2330	0.9725
18.5	601.699	601.895	737.725	737.754	0.9698	0.2437	−0.2437	0.9698
19.5	666.866	667.069	804.524	804.558	0.9672	0.2542	−0.2542	0.9672
20.5	735.368	735.578	874.720	874.759	0.9644	0.2643	−0.2643	0.9644
21.5	807.203	807.421	948.309	948.353	0.9617	0.2742	−0.2742	0.9617
22.5	882.372	882.596	1,025.286	1,025.336	0.9589	0.2838	−0.2838	0.9589
23.5	960.871	961.101	1,105.648	1,105.703	0.9561	0.2931	−0.2931	0.9561

Table 3.3 (*cont.*)

J	F_{1e}	F_{1f}	F_{2e}	F_{2f}	$S_{1/2,N+1/2}$	$S_{3/2,N+1/2}$	$S_{1/2,N-1/2}$	$S_{3/2,N-1/2}$
24.5	1,042.699	1,042.934	1,189.389	1,189.450	0.9533	0.3021	−0.3021	0.9533
26.5	1,216.332	1,216.578	1,366.995	1,367.069	0.9476	0.3194	−0.3194	0.9476
27.5	1,308.133	1,308.383	1,460.850	1,460.932	0.9448	0.3276	−0.3276	0.9448
28.5	1,403.253	1,403.508	1,558.067	1,558.156	0.9420	0.3356	−0.3356	0.9420
29.5	1,501.690	1,501.949	1,658.641	1,658.739	0.9392	0.3434	−0.3434	0.9392
30.5	1,603.441	1,603.703	1,762.568	1,762.673	0.9364	0.3508	−0.3508	0.9364
31.5	1,708.503	1,708.768	1,869.842	1,869.956	0.9337	0.3581	−0.3581	0.9337
32.5	1,816.872	1,817.141	1,980.458	1,980.582	0.9310	0.3651	−0.3651	0.9310
33.5	1,928.545	1,928.817	2,094.412	2,094.545	0.9283	0.3719	−0.3719	0.9283
34.5	2,043.519	2,043.793	2,211.698	2,211.840	0.9256	0.3785	−0.3785	0.9256
35.5	2,161.791	2,162.066	2,332.310	2,332.462	0.9229	0.3849	−0.3849	0.9229
36.5	2,283.355	2,283.632	2,456.243	2,456.406	0.9203	0.3911	−0.3911	0.9203
37.5	2,408.208	2,408.487	2,583.493	2,583.665	0.9178	0.3971	−0.3971	0.9178
38.5	2,536.346	2,536.626	2,714.052	2,714.235	0.9152	0.4029	−0.4029	0.9152
39.5	2,667.765	2,668.045	2,847.915	2,848.110	0.9127	0.4085	−0.4085	0.9127

Table 3.4 Spectroscopic constants (MHz) for the rotational levels of the $X^2\Pi(\mathrm{v}=0)$ vibrational level of OH. Spectroscopic constants extracted from Brown et al. (1978). Constants can be converted to units of cm^{-1} by dividing by $c = 2.99792458 \times 10^4 \mathrm{cm}/\mu\mathrm{s}$

T_v	0.00000	A_v	−4,169,030	p_v	7,051.5598
B_v	555,538	γ_v	−3,690.5	$p_{D\mathrm{v}}$	−1.5570
D_v	57.4	$A_{D\mathrm{v}}$	0.000	q_v	−1,159.5557
H_v	0.0044			$q_{D\mathrm{v}}$	0.441884

The wavefunction for a positive parity state with even ρ is zero. For a negative parity state with even ρ, we obtain

$$\begin{aligned}|F_2;\mathrm{v};JM;-\rangle &= -\frac{1+(-1)^\rho}{4\sqrt{2}}\left[\left|0;\frac{1}{2}-\frac{1}{2};\mathrm{v};J-\frac{1}{2}M\right\rangle-(-1)^\rho\left|0;\frac{1}{2}\frac{1}{2};\mathrm{v};J\frac{1}{2}M\right\rangle\right]\\ &\quad\times\frac{1+(-1)^\rho}{4\sqrt{2}}\left[\left|0;\frac{1}{2}\frac{1}{2};\mathrm{v};J\frac{1}{2}M\right\rangle-(-1)^\rho\left|0;\frac{1}{2}-\frac{1}{2};\mathrm{v};J-\frac{1}{2}M\right\rangle\right]\\ &=\frac{1}{\sqrt{2}}\left|0;\frac{1}{2}\frac{1}{2};\mathrm{v};J\frac{1}{2}M\right\rangle-\frac{1}{\sqrt{2}}\left|0;\frac{1}{2}-\frac{1}{2};\mathrm{v};J-\frac{1}{2}M\right\rangle.\end{aligned} \tag{3.209}$$

For $^2\Sigma^+$ electronic levels, all F_2 levels are f levels; for $^2\Sigma^-$ electronic levels, all F_2 levels are e levels.

For F_1 levels, the rotational energy is given by

$$\begin{aligned}\varepsilon_R &= \langle F_1;\mathrm{v};JM;\pm|\hat{H}_R|F_1;\mathrm{v};JM;\pm\rangle = \frac{1}{2}\left\{\left\langle 0;\frac{1}{2}\frac{1}{2};\mathrm{v};J\frac{1}{2}M_J\middle|\hat{H}_{R1}\middle|0;\frac{1}{2}\frac{1}{2};\mathrm{v};J\frac{1}{2}M_J\right\rangle\right.\\ &\quad\left.+\left\langle 0;-\frac{1}{2}-\frac{1}{2};\mathrm{v};J-\frac{1}{2}M_J\middle|\hat{H}_{R1}\middle|0;-\frac{1}{2}-\frac{1}{2};\mathrm{v};J-\frac{1}{2}M_J\right\rangle\right\}\\ &\quad+\frac{1}{2}\left\{\left\langle 0;-\frac{1}{2}-\frac{1}{2};\mathrm{v};J-\frac{1}{2}M_J\middle|\hat{H}_{R2}\middle|0;\frac{1}{2}\frac{1}{2};\mathrm{v};J\frac{1}{2}M_J\right\rangle\right.\\ &\quad\left.+\left\langle 0;\frac{1}{2}\frac{1}{2};\mathrm{v};J\frac{1}{2}M_J\middle|\hat{H}_{R2}\middle|0;-\frac{1}{2}-\frac{1}{2};\mathrm{v};J-\frac{1}{2}M_J\right\rangle\right\}\\ &=\frac{1}{2}B_\mathrm{v}\left[J(J+1)-2\left(\frac{1}{2}\right)\left(\frac{1}{2}\right)+\frac{1}{2}\left(\frac{3}{2}\right)\right]+\frac{1}{2}B_\mathrm{v}\left[J(J+1)-2\left(-\frac{1}{2}\right)\left(-\frac{1}{2}\right)+\frac{1}{2}\left(\frac{3}{2}\right)\right]\\ &\quad-\frac{1}{2}B_\mathrm{v}\left[J(J+1)-\left(\frac{1}{2}\right)\left(-\frac{1}{2}\right)\right]^{1/2}\left[\frac{1}{2}\left(\frac{3}{2}\right)-\left(\frac{1}{2}\right)\left(-\frac{1}{2}\right)\right]^{1/2}\\ &\quad-\frac{1}{2}B_\mathrm{v}\left[J(J+1)-\left(-\frac{1}{2}\right)\left(\frac{1}{2}\right)\right]^{1/2}\left[\frac{1}{2}\left(\frac{3}{2}\right)-\left(-\frac{1}{2}\right)\left(\frac{1}{2}\right)\right]^{1/2}\\ &=B_\mathrm{v}\left[J(J+1)+\frac{1}{4}\right]-B_\mathrm{v}\left[J(J+1)+\frac{1}{4}\right]^{1/2}=B_\mathrm{v}\left(J+\frac{1}{2}\right)\left[\left(J+\frac{1}{2}\right)-1\right],\end{aligned} \tag{3.210}$$

where we have used Eqs. (3.81), (3.83), and (3.84). A similar analysis for F_2 levels results in the following expression:

Table 3.5 Level energies (cm^{-1}) and Hund's case (a)-basis wavefunction coefficients for the $X^2\Pi(v = 0)$ vibrational level of OH. The Hund's case (a) coefficients $S_{\Omega,J}$ are listed for the *e* levels

J	F_{1e}	F_{1f}	F_{2e}	F_{2f}	$S_{1/2,N+1/2}$	$S_{3/2,N+1/2}$	$S_{1/2,N-1/2}$	$S_{3/2,N-1/2}$
0.5			126.297	126.455			1.0000	0.0000
1.5	0.000	0.055	187.498	187.758	0.1747	0.9846	0.9846	−0.1747
2.5	83.720	83.921	288.776	289.048	0.2665	0.9638	0.9638	−0.2665
3.5	201.924	202.371	429.282	429.465	0.3367	0.9416	0.9416	−0.3367
4.5	355.107	355.900	608.199	608.193	0.3911	0.9203	0.9203	−0.3911
6.5	767.449	769.203	1,078.498	1,077.842	0.4671	0.8842	0.8842	−0.4671
7.5	1,026.708	1,029.063	1,368.693	1,367.590	0.4939	0.8695	0.8695	−0.4939
8.5	1,321.210	1,324.240	1,694.865	1,693.240	0.5157	0.8568	0.8568	−0.5157
9.5	1,650.720	1,654.493	2,056.487	2,054.271	0.5336	0.8457	0.8457	−0.5336
10.5	2,014.929	2,019.507	2,453.029	2,450.158	0.5485	0.8361	0.8361	−0.5485
11.5	2,413.459	2,418.902	2,883.942	2,880.356	0.5611	0.8277	0.8277	−0.5611
12.5	2,845.879	2,852.239	3,348.654	3,344.299	0.5718	0.8204	0.8204	−0.5718
13.5	3,311.707	3,319.033	3,846.570	3,841.395	0.5810	0.8139	0.8139	−0.5810
14.5	3,810.414	3,818.749	4,377.064	4,371.024	0.5890	0.8081	0.8081	−0.5890
15.5	4,341.428	4,350.810	4,939.481	4,932.537	0.5960	0.8030	0.8030	−0.5960
16.5	4,904.141	4,914.601	5,533.140	5,525.257	0.6022	0.7984	0.7984	−0.6022
17.5	5,497.904	5,509.469	6,157.329	6,148.479	0.6076	0.7942	0.7942	−0.6076
18.5	6,122.037	6,134.727	6,811.310	6,801.471	0.6124	0.7905	0.7905	−0.6124
19.5	6,775.829	6,789.656	7,494.317	7,483.474	0.6168	0.7872	0.7872	−0.6168
20.5	7,458.537	7,473.507	8,205.563	8,193.708	0.6206	0.7841	0.7841	−0.6206
21.5	8,169.393	8,185.504	8,944.236	8,931.366	0.6241	0.7813	0.7813	−0.6241
22.5	8,907.603	8,924.847	9,709.504	9,695.626	0.6273	0.7788	0.7788	−0.6273
23.5	9,672.353	9,690.713	10,500.515	10,485.643	0.6301	0.7765	0.7765	−0.6301
24.5	10,462.808	10,482.257	11,316.402	11,300.560	0.6327	0.7744	0.7744	−0.6327
26.5	12,117.407	12,138.924	13,019.272	13,001.589	0.6373	0.7707	0.7707	−0.6373
27.5	12,979.808	13,002.283	13,904.460	13,885.926	0.6392	0.7690	0.7690	−0.6392
28.5	13,864.428	13,887.801	14,810.944	14,791.617	0.6410	0.7675	0.7675	−0.6410
29.5	14,770.376	14,794.572	15,737.813	15,717.763	0.6427	0.7661	0.7661	−0.6427
30.5	15,696.754	15,721.691	16,684.157	16,663.463	0.6442	0.7649	0.7649	−0.6442

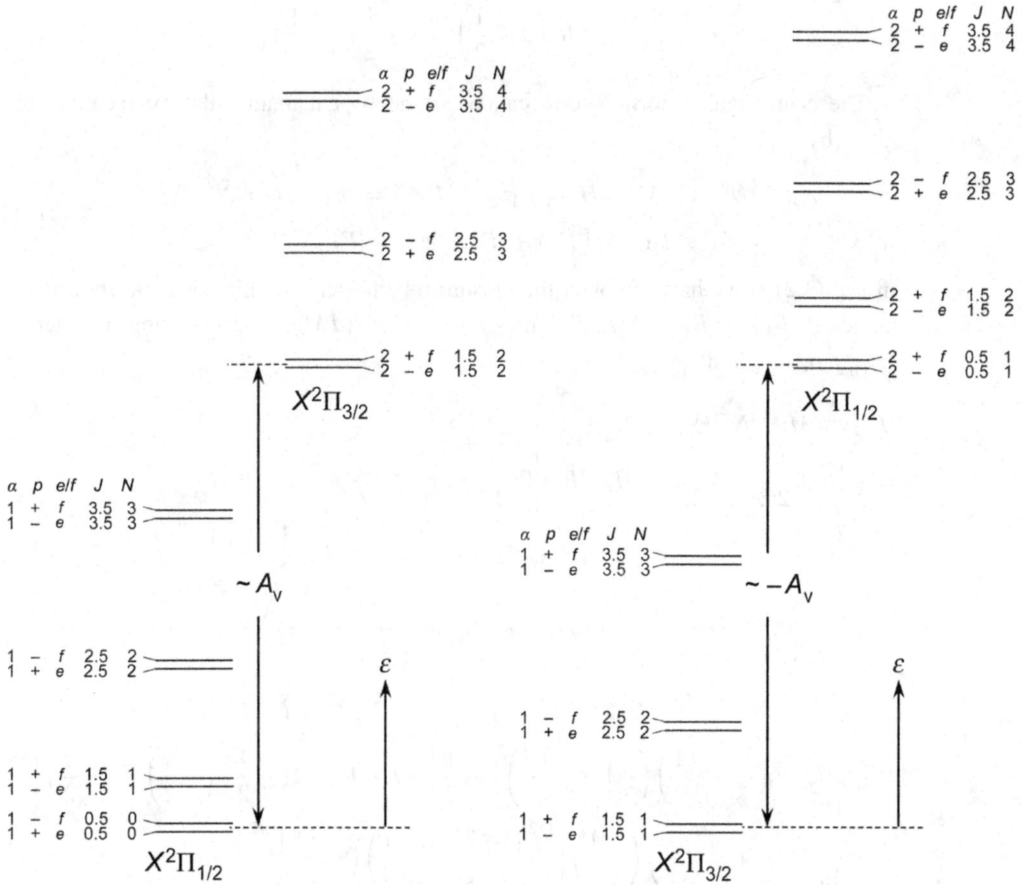

Figure 3.7 Rotational energy level structure in the vibrational bands of $^2\Pi$ electronic levels. The structure on the left is for $A_v > 0$ and the structure on the right is for $A_v < 0$. The energy level spacing as depicted is only approximate.

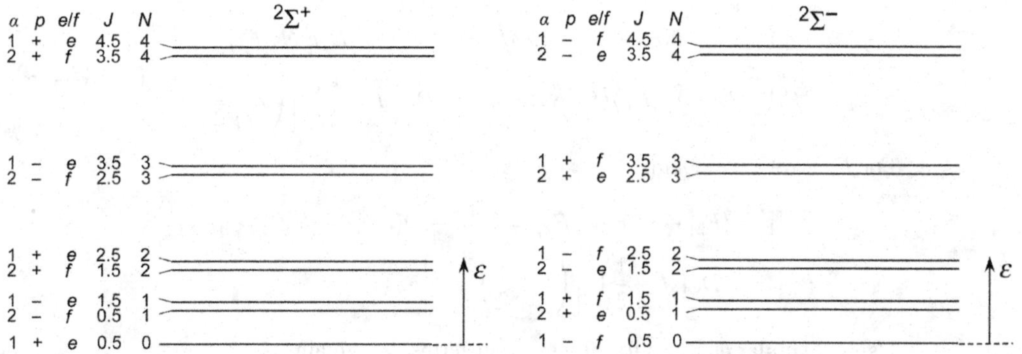

Figure 3.8 Rotational energy level structure in the vibrational bands of $^2\Sigma^{\pm}$ electronic levels. The energy level spacing as depicted is only approximate.

$$\varepsilon_R = B_v\left(J+\frac{1}{2}\right)\left[\left(J+\frac{1}{2}\right)+1\right]. \tag{3.211}$$

The centrifugal distortion term can be found by calculating the matrix element given by

$$\begin{aligned}\varepsilon_{CD1} &= \langle F_1; v; JM; \pm|\hat{H}_{CD1}|F_1; v; JM; \pm\rangle = \langle F_1| - D(r)\hat{N}^4|F_1\rangle \\ &= \langle F_1| - D(r)\hat{N}^2[|F_1\rangle + |F_2\rangle][\langle F_1| + \langle F_2|]\hat{N}^2|F_1\rangle.\end{aligned} \tag{3.212}$$

In Eq. (3.212) we have invoked the closure relation and we are using the shorthand notation $|F_1\rangle = |F_1; v; JM; \pm\rangle$ and $|F_2\rangle = |F_2; v; JM; \pm\rangle$. Evaluating the term $\langle F_1|\hat{N}^2|F_2\rangle$, we obtain

$$\begin{aligned}&\langle F_1; v; JM; \pm|\hat{N}^2|F_2; v; JM; \pm\rangle \\ &= \frac{1}{2}\left\{\left\langle 0;\frac{1}{2}\frac{1}{2}; v; J\frac{1}{2}M_J\right|(\hat{H}_{R1}/B_v)\left|0;\frac{1}{2}\frac{1}{2}; v; J\frac{1}{2}M_J\right\rangle\right. \\ &\left.-\left\langle 0;-\frac{1}{2}-\frac{1}{2}; v; J-\frac{1}{2}M_J\right|(\hat{H}_{R1}/B_v)\left|0;-\frac{1}{2}-\frac{1}{2}; v; J-\frac{1}{2}M_J\right\rangle\right\} \\ &+\frac{1}{2}\left\{\left\langle 0;-\frac{1}{2}-\frac{1}{2}; v; J-\frac{1}{2}M_J\right|(\hat{H}_{R2}/B_v)\left|0;\frac{1}{2}\frac{1}{2}; v; J\frac{1}{2}M_J\right\rangle\right. \\ &\left.-\left\langle 0;\frac{1}{2}\frac{1}{2}; v; J\frac{1}{2}M_J\right|(\hat{H}_{R2}/B_v)\left|0;-\frac{1}{2}-\frac{1}{2}; v; J-\frac{1}{2}M_J\right\rangle\right\} \\ &= \frac{1}{2}\left[J(J+1)-2\left(\frac{1}{2}\right)\left(\frac{1}{2}\right)+\frac{1}{2}\left(\frac{3}{2}\right)\right]-\frac{1}{2}\left[J(J+1)-2\left(-\frac{1}{2}\right)\left(-\frac{1}{2}\right)+\frac{1}{2}\left(\frac{3}{2}\right)\right] \\ &+\frac{1}{2}\left[J(J+1)-\left(\frac{1}{2}\right)\left(-\frac{1}{2}\right)\right]^{1/2}\left[\frac{1}{2}\left(\frac{3}{2}\right)-\left(\frac{1}{2}\right)\left(-\frac{1}{2}\right)\right]^{1/2} \\ &-\frac{1}{2}\left[J(J+1)-\left(-\frac{1}{2}\right)\left(\frac{1}{2}\right)\right]^{1/2}\left[\frac{1}{2}\left(\frac{3}{2}\right)-\left(-\frac{1}{2}\right)\left(\frac{1}{2}\right)\right]^{1/2} = 0.\end{aligned} \tag{3.213}$$

Consequently, we obtain

$$\begin{aligned}\varepsilon_{CD1} &= \langle F_1; v; JM; \pm|\hat{H}_{CD1}|F_1; v; JM; \pm\rangle = \langle F_1| - D(r)\hat{N}^4|F_1\rangle \\ &= \langle F_1| - D(r)\hat{N}^2|F_1\rangle\langle F_1|\hat{N}^2|F_1\rangle = -D_v\left(J+\frac{1}{2}\right)^2\left[\left(J+\frac{1}{2}\right)-1\right]^2.\end{aligned} \tag{3.214}$$

For F_2 levels, we obtain

$$\begin{aligned}\varepsilon_{CD1} &= \langle F_2| - D(r)\hat{N}^4|F_2\rangle = \langle F_2| - D(r)\hat{N}^2|F_2\rangle\langle F_2|\hat{N}^2|F_2\rangle \\ &= -D_v\left(J+\frac{1}{2}\right)^2\left[\left(J+\frac{1}{2}\right)+1\right]^2.\end{aligned} \tag{3.215}$$

Using a similar analysis for higher-order terms, we obtain

$$\varepsilon_R = B_v w(w\pm 1) - D_v w^2(w\pm 1)^2 + H_v w^3(w\pm 1)^3 - L_v w^4(w\pm 1)^4 + \cdots, \tag{3.216}$$

where the + sign applies for F_2 levels, the − sign applies for F_1 levels, and $w = \left(J + \frac{1}{2}\right) = \sqrt{z}$.

For F_1 levels, the spin–rotation energy is given by

$$\begin{aligned}
\varepsilon_{SR} &= \langle F_1; \mathrm{v}; JM; \pm | \hat{H}_{SR} | F_1; \mathrm{v}; JM; \pm \rangle \\
&= \frac{1}{2}\left\{\left\langle 0; \frac{1}{2}\frac{1}{2}; \mathrm{v}; J\frac{1}{2}M_J \middle| \hat{H}_{SR1} \middle| 0; \frac{1}{2}\frac{1}{2}; \mathrm{v}; J\frac{1}{2}M_J \right\rangle\right. \\
&\quad \left. + \left\langle 0; -\frac{1}{2}-\frac{1}{2}; \mathrm{v}; J-\frac{1}{2}M_J \middle| \hat{H}_{SR1} \middle| 0; -\frac{1}{2}-\frac{1}{2}; \mathrm{v}; J-\frac{1}{2}M_J \right\rangle\right\} \\
&\quad + \frac{1}{2}\left\{\left\langle 0; -\frac{1}{2}-\frac{1}{2}; \mathrm{v}; J-\frac{1}{2}M_J \middle| \hat{H}_{SR2} \middle| 0; \frac{1}{2}\frac{1}{2}; \mathrm{v}; J\frac{1}{2}M_J \right\rangle\right. \\
&\quad \left. + \left\langle 0; \frac{1}{2}\frac{1}{2}; \mathrm{v}; J\frac{1}{2}M_J \middle| \hat{H}_{SR2} \middle| 0; -\frac{1}{2}-\frac{1}{2}; \mathrm{v}; J-\frac{1}{2}M_J \right\rangle\right\} \\
&= \frac{1}{2}\gamma_{\mathrm{v}}\left[\left(\frac{1}{2}\right)\left(\frac{1}{2}\right) - \frac{1}{2}\left(\frac{3}{2}\right)\right] + \frac{1}{2}\gamma_{\mathrm{v}}\left[\left(-\frac{1}{2}\right)\left(-\frac{1}{2}\right) - \frac{1}{2}\left(\frac{3}{2}\right)\right] \\
&\quad + \frac{1}{4}\gamma_{\mathrm{v}}\left[J(J+1) - \left(\frac{1}{2}\right)\left(-\frac{1}{2}\right)\right]^{1/2}\left[\frac{1}{2}\left(\frac{3}{2}\right) - \left(\frac{1}{2}\right)\left(-\frac{1}{2}\right)\right]^{1/2} \\
&\quad + \frac{1}{4}\gamma_{\mathrm{v}}\left[J(J+1) - \left(-\frac{1}{2}\right)\left(\frac{1}{2}\right)\right]^{1/2}\left[\frac{1}{2}\left(\frac{3}{2}\right) - \left(-\frac{1}{2}\right)\left(\frac{1}{2}\right)\right]^{1/2} \\
&= -\frac{1}{2}\gamma_{\mathrm{v}} - \frac{1}{2}\gamma_{\mathrm{v}}\left[J(J+1) + \frac{1}{4}\right]^{1/2} = -\frac{1}{2}\gamma_{\mathrm{v}}\left[1 - \left(J + \frac{1}{2}\right)\right],
\end{aligned} \tag{3.217}$$

where we have used Eqs. (3.102), and (3.104). Using a similar analysis for F_2 levels, we obtain

$$\varepsilon_{SR} = -\frac{1}{2}\gamma_{\mathrm{v}}\left[1 + \left(J + \frac{1}{2}\right)\right]. \tag{3.218}$$

An analysis of centrifugal distortion effects similar to that for the rotational energy results in

$$\varepsilon_{SR} = -\frac{1}{2}(1 \pm w)\left[\gamma_{\mathrm{v}} + \gamma_{D\mathrm{v}} w(w \pm 1) + \gamma_{H\mathrm{v}} w^2 (w \pm 1)^2 + \cdots\right]. \tag{3.219}$$

The energy level structure of the $A^2\Sigma^+(\mathrm{v}=0)$ vibrational level of NO is detailed in Table 3.6. Note the very small differences in energy between F_1 and F_2 levels with the same J due to the small value of the spin-rotation constant.

3.5.3 Triplet Electronic Levels

Triplet Electronic Levels with $\Lambda > 0$

The wavefunctions for the vibration-rotation levels for the ^3X electronic levels with $\Lambda > 0$ are again assumed to be intermediate between case (a) and case (b). The wavefunctions are given by

Table 3.6 Level energies (cm^{-1}) for the $A^2\Sigma^+(\mathrm{v}=0)$ vibrational level of NO. The spectroscopic constants used for the calculation of the term energies were $T_0 = 44,198.943$, $B_\mathrm{v} = 1.9862594$, $D_\mathrm{v} = 5.643 \times 10^{-6}$, and $\gamma_\mathrm{v} = -2.6802 \times 10^{-3}$, all in cm^{-1} (Paul, 1997)

J	F_{1e}	F_{2f}	J	F_{1e}	F_{2f}
0.5	44,198.943	44,202.918	20.5	45,032.150	45,115.420
1.5	44,202.914	44,210.864	21.5	45,115.362	45,202.576
2.5	44,210.858	44,222.783	22.5	45,202.516	45,293.671
3.5	44,222.773	44,238.673	23.5	45,293.608	45,388.701
4.5	44,238.661	44,258.534	24.5	45,388.635	45,487.662
5.5	44,258.519	44,282.365	25.5	45,487.594	45,590.552
6.5	44,282.348	44,310.167	26.5	45,590.481	45,697.367
7.5	44,310.146	44,341.936	27.5	45,697.294	45,808.104
8.5	44,341.914	44,377.674	28.5	45,808.027	45,922.758
9.5	44,377.649	44,417.378	29.5	45,922.679	46,041.325
10.5	44,417.350	44,461.047	30.5	46,041.243	46,163.802
11.5	44,461.016	44,508.680	31.5	46,163.718	46,290.184
12.5	44,508.646	44,560.274	32.5	46,290.097	46,420.468
13.5	44,560.238	44,615.829	33.5	46,420.378	46,554.648
14.5	44,615.790	44,675.342	34.5	46,554.555	46,692.719
15.5	44,675.300	44,738.811	35.5	46,692.624	46,834.678
16.5	44,738.767	44,806.234	36.5	46,834.580	46,980.519
17.5	44,806.187	44,877.609	37.5	46,980.419	47,130.238
18.5	44,877.560	44,952.934	38.5	47,130.135	47,283.828
19.5	44,952.881	45,032.205	39.5	47,283.723	47,441.286

$$\begin{aligned}|F_1;\mathrm{v};JM;\pm\rangle &= S_{-1,N+1}\left|{}^3X_{\Lambda-1};\Lambda;\mathrm{v};JM;\pm\right\rangle + S_{0,N+1}\left|{}^3X_{\Lambda};\Lambda;\mathrm{v};JM;\pm\right\rangle \\ &\quad + S_{1,N+1}\left|{}^3X_{\Lambda+1};\Lambda;\mathrm{v};JM;\pm\right\rangle,\end{aligned} \tag{3.220}$$

$$\begin{aligned}|F_2;\mathrm{v};JM;\pm\rangle &= S_{-1,N}\left|{}^3X_{\Lambda-1};\Lambda;\mathrm{v};JM;\pm\right\rangle + S_{0,N}\left|{}^3X_{\Lambda};\Lambda;\mathrm{v};JM;\pm\right\rangle \\ &\quad + S_{1,N}\left|{}^3X_{\Lambda+1};\Lambda;\mathrm{v};JM;\pm\right\rangle,\end{aligned} \tag{3.221}$$

$$\begin{aligned}|F_3;\mathrm{v};JM;\pm\rangle &= S_{-1,N-1}\left|{}^3X_{\Lambda-1};\Lambda;\mathrm{v};JM;\pm\right\rangle + S_{0,N-1}\left|{}^3X_{\Lambda};\Lambda;\mathrm{v};JM;\pm\right\rangle \\ &\quad + S_{1,N-1}\left|{}^3X_{\Lambda+1};\Lambda;\mathrm{v};JM;\pm\right\rangle,\end{aligned} \tag{3.222}$$

where

$$\begin{aligned}\left|{}^3X_{\Lambda-1};\Lambda;\mathrm{v};JM;\pm\right\rangle &= \frac{1}{\sqrt{2}}\Big[|\Lambda;S=1\,\Sigma=-1;\mathrm{v};J\,\Omega=\Lambda-1\,M\rangle \\ &\quad \pm(-1)^{J-1+s}|-\Lambda;1\,1;\mathrm{v};J-\Lambda+1\,M\rangle\Big].\end{aligned} \tag{3.223}$$

$$\left|{}^3X_{\Lambda};\Lambda;\mathrm{v};JM;\pm\right\rangle = \frac{1}{\sqrt{2}}\Big[|\Lambda;1\,0;\mathrm{v};J\,\Lambda\,M\rangle \pm(-1)^{J-1+s}|-\Lambda;1\,0;\mathrm{v};J-\Lambda\,M\rangle\Big]. \tag{3.224}$$

$$|^{3}X_{\Lambda+1};\Lambda;\mathrm{v};JM;\pm\rangle = \frac{1}{\sqrt{2}}\left[|\Lambda;11;\mathrm{v};J\Lambda+1M\rangle \pm (-1)^{J-1+s}|-\Lambda;1-1;\mathrm{v};J-(\Lambda+1)M\rangle\right]. \tag{3.225}$$

The requirement that the wavefunctions in Eqs. (3.220)–(3.222) are normalized results in the following three equations:

$$1 = S^2_{\Lambda-1,N+1} + S^2_{\Lambda,N+1} + S^2_{\Lambda+1,N+1}. \tag{3.226}$$

$$1 = S^2_{\Lambda-1,N} + S^2_{\Lambda,N} + S^2_{\Lambda+1,N}. \tag{3.227}$$

$$1 = S^2_{\Lambda-1,N-1} + S^2_{\Lambda,N-1} + S^2_{\Lambda+1,N-1}. \tag{3.228}$$

The requirement that the wavefunctions in Eqs. (3.220)–(3.222) are orthogonal results in the following three equations:

$$0 = S_{\Lambda-1,N+1}S_{\Lambda-1,N} + S_{\Lambda,N+1}S_{\Lambda,N} + S_{\Lambda+1,N+1}S_{\Lambda+1,N}. \tag{3.229}$$

$$0 = S_{\Lambda-1,N-1}S_{\Lambda-1,N} + S_{\Lambda,N-1}S_{\Lambda,N} + S_{\Lambda+1,N-1}S_{\Lambda+1,N}. \tag{3.230}$$

$$0 = S_{\Lambda-1,N+1}S_{\Lambda-1,N-1} + S_{\Lambda,N+1}S_{\Lambda,N-1} + S_{\Lambda+1,N+1}S_{\Lambda+1,N-1}. \tag{3.231}$$

Following the same procedures as in Section 3.5.2, we obtain the following sets of equations in matrix form for the energies and Hund's case (a) coefficients for the wavefunctions:

$$\begin{bmatrix} H_{11}-\varepsilon_1 & H_{12} & H_{13} \\ H_{21} & H_{22}-\varepsilon_1 & H_{23} \\ H_{31} & H_{32} & H_{33}-\varepsilon_1 \end{bmatrix} \begin{bmatrix} S_{\Lambda-1,N-1} \\ S_{\Lambda,N-1} \\ S_{\Lambda+1,N-1} \end{bmatrix} = \begin{bmatrix} 0 \\ 0 \\ 0 \end{bmatrix} \tag{3.232}$$

$$\begin{bmatrix} H_{11}-\varepsilon_2 & H_{12} & H_{13} \\ H_{21} & H_{22}-\varepsilon_2 & H_{23} \\ H_{31} & H_{32} & H_{33}-\varepsilon_2 \end{bmatrix} \begin{bmatrix} S_{\Lambda-1,N} \\ S_{\Lambda,N} \\ S_{\Lambda+1,N} \end{bmatrix} = \begin{bmatrix} 0 \\ 0 \\ 0 \end{bmatrix} \tag{3.233}$$

and

$$\begin{bmatrix} H_{11}-\varepsilon_3 & H_{12} & H_{13} \\ H_{21} & H_{22}-\varepsilon_3 & H_{23} \\ H_{31} & H_{32} & H_{33}-\varepsilon_3 \end{bmatrix} \begin{bmatrix} S_{\Lambda-1,N+1} \\ S_{\Lambda,N+1} \\ S_{\Lambda+1,N+1} \end{bmatrix} = \begin{bmatrix} 0 \\ 0 \\ 0 \end{bmatrix}. \tag{3.234}$$

The matrix equations represented in Eqs. (3.232)–(3.234) constitute a set of nine equations. Together with Eqs. (3.226)–(3.231), this constitutes a set of 15 equations in 12 unknowns.

In Eqs. (3.232)–(3.234), the density matrix elements H_{ij} are given by

$$
\begin{aligned}
H_{11} &= \langle {}^3X_{\Lambda-1};\Lambda;\mathrm{v};JM;\pm|\hat{H}_R+\hat{H}_{SO}+\hat{H}_{SR}+\hat{H}_{SS}+\hat{H}_{CD}+\hat{H}_{SD}|{}^3X_{\Lambda-1};\Lambda;\mathrm{v};JM;\pm\rangle \\
&= \frac{1}{2}\Big[\langle\Lambda;S\Sigma_-;\mathrm{v};J\Omega_-M|\hat{H}'|\Lambda;S\Sigma_-;\mathrm{v};J\Omega_-M\rangle \\
&\quad +(-1)^{2p}\langle-\Lambda;S-\Sigma_-;\mathrm{v};J-\Omega_-M|\hat{H}'|-\Lambda;S-\Sigma_-1;\mathrm{v};J-\Omega_-M\rangle \\
&\quad \pm(-1)^{p}\langle-\Lambda;S-\Sigma_-;\mathrm{v};J-\Omega_-M|\hat{H}'|\Lambda;S\Sigma_-;\mathrm{v};J\Omega_-M\rangle \\
&\quad \pm(-1)^{p}\langle\Lambda;S\Sigma_-;\mathrm{v};J\Omega_-M|\hat{H}'|-\Lambda;S-\Sigma_-;\mathrm{v};J-\Omega_-M\rangle\Big] \\
&= \frac{1}{2}\Big[\langle\Lambda;S\Sigma_-;\mathrm{v};J\Omega_-M|\hat{H}'|\Lambda;S\Sigma_-;\mathrm{v};J\Omega_-M\rangle \\
&\quad +\langle-\Lambda;S-\Sigma_-;\mathrm{v};J-\Omega_-M|\hat{H}'|-\Lambda;S-\Sigma_-1;\mathrm{v};J-\Omega_-M\rangle\Big] \\
&= B_\mathrm{v}[J(J+1)+S(S+1)-2\Omega_-\Sigma_-]+\frac{1}{2}A_\mathrm{v}[\Lambda\Sigma_-+(-\Lambda)(-\Sigma_-)] \\
&\quad +\frac{1}{2}\gamma_\mathrm{v}[\Sigma_-\Omega_-+(-\Sigma_-)(-\Omega_-)-2S(S+1)]+2\lambda_\mathrm{v}\left[\Sigma_-^2-\frac{1}{3}S(S+1)\right] \\
&\quad -D_\mathrm{v}\Big\{[J(J+1)+S(S+1)-2\Omega_-\Sigma_-]^2+2\left[J(J+1)-\Omega_-^2\right]\left[S(S+1)-\Sigma_-^2\right] \\
&\quad +\Omega_-\Sigma_-+(-\Omega_-)(-\Sigma_-)\Big\}+\frac{1}{2}A_{D\mathrm{v}}[\Lambda\Sigma_-+(-\Lambda)(-\Sigma_-)] \\
&\quad \times[J(J+1)+S(S+1)-2\Omega_-\Sigma_-].
\end{aligned}
\tag{3.235}
$$

Substituting $\Omega_-=\Lambda-1$ and $\Sigma_-=-1$, we obtain

$$
\begin{aligned}
H_{11} &= B_\mathrm{v}[J(J+1)+2\Lambda]-A_\mathrm{v}\Lambda+\gamma_\mathrm{v}(-\Lambda-1)+2\lambda_\mathrm{v}/3 \\
&\quad -D_\mathrm{v}\Big\{[J(J+1)+2+2(\Lambda-1)]^2+2\left[J(J+1)-(\Lambda-1)^2\right]-2(\Lambda-1)\Big\} \\
&\quad -A_{D\mathrm{v}}\Lambda[J(J+1)+2\Lambda].
\end{aligned}
\tag{3.236}
$$

$$
\begin{aligned}
H_{22} &= \langle {}^3X_{\Lambda};\Lambda;\mathrm{v};JM;\pm|\hat{H}_R+\hat{H}_{SO}+\hat{H}_{SR}+\hat{H}_{SS}+\hat{H}_{CD}+\hat{H}_{SD}|{}^3X_{\Lambda};\Lambda;\mathrm{v};JM;\pm\rangle \\
&= \frac{1}{2}\Big[\langle\Lambda;S\Sigma_0;\mathrm{v};J\Omega_0M|\hat{H}'|\Lambda;S\Sigma_0;\mathrm{v};J\Omega_0M\rangle \\
&\quad +\langle-\Lambda;S\Sigma_0;\mathrm{v};J-\Omega_0M|\hat{H}'|-\Lambda;S\Sigma_0;\mathrm{v};J-\Omega_0M\rangle\Big] \\
&= B_\mathrm{v}[J(J+1)+S(S+1)-2\Omega_0\Sigma_0]+\frac{1}{2}\gamma_\mathrm{v}[-2S(S+1)]+2\lambda_\mathrm{v}\left[-\frac{1}{3}S(S+1)\right] \\
&\quad -D_\mathrm{v}\Big\{[J(J+1)+S(S+1)-2\Omega_0\Sigma_0]^2+2\left[J(J+1)-\Omega_0^2\right]\left[S(S+1)-\Sigma_0^2\right]+2\Omega_0\Sigma_0\Big\} \\
&\quad +A_{D\mathrm{v}}\Lambda\Sigma_0[J(J+1)+S(S+1)-2\Omega_0\Sigma_0] \\
&= B_\mathrm{v}[J(J+1)+2]-2\gamma_\mathrm{v}-4\lambda_\mathrm{v}/3-D_\mathrm{v}\Big\{[J(J+1)+2]^2+4\left[J(J+1)-\Lambda^2\right]\Big\}.
\end{aligned}
\tag{3.237}
$$

$$
\begin{aligned}
H_{33} &= \left\langle {}^3X_{\Lambda+1};\Lambda;\mathrm{v};JM;\pm\middle|\hat{H}_R+\hat{H}_{SO}+\hat{H}_{SR}+\hat{H}_{SS}+\hat{H}_{CD}+\hat{H}_{SD}\middle|{}^3X_{\Lambda+1};\Lambda;\mathrm{v};JM;\pm\right\rangle \\
&= \frac{1}{2}\left[\langle\Lambda;S\Sigma_+;\mathrm{v};J\Omega_+M|\hat{H}|\Lambda;S\Sigma_+;\mathrm{v};J\Omega_+M\rangle\right. \\
&\quad \left.+\langle-\Lambda;S\Sigma_+;\mathrm{v};J-\Omega_+M|\hat{H}|-\Lambda;S-\Sigma_+;\mathrm{v};J-\Omega_+M\rangle\right] \\
&= B_\mathrm{v}[J(J+1)+S(S+1)-2\Omega_+\Sigma_+]+(A_\mathrm{v}/2)[\Lambda\Sigma_++(-\Lambda)(-\Sigma_+)] \\
&\quad +(\gamma_\mathrm{v}/2)[\Sigma_+\Omega_++(-\Sigma_+)(-\Omega_+)-2S(S+1)]+2\lambda_\mathrm{v}\left[\Sigma_+^2-\frac{1}{3}S(S+1)\right] \\
&\quad -D_\mathrm{v}\Big\{[J(J+1)+S(S+1)-2\Omega_+\Sigma_+]^2+2\left[J(J+1)-\Omega_+^2\right]\left[S(S+1)-\Sigma_+^2\right] \\
&\quad +\Omega_+\Sigma_++(-\Omega_+)(-\Sigma_+)\Big\}+(A_{D\mathrm{v}}/2)[\Lambda\Sigma_++(-\Lambda)(-\Sigma_+)] \\
&\quad \times[J(J+1)+S(S+1)-2\Omega_+\Sigma_+] \\
&= B_\mathrm{v}[J(J+1)-2\Lambda]+A_\mathrm{v}\Lambda+\gamma_\mathrm{v}(\Lambda-1)+2\lambda_\mathrm{v}/3 \\
&\quad -D_\mathrm{v}\Big\{[J(J+1)+2-2(\Lambda+1)]^2+2\left[J(J+1)-(\Lambda+1)^2\right]+2(\Lambda+1)\Big\} \\
&\quad +A_{D\mathrm{v}}\Lambda[J(J+1)-2\Lambda].
\end{aligned}
\tag{3.238}
$$

$$
\begin{aligned}
H_{21} &= \left\langle {}^3X_{\Lambda};\Lambda;\mathrm{v};JM;\pm\middle|\hat{H}_R+\hat{H}_{SO}+\hat{H}_{SR}+\hat{H}_{SS}+\hat{H}_{CD}+\hat{H}_{SD}\middle|{}^3X_{\Lambda};\Lambda;\mathrm{v};JM;\pm\right\rangle \\
&= \frac{1}{2}\left[\langle\Lambda;S\Sigma_0;\mathrm{v};J\Omega_0M|\hat{H}|\Lambda;S\Sigma_-;\mathrm{v};J\Omega_-M\rangle\right. \\
&\quad \left.+\langle-\Lambda;S\Sigma_0;\mathrm{v};J-\Omega_0M|\hat{H}|-\Lambda;S-\Sigma_-;\mathrm{v};J-\Omega_-M\rangle\right] \\
&= -(B_\mathrm{v}/2)[J(J+1)-\Omega_-(\Omega_-+1)]^{1/2}[S(S+1)-\Sigma_-(\Sigma_-+1)]^{1/2} \\
&\quad -(B_\mathrm{v}/2)[J(J+1)-(-\Omega_-)(-\Omega_--1)]^{1/2}[S(S+1)-(-\Sigma_-)(-\Sigma_--1)]^{1/2} \\
&\quad +(\gamma_\mathrm{v}/4)[J(J+1)-\Omega_-(\Omega_-+1)]^{1/2}[S(S+1)-\Sigma_-(\Sigma_-+1)]^{1/2} \\
&\quad +(\gamma_\mathrm{v}/4)[J(J+1)-(-\Omega_-)(-\Omega_--1)]^{1/2}[S(S+1)-(-\Sigma_-)(-\Sigma_--1)]^{1/2} \\
&\quad +D_\mathrm{v}[J(J+1)-\Omega_-(\Omega_-+1)]^{1/2}[S(S+1)-\Sigma(\Sigma_-+1)]^{1/2} \\
&\quad \times[J(J+1)+S(S+1)-4\Omega_-\Sigma_--2(\Omega_-+\Sigma_-)-2] \\
&\quad +D_\mathrm{v}[J(J+1)-(-\Omega_-)(-\Omega_--1)]^{1/2}[S(S+1)-(-\Sigma_-)(-\Sigma_--1)]^{1/2} \\
&\quad \times[J(J+1)+S(S+1)-4(-\Omega_-)(-\Sigma_-)+2(-\Omega_--\Sigma_-)-2] \\
&\quad -(A_{D\mathrm{v}}/2)\,\Lambda\left(\Sigma_-+\frac{1}{2}\right)[J(J+1)-\Omega_-(\Omega_-+1)]^{1/2}[S(S+1)-\Sigma_-(\Sigma_-+1)]^{1/2} \\
&\quad -(A_{D\mathrm{v}}/2)\,(-\Lambda)\left(-\Sigma_--\frac{1}{2}\right)[J(J+1)-(-\Omega_-)(-\Omega_--1)]^{1/2} \\
&\quad \times[S(S+1)-(-\Sigma_-)(-\Sigma_--1)]^{1/2} \\
&= \left\{-B_\mathrm{v}+\frac{\gamma_\mathrm{v}}{2}+\frac{A_{D\mathrm{v}}}{2}\Lambda+2D_\mathrm{v}[J(J+1)+4(\Lambda-1)+2]\right\} \\
&\quad \times[2J(J+1)-2\Lambda(\Lambda-1)]^{1/2}.
\end{aligned}
\tag{3.239}
$$

$$
\begin{aligned}
H_{32} &= \langle {}^3X_{\Lambda+1}; \Lambda; \mathrm{v}; JM; \pm | \hat{H}_R + \hat{H}_{SO} + \hat{H}_{SR} + \hat{H}_{SS} + \hat{H}_{CD} + \hat{H}_{SD} | {}^3X_{\Lambda}; \Lambda; \mathrm{v}; JM; \pm \rangle \\
&= \frac{1}{2} \Big[\langle \Lambda; S\Sigma_+; \mathrm{v}; J\Omega_+ M | \hat{H} | \Lambda; S\Sigma_0; \mathrm{v}; J\Omega_0 M \rangle \\
&\quad + (-1)^{2p} \langle -\Lambda; S-\Sigma_+; \mathrm{v}; J-\Omega_+ M | \hat{H} | -\Lambda; S-\Sigma_0; \mathrm{v}; J-\Omega_0 M \rangle \Big] \\
&= -(B_\mathrm{v}/2)[J(J+1) - \Omega_0(\Omega_0+1)]^{1/2}[S(S+1) - \Sigma_0(\Sigma_0+1)]^{1/2} \\
&\quad -(B_\mathrm{v}/2)[J(J+1) - (-\Omega_0)(-\Omega_0-1)]^{1/2}[S(S+1) - (-\Sigma_0)(-\Sigma_0-1)]^{1/2} \\
&\quad +(\gamma_\mathrm{v}/4)[J(J+1) - \Omega_0(\Omega_0+1)]^{1/2}[S(S+1) - \Sigma_0(\Sigma_0+1)]^{1/2} \\
&\quad +(\gamma_\mathrm{v}/4)[J(J+1) - (-\Omega_0)(-\Omega_0-1)]^{1/2}[S(S+1) - (-\Sigma_0)(-\Sigma_0-1)]^{1/2} \\
&\quad +D_\mathrm{v}[J(J+1) - \Omega_0(\Omega_0+1)]^{1/2}[S(S+1) - \Sigma_0(\Sigma_0+1)]^{1/2} \\
&\quad \times[J(J+1) + S(S+1) - 4\Omega_0\Sigma_0 - 2(\Omega_0+\Sigma_0) - 1] \\
&\quad +D_\mathrm{v}[J(J+1) - (-\Omega_0)(-\Omega_0-1)]^{1/2}[S(S+1) - (-\Sigma_0)(-\Sigma_0-1)]^{1/2} \\
&\quad \times[J(J+1) + S(S+1) - 4(-\Omega_0)(-\Sigma_0) + 2(-\Omega_0-\Sigma_0) - 1] \\
&\quad -(A_{D\mathrm{v}}/2)\,\Lambda\left(\Sigma_0 + \frac{1}{2}\right)[J(J+1) - \Omega_0(\Omega_0+1)]^{1/2}[S(S+1) - \Sigma_0(\Sigma_0+1)]^{1/2} \\
&\quad -(A_{D\mathrm{v}}/2)\,(-\Lambda)\left(-\Sigma_0 - \frac{1}{2}\right)[J(J+1) + \Omega_0(-\Omega_0-1)]^{1/2} \\
&\quad \times[S(S+1) + \Sigma_0(-\Sigma_0-1)]^{1/2} \\
&= \left\{ -B_\mathrm{v} + \frac{\gamma_\mathrm{v}}{2} - \frac{A_{D\mathrm{v}}}{2}\Lambda + 2D_\mathrm{v}[J(J+1) - 2\Lambda + 2] \right\}[2J(J+1) - 2\Lambda(\Lambda+1)]^{1/2}.
\end{aligned}
\tag{3.240}
$$

$$
\begin{aligned}
H_{31} &= \langle {}^3X_{\Lambda+1}; \Lambda; \mathrm{v}; JM; \pm | \hat{H}_R + \hat{H}_{SO} + \hat{H}_{SR} + \hat{H}_{SS} + \hat{H}_{CD} + \hat{H}_{SD} | {}^3X_{\Lambda-1}; \Lambda; \mathrm{v}; JM; \pm \rangle \\
&= \frac{1}{2} \Big[\langle \Lambda; S\Sigma_+; \mathrm{v}; J\Omega_+ M | \hat{H} | \Lambda; S\Sigma_-; \mathrm{v}; J\Omega_- M \rangle \\
&\quad + (-1)^{2p} \langle -\Lambda; S-\Sigma_+; \mathrm{v}; J-\Omega_+ M_J | \hat{H} | -\Lambda; S-\Sigma_-; \mathrm{v}; J-\Omega_- M_J \rangle \Big] \\
&= -(D_\mathrm{v}/2)[J(J+1) - \Omega_-(\Omega_-+1)]^{1/2}[J(J+1) - (\Omega_-+1)(\Omega_-+2)]^{1/2} \\
&\quad \times[S(S+1) - \Sigma_-(\Sigma_-+1)]^{1/2}[S(S+1) - (\Sigma_-+1)(\Sigma_-+2)]^{1/2} \\
&\quad -(D_\mathrm{v}/2)[J(J+1) - (-\Omega_-)(-\Omega_- - 1)]^{1/2}[J(J+1) - (-\Omega_- - 1)(-\Omega_- - 2)]^{1/2} \\
&\quad \times[S(S+1) - (-\Sigma_-)(-\Sigma_- - 1)]^{1/2}[S(S+1) - (-\Sigma_- - 1)(-\Sigma_- - 2)]^{1/2} \\
&= -2D_\mathrm{v}[J(J+1) - \Lambda(\Lambda-1)]^{1/2}[J(J+1) - \Lambda(\Lambda+1)]^{1/2}.
\end{aligned}
\tag{3.241}
$$

The density matrix element for the lambda-doubling operator between two states characterized by $\Omega = \Lambda - 1 = 0$ is obtained using Eq. (3.136),

$$
\begin{aligned}
H_{11\Lambda\pm} &= \left\langle {}^3\Pi_0;\Lambda;\mathrm{v};JM;\pm\middle|\hat{H}_{LD}\middle|{}^3\Pi_0;\Lambda;\mathrm{v};JM;\pm\right\rangle \\
&= \frac{1}{2}\left[\pm(-1)^{\rho}\langle -1;S1\,;\mathrm{v};J0M|\hat{H}_{LD}|\,1;S-1;\mathrm{v};J0M\rangle\right. \\
&\quad \left.\pm(-1)^{\rho}\langle 1;S-1;\mathrm{v};J0M|\hat{H}_{LD}|-1;S1;\mathrm{v};J0M\rangle\right] \\
&= \pm\frac{1}{2}(-1)^{\rho}\left(\frac{q_{\mathrm{v}}}{2}+\frac{p_{\mathrm{v}}}{2}+\frac{o_{\mathrm{v}}}{2}\right)[1(2)-(-1)(-1+1)]^{1/2}[1(2)-(-1+1)(-1+2)]^{1/2} \\
&\quad \pm\frac{1}{2}(-1)^{\rho}\left(\frac{q_{\mathrm{v}}}{2}+\frac{p_{\mathrm{v}}}{2}+\frac{o_{\mathrm{v}}}{2}\right)[1(2)-(1)(1-1)]^{1/2}[1(2)-(1-1)(-1+2)]^{1/2} \\
&= \pm(-1)^{J-1}(q_{\mathrm{v}}+p_{\mathrm{v}}+o_{\mathrm{v}}) = \mp(-1)^{J}(q_{\mathrm{v}}+p_{\mathrm{v}}+o_{\mathrm{v}}).
\end{aligned}
\tag{3.242}
$$

The density matrix element for the lambda-doubling operator between two states characterized by $\Omega=\Lambda=1$ is obtained using Eq. (3.134),

$$
\begin{aligned}
H_{22\Lambda\pm} &= \left\langle {}^3\Pi_1;\Lambda;\mathrm{v};JM;\pm\middle|\hat{H}_{LD}\middle|{}^3\Pi_1;\Lambda;\mathrm{v};JM;\pm\right\rangle \\
&= \frac{1}{2}\left[\pm(-1)^{\rho}\langle -1;S0\,;\mathrm{v};J-1M|\hat{H}_{LD}|\,1;S0;\mathrm{v};J1M\rangle\right. \\
&\quad \left.\pm(-1)^{\rho}\langle 1;S0;\mathrm{v};J1M|\hat{H}_{LD}|-1;S0;\mathrm{v};J1M\rangle\right] \\
&= \pm(-1)^{\rho}\frac{q_{\mathrm{v}}}{4}[J(J+1)-1(1-1)]^{1/2}[J(J+1)-(1-1)(1-2)]^{1/2} \\
&\quad \pm(-1)^{\rho}\frac{q_{\mathrm{v}}}{4}[J(J+1)-(-1)(-1+1)]^{1/2}[J(J+1)-(-1)(-1+1)(-1+2)]^{1/2} \\
&= \pm(-1)^{J-1}\frac{q_{\mathrm{v}}}{2}J(J+1) = \mp(-1)^{J}\frac{q_{\mathrm{v}}}{2}J(J+1).
\end{aligned}
\tag{3.243}
$$

The density matrix element for the lambda-doubling operator between two states characterized by $\Omega=\Lambda+1=2$ is given by

$$
\begin{aligned}
H_{33\Lambda\pm} &= \left\langle {}^3\Pi_2;\Lambda;\mathrm{v};JM_J;\pm\middle|\hat{H}_{LD}\middle|{}^3\Pi_2;\Lambda;\mathrm{v};JM_J;\pm\right\rangle \\
&= \frac{1}{2}\left[\pm(-1)^{\rho}\langle -1;S-1\,;\mathrm{v};J-2M_J|\hat{H}_{LD}|\,1;S1;\mathrm{v};J2M_J\rangle\right. \\
&\quad \left.\pm(-1)^{\rho}\langle 1;S1;\mathrm{v};J2M_J|\hat{H}_{LD}|-1;S-1;\mathrm{v};J-2M_J\rangle\right] = 0.
\end{aligned}
\tag{3.244}
$$

The density matrix element for the lambda-doubling operator between two states characterized by $\Omega=\Lambda-1=0$ and $\Omega=\Lambda=1$ is obtained using Eq. (3.135),

$$
\begin{aligned}
H_{21\Lambda\pm} &= \left\langle {}^3\Pi_1;\Lambda;\mathrm{v};JM;\pm\middle|\hat{H}_{LD}\middle|{}^3\Pi_0;\Lambda;\mathrm{v};JM;\pm\right\rangle \\
&= \frac{1}{2}\left[\pm(-1)^{\rho}\langle -1;S0;\mathrm{v};J-1M|\hat{H}_{LD}|\,1;S-1;\mathrm{v};J0M\rangle\right. \\
&\quad \left.\pm(-1)^{\rho}\langle 1;S0;\mathrm{v};J1M|\hat{H}_{LD}|-1;S1;\mathrm{v};J0M\rangle\right. \\
&= \mp\frac{1}{2}(-1)^{\rho}\left(q_{\mathrm{v}}+\frac{p_{\mathrm{v}}}{2}\right)[J(J+1)-0(0-1)]^{1/2}[S(S+1)-(-1)(-1+1)]^{1/2} \\
&\quad \mp\frac{1}{2}(-1)^{\rho}\left(q_{\mathrm{v}}+\frac{p_{\mathrm{v}}}{2}\right)[J(J+1)-0(0+1)]^{1/2}[S(S+1)-1(1-1)]^{1/2} \\
&= \mp\sqrt{2}(-1)^{J-1}\left(q_{\mathrm{v}}+\frac{p_{\mathrm{v}}}{2}\right)\sqrt{J(J+1)} = \pm\sqrt{2}(-1)^{J}\left(q_{\mathrm{v}}+\frac{p_{\mathrm{v}}}{2}\right)\sqrt{J(J+1)}.
\end{aligned}
\tag{3.245}
$$

The density matrix element for the lambda-doubling operator between two states characterized by $\Omega = \Lambda = 1$ and $\Omega = \Lambda + 1 = 2$ is given by

$$\begin{aligned} H_{32\Lambda\pm} &= \left\langle {}^3\Pi_2; \Lambda; \mathrm{v}; JM; \pm \middle| \hat{H}_{LD} \middle| {}^3\Pi_1; \Lambda; \mathrm{v}; JM; \pm \right\rangle \\ &= \frac{1}{2}\left[\pm(-1)^p \langle -1; S-1; \mathrm{v}; J-2M | \hat{H}_{LD} | 1; S0; \mathrm{v}; J1M \rangle \right. \\ &\left. \pm(-1)^p \langle 1; S1; \mathrm{v}; J2M | \hat{H}_{LD} | -1; S0; \mathrm{v}; J-1M \rangle \right] = 0. \end{aligned} \tag{3.246}$$

The density matrix element for the lambda-doubling operator between two states characterized by $\Omega = \Lambda - 1 = 0$ and $\Omega = \Lambda + 1 = 2$ is obtained using Eq. (3.134),

$$\begin{aligned} H_{31\Lambda\pm} &= \left\langle {}^3\Pi_2; \Lambda; \mathrm{v}; JM; \pm \middle| \hat{H}_{LD} \middle| {}^3\Pi_0; \Lambda; \mathrm{v}; JM; \pm \right\rangle \\ &= \frac{1}{2}\left[\pm(-1)^p \langle -1; S-1; \mathrm{v}; J-2M | \hat{H}_{LD} | 1; S-1; \mathrm{v}; J0M \rangle \right. \\ &\left. \pm(-1)^p \langle 1; S1; \mathrm{v}; J2M | \hat{H}_{LD} | -1; S1; \mathrm{v}; J0M \rangle \right] \\ &= \pm\frac{1}{2}(-1)^p \frac{q_\mathrm{v}}{2} [J(J+1) - 0(0-1)]^{1/2} [J(J+1) - (0-1)(0-2)]^{1/2} \\ &\pm\frac{1}{2}(-1)^p \frac{q_\mathrm{v}}{2} [J(J+1) - 0(0+1)]^{1/2} [J(J+1) - (0+1)(0+2)]^{1/2} \\ &= \pm(-1)^{J-1} \frac{q_\mathrm{v}}{2} \sqrt{J(J+1)[J(J+1)-2]} \\ &= \mp(-1)^J \frac{q_\mathrm{v}}{2} \sqrt{J(J+1)[J(J+1)-2]}. \end{aligned} \tag{3.247}$$

The structures of the $X^3\Sigma^-$ and $A^3\Pi$ electronic levels for the NH molecule are discussed in detail by Brazier et al. (1986). The rotational structure for the $A^3\Pi$ electronic level of NH is depicted schematically in Figure 3.9. The density matrix elements for a $^3\Pi$ electronic level including higher-order terms are given by (Brazier et al., 1986)

$$\begin{aligned} H_{11e/f} &= B_\mathrm{v}(x+2) - A_\mathrm{v} - A_{D\mathrm{v}}(x+2) - D_\mathrm{v}(x^2+6x+4) \\ &+H_\mathrm{v}(x^3+12x^2+24x+8) + L_\mathrm{v}(x^4+20x^3+80x^2+72x+16) \\ &+M_\mathrm{v}(x^5+30x^4+200x^3+344x^2+192x+32) - 2\gamma_\mathrm{v} - \gamma_{D\mathrm{v}}(3x+4) \\ &+2\lambda_\mathrm{v}/3 + 2\lambda_{D\mathrm{v}}(x+2)/3 \mp o_\mathrm{v} \mp o_{D\mathrm{v}}(x+2) \mp o_{H\mathrm{v}}\sqrt{x(x-2)} \\ &\mp p_\mathrm{v} \mp 2p_{D\mathrm{v}}(x+1) \mp p_{H\mathrm{v}}(3x^2+10x+4) \mp q_\mathrm{v} \mp q_{D\mathrm{v}}(3x+2) \mp q_{H\mathrm{v}}(6x^2+12x+4) \\ &\mp q_{L\mathrm{v}}(10x^3+40x^2+36x+8), \end{aligned} \tag{3.248}$$

$$\begin{aligned} H_{22e/f} &= B_\mathrm{v}(x+2) - D_\mathrm{v}(x^2+8x) + H_\mathrm{v}(x^3+18x^2+16x) \\ &+L_\mathrm{v}(x^4+32x^3+80x^2+32x) + M_\mathrm{v}(x^5+50x^4+240x^3+256x^2+64x) \\ &-2\gamma_\mathrm{v} - \gamma_{D\mathrm{v}}(4x+2) - 4\lambda_{D\mathrm{v}}(x+2)/3 \mp p_{D\mathrm{v}}x \mp 2p_{H\mathrm{v}}x(x+2) \\ &\mp q_\mathrm{v}x/2 \mp q_{D\mathrm{v}}x(x+6)/2 \mp q_{H\mathrm{v}}(x^3/2+8x^2+8x) \\ &\mp q_{L\mathrm{v}}(x^4/2+15x^3+40x^2+16x), \end{aligned} \tag{3.249}$$

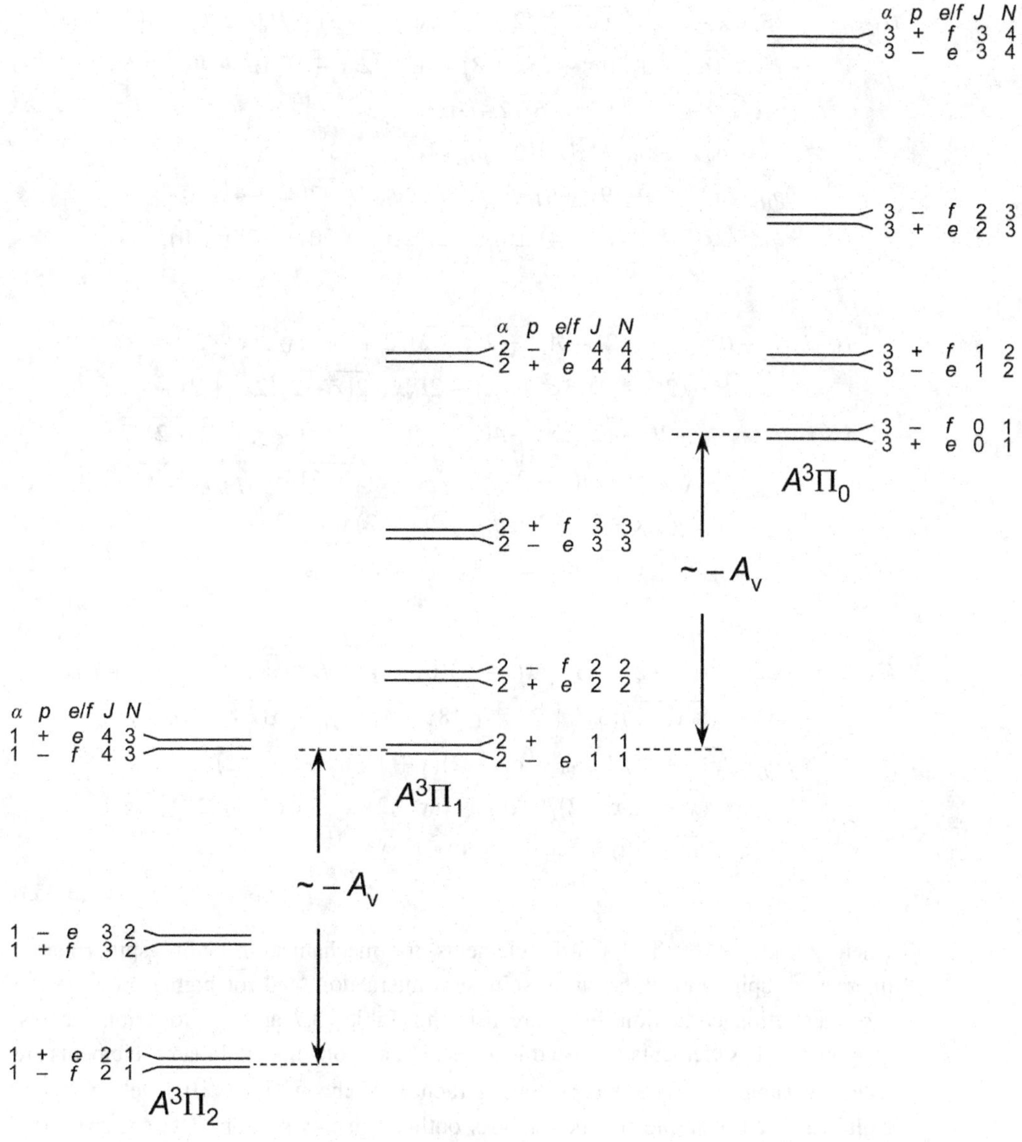

Figure 3.9 Rotational energy level structure in the vibrational bands of $^3\Pi$ electronic levels. The energy level spacing as depicted is only approximate. The energy ordering of the Λ-doubled levels is appropriate for the $A^3\Pi(\mathrm{v}=0)$ level of NH.

$$\begin{aligned} H_{33e/f} &= B_\mathrm{v}(x-2) + A_\mathrm{v} + A_{D\mathrm{v}}(x-2) - D_\mathrm{v}(x^2-2x) + H_\mathrm{v}(x^3-4x) \\ &+L_\mathrm{v}(x^4+4x^3-8x^2-8x) + M_\mathrm{v}(x^5+10x^4-40x^2-16x) + \gamma_{D\mathrm{v}}(-x+2) \\ &+2\lambda_\mathrm{v}/3 + 2\lambda_{D\mathrm{v}}(x-2)/3 \mp q_{H\mathrm{v}}x(x-2) \mp q_{L\mathrm{v}}(3x^3-4x^2-4x), \end{aligned} \tag{3.250}$$

$$\begin{aligned}H_{12e/f} = & -2B_v\sqrt{x} - A_{Dv}\sqrt{2(x-2)}/2 + 2D_v x\sqrt{2(x-2)} - H_v\sqrt{2(x-2)}\left(3x^2+4x\right) \\ & -2L_v\sqrt{2x}\left(2x^3+20x^2+28x+8\right) - M_v\sqrt{2(x-2)}\left(5x^4+40x^3+56x^2+16x\right) \\ & +\gamma_v\sqrt{x/2} + \gamma_{Dv}\sqrt{2x}(x+6)/2 + \lambda_{Dv}\sqrt{2x}/3 \\ & \pm o_{Dv}\sqrt{x/2} \pm o_{Hv}\sqrt{2x}(x+2) \mp p_{Dv}\sqrt{x/2}(x+3) \\ & \pm p_{Hv}\sqrt{x/2}\left(x^2+19x+6\right) \pm q_v\sqrt{2x} \pm q_{Dv}\sqrt{x/2}(3x+4) \\ & \pm q_{Hv}\sqrt{2x}\left(2x^2+18x+4\right) \pm q_{Lv}\sqrt{2x}\left(2.5x^3+20x^2+28x+8\right),\end{aligned} \tag{3.251}$$

$$\begin{aligned}H_{23e/f} = & -B_v\sqrt{2(x-2)} - A_{Dv}\sqrt{2(x-2)}/2 + 2D_v x\sqrt{2(x-2)} \\ & -H_v\sqrt{2(x-2)}\left(3x^2+4x\right) - 2L_v x\sqrt{2(x-2)}\left(2x^2+8x+4\right) \\ & -M_v\sqrt{2(x-2)}\left(5x^4+40x^3+56x^2+16x\right) + \gamma_v\sqrt{(x-2)/2} \\ & +\gamma_{Dv}\sqrt{2(x-2)}(x+2)/2 + \lambda_{Dv}\sqrt{2(x-2)}/3 \pm p_{Hv}\sqrt{2(x-2)}\,x/2 \\ & \pm q_{Dv}\sqrt{2(x-2)}\,x/2 \pm q_{Hv}\sqrt{2(x-2)}\,x(x+2) \\ & \pm q_{Lv}\sqrt{2(x-2)}\left(1.5x^3+8x^2+4x\right),\end{aligned} \tag{3.252}$$

$$\begin{aligned}H_{13e/f} = & -2D_v\sqrt{x(x-2)} + H_v\sqrt{x(x-2)}(6x+4) + L_v\sqrt{x(x-2)}\left(12x^2+24x+8\right) \\ & +4M_v\sqrt{x(x-2)}\left(5x^3+20x^2+18x+4\right) - \gamma_{Dv}\sqrt{x(x-2)} \mp o_{Hv}\sqrt{x(x-2)} \\ & \mp p_{Dv}\sqrt{x(x-2)}/2 \mp p_{Hv}\sqrt{x(x-2)}(x+1) \mp q_v\sqrt{x(x-2)}/2 \\ & \mp q_{Dv}\sqrt{x(x-2)}(x+2)/2 \mp q_{Hv}\sqrt{x(x-2)}\left(x^2+6x+4\right)/2 \\ & \mp q_{Lv}\sqrt{x(x-2)}\left(0.5x^3+6x^2+12x+4\right),\end{aligned} \tag{3.253}$$

where $x = J(J+1)$. The matrix elements for mechanical rotation, spin–rotation interaction, spin–orbit interaction, spin–spin interaction, and for higher-order centrifugal distortion correction terms are listed in Tables 3.7 and 3.8 for triplet levels. Diagonal matrix elements are listed in Table 3.7, and off-diagonal matrix elements are listed in Table 3.8. The energy level structure of the $A^3\Pi(\mathrm{v}=0)$ level of NH is included as an example for the analysis outlined in this section. The spectroscopic constants for the $A^3\Pi(\mathrm{v}=0)$ level are listed in Table 3.9, the level energies are listed in Table 3.10, and the basis wavefunction coefficients are listed in Table 3.11.

$^3\Sigma$ Electronic Levels

The basis sets for the wavefunctions of the vibration-rotation levels for $^3\Sigma^\pm$ electronic levels of diatomic molecules are discussed in detail by Hougen (1970). As depicted in Figure 3.10, for $^3\Sigma^+$ electronic levels, all F_1 and F_3 levels are *f* levels and all F_2 levels are *e* levels; for $^3\Sigma^-$ electronic levels, all F_1 and F_3 levels are *e* levels and all F_2 levels are *f* levels The basis wavefunctions for vibration-rotation states in $^3\Sigma^+$ electronic levels are given by

Table 3.7 Diagonal matrix elements, for triplet electronic levels with $\Lambda \geq 1$, for mechanical rotation, spin–rotation interaction, spin–orbit interaction, spin-spin interaction, and for higher-order centrifugal distortion correction terms. The parameter $x = J(J+1)$

Energy interaction	H_{11}	H_{22}	H_{33}
Rotational	$B_v(x+2\Lambda)$	$B_v(x+2\Lambda)$	$B_v(x-2\Lambda)$
Rot CD1	$-D_v[x^2+(2+4\Lambda)x+2\Lambda(\Lambda+1)]$	$-D_v[x^2+8x+4(1-\Lambda^2)]$	$-D_v[x^2+(2-4\Lambda)x+2\Lambda(\Lambda-1)]$
Rot CD2	$H_v[x^3+(6\Lambda+6)x^2+(6\Lambda^2+14\Lambda+4)x$	$H_v[x^3+18x^2+(-12\Lambda^2$	$H_v[x^3+(-6\Lambda+6)x^2$
	$+4\Lambda(\Lambda+1)]$	$+28)x-8\Lambda(1-\Lambda)]$	$+(6\Lambda^2-14\Lambda+4)x]$
Spin–rotation	$-\gamma_v(\Lambda+1)$	$-2\gamma_v$	$\gamma_v(\Lambda-1)$
SR CD1	$-\gamma_{Dv}[(\Lambda+2)x+\Lambda(\Lambda+3)]$	$-\gamma_{Dv}[4x-2\Lambda^2+4]$	$-\gamma_{Dv}[(\Lambda-2)x+\Lambda(\Lambda-3)]$
Spin–Orbit	$-A_v\Lambda$	0	$A_v\Lambda$
SO CD1	$-A_{Dv}\Lambda(x+2\Lambda)$	0	$A_{Dv}\Lambda(x-2\Lambda)$
Spin–spin	$2\lambda_v/3$	$-4\lambda_v/3$	$2\lambda_v/3$
SS CD1	$\lambda_{Dv}[2(x+2\Lambda)/3]$	$\lambda_{Dv}[-4(x+2)/3]$	$\lambda_{Dv}[2(x-2\Lambda)/3]$

Table 3.8 Off-diagonal matrix elements for triplet electronic levels for mechanical rotation, spin–rotation interaction, spin–orbit interaction, spin-spin interaction, and for higher-order centrifugal distortion correction terms. The parameter $x = J(J+1)$

Energy interaction	$H_{12} = H_{21}$	$H_{23} = H_{32}$	$H_{13} = H_{31}$
Rotational	$-B_v[2x - 2\Lambda(\Lambda-1)]^{1/2}$	$-B_v[2x - 2\Lambda(\Lambda+1)]^{1/2}$	0
Rot CD1	$2D_v[2x - 2\Lambda(\Lambda-1)]^{1/2}$ $\times[x + (\Lambda+1)]$	$2D_v[2x - 2\Lambda(\Lambda+1)]^{1/2}$ $\times[x - (\Lambda-1)]$	$-2D_v\left[x^2 - 2x\Lambda^2 + \Lambda^2(\Lambda^2-1)\right]^{1/2}$
Rot CD2	$-H_v\left\{[2x - 2\Lambda(\Lambda-1)]^{1/2} \times\left[3x^2 + (6\Lambda+9)x + \Lambda^2 + 5\Lambda + 4\right] + [2x - 2\Lambda(\Lambda+1)]^{1/2} \times\left[x^2 - 2x\Lambda^2 + \Lambda^2(\Lambda^2-1)\right]^{1/2}\right\}$	$-H_v\left\{[2x - 2\Lambda(\Lambda+1)]^{1/2} \times\left[3x^2 + (-6\Lambda+9)x + \Lambda^2 - 5\Lambda + 4\right] + [2x - 2\Lambda(\Lambda-1)]^{1/2} \times\left[x^2 - 2x\Lambda^2 + \Lambda^2(\Lambda^2-1)\right]^{1/2}\right\}$	$H_v\left[x^2 - 2x\Lambda^2 + \Lambda^2(\Lambda^2-1)\right]^{1/2} \times(6x+4)$
Spin–rotation	$\frac{1}{2}\gamma_v[2x - 2\Lambda(\Lambda-1)]^{1/2}$	$\frac{1}{2}\gamma_v[2x - 2\Lambda(\Lambda+1)]^{1/2}$	0
SR CD1	$\frac{1}{2}\gamma_{Dv}\sqrt{2x - 2\Lambda(\Lambda-1)}$ $\times[x + 2\Lambda + 4]$	$\frac{1}{2}\gamma_{Dv}\sqrt{2x - 2\Lambda(\Lambda+1)}$ $\times[x + 4 - 2\Lambda]$	$-\gamma_{Dv}\{[x - \Lambda(\Lambda+1)] \times[x - \Lambda(\Lambda-1)]\}^{1/2}$
Spin–orbit	0	0	0
SO CD1	$\frac{1}{2}A_{Dv}\Lambda[2x-2\Lambda(1-\Lambda)]^{1/2}$	$-\frac{1}{2}A_{Dv}\Lambda[2x-2\Lambda(1+\Lambda)]^{1/2}$	0
Spin–spin	0	0	0
SS CD1	$\lambda_{Dv}\sqrt{2x - 2\Lambda(\Lambda-1)}/3$	$\lambda_{Dv}\sqrt{2x - 2\Lambda(\Lambda+1)}/3$	0

Table 3.9 Spectroscopic constants (cm^{-1}) for the rotational levels of the $A^3\Pi(v = 0)$ vibrational level of the NH radical. Spectroscopic constants extracted from Ram and Bernath (2010)

T_v	29,761.182869	$M_v \times 10^{15}$	−2.5332	λ_v	−0.201511	$q_{Dv} \times 10^5$	1.38538
B_v	16.32147263	A_v	−34.619342	$p_v \times 10^2$	5.52344	$q_{Hv} \times 10^9$	−1.9159
$D_v \times 10^3$	1.7897015	$A_{Dv} \times 10^5$	−8.14	$p_{Dv} \times 10^5$	−1.7785	$q_{Lv} \times 10^{13}$	2.275
$H_v \times 10^7$	1.07610	$\gamma_v \times 10^2$	2.97879	$p_{Hv} \times 10^9$	2.107	o_v	1.284426
$L_v \times 10^{11}$	−1.6904	$\gamma_{Dv} \times 10^6$	−5.2496	$q_v \times 10^2$	−3.159263	$o_{Dv} \times 10^6$	−1.3100
						$o_{Hv} \times 10^8$	−3.777

Table 3.10 Level energies (cm^{-1}) for the $A^3\Pi(v = 0)$ vibrational level of the NH radical

J	F_{1e}	F_{2e}	F_{3e}	F_{1f}	F_{2f}	F_{3f}
0			29,826.94			29,829.55
1		29,806.61	29,879.62		29,807.28	29,881.51
2	29,770.60	29,866.11	29,971.61	29,770.59	29,867.16	29,973.00
3	29,842.29	29,960.72	30,098.49	29,842.20	29,962.05	30,099.48
4	29,943.94	30,089.07	30,258.77	29,943.70	30,090.69	30,259.38
5	30,076.67	30,250.30	30,451.73	30,076.19	30,252.24	30,451.94
6	30,240.87	30,443.85	30,676.86	30,240.08	30,446.17	30,676.64
7	30,436.64	30,669.31	30,933.71	30,435.46	30,672.06	30,933.02
8	30,663.88	30,926.31	31,221.87	30,662.26	30,929.54	31,220.66
9	30,922.39	31,214.47	31,540.91	30,920.26	31,218.23	31,539.15
10	31,211.89	31,533.38	31,890.38	31,209.20	31,537.71	31,888.02
11	31,532.03	31,882.61	32,269.80	31,528.73	31,887.57	32,266.80
12	31,882.43	32,261.71	32,678.65	31,878.47	32,267.32	32,674.98
13	32,262.64	32,670.16	33,116.40	32,257.98	32,676.48	33,112.02
14	32,672.18	33,107.44	33,582.45	32,666.78	33,114.49	33,577.33
15	33,110.53	33,572.97	34,076.21	33,104.35	33,580.79	34,070.32
16	33,577.12	34,066.15	34,597.01	33,570.13	34,074.77	34,590.33
17	34,071.35	34,586.35	35,144.17	34,063.53	34,595.78	35,136.69
18	34,592.59	35,132.88	35,716.99	34,583.91	35,143.14	35,708.69
19	35,140.14	35,705.04	36,314.71	35,130.60	35,716.14	36,305.58
20	35,713.32	36,302.08	36,936.56	35,702.89	36,314.03	36,926.59
21	36,311.37	36,923.24	37,581.71	36,300.05	36,936.05	37,570.91
22	36,933.52	37,567.71	38,249.33	36,921.31	37,581.36	38,237.70
23	37,578.95	38,234.65	38,938.53	37,565.86	38,249.14	38,926.09
24	38,246.83	38,923.18	39,648.40	38,232.86	38,938.50	39,635.16
25	38,936.28	39,632.41	40,378.01	38,921.45	39,648.54	40,363.99
26	39,646.39	40,361.40	41,126.38	39,630.72	40,378.31	41,111.61
27	40,376.22	41,109.17	41,892.49	40,359.74	41,126.83	41,877.00
28	41,124.81	41,874.73	42,675.29	41,107.54	41,893.11	42,659.13
29	41,891.13	42,657.04	43,473.72	41,873.12	42,676.10	43,456.91
30	42,674.16	43,455.02	44,286.63	42,655.44	43,474.71	44,269.24

$$|^3\Sigma^+_{1e}; v; JM\rangle = \frac{1}{\sqrt{2}}[|0^+; 11; v; J\,1\,M\rangle - |0^+; 1-1; v; J-1\,M\rangle]. \tag{3.254}$$

$$\left|^3\Sigma^+_{0f}; v; JM\right\rangle = |0^+; 10; v; J\,0\,M\rangle. \tag{3.255}$$

$$\left|^3\Sigma^+_{1f}; v; JM\right\rangle = \frac{1}{\sqrt{2}}[|0^+; 11; v; J\,1\,M\rangle + |0^+; 1-1; v; J-1\,M\rangle]. \tag{3.256}$$

For $^3\Sigma^-$ electronic levels, the basis wavefunctions are given by

$$|^3\Sigma^-_{1e}; v; JM\rangle = \frac{1}{\sqrt{2}}[|0^-; 11; v; J\,1\,M\rangle + |0^-; 1-1; v; J-1\,M\rangle]. \tag{3.257}$$

$$|^3\Sigma^-_{0e}; v; JM\rangle = |0^-; 10; v; J\,0\,M\rangle. \tag{3.258}$$

Table 3.11 Hund's case (a)-basis wavefunction coefficients for the $A^3\Pi(\mathrm{v}=0)$ vibrational level of the NH radical. The Hund's case (a) coefficients $S_{\Omega,J}$ are listed for the *e* levels

J	$S_{0,N-1}$	$S_{1,N-1}$	$S_{2,N-1}$	$S_{0,N}$	$S_{1,N}$	$S_{2,N}$	$S_{0,N+1}$	$S_{1,N+1}$	$S_{2,N+1}$
0	1.000								
1	0.8516	−0.5242		0.5242	0.8516				
2	0.7590	−0.6306	0.1618	0.6335	0.6581	−0.4069	0.1502	0.4113	0.8990
3	0.7033	−0.6713	0.2341	0.6715	0.5192	−0.5287	0.2334	0.5290	0.8159
4	0.6668	−0.6900	0.2817	0.6872	0.4229	−0.5907	0.2884	0.5875	0.7561
5	0.6411	−0.6996	0.3153	0.6948	0.3547	−0.6257	0.3259	0.6202	0.7135
6	0.6223	−0.7050	0.3401	0.6989	0.3047	−0.6470	0.3525	0.6404	0.6824
7	0.6078	−0.7082	0.3591	0.7014	0.2668	−0.6609	0.3722	0.6536	0.6589
8	0.5965	−0.7101	0.3741	0.7030	0.2373	−0.6704	0.3873	0.6629	0.6407
9	0.5873	−0.7113	0.3863	0.7041	0.2136	−0.6772	0.3992	0.6697	0.6263
10	0.5797	−0.7120	0.3962	0.7049	0.1943	−0.6822	0.4087	0.6748	0.6145
11	0.5734	−0.7124	0.4046	0.7055	0.1783	−0.6860	0.4166	0.6787	0.6048
12	0.5680	−0.7127	0.4116	0.7059	0.1647	−0.6889	0.4231	0.6819	0.5967
13	0.5634	−0.7128	0.4177	0.7062	0.1532	−0.6912	0.4287	0.6844	0.5897
14	0.5594	−0.7129	0.4229	0.7065	0.1433	−0.6931	0.4334	0.6865	0.5838
15	0.5560	−0.7128	0.4275	0.7067	0.1346	−0.6946	0.4376	0.6883	0.5786
16	0.5529	−0.7128	0.4316	0.7069	0.1270	−0.6958	0.4411	0.6898	0.5741
17	0.5502	−0.7127	0.4351	0.7070	0.1204	−0.6969	0.4443	0.6911	0.5701
18	0.5478	−0.7126	0.4383	0.7072	0.1144	−0.6977	0.4471	0.6922	0.5666
19	0.5456	−0.7125	0.4412	0.7073	0.1091	−0.6985	0.4496	0.6931	0.5634
20	0.5437	−0.7124	0.4437	0.7073	0.1043	−0.6991	0.4518	0.6940	0.5606
21	0.5419	−0.7123	0.4461	0.7074	0.1000	−0.6997	0.4538	0.6947	0.5581
22	0.5403	−0.7122	0.4482	0.7075	0.09613	−0.7002	0.4556	0.6954	0.5558
23	0.5389	−0.7121	0.4501	0.7075	0.09259	−0.7006	0.4572	0.6960	0.5537
24	0.5376	−0.7119	0.4518	0.7075	0.08938	−0.7010	0.4587	0.6965	0.5518
25	0.5364	−0.7118	0.4534	0.7076	0.08645	−0.7013	0.4600	0.6970	0.5500
26	0.5353	−0.7117	0.4549	0.7076	0.08377	−0.7016	0.4613	0.6974	0.5485
27	0.5343	−0.7116	0.4562	0.7076	0.08131	−0.7019	0.4624	0.6978	0.5470
28	0.5334	−0.7115	0.4574	0.7076	0.07906	−0.7022	0.4634	0.6982	0.5457
29	0.5326	−0.7114	0.4585	0.7076	0.07700	−0.7024	0.4644	0.6985	0.5444
30	0.5318	−0.7113	0.4595	0.7076	0.07511	−0.7026	0.4653	0.6988	0.5433

$$\left|{}^3\Sigma^-_{1f};\mathrm{v};JM\right\rangle = \frac{1}{\sqrt{2}}\left[|0^-;11;\mathrm{v};J1M\rangle - |0^-;1-1;\mathrm{v};J-1M\rangle\right]. \tag{3.259}$$

The vibration-rotation state wavefunctions for ${}^3\Sigma^+$ electronic levels are given by

$$|F_1;\mathrm{v};JM;f\rangle = S_{1f,N+1}\left|{}^3\Sigma^+_{1f};\mathrm{v};JM\right\rangle + S_{0f,N+1}\left|{}^3\Sigma^+_{0f};\mathrm{v};JM\right\rangle \tag{3.260}$$

$$|F_2;\mathrm{v};JM;e\rangle = S_{1e,N}\left|{}^3\Sigma^+_{1e};\mathrm{v};JM\right\rangle \tag{3.261}$$

$$|F_3;\mathrm{v};JM;f\rangle = S_{1f,N-1}\left|{}^3\Sigma^+_{1f};\mathrm{v};JM\right\rangle + S_{0f,N-1}\left|{}^3\Sigma^+_{0f};\mathrm{v};JM\right\rangle \tag{3.262}$$

and for ${}^3\Sigma^-$ electronic levels they are given by

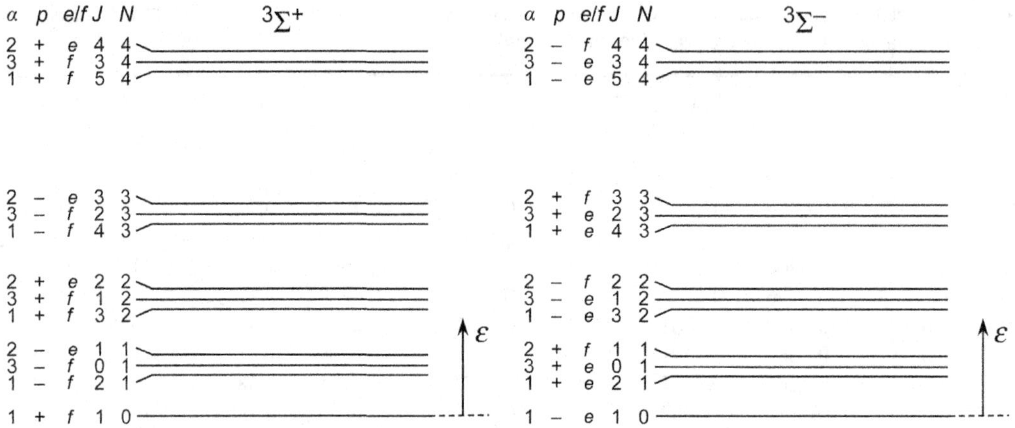

Figure 3.10 Rotational energy level structure in the vibrational bands of $^3\Sigma^{\pm}$ electronic levels. The energy level spacing as depicted is only approximate.

$$|F_1; \mathrm{v}; JM; e\rangle = S_{1e,N+1}\left|{}^3\Sigma^-_{1e}; \mathrm{v}; JM\right\rangle + S_{0e,N+1}\left|{}^3\Sigma^-_{0e}; \mathrm{v}; JM\right\rangle. \tag{3.263}$$

$$|F_2; \mathrm{v}; JM; f\rangle = S_{1f,N}\left|{}^3\Sigma^-_{1f}; \mathrm{v}; JM\right\rangle. \tag{3.264}$$

$$|F_3; \mathrm{v}; JM; e\rangle = S_{1e,N-1}\left|{}^3\Sigma^-_{1e}; \mathrm{v}; JM\right\rangle + S_{0e,N-1}\left|{}^3\Sigma^-_{0e}; \mathrm{v}; JM\right\rangle. \tag{3.265}$$

Note that the vibration-rotation state wavefunctions and the associated basis wavefunctions must have the same parity.

To determine the energies $\varepsilon_1, \varepsilon_2,$ and ε_3 of the vibration-rotation states and the values of the basis state coefficients $S_{0e/f,N+1}, S_{1e/f,N+1}, S_{0f/e,N}, S_{1e/f,N-1},$ and $S_{0e/f,N-1}$, we follow a procedure similar to that described in the last section for $^2\Pi$ vibration-rotation states. The $F_1, F_2,$ and F_3 wavefunctions must be normalized, and so

$$S^2_{1e,N+1} + S^2_{0e,N+1} = 1. \tag{3.266}$$

$$S_{1f,N} = 1. \tag{3.267}$$

$$S^2_{1e,N-1} + S^2_{0e,N-1} = 1. \tag{3.268}$$

In addition, the F_1 and F_3 wavefunctions, which have the same parity, must be orthogonal. Therefore,

$$S_{1e,N+1}S_{1e,N-1} + S_{0e,N+1}S_{0e,N-1} = 0. \tag{3.269}$$

The operation of the operator $\hat{H}'$ on the F_1 ket for $^3\Sigma^-$ electronic levels gives the product of the energy eigenvalue and the original ket,

$$\begin{aligned}\hat{H}'|F_1; \mathrm{v}; JM; f\rangle = \varepsilon_1|F_1; \mathrm{v}; JM; f\rangle = \varepsilon_1 S_{1e,N+1}\left|\Sigma^-_{1e}; \mathrm{v}; JM\right\rangle \\ +\varepsilon_1 S_{0e,N+1}\left|\Sigma^-_{0e}; \mathrm{v}; JM\right\rangle.\end{aligned} \tag{3.270}$$

The density matrix elements that result from using the basis wavefunctions as the bra in Eq. (3.270) gives

$$\begin{aligned}\langle \Sigma_{1e}^-; \mathrm{v}; JM|\hat{H}'|F_1; \mathrm{v}; JM; f\rangle &= \varepsilon_1 S_{1e,N+1} = S_{1e,N+1}\langle \Sigma_{1e}^-; \mathrm{v}; JM|H'|\Sigma_{1e}^-; \mathrm{v}; JM\rangle \\ &\quad + S_{0e,N+1}\langle \Sigma_{1e}^-; \mathrm{v}; JM|H'|\Sigma_{0e}^-; \mathrm{v}; JM\rangle \\ &= S_{1e,N+1}H_{11} + S_{0e,N+1}H_{12}.\end{aligned} \tag{3.271}$$

$$\begin{aligned}\langle \Sigma_{0e}^-; \mathrm{v}; JM|\hat{H}'|F_1; \mathrm{v}; JM; f\rangle &= \varepsilon_1 S_{0e,N+1} = S_{1e,N+1}\langle \Sigma_{0e}^-; \mathrm{v}; JM|H'|\Sigma_{1e}^-; \mathrm{v}; JM\rangle \\ &\quad + S_{0e,N+1}\langle \Sigma_{0e}^-; \mathrm{v}; JM|H'|\Sigma_{0e}^-; \mathrm{v}; JM\rangle \\ &= S_{1e,N+1}H_{21} + S_{0e,N+1}H_{22}.\end{aligned} \tag{3.272}$$

Repeating the procedure for the F_2 and F_3 kets, we obtain

$$\left\langle \Sigma_{1f}^-; \mathrm{v}; JM\middle|\hat{H}'\middle|F_2; \mathrm{v}; JM; f\right\rangle = \varepsilon_2 = \left\langle \Sigma_{1f}^-; \mathrm{v}; JM\middle|H'\middle|\Sigma_{1f}^-; \mathrm{v}; JM\right\rangle = H_{33}. \tag{3.273}$$

$$\begin{aligned}\langle \Sigma_{1e}^-; \mathrm{v}; JM|\hat{H}'|F_3; \mathrm{v}; JM; f\rangle &= \varepsilon_3 S_{1e,N-1} = S_{1e,N-1}\langle \Sigma_{1e}^-; \mathrm{v}; JM|H'|\Sigma_{1e}^-; \mathrm{v}; JM\rangle \\ &\quad + S_{0e,N-1}\langle \Sigma_{1e}^-; \mathrm{v}; JM|H'|\Sigma_{0e}^-; \mathrm{v}; JM\rangle \\ &= S_{1e,N-1}H_{11} + S_{0e,N-1}H_{12}.\end{aligned} \tag{3.274}$$

$$\begin{aligned}\langle \Sigma_{0e}^-; \mathrm{v}; JM_J|\hat{H}'|F_3; \mathrm{v}; JM_J; f\rangle &= \varepsilon_3 S_{0e,N-1} = S_{1e,N-1}\langle \Sigma_{0e}^-; \mathrm{v}; JM_J|H'|\Sigma_{1e}^-; \mathrm{v}; JM_J\rangle \\ &\quad + S_{0e,N-1}\langle \Sigma_{0e}^-; \mathrm{v}; JM_J|H'|\Sigma_{0e}^-; \mathrm{v}; JM_J\rangle \\ &= S_{1e,N-1}H_{21} + S_{0e,N-1}H_{22}.\end{aligned} \tag{3.275}$$

As is the case for the $^2\Pi$, Eqs. (3.271), (3.272), (3.274), and (3.275) can be written in matrix form as

$$\begin{bmatrix} H_{11} - \varepsilon_1 & H_{12} \\ H_{21} & H_{22} - \varepsilon_1 \end{bmatrix} \begin{bmatrix} S_{1e,N+1} \\ S_{0e,N+1} \end{bmatrix} = \begin{bmatrix} 0 \\ 0 \end{bmatrix} \tag{3.276}$$

and

$$\begin{bmatrix} H_{11} - \varepsilon_3 & H_{12} \\ H_{21} & H_{22} - \varepsilon_3 \end{bmatrix} \begin{bmatrix} S_{1e,N-1} \\ S_{0e,N-1} \end{bmatrix} = \begin{bmatrix} 0 \\ 0 \end{bmatrix}. \tag{3.277}$$

As discussed in Zare (1988), the system of two equations will have a solution only if the determinant $(H_{11} - \varepsilon)(H_{22} - \varepsilon) - H_{12}H_{21}$ of the secular matrix $\begin{bmatrix} H_{11} - \varepsilon & H_{12} \\ H_{21} & H_{22} - \varepsilon \end{bmatrix}$ is zero,

$$\varepsilon^2 - \varepsilon(H_{11} + H_{22}) + H_{11} + H_{22} - H_{12}^2 = 0. \tag{3.278}$$

Solving Eq. (3.175) for ε we obtain

$$\varepsilon = \left(\frac{H_{11} + H_{22}}{2}\right) \pm \left[\left(\frac{H_{11} - H_{22}}{2}\right)^2 + H_{12}^2\right]^{1/2}. \tag{3.279}$$

The energy ε_1 associated with the F_1 levels is lower than the energy associated with the F_3 levels, so

$$\varepsilon_1 = \left(\frac{H_{11}+H_{22}}{2}\right) - \left[\left(\frac{H_{11}-H_{22}}{2}\right)^2 + H_{12}^2\right]^{1/2} \tag{3.280}$$

and

$$\varepsilon_3 = \left(\frac{H_{11}+H_{22}}{2}\right) + \left[\left(\frac{H_{11}-H_{22}}{2}\right)^2 + H_{12}^2\right]^{1/2}. \tag{3.281}$$

Solving for the case (a) coefficients, we obtain

$$S_{1e,N+1} = \left[\frac{H_{12}^2}{(\varepsilon_1 - H_{11})^2 + H_{12}^2}\right]^{1/2}. \tag{3.282}$$

$$S_{0e,N+1} = \left[\frac{(\varepsilon_1 - H_{11})^2}{(\varepsilon_1 - H_{11})^2 + H_{12}^2}\right]^{1/2}. \tag{3.283}$$

$$S_{1e,N-1} = -\left[\frac{H_{12}^2}{(\varepsilon_3 - H_{11})^2 + H_{12}^2}\right]^{1/2}. \tag{3.284}$$

$$S_{0e,N+1} = \left[\frac{(\varepsilon_3 - H_{11})^2}{(\varepsilon_3 - H_{11})^2 + H_{12}^2}\right]^{1/2}. \tag{3.285}$$

In Eqs. (3.271)–(3.285), the density matrix elements H_{ij} are given by

$$\begin{aligned}
H_{11} &= \left\langle \Sigma_{1e}^-; \mathrm{v}; JM \middle| \hat{H}_R + \hat{H}_{SO} + \hat{H}_{SR} + \hat{H}_{SS} + \hat{H}_{CD} + \hat{H}_{SD} \middle| \Sigma_{1e}^-; \mathrm{v}; JM \right\rangle \\
&= \frac{1}{2}\left[\langle 0;11;\mathrm{v};J1M|\hat{H}'|0;11;\mathrm{v};J1M\rangle + \langle 0;1-1;\mathrm{v};J-1M|\hat{H}'|0;1-1;\mathrm{v};J-1M\rangle\right] \\
&\quad + \frac{1}{2}\left[\langle 0;11;\mathrm{v};J1M|\hat{H}'|0;1-1;\mathrm{v};J-1M\rangle + \langle 0;1-1;\mathrm{v};J-1M|\hat{H}'|0;11;\mathrm{v};J1M\rangle\right] \\
&= B_\mathrm{v}[J(J+1) + S(S+1) - 2\Omega\Sigma] + \frac{1}{2}\gamma_\mathrm{v}[2\Omega\Sigma - 2S(S+1)] \\
&\quad + 2\lambda_\mathrm{v}\left[\Sigma^2 - \frac{1}{3}S(S+1)\right] - D_\mathrm{v}\Big\{[J(J+1) + S(S+1) - 2\Omega\Sigma]^2 \\
&\quad + 2\left[J(J+1) - \Omega^2\right]\left[S(S+1) - \Sigma^2\right] + 2\Omega\Sigma\Big\} - D_\mathrm{v}\Big\{[J(J+1) - \Omega(\Omega+1)]^{1/2} \\
&\quad \Big\{\times[J(J+1) - (\Omega+1)(\Omega+2)]^{1/2}[S(S+1) - \Sigma(\Sigma+1)]^{1/2}[S(S+1) - (\Sigma+1)(\Sigma+2)]^{1/2} \\
&= B_\mathrm{v}J(J+1) - \gamma_\mathrm{v} + 2\lambda_\mathrm{v}/3 - D_\mathrm{v}\Big\{[J(J+1)]^2 + 2J(J+1)\Big\} - 2D_\mathrm{v}J(J+1) \\
&= B_\mathrm{v}J(J+1) - \gamma_\mathrm{v} + 2\lambda_\mathrm{v}/3 - D_\mathrm{v}\Big\{[J(J+1)]^2 + 4J(J+1)\Big\}.
\end{aligned} \tag{3.286}$$

$$
\begin{aligned}
H_{22} &= \langle \Sigma_{0e}^{-}; \mathrm{v}; JM | \hat{H}_R + \hat{H}_{SR} + \hat{H}_{SS} + \hat{H}_{CD} | \Sigma_{0e}^{-}; \mathrm{v}; JM \rangle \\
&= \langle 0; S0; \mathrm{v}; J0M | \hat{H}' | 0; S0; \mathrm{v}; J0M \rangle \\
&= B_{\mathrm{v}}[J(J+1) + S(S+1) - 2\Omega\Sigma] + \frac{1}{2}\gamma_{\mathrm{v}}[2\Omega\Sigma - 2S(S+1)] + 2\lambda_{\mathrm{v}}\left[\Sigma^2 - \frac{1}{3}S(S+1)\right] \\
&\quad - D_{\mathrm{v}}\left\{[J(J+1) + S(S+1) - 2\Omega\Sigma]^2 + 2\left[J(J+1) - \Omega^2\right]\left[S(S+1) - \Sigma^2\right] + 2\Omega\Sigma\right\} \\
&= B_{\mathrm{v}}[J(J+1) + 2] - 2\gamma_{\mathrm{v}} - 4\lambda_{\mathrm{v}}/3 - D_{\mathrm{v}}\left\{[J(J+1)]^2 + 8J(J+1) + 4\right\}.
\end{aligned}
\tag{3.287}
$$

$$
\begin{aligned}
H_{33} &= \left\langle \Sigma_{1f}^{-}; \mathrm{v}; JM \middle| \hat{H}_R + \hat{H}_{SO} + \hat{H}_{SR} + \hat{H}_{SS} + \hat{H}_{CD} + \hat{H}_{SD} \middle| \Sigma_{1f}^{-}; \mathrm{v}; JM \right\rangle \\
&= \frac{1}{2}\left[\langle 0; 11; \mathrm{v}; J1M | \hat{H}' | 0; 11; \mathrm{v}; J1M \rangle + \langle 0; 1-1; \mathrm{v}; J-1M | \hat{H}' | 0; 1-1; \mathrm{v}; J-1M \rangle\right] \\
&\quad - \frac{1}{2}\left[\langle 0; 11; \mathrm{v}; J1M | \hat{H}' | 0; 1-1; \mathrm{v}; J-1M \rangle + \langle 0; 1-1; \mathrm{v}; J-1M | \hat{H}' | 0; 11; \mathrm{v}; J1M \rangle\right] \\
&= B_{\mathrm{v}}[J(J+1) + S(S+1) - 2\Omega\Sigma] + \frac{1}{2}\gamma_{\mathrm{v}}[2\Omega\Sigma - 2S(S+1)] \\
&\quad + 2\lambda_{\mathrm{v}}\left[\Sigma^2 - \frac{1}{3}S(S+1)\right] - D_{\mathrm{v}}\left\{[J(J+1) + S(S+1) - 2\Omega\Sigma]^2\right. \\
&\quad \left. + 2\left[J(J+1) - \Omega^2\right]\left[S(S+1) - \Sigma^2\right] + 2\Omega\Sigma\right\} + D_{\mathrm{v}}\left\{[J(J+1) - \Omega(\Omega+1)]^{1/2}\right. \\
&\quad \left\{ \times [J(J+1) - (\Omega+1)(\Omega+2)]^{1/2}[S(S+1) - \Sigma(\Sigma+1)]^{1/2}[S(S+1) - (\Sigma+1)(\Sigma+2)]^{1/2} \right. \\
&= B_{\mathrm{v}}J(J+1) - \gamma_{\mathrm{v}} + 2\lambda_{\mathrm{v}}/3 - D_{\mathrm{v}}\left\{[J(J+1)]^2 + 2J(J+1)\right\} + 2D_{\mathrm{v}}J(J+1) \\
&= B_{\mathrm{v}}J(J+1) - \gamma_{\mathrm{v}} + 2\lambda_{\mathrm{v}}/3 - D_{\mathrm{v}}[J(J+1)]^2.
\end{aligned}
\tag{3.288}
$$

$$
\begin{aligned}
H_{12} &= \langle \Sigma_{0e}^{-}; \mathrm{v}; JM | \hat{H}_R + \hat{H}_{SR} + \hat{H}_{SS} + \hat{H}_{CD} | \Sigma_{1e}^{-}; \mathrm{v}; JM \rangle \\
&= \frac{1}{\sqrt{2}}\left[\langle 0; 10; \mathrm{v}; J0M | \hat{H}' | 0; 11; \mathrm{v}; J1M \rangle + \langle 0; 10; \mathrm{v}; J0M | \hat{H}' | 0; 1-1; \mathrm{v}; J-1M \rangle\right] \\
&= -\left(B_{\mathrm{v}}/\sqrt{2}\right)[J(J+1) - \Omega_+(\Omega_+ - 1)]^{1/2}[S(S+1) - \Sigma_+(\Sigma_+ - 1)]^{1/2} \\
&\quad - \left(B_{\mathrm{v}}/\sqrt{2}\right)[J(J+1) - \Omega_-(\Omega_- + 1)]^{1/2}[S(S+1) - \Sigma_-(\Sigma_- + 1)]^{1/2} \\
&\quad + \left(\gamma_{\mathrm{v}}/2\sqrt{2}\right)[J(J+1) - \Omega_+(\Omega_+ - 1)]^{1/2}[S(S+1) - \Sigma_+(\Sigma_+ - 1)]^{1/2} \\
&\quad + \left(\gamma_{\mathrm{v}}/2\sqrt{2}\right)[J(J+1) - \Omega_-(\Omega_- + 1)]^{1/2}[S(S+1) - \Sigma_-(\Sigma_- + 1)]^{1/2} \\
&\quad + \left(2D_{\mathrm{v}}/\sqrt{2}\right)[J(J+1) - \Omega_+(\Omega_+ - 1)]^{1/2}[S(S+1) - \Sigma_+(\Sigma_+ - 1)]^{1/2} \\
&\quad \times [J(J+1) + S(S+1) - 2\Omega_+\Sigma_+ + (\Omega_+ + \Sigma_+) - 1] \\
&\quad + \left(2D_{\mathrm{v}}/\sqrt{2}\right)[J(J+1) - \Omega_-(\Omega_- + 1)]^{1/2}[S(S+1) - \Sigma_-(\Sigma_- + 1)]^{1/2} \\
&\quad \times [J(J+1) + S(S+1) - 2\Omega_-\Sigma_- - (\Omega_- + \Sigma_-) - 1] \\
&= \{-2B_{\mathrm{v}} + \gamma_{\mathrm{v}} + 4D_{\mathrm{v}}[J(J+1) + 1]\}[J(J+1)]^{1/2}.
\end{aligned}
\tag{3.289}
$$

Including higher-order spectroscopic corrections for centrifugal terms, again tabulated by Brazier et al. (1986), the density matrix elements are given by

$$\begin{aligned} H_{11} = {} & B_v x - D_v\left(x^2 + 4x\right) + H_v\left(x^3 + 12x^2 + 8x\right) + L_v\left(x^4 + 24x^3 + 48x^2 + 16x\right) \\ & + M_v\left(x^5 + 40x^4 + 160x^3 + 144x^2 + 32x\right) - \gamma_v - 3\gamma_{Dv}x - \gamma_{Hv}\left(5x^2 + x\right) \\ & + 2\lambda_v/3 + 2\lambda_{Dv}x/3, \end{aligned} \tag{3.290}$$

$$\begin{aligned} H_{22} = {} & B_v(x + 2) - D_v\left(x^2 + 8x + 4\right) + H_v\left(x^3 + 18x^2 + 28x + 8\right) \\ & + L_v\left(x^4 + 32x^3 + 104x^2 + 80x + 16\right) \\ & + M_v\left(x^5 + 50x^4 + 280x^3 + 416x^2 + 208x + 32\right) \\ & - 2\gamma_v - 4\gamma_{Dv}(x + 1) - \gamma_{Hv}\left(6x^2 - 20x - 8\right) - 4\lambda_v/3 - 4\lambda_{Dv}(x + 2)/3, \end{aligned} \tag{3.291}$$

$$H_{33} = B_v x - D_v x^2 + H_v x^3 + L_v x^4 + M_v x^5 - \gamma_v - \gamma_{Dv}x - \gamma_{Hv}x^2 + 2\lambda_v/3 + 2\lambda_{Dv}x/3, \tag{3.292}$$

$$\begin{aligned} H_{12} = {} & -2B_v\sqrt{x} + 4D_v\sqrt{x}(x + 1) - H_v\sqrt{x}\left(6x^2 + 20x + 8\right) \\ & - 4L_v\sqrt{x}\left(2x^3 + 14x^2 + 16x + 4\right) \\ & - 2M_v\sqrt{x}\left(5x^4 + 60x^3 + 136x^2 + 88x + 16\right) \\ & + \gamma_v\sqrt{x} + \gamma_{Dv}\sqrt{x}(x + 4) - \gamma_{Hv}\sqrt{x}\left(x^2 + 12x + 8\right) + 2\lambda_{Dv}\sqrt{x}/3, \end{aligned} \tag{3.293}$$

where $x = J(J + 1)$.

The energy level structure of the $X^3\Sigma^-(v = 0)$ level of NH is included as an example for the analysis outlined in this section. The spectroscopic constants for the $X^3\Sigma^-(v = 0)$ level are listed in Table 3.12, and the level energies and the basis wavefunction coefficients are listed in Table 3.13.

Table 3.12 Spectroscopic constants (cm^{-1}) for the rotational levels of the $X^3\Sigma^-(v = 0)$ vibrational level of the NH radical. Spectroscopic constants extracted from Ram and Bernath (2010)

T_v	0.00	$H_v \times 10^7$	1.238065	$\gamma_v \times 10^2$	−5.485506	λ_v	0.91989675
B_v	16.343275263	$L_v \times 10^{11}$	−1.45460	$\gamma_{Dv} \times 10^5$	1.51582	$\lambda_{Dv} \times 10^7$	3.436
$D_v \times 10^3$	1.7028445	$M_v \times 10^{16}$	6.9165	$\gamma_{Hv} \times 10^9$	−1.2744		

Table 3.13 Energies (cm^{-1}) and Hund's case (a) coefficients for the rotational levels of the $X^3\Sigma^-(v=0)$ vibrational level of the NH radical

J	F_{1e}	F_{2f}	F_{3e}	$S_{0e,N+1}$	$S_{1e,N+1}$	$S_{1f,N}$	$S_{0e,N-1}$	$S_{1e,N-1}$
0			31.56				1.0000	0.0000
1	−0.01	33.35	97.57	0.5846	0.8113	1.0000	0.8113	−0.5846
2	32.49	98.67	195.63	0.6368	0.7710	1.0000	0.7710	−0.6368
3	97.71	196.54	326.06	0.6577	0.7532	1.0000	0.7532	−0.6577
4	195.51	326.85	488.75	0.6691	0.7432	1.0000	0.7432	−0.6691
5	325.76	489.44	683.50	0.6762	0.7367	1.0000	0.7367	−0.6762
6	488.29	684.09	910.07	0.6811	0.7322	1.0000	0.7322	−0.6811
7	682.90	910.57	1,168.19	0.6847	0.7289	1.0000	0.7289	−0.6847
8	909.34	1,168.60	1,457.54	0.6874	0.7263	1.0000	0.7263	−0.6874
9	1,167.34	1,457.86	1,777.76	0.6895	0.7243	1.0000	0.7243	−0.6895
10	1,456.57	1,777.99	2,128.46	0.6913	0.7226	1.0000	0.7226	−0.6913
11	1,776.67	2,128.59	2,509.20	0.6927	0.7212	1.0000	0.7212	−0.6927
12	2,127.26	2,509.24	2,919.53	0.6939	0.7201	1.0000	0.7201	−0.6939
13	2,507.90	2,919.47	3,358.94	0.6949	0.7191	1.0000	0.7191	−0.6949
14	2,918.12	3,358.78	3,826.89	0.6958	0.7182	1.0000	0.7182	−0.6958
15	3,357.42	3,826.63	4,322.83	0.6966	0.7175	1.0000	0.7175	−0.6966
16	3,825.28	4,322.46	4,846.15	0.6973	0.7168	1.0000	0.7168	−0.6973
17	4,321.12	4,845.68	5,396.23	0.6979	0.7162	1.0000	0.7162	−0.6979
18	4,844.35	5,395.65	5,972.40	0.6984	0.7157	1.0000	0.7157	−0.6984
19	5,394.33	5,971.71	6,573.98	0.6989	0.7152	1.0000	0.7152	−0.6989
20	5,970.42	6,573.18	7,200.26	0.6994	0.7147	1.0000	0.7147	−0.6994
21	6,571.92	7,199.35	7,850.50	0.6998	0.7143	1.0000	0.7143	−0.6998
22	7,198.12	7,849.48	8,523.93	0.7002	0.7140	1.0000	0.7140	−0.7002
23	7,848.28	8,522.79	9,219.76	0.7005	0.7136	1.0000	0.7136	−0.7005
24	8,521.64	9,218.51	9,937.19	0.7008	0.7133	1.0000	0.7133	−0.7008
25	9,217.41	9,935.82	10,675.37	0.7011	0.7130	1.0000	0.7130	−0.7011
26	9,934.77	10,673.88	11,433.46	0.7014	0.7128	1.0000	0.7128	−0.7014
27	10,672.89	11,431.85	12,210.57	0.7017	0.7125	1.0000	0.7125	−0.7017
28	11,430.92	12,208.84	13,005.82	0.7019	0.7123	1.0000	0.7123	−0.7019
29	12,207.98	13,003.96	13,818.28	0.7022	0.7120	1.0000	0.7120	−0.7022
30	13,003.17	13,816.30	14,647.02	0.7024	0.7118	1.0000	0.7118	−0.7024

4 Quantum Mechanical Analysis of the Interaction of Laser Radiation with Electric Dipole Resonances

4.1 Introduction

The subject of this chapter is the quantum mechanical analysis of the interaction of electromagnetic radiation with atomic transitions. The analysis is based on the Schrödinger wave equation (SWE), and in the first section the gauge-invariant form of the external electromagnetic field is introduced. In Section 4.3 the electric dipole interaction and the long-wavelength approximation for the analysis of this interaction are discussed. The perturbative analysis of both single-photon and two-photon electric dipole interactions is subject of Section 4.4, and density matrix analysis is introduced in Section 4.5. The interaction of radiation with the resonances of atomic hydrogen is then discussed. The analysis is performed for resonances in both the uncoupled (n, l, m_l, m_s) and coupled (J, m_J) representations. In the last section of the chapter, the radiative interactions for multielectron atoms are discussed. The Wigner–Eckart theorem and selection rules for transitions between levels characterized by LS coupling are developed. The effect of hyperfine splitting on radiative transitions is also briefly discussed.

4.2 The Schrödinger Wave Equation for Charged Particles in External Electromagnetic Fields

The Schrödinger wave equation (SWE) is the fundamental equation of nonrelativistic wave mechanics and is the basis for our study of the interaction of laser radiation with atomic and molecular resonances, and the objective of this section is the development of the SWE for the interaction of atoms or molecules with an external electric field. The effect of the external electric field is to cause transitions from some initial quantum state $\Psi_i(\boldsymbol{r}, t_i)$ to some final quantum state $\Psi_f(\boldsymbol{r}, t_f)$. It is assumed that the electric field is weak enough that the quantum states of the atom or molecule, calculated using methods described in Chapters 2 and 3, respectively, are still valid. Consequently, the quantum states at both the initial and final conditions can be expressed as linear combinations of the unperturbed, stationary quantum states $u_k(\boldsymbol{r})$ of the molecule,

$$\Psi(\boldsymbol{r}, t) = \sum_k a_k(t)\, u_k(\boldsymbol{r}). \tag{4.1}$$

The time-dependent coefficients $a_k(t)$ contain the time-dependent exponential term $\exp(-i\varepsilon_k t/\hbar)$, where ε_k is the eigenenergy of the basis set quantum state k. The effect of the interaction of an external electromagnetic field with the atom or molecule is captured in the time-dependent coefficients $a_k(t)$. The products $a_k a_k^*$ and $a_k a_j^*$ are proportional to the population of quantum state k and the coherence between quantum states k and j, respectively, as will be discussed in detail. However, the Hamiltonian for the interaction of the external electromagnetic field with the atom or molecule must be developed in a gauge-invariant manner so that the meaning of the time-dependent coefficients $a_k(t)$ is preserved (Kobe & Smirl, 1978; Lamb et al., 1987; Scully & Zubairy, 1997; Yang, 1976).

The SWE can be written in deceptively compact form as

$$\hat{H}\Psi = \hat{\varepsilon}\Psi = i\hbar\frac{\partial \Psi}{\partial t}, \tag{4.2}$$

where $\hat{H}$ is the quantum mechanical Hamiltonian operator and $\hat{\varepsilon} = i\frac{\partial}{\partial t}$ is the energy operator. The classical Hamiltonian is a good starting point for the determination of the quantum mechanical Hamiltonian operator for a given physical system. However, the quantum mechanical Hamiltonian operator must be Hermitian, and it is not always straightforward to determine the correct form for $\hat{H}$. This issue is discussed in depth by Slater (1960a). The classical Hamiltonian for a single charged particle in an external electromagnetic field is given by

$$H = \frac{(\boldsymbol{p} - q\boldsymbol{A})^2}{2m} + q\phi. \tag{4.3}$$

The external electromagnetic field is specified in term of the vector potential $\boldsymbol{A}$ and the scalar potential ϕ,

$$\boldsymbol{E}(\boldsymbol{r}, t) = -\nabla\phi(\boldsymbol{r}, t) - \frac{\partial \boldsymbol{A}(\boldsymbol{r}, t)}{\partial t} \qquad \boldsymbol{B}(\boldsymbol{r}, t) = \nabla \times \boldsymbol{A}(\boldsymbol{r}, t). \tag{4.4}$$

The potential can undergo a gauge transformation characterized by a real differentiable function $\Lambda(\boldsymbol{r}, t)$,

$$\boldsymbol{A}'(\boldsymbol{r}, t) = \boldsymbol{A}(\boldsymbol{r}, t) + \nabla\Lambda(\boldsymbol{r}, t) \qquad \phi'(\boldsymbol{r}, t) = \phi(\boldsymbol{r}, t) - \frac{\partial \Lambda(\boldsymbol{r}, t)}{\partial t}. \tag{4.5}$$

The electric and magnetic fields are unchanged by the gauge transformation, as is evident by comparison of

$$\boldsymbol{E}(\boldsymbol{r}, t) = -\nabla\phi'(\boldsymbol{r}, t) - \frac{\partial \boldsymbol{A}'(\boldsymbol{r}, t)}{\partial t} \qquad \boldsymbol{B}(\boldsymbol{r}, t) = \nabla \times \boldsymbol{A}'(\boldsymbol{r}, t). \tag{4.6}$$

with Eq. (4.4). However, if the Hamiltonian for the field interaction is not developed in a gauge-invariant manner, the calculation of the resulting state population coefficients $a_k(t)$ will be in error.

The so-called "conventional" approach (Yang, 1976) to the quantum mechanical analysis of the interaction of an external electromagnetic field with atomic or molecular resonances is to define the Hamiltonian operator as

$$\hat{H} = \frac{1}{2m}\left[\hat{\boldsymbol{p}} - q\hat{\boldsymbol{A}}(\boldsymbol{r}, t)\right]^2 + q\phi(\boldsymbol{r}, t) + U(\boldsymbol{r}). \tag{4.7}$$

In this case the potential function $U(\boldsymbol{r})$ represents a non-time-dependent force that is responsible for the stable structure of atom or molecule. In applying Eq. (4.7) to the electron in atomic hydrogen, for example, this would be the electromagnetic attractive potential due to the proton, and the vector $\boldsymbol{r}$ would extend from the proton to the electron. A more general equation for multielectron atoms and molecules will be discussed in the next section. Expanding the first term in Eq. (4.7), we obtain

$$\hat{H} = \frac{\hat{\boldsymbol{p}}^2}{2m} - \frac{q}{2m}\left[\hat{\boldsymbol{p}}\bullet\hat{\boldsymbol{A}}(\boldsymbol{r}, t) + \hat{\boldsymbol{A}}(\boldsymbol{r}, t)\bullet\hat{\boldsymbol{p}}\right] + \frac{q\hat{\boldsymbol{A}}^2(\boldsymbol{r}, t)}{2m} + q\phi(\boldsymbol{r}, t) + U(\boldsymbol{r}). \tag{4.8}$$

In the semiclassical treatment of the interaction of radiation with atoms and molecules, we regard the electromagnetic field as a classical field, and the vector potential operator $\hat{\boldsymbol{A}}(\boldsymbol{r}_j) = \boldsymbol{A}(\boldsymbol{r}_j)$ is thus a multiplicative operator. It can be shown (Schiff, 1968) that

$$[\hat{\boldsymbol{p}}, \boldsymbol{A}(\boldsymbol{r}, t)] = \hat{\boldsymbol{p}}\bullet\boldsymbol{A}(\boldsymbol{r}, t) - \boldsymbol{A}(\boldsymbol{r}, t)\bullet\hat{\boldsymbol{p}} = -i\hbar\nabla\bullet\boldsymbol{A}(\boldsymbol{r}, t). \tag{4.9}$$

Substituting for $\hat{\boldsymbol{p}}\bullet\boldsymbol{A}(\boldsymbol{r}, t)$ in Eq. (4.7) using Eq. (4.9) and simplifying, we obtain

$$\hat{H} = \frac{\hat{\boldsymbol{p}}^2}{2m} - \frac{q\boldsymbol{A}(\boldsymbol{r}, t)\bullet\hat{\boldsymbol{p}}}{m} + \frac{i\hbar q\nabla\bullet\boldsymbol{A}(\boldsymbol{r}, t)}{2m} + \frac{q\boldsymbol{A}^2(\boldsymbol{r}, t)}{2m} + q\phi(\boldsymbol{r}, t) + U(\boldsymbol{r}). \tag{4.10}$$

For the interaction of atoms and molecules with an external field, the term that is quadratic in $\boldsymbol{A}$ is very small compared to the term that is linear in $\boldsymbol{A}$, and we can neglect this term. As noted by Scully and Zubairy (1997), at this stage a choice of gauge is now usually made such that $\phi(\boldsymbol{r}, t) = 0$ and $\nabla\bullet\boldsymbol{A}(\boldsymbol{r}, t) = 0$. The result is

$$\hat{H} = \frac{\hat{\boldsymbol{p}}^2}{2m} + U(\boldsymbol{r}) - \frac{q\boldsymbol{A}(\boldsymbol{r}, t)\bullet\hat{\boldsymbol{p}}}{m} = \hat{H}_B^0 + \hat{H}_R^g, \tag{4.11}$$

where

$$\hat{H}_B^0 = \frac{\hat{\boldsymbol{p}}^2}{2m} + U(\boldsymbol{r}) \tag{4.12}$$

is the Hamiltonian for the unperturbed system and

$$\hat{H}_R^g = -\frac{q\boldsymbol{A}(\boldsymbol{r}, t)\bullet\hat{\boldsymbol{p}}}{m} \tag{4.13}$$

is the residual Hamiltonian that describes the interaction of the atom or molecule with the external field (Yang, 1976). Although this procedure for describing the interaction of the external field is widely used (Slater, 1960a; Struve, 1988), the calculations of transition rates are not gauge invariant in this formulation, and the quantum state coefficients $a_k(t)$ for the stationary state Hamiltonian no longer give the correct occupation probabilities $a_k a_k^*$ for the atomic quantum states calculated in the absence of the external electromagnetic field. The problem was first pointed out by Lamb (1952).

A gauge-invariant formulation of the radiative interaction was outlined by Yang (1976). In this formulation, the Hamiltonian that is used to calculate the stationary states is written as

$$\hat{H}_B = \frac{1}{2m}[\hat{\boldsymbol{p}} - q\boldsymbol{A}(\boldsymbol{r}, t)]^2 + U(\boldsymbol{r}). \tag{4.14}$$

The full Hamiltonian is still given by Eq. (4.7), but now the residual Hamiltonian is given by

$$\hat{H}_R = q\phi(\boldsymbol{r}, t). \tag{4.15}$$

The basis Hamiltonian satisfies the following eigenvalue equation:

$$\hat{H}_B \psi_k(\boldsymbol{r}, t) = \varepsilon_k(t)\psi_k(\boldsymbol{r}, t), \tag{4.16}$$

where both the basis state energies ε_k and the basis state wavefunctions ψ_k are in general functions of time because of the time dependence of the vector potential in Eq. (4.14). The wavefunction for the system is expressed in terms of the basis wavefunctions,

$$\Psi(\boldsymbol{r}, t) = \sum_k a_k(t)\psi_k(\boldsymbol{r}, t). \tag{4.17}$$

The basis wavefunctions form a complete set, and the eigenstates are orthonormal,

$$\langle \psi_j(\boldsymbol{r}, t) | \psi_k(\boldsymbol{r}, t) \rangle = \delta_{jk}. \tag{4.18}$$

In terms of the Hamiltonian in Eq. (4.7), we can write the SWE as

$$\left\{ \frac{1}{2m}[\hat{\boldsymbol{p}} - q\boldsymbol{A}]^2 + qU \right\}\Psi + q\phi\Psi = i\hbar\frac{\partial\Psi}{\partial t}. \tag{4.19}$$

Substituting Eq. (4.17) into Eq. (4.19) and using the results of Eq. (4.16), we obtain

$$\sum_k a_k(t)\varepsilon_k(t)\psi_k(\boldsymbol{r}, t) + q\phi\sum_k a_k(t)\psi_k(\boldsymbol{r}, t) = i\hbar\frac{\partial}{\partial t}\sum_k a_k(t)\psi_k(\boldsymbol{r}, t). \tag{4.20}$$

Multiplying the left- and right-hand sides of Eq. (4.20) by $\psi_j^*(\boldsymbol{r}, t)$ and integrating over all space, and using Eq. (4.18), we obtain

$$a_j(t)\varepsilon_j(t) + \sum_k a_k(t)\langle \psi_j(\boldsymbol{r}, t) | q\phi\psi_k(\boldsymbol{r}, t) \rangle = i\hbar\frac{da_j(t)}{dt} + \sum_k a_k(t)\left\langle \psi_j(\boldsymbol{r}, t) \middle| i\hbar\frac{\partial}{\partial t}\psi_k(\boldsymbol{r}, t) \right\rangle. \tag{4.21}$$

Rearranging, we obtain

$$\frac{da_j(t)}{dt} = \frac{1}{i\hbar}\left\{a_j(t)\varepsilon_j(t) + \sum_k a_k(t)\left\langle \psi_j(\boldsymbol{r},t)\left|\left[q\phi - i\hbar\frac{\partial}{\partial t}\right]\right|\psi_k(\boldsymbol{r},t)\right\rangle\right\}. \tag{4.22}$$

Following the treatment of Yang (1976) and Kobe and Smirl (1978), we can show that the coefficients $a_j(t)$ are "unchanged under a gauge transformation." For a gauge transformation of the type indicated by Eq. (4.5), the wavefunction in the new gauge is given by

$$\Psi'(\boldsymbol{r},t) = \exp\left[i\frac{q\Lambda(\boldsymbol{r},t)}{\hbar}\right]\Psi(\boldsymbol{r},t). \tag{4.23}$$

Substituting for $\Psi(\boldsymbol{r},t)$ in Eq. (4.19) using Eq. (4.23), we obtain, after a great deal of operator algebra,

$$\left\{\frac{1}{2m}[\hat{\boldsymbol{p}} - q\boldsymbol{A}']^2 + U\right\}\Psi' + q\phi'\Psi' = i\hbar\frac{\partial\Psi'}{\partial t}, \tag{4.24}$$

where

$$\boldsymbol{A}'(\boldsymbol{r},t) = \boldsymbol{A}(\boldsymbol{r},t) + \nabla\Lambda(\boldsymbol{r},t) \tag{4.25}$$

and

$$\phi'(\boldsymbol{r},t) = \phi(\boldsymbol{r},t) + \frac{\partial\Lambda(\boldsymbol{r},t)}{\partial t}. \tag{4.26}$$

Note that it is much easier to show that substitution of Eqs. (4.23), (4.25), and (4.26) into Eq. (4.24) results in Eq. (4.19) than to derive Eq. (4.24). Expanding $\Psi'(\boldsymbol{r},t)$ in terms of the basis states resulting from the transformed basis Hamiltonian

$$H'_B = \frac{1}{2m}[\hat{\boldsymbol{p}} - q\boldsymbol{A}']^2 + U, \tag{4.27}$$

we obtain:

$$\begin{aligned}\Psi'(\boldsymbol{r},t) &= \sum_k a'_k(t)\psi'_k(\boldsymbol{r},t) = \exp\left(i\frac{q\Lambda}{\hbar}\right)\Psi(\boldsymbol{r},t)\\ &= \sum_k a_k(t)\exp\left(i\frac{q\Lambda}{\hbar}\right)\psi_k(\boldsymbol{r},t) = \sum_k a_k(t)\psi'_k(\boldsymbol{r},t).\end{aligned} \tag{4.28}$$

The gauge-transformed energy operator $\hat{\varepsilon}'$ is defined by

$$\hat{\varepsilon}' = \exp\left[i\frac{q\Lambda(\boldsymbol{r},t)}{\hbar}\right]\hat{\varepsilon}\exp\left[-i\frac{q\Lambda(\boldsymbol{r},t)}{\hbar}\right]. \tag{4.29}$$

Operating on the basis state wavefunctions in the transformed gauge,

$$\hat{\varepsilon}'\psi'_k = \left[\exp\left(i\frac{q\Lambda}{\hbar}\right)\hat{\varepsilon}\exp\left(-i\frac{q\Lambda}{\hbar}\right)\right]\left[\exp\left(i\frac{q\Lambda}{\hbar}\right)\psi_k\right] = \exp\left(i\frac{q\Lambda}{\hbar}\right)\hat{\varepsilon}\psi_k = \varepsilon_k\psi'_k, \tag{4.30}$$

we can show that the eigenenergies of the stationary state wavefunctions are unchanged by the gauge transformation. Now substituting Eq. (4.28) into Eq. (4.24) and performing the same operations that were used to derive Eq. (4.22), we obtain

$$\frac{da_j(t)}{dt} = \frac{1}{i\hbar}\left\{ a_j(t)\varepsilon_j(t) + \sum_k a_k(t)\left\langle \psi_j'(\boldsymbol{r},t)\left|\left[q\phi' - i\hbar\frac{\partial}{\partial t}\right]\right|\psi_k'(\boldsymbol{r},t)\right\rangle\right\}. \tag{4.31}$$

The matrix element has the same value in both gauges,

$$\begin{aligned}\left\langle \psi_j'\left|\left[q\phi' - i\hbar\frac{\partial}{\partial t}\right]\right|\psi_k'\right\rangle &= \left\langle \psi_j\left|\exp\left(+i\frac{q\Lambda}{\hbar}\right)\left[q\left(\phi - \frac{\partial\Lambda}{\partial t}\right) - i\hbar\frac{\partial}{\partial t}\right]\exp\left(+i\frac{q\Lambda}{\hbar}\right)\psi_k\right.\right\rangle \\ &= \left\langle \psi_j\left|\exp\left(+i\frac{q\Lambda}{\hbar}\right)\left[q\phi\psi_k - q\frac{\partial\Lambda}{\partial t}\psi_k + q\frac{\partial\Lambda}{\partial t}\psi_k - i\hbar\frac{\partial\psi_k}{\partial t}\right]\exp\left(+i\frac{q\Lambda}{\hbar}\right)\right.\right\rangle \\ &= \left\langle \psi_j\left|\left[q\phi - i\hbar\frac{\partial}{\partial t}\right]\right|\psi_k\right\rangle.\end{aligned} \tag{4.32}$$

Consequently, the time development of the coefficients $a_j(t)$ will be gauge invariant, and the occupation probabilities $a_j^* a_j$ and coherence terms $a_j^* a_k$ will have the same values in every gauge.

4.3 The Electric Dipole Interaction and the Long Wavelength Approximation

For an atom or molecule with multiple charged electrons and protons, Eq. (4.19) can be rewritten as

$$\begin{aligned}&\sum_{j=1}^{N}\left\{\frac{1}{2m_j}\left[\hat{\boldsymbol{p}}_j - \boldsymbol{A}(\boldsymbol{r}_j,t)\right]^2 + q_j\phi(\boldsymbol{r}_j,t)\right\}\Psi(\boldsymbol{r}_1,\ldots\boldsymbol{r}_N,t) \\ &\qquad + U(\boldsymbol{r}_1,q_1,\ldots\boldsymbol{r}_N,q_N)\Psi(\boldsymbol{r}_1,\ldots\boldsymbol{r}_N,t) = i\hbar\frac{\partial\Psi(\boldsymbol{r}_1,\ldots\boldsymbol{r}_N,t)}{\partial t}.\end{aligned} \tag{4.33}$$

The potential function $U(\boldsymbol{r}_1,q_1,\ldots\boldsymbol{r}_N,q_N)$ results from the attractive and repulsive electrical forces due to the multiple electrons and protons in the atom or molecule. For electric dipole interactions of atoms or molecules with the external field, we can assume that the magnetic field is negligible, $\boldsymbol{B}(\boldsymbol{r},t) \cong 0$. Therefore, we can choose a gauge where the vector potential $\boldsymbol{A}(\boldsymbol{r},t) = 0$, and the electric field is then calculated from the time-dependent scalar potential for the external field,

$$\boldsymbol{E}(\boldsymbol{r},t) = -\nabla\phi(\boldsymbol{r},t). \tag{4.34}$$

The scalar potential can be expressed in terms of the electric field as

$$\phi(\boldsymbol{r},t) = -\int_0^{\boldsymbol{r}} \boldsymbol{E}(\boldsymbol{r}_1,t)\bullet d\boldsymbol{r}_1. \tag{4.35}$$

For an atom or molecule illuminated with electromagnetic radiation in the ultraviolet, visible, or infrared, the wavelength of the radiation will be large compared to the dimensions of the atom or molecule, and the long wavelength approximation can be used (Scully & Zubairy, 1997). The electric field can be written as

$$\boldsymbol{E}(\boldsymbol{r},t)=\frac{1}{2}\boldsymbol{E}_0(t)\exp(i\boldsymbol{k}\bullet\boldsymbol{r}-i\omega t)+\frac{1}{2}\boldsymbol{E}_0^*(t)\exp(-i\boldsymbol{k}\bullet\boldsymbol{r}+i\omega t). \tag{4.36}$$

The spatial phase terms $\exp(i\boldsymbol{k}\bullet\boldsymbol{r})$ and $\exp(-i\boldsymbol{k}\bullet\boldsymbol{r})$ will vary only slightly over the dimensions of the atom or molecule, and the electric field can be written as

$$\boldsymbol{E}(\boldsymbol{r},t)=\frac{1}{2}\boldsymbol{E}(t)\exp(i\boldsymbol{k}\bullet\boldsymbol{r}_0)\exp[i\boldsymbol{k}\bullet(\boldsymbol{r}-\boldsymbol{r}_0)]+\frac{1}{2}\boldsymbol{E}^*(t)\exp(i\boldsymbol{k}\bullet\boldsymbol{r}_0)\exp[i\boldsymbol{k}\bullet(\boldsymbol{r}-\boldsymbol{r}_0)], \tag{4.37}$$

where $\boldsymbol{E}(t)=\boldsymbol{E}_0(t)\exp(-i\omega t)$; for pulsed lasers, the amplitude term $\boldsymbol{E}_0(t)$ will also be a function of time. The term $\boldsymbol{k}\bullet(\boldsymbol{r}-\boldsymbol{r}_0)\ll 1$, and in this limit we can expand the exponentials in Eq. (4.37) to give

$$\begin{aligned}\boldsymbol{E}(\boldsymbol{r},t)=&\frac{1}{2}\boldsymbol{E}(t)\exp(i\boldsymbol{k}\bullet\boldsymbol{r}_0)\left[1+i\boldsymbol{k}\bullet\boldsymbol{r}'+\frac{1}{2}(i\boldsymbol{k}\bullet\boldsymbol{r}')^2+\cdots\right]\\&+\frac{1}{2}\boldsymbol{E}^*(t)\exp(-i\boldsymbol{k}\bullet\boldsymbol{r}_0)\left[1-i\boldsymbol{k}\bullet\boldsymbol{r}'+\frac{1}{2}(i\boldsymbol{k}\bullet\boldsymbol{r}')^2+\cdots\right].\end{aligned} \tag{4.38}$$

where $\boldsymbol{r}'=\boldsymbol{r}-\boldsymbol{r}_0$. For the electric dipole interaction, only the first term in the exponential expansion is retained; the second and third terms give the magnetic dipole and electric quadrupole interactions, respectively (Struve, 1988). Substituting Eq. (4.38) into Eq. (4.35) and integrating,

$$\begin{aligned}\phi(\boldsymbol{r}_j,t)&=-\frac{1}{2}\boldsymbol{E}(t)\exp(i\boldsymbol{k}\bullet\boldsymbol{r}_0)\bullet\int_0^{\boldsymbol{r}_j'}[1]\,d\boldsymbol{r}_1'-\frac{1}{2}\boldsymbol{E}^*(t)\exp(-i\boldsymbol{k}\bullet\boldsymbol{r}_0)\bullet\int_0^{\boldsymbol{r}_j'}[1]\,d\boldsymbol{r}_1'\\&=-\frac{1}{2}\boldsymbol{E}(t)\bullet\boldsymbol{r}_j'\exp(i\boldsymbol{k}\bullet\boldsymbol{r}_0)-\frac{1}{2}\boldsymbol{E}^*(t)\bullet\boldsymbol{r}_j'\exp(-i\boldsymbol{k}\bullet\boldsymbol{r}_0).\end{aligned} \tag{4.39}$$

If we set $\boldsymbol{r}_0=0$, then $\boldsymbol{r}_j'=\boldsymbol{r}_j$ and

$$\phi(\boldsymbol{r}_j,t)=-\frac{1}{2}[\boldsymbol{E}(t)+\boldsymbol{E}^*(t)]\bullet\boldsymbol{r}_j. \tag{4.40}$$

Substituting Eq. (4.40) into Eq. (4.33), we obtain

$$\begin{aligned}&\left\{\left[\sum_{j=1}^{N}\frac{\hat{\boldsymbol{p}}_j^2}{2m_j}+U(\boldsymbol{r}_1,q_1,\ldots\boldsymbol{r}_N,q_N)\right]-\left(\sum_{j=1}^{N}q_j\boldsymbol{r}_j\right)\bullet\frac{[\boldsymbol{E}(t)+\boldsymbol{E}^*(t)]}{2}\right\}\Psi(\boldsymbol{r}_1,\ldots\boldsymbol{r}_N,t)\\&\quad=i\hbar\frac{\partial\Psi(\boldsymbol{r}_1,\ldots\boldsymbol{r}_N,t)}{\partial t}.\end{aligned} \tag{4.41}$$

A major advantage of choosing the gauge with $\boldsymbol{A}(\boldsymbol{r},t)=0$ is that the basis state wavefunctions are no longer time dependent, except for the term $\exp(-i\varepsilon_k t/\hbar)$, which results from the separation of variables solution. In the absence of an external electric field, Eq. (4.41) becomes

$$\left[\sum_{j=1}^{N}\frac{\hat{p}_j^2}{2m_j}+U(\boldsymbol{r}_1,q_1,\ldots\boldsymbol{r}_N,q_N)\right]\Psi(\boldsymbol{r}_1,\ldots\boldsymbol{r}_N,t)=i\hbar\frac{\partial\Psi(\boldsymbol{r}_1,\ldots\boldsymbol{r}_N,t)}{\partial t}. \tag{4.42}$$

It can easily be demonstrated, using separation of variables techniques, that Eq. (4.42) has solutions of the form

$$\Psi_k(\boldsymbol{r}_1,\ldots\boldsymbol{r}_N,t)=u_k(\boldsymbol{r}_1,\ldots\boldsymbol{r}_N)\exp(-i\varepsilon_k t/\hbar), \tag{4.43}$$

where the function Ψ_k is the eigenfunction associated with the quantum number n, u_k is the time-independent eigenfunction component, and ε_k is the energy of the eigenstate. We assume that any wavefunction $\Psi(\boldsymbol{r},t)$ can be written as the sum of the eigenstates Ψ_k,

$$\Psi(\boldsymbol{r}_1,\ldots\boldsymbol{r}_N,t)=\sum_k c_k(t)\Psi_k(\boldsymbol{r}_1,\ldots\boldsymbol{r}_N,t)=\sum_k c_k(t)u_k(\boldsymbol{r}_1,\ldots\boldsymbol{r}_N)\exp(-i\varepsilon_k t/\hbar), \tag{4.44}$$

i.e., the eigenstates form a complete set. The slowly varying coefficient $c_k(t)$ is used instead of instead of $a_k(t)$ to remove the component that oscillates at optical frequencies. The eigenstates again form an orthonormal set,

$$\langle\Psi_k|\Psi_m\rangle=\langle u_k|u_m\rangle=\delta_{km}. \tag{4.45}$$

Considering now the case where the external field is nonzero, we rewrite Eq. (4.41) as

$$\hat{H}_B\Psi(\boldsymbol{r}_1,\ldots\boldsymbol{r}_N,t)+\hat{V}\Psi(\boldsymbol{r}_1,\ldots\boldsymbol{r}_N,t)=i\hbar\frac{\partial\Psi(\boldsymbol{r}_1,\ldots\boldsymbol{r}_N,t)}{\partial t}, \tag{4.46}$$

where

$$\hat{H}_B=\sum_{j=1}^{N}\frac{\hat{p}_j^2}{2m_j}+U(\boldsymbol{r}_1,q_1,\ldots\boldsymbol{r}_N,q_N) \tag{4.47}$$

and

$$\hat{V}=\hat{V}(\boldsymbol{r}_1,\ldots\boldsymbol{r}_N,t)=-\left(\sum_{j=1}^{N}q_j\boldsymbol{r}_j\right)\bullet\frac{[\boldsymbol{E}(t)+\boldsymbol{E}^*(t)]}{2}, \tag{4.48}$$

where the symbol $\hat{V}$ is commonly used for the radiative interaction operator.

4.4 Perturbation Analysis of the Interaction of Radiation with Resonant Transitions

The perturbation analysis of the interaction of electromagnetic radiation with resonant transitions in atoms and molecules will now be discussed. Again, it will be assumed that the strength of the external electromagnetic field is low enough that the quantum states of the atom or molecule can still be represented by the field-free quantum states.

The perturbation analysis will be performed for both single-photon and two-photon resonances. The two-photon resonances include the important case of Raman resonances.

4.4.1 Perturbation Analysis for Single-Photon Resonances

The Hamiltonian operator $\hat{H}$ is written in Eq. (4.46) as the sum of a time-independent basis term $\hat{H}_B$, which gives the energy levels in the absence of an external field, and a time-dependent interaction operator $\hat{V}(\boldsymbol{r},t)$, which accounts for the electric dipole interaction of the atom or molecule with the external laser field (for compactness of notation in this section, the symbol $\boldsymbol{r}$ is meant to imply $\boldsymbol{r}_1, \ldots \boldsymbol{r}_N$). The time-dependent interaction term $\hat{V}(\boldsymbol{r},t)$ is assumed to be weak compared to $\hat{H}_B$. Following Boyd (2008) we write the Hamiltonian as the sum of the time-independent term and the product of the interaction term and a perturbation parameter:

$$\hat{H} = \hat{H}_B + \lambda \hat{V}(t). \tag{4.49}$$

The SWE becomes

$$\hat{H}_B(\boldsymbol{r})\Psi(\boldsymbol{r},t) + \lambda \hat{V}(\boldsymbol{r},t)\Psi(\boldsymbol{r},t) = i\hbar \frac{\partial \Psi(\boldsymbol{r},t)}{\partial t}. \tag{4.50}$$

The wavefunction for the system is then written as a power series in the perturbation parameter λ:

$$\Psi(\boldsymbol{r},t) = \Psi^{(0)}(\boldsymbol{r},t) + \lambda \Psi^{(1)}(\boldsymbol{r},t) + \lambda^2 \Psi^{(2)}(\boldsymbol{r},t) + \cdots. \tag{4.51}$$

The wavefunctions $\Psi^{(n)}(\boldsymbol{r},t)$ in the power series are expressed in terms of the basis wavefunctions $|u_k(\boldsymbol{r})\rangle$ and time-dependent coefficients $c_k^{(n)}(t)$:

$$\Psi^{(n)}(\boldsymbol{r},t) = \sum_k c_k^{(n)}(t) \exp(-i\varepsilon_k t/\hbar) u_k(\boldsymbol{r}). \tag{4.52}$$

Substituting Eqs. (4.49) and (4.51) into Eq. (4.52), we obtain:

$$\begin{aligned} &\hat{H}_B(\boldsymbol{r})\Psi^{(0)}(\boldsymbol{r},t) + \lambda H_B(\boldsymbol{r})\Psi^{(1)}(\boldsymbol{r},t) + \lambda^2 H_B(\boldsymbol{r})\Psi^{(2)}(\boldsymbol{r},t) + \cdots \\ &\quad + \lambda \hat{V}(\boldsymbol{r},t)\Psi^{(0)}(\boldsymbol{r},t) + \lambda^2 \hat{V}(\boldsymbol{r},t)\Psi^{(1)}(\boldsymbol{r},t) + \lambda^3 \hat{V}(\boldsymbol{r},t)\Psi^{(2)}(\boldsymbol{r},t) + \cdots \\ &\quad = i\hbar \left[\frac{\partial \Psi^{(0)}(\boldsymbol{r},t)}{\partial t} + \lambda \frac{\partial \Psi^{(1)}(\boldsymbol{r},t)}{\partial t} + \lambda^2 \frac{\partial \Psi^{(2)}(\boldsymbol{r},t)}{\partial t} + \cdots \right]. \end{aligned} \tag{4.53}$$

Requiring that terms with the same power n of λ on the LHS and RHS of Eq. (4.53) be equal, we obtain

$$\hat{H}_B(\boldsymbol{r})\Psi^{(0)}(\boldsymbol{r},t) = i\hbar \frac{\partial \Psi^{(0)}(\boldsymbol{r},t)}{\partial t} \tag{4.54}$$

for $n = 0$ and

$$\hat{H}_B(\boldsymbol{r})\Psi^{(n)}(\boldsymbol{r},t) + \hat{V}\Psi^{(n-1)}(\boldsymbol{r},t) = i\hbar\frac{\partial\Psi^{(n)}(\boldsymbol{r},t)}{\partial t} \tag{4.55}$$

for $n > 0$. Substituting Eq. (4.52) into Eq. (4.55), we obtain

$$\begin{aligned}\sum_k \varepsilon_k c_k^{(n)}(t)\exp(-i\varepsilon_k t/\hbar)u_k(\boldsymbol{r}) + \hat{V}(\boldsymbol{r},t)\sum_k c_k^{(n-1)}(t)\exp(-i\varepsilon_k t/\hbar)u_k(\boldsymbol{r})\\ = i\hbar\sum_k \dot{c}_k^{(n)}(t)\exp(-i\varepsilon_k t/\hbar)u_k(\boldsymbol{r}) + \sum_k \varepsilon_k c_k^{(n)}(t)\exp(-i\varepsilon_k t/\hbar)u_k(\boldsymbol{r}).\end{aligned} \tag{4.56}$$

Two of these terms obviously cancel. We now multiply both sides of the equation by $-\frac{i}{\hbar}u_m^*(\boldsymbol{r})$ and integrate over all space. Using the orthonormality relation $\langle u_m(\boldsymbol{r})|u_k(\boldsymbol{r})\rangle = \delta_{km}$, we obtain

$$\begin{aligned}\frac{dc_m^{(n)}(t)}{dt} &= -\frac{i}{\hbar}\sum_k c_k^{(n-1)}(t)\langle u_m(\boldsymbol{r})|\hat{V}(\boldsymbol{r},t)|u_k(\boldsymbol{r})\rangle\exp(-i\omega_{km}t)\\ &= -\frac{i}{\hbar}\sum_k c_k^{(n-1)}(t)\,V_{mk}\exp(i\omega_{mk}t),\end{aligned} \tag{4.57}$$

where $\omega_{mk} = (\varepsilon_m - \varepsilon_k)/\hbar$ and

$$\begin{aligned}V_{mk} &= \langle u_m(\boldsymbol{r})|\hat{V}(\boldsymbol{r},t)|u_k(\boldsymbol{r})\rangle = \langle u_m(\boldsymbol{r})| - \left(\sum_{j=1}^{N} q_j\boldsymbol{r}_j\right)\bullet\frac{[\boldsymbol{E}(t)+\boldsymbol{E}^*(t)]}{2}|u_k(\boldsymbol{r})\rangle\\ &= -\langle u_m(\boldsymbol{r})|\left(\sum_{j=1}^{N} q_j\boldsymbol{r}_j\right)|u_k(\boldsymbol{r})\rangle\bullet\frac{[\boldsymbol{E}(t)+\boldsymbol{E}^*(t)]}{2} = -\frac{\boldsymbol{\mu}_{mk}\bullet[\boldsymbol{E}(t)+\boldsymbol{E}^*(t)]}{2},\end{aligned} \tag{4.58}$$

where $\boldsymbol{\mu}_{mk}$ is the electric dipole matrix element between states m and k.

The zeroth-order terms $c_k^{(0)}$ are only nonzero for states that are populated in the absence of an external electromagnetic field, and in this perturbative treatment they are not functions of time. We will designate those states with nonzero initial population with the subscript g. The equation for the first-order term $c_m^{(1)}(t)$ is given by

$$\frac{dc_m^{(1)}(t)}{dt} = -\frac{i}{\hbar}\sum_g c_g^{(0)}\exp\left(i\omega_{mg}t\right)V_{mg}. \tag{4.59}$$

Equation (4.59) can be solved by direct integration from $t_1 = 0$ to $t_1 = t$:

$$c_m^{(1)}(t) = -\frac{i}{\hbar}\int_0^t \sum_g c_g^{(0)}\exp\left(i\omega_{mg}t_1\right)V_{mg}\,dt_1. \tag{4.60}$$

If state m is initially unpopulated, it will be populated only if it is coupled to a state or states with nonzero population prior to the application of the external field; that is, the coefficients $c_g^{(0)}$ and the coupling term V_{mg} must be nonzero for some states g. Substituting for V_{mg} using Eq. (4.58), we obtain

$$c_m^{(1)}(t) = -\frac{i}{\hbar}\sum_g \frac{c_g^{(0)}}{2}\left\{\boldsymbol{E}_0\boldsymbol{\cdot}\boldsymbol{\mu}_{mg}\int_0^t \exp\left[-i(\omega-\omega_{mg})t_1\right]dt_1 \right.$$
$$\left. +\boldsymbol{E}_0^*\boldsymbol{\cdot}\boldsymbol{\mu}_{mg}\int_0^t \exp\left[+i(\omega+\omega_{mg})t_1\right]dt_1\right\}, \tag{4.61}$$

where in writing Eq. (4.61) we have assumed the electric field amplitude is zero until $t = 0$ and constant thereafter. Performing the time integrals on the RHS of Eq. (4.61) after this substitution, we obtain

$$c_m^{(1)}(t) = \frac{1}{\hbar}\sum_g \frac{c_g^{(0)}}{2}\left\{\boldsymbol{E}_0\boldsymbol{\cdot}\boldsymbol{\mu}_{mg}\frac{\exp\left[i(\omega_{mg}-\omega)t\right]-1}{(\omega_{mg}-\omega)}\right.$$
$$\left. -\boldsymbol{E}_0^*\boldsymbol{\cdot}\boldsymbol{\mu}_{mg}\frac{\exp\left[i(\omega_{mg}+\omega)t\right]-1}{(\omega_{mg}+\omega)}\right\}. \tag{4.62}$$

Consider the case where $\omega_{mg} > 0$, and at this point regard the frequency of the laser field as a positive quantity, $\omega > 0$. The denominator for the first term on the RHS will be close to zero for $\omega \cong \omega_{mg}$; the first term exhibits resonant behavior for $\omega \cong \omega_{mg}$ and will be much larger than the second term. The first term therefore applies for the process of stimulated absorption. If $\omega_{mg} < 0$, so that state m has lower energy than state g, the second term will exhibit resonant behavior for $\omega \cong -\omega_{mg}$. The second term on the RHS thus applies for the process of stimulated emission.

Consider the process of absorption from an initially occupied state g to a state m of higher energy. The occupation probability for state m is given by the product

$$c_m^{(1)}(t)c_m^{(1)*}(t) = \left|c_m^{(1)}(t)\right|^2 = \frac{1}{\hbar^2}\sum_g \left|\boldsymbol{E}_0\boldsymbol{\cdot}\boldsymbol{\mu}_{mg}\right|^2\left|c_g^{(0)}\right|^2\frac{1-\cos\left[(\omega_{mg}-\omega)t\right]}{(\omega_{mg}-\omega)^2}. \tag{4.63}$$

Using the trigonometric relation $1 - \cos 2\theta = 2\sin^2\theta$, we can rewrite Eq. (4.63) as

$$\left|c_m^{(1)}(t)\right|^2 = \frac{2}{\hbar^2}\sum_g \left|\boldsymbol{E}_0\boldsymbol{\cdot}\boldsymbol{\mu}_{mg}\right|^2\left|c_g^{(0)}\right|^2\frac{\sin^2\left[\frac{(\omega_{mg}-\omega)t}{2}\right]}{(\omega_{mg}-\omega)^2}. \tag{4.64}$$

In the limit of time $t \gg \omega_{mg}^{-1}$, the $\sin^2$ term approaches the delta function,

$$\lim_{t\gg\omega_{mg}^{-1}}\frac{\sin^2\left[\frac{(\omega_{mg}-\omega)t}{2}\right]}{(\omega_{mg}-\omega)^2} = \pi t\delta(\omega_{mg}-\omega), \tag{4.65}$$

where

$$\int_0^\infty f(\omega)\delta(\omega_{mg}-\omega)\,d\omega = f(\omega_{mg}). \tag{4.66}$$

The occupation probability $\left|c_m^{(1)}(t)\right|^2$ will be significantly greater than zero only if $\omega = \omega_{mg}$. In this case Eq. (4.64) reduces to

$$\left|c_m^{(1)}(t)\right|^2 = \frac{2\pi t}{\hbar^2}\sum_g \left|\boldsymbol{E}_0(t)\boldsymbol{\cdot}\boldsymbol{\mu}_{mg}\right|^2 \left|c_g^{(0)}\right|^2 \delta\left(\omega_{mg}-\omega\right). \tag{4.67}$$

Equation (4.67) would apply if the energies of both states m and g were precisely defined. However, the energy will be precisely defined only if the lifetime of the state is infinite. The process of spontaneous emission from state m will introduce some uncertainty in the energy of state m and consequent broadening of the transition. Collisional dephasing will also cause the transition to occur over a spread of frequencies rather than at a single infinitely narrow frequency. Equation (4.67) can be rewritten as

$$\left|c_m^{(1)}(t)\right|^2 d\omega = \frac{2\pi t}{\hbar^2}\sum_g \left|\boldsymbol{E}_0(t)\boldsymbol{\cdot}\boldsymbol{\mu}_{mg}\right|^2 \left|c_g^{(0)}\right|^2 g\left(\omega_{mg}-\omega\right)d\omega, \tag{4.68}$$

where for homogenous broadening the normalized lineshape function $g\left(\omega_{mg}-\omega\right)$ is given by the Lorentzian function

$$g\left(\omega_{mg}-\omega\right) = \frac{1}{\pi}\frac{\gamma/2}{\left(\omega-\omega_{mg}\right)^2 + (\gamma/2)^2}, \tag{4.69}$$

and

$$\int_0^\infty g\left(\omega_{mg}-\omega\right)d\omega = 1. \tag{4.70}$$

Lineshape broadening processes will be discussed in more detail in Chapter 6.

4.4.2 Perturbation Analysis for Multi-Photon Resonances

Now consider the case where the atom or molecule is subjected to a laser field with two different frequencies ω_1 and ω_2. In this case, the first-order coefficient $c_m^{(1)}(t)$ will be given by

$$c_m^{(1)}(t) = \frac{1}{\hbar}\sum_g c_g^{(0)}\left(\boldsymbol{Z}_1 + \boldsymbol{Z}_2\right)\boldsymbol{\cdot}\boldsymbol{\mu}_{mg}, \tag{4.71}$$

where

$$\boldsymbol{Z}_i = \frac{\boldsymbol{E}_{0i}}{2}\frac{\exp\left[i\left(\omega_{mg}-\omega_i\right)t\right]-1}{\left(\omega_{mg}-\omega_i\right)} - \frac{\boldsymbol{E}_{0i}^*}{2}\frac{\exp\left[i\left(\omega_{mg}+\omega_i\right)t\right]-1}{\left(\omega_{mg}+\omega_i\right)}. \tag{4.72}$$

At this point the notation explicitly indicating the time dependence of the electric field amplitude will be dropped to simplify the notation somewhat. The time derivative of the second-order coefficient $c_k^{(2)}(t)$ is given by

$$\frac{dc_k^{(2)}(t)}{dt} = -\frac{i}{\hbar}\sum_m c_m^{(1)}(t)\,V_{km}\exp\left(i\omega_{km}t\right). \tag{4.73}$$

Substituting for $c_m^{(1)}(t)$ using Eq. (4.62), we obtain

$$\frac{dc_k^{(2)}(t)}{dt} = -\frac{i}{\hbar^2}\sum_g\sum_m V_{km}\exp(i\omega_{km}t)c_g^{(0)}\sum_i\left\{\frac{\mathbf{E}_{0i}\boldsymbol{\cdot}\boldsymbol{\mu}_{mg}}{2}\frac{\exp\left[i\left(\omega_{mg}-\omega_i\right)t\right]-1}{\left(\omega_{mg}-\omega_i\right)}\right.$$
$$\left.-\frac{\mathbf{E}_{0i}\boldsymbol{\cdot}\boldsymbol{\mu}_{mg}}{2}\frac{\exp\left[i\left(\omega_{mg}+\omega_i\right)t\right]-1}{\left(\omega_{mg}+\omega_i\right)}\right\}, \tag{4.74}$$

where the index i can take on the values 1 or 2. Substituting for V_{km} using Eq. (4.58), again modifying the expression for two laser frequencies, we obtain

$$\frac{dc_k^{(2)}(t)}{dt} = -\frac{i}{\hbar^2}\sum_g\sum_m c_g^{(0)}\left\{\sum_j\left[\frac{\exp(-i\omega_j t)}{2}\mathbf{E}_{0j}\boldsymbol{\cdot}\boldsymbol{\mu}_{km}+\frac{\exp(+i\omega_j t)}{2}\mathbf{E}_{0j}^*\boldsymbol{\cdot}\boldsymbol{\mu}_{km}\right]\right\}$$
$$\left\{\sum_i\left[\frac{\mathbf{E}_{0i}\boldsymbol{\cdot}\boldsymbol{\mu}_{mg}}{2}\frac{\exp\left[i\left(\omega_{kg}-\omega_i\right)t\right]-1}{\left(\omega_{mg}-\omega_i\right)}-\frac{\mathbf{E}_{0i}^*\boldsymbol{\cdot}\boldsymbol{\mu}_{mg}}{2}\frac{\exp\left[i\left(\omega_{kg}+\omega_i\right)t\right]-1}{\left(\omega_{mg}+\omega_i\right)}\right]\right\}, \tag{4.75}$$

where the index j can take on values 1 or 2, and we have used the relation $\omega_{kg}=\omega_{km}+\omega_{mg}$ to simplify Eq. (4.75). Writing out the cross terms in Eq. (4.75) explicitly, we obtain

$$\frac{dc_k^{(2)}(t)}{dt} = -\frac{i}{\hbar^2}\sum_g\sum_m c_g^{(0)}$$
$$\times\left\{\sum_{i,j}\frac{\left[\mathbf{E}_{0j}\boldsymbol{\cdot}\boldsymbol{\mu}_{km}\right]\left[\mathbf{E}_{0i}\boldsymbol{\cdot}\boldsymbol{\mu}_{mg}\right]}{4}\left(\frac{\exp\left[i\left(\omega_{kg}-\omega_i-\omega_j\right)t\right]-\exp\left(-i\omega_j t\right)}{\left(\omega_{mg}-\omega_i\right)}\right)\right.$$
$$+\sum_{i,j}\frac{\left[\mathbf{E}_{0j}^*\boldsymbol{\cdot}\boldsymbol{\mu}_{km}\right]\left[\mathbf{E}_{0i}\boldsymbol{\cdot}\boldsymbol{\mu}_{mg}\right]}{4}\left(\frac{\exp\left[i\left(\omega_{kg}-\omega_i+\omega_j\right)t\right]-\exp\left(i\omega_j t\right)}{\left(\omega_{mg}-\omega_i\right)}\right)$$
$$-\sum_{i,j}\frac{\left[\mathbf{E}_{0j}\boldsymbol{\cdot}\boldsymbol{\mu}_{km}\right]\left[\mathbf{E}_{0i}^*\boldsymbol{\cdot}\boldsymbol{\mu}_{mg}\right]}{4}\left(\frac{\exp\left[i\left(\omega_{kg}+\omega_i-\omega_j\right)t\right]-\exp\left(-i\omega_j t\right)}{\left(\omega_{mg}-\omega_i\right)}\right)$$
$$\left.-\sum_{i,j}\frac{\left[\mathbf{E}_{0j}^*\boldsymbol{\cdot}\boldsymbol{\mu}_{km}\right]\left[\mathbf{E}_{0i}^*\boldsymbol{\cdot}\boldsymbol{\mu}_{mg}\right]}{4}\left(\frac{\exp\left[i\left(\omega_{kg}+\omega_i+\omega_j\right)t\right]-\exp\left(i\omega_j t\right)}{\left(\omega_{mg}-\omega_i\right)}\right)\right\}. \tag{4.76}$$

Once again we will assume that the laser frequencies are positive, $\omega_1>0$ and $\omega_2>0$. It should be apparent from the preceding analysis of the first-order coefficient $c_m^{(1)}(t)$ that when we integrate Eq. (4.76), only those terms having an argument for the imaginary exponential that is near zero will contribute a significant change in $c_k^{(2)}(t)$. As we will show, the first and fourth terms on the RHS give rise to two-photon absorption or two-photon emission, depending on the sign of ω_{kg}, and the second and third terms give rise to Stokes Raman scattering or anti-Stokes Raman scattering, again depending on the sign of ω_{kg}.

Perturbation Analysis for Two-Photon Absorption

Consider the first term on the RHS of Eq. (4.76). Assume that only one laser field is present $[\boldsymbol{E}_{02} = 0]$, and neglect the term $\exp(-i\omega_j t)$, which does not give rise to resonant behavior. Equation (4.76) becomes

$$\frac{dc_k^{(2)}(t)}{dt} = -\frac{i}{\hbar^2}\sum_g\sum_m \frac{c_g^{(0)}}{4(\omega_{mg} - \omega_1)}[\boldsymbol{E}_{01}\bullet\boldsymbol{\mu}_{km}]\left[\boldsymbol{E}_{01}\bullet\boldsymbol{\mu}_{mg}\right]\exp\left[i(\omega_{kg} - 2\omega_1)t\right]. \quad (4.77)$$

Integrating Eq. (4.75),we obtain

$$c_k^{(2)}(t) = \sum_g\sum_m \frac{c_g^{(0)}[\boldsymbol{E}_{01}\bullet\boldsymbol{\mu}_{km}]\left[\boldsymbol{E}_{01}\bullet\boldsymbol{\mu}_{mg}\right]}{4\hbar^2(\omega_{mg} - \omega_1)}\left\{\frac{1 - \exp\left[i(\omega_{kg} - 2\omega_1)t\right]}{(\omega_{kg} - 2\omega_1)}\right\}. \quad (4.78)$$

The occupation probability $\left|c_k^{(2)}(t)\right|^2$ is given by

$$\begin{aligned}\left|c_k^{(2)}(t)\right|^2 &= \sum_g\sum_m \frac{\left|c_g^{(0)}\right|^2\left|\boldsymbol{E}_{01}\bullet\boldsymbol{\mu}_{km}\right|^2\left|\boldsymbol{E}_{01}\bullet\boldsymbol{\mu}_{mg}\right|^2}{16\hbar^4(\omega_{mg} - \omega_1)^2}\left\{\frac{1 - \cos\left[(\omega_{kg} - 2\omega_1)t\right]}{(\omega_{kg} - 2\omega_1)^2}\right\} \\ &= \sum_g\sum_m \frac{\left|c_g^{(0)}\right|^2\left|\boldsymbol{E}_{01}\bullet\boldsymbol{\mu}_{km}\right|^2\left|\boldsymbol{E}_{01}\bullet\boldsymbol{\mu}_{mg}\right|^2}{8\hbar^4(\omega_{mg} - \omega_1)^2}\left\{\frac{\sin^2\left[(\omega_{kg} - 2\omega_1)t/2\right]}{(\omega_{kg} - 2\omega_1)^2}\right\}.\end{aligned} \quad (4.79)$$

As pointed out by Lamb et al. (1987), derivation of this formula using the $\boldsymbol{p}\bullet\boldsymbol{A}$ gauge results in the same formula except that the terms inside the summation signs are multiplied by β_{kg}, where

$$\beta_{kg} = \left(\frac{\omega_{km}\omega_{mg}}{\omega_1^2}\right)^2. \quad (4.80)$$

The presence or absence of this term in the two-photon absorption cross section is discussed in detail in a number of papers, including Peticolas et al. (1965), Forney et al. (1977), Bassani et al. (1977), and Lamb et al. (1987). Bassani et al. (1977) show that for atomic hydrogen, the two-photon absorption cross section calculated in the two different formulae is exactly the same regardless of whether $\beta_{kg} = 1$ or $\beta_{kg} = (\omega_{km}\omega_{mg}/\omega_1^2)^2$, as long as the sum is performed over the complete set of intermediate states m. However, the contribution of the individual intermediate levels is very different for the two formulae.

Returning again to Eq. (4.79), for $t \gg \omega_{kg}^{-1}$, the $\sin^2$ term approaches the delta function, and we can write

$$\left|c_k^{(2)}(t)\right|^2 = \sum_g\sum_m \frac{\left|c_g^{(0)}\right|^2\left|\boldsymbol{E}_{01}\bullet\boldsymbol{\mu}_{km}\right|^2\left|\boldsymbol{E}_{01}\bullet\boldsymbol{\mu}_{mg}\right|^2}{8\hbar^4(\omega_{mg} - \omega_1)^2}\pi t\delta(\omega_{kg} - 2\omega_1). \quad (4.81)$$

Perturbation Analysis for Raman Scattering

Now consider the second term on the RHS of Eq. (4.76), and assume that $\omega_{kg} > 0$ and that $\omega_1 > \omega_2 > 0$. In this case the term $\exp\left[i(\omega_{kg} - \omega_i + \omega_j)t\right]$ will give rise to resonant behavior when $\omega_{kg} = \omega_1 - \omega_2$. Substituting $\omega_i = \omega_1$ and $\omega_j = \omega_2$, we obtain

$$\frac{dc_k^{(2)}(t)}{dt} = -\frac{i}{\hbar^2}\sum_g\sum_m c_g^{(0)} \frac{\left[\boldsymbol{E}_{02}^*\bullet\boldsymbol{\mu}_{km}\right]\left[\boldsymbol{E}_{01}\bullet\boldsymbol{\mu}_{mg}\right]}{4} \frac{\exp\left[i(\omega_{kg} - \omega_1 + \omega_2)t\right]}{\left(\omega_{mg} - \omega_1\right)}. \tag{4.82}$$

Following the same procedure as for two-photon absorption, we obtain

$$\left|c_k^{(2)}(t)\right|^2 = \sum_g\sum_m \frac{\left|c_g^{(0)}\right|^2\left|\boldsymbol{E}_{02}^*\bullet\boldsymbol{\mu}_{km}\right|^2\left|\boldsymbol{E}_{01}\bullet\boldsymbol{\mu}_{mg}\right|^2}{8\hbar^4\left(\omega_{mg} - \omega_1\right)^2}\pi t\delta\left(\omega_{kg} - \omega_1 + \omega_2\right). \tag{4.83}$$

The expression given in Eq. (4.83) describes the population of upper state k due to stimulated Raman scattering (SRS). During the SRS process, a pump photon of frequency ω_1 is destroyed (annihilated), and a Stokes photon of frequency ω_2 is created; the appearance of the complex conjugate of the electric field amplitude $\boldsymbol{E}_{02}^*$ is consistent with the creation of a Stokes photon.

4.5 The Density Matrix and the Quantum Mechanical Liouville Equation

Our analysis of the interaction of laser radiation with single- and two-photon resonances in atoms and molecules is based on the quantum mechanical Liouville equation. In the Liouville equation, populations of quantum states and the coherences between these quantum states are expressed in terms of the density matrix. In this section we introduce the density matrix and develop the Liouville equation from the SWE. The treatment that is given here is consistent with the very detailed treatment of the density matrix found in Weissbluth (1989). We start with the SWE and use Dirac's bra and ket notation, which is especially useful for this application. The SWE is given by

$$i\hbar\frac{\partial}{\partial t}|\Psi(\boldsymbol{r},t)\rangle = \hat{H}|\Psi(\boldsymbol{r},t)\rangle, \tag{4.84}$$

where the Hamiltonian operator $\hat{H}$ is the sum of a time-independent term $\hat{H}_0$, which gives the energy levels in the absence of an external field, and a time-dependent interaction operator $\hat{V}(t)$, which accounts for the electric dipole interaction of the atom or molecule with the external laser field:

$$\hat{H} = \hat{H}_B + \hat{V}(t). \tag{4.85}$$

The wavefunction for the system is expressed as the linear superposition of basis wavefunctions $|u_k(\boldsymbol{r})\rangle$:

$$|\Psi(\boldsymbol{r},t)\rangle = \sum_m a_m(t)|u_m(\boldsymbol{r})\rangle, \tag{4.86}$$

where $a_m(t)$ is the time-dependent coefficient introduced in Eq. (4.1), and $a_m(t) = c_m(t)\exp(-i\varepsilon_m t/\hbar)$. Operating with $\hat{H}_B$ on the normalized basis wavefunctions $|u_m(\boldsymbol{r})\rangle$ results in

$$\hat{H}_B|u_m(\boldsymbol{r})\rangle = \varepsilon_m|u_m(\boldsymbol{r})\rangle, \tag{4.87}$$

where ε_m is the energy of the quantum state. Substituting Eqs. (4.85) and (4.86) into Eq. (4.84) and using Eq. (4.87), we obtain:

$$i\hbar\sum_m \dot{a}_m(t)|u_m(\boldsymbol{r})\rangle = \sum_m \varepsilon_m a_m(t)|u_m(\boldsymbol{r})\rangle + \hat{V}(t)\sum_m a_m(t)|u_m(\boldsymbol{r})\rangle. \tag{4.88}$$

We can now multiply both sides of the equation by $-\frac{i}{\hbar}\langle u_k(\boldsymbol{r})|$ and integrate over all space to obtain

$$\dot{a}_k(t) = -\frac{i}{\hbar}\left[\varepsilon_k a_k(t) + \sum_m a_m(t)\langle u_k(\boldsymbol{r})|\hat{V}(t)|u_m(\boldsymbol{r})\rangle\right] = -\frac{i}{\hbar}\left[\varepsilon_k a_k(t) + \sum_m a_m(t)\, V_{km}\right]. \tag{4.89}$$

In deriving Eq. (4.89) we have used the orthonormality property of the basis wavefunctions, $\langle u_k(\boldsymbol{r})|u_m(\boldsymbol{r})\rangle = \delta_{km}$. The rate of change of the amplitude coefficient for quantum state k due to transitions from state m is proportional to the amplitude coefficient of state m and the energy interaction term V_{km}. The density matrix element ρ_{jk} is now introduced:

$$\rho_{jk} = a_j a_k^*. \tag{4.90}$$

The time derivative of the density matrix is given by

$$\begin{aligned}\dot{\rho}_{jk} = \dot{a}_j a_k^* + a_j \dot{a}_k^* &= -\frac{i}{\hbar}\left(\varepsilon_j a_j + \sum_m a_m V_{jm}\right)a_k^* + \frac{i}{\hbar}a_j\left(\varepsilon_k a_k^* + \sum_m a_k^* V_{km}^*\right)\\ &= -\frac{i}{\hbar}\rho_{jk}(\varepsilon_j - \varepsilon_k) - \frac{i}{\hbar}\left(\sum_m V_{jm}\rho_{mk} - \sum_m \rho_{jm} V_{mk}\right),\end{aligned} \tag{4.91}$$

where we have used the relation $V_{mk} = V_{km}^*$. For the case where $j = k$, we obtain

$$\dot{\rho}_{jj} = -\frac{i}{\hbar}\left(\sum_m V_{jm}\rho_{mj} - \sum_m \rho_{jm} V_{mj}\right). \tag{4.92}$$

The off-diagonal density matrix element ρ_{jk} is proportional to the degree of coherence that exists between two quantum states j and k, while the diagonal density matrix element ρ_{jj} is proportional to the population of quantum state j. Defining the matrices for the Hamiltonian as

$$\hat{H} = \begin{bmatrix} \varepsilon_1 & 0 & 0 & \cdots \\ 0 & \varepsilon_2 & 0 & \cdots \\ 0 & 0 & \varepsilon_3 & 0 \\ \vdots & \vdots & 0 & \ddots \end{bmatrix} + \begin{bmatrix} V_{11} & V_{12} & V_{13} & \cdots \\ V_{21} & V_{22} & V_{23} & \cdots \\ V_{31} & V_{32} & V_{33} & \cdots \\ \vdots & \vdots & \vdots & \ddots \end{bmatrix}, \tag{4.93}$$

we can write the density matrix equations in compact form as

$$\dot{\rho} = -\frac{i}{\hbar}[\hat{H},\rho] = -\frac{i}{\hbar}(\hat{H}\rho - \rho\hat{H}) \equiv -\frac{i}{\hbar}\hat{L}\rho, \tag{4.94}$$

where $\hat{L}$ is the Liouville operator, and Eq. (4.94) is the quantum mechanical Liouville equation (Weissbluth, 1989).

The density matrix is introduced to describe assemblies of atoms (molecules). Consider an assembly of atoms where half of the atoms are in quantum state j and half are in quantum state k. For an assembly in thermodynamic equilibrium, the time-dependent coefficients a_j and a_k include random phase factors $\exp(i\varphi_j)$ and $\exp(i\varphi_k)$ that in general are different for each atom in the assembly. Consequently, for an assembly in thermodynamic equilibrium, the off-diagonal density matrix elements are zero. When a system in equilibrium is exposed to an external disturbance such as an irradiating laser field, coherences among the quantum states may be established, and the state populations may depart from the values at equilibrium in the absence of the external disturbance. Once the disturbance is removed, the system returns to thermodynamic equilibrium. In the density matrix formulation, this return to equilibrium is described by introducing phenomenological decay constants. At this point, we introduce coherence decay processes and population transfer processes. Eqs. (4.91) and (4.92) are modified as follows:

$$\dot{\rho}_{jk} = -\frac{i}{\hbar}\rho_{jk}(\varepsilon_j - \varepsilon_k) - \frac{i}{\hbar}\left(\sum_m V_{jm}\rho_{mk} - \sum_m \rho_{jm}V_{mk}\right) - \gamma_{jk}\rho_{jk}. \tag{4.95}$$

$$\dot{\rho}_{jj} = -\frac{i}{\hbar}\left(\sum_m V_{jm}\rho_{mj} - \sum_m \rho_{jm}V_{mj}\right) - \sum_m \Gamma_{jm}\rho_{jj} + \sum_m \Gamma_{mj}\rho_{mm}. \tag{4.96}$$

4.6 Density Matrix Elements for Electric Dipole Interactions: The Single-Electron Hydrogen Atom

Now the treatment is specialized to the interaction of monochromatic laser radiation with single-photon atomic resonances. The strongest interaction of the laser field with atomic and molecular resonances occurs typically via the electric dipole interaction. The energy associated with this interaction is given by

$$V_{km}(\boldsymbol{r}_0,t) = \left\langle \phi_k \middle| -\sum_i e(\boldsymbol{r}_{ei} - \boldsymbol{r}_{nuc})\boldsymbol{\cdot}\boldsymbol{E}(\boldsymbol{r}_0,t) \middle| \phi_m \right\rangle, \tag{4.97}$$

where $\boldsymbol{r}_{ei}$ is the coordinate of the ith electron in the atom, $\boldsymbol{r}_{nuc}$ is the position of the atomic nucleus, and $\boldsymbol{E}(\boldsymbol{r}_0,t)$ is the electric field at the position $\boldsymbol{r}_0 = \boldsymbol{r}_{nuc}$. It is assumed that the nucleus is too massive to respond to the rapidly oscillating electric field and that the variation in the electric field over the dimensions of the atom is negligible. With these assumptions, we obtain

$$V_{km}(\boldsymbol{R},t) = -\left\langle \phi_k \middle| \sum_i e(\boldsymbol{r}_{ei} - \boldsymbol{r}_0) \middle| \phi_m \right\rangle \boldsymbol{\cdot}\boldsymbol{E}(\boldsymbol{r}_0,t) = -\boldsymbol{\mu}_{km}\boldsymbol{\cdot}\boldsymbol{E}(\boldsymbol{r}_0,t). \tag{4.98}$$

4.6.1 Density Matrix Elements in the Uncoupled and Coupled Representations

We now calculate the density matrix elements for the hydrogen atom in both uncoupled and coupled representations. The analysis of the hydrogen atom structure with the SWE results in the quantization of the orbital angular momentum $\boldsymbol{L}$. The spin $\boldsymbol{S}$ of the electron arises in the course of the solution of the Dirac equation for the structure of the hydrogen atom. In the Dirac analysis of the hydrogen atom, the orbital angular momentum $\boldsymbol{L}$ and spin angular momentum $\boldsymbol{S}$ combine to yield the total angular momentum $\boldsymbol{J}$. Quantum states in the hydrogen atom with the same principal quantum number n but different values of $\boldsymbol{J}$ have slightly different energies.

The Dirac analysis of the hydrogen atom structure is complicated. As a first approximation that is suitable for the calculation of electric dipole moment matrix elements, the wavefunction can be represented as the product of the spatially dependent Schrödinger wavefunctions and a spin function. In the uncoupled representation, the wavefunction of the electron can be represented as (Weissbluth, 1978)

$$\psi(\lambda) = u(\boldsymbol{r})\xi(m_s) = R_{nl}(r)Y_{lm_l}(\theta,\varphi)\xi(m_s), \tag{4.99}$$

where $\boldsymbol{r} = \boldsymbol{r}_e - \boldsymbol{r}_{nuc}$. The angular ($Y_{lm_l}$) and radial ($R_{nl}$) components of the hydrogen wavefunctions are listed in Appendices A1 and A2, respectively, for $l \leq 2$. For the spin-1/2 electron in the hydrogen atom, $m_s = \pm\frac{1}{2}$. The spin function $\xi(m_s)$ is a matrix with the values

$$\xi\left(m_s = +\frac{1}{2}\right) = |\alpha\rangle = \begin{pmatrix}1\\0\end{pmatrix} \quad \xi\left(m_s = -\frac{1}{2}\right) = |\beta\rangle = \begin{pmatrix}0\\1\end{pmatrix}. \tag{4.100}$$

The wavefunction $u(\boldsymbol{r})$ is the solution of the SWE for the hydrogen atom. The Dirac equation takes into account relativistic effects, and the electron spin arises naturally in the solution of the equation. The Dirac equation also accounts correctly for spin–orbit splitting of the levels with the same value of n but different values of l but does not account for the Lamb shift. However, these energy splittings can be regarded as a perturbation on the Schrödinger equation solution (Weissbluth, 1978), and we can account for the effects of electron spin by multiplying $u(\boldsymbol{r})$ by the spin function $\xi(m_s)$. The function $\psi(\lambda)$ is referred to as a spin orbital, and when we calculate matrix elements, we must integrate over the spatial coordinates and sum over the spin coordinates.

The electric dipole moment matrix element $\boldsymbol{\mu}_{km}$ is given by

$$\begin{aligned}\boldsymbol{\mu}_{km} &= -\langle u_k(\boldsymbol{r})\xi_k(m_s')|e\boldsymbol{r}|u_m(\boldsymbol{r})\xi_m(m_s)\rangle \\ &= -e\langle u_k(\boldsymbol{r})|\boldsymbol{r}|u_m(\boldsymbol{r})\rangle\langle\xi_k(m_s')|\xi_m(m_s)\rangle = -e\langle u_k(\boldsymbol{r})|\boldsymbol{r}|u_m(\boldsymbol{r})\rangle\,\delta_{m_s',m_s},\end{aligned} \tag{4.101}$$

where m_s' and m_s are the spin projection quantum numbers for states k and m, respectively. For the single-photon matrix elements, the sum over the spin coordinates leads immediately to the selection rule that $\Delta m_s = 0$. The spatially dependent term in Eq. (4.101) is evaluated as

$$\langle u_k(\boldsymbol{r})|\boldsymbol{r}|u_m(\boldsymbol{r})\rangle = \langle R_{n_k,l_k}(r)Y_{l_k,m_{lk}}(\theta,\varphi)|\boldsymbol{r}|R_{n_m,l_m}(r)Y_{l_m,m_{lm}}(\theta,\varphi)\rangle, \tag{4.102}$$

where again the primed quantum numbers n', l', m'_l are associated with state k, and the unprimed quantum numbers (n, l, m_l) are associated with state m. But we can also write $\boldsymbol{r}$ in terms of spherical harmonics (Weissbluth, 1978):

$$\begin{aligned}\boldsymbol{r} &= r(\sin\theta\cos\varphi\hat{\boldsymbol{x}} + \sin\theta\sin\varphi\hat{\boldsymbol{y}} + \cos\theta\hat{\boldsymbol{z}}) \\ &= r\left[\sqrt{\frac{2\pi}{3}}(-Y_{11} + Y_{1-1})\hat{\boldsymbol{x}} + \sqrt{\frac{2\pi}{3}}i(Y_{11} + Y_{1-1})\hat{\boldsymbol{y}} + \sqrt{\frac{4\pi}{3}}Y_{10}\hat{\boldsymbol{z}}\right].\end{aligned} \tag{4.103}$$

We can now write the electric dipole moment matrix element in terms of integrals over the radial and angular coordinates,

$$\begin{aligned}\langle u_k(\boldsymbol{r})|\boldsymbol{r}|u_m(\boldsymbol{r})\rangle &= \langle n'l'm'_l s'm'_s|\boldsymbol{r}|nlm_l sm_s\rangle \\ &= \delta_{m'_s m_s}\left[\int_0^\infty R^*_{n',l'}(r)\,R_{n,l}(r)\,r^3\,dr\right] \\ &\times\sqrt{\frac{4\pi}{3}}\left[-\frac{(\hat{\boldsymbol{x}} - i\hat{\boldsymbol{y}})}{\sqrt{2}}\int_0^{2\pi}d\varphi\int_0^\pi \sin\theta d\theta\, Y^*_{l',m'_l}(\theta,\varphi)Y_{11}(\theta,\varphi)Y_{l,m_l}(\theta,\varphi)\right. \\ &+\frac{(\hat{\boldsymbol{x}} + i\hat{\boldsymbol{y}})}{\sqrt{2}}\int_0^{2\pi}d\varphi\int_0^\pi \sin\theta d\theta\, Y^*_{l',m'_l}(\theta,\varphi)Y_{1-1}(\theta,\varphi)Y_{l,m_l}(\theta,\varphi) \\ &\left.+\hat{\boldsymbol{z}}\int_0^{2\pi}d\varphi\int_0^\pi \sin\theta d\theta\, Y^*_{l',m'_l}(\theta,\varphi)Y_{10}(\theta,\varphi)Y_{l,m_l}(\theta,\varphi)\right].\end{aligned} \tag{4.104}$$

The radial components of the Schrödinger solution for the one-electron atomic hydrogen wavefunctions are shown in Appendix A2 for $n \leq 3$. The angular integrals of the spherical harmonics will be given in terms of *3j* symbols,

$$\begin{aligned}&\int_0^{2\pi}d\varphi\int_0^\pi \sin\theta d\theta\, Y^*_{l'm'}(\theta,\varphi)Y_{LM}(\theta,\varphi)Y_{lm}(\theta,\varphi) \\ &= (-1)^{m'}\sqrt{\frac{(2l'+1)(2L+1)(2l+1)}{4\pi}}\begin{pmatrix} l' & L & l \\ -m' & M & m\end{pmatrix}\begin{pmatrix} l' & L & l \\ 0 & 0 & 0\end{pmatrix}.\end{aligned} \tag{4.105}$$

The properties of both the *3j* symbols and the related Clebsch–Gordan coefficients are covered in great detail in a number of excellent texts (e.g., Edmonds, 1960; Rose, 1957; Zare, 1988). For electric dipole interactions, the most important properties of the *3j* symbols are summarized in Appendix A3. Note that the *3j* symbol $\begin{pmatrix} l' & L & l \\ 0 & 0 & 0\end{pmatrix}$ is zero unless $l' + L + l$ is an even number. For the electric dipole interaction, $L = 1$, which leads immediately to the conclusion that, for allowed transitions, $l' \neq l$.

In the coupled representation, the orbital angular momentum $\boldsymbol{L}$ and spin angular momentum $\boldsymbol{S}$ of the electron combine to give the total electron angular momentum $\boldsymbol{J} = \boldsymbol{L} + \boldsymbol{S}$. The projection of the quantum number $\boldsymbol{J}$ on the z-axis is labeled m_J. The transformation between the uncoupled representation and the coupled representation is determined using Clebsch–Gordan coefficients,

$$|Jm_J\rangle = \sum_{m_l m_s} |lm_l, sm_s\rangle\langle lm_l, sm_s|Jm_J\rangle. \tag{4.106}$$

$$|lm_l, sm_s\rangle = \sum_{Jm_J} |Jm_J\rangle\langle Jm_J|lm_l, sm_s\rangle. \tag{4.107}$$

The terms $\langle lm_l,sm_s|Jm_J\rangle$ and $\langle Jm_J|lm_l,sm_s\rangle$ are the Clebsch–Gordan (C–G) coefficients; the C–G coefficients in Eqs. (2.84) and (2.85) are zero unless $m_J = m_l + m_s$. The C–G coefficients are real quantities and are related to the *3j* symbols by the following formula (Weissbluth, 1978):

$$\begin{pmatrix} j_1 & j_2 & j \\ m_1 & m_2 & m \end{pmatrix} = \frac{(-1)^{j_1-j_2-m}}{\sqrt{2j+1}} \langle j_1 m_1, j_2 m_2 | j - m\rangle. \tag{4.108}$$

Using Eq. (2.86), we can immediately write

$$\langle lm_l, sm_s|Jm_J\rangle = (-1)^{s-l-m_J}\sqrt{2J+1}\begin{pmatrix} l & s & J \\ m_l & m_s & -m_J \end{pmatrix}. \tag{4.109}$$

In the uncoupled representation, the electric dipole density matrix elements are given by

$$\boldsymbol{\mu}_{km} = -e\langle \phi_k(\mathbf{r})\xi_k(m_s')|\mathbf{r}|\phi_m(\mathbf{r})\xi_m(m_s)\rangle = -e\langle l'm'_l s'm'_s|\mathbf{r}|lm_l sm_s\rangle\delta_{m_s',m_s}. \tag{4.110}$$

In the coupled representation, we wish to find the following density matrix elements:

$$\boldsymbol{\mu}_{pq} = -e\langle J'm'_J|\mathbf{r}|Jm_J\rangle, \tag{4.111}$$

where the states p and q quantum numbers sets $n'l's'J'm'_J$ and $nlsJm_J$, respectively; the quantum numbers nls have been suppressed in the notation for Eq. (4.111) for clarity. Using Eq. (2.84) we can write

$$\begin{aligned}\boldsymbol{\mu}_{pq} &= -e\langle J'm'_J|\mathbf{r}|Jm_J\rangle \\ &= -e\left(\sum_{m'_l m'_s}\langle l'm'_l, s'm'_s|J'm'_J\rangle\langle l'm'_l, s'm'_s|\right)\mathbf{r}\left(\sum_{m_l m_s}|lm_l, sm_s\rangle\langle lm_l, sm_s|Jm_J\rangle\right) \\ &= -e\sum_{m'_l m'_s m_l m_s}\langle l'm'_l, s'm'_s|\mathbf{r}|lm_l, sm_s\rangle\langle l'm'_l, s'm'_s|J'm'_J\rangle\langle lm_l, sm_s|Jm_J\rangle.\end{aligned} \tag{4.112}$$

Thus, we can determine the electric dipole moment matrix elements for the coupled representation from the electric dipole moment matrix elements for the uncoupled representation and the appropriate C–G coefficients.

4.6.2 Calculation of Electric Dipole Moment Matrix Elements for the 1*s*–2*p* Transition of Atomic Hydrogen in the Uncoupled Representation

In the uncoupled representation, the electric dipole moment matrix elements are given by Eq. (4.104). For the 1*s*–2*p* transition,

$$1s \ level: \quad n = 1, \ l = 0, \ m_l = 0, \ s = \frac{1}{2}, \ m_s = \pm\frac{1}{2}.$$

$$2p \ level: \quad n' = 2, \ l' = 1, \ m_l' = -1, 0, +1, \ s = \frac{1}{2}, \ m_s = \pm\frac{1}{2}.$$

The radial components of the wavefunction are given by

$$1s \ level: \quad R_{10} = \frac{2\exp(-r/a_0)}{a_0^{3/2}}.$$

$$2p \ level: \quad R_{21} = \frac{r\exp(-r/2a_0)}{\sqrt{24}a_0^{5/2}}.$$

Using the definite integral relation (Beyer, 1978)

$$\int_0^\infty x^n \exp(-bx)dx = \frac{n!}{b^{n+1}} \quad (b > 0, \ n \text{ is a positive integer}), \tag{4.113}$$

we obtain

$$\begin{aligned} \int_0^\infty R_{21}^*(r) R_{10}(r) r^3 \, dr &= \frac{\sqrt{2}}{\sqrt{3}a_0^4} \int_0^\infty r^4 \exp\left(-\frac{3r}{2a_0}\right) dr \\ &= \frac{\sqrt{2}}{\sqrt{3}a_0^4} \frac{4!}{(3/2a_0)^5} = \frac{128\sqrt{2}}{81\sqrt{3}} a_0 = 1.290\, a_0. \end{aligned} \tag{4.114}$$

Similarly, we can show that for the $1s$–$3p$ transition,

$$\int_0^\infty R_{31}^*(r) R_{10}(r) r^3 \, dr = 0.5167 a_0; \tag{4.115}$$

for the $2s$–$3p$ transition,

$$\int_0^\infty R_{31}^*(r) R_{20}(r) r^3 \, dr = 3.065\, a_0; \tag{4.116}$$

for the $2p$–$3s$ transition,

$$\int_0^\infty R_{30}^*(r) R_{21}(r) r^3 \, dr = 0.938 a_0; \tag{4.117}$$

and for the $2p$-$3d$ transition,

$$\int_0^\infty R_{32}^*(r) R_{21}(r) r^3 \, dr = 4.748 a_0. \tag{4.118}$$

A much more complete listing of the square of the radial integrals is provided as table 13 in Bethe and Salpeter (1957).

Now the integrations over the angular variables will be performed. Using Eq. (4.105), the first angular integral in Eq. (4.104) becomes

$$
\begin{aligned}
&-\frac{(\hat{x}-i\hat{y})}{\sqrt{2}}\int_0^{2\pi}d\varphi\int_0^{\pi}\sin\theta d\theta Y^*_{1m'_l}(\theta,\varphi)Y_{11}(\theta,\varphi)Y_{00}(\theta,\varphi)\\
&=-\frac{(\hat{x}-i\hat{y})}{\sqrt{2}}(-1)^{m'_l}\sqrt{\frac{(3)(3)(1)}{4\pi}}\begin{pmatrix}1 & 1 & 0\\ -m'_l & 1 & 0\end{pmatrix}\begin{pmatrix}1 & 1 & 0\\ 0 & 0 & 0\end{pmatrix}.
\end{aligned}
\tag{4.119}
$$

We need to manipulate the last $3j$ symbol into a form that can be evaluated using the formulae in Appendix A3. From Eq. (A3.2), we find

$$
\begin{aligned}
\begin{pmatrix}1 & 1 & 0\\ 0 & 0 & 0\end{pmatrix}&=(-1)^{1+1+0}\begin{pmatrix}1 & 0 & 1\\ 0 & 0 & 0\end{pmatrix}=\begin{pmatrix}1 & 0 & 1\\ 0 & 0 & 0\end{pmatrix}\\
&=\begin{pmatrix}j+1 & j & 1\\ m & -m & 0\end{pmatrix}\quad \text{for } j=m=0.
\end{aligned}
\tag{4.120}
$$

This $3j$ symbol can be evaluated from Eq. (A3.5),

$$
\begin{pmatrix}j+1 & j & 1\\ m & -m & 0\end{pmatrix}=(-1)^{j-m-1}\sqrt{\frac{(j+m+1)(j-m+1)}{(2j+3)(j+1)(2j+1)}}=-\frac{1}{\sqrt{3}}=\begin{pmatrix}1 & 1 & 0\\ 0 & 0 & 0\end{pmatrix}.
\tag{4.121}
$$

Thus Eq. (4.119) reduces to

$$
\begin{aligned}
&-\frac{(\hat{x}-i\hat{y})}{\sqrt{2}}\int_0^{2\pi}d\varphi\int_0^{\pi}\sin\theta d\theta Y^*_{1m'_l}(\theta,\varphi)Y_{11}(\theta,\varphi)Y_{00}(\theta,\varphi)\\
&=+\frac{(\hat{x}-i\hat{y})}{\sqrt{2}}\sqrt{\frac{3}{4\pi}}(-1)^{m'_l}\begin{pmatrix}1 & 1 & 0\\ -m'_l & 1 & 0\end{pmatrix}.
\end{aligned}
\tag{4.122}
$$

The remaining $3j$ symbol in Eq. (4.122) is nonzero only for $m'_l=1$. The $3j$ symbol is evaluated again using relations from Appendix A3,

$$
\begin{aligned}
\begin{pmatrix}1 & 1 & 0\\ -m'_\ell & 1 & 0\end{pmatrix}&=\begin{pmatrix}1 & 1 & 0\\ -1 & 1 & 0\end{pmatrix}=(-1)^2\begin{pmatrix}1 & 0 & 1\\ -1 & 0 & 1\end{pmatrix}\\
&=\begin{pmatrix}j+1 & j & 1\\ m & -(m+1) & 1\end{pmatrix}\quad \text{for } j=0,\ m=-1.
\end{aligned}
\tag{4.123}
$$

From Eq. (A3.4), we obtain

$$
\begin{aligned}
\begin{pmatrix}j+1 & j & 1\\ m & -(m+1) & 1\end{pmatrix}&=(-1)^{j-m-1}\sqrt{\frac{(j-m)(j-m+1)}{(2j+3)(2j+2)(2j+1)}}\\
&=(-1)^0\sqrt{\frac{(1)(2)}{(3)(2)(1)}}=\frac{1}{\sqrt{3}}=\begin{pmatrix}1 & 1 & 0\\ -1 & 1 & 0\end{pmatrix}.
\end{aligned}
\tag{4.124}
$$

Thus Eq. (4.122) reduces to

$$+\frac{(\hat{x}-i\hat{y})}{\sqrt{2}}\sqrt{\frac{3}{4\pi}}(-1)^{-1}\begin{pmatrix}1 & 1 & 0\\ 1 & 1 & 0\end{pmatrix}=-\frac{(\hat{x}-i\hat{y})}{\sqrt{2}}\sqrt{\frac{1}{4\pi}}. \tag{4.125}$$

The second integral over angular variables in Eq. (4.104) is given by

$$\begin{aligned}
&+\frac{(\hat{x}+i\hat{y})}{\sqrt{2}}\int_0^{2\pi}d\varphi\int_0^{\pi}\sin\theta d\theta Y^*_{1m_l'}(\theta,\varphi)Y_{1-1}(\theta,\varphi)Y_{00}(\theta,\varphi)\\
&=+\frac{(\hat{x}+i\hat{y})}{\sqrt{2}}(-1)^{m_l'}\frac{3}{\sqrt{4\pi}}\begin{pmatrix}1 & 1 & 0\\ -m_l' & -1 & 0\end{pmatrix}\begin{pmatrix}1 & 1 & 0\\ 0 & 0 & 0\end{pmatrix}\\
&=-\frac{(\hat{x}+i\hat{y})}{\sqrt{2}}(-1)^{m_l'}\sqrt{\frac{3}{4\pi}}\begin{pmatrix}1 & 1 & 0\\ -m_l' & -1 & 0\end{pmatrix},
\end{aligned} \tag{4.126}$$

where we have used Eq. (4.121) to simplify the expression. The *3j* symbol in Eq. (4.126) will be nonzero only for $m_l'=-1$. The *3j* symbol is given by

$$\begin{aligned}
\begin{pmatrix}1 & 1 & 0\\ -m_l' & -1 & 0\end{pmatrix}&=\begin{pmatrix}1 & 1 & 0\\ 1 & -1 & 0\end{pmatrix}=\begin{pmatrix}1 & 0 & 1\\ -1 & 0 & 1\end{pmatrix}\\
&=\begin{pmatrix}j+1 & j & 1\\ m & -(m+1) & 1\end{pmatrix} \quad \textit{for } j=0,\ m=-1.
\end{aligned} \tag{4.127}$$

This last *3j* symbol is the same as that evaluated in Eq. (4.124). Thus we obtain

$$\begin{pmatrix}1 & 1 & 0\\ 1 & -1 & 0\end{pmatrix}=\begin{pmatrix}1 & 0 & 1\\ -1 & 0 & 1\end{pmatrix}=\frac{1}{\sqrt{3}}. \tag{4.128}$$

Substituting Eq. (4.128) into Eq. (4.126) and simplifying, we obtain

$$\begin{aligned}
&+\frac{(\hat{x}+i\hat{y})}{\sqrt{2}}\int_0^{2\pi}d\varphi\int_0^{\pi}\sin\theta d\theta Y^*_{1-1}(\theta,\varphi)Y_{1-1}(\theta,\varphi)Y_{00}(\theta,\varphi)\\
&=+\frac{(\hat{x}+i\hat{y})}{\sqrt{2}}\sqrt{\frac{1}{4\pi}}.
\end{aligned} \tag{4.129}$$

A similar analysis for the third integral over angular variables in Eq. (4.104) gives us

$$\begin{aligned}
&+\hat{z}\int_0^{2\pi}d\varphi\int_0^{\pi}\sin\theta d\theta Y^*_{1m_l'}(\theta,\varphi)Y_{10}(\theta,\varphi)Y_{00}(\theta,\varphi)\Big]\\
&=+\hat{z}\int_0^{2\pi}d\varphi\int_0^{\pi}\sin\theta d\theta Y^*_{10}(\theta,\varphi)Y_{10}(\theta,\varphi)Y_{00}(\theta,\varphi)\Big]\\
&=+\hat{z}(-1)^0\frac{3}{\sqrt{4\pi}}\begin{pmatrix}1 & 1 & 0\\ 0 & 0 & 0\end{pmatrix}\begin{pmatrix}1 & 1 & 0\\ 0 & 0 & 0\end{pmatrix}=+\hat{z}\frac{1}{\sqrt{4\pi}}.
\end{aligned} \tag{4.130}$$

Substituting Eqs. (4.125), (4.129), and (4.130) sequentially into Eq. (4.104), we obtain the following three nonzero density matrix elements for the 1*s*–2*p* transition:

$$\begin{aligned}\langle n'l'm'_l s'm'_s|\boldsymbol{r}|nlm_l sm_s\rangle &= \left\langle 211\frac{1}{2}m'_s\middle|\boldsymbol{r}\middle|100\frac{1}{2}m_s\right\rangle \\ &= (1.290\,a_0)\left(-\frac{(\hat{\boldsymbol{x}}-i\hat{\boldsymbol{y}})}{\sqrt{6}}\right)\delta_{m'_s m_s}.\end{aligned} \tag{4.131}$$

$$\left\langle 21-1\frac{1}{2}m'_s\middle|\boldsymbol{r}\middle|100\frac{1}{2}m_s\right\rangle = (1.290\,a_0)\left(+\frac{(\hat{\boldsymbol{x}}+i\hat{\boldsymbol{y}})}{\sqrt{6}}\right)\delta_{m'_s m_s}. \tag{4.132}$$

$$\left\langle 210\frac{1}{2}m'_s\middle|\boldsymbol{r}\middle|100\frac{1}{2}m_s\right\rangle = (1.290\,a_0)\left(+\frac{\hat{\boldsymbol{z}}}{\sqrt{3}}\right)\delta_{m'_s m_s}. \tag{4.133}$$

4.6.3 Calculation of Electric Dipole Moment Matrix Elements for $^2S_{1/2}$–$^2P_{3/2,1/2}$ Transitions in Atomic Hydrogen in the Coupled Representation

In the coupled representation, the electric dipole moment matrix elements are calculated using Eq. (4.112). In the coupled representation, the orbital angular momentum $\boldsymbol{L}$ and the spin angular momentum $\boldsymbol{S}$ combine to give the total angular momentum $\boldsymbol{J}=\boldsymbol{L}+\boldsymbol{S}$. For the 1$s$ level, the resultant states are

$$\begin{aligned}&1s\ level:\quad n=1, l=0, m_l=0, s=\frac{1}{2}, m_s=\pm\frac{1}{2}\\ &\Rightarrow J=\frac{1}{2}, m_J=\pm\frac{1}{2}.\\ &2p\ level:\quad n=2, l=1, m_l=-1, 0, +1,\quad s=\frac{1}{2}, m_s=\pm\frac{1}{2}\\ &\Rightarrow J=\frac{1}{2}, m_J=\pm\frac{1}{2}\ \text{and}\ J=\frac{3}{2},\quad m_J=-\frac{3}{2}, -\frac{1}{2}, +\frac{1}{2}, +\frac{3}{2}.\end{aligned}$$

To calculate the electric dipole moment matrix elements for the coupled representation, we need to calculate the Clebsch–Gordan coefficients $\langle lm_l, sm_s|Jm_J\rangle$. These are given by Eq. (2.88). For the $J=\frac{1}{2}, m_J=+\frac{1}{2}$ quantum state in the 1s level, the Clebsch–Gordan coefficient for $m_s=-\frac{1}{2}$ is given by

$$\langle lm_l, sm_s, Jm_J\rangle = \left\langle 00\frac{1}{2}-\frac{1}{2}\middle|\frac{1}{2}\frac{1}{2}\right\rangle = (-1)^{\frac{1}{2}-0-\frac{1}{2}}\sqrt{2\left(\frac{1}{2}\right)+1}\begin{pmatrix}0 & \frac{1}{2} & \frac{1}{2}\\ 0 & -\frac{1}{2} & -\frac{1}{2}\end{pmatrix} = 0. \tag{4.134}$$

The $3j$ symbol in Eq. (4.134) is zero because the sum of the coefficients in the lower row is nonzero. For $m_s=+\frac{1}{2}$, we obtain

$$\begin{aligned}\langle lm_l, sm_s, Jm_J\rangle &= \left\langle 00\frac{1}{2}\frac{1}{2}\middle|\frac{1}{2}\frac{1}{2}\right\rangle = (-1)^{\frac{1}{2}-0-\frac{1}{2}}\sqrt{2\left(\frac{1}{2}\right)+1}\begin{pmatrix}0 & \frac{1}{2} & \frac{1}{2}\\ 0 & \frac{1}{2} & -\frac{1}{2}\end{pmatrix}\\ &= \sqrt{2}\begin{pmatrix}\frac{1}{2} & 0 & \frac{1}{2}\\ -\frac{1}{2} & 0 & \frac{1}{2}\end{pmatrix} = \sqrt{2}\begin{pmatrix}j+\frac{1}{2} & j & \frac{1}{2}\\ m & -\left(m+\frac{1}{2}\right) & \frac{1}{2}\end{pmatrix}\\ &\quad \text{for } j=0,\ m=-\frac{1}{2}.\end{aligned} \tag{4.135}$$

From Eq. (A3.8) in Appendix A3, we obtain

$$\begin{pmatrix} \frac{1}{2} & 0 & \frac{1}{2} \\ -\frac{1}{2} & 0 & \frac{1}{2} \end{pmatrix} = (-1)^{0+\frac{1}{2}-\frac{1}{2}} \sqrt{\frac{0+\frac{1}{2}+\frac{1}{2}}{(2)(1)}} = \frac{1}{\sqrt{2}}. \tag{4.136}$$

Substituting into Eq. (4.135), we obtain

$$\left\langle 00\frac{1}{2}\frac{1}{2} \middle| \frac{1}{2}\frac{1}{2} \right\rangle = \sqrt{2}\frac{1}{\sqrt{2}} = 1. \tag{4.137}$$

Thus, the wavefunction in the coupled representation is given by

$$\begin{aligned} |Jm_J\rangle &= \left|\frac{1}{2}\frac{1}{2}\right\rangle = \sum_{m_l m_s} |lm_l, sm_s\rangle \langle lm_l, sm_s | Jm_J\rangle \\ &= \left|00\frac{1}{2} - \frac{1}{2}\right\rangle (0) + \left|00\frac{1}{2}\frac{1}{2}\right\rangle (1) = \left|00\frac{1}{2}\frac{1}{2}\right\rangle. \end{aligned} \tag{4.138}$$

A similar analysis for the $J = \frac{1}{2}, m_J = -\frac{1}{2}$ quantum state in the 1s level gives us

$$\left\langle 00\frac{1}{2} - \frac{1}{2} \middle| \frac{1}{2} - \frac{1}{2} \right\rangle = 1 \quad \left\langle 00\frac{1}{2} + \frac{1}{2} \middle| \frac{1}{2} - \frac{1}{2} \right\rangle = 0 \quad \left|\frac{1}{2} - \frac{1}{2}\right\rangle = \left|00\frac{1}{2} - \frac{1}{2}\right\rangle. \tag{4.139}$$

The transformation from the uncoupled to the coupled representation is more complicated for the 2p level. For the $J = \frac{1}{2}, m_J = +\frac{1}{2}$ quantum state in the 2p level, the Clebsch–Gordan coefficient for $m_l = 1, m_s = -\frac{1}{2}$ and for $m_l = 0, m_s = +\frac{1}{2}$ will be the only nonzero coefficients. The Clebsch–Gordan coefficient for $m_l = 1, m_s = -\frac{1}{2}$ is given by

$$\begin{aligned} \langle lm_l, sm_s | Jm_J\rangle &= \left\langle 11\frac{1}{2} - \frac{1}{2} \middle| \frac{1}{2}\frac{1}{2} \right\rangle = (-1)^{\frac{1}{2}-1-\frac{1}{2}} \sqrt{2\left(\frac{1}{2}\right)+1} \begin{pmatrix} 1 & \frac{1}{2} & \frac{1}{2} \\ 1 & -\frac{1}{2} & -\frac{1}{2} \end{pmatrix} \\ &= -\sqrt{2} \begin{pmatrix} \frac{1}{2} & \frac{1}{2} & 1 \\ -\frac{1}{2} & -\frac{1}{2} & 1 \end{pmatrix} = -\sqrt{2} \begin{pmatrix} j & j & 1 \\ m & -(m+1) & 1 \end{pmatrix} \\ &\quad \text{for } j = 0, \ m = -\frac{1}{2}. \end{aligned} \tag{4.140}$$

From Eq. (A3.7) in Appendix A3, we obtain

$$\begin{pmatrix} \frac{1}{2} & \frac{1}{2} & 1 \\ -\frac{1}{2} & -\frac{1}{2} & 1 \end{pmatrix} = (-1)^{\frac{1}{2}+\frac{1}{2}} \sqrt{\frac{\left(\frac{1}{2}+\frac{1}{2}\right)\left(\frac{1}{2}-\frac{1}{2}+1\right)}{\left[2\left(\frac{1}{2}\right)+1\right]\left[2\left(\frac{1}{2}\right)\right]\left[\frac{1}{2}+1\right]}} = -\frac{1}{\sqrt{3}}. \tag{4.141}$$

Substituting into Eq. (4.140), we obtain

$$\left\langle 11\frac{1}{2}-\frac{1}{2}\middle|\frac{1}{2}\frac{1}{2}\right\rangle = \sqrt{\frac{2}{3}}. \tag{4.142}$$

The Clebsch–Gordan coefficient for $m_l = 0, m_s = +\frac{1}{2}$ is given by

$$\begin{aligned}\langle lm_l, sm_s|Jm_J\rangle &= \left\langle 10\frac{1}{2}\frac{1}{2}\middle|\frac{1}{2}\frac{1}{2}\right\rangle = (-1)^{\frac{1}{2}-1-\frac{1}{2}}\sqrt{2\left(\frac{1}{2}\right)+1}\begin{pmatrix}1 & \frac{1}{2} & \frac{1}{2}\\ 0 & \frac{1}{2} & -\frac{1}{2}\end{pmatrix}\\ &= -\sqrt{2}\begin{pmatrix}\frac{1}{2} & \frac{1}{2} & 1\\ \frac{1}{2} & -\frac{1}{2} & 0\end{pmatrix} = -\sqrt{2}\begin{pmatrix}j & j & 1\\ m & -m & 1\end{pmatrix}\\ &\text{for } j=\frac{1}{2},\ m=\frac{1}{2}.\end{aligned} \tag{4.143}$$

From Eq. (A3.6) in Appendix A3, we obtain

$$\begin{pmatrix}\frac{1}{2} & \frac{1}{2} & 1\\ \frac{1}{2} & -\frac{1}{2} & 0\end{pmatrix} = (-1)^{\frac{1}{2}-\frac{1}{2}}\frac{\frac{1}{2}}{\sqrt{\left[2\left(\frac{1}{2}\right)+1\right]\left(\frac{1}{2}\right)\left[\frac{1}{2}+1\right]}} = \frac{1}{\sqrt{6}}. \tag{4.144}$$

Substituting into Eq. (4.140), we obtain

$$\left\langle 1\frac{1}{2}0\frac{1}{2}\middle|1\frac{1}{2}\frac{1}{2}\frac{1}{2}\right\rangle = -\frac{1}{\sqrt{3}}. \tag{4.145}$$

Therefore, for the $J = \frac{1}{2}, m_J = +\frac{1}{2}$ quantum state in the $2p$ level, we can write

$$\begin{aligned}|Jm_J\rangle &= \left|\frac{1}{2}\frac{1}{2}\right\rangle = \sum_{m_l m_s}|lm_l, sm_s\rangle\langle lm_l, sm_s|Jm_J\rangle\\ &= -\frac{1}{\sqrt{3}}\left|1\frac{1}{2}0\frac{1}{2}\right\rangle + \sqrt{\frac{2}{3}}\left|1\frac{1}{2}1-\frac{1}{2}\right\rangle.\end{aligned} \tag{4.146}$$

For this quantum state in the coupled representation, therefore, we can write the wavefunction as the sum of two wavefunctions in the uncoupled representation. The Clebsch–Gordan coefficients for the $2p$ quantum states in the coupled representation are listed in Appendix A2. For the other quantum states in the coupled representation, we can write

$$|Jm_J\rangle = \left|\frac{1}{2}-\frac{1}{2}\right\rangle = \frac{1}{\sqrt{3}}\left|10\frac{1}{2}-\frac{1}{2}\right\rangle - \sqrt{\frac{2}{3}}\left|1-1\frac{1}{2}\frac{1}{2}\right\rangle \tag{4.147}$$

$$\left|\frac{3}{2}-\frac{3}{2}\right\rangle = \left|1-1\frac{1}{2}-\frac{1}{2}\right\rangle \tag{4.148}$$

$$\left|\frac{3}{2} - \frac{1}{2}\right\rangle = \frac{1}{\sqrt{3}}\left|1 - 1\frac{1}{2}\frac{1}{2}\right\rangle + \sqrt{\frac{2}{3}}\left|10\frac{1}{2} - \frac{1}{2}\right\rangle \tag{4.149}$$

$$\left|\frac{3}{2}\frac{1}{2}\right\rangle = \frac{1}{\sqrt{3}}\left|11\frac{1}{2} - \frac{1}{2}\right\rangle + \sqrt{\frac{2}{3}}\left|10\frac{1}{2}\frac{1}{2}\right\rangle \tag{4.150}$$

$$\left|\frac{3}{2}\frac{3}{2}\right\rangle = \left|11\frac{1}{2}\frac{1}{2}\right\rangle. \tag{4.151}$$

For the coupled representation, the wavefunctions listed in Eqs. (4.146)–(4.151) are normalized and can be written as sums of functions in the form given in Eq. (4.99). The electric dipole moment matrix elements are found using Eq. (4.112). For the $J = \frac{1}{2}, m_J = +\frac{1}{2}$ quantum state in the $2p$ level, the matrix element for coupling with the $J = \frac{1}{2}, m_J = +\frac{1}{2}$ quantum state in the $1s$ level is given by

$$\begin{aligned}\boldsymbol{\mu} &= -e\langle J'm'_J|\boldsymbol{r}|Jm_J\rangle = -e\left\langle\frac{1}{2}\frac{1}{2}\middle|\boldsymbol{r}\middle|\frac{1}{2}\frac{1}{2}\right\rangle \\ &= -e\sum_{m'_l m'_s m_l m_s}\langle l'm'_l, s'm'_s|\boldsymbol{r}|lm_l, sm_s\rangle\langle l'm'_l, s'm'_s|J'm'_J\rangle\langle lm_l, sm_s|Jm_J\rangle \\ &= -e\left\langle 10\frac{1}{2}\frac{1}{2}\middle|\boldsymbol{r}\middle|00\frac{1}{2}\frac{1}{2}\right\rangle\left(-\frac{1}{\sqrt{3}}\right) = \left(\frac{1.290\,ea_0}{3}\right)\hat{\boldsymbol{z}},\end{aligned} \tag{4.152}$$

where we have substituted using Eq. (4.133). The $J = \frac{1}{2}, m_J = +\frac{1}{2}$ quantum state in the $2p$ level is also coupled with the $J = \frac{1}{2}, m_J = -\frac{1}{2}$ quantum state in the $1s$ level:

$$\begin{aligned}\boldsymbol{\mu} &= -e\langle J'm'_J|\boldsymbol{r}|Jm_J\rangle = -e\left\langle\frac{1}{2}\frac{1}{2}\middle|\boldsymbol{r}\middle|\frac{1}{2} - \frac{1}{2}\right\rangle \\ &= -e\sum_{m'_l m'_s m_l m_s}\langle l'm'_l, s'm'_s|\boldsymbol{r}|lm_l, sm_s\rangle\langle l'm'_l, s'm'_s|J'm'_J\rangle\langle lm_l, sm_s|Jm_J\rangle \\ &= -e\left\langle 11\frac{1}{2} - \frac{1}{2}\middle|\boldsymbol{r}\middle|00\frac{1}{2} - \frac{1}{2}\right\rangle\left(\sqrt{\frac{2}{3}}\right) = \left(\frac{1.290\,ea_0}{3}\right)\left(\frac{\hat{\boldsymbol{x}} - i\hat{\boldsymbol{y}}}{\sqrt{2}}\right).\end{aligned} \tag{4.153}$$

The electric dipole moment matrix elements for the $1s$–$2p$ transitions for the coupled representation are listed in Table A2.2.

4.6.4 Spontaneous Emission Coefficients for the Hydrogen Atom

In this section we develop expressions for the spontaneous emission coefficients for $2p \rightarrow 1s$ transitions in terms of the electric dipole moment density matrix elements discussed in the previous section. The primary reason for developing these expressions is that single-photon transition strengths are most frequently tabulated in terms of the Einstein coefficient for spontaneous emission or, equivalently, the lifetime of the excited level of the transition. Although care must be exercised in the determination of exactly which upper and lower levels are coupled, a great advantage of tabulating transition strengths in terms of spontaneous emission coefficients and/or radiative lifetimes is that the units for these quantities are clearly s^{-1} and s, respectively.

Theoretical analysis of the process of spontaneous emission requires consideration of the interaction of the excited atom or molecule with the quantized electromagnetic field (Craig & Tirunamachandran, 1984; Marcuse, 1980). This analysis is outside the scope of our present discussion. The total rate of spontaneous emission for a transition between upper state b and lower state a is given by

$$A_{ba} = \frac{\omega_{ba}^3}{3\pi\varepsilon_0\hbar c^3}|\boldsymbol{\mu}_{ba}|^2. \tag{4.154}$$

In Eq. (4.154), $\omega_{ba} = (\varepsilon_b - \varepsilon_a)/\hbar$, where ε_b and ε_a are the energies for quantum states b and a, respectively. For the $J_b = \frac{1}{2}, m_b = \frac{1}{2}$ quantum state in the $2p$ level, the Einstein coefficient for spontaneous emission to the $J_a = \frac{1}{2}, m_a = \frac{1}{2}$ quantum state in the $1s$ level is given by

$$\begin{aligned} A_{ba} &= \frac{\omega_{ba}^3}{3\pi\varepsilon_0\hbar c^3}\left(\frac{1.290 e a_0}{3}\right)^2 \hat{\mathbf{z}}\bullet\hat{\mathbf{z}} = \frac{1.664\,\omega_{ba}^3 e^2 a_0^2}{27\pi\,\varepsilon_0\hbar\,c^3} \\ &= \frac{1.664\left(1.54948\times10^{16}\ \mathrm{s}^{-1}\right)^3\left(1.6022\times10^{-19}\ \mathrm{C}\right)^2\left(5.29177\times10^{-11}\ \mathrm{m}\right)^2}{27\pi\left(8.8542\times10^{-12}\ \frac{\mathrm{C}^2}{\mathrm{J-m}}\right)\left(1.05457\times10^{-34}\ J-s\right)\left(2.99792\times10^8\ \frac{\mathrm{m}}{\mathrm{s}}\right)^3} \\ &= 2.085\times10^8\ \mathrm{s}^{-1}. \end{aligned} \tag{4.155}$$

This state can also undergo radiative decay to the $J_a = \frac{1}{2}, m_a = -\frac{1}{2}$ quantum state in the $1s$ level. The spontaneous emission coefficient for that transition is given by

$$A_{ba} = \frac{\omega_{ba}^3}{3\pi\varepsilon_0\hbar c^3}\left(\frac{1.290\sqrt{2}\,e a_0}{3}\right)^2\left(\frac{\hat{\mathbf{x}}+i\hat{\mathbf{y}}}{\sqrt{2}}\right)\bullet\left(\frac{\hat{\mathbf{x}}-i\hat{\mathbf{y}}}{\sqrt{2}}\right) = \frac{3.328\,\omega_{ba}^3 e^2 a_0^2}{27\pi\,\varepsilon_0\hbar\,c^3} = 4.170\times10^8\ s^{-1}. \tag{4.156}$$

For the $J_b = \frac{3}{2}, m_b = \frac{1}{2}$ quantum state in the $2p$ level, the Einstein coefficient for spontaneous emission to the $J_a = \frac{1}{2}, m_a = \frac{1}{2}$ quantum state in the $1s$ level is given by

$$A_{ba} = \frac{\omega_{ba}^3}{3\pi\varepsilon_0\hbar c^3}\left(-\frac{1.290\sqrt{2}\,e a_0}{3}\right)^2 \hat{\mathbf{z}}\bullet\hat{\mathbf{z}} = \frac{3.328\,\omega_{ba}^3 e^2 a_0^2}{27\pi\,\varepsilon_0\hbar\,c^3} = 4.170\times10^8\ \mathrm{s}^{-1}. \tag{4.157}$$

The $J_b = \frac{3}{2}, m_b = \frac{1}{2}$ quantum state in the $2p$ level can also undergo radiative decay to the $J_a = \frac{3}{2}, m_a = -\frac{1}{2}$ quantum state in the $1s$ level; the spontaneous emission coefficient for this transition is given by

$$A_{ba} = \frac{\omega_{ba}^3}{3\pi\varepsilon_0\hbar c^3}\left(-\frac{1.290 e a_0}{3}\right)^2\left(\frac{\hat{\mathbf{x}}+i\hat{\mathbf{y}}}{\sqrt{2}}\right)\bullet\left(\frac{\hat{\mathbf{x}}-i\hat{\mathbf{y}}}{\sqrt{2}}\right) = \frac{1.664\,\omega_{ba}^3 e^2 a_0^2}{27\pi\,\varepsilon_0\hbar\,c^3} = 2.085\times10^8\ \mathrm{s}^{-1}. \tag{4.158}$$

The total spontaneous emission coefficients for both the $J_b = \frac{1}{2}, m_b = \frac{1}{2}$ and $J_b = \frac{3}{2}, m_b = \frac{1}{2}$ quantum states in the $2p$ level are given by

$$A_{bA} = \sum_a A_{ba} = \frac{\omega_{ba}^3}{9\pi\varepsilon_0\hbar c^3}(1.290ea_0)^2 = 6.255\times 10^8\ \text{s}^{-1}. \tag{4.159}$$

The state-to-state and total spontaneous emission coefficients for radiative decay of the $2p$ quantum states to the $1s$ quantum states are listed in Table A2.3. Note that the total spontaneous emission coefficient is that same for all quantum states in the $2p$ level. In general, for any energy level characterized by a quantum number J_b, the total spontaneous emission coefficient A_{bA} to any lower level A must be the same for all quantum states b in the upper level. If this were not the case, an isotropic population distribution in the upper level B (equal population in each m_b quantum state) would become anisotropic as a result of radiative decay. We can also note that the total spontaneous emission coefficient is the same for the quantum states in the $J_b = \frac{1}{2}$ and $J_b = \frac{3}{2}$ levels; this is a special case for the one-electron hydrogen atom and does not hold in general for multielectron atoms and molecules.

For either of the $J_b = \frac{1}{2}$ and $J_b = \frac{3}{2}$ $2p$ energy levels, we can define a quantum-state-averaged spontaneous emission coefficient A_{BA} given by

$$A_{BA} = \frac{1}{2J_B+1}\sum_{m_J=-J_b}^{m_J=J_b} A_{bA} = A_{bA} = \frac{\omega_{ba}^3}{9\pi\varepsilon_0\hbar c^3}(1.290ea_0)^2 = 6.255\times 10^8\,\text{s}^{-1}. \tag{4.160}$$

4.7 Density Matrix Elements for Electric Dipole Interactions: Multielectron Atoms and Molecules

The results of the single-electron hydrogen atom analysis of Section 4.6 can be generalized for radiative transitions for multielectron atoms and molecules so long as the transitions occur between energy levels with well-defined values of the total angular momentum $\boldsymbol{J}$ (or $\boldsymbol{F}$ if the hyperfine interaction, resulting in the coupling of the nuclear spin $\boldsymbol{I}$ with $\boldsymbol{J}$, is included). The Wigner–Eckart theorem is a key step in this generalization. The Wigner–Eckart theorem allows the separation of the electric dipole moment matrix elements into a reduced density matrix element that does not depend on the projection quantum numbers M and M', and a $3j$ symbol that contains the information carried by these projection quantum numbers.

4.7.1 The Wigner–Eckart Theorem and Reduced Density Matrix Elements

The Wigner–Eckart theorem is given by (Rose, 1957; Weissbluth, 1978)

$$\langle \alpha JM|\boldsymbol{T}_{kq}|\alpha' J'M'\rangle = (-1)^{J-M}\begin{pmatrix} J & k & J' \\ -M & q & M' \end{pmatrix}\langle \alpha J||\boldsymbol{T}_k||\alpha' J'\rangle. \tag{4.161}$$

In Eq. (4.161), the quantity $\boldsymbol{T}_{kq}$ is the qth component of an irreducible spherical tensor operator of rank k, and there are $2k+1$ values of q ranging from $-k$ to $+k$. The index α indicates all quantum numbers other than J and M that characterize the state of the atom or molecule. The reduced matrix element $\langle \alpha J \| \boldsymbol{T}_k \| \alpha' J' \rangle$ is independent of M, M', and q.

The properties of the irreducible spherical tensor operators are discussed in detail in a number of excellent texts. The electric dipole moment operator can be represented as an irreducible tensor operator of rank $k = 1$. An irreducible spherical tensor operator has components of the form $T_{kq} = f(r) Y_{kq}(\theta, \varphi)$, where $f(r)$ is a function only of the radial coordinate r, and $Y_{kq}(\theta, \varphi)$ is a spherical harmonic function. Consider the vector $\boldsymbol{A}$ in the Cartesian representation:

$$\boldsymbol{A} = A_x \hat{\boldsymbol{x}} + A_y \hat{\boldsymbol{y}} + A_z \hat{\boldsymbol{z}}. \tag{4.162}$$

In the spherical basis, this vector is written as (Sobelman, 1992; Weissbluth, 1978)

$$\boldsymbol{A} = A_{-1} \hat{\boldsymbol{e}}^*_{-1} + A_0 \hat{\boldsymbol{e}}_0 + A_{+1} \hat{\boldsymbol{e}}^*_{+1} = \sum_q A_q \hat{\boldsymbol{e}}^*_q \quad (q = -1,\, 0,\, +1), \tag{4.163}$$

where

$$A_{+1} = -\frac{A_x + iA_y}{\sqrt{2}} \tag{4.164}$$

$$A_0 = A_z \tag{4.165}$$

$$A_{-1} = \frac{A_x - iA_y}{\sqrt{2}} \tag{4.166}$$

and the unit vectors are written as

$$\hat{\boldsymbol{e}}_{+1} = -\frac{\hat{\boldsymbol{x}} + i\hat{\boldsymbol{y}}}{\sqrt{2}}. \tag{4.167}$$

$$\hat{\boldsymbol{e}}_0 = \hat{\boldsymbol{z}}. \tag{4.168}$$

$$\hat{\boldsymbol{e}}_{-1} = \frac{\hat{\boldsymbol{x}} - i\hat{\boldsymbol{y}}}{\sqrt{2}}. \tag{4.169}$$

From Eq. (4.103), for an atom with only a single electron and with the origin of the coordinate system at the nucleus, we can write

$$\mu_x = -er\sqrt{\frac{4\pi}{3}} \frac{(-Y_{11} + Y_{1-1})}{\sqrt{2}}. \tag{4.170}$$

$$\mu_y = -ier\sqrt{\frac{4\pi}{3}} \frac{(Y_{11} + Y_{1-1})}{\sqrt{2}}. \tag{4.171}$$

$$\mu_z = -er\sqrt{\frac{4\pi}{3}} Y_{10}. \tag{4.172}$$

For a multielectron atom we would substitute $-e\sum_i \boldsymbol{r}_i$ for $-e\boldsymbol{r}$, but the basic results will be unchanged. Using Eqs. (4.164)–(4.166), we can write

$$\mu_{+1} = -\frac{\mu_x + i\mu_y}{\sqrt{2}} = er\sqrt{\frac{4\pi}{3}}\frac{-2Y_{11}}{2} = -er\sqrt{\frac{4\pi}{3}}Y_{11}. \tag{4.173}$$

$$\mu_0 = -er\sqrt{\frac{4\pi}{3}}Y_{10}. \tag{4.174}$$

$$\mu_{-1} = \frac{\mu_x - i\mu_y}{\sqrt{2}} = -er\sqrt{\frac{4\pi}{3}}\frac{2Y_{1-1}}{2} = -er\sqrt{\frac{4\pi}{3}}Y_{1-1}. \tag{4.175}$$

Using Eq. (4.163), we find that the irreducible spherical tensor representation of the electric dipole moment operator is given by

$$\boldsymbol{T}_{11} = \boldsymbol{\mu}_{11} = \mu_{+1}\hat{e}^*_{+1} = er\sqrt{\frac{4\pi}{3}}Y_{11}\frac{\hat{x} - iy}{\sqrt{2}}. \tag{4.176}$$

$$\boldsymbol{T}_{10} = \boldsymbol{\mu}_{10} = \mu_0\,\hat{e}^*_0 = -er\sqrt{\frac{4\pi}{3}}Y_{10}\hat{z}. \tag{4.177}$$

$$\boldsymbol{T}_{1-1} = \boldsymbol{\mu}_{1-1} = -er\sqrt{\frac{4\pi}{3}}Y_{1-1}\frac{\hat{x} + iy}{\sqrt{2}}. \tag{4.178}$$

The Wigner–Eckart theorem is very useful for separating the geometrical aspects of matrix elements. For electric dipole radiation, substitution of the index $k = 1$ in Eq. (4.161) results in

$$\langle \alpha JM|\mu_{1q}|\alpha' J'M'\rangle = (-1)^{J-M}\begin{pmatrix} J & 1 & J' \\ -M & q & M' \end{pmatrix}\langle \alpha J\|\boldsymbol{\mu}_1\|\alpha' J'\rangle. \tag{4.179}$$

From the properties of the *3j* symbols we note immediately that the electric dipole density matrix elements will be zero unless

$$J' = J, J \pm 1 \tag{4.180}$$

$$0 = -M + q + M' \quad \Rightarrow \quad M' = M - q. \tag{4.181}$$

The vector components of the electric dipole density matrix elements are then given by

$$\langle \alpha JM|\boldsymbol{\mu}_1|\alpha' J'M'\rangle = \sum_q \hat{e}^*_q\langle \alpha JM|\mu_{1q}|\alpha' J'M'\rangle. \tag{4.182}$$

The solution of Eq. (4.179) for different values of J' and M' is straightforward, and all three components of the electric dipole density matrix for $J' = J, J \pm 1$ are tabulated in Appendix A3, along with the results. For example, the $q = -1$ component for a transition where $J' = J + 1$ is given by

$$\hat{e}^*_{-1}\langle \alpha JM|\mu_{1-1}|\alpha' J+1M'\rangle = \hat{e}^*_{-1}(-1)^{J-M}\begin{pmatrix} J & 1 & J' \\ -M & -1 & M' \end{pmatrix}\langle \alpha J\|\boldsymbol{\mu}_1\|\alpha' J'\rangle. \quad (4.183)$$

The $3j$ symbol will only be nonzero for $M' = M+1$. Therefore, we can write

$$\begin{aligned}\hat{e}^*_{-1}\langle \alpha JM|\mu_{1-1}|\alpha' J+1M+1\rangle &= \hat{e}^*_{-1}(-1)^{J-M}\begin{pmatrix} J & 1 & J' \\ -M & -1 & M+1 \end{pmatrix}\langle \alpha J\|\boldsymbol{\mu}_1\|\alpha' J'\rangle \\ &= \hat{e}^*_{-1}(-1)^{J-M+2(J+1+J')}\begin{pmatrix} J & J+1 & 1 \\ M & -(M+1) & 1 \end{pmatrix}\langle \alpha J\|\boldsymbol{\mu}_1\|\alpha' J'\rangle \\ &= \hat{e}^*_{-1}(-1)^{J-M}\begin{pmatrix} J & J+1 & 1 \\ M & -(M+1) & 1 \end{pmatrix}\langle \alpha J\|\boldsymbol{\mu}_1\|\alpha' J'\rangle.\end{aligned} \quad (4.184)$$

The $3j$ symbol is then evaluated from Eq. (A3.16):

$$\begin{pmatrix} J & J+1 & 1 \\ M & -(M+1) & 1 \end{pmatrix} = (-1)^{3J+M}\sqrt{\frac{(J+M+1)(J+M+1)}{(2J+3)(2J+2)(2J+1)}}. \quad (4.185)$$

Substituting Eq. (4.185) into Eq. (4.184) and noting that $4J$ will always be an even number, we obtain

$$\hat{e}^*_{-1}\langle \alpha JM|\mu_{1-1}|\alpha' J+1M+1\rangle = \hat{e}^*_{-1}\sqrt{\frac{(J+M+1)(J+M+1)}{(2J+3)(2J+2)(2J+1)}}\langle \alpha J\|\boldsymbol{\mu}_1\|\alpha' J'\rangle. \quad (4.186)$$

Before we develop a general expression for the reduced density matrix element, consider the relation between the reduced matrix elements $\langle \alpha J\|\boldsymbol{\mu}_1\|\alpha' J'\rangle$ and $\langle \alpha' J'\|\boldsymbol{\mu}_1\|\alpha J\rangle$. Let's consider the case where $J' = J+1$. From Eq. (4.182), we can write

$$\begin{aligned}\langle \alpha' J'M'|\boldsymbol{\mu}_1|\alpha JM\rangle &= \sum_q \hat{e}^*_q\langle \alpha' J'M'|\mu_{1q}|\alpha JM\rangle \\ &= \langle \alpha JM|\boldsymbol{\mu}|\alpha' J'M'\rangle^* = \sum_q \hat{e}_q\langle \alpha JM|\mu_{1q}|\alpha' J'M'\rangle.\end{aligned} \quad (4.187)$$

Selecting the $q=0$ component of Eq. (4.187), we obtain

$$\hat{z}\langle \alpha' J'M'|\mu_{10}|\alpha JM\rangle = \hat{z}\langle \alpha JM|\mu_{10}|\alpha' J'M'\rangle. \quad (4.188)$$

The $q=0$ component was selected because it is the only term that involves the unit vector $\hat{z}$. Using the Wigner–Eckart theorem (Eq. (4.161)) to evaluate the terms in Eq. (4.188), we obtain

$$\begin{aligned}\langle \alpha JM|\mu_{10}|\alpha' J'M'\rangle &= (-1)^{J-M}\begin{pmatrix} J & 1 & J' \\ -M & 0 & M \end{pmatrix}\langle \alpha J\|\boldsymbol{\mu}_1\|\alpha' J'\rangle \\ = \langle \alpha' J'M'|\mu_{10}|\alpha JM\rangle &= (-1)^{J'-M}\begin{pmatrix} J' & 1 & J \\ -M & 0 & M \end{pmatrix}\langle \alpha' J'\|\boldsymbol{\mu}_1\|\alpha J\rangle,\end{aligned} \quad (4.189)$$

where we have used the relation $M' = M$ for $q = 0$. We can rearrange Eq. (4.189) to obtain

$$\langle \alpha' J' \| \boldsymbol{\mu}_1 \| \alpha J \rangle = \frac{(-1)^{J-M} \begin{pmatrix} J & 1 & J' \\ -M & 0 & M \end{pmatrix}}{(-1)^{J'-M} \begin{pmatrix} J' & 1 & J \\ -M & 0 & M \end{pmatrix}} \langle \alpha J \| \boldsymbol{\mu}_1 \| \alpha' J' \rangle. \tag{4.190}$$

But the $3j$ symbol in the denominator can be rewritten as

$$\begin{pmatrix} J' & 1 & J \\ -M & 0 & M \end{pmatrix} = (-1)^{J'+J+1} \begin{pmatrix} J & 1 & J' \\ M & 0 & -M \end{pmatrix} = (-1)^{2(J'+J+1)} \begin{pmatrix} J & 1 & J' \\ -M & 0 & M \end{pmatrix}. \tag{4.191}$$

Substituting Eq. (4.191) into Eq. (4.190), we obtain

$$\langle \alpha' J' \| \boldsymbol{\mu}_1 \| \alpha J \rangle = (-1)^{J-J'} \langle \alpha J \| \boldsymbol{\mu}_1 \| \alpha' J' \rangle. \tag{4.192}$$

We now develop general expressions for the reduced density matrix element for electric dipole interaction in terms of the spontaneous emission rate coefficient from the upper energy level to the lower energy level. The rate coefficient for spontaneous emission from an upper quantum state b to a lower quantum state a is given by

$$A_{ba} = \frac{\omega_{ba}^3}{3\pi\varepsilon_0 \hbar c^3} |\boldsymbol{\mu}_{ba}|^2. \tag{4.193}$$

The rate coefficient for spontaneous emission from an upper energy level B to a lower energy level A is given by

$$A_{BA} = \frac{1}{2J'+1} \sum_{M'=-J'}^{M'=J'} A_{bA} = \frac{1}{2J'+1} \sum_{M'=-J'}^{M'=J'} \sum_{M=-J}^{M=-J} A_{ba}, \tag{4.194}$$

where in Eq. (4.194), when $J = J_A$, $J' = J_B$, $M = M_a$, and $M' = M_b$, we can write

$$\begin{aligned} A_{J'J} &= \frac{1}{2J'+1} \left(\frac{\omega_{J'J}^3}{3\pi\varepsilon_0 \hbar c^3} \right) \sum_{M'=-J'}^{M'=+J'} \sum_{M=-J}^{M=+J} |\langle \alpha' J' M' | \boldsymbol{\mu}_1 | \alpha J M \rangle|^2 \\ &= \frac{1}{2J'+1} \left(\frac{\omega_{J'J}^3}{3\pi\varepsilon_0 \hbar c^3} \right) \sum_{M'=-J'}^{M'=+J'} \sum_{q=-1}^{q=+1} \left| \hat{e}_q^* \langle \alpha' J' M' | \mu_{1q} | \alpha J M \rangle \right|^2. \end{aligned} \tag{4.195}$$

For $J' = J$ we can write,

$$\begin{aligned} A_{J'J} = \frac{1}{2J'+1} \left(\frac{\omega_{J'J}^3}{3\pi\varepsilon_0 \hbar c^3} \right) \sum_{M'=-J'}^{M'=+J'} \Big[\hat{e}_{-1}^* \bullet \hat{e}_{-1} \langle \alpha' J' M' | \mu_{1-1} | \alpha J' M' + 1 \rangle^2 \\ + \hat{e}_0^* \bullet \hat{e}_0 \langle \alpha' J' M' | \mu_{10} | \alpha J' M' \rangle^2 + \hat{e}_{+1}^* \bullet \hat{e}_{+1} \langle \alpha' J' M' | \mu_{1+1} | \alpha J' M' - 1 \rangle^2 \Big]. \end{aligned} \tag{4.196}$$

Using Eqs. (A7.4)–(A7.6) from Appendix A7, we obtain

$$\begin{aligned} A_{J'J} &= \frac{1}{2J'+1}\left(\frac{\omega_{J'J}^3}{3\pi\varepsilon_0\hbar c^3}\right)|\langle\alpha' J'\|\boldsymbol{\mu}_1\|\alpha J'\rangle|^2 \sum_{M'=-J'}^{M'=+J'}\left[\frac{(J'+M'+1)(J'-M')}{(2J'+1)(2J')(J'+1)}\right. \\ &\left. + \frac{(M')^2}{(2J'+1)J'(J'+1)} + \frac{(J'-M'+1)(J'+M')}{(2J'+1)(2J')(J'+1)}\right] \\ &= \frac{1}{2J+1}\left(\frac{\omega_{J'J}^3}{3\pi\varepsilon_0\hbar c^3}\right)|\langle\alpha' J'\|\boldsymbol{\mu}_1\|\alpha J'\rangle|^2 \sum_{M'=-J'}^{M'=+J'}\left[\frac{2(J')^2+2J'}{(2J'+1)(2J')(J'+1)}\right] \\ &= \frac{1}{2J'+1}\left(\frac{\omega_{J'J}^3}{3\pi\varepsilon_0\hbar c^3}\right)|\langle\alpha' J'\|\boldsymbol{\mu}_1\|\alpha J'\rangle|^2 \sum_{M'=-J'}^{M'=+J'}\left[\frac{1}{(2J'+1)}\right]. \end{aligned} \tag{4.197}$$

Performing the summation from $M' = -J'$ to $M' = +J'$, we obtain

$$A_{J'J} = \frac{1}{2J'+1}\left(\frac{\omega_{J'J}^3}{3\pi\varepsilon_0\hbar c^3}\right)|\langle\alpha' J'\|\boldsymbol{\mu}_1\|\alpha J'\rangle|^2. \tag{4.198}$$

Solving for the reduced density matrix element gives us

$$\langle\alpha' J'\|\boldsymbol{\mu}_1\|\alpha J'\rangle = \pm\sqrt{\frac{A_{J'J}(2J'+1)3\pi\varepsilon_0\hbar c^3}{\omega_{J'J}^3}}. \tag{4.199}$$

A similar analysis for $J = J'+1$ and $J = J'-1$ results in the same functional relations:

$$\langle\alpha' J'\|\boldsymbol{\mu}_1\|\alpha J'+1\rangle = \pm\sqrt{\frac{A_{J'J}(2J'+1)3\pi\varepsilon_0\hbar c^3}{\omega_{J'J}^3}}. \tag{4.200}$$

$$\langle\alpha' J'\|\boldsymbol{\mu}_1\|\alpha J'-1\rangle = \pm\sqrt{\frac{A_{J'J}(2J'+1)3\pi\varepsilon_0\hbar c^3}{\omega_{J'J}^3}}. \tag{4.201}$$

The selection of sign for the reduced matrix element is arbitrary, except that from Eq. (4.192) we note that $\langle\alpha' J'\|\boldsymbol{\mu}_1\|\alpha J'-1\rangle$ and $\langle\alpha' J'\|\boldsymbol{\mu}_1\|\alpha J'+1\rangle$ must have opposite signs. For consistency with Condon and Shortley (1951) we choose

$$\langle\alpha' J'\|\boldsymbol{\mu}_1\|\alpha J'+1\rangle = -\sqrt{\frac{A_{J'J}(2J'+1)3\pi\varepsilon_0\hbar c^3}{\omega_{J'J}^3}}. \tag{4.202}$$

$$\langle\alpha' J'\|\boldsymbol{\mu}_1\|\alpha J'\rangle = +\sqrt{\frac{A_{J'J}(2J'+1)3\pi\varepsilon_0\hbar c^3}{\omega_{J'J}^3}}. \tag{4.203}$$

$$\langle\alpha' J'\|\boldsymbol{\mu}_1\|\alpha J'-1\rangle = +\sqrt{\frac{A_{J'J}(2J'+1)3\pi\varepsilon_0\hbar c^3}{\omega_{J'J}^3}}. \tag{4.204}$$

4.7.2 Effects of Hyperfine Splitting

When the nuclear spin $\boldsymbol{I}$ of the nucleus of an atom or the one of the nuclei of a molecule is nonzero, there is an interaction between the electromagnetic fields of the electrons and the magnetic field of the nucleus. This results in the splitting of the electron energy level $\boldsymbol{J}$ into $2I+1$ different levels when $J \geq I$, and $2J+1$ levels when $I \geq J$. The total angular momentum $\boldsymbol{F}$ is defined as a result of this interaction:

$$\boldsymbol{F} = \boldsymbol{I} + \boldsymbol{J} \qquad F = I + J,\ I + J - 1, \ldots |I - J|. \tag{4.205}$$

The hyperfine splitting is small and is frequently neglected except for high-resolution spectroscopic measurements.

The effect of the hyperfine splitting on the electric dipole moment density matrix elements can be analyzed using the following relation for irreducible tensor operators (Sobelman, 1992):

$$\begin{aligned}\langle \alpha J_1 J_2 J_3 \| \boldsymbol{T}_k \| \alpha' J_1' J_2 J_3' \rangle &= (-1)^{J_1 + J_2 + J_3' + k} \langle \alpha J_1 \| \boldsymbol{T}_k \| \alpha' J_1' \rangle \\ &\quad \times \sqrt{(2J_3 + 1)(2J_3' + 1)} \begin{Bmatrix} J_1 & J_3 & J_2 \\ J_3' & J_1' & k \end{Bmatrix}.\end{aligned} \tag{4.206}$$

In Eq. (4.206), $\boldsymbol{J}_1$ and $\boldsymbol{J}_2$ are angular momenta that couple to give the resultant angular momentum $\boldsymbol{J}_3$ and it is assumed that the irreducible tensor operator $\boldsymbol{T}_{kq}$ commutes with $\boldsymbol{J}_2$. The quantity $\begin{Bmatrix} J_1 & J_3 & J_2 \\ J_3' & J_1' & k \end{Bmatrix}$ is a *6j* symbol. Pertinent expressions for evaluation of *6j* symbols are listed in Appendix A4.

The effects of hyperfine splitting are now analyzed by substituting J, I, and F for J_1, J_2, and J_3, respectively, and by replacing the general irreducible tensor operator $\boldsymbol{T}_k$ with the electric dipole operator $\boldsymbol{\mu}_1$ in Eq. (4.206):

$$\langle \alpha JIF \| \boldsymbol{\mu}_1 \| \alpha' J' I F' \rangle = (-1)^{J+I+F'+1} \langle \alpha J \| \boldsymbol{\mu}_1 \| \alpha' J' \rangle \sqrt{(2F+1)(2F'+1)} \begin{Bmatrix} J & F & I \\ F' & J' & 1 \end{Bmatrix}. \tag{4.207}$$

It is evident from Eq. (4.207) that, as expected, the nuclear spin I is not affected by the electric dipole interaction. We can now substitute Eq. (4.207) into Eq. (4.179) to obtain a general expression for the electric dipole moment matrix elements for a transition between the quantum states (F, M_F) and (F', M_F'):

$$\langle \alpha F M_F | \mu_{1q} | \alpha' F' M'_F \rangle = (-1)^{F - M_F} \begin{pmatrix} F & 1 & F' \\ -M_F & q & M_F' \end{pmatrix} \langle \alpha JIF \| \boldsymbol{\mu}_1 \| \alpha' J' I F' \rangle. \tag{4.208}$$

The effects of hyperfine splitting are generally ignored in optical spectroscopy. The hyperfine levels are in general so close together in energy that it is very difficult or impossible to resolve the different hyperfine levels. The total spontaneous emission coefficient or the absorption coefficient for a level J will not be affected by the hyperfine splitting, as shown in Appendix A8. Consequently it is usually ignored except in experiments where special measures are taken to resolve the hyperfine structure (Goldsmith & Rahn, 1988).

4.7.3 Selection Rules for *LS* Coupling

The analysis presented in Sections 4.7.1 and 4.7.2 allows the calculation of the electric dipole moment matrix elements for transitions $JM \rightarrow J'M'$ or $FM_F \rightarrow F'M'_F$ provided that the spontaneous mission rate coefficient $A_{J'J}$ is known. In many cases, however, we can obtain additional information concerning the ratios of spontaneous emission rate coefficients for transitions that share common integrals for the non-angular variables in the determination of the electric dipole moment matrix elements. For example, the relative strength of the transitions between the different multiplet terms in atoms with *LS* coupling can be calculated from the angular momentum coupling algebra developed in Sections 4.7.1–4.7.2. For such transitions, Eq. (4.179) can be written as

$$\langle \alpha LSJM|\mu_{1q}|\alpha' L'S'J'M'\rangle = (-1)^{J-M}\begin{pmatrix} J & 1 & J' \\ -M & q & M' \end{pmatrix}\langle \alpha LSJ\|\mu_1\|\alpha' L'S'J'\rangle. \quad (4.209)$$

Equation (4.206) is then used to develop the following expression for the reduced matrix element in Eq. (4.209):

$$\langle \alpha LSJ\|\boldsymbol{\mu}_1\|\alpha' L'S'J'\rangle = \delta_{SS'}(-1)^{L+S+J'+1}\langle \alpha L\|\boldsymbol{\mu}_1\|\alpha' L'\rangle\sqrt{(2J+1)(2J'+1)}\begin{Bmatrix} L & J & S \\ J' & L' & 1 \end{Bmatrix}. \quad (4.210)$$

Selection rules for *LS* radiative transitions can be developed based on the properties of the *6j* symbol in Eq. (4.210). For the general *6j* symbol $\begin{Bmatrix} j_1 & j_2 & j_3 \\ k_1 & k_2 & k_3 \end{Bmatrix}$, the value of the *6j* symbol will be nonzero only if all of the triangle conditions $\delta(j_1j_2j_3), \delta(j_1k_2k_3), \delta(k_1j_2k_3)$, and $\delta(k_1k_2j_3)$ are satisfied (Cowan, 1981). The triangle condition $\delta(j_1j_2j_3)$ is satisfied if

$$\begin{aligned} j_1 + j_2 &\geq j_3 \\ j_1 + j_3 &\geq j_2 \\ j_2 + j_3 &\geq j_1. \end{aligned} \quad (4.211)$$

Application of the triangle condition to the *6j* symbol in Eq. (4.210) leads to the selection rules

$$\begin{aligned} &L' - L = \pm 1 \quad \text{or } 0 \\ &L' = 0 \rightarrow L = 0 \quad \text{not allowed} \end{aligned} \quad (4.212)$$

in addition to the selection rule that $S = S'$.

4.7.4 Parity Selection Rule for Single-Photon Electric Dipole Transitions

For *LS* electric dipole transitions where only a single electron is involved, an additional more restrictive selection rule is that $L' - L = \pm 1$ (Cowan, 1981). This selection rule is consistent with the much more rigorous selection rule that the parity of the initial and final states must be different for allowed single-photon electric dipole transitions. As discussed in Chapter 2, the parity of a wavefunction $\psi(\boldsymbol{r})$ is even if $\psi(\boldsymbol{r}) = \psi(-\boldsymbol{r})$ and odd if $\psi(\boldsymbol{r}) = -\psi(-\boldsymbol{r})$. The single-photon electric dipole moment matrix element involves the integral

$$\langle \Psi(\boldsymbol{r}) | \boldsymbol{\mu}_1 | \Psi'(\boldsymbol{r}) \rangle = -\sum_i e \int\int\int_{-\infty}^{+\infty} \psi^*(\boldsymbol{r}_i) \boldsymbol{r}_i \psi'(\boldsymbol{r}_i) dx_i dy_i dz_i. \tag{4.213}$$

The integral for the ith electron in the configuration is given by

$$\langle \psi(\boldsymbol{r}) | \boldsymbol{\mu}_1 | \psi'(\boldsymbol{r}) \rangle = -e \int\int\int_{-\infty}^{+\infty} \psi^*(\boldsymbol{r}_i) \boldsymbol{r}_i \psi'(\boldsymbol{r}_i) dx_i dy_i dz_i. \tag{4.214}$$

Now consider the effect of replacing $\boldsymbol{r}$ by $-\boldsymbol{r}$ in Eq. (4.214). Since the integral is over all space, the value of the integral will be unchanged by replacing $\boldsymbol{r}$ by $-\boldsymbol{r}$. Therefore

$$\langle \psi(\boldsymbol{r}) | \boldsymbol{\mu}_1 | \psi'(\boldsymbol{r}) \rangle = -e \int\int\int_{+\infty}^{-\infty} \psi^*(-\boldsymbol{r}_i)(-\boldsymbol{r}_i)\psi'(-\boldsymbol{r}_i)(-dx_i)(-dy_i)(-dz_i). \tag{4.215}$$

The parity of the wavefunctions is given by p and p', such that $\psi(-\boldsymbol{r}_i) = (-1)^p \psi(\boldsymbol{r}_i)$ and $\psi'(-\boldsymbol{r}_i) = (-1)^{p'} \psi'(\boldsymbol{r}_i)$, where p and p' will be 1 for odd parity wavefunctions and 0 for even parity wavefunctions. Substituting into Eq. (4.215) and inverting the order of integration, we obtain

$$\begin{aligned} \langle \psi(\boldsymbol{r}) | \boldsymbol{\mu}_1 | \psi'(\boldsymbol{r}) \rangle &= (-1)^{p+1+p'} \left[-e \int\int\int_{-\infty}^{+\infty} \psi^*(\boldsymbol{r}_i) \boldsymbol{r}_i \psi'(\boldsymbol{r}_i) dx_i dy_i dz_i \right] \\ &= (-1)^{p+1+p'} \langle \psi(\boldsymbol{r}) | \boldsymbol{\mu}_1 | \psi'(\boldsymbol{r}) \rangle. \end{aligned} \tag{4.216}$$

This implies that unless $p \neq p'$, the value of the integral $\langle \psi(\boldsymbol{r}) | \boldsymbol{\mu}_1 | \psi'(\boldsymbol{r}) \rangle$ must be zero, and the single-photon electric dipole transition between $\psi(\boldsymbol{r})$ and $\psi'(\boldsymbol{r})$ is not allowed. If only a single electron is allowed in the transition, then the angular momentum quantum number l must change for that electron.

4.7.5 Intensity of Multiplet Terms for *LS* Coupling

The use of Eq. (4.210) to calculate relative strengths for the different J levels of a multiplet transition is now demonstrated for the ^3P–^3D transitions in atomic oxygen. The levels considered are depicted in Figure 4.1. The ground term is the ^3P multiplet with level energies of 0.000, 158.265, and 226.977 cm^{-1} for J values of 2, 1, and 0, respectively. The excited term is the ^3D multiplet with level energies of 97,488.378, 97,488.448, and 97,488.538 cm^{-1} for J values of 1, 2, and 3, respectively (Kramida et al., 2022). The relative strengths of the transitions between the different J levels are calculated from Eq. (4.210). For the $^3D_3 \rightarrow {}^3P_2$ transition, for example, the reduced density matrix element is given by

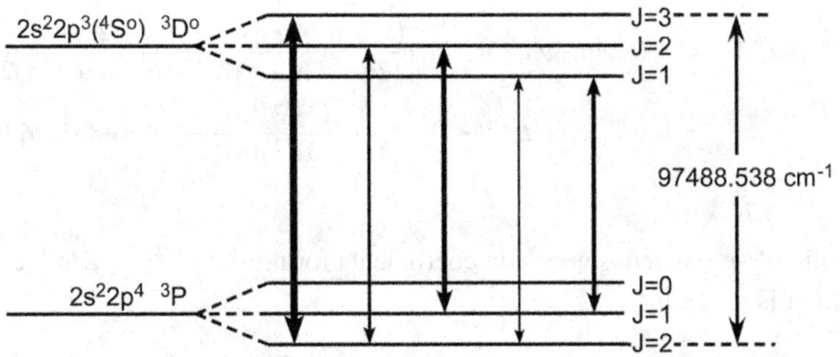

Figure 4.1 Energy level diagram for the $^3D^0 - {}^3P$ transitions in atomic oxygen. The energy levels splittings are not to scale. The width of the arrow for each transition are roughly proportional to the value of $A_{j'j}$.

$$\begin{aligned}
\langle\alpha LSJ\|\boldsymbol{\mu}_1\|\alpha' L'SJ'\rangle &= \langle\alpha 112\|\boldsymbol{\mu}_1\|\alpha' 213\rangle \\
&= (-1)^6\langle\alpha L\|\boldsymbol{\mu}_1\|\alpha' L'\rangle\sqrt{(5)(7)}\begin{Bmatrix} 1 & 2 & 1 \\ 3 & 2 & 1 \end{Bmatrix} \\
&= \langle\alpha L\|\boldsymbol{\mu}_1\|\alpha' L'\rangle\sqrt{35}\begin{Bmatrix} 3 & 2 & 1 \\ 1 & 2 & 1 \end{Bmatrix} \\
&= \langle\alpha L\|\boldsymbol{\mu}_1\|\alpha' L'\rangle\sqrt{35}\begin{Bmatrix} 3 & 1 & 2 \\ 1 & 1 & 2 \end{Bmatrix}.
\end{aligned} \tag{4.217}$$

The $6j$ symbol can be evaluated using Eq. (A4.8), with $a = 3$, $b = 1$, $c = 2$, and $s = 6$:

$$\begin{aligned}
\langle\alpha 112\|\boldsymbol{\mu}_1\|\alpha' 213\rangle &= \langle\alpha L\|\boldsymbol{\mu}_1\|\alpha' L'\rangle\sqrt{35} \\
&\quad\times(-1)^s\left[\frac{(s-2b-1)(s-2b)(s-2c+1)(s-2c+2)}{(2b+1)(2b+2)(2b+3)(2c-1)2c(2c+1)}\right]^{1/2} \\
&= \langle\alpha L\|\boldsymbol{\mu}_1\|\alpha' L'\rangle\sqrt{35}(-1)^6\left[\frac{(3)(4)(3)(4)}{(3)(4)(5)(3)(4)(5)}\right]^{1/2} \\
&= \langle\alpha L\|\boldsymbol{\mu}_1\|\alpha' L'\rangle\sqrt{\frac{7}{5}}.
\end{aligned} \tag{4.218}$$

For the $^3D_2 \to {}^3P_2$ transition, the reduced density matrix element is given by

$$\begin{aligned}
\langle\alpha LSJ\|\boldsymbol{\mu}_1\|\alpha' L'SJ'\rangle &= \langle\alpha 112\|\boldsymbol{\mu}_1\|\alpha' 212\rangle \\
&= (-1)^5\langle\alpha L\|\boldsymbol{\mu}_1\|\alpha' L'\rangle\sqrt{(5)(5)}\begin{Bmatrix} 1 & 2 & 1 \\ 2 & 2 & 1 \end{Bmatrix} \\
&= -\langle\alpha L\|\boldsymbol{\mu}_1\|\alpha' L'\rangle 5\begin{Bmatrix} 2 & 2 & 1 \\ 1 & 2 & 1 \end{Bmatrix} \\
&= -\langle\alpha L\|\boldsymbol{\mu}_1\|\alpha' L'\rangle 5\begin{Bmatrix} 2 & 1 & 2 \\ 1 & 1 & 2 \end{Bmatrix}.
\end{aligned} \tag{4.219}$$

The $6j$ symbol can again be evaluated using Eq. (A4.8), with $a = 2$, $b = 1$, $c = 2$, and $s = 5$:

$$\langle\alpha 112\|\boldsymbol{\mu}_1\|\alpha' 212\rangle = -\langle\alpha L\|\boldsymbol{\mu}_1\|\alpha' L'\rangle 5(-1)^s \left[\frac{(s-2b-1)(s-2b)(s-2c+1)(s-2c+2)}{(2b+1)(2b+2)(2b+3)(2c-1)2c(2c+1)}\right]^{1/2}$$
$$= -\langle\alpha L\|\boldsymbol{\mu}_1\|\alpha' L'\rangle 5(-1)^5 \left[\frac{(2)(3)(2)(3)}{(3)(4)(5)(3)(4)(5)}\right]^{1/2} = +\langle\alpha L\|\boldsymbol{\mu}_1\|\alpha' L'\rangle \frac{1}{2}. \tag{4.220}$$

The ratio of spontaneous emission coefficients for the ${}^3D_3 \rightarrow {}^3P_2$ and the ${}^3D_2 \rightarrow {}^3P_2$ transition is given by

$$\frac{A_{JJ'}\left({}^3D_3 \rightarrow {}^3P_2\right)}{A_{JJ'}\left({}^3D_2 \rightarrow {}^3P_2\right)} = \frac{\left\{\dfrac{|\langle\alpha 213\|\boldsymbol{\mu}_1\|\alpha' 112\rangle|^2}{[2(3)+1]}\right\}}{\left\{\dfrac{|\langle\alpha 212\|\boldsymbol{\mu}_1\|\alpha' 112\rangle|^2}{[2(2)+1]}\right\}} = \frac{\left[\dfrac{|\langle\alpha L\|\boldsymbol{\mu}_1\|\alpha' L'\rangle|^2\left(\sqrt{7/5}\right)^2}{7}\right]}{\left[\dfrac{|\langle\alpha L\|\boldsymbol{\mu}_1\|\alpha' L'\rangle|^2 (1/2)^2}{5}\right]} = 4. \tag{4.221}$$

From the NIST database, the ratio is

$$\frac{A_{JJ'}\left({}^3D_3 \rightarrow {}^3P_2\right)}{A_{JJ'}\left({}^3D_2 \rightarrow {}^3P_2\right)} = \frac{7.66 \times 10^7\ \mathrm{s}^{-1}}{1.91 \times 10^7\ \mathrm{s}^{-1}} = 4.01. \tag{4.222}$$

Given the values of $A_{J'J}$ for the transitions, the values of the reduced matrix elements $\langle\alpha L\|\boldsymbol{\mu}_1\|\alpha' L'\rangle$ and $\langle\alpha LSJ\|\boldsymbol{\mu}_1\|\alpha' L'SJ'\rangle$ can be calculated from Eqs. (4.202)–(4.204). For the ${}^3D_3 \rightarrow {}^3P_2$ transition,

$$\langle\alpha L'SJ'\|\boldsymbol{\mu}_1\|\alpha LSJ'-1\rangle = (-1)^{J'-J}\langle\alpha LSJ\|\boldsymbol{\mu}_1\|\alpha' L'S'J+1\rangle = +\sqrt{\frac{A_{J'J}(2J'+1)3\pi\varepsilon_0\hbar c^3}{\omega_{J'J}^3}}$$
$$= +\sqrt{\frac{\left(7.66\times 10^7\ s^{-1}\right)(7)3\pi\left(8.8542\times 10^{-12}\ \frac{\mathrm{C}^2}{\mathrm{Jm}}\right)\left(1.0546\times 10^{-34}\ Js\right)\left(2.9979\times 10^8\ \frac{\mathrm{m}}{\mathrm{s}}\right)^3}{\left[2\pi\left(2.9979\times 10^{10}\ \frac{\mathrm{cm}}{\mathrm{s}}\right)\left(97{,}488.538\ \mathrm{cm}^{-1}\right)\right]^3}}$$
$$= +4.53\times 10^{-30}\ \mathrm{Cm}. \tag{4.223}$$

We can then calculate $\langle\alpha L\|\boldsymbol{\mu}_1\|\alpha' L'\rangle$ from Eq. (4.218),

$$\langle\alpha L\|\boldsymbol{\mu}_1\|\alpha' L'\rangle = \sqrt{\frac{5}{7}}\langle\alpha 112\|\boldsymbol{\mu}_1\|\alpha' 213\rangle = 3.83\times 10^{-30}\ \mathrm{Cm}. \tag{4.224}$$

Calculation of the values of the reduced density matrix elements $\langle\alpha LSJ\|\boldsymbol{\mu}_1\|\alpha' L'SJ'\rangle$ from Eq. (4.210) and the spontaneous emission coefficients from

$$A_{J'J} = \frac{\langle\alpha L'SJ'\|\boldsymbol{\mu}_1\|\alpha LSJ\rangle^2\omega_{J'J}^3}{(2J'+1)3\pi\varepsilon_0\hbar c^3} \tag{4.225}$$

Table 4.1 Values of the reduced matrix elements $\langle \alpha LSJ \| \mu_1 \| \alpha' L'SJ' \rangle$ and spontaneous emission coefficients

Transition	$\tilde{\nu}_{J'J}$ (cm^{-1})	$\begin{Bmatrix} L & J & S \\ J' & L' & 1 \end{Bmatrix}$	$\langle \alpha LSJ \| \mu_1 \| \alpha' L'SJ' \rangle$ (Cm)	$A_{J'J}$ (s^{-1}) Eq. (4.225)	$A_{J'J}$ (s^{-1}) NIST
$^3D_3 \to {}^3P_2$	97,488.538	0.20000	4.53×10^{-30}	7.66×10^7	7.66×10^7
$^3D_2 \to {}^3P_2$	97,488.448	−0.10000	1.92×10^{-30}	1.92×10^7	1.91×10^7
$^3D_2 \to {}^3P_1$	97,330.183	−0.22361	3.32×10^{-30}	5.72×10^7	5.71×10^7
$^3D_1 \to {}^3P_2$	97,488.378	0.03333	4.94×10^{-31}	2.13×10^6	2.11×10^6
$^3D_1 \to {}^3P_1$	97,330.113	0.16667	1.92×10^{-30}	3.18×10^7	3.17×10^7
$^3D_1 \to {}^3P_0$	97,261.401	0.33333	2.21×10^{-30}	4.23×10^7	4.22×10^7

for the $^3D_2 \to {}^3P_2$, $^3D_2 \to {}^3P_1$, $^3D_1 \to {}^3P_2$, $^3D_1 \to {}^3P_1$, and $^3D_1 \to {}^3P_0$ transitions is left as an exercise for the reader. The results of the calculations and the values of the spontaneous emission coefficients for these transitions from the NIST tables are shown in Table 4.1 (Kramida, 2022).

4.7.6 Deviations from the Selection Rules for *LS* Coupling

For light atoms such as atomic oxygen, the *LS* coupling model is accurate in describing the energy level structure, selection rules, and radiative transition strengths except for electron configurations with highly excited electrons, where pair coupling occurs. For heavier atoms and for some excited electron configurations, the *LS* coupling model may not apply, or the wavefunction for some levels may be a combination of two or more *LS* coupling basis states. In atomic carbon, for example, numerous levels with energies above 70,000 cm^{-1} are listed in Kramida et al. (2022) as combinations of different *LS* coupling basis states and in most cases different electron configurations. The $J = 0$ level at 75,254.0017 cm^{-1} is listed as 93% $^3P^o_0\ 2s2p^3$ and 7% $^3P^o_0\ 2s^22p3d$. The $J = 2$ level at 78,199.09144 cm^{-1} is listed as 86% $^3F^o_2\ 2s^22p3d$ and 13% $^3D^o_2\ 2s^22p3d$.

The mixed configurations in heavier atoms can lead to some seemingly strange violations of selection rules. For example, a very well-known transition in atomic spectroscopy is the $^1S_0 - {}^3P^o_1$ transition at 253.652 nm in atomic mercury, for which $A = 8.4 \times 10^6\ s^{-1}$. In this case the wavefunction for the $^3P^o_1$ level also contains significant contributions from a $^1P^o_1$ *LS* coupling basis function, and this contribution from the $^1P^o_1$ basis function accounts for the strength of the transition.

5 Quantum Mechanical Analysis of Single-Photon Electric Dipole Resonances for Diatomic Molecules

5.1 Introduction

Evaluation of the electric dipole moment matrix elements for diatomic molecules is complicated by the different energy storage modes of the molecule. The electric dipole moment matrix element for diatomic molecules is given by

$$\begin{aligned}\boldsymbol{\mu}_{ba} &= \langle\phi_b|\hat{\boldsymbol{\mu}}|\phi_a\rangle = e\langle\phi_b|\sum_j\left(Z_j\boldsymbol{R}_{nj} - \boldsymbol{R}_{CM}\right) - \sum_i\left(\boldsymbol{R}_{ei} - \boldsymbol{R}_{CM}\right)|\phi_a\rangle \\ &= e\langle\eta_B S_B \mathrm{v}_B R_b|\sum_j Z_j\boldsymbol{r}_{nj} - \sum_i \boldsymbol{r}_{ei}|\eta_A S_A \mathrm{v}_A R_a\rangle,\end{aligned} \tag{5.1}$$

where $\boldsymbol{R}_{nj}$ and $\boldsymbol{R}_{ei}$ are the coordinates of the jth nucleus and ith electron, respectively, in the laboratory frame; $\boldsymbol{r}_{nj}$ and $\boldsymbol{r}_{ei}$ are the coordinates of the jth nucleus and ith electron, respectively, referenced to the center of mass; Z_j is the atomic number of the jth nucleus; and $\boldsymbol{R}_{CM}$ is the center of mass of the molecule. The symbols η_B, v_B, and R_b represent the electronic, vibrational, and rotational parts of the wavefunction for state b; and η_A, v_A, and R_a represent the electronic, vibrational, and rotational parts of the wavefunction for state a. The quantum states b and a are included in the energy levels B and A, respectively, and we will assume that $\varepsilon_b = \varepsilon_B$ for all degenerate states b and $\varepsilon_a = \varepsilon_A$ for all degenerate states a. We will also assume that $\varepsilon_B > \varepsilon_A$. The total spins S_B and S_A of the electron clouds for states b and a are denoted separately because the spin is not affected by the spatial operators, and the dipole moment will be zero unless $S_B = S_A$.

The evaluation of the electric dipole moment matrix elements is more complicated than for atomic species because of the rotation of the molecule around its center of mass and because of the vibration of the molecule about the equilibrium internuclear spacing. The properties of the electric dipole moment of the molecule are determined primarily by the characteristics of the electron cloud that binds the two nuclei together, and these properties can be understood by considering a reference frame fixed to the molecule. However, in the laboratory frame the molecule is rotating; therefore, to calculate the interactions of the molecule with a laser, for example, the response of the molecule must be averaged over all possible orientations of the molecule. Using irreducible spherical tensors greatly simplifies the orientation averaging of the molecular response.

The Born–Oppenheimer approximation will be invoked to initially account for the effect of the electronic, vibrational, and rotational modes of the molecule. Then corrections will be applied to account for the coupling and interactions of the different modes. In a space-fixed reference frame, the pth component of the electric dipole moment matrix element is given by (Brown & Carrington, 2003)

$$\begin{aligned}\mu_{ba,p} &= \langle \eta_B S_B \mathrm{v}_B R_b | \hat{\mu}_p | \eta_A S_A \mathrm{v}_A R_a \rangle = \langle \eta_B S_B \mathrm{v}_B R_b | \sum_q \hat{\mu}_{1q} D^{1*}_{pq}(\omega) | \eta_A S_A \mathrm{v}_A R_a \rangle \\ &= \sum_q \langle \eta_B \mathrm{v}_B | \hat{\mu}_{1q} | \eta_A \mathrm{v}_A \rangle \langle R_b | D^{1*}_{pq}(\omega) | R_a \rangle \delta_{S_B S_A}.\end{aligned} \tag{5.2}$$

In Eq. (5.2), averaging over all possible molecular orientations is accomplished by summing over the $2k+1$ components of the product of the molecule-fixed electric dipole moment operator $\hat{\mu}_{1q}$ and the rotation operator $D^{1*}_{pq}(\omega)$. The rotation operator $D^{1*}_{pq}(\omega)$ rotates the molecule-fixed spatial component q into the space-fixed component p and depends only on the rotational part of the wavefunction. The molecule-fixed dipole moment $\hat{\mu}_{1q}$ does not depend on the rotational part of the wavefunction, except indirectly due to the effects of centrifugal distortion, as will be discussed in detail later.

Evaluation of Eq. (5.2) will depend on whether the transitions are pure rotational, where initial and final rotational levels are different but the initial and final vibrational and electronic levels are the same; vibrational, where the initial and final vibrational levels are different but the initial and final electronic levels are the same; or electronic, where the initial and final electronic levels are different. The pure rotational and vibrational levels will be considered first.

5.1.1 Pure Rotational and Vibrational Transitions

Transitions within the same electronic level $\eta_a = \eta_b = \eta$ will be considered in this section. The first step in the evaluation is the integration of the electronic part of the wave function for fixed internuclear separation, as given by

$$R_{e,q}(r) = e\int_0^\infty \psi_e(r, r_{e,qi}) \left\{ \sum_j Z_j r_{n,qj} - \sum_i r_{e,qi} \right\} \psi_e(r, r_{e,qi}) d\tau. \tag{5.3}$$

In Eq. (5.3), r (m) is the internuclear distance and $R_{e,q}(r)$ is the qth component of the electronic transition moment (C-m), $\psi_e(r, r_{e,qi})$ is the electronic wavefunction (usually the ground electronic wavefunction), Z_j and $r_{n,qj}$ (m) are the atomic number and coordinate for the jth nucleus, $r_{e,qi}$ (m) is the coordinate for the ith electron, and the integration is performed over all space. The coordinates $r_{n,qj}$ and $r_{e,qi}$ are referenced to the center of mass of the molecule in a molecule-fixed reference frame.

Note that the integration in Eq. (5.3) is performed for a fixed value of the internuclear distance r. Due to the symmetry of the electron cloud about the internuclear axis, the only nonzero component of the electronic transition moment will be the $q = 0$ component; this will be the case for vibrational transitions as well. For pure rotational and vibrational transitions, we will use the notation $\hat{\mu}_{1q=0} = R_{e,0}(r)$. The dipole moment operator $\hat{\mu}_{1q=0}$ is a function of the internuclear separation, and a Taylor's series expansion about the equilibrium internuclear separation $Q = r - r_e$ can be performed (Brown & Carrington, 2003),

$$\hat{\mu}_{1q=0} = (\mu_{q=0})_0 + \left(\frac{\partial \mu_{q=0}}{\partial r}\right) Q + \frac{1}{2!}\left(\frac{\partial^2 \mu_{q=0}}{\partial r^2}\right) Q^2 + \frac{1}{3!}\left(\frac{\partial^3 \mu_{q=0}}{\partial r^3}\right) Q^3 + \cdots. \tag{5.4}$$

Note that in Eq. (5.4) the superscript indicating the tensor rank $k = 1$ of the dipole moment operator has been omitted; henceforth we will in general omit the superscript for the dipole moment operator unless needed for clarity. For pure rotational transitions, $\mathrm{v}_A = \mathrm{v}_B = \mathrm{v}$, and the molecule-fixed electric dipole moment component $\langle \eta \mathrm{v} | \hat{\mu}_{1q} | \eta \mathrm{v} \rangle$ must be evaluated. For pure rotational transitions, considering only the first two terms in Eq. (5.4), we obtain

$$\langle \eta \mathrm{v} | \hat{\mu}_{1q} | \eta \mathrm{v} \rangle = \langle \mathrm{v} | \hat{\mu}_{1q=0} | \mathrm{v} \rangle = (\mu_{q=0})_0 \langle \mathrm{v} | \mathrm{v} \rangle + \left(\frac{\partial \mu_{q=0}}{\partial r}\right) \langle \mathrm{v} | Q | \mathrm{v} \rangle = (\mu_{q=0})_0 \tag{5.5}$$

because $\langle \mathrm{v} | Q | \mathrm{v} \rangle = 0$. For vibrational transitions, where $\mathrm{v}_A \neq \mathrm{v}_B$, the molecule-fixed electric dipole moment component $\langle \eta \mathrm{v}_B | \hat{\mu}_{1q} | \eta \mathrm{v}_A \rangle$ must be evaluated. Again considering only the first two terms, and assuming that $\mathrm{v}_B = \mathrm{v}_A + 1$, we obtain

$$\begin{aligned} \langle \eta \mathrm{v}_B | \hat{\mu}_{1q} | \eta \mathrm{v}_A \rangle &= \langle \mathrm{v}_A + 1 | \hat{\mu}_{1q=0} | \mathrm{v}_A \rangle = (\mu_{q=0})_0 \langle \mathrm{v}_A + 1 | \mathrm{v}_A \rangle + \left(\frac{\partial \mu_{q=0}}{\partial r}\right) \langle \mathrm{v}_A + 1 | Q | \mathrm{v}_A \rangle \\ &= \left(\frac{\partial \mu_{q=0}}{\partial r}\right) \langle \mathrm{v}_A + 1 | Q | \mathrm{v}_A \rangle. \end{aligned} \tag{5.6}$$

The effects of anharmonicity on the intensity of vibrational overtone transitions are discussed in more detail by Brown and Carrintgton (2003).

At this point we will turn our attention to the part of the electric dipole moment matrix element that involves the integration of the rotational wavefunctions and the rotation operator. The pth component of electric dipole moment matrix element is given by

$$\begin{aligned} \mu_{ba,p} &= \langle \eta \mathrm{v} R_b | \hat{\mu}_p | \eta \mathrm{v} R_a \rangle = \langle \eta \mathrm{v} | \hat{\mu}_q | \eta \mathrm{v} \rangle \langle R_b | D^{1*}_{p0}(\omega) | R_a \rangle \\ &= (\mu_{q=0})_0 \langle R_b | D^{1*}_{p0}(\omega) | R_a \rangle \end{aligned} \tag{5.7}$$

for pure rotational transitions and

$$\begin{aligned} \mu_{ba,p} &= \langle \eta \mathrm{v}_B R_b | \hat{\mu}_p | \eta \mathrm{v}_A R_a \rangle = \langle \eta \mathrm{v}_B | \hat{\mu}_q | \eta_A \mathrm{v} \rangle \langle R_b | D^{1*}_{p0}(\omega) | R_a \rangle \\ &= \left(\frac{\partial \hat{\mu}_{q=0}}{\partial r}\right) \langle \mathrm{v}_A + 1 | Q | \mathrm{v}_A \rangle \langle R_b | D^{1*}_{p0}(\omega) | R_a \rangle \end{aligned} \tag{5.8}$$

for vibrational transitions.

For a Hund's case (a) molecule, the quantum numbers Λ, Σ, and Ω are well defined, and the wavefunction for a case (a) molecule is given by (Brown & Carrington, 2003)

$$|J\Omega M\rangle = \sqrt{(2J+1)/8\pi^2} D^{J*}_{M\Omega}(\omega). \tag{5.9}$$

The general form of the second term on the RHS of Eq. (5.7) can be written

$$\begin{aligned} \langle R_b|D^{k*}_{pq}(\omega)|R_a\rangle &= \langle J_B\Omega_B M_b|D^{k*}_{pq}(\omega)|J_A\Omega_A M_a\rangle \\ &= \frac{\sqrt{(2J_B+1)(2J_A+1)}}{8\pi^2}\int_\omega D^{J_B}_{M_b\Omega_B}(\omega)D^{k*}_{pq}(\omega)D^{J_A*}_{M_a\Omega_A}(\omega)d\omega. \end{aligned} \tag{5.10}$$

But we know that

$$D^{k*}_{pq}(\omega) = (-1)^{p-q}D^k_{-p,-q}(\omega) \tag{5.11}$$

and

$$D^{J_A*}_{M_a\Omega_A}(\omega) = (-1)^{M_a-\Omega_A}D^{J_A}_{-M_a,-\Omega_A}(\omega). \tag{5.12}$$

Rewriting the integral in Eq. (5.10) using Eqs. (5.11) and (5.12), we obtain

$$\begin{aligned} &\int_\omega D^{J_B}_{M_b\Omega_B}(\omega)D^{k*}_{p0}(\omega)D^{J_A*}_{M_a\Omega_A}(\omega)d\omega \\ &= (-1)^{p-q}(-1)^{M_a-\Omega_A}\int_\omega D^{J_B}_{M_b\,\Omega_B}(\omega)D^k_{-p,-q}(\omega)D^{J_A}_{-M_a,-\Omega_A}(\omega)d\omega. \end{aligned} \tag{5.13}$$

The integral term can now be evaluated using

$$\begin{aligned} &\int_\omega D^{J_B}_{M_b\,\Omega_B}(\omega)D^k_{-p,-q}(\omega)\,D^{J_A}_{-M_a,-\Omega_A}(\omega)d\omega \\ &= 8\pi^2\begin{pmatrix} J_B & k & J_A \\ M_b & -p & -M_a \end{pmatrix}\begin{pmatrix} J_B & k & J_A \\ \Omega_B & -q & -\Omega_A \end{pmatrix} \\ &= 8\pi^2(-1)^{2(J_B+k+J_A)}\begin{pmatrix} J_B & k & J_A \\ -M_b & p & M_a \end{pmatrix}\begin{pmatrix} J_B & k & J_A \\ -\Omega_B & q & \Omega_A \end{pmatrix} \\ &= 8\pi^2\begin{pmatrix} J_B & k & J_A \\ -M_b & p & M_a \end{pmatrix}\begin{pmatrix} J_B & k & J_A \\ -\Omega_B & q & \Omega_A \end{pmatrix}. \end{aligned} \tag{5.14}$$

The $3j$ symbols are subject to the restriction that they are nonzero only when $|J_B - J_A| \leq k$. Substituting Eqs. (5.13) and (5.14) into Eq. (5.10) we obtain

$$\begin{aligned} &\langle J_B\Omega_B M_b|D^{k*}_{pq}(\omega)|J_A\Omega_A M_a\rangle \\ &= (-1)^{p-q}(-1)^{M_a-\Omega_A}\sqrt{(2J_B+1)(2J_A+1)}\begin{pmatrix} J_B & k & J_A \\ -M_b & p & M_a \end{pmatrix}\begin{pmatrix} J_B & k & J_A \\ -\Omega_B & q & \Omega_A \end{pmatrix} \\ &= (-1)^{M_b-\Omega_B}\sqrt{(2J_B+1)(2J_A+1)}\begin{pmatrix} J_A & J_B & k \\ M_a & -M_b & p \end{pmatrix}\begin{pmatrix} J_A & J_B & k \\ \Omega_A & -\Omega_B & q \end{pmatrix}. \end{aligned} \tag{5.15}$$

The second term on the RHS of Eq. (5.8) is thus given by

$$\langle R_b|D^{1*}_{p0}(\omega)|R_a\rangle = \langle J_B\Omega_B M_b|D^{1*}_{p0}(\omega)|J_A\Omega_A M_a\rangle$$
$$= (-1)^{M_b-\Omega_B}\sqrt{(2J_B+1)(2J_A+1)}\begin{pmatrix} J_A & J_B & k \\ M_a & -M_b & p \end{pmatrix}\begin{pmatrix} J_A & J_B & k \\ \Omega_A & -\Omega_B & 0 \end{pmatrix}. \tag{5.16}$$

The *3j* symbol $\begin{pmatrix} J_B & 1 & J_A \\ -\Omega_B & 0 & \Omega_A \end{pmatrix}$ will be zero unless $\Omega_B = \Omega_A$, resulting in the selection rule $\Delta\Omega = 0$ for case (a) molecules. When $\Omega = \Omega_B = \Omega_A = 0$, the same *3j* symbol will be zero unless $J_B = J_A \pm 1$; $\Delta J = 0$ transitions are not allowed for $\Omega = 0$. The first *3j* symbol will be zero unless $p + M_a = M_b$. The final expression for Hund's case (a) molecules is thus given by

$$\mu_{ba,p} = \langle\eta\mathrm{v}|\hat{\mu}_{q=0}|\eta\mathrm{v}\rangle(-1)^{M_b-\Omega}\sqrt{(2J_B+1)(2J_A+1)} \times \begin{pmatrix} J_B & 1 & J_A \\ -M_b & p & M_a \end{pmatrix}\begin{pmatrix} J_B & 1 & J_A \\ -\Omega & 0 & \Omega \end{pmatrix}. \tag{5.17}$$

The expression given in Eq. (5.17) is not rigorously correct, however, because the Hund's case (a) wavefunction $|\eta\Lambda^s;\mathrm{v};S\Sigma;J\Omega M\rangle$ does not have definite parity, as discussed in Chapter 3. In addition, for many light molecules of importance in combustion and in plasmas, the molecular electronic levels are either Hund's case (b), in the case of Σ levels with $\Lambda = 0$, or intermediate between Hund's case (a) and (b). In both cases, a basis set of Hund's case (a) wavefunctions $|\eta\Lambda^s;\mathrm{v};S\Sigma;J\Omega M\rangle$ can be used to analyze molecular structure, as shown in Chapter 3, or to calculate electric dipole moment matrix elements. The calculation of the electric dipole moment matrix elements for transitions between states that reside in different electronic levels will be discussed in detail in this chapter. Pure rotational and vibrational transitions can be regarded as degenerate categories of these electronic transitions, and detailed analysis of pure rotational and vibrational transitions will be deferred at this point.

5.1.2 Electronic Transitions

The evaluation of the molecule-fixed electric dipole moment matrix element $\langle\eta_b\mathrm{v}_b|\hat{\mu}_{1q}|\eta_a\mathrm{v}_a\rangle$, or transition dipole moment (TDM) for short, in Eq. (5.2) is more complicated for transitions coupling states in different electronic levels. The electronic wave functions will depend on the internuclear separation, and therefore the electronic transition dipole moment function must be calculated for different fixed values of the internuclear separation, as shown in the following equation:

$$R_{e,q}(r) = e\int_0^\infty \psi'_e(r,r_{e,qi})\left\{\sum_j Z_j r_{n,qj} - \sum_i r_{e,qi}\right\}\psi''_e(r,r_{e,qi})d\tau. \tag{5.18}$$

In Eq. (5.18), $\psi_e{}'(r,r_{e,qi})$ and $\psi''_e(r,r_{e,qi})$ are the electronic wavefunctions for the excited and ground electronic levels, respectively, Z_j and $r_{n,qj}$ (m) are the atomic number and coordinate for the jth nucleus, $r_{e,qi}$(m) is the coordinate for the ith electron, and the integration is performed over all space. The coordinates $r_{n,qj}$ and $r_{e,qi}$ are referenced to the center of mass of the molecule in a molecule-fixed reference frame. The irreducible spherical coordinates for $q = 0, +1, \text{and} -1$ are given by

$$r_{e,0i} = z_{e,i}. \tag{5.19}$$

$$r_{e,+1i} = \frac{1}{\sqrt{2}}\left(x_{e,i} + iy_{e,i}\right). \tag{5.20}$$

$$r_{e,-1i} = \frac{1}{\sqrt{2}}\left(x_{e,i} - iy_{e,i}\right). \tag{5.21}$$

The electronic TDM can be calculated using *ab initio* electronic wavefunctions, but it is usually assumed to have a particular functional dependence on the internuclear separation r. The electronic transition moment can also be determined from ratios of band intensities for different vibrational transitions provided that the vibration-rotation wavefunctions are known. The electric dipole moment does not necessarily lie along the internuclear axis for electronic transitions in diatomic molecules; therefore, $\hat{\mu}_{q=0}$, $\hat{\mu}_{q=+1}$, and $\hat{\mu}_{q=-1}$ can all contribute to the electric dipole moment matrix elements.

The evaluation of the integral $\langle \eta_B \mathrm{v}_B | \hat{\mu}_{1q} | \eta_A \mathrm{v}_A \rangle$, also known as the band strength or band intensity, is then given by (Bernath, 2016; Luque & Crosley, 1998)

$$\tilde{\mu}_{\mathrm{v}'\mathrm{v}''} = \left| \int_0^\infty \psi_{\mathrm{v}'}(r) \boldsymbol{R}_e(r) \psi_{\mathrm{v}''}(r) dr \right|^2 = \sum_q \left| \int_0^\infty \psi_{\mathrm{v}'}(r) R_{e,q}(r) \psi_{\mathrm{v}''}(r) dr \right|^2. \tag{5.22}$$

However, as will be discussed in detail in Section 5.4, the vibrational wavefunction will be altered by centrifugal distortion for high values of the angular momentum quantum number J. Consequently, Eq. (5.22) can be rewritten as

$$\tilde{\mu}_{\mathrm{v}'\mathrm{v}''}^{J'J''} = \left| \int_0^\infty \psi_{\mathrm{v}'J'}(r) \boldsymbol{R}_e(r) \psi_{\mathrm{v}''J''}(r) dr \right|^2. \tag{5.23}$$

The spontaneous emission coefficient $A_{J'J''}$ (s^{-1}) is given by (Bernath, 2016; Luque & Crosley, 1996a, 1996b, 1998, 1999b; Noda & Zare, 1982)

$$A_{J'J''} = \frac{16\pi^3 \nu_{J'J''}^3 \tilde{\mu}_{\mathrm{v}'\mathrm{v}''}^{J'J''} S_{J'}^{\Delta J}}{3h\varepsilon_0 c^3 (2J'+1)} = \frac{16\pi^3 \tilde{\nu}_{J'J''}^3 \tilde{\mu}_{\mathrm{v}'\mathrm{v}''}^{J'J''} S_{J'}^{\Delta J}}{3h\varepsilon_0 (2J'+1)} = \frac{2\omega_{J'J''}^3 \tilde{\mu}_{\mathrm{v}'\mathrm{v}''}^{J'J''} S_{J'}^{\Delta J}}{3h\varepsilon_0 c^3 (2J'+1)}, \tag{5.24}$$

where $S_{J'}^{\Delta J}$ is the Hönl–London factor, also commonly termed the rotational line strength for emission, $\nu_{J'J''}$ and $\tilde{\nu}_{J'J''}$ are the transition frequencies in Hz and cm^{-1}, respectively, and $\omega_{J'J''}$ is the angular frequency of the transition in s^{-1}. The dimensionless oscillator strength for absorption is given by

$$f_{J''J'} = \frac{8\pi^2 m_e \nu_{J'J''} p_{\mathrm{v}''J''}^{\mathrm{v}'J'} S_{J''}^{\Delta J}}{3he^2(2J''+1)} = \frac{8\pi^2 m_e \tilde{\nu}_{J'J''} p_{\mathrm{v}''J''}^{\mathrm{v}'J'} S_{J''}^{\Delta J}}{3he^2 c(2J''+1)} = \frac{4\pi m_e \omega_{J'J''} p_{\mathrm{v}''J''}^{\mathrm{v}'J'} S_{J''}^{\Delta J}}{3he^2(2J''+1)}. \tag{5.25}$$

The Hönl–London factors obey the following sum rules (Bernath, 2016):

$$\sum_{J''} S_{J''}^{\Delta J} = (2 - \delta_{0,\Lambda'}\delta_{0,\Lambda''})(2S+1)(2J''+1). \tag{5.26}$$

$$\sum_{J'} S_{J'}^{\Delta J} = (2 - \delta_{0,\Lambda'}\delta_{0,\Lambda''})(2S+1)(2J'+1). \tag{5.27}$$

In the limit where the vibration rotation interaction is negligible, the spontaneous emission coefficient for a rotational transition can be written as (Bernath, 2016; Nicholls & Stewart, 1962)

$$A_{J'J''} = \frac{16\pi^3 \tilde{\nu}_{J'J''}^3 |\boldsymbol{R}_e(\bar{r})|^2 q_{\mathrm{v}'\mathrm{v}''} S_{J'}^{\Delta J}}{3h\varepsilon_0(2J'+1)}, \tag{5.28}$$

where the $|\boldsymbol{R}_e(\bar{r})|^2$ is defined by

$$|\boldsymbol{R}_e(\bar{r})|^2 = \frac{\tilde{\mu}_{\mathrm{v}'\mathrm{v}''}}{q_{\mathrm{v}'\mathrm{v}''}} = \frac{\left|\int_0^\infty \psi_{\mathrm{v}'}(r)\boldsymbol{R}_e(r)\,\psi_{\mathrm{v}''}(r)dr\right|^2}{\left|\int_0^\infty \psi_{\mathrm{v}'}(r)\,\psi_{\mathrm{v}''}(r)dr\right|^2} \tag{5.29}$$

and the Franck–Condon factor $q_{\mathrm{v}'\mathrm{v}''}$ is given by

$$q_{\mathrm{v}'\mathrm{v}''} = \left|\int \psi_{\mathrm{v}'}\psi_{\mathrm{v}''}\,dr\right|^2. \tag{5.30}$$

The Franck–Condon factors obey the sum rules (Bernath, 2016)

$$\sum_{\mathrm{v}'} q_{\mathrm{v}'\mathrm{v}''} = \sum_{\mathrm{v}''} q_{\mathrm{v}'\mathrm{v}''} = 1. \tag{5.31}$$

The band spontaneous emission coefficient is given by (Luque & Crosley, 1998)

$$A_{\mathrm{v}'\mathrm{v}''} = \frac{64\pi^4 \tilde{\nu}_{\mathrm{v}'\mathrm{v}''}^3}{3h} \cdot \frac{(2-\delta_{0,\Lambda''+\Lambda'})}{(2-\delta_{0,\Lambda'})}\tilde{\mu}_{\mathrm{v}'\mathrm{v}''}. \tag{5.32}$$

5.2 Singlet Electronic Transitions

5.2.1 $^1\Sigma$–$^1\Sigma$ Transitions

Electronic transitions between singlet levels ($S = 0$) will be considered first. The first transitions that will be discussed are transitions between rotational levels contained within ground and excited $^1\Sigma^\pm$ electronic levels. The wavefunctions for states a in level A and states b in level B are given by

$$\begin{aligned}|^1\Sigma^\pm;0^s;\mathrm{v}_A;J_AM_a;\pm\rangle = \frac{1}{2}\big[&|\Lambda_A = 0^\pm; S_A = 0\Sigma_A = 0; \mathrm{v}_A; J_A\Omega_A = 0M_a\rangle \\ &\pm(-1)^{J_A+s}|0^\pm;00;\,\mathrm{v}_A;J_A0M_a\rangle\big].\end{aligned} \tag{5.33}$$

$$\left|{}^1\Sigma^\pm;0^s;\mathrm{v}_B;J_BM_b;\pm\right\rangle=\frac{1}{2}\left[\left|0^\pm;00;\mathrm{v}_B;J_B0M_B\right\rangle\pm(-1)^{J_B+s}\left|0^\pm;00;\,\mathrm{v}_B;J_B0M_b\right\rangle\right].\tag{5.34}$$

For ${}^1\Sigma^+$ electronic levels, $s=0$ and the positive- and negative-parity states will have even and odd J, respectively, whereas for ${}^1\Sigma^-$ electronic levels, $s=1$ and the positive- and negative-parity states will have odd and even J, respectively. From Eq. (5.2), the dipole moment for transitions between a ${}^1\Sigma^\pm$ lower state and a ${}^1\Sigma^\pm$ upper state will be given by

$$\begin{aligned}
\mu_{ba,p}&=\left\langle{}^1\Sigma^\pm;\mathrm{v}_B;J_BM_b;\pm\right|\sum_q\hat{\mu}_qD^{1*}_{pq}(\omega)\left|{}^1\Sigma^\pm;0^\pm;\mathrm{v}_A;J_AM_a;\pm\right\rangle\\
&=\frac{1}{4}\Bigg\{\left\langle0^\pm;00;\mathrm{v}';J'0M'\right|\sum_q\hat{\mu}_qD^{1*}_{pq}(\omega)\left|0^\pm;00;\mathrm{v};J0M\right\rangle\\
&\quad(\pm)(-1)^{J+s}\left\langle0^\pm;00;\mathrm{v}';J'0M'\right|\sum_q\hat{\mu}_qD^{1*}_{pq}(\omega)\left|0^\pm;00;\mathrm{v};J0M\right\rangle\\
&\quad(\pm)'(-1)^{J'+s'}\left\langle0^\pm;00;\mathrm{v}';J'0M'\right|\sum_q\hat{\mu}_qD^{1*}_{pq}(\omega)\left|0^\pm;00;\mathrm{v};J0M\right\rangle\\
&\quad(\pm)(\pm)'(-1)^{J+s+J'+s'}\left\langle0^\pm;00;\mathrm{v}';J'0M'\right|\sum_q\hat{\mu}_qD^{1*}_{pq}(\omega)\left|0^\pm;\mathrm{v};00;J0M\right\rangle\Bigg\}\\
&=\frac{1}{4}\sum_q\Big\{\left\langle0^\pm;00;\mathrm{v}';\right|\hat{\mu}_q\left|0^\pm;00;\mathrm{v}\right\rangle\left\langle J'0M'\right|D^{1*}_{pq}(\omega)\left|J0M\right\rangle\\
&\quad\pm(-1)^{J+s}\left\langle0^\pm;00;\mathrm{v}';\right|\hat{\mu}_q\left|0^\pm;00;\mathrm{v}\right\rangle\left\langle J'0M'\right|D^{1*}_{pq}(\omega)\left|J0M\right\rangle\\
&\quad\pm(-1)^{J'+s'}\left\langle0^\pm;00;\mathrm{v}';\right|\hat{\mu}_q\left|0^\pm;00;\mathrm{v}\right\rangle\left\langle J'0M'\right|D^{1*}_{pq}(\omega)\left|J0M\right\rangle\\
&\quad(\pm)(\pm)'(-1)^{J+s+J'+s'}\left\langle0^\pm;00;\mathrm{v}';\right|\hat{\mu}_q\left|0^\pm;00;\mathrm{v}\right\rangle\left\langle J'0M'\right|D^{1*}_{pq}(\omega)\left|J0M\right\rangle\Big\}.
\end{aligned}\tag{5.35}$$

In Eq. (5.35) we have made the substitutions $J=J_A, M=M_a, \mathrm{v}=\mathrm{v}_A, J'=J_B$, $M'=M_b$, and $\mathrm{v}'=\mathrm{v}_B$. Simplifying Eq. (5.35) and using the relation (Brown & Carrington, 2003, p. 167)

$$\begin{aligned}
&\left\langle J'\Omega'M'\right|D^{1*}_{pq}(\omega)\left|J\Omega M\right\rangle\\
&=(-1)^{M'-\Omega'}\sqrt{(2J'+1)(2J+1)}\begin{pmatrix}J'&1&J\\-M'&p&M\end{pmatrix}\begin{pmatrix}J'&1&J\\-\Omega'&q&\Omega\end{pmatrix}\\
&=(-1)^{M'-\Omega'}\sqrt{(2J'+1)(2J+1)}\begin{pmatrix}J&J'&1\\M&-M'&p\end{pmatrix}\begin{pmatrix}J&J'&1\\\Omega&-\Omega'&q\end{pmatrix},
\end{aligned}\tag{5.36}$$

we obtain

$$\begin{aligned}\mu_{ba,p} &= \frac{\tilde{\mu}_{0^\pm 0^\pm,0}}{4}\langle 0^\pm;00;\mathrm{v}';|\hat{\mu}_0|0^\pm;00;\mathrm{v}\rangle\langle J'0M'|D^{1*}_{p0}(\omega)|J0M\rangle \\ &\quad\times\left[1+(\pm)(-1)^{J+s}+(\pm)'(-1)^{J'+s'}+(\pm)(\pm)'(-1)^{J+s+J'+s'}\right] \\ &= \frac{\tilde{\mu}_{0^\pm 0^\pm,0}}{4}\left\{(-1)^{M'}\sqrt{(2J'+1)(2J+1)}\begin{pmatrix} J' & 1 & J \\ -M' & p & M\end{pmatrix}\begin{pmatrix} J' & 1 & J \\ 0 & 0 & 0\end{pmatrix}\right. \\ &\quad\times\left[1+(\pm)(-1)^{J+s}+(\pm)'(-1)^{J'+s'}+(\pm)(\pm)'(-1)^{J+s+J'+s'}\right].\end{aligned} \tag{5.37}$$

Consider first the molecule-fixed dipole moment term $\tilde{\mu}_{0^\pm 0^\pm,0} = \langle 0^\pm;00;\mathrm{v}';|\hat{\mu}_0|0^\pm;00;\mathrm{v}\rangle$. Let us assume that the ground and excited electronic levels are $^1\Sigma^+$ and $^1\Sigma^-$ levels, respectively. The results of applying the symmetry operator σ_V, corresponding to reflection about the internuclear axis in the *xz* plane, are given by (Hougen, 1970)

$$\sigma_v|0^+;00;\mathrm{v}\rangle = +|0^+;00;\mathrm{v}\rangle. \tag{5.38}$$

$$\sigma_v\hat{\mu}_0 = \hat{\mu}_0. \tag{5.39}$$

$$\sigma_v\langle 0^-;00;\mathrm{v}| = -\langle 0^-;00;\mathrm{v}|. \tag{5.40}$$

In this case then

$$\begin{aligned}\langle 0^-;00;\mathrm{v}';|\hat{\mu}_0|0^+;00;\mathrm{v}\rangle &= \sigma_V\langle 0^-;00;\mathrm{v}';|\hat{\mu}_0|0^+;00;\mathrm{v}\rangle \\ &= \int_\tau \sigma_V\left(\Psi^*_{0^-}\hat{\mu}_0\Psi_{0^+}\right)d\tau \\ &= -\langle 0^-;00;\mathrm{v}';|\hat{\mu}_0|0^+;00;\mathrm{v}\rangle.\end{aligned} \tag{5.41}$$

The value of the integral used to calculate the molecule-fixed electric dipole matrix element must be unchanged by the σ_V symmetry operation because the integration is carried out over all space. The only way that Eq. (5.41) can therefore be satisfied is if $\langle 0^-;00;\mathrm{v}';|\hat{\mu}_0|0^+;00;\mathrm{v}\rangle = 0$; electric dipole transitions are not allowed between states in $^1\Sigma^+$ and $^1\Sigma^-$ electronic levels.

Consider now the case where both the ground and the excited levels are in $^1\Sigma^+$ electronic levels ($s = s' = 0$), and the ground and the excited levels are positive- and negative-parity levels, respectively $\left[(\pm) = +, (\pm)' = -\right]$. In this case Eq. (5.37) reduces to

$$\begin{aligned}\mu_{ba,p} &= \frac{\tilde{\mu}_{0^+0^+,0}}{4}\left\{(-1)^{M'}\sqrt{(2J'+1)(2J+1)}\begin{pmatrix} J' & 1 & J \\ -M' & p & M\end{pmatrix}\begin{pmatrix} J' & 1 & J \\ 0 & 0 & 0\end{pmatrix}\right. \\ &\quad\left.\times\left[1+(-1)^J-(-1)^{J'}-(-1)^{J+J'}\right]\right\}.\end{aligned} \tag{5.42}$$

In this case all positive-parity ground levels have even J, all negative-parity upper levels have odd J', and $\left[1+(-1)^J-(-1)^{J'}-(-1)^{J+J'}\right] = 4$. Because J is even and

J' is odd, $J' + 1 + J$ is an even number; hence the *3j* symbol $\begin{pmatrix} J' & 1 & J \\ 0 & 0 & 0 \end{pmatrix}$ will be nonzero. If, on the other hand, the upper rotational level is a positive-parity level, then Eq. (5.37) becomes

$$\mu_{ba,p} = \frac{\tilde{\mu}_{0^+0^+,0}}{4} \left\{ (-1)^{p+M} \sqrt{(2J'+1)(2J+1)} \begin{pmatrix} J' & 1 & J \\ -M' & p & M \end{pmatrix} \begin{pmatrix} J' & 1 & J \\ 0 & 0 & 0 \end{pmatrix} \right. \\ \times \left[1 + (-1)^J + (-1)^{J'} + (-1)^{J+J'} \right]. \tag{5.43}$$

In this case both J and J' will be even numbers, $J' + 1 + J$ is an odd number, and the *3j* symbol $\begin{pmatrix} J' & 1 & J \\ 0 & 0 & 0 \end{pmatrix}$ will be zero. Therefore, transitions between levels with the same parity are forbidden, as is the case for all single-photon electric dipole transitions.

For allowed transitions, Eq. (5.42) reduces to

$$\mu_{ba,p} = (-1)^{M'} \sqrt{(2J'+1)(2J+1)} \begin{pmatrix} J' & 1 & J \\ -M' & p & M \end{pmatrix} \begin{pmatrix} J' & 1 & J \\ 0 & 0 & 0 \end{pmatrix} \tilde{\mu}_{0^+0^+,0}. \tag{5.44}$$

The modulus squared of the dipole moment is thus given by

$$\mu_{ba,p}\mu^*_{ba,p} = (2J'+1)(2J+1) \begin{pmatrix} J' & 1 & J \\ -M' & p & M \end{pmatrix}^2 \begin{pmatrix} J & J' & 1 \\ 0 & 0 & 0 \end{pmatrix}^2 \tilde{\mu}^2_{0^+0^+,0}. \tag{5.45}$$

The Hönl–London factor $S_{J''}^{\Delta J}$, also commonly termed the rotational line strength, is given by (Kovacs, 1969; Zare, 1988)

$$S_{J''}^{\Delta J} = S_{J'}^{\Delta J} = \sum_p \sum_{M'} \sum_M \left| \left\langle J'M' \left| \sum_q D^{1*}_{pq} \right| JM \right\rangle \right|^2 = 3 \sum_{M'} \sum_M \left| \left\langle J'M' \left| \sum_q D^{1*}_{0q} \right| JM \right\rangle \right|^2 \\ = 3 \sum_{M'} \sum_M \frac{\mu_{ba,0}\mu^*_{ba,0}}{\tilde{\mu}^2_{0^+0^+,0}}. \tag{5.46}$$

From Appendix A3, we obtain

$$\sum_{-M'} \sum_M \begin{pmatrix} J' & 1 & J \\ -M' & p & M \end{pmatrix}^2 = \sum_{-M'} \sum_M \begin{pmatrix} J' & 1 & J \\ -M' & p & M \end{pmatrix}^2 = \frac{1}{2k+1} = \frac{1}{2(1)+1} = \frac{1}{3}. \tag{5.47}$$

Substituting Eq. (5.45) into Eq. (5.46) and using Eq. (5.47), we obtain

$$S_{J''}^{\Delta J} = (2J'+1)(2J+1) \begin{pmatrix} J & J' & 1 \\ 0 & 0 & 0 \end{pmatrix}^2. \tag{5.48}$$

The $3j$ symbol in Eq. (5.48) is in a form where it can now be easily evaluated using formulae from Appendix A3. For a $Q(J)$ transition, $J' = J$, we obtain $S_{J''}^{\Delta J} = 0$ because $J' + 1 + J$ is an odd number. For a $P(J)$ transition, $J' = J - 1$, we obtain

$$S_{J''}^{\Delta J} = (2J-1)(2J+1)\frac{2J^2}{(2J+1)2J(2J-1)} = J. \tag{5.49}$$

For an $R(J)$ transition, $J' = J + 1$, and

$$S_{J''}^{\Delta J} = (2J+3)(2J+1)\frac{2(J+1)^2}{(2J+3)(2J+2)(2J+1)} = J + 1. \tag{5.50}$$

The sum of the rotational line strengths for the $Q(J), P(J)$, and $R(J)$ transitions is $2J + 1$.

5.2.2 $^1\Sigma$–$^1\Pi$ Transitions

In this section we will consider radiative transitions between a Zeeman state a in a rotational level A in a lower $^1\Sigma^+$ electronic level and a Zeeman state b in a rotational level B with positive parity in a $^1\Pi$ upper electronic level. The wavefunction for the Zeeman states for the $^1\Sigma$ level is given by Eq. (5.33) and for the $^1\Pi$ levels by

$$\left|^1\Pi; \mathrm{v}_B; J_B M_b; \pm\right\rangle = \frac{1}{\sqrt{2}}\left[|1;00;\mathrm{v}';J'1M'\rangle \pm (-1)^{J_B}|-1;00;\mathrm{v}';J'-1M'\rangle\right], \tag{5.51}$$

where again $J' = J_B, M' = M_B$, and $\mathrm{v}' = \mathrm{v}_B$. From Eq. (5.2), the dipole moment for transitions between a negative-parity $^1\Sigma^\pm$ lower state and a positive-parity $^1\Pi$ upper state will be given by

$$\begin{aligned}
\mu_{ba,p} &= \left\langle ^1\Pi; \mathrm{v}_B; J'M'; + \right| \sum_q \hat{\mu}_q D_{pq}^{1*}(\omega) \left| ^1\Sigma^\pm; 0^\pm; \mathrm{v}_A; JM; - \right\rangle \\
&= \frac{1}{2\sqrt{2}} \Bigg\{ \langle 1;\ 00;\mathrm{v}';J'1M'| \sum_q \hat{\mu}_q D_{pq}^{1*}(\omega) |0^\pm;00;\mathrm{v};\ J0M\rangle \\
&\quad - (-1)^{J+s}\langle 1;00;\mathrm{v}';J'1M'| \sum_q \hat{\mu}_q D_{pq}^{1*}(\omega) |0^\pm;00;\mathrm{v};\ J0M\rangle \\
&\quad + (-1)^{J'}\langle -1;00;\mathrm{v}';J'-1M'| \sum_q \hat{\mu}_q D_{pq}^{1*}(\omega) |0^\pm;00;\mathrm{v};\ J0M\rangle \\
&\quad - (-1)^{J+s+J'}\langle -1;00;\mathrm{v}';J'-1M'| \sum_q \hat{\mu}_q D_{pq}^{1*}(\omega) |0^\pm;\mathrm{v};00;J0M\rangle \Bigg\} \\
&= \frac{1}{2\sqrt{2}} \sum_q \Big\{ \langle 1;00;\mathrm{v}'|\hat{\mu}_q|0^\pm;00;\mathrm{v}\rangle \langle J'1M'|D_{pq}^{1*}(\omega)|J0M\rangle \\
&\quad - (-1)^{J+s}\langle 1;00;\mathrm{v}'|\hat{\mu}_q|0^\pm;00;\mathrm{v}\rangle \langle J'1M'|D_{pq}^{1*}(\omega)|J0M\rangle \\
&\quad + (-1)^{J'}\langle -1;00;\mathrm{v}'|\hat{\mu}_q|0^\pm;00;\mathrm{v}\rangle \langle J'-1M'|D_{pq}^{1*}(\omega)|J0M\rangle \\
&\quad - (-1)^{J+s+J'}\langle -1;00;\mathrm{v}'|\hat{\mu}_q|0^\pm;00;\mathrm{v}\rangle \langle J'-1M'|D_{pq}^{1*}(\omega)|J0M\rangle \Big\},
\end{aligned} \tag{5.52}$$

where $J = J_A, J' = J_B, M = M_A,$ and $M' = M_B$. The rotational density matrix elements can be evaluated using Eq. (5.36). Using this expression, simplifying, and using the fact that the 3j symbol $\begin{pmatrix} J' & 1 & J \\ -\Omega' & q & \Omega \end{pmatrix}$ will be zero unless the sum of the coefficients in the bottom row is zero, we obtain

$$\mu_{ba,p} = \frac{\sqrt{(2J'+1)(2J+1)}}{2\sqrt{2}} \begin{pmatrix} J' & 1 & J \\ -M' & p & M \end{pmatrix} (-1)^{M'} \\ \left\{ (-1)^{-1} \begin{pmatrix} J' & 1 & J \\ -1 & 1 & 0 \end{pmatrix} \tilde{\mu}_{+1} \left[1 - (-1)^{J+s}\right] \right. \\ \left. +(-1)^{J'+1} \begin{pmatrix} J' & 1 & J \\ 1 & -1 & 0 \end{pmatrix} \tilde{\mu}_{-1} \left[1 - (-1)^{J+s}\right] \right\}, \tag{5.53}$$

where

$$\tilde{\mu}_{+1} = \langle 1;00;\mathrm{v}'|\hat{\mu}_{+1}|0^{\pm};00;\mathrm{v}\rangle = \langle 1;00;\mathrm{v}'|\hat{\mu}_x + i\hat{\mu}_y|0^{\pm};00;\mathrm{v}\rangle \tag{5.54}$$

and

$$\tilde{\mu}_{-1} = \langle 1;00;\mathrm{v}'|\hat{\mu}_{-1}|0^{\pm};00;\ \mathrm{v}\rangle = \langle 1;00;\mathrm{v}'|\hat{\mu}_x - i\hat{\mu}_y|0^{\pm};00;\ \mathrm{v}\rangle. \tag{5.55}$$

For negative-parity states, the rotational quantum number J will be odd and $s = 0$ for ground $^1\Sigma^+$ electronic levels, and J will be even and $s = 1$ for ground $^1\Sigma^-$ electronic levels. In both cases, the term $\left[1 - (-1)^{J+s}\right] = 2$. In addition, the 3j symbols will be zero unless the sum of the coefficients in the bottom row is zero. Therefore, we can write

$$\mu_{ba,p} = -\frac{\sqrt{(2J'+1)(2J+1)}}{\sqrt{2}} \begin{pmatrix} J' & 1 & J \\ -M' & p & M \end{pmatrix} (-1)^{M'} \sum_q \left\{ \begin{pmatrix} J' & 1 & J \\ -1 & 1 & 0 \end{pmatrix} \tilde{\mu}_{+1} \right. \\ \left. +(-1)^{J'} \begin{pmatrix} J' & 1 & J \\ 1 & -1 & 0 \end{pmatrix} \tilde{\mu}_{-1} \right\}. \tag{5.56}$$

Performing an even permutation of the columns of the 3j symbols and using the relation

$$\begin{pmatrix} J & J' & 1 \\ \Omega & -\Omega' & -1 \end{pmatrix} = (-1)^{J+J'+1} \begin{pmatrix} J & J' & 1 \\ -\Omega & \Omega' & 1 \end{pmatrix}, \tag{5.57}$$

we obtain

$$\mu_{ba,p} = -\frac{\sqrt{(2J+1)(2J'+1)}}{\sqrt{2}} \begin{pmatrix} J' & 1 & J \\ -M' & p & M \end{pmatrix} (-1)^{M'} \begin{pmatrix} J & J' & 1 \\ 0 & -1 & 1 \end{pmatrix} \\ \times \left[\tilde{\mu}_{+1} + (-1)^{2J'+J+1} \tilde{\mu}_{-1}\right]. \tag{5.58}$$

Because $2J'$ is always an even number, $(-1)^{2J'+J+1} = (-1)^{J+1}$.

The electronic transition dipole moment (TDM) factors $\tilde{\mu}_{+1}$, $\tilde{\mu}_{-1}$, and $\tilde{\mu}_0$, where

$$\tilde{\mu}_0 = \langle \eta_B \mathrm{v}_B | \hat{\mu}_0 | \eta_B \mathrm{v}_B \rangle = \langle \eta_B \mathrm{v}_B | \hat{\mu}_z | \eta_A \mathrm{v}_A \rangle \tag{5.59}$$

are discussed in great detail by Hougen (1970) and Whiting and Nicholls (1974). Based on the sequential application of a symmetry operation involving reflection through a "plane containing the internuclear axis" (Whiting & Nicholls, 1974) and a time reversal operation, the factors $\tilde{\mu}_{+1}$, $\tilde{\mu}_{-1}$, and $\tilde{\mu}_0$ must be purely real or purely imaginary, and can always be considered purely real.

Application of the symmetry operator σ_v, representing reflection of coordinates across the xz plane that contains the internuclear axis, will leave the value of the dipole moment matrix element unchanged because the dipole moment matrix element is calculated by performing an integration over all coordinates (Hougen, 1970). Applying the symmetry operator σ_v we obtain

$$\sigma_v \big| 0^{\pm}; 00; \mathrm{v} \big\rangle = \pm \big| 0^{\pm}; 00; \mathrm{v} \big\rangle. \tag{5.60}$$

$$\sigma_v \left(\hat{\mu}_x \pm i\hat{\mu}_y \right) = \sigma_v \hat{\mu}_{\pm 1} = \hat{\mu}_x \mp i\hat{\mu}_y = \sigma_v \hat{\mu}_{\mp 1}. \tag{5.61}$$

For the $^1\Pi$ electronic level, the result of the symmetry operation is given by (Hougen, 1970)

$$\sigma_v \langle 1; 00; \mathrm{v}' | = (-1)^{L-\Lambda} (-1)^{S-\Sigma} \langle -1; 00; \mathrm{v}' |. \tag{5.62}$$

The quantum number L is not well defined. Choosing $L = 1$, we obtain

$$\sigma_v \langle 1; 00; \mathrm{v}' | = \langle -1; 00; \mathrm{v}' |. \tag{5.63}$$

Therefore, when the ground electronic level is a $^1\Sigma^+$level,

$$\begin{aligned} \tilde{\mu}_{+1} &= \sigma_v \left(\tilde{\mu}_{BA.+1} \right) = \sigma_v \left(\langle \mathrm{v}' | \langle 1; 00 | \hat{\mu}_x \pm i\hat{\mu}_y | 0^+; 00 \rangle | \mathrm{v} \rangle \right) \\ &= + \langle \mathrm{v}' | \langle -1; 00 | \hat{\mu}_x \mp i\hat{\mu}_y | 0^+; 00 \rangle | \mathrm{v} \rangle = \tilde{\mu}_{-1}. \end{aligned} \tag{5.64}$$

When the ground electronic level is a $^1\Sigma^-$level, then $\tilde{\mu}_{-1} = -\tilde{\mu}_{+1}$. Consequently, Eq. (5.58) can be written as

$$\mu_{ba,p} = -\frac{\sqrt{(2J+1)(2J'+1)}}{\sqrt{2}} \begin{pmatrix} J' & 1 & J \\ -M' & p & M \end{pmatrix} (-1)^{M'} \begin{pmatrix} J & J' & 1 \\ 0 & -1 & 1 \end{pmatrix} \tilde{\mu}_{+1} \left[1 \pm (-1)^{J+1} \right], \tag{5.65}$$

where the + and – signs in the last term on the RHS of Eq. (5.65) apply for $^1\Sigma^+$ and $^1\Sigma^-$ ground electronic levels, respectively. For negative-parity levels in $^1\Sigma^+$ and $^1\Sigma^-$ ground electronic levels, the rotational quantum numbers J will be odd and even, respectively, and so in both cases $\left[1 \pm (-1)^{J+1}\right] = 2$. The modulus squared of the dipole moment is thus given by

$$\mu_{ba,p}\mu^*_{ba,p} = 2(2J'+1)(2J+1)\begin{pmatrix} J' & 1 & J \\ -M' & p & M \end{pmatrix}^2 \begin{pmatrix} J & J' & 1 \\ 0 & -1 & 1 \end{pmatrix}^2 \tilde{\mu}^2_{+1}. \tag{5.66}$$

The Hönl–London factor $S^{\Delta J}_{J''}$ is given by (Kovacs, 1969; Zare, 1988)

$$\begin{aligned} S^{\Delta J}_{J''} = S^{\Delta J}_{J'} &= \sum_p \sum_{M'} \sum_M \left| \left\langle J'M' \left| \sum_q D^{1*}_{pq} \right| JM \right\rangle \right|^2 = 3 \sum_{M'} \sum_M \left| \left\langle J'M' \left| \sum_q D^{1*}_{0q} \right| JM \right\rangle \right|^2 \\ &= 3 \sum_{M'} \sum_M \frac{\mu_{ba,0}\mu^*_{ba,0}}{2\tilde{\mu}^2_1}. \end{aligned} \tag{5.67}$$

Substituting Eq. (5.66) into Eq. (5.67) and using Eq. (5.47), we obtain

$$S^{\Delta J}_{J''} = (2J'+1)(2J+1)\begin{pmatrix} J & J' & 1 \\ 0 & -1 & 1 \end{pmatrix}^2. \tag{5.68}$$

The *3j* symbol in Eq. (5.68) is in a form where it can now be evaluated using formulae from Appendix A3. For a *Q*(*J*) transition, $J' = J$, we obtain

$$S^{\Delta J}_{J''} = (2J+1)^2 \frac{2J(J+1)}{(2J+2)(2J+1)2J} = \frac{2J+1}{2}. \tag{5.69}$$

For a *P*(*J*) transition, $J' = J - 1$, we obtain

$$S^{\Delta J}_{J''} = (2J-1)(2J+1)\frac{J(J-1)}{(2J+1)2J(2J-1)} = \frac{J-1}{2}. \tag{5.70}$$

For an *R*(*J*) transition, $J' = J + 1$, and

$$S^{\Delta J}_{J''} = (2J+3)(2J+1)\frac{(J+1)(J+2)}{(2J+3)(2J+2)(2J+1)} = \frac{J+2}{2}. \tag{5.71}$$

The sum of the rotational line strengths for the *Q*(*J*), *P*(*J*), and *R*(*J*) transitions is $2J + 1$. This sum rule can be derived directly from Eq. (5.68) by applying Eq. (A3.4) from Appendix A3,

$$\sum_{J'} S^{\Delta J}_{J''} = \sum_{J'} (2J'+1)(2J+1)\begin{pmatrix} J & J' & 1 \\ 0 & -1 & 1 \end{pmatrix}^2 = (2J+1). \tag{5.72}$$

Similarly,

$$\sum_J S_{J'}^{\Delta J} = \sum_J (2J'+1)(2J+1)\begin{pmatrix} J & J' & 1 \\ 0 & -1 & 1 \end{pmatrix}^2 = (2J'+1). \tag{5.73}$$

5.2.3 Singlet Electronic Transitions with $\Lambda \geq 1$, $\Lambda' \geq 1$

In this section, we will consider transitions where both electronic levels have nonzero values of the orbital angular momentum projection quantum number Λ. The wavefunctions for states a and b in levels A and B, respectively, are given by

$$|{}^1A;\mathrm{v}_A;J_AM_a;\pm\rangle = \frac{1}{\sqrt{2}}\left[|\Lambda;\ 00;\mathrm{v}';\ J'\Lambda M'\rangle \pm (-1)^{J'}|-\Lambda;\ 00;\mathrm{v}';\ J'-\Lambda M'\rangle\right]. \tag{5.74}$$

$$|{}^1B;\mathrm{v}_B;J_BM_b;\pm\rangle = \frac{1}{\sqrt{2}}\left[|\Lambda';\ 00;\mathrm{v}';\ J'\Lambda' M'\rangle \pm (-1)^{J'}|-\Lambda';\ 00;\mathrm{v}';\ J'-\Lambda' M'\rangle\right]. \tag{5.75}$$

The dipole moment matrix element for transitions between a negative-parity lower state a and an upper state b is given by

$$\begin{aligned}
\mu_{ba,p} &= \langle {}^1B;\mathrm{v}_B;J'M';+|\sum_q \hat{\mu}_q D_{pq}^{1*}(\omega)|{}^1A;\mathrm{v}_A;JM;-\rangle \\
&= \frac{1}{2}\sum_q \Big\{\langle\Lambda';00;\mathrm{v}'|\hat{\mu}_q|\Lambda;00;\mathrm{v}\rangle\langle J'\Lambda'M'|D_{pq}^{1*}(\omega)|J\Lambda M\rangle \\
&\quad -(-1)^J\langle\Lambda';00;\mathrm{v}'|\hat{\mu}_q|-\Lambda;00;\mathrm{v}\rangle\langle J'\Lambda'M'|D_{pq}^{1*}(\omega)|J-\Lambda M\rangle \\
&\quad +(-1)^{J'}\langle-\Lambda';00;\mathrm{v}'|\hat{\mu}_q|\Lambda;00;\mathrm{v}\rangle\langle J'-\Lambda'M'|D_{pq}^{1*}(\omega)|J\Lambda M\rangle \\
&\quad -(-1)^{J+J'}\langle-\Lambda';00;\mathrm{v}'|\hat{\mu}_q|-\Lambda;00;\mathrm{v}\rangle\langle J'-\Lambda'M'|D_{pq}^{1*}(\omega)|J-\Lambda M\rangle\Big\},
\end{aligned} \tag{5.76}$$

where $J = J_A, J' = J_B, M = M_A,$ and $M' = M_B$. The rotational density matrix elements can be evaluated using Eq. (5.36). Using this expression, simplifying, and using the fact that the *3j* symbol $\begin{pmatrix} J' & 1 & J \\ -\Omega' & q & \Omega \end{pmatrix}$ will be zero unless the sum of the coefficients in the bottom row is zero, we obtain

$$\begin{aligned}
\mu_{ba,p} &= \frac{\sqrt{(2J'+1)(2J+1)}}{2}\begin{pmatrix} J' & 1 & J \\ -M' & p & M \end{pmatrix}(-1)^{M'}\Big\{(-1)^{-\Lambda'}\begin{pmatrix} J' & 1 & J \\ -\Lambda' & \Delta\Lambda & \Lambda \end{pmatrix}\tilde{\mu}_{\Delta\Lambda} \\
&\quad -(-1)^{J-\Lambda'}\begin{pmatrix} J' & 1 & J \\ -\Lambda' & \Lambda+\Lambda' & -\Lambda \end{pmatrix}\tilde{\mu}_{\Lambda+\Lambda'} + (-1)^{J'+\Lambda'}\begin{pmatrix} J' & 1 & J \\ \Lambda' & -\Lambda-\Lambda' & \Lambda \end{pmatrix}\tilde{\mu}_{-\Lambda-\Lambda'} \\
&\quad -(-1)^{J'+J+\Lambda'}\begin{pmatrix} J' & 1 & J \\ \Lambda' & -\Delta\Lambda & -\Lambda \end{pmatrix}\tilde{\mu}_{-\Delta\Lambda}\Big\}.
\end{aligned} \tag{5.77}$$

However, the 3*j* symbols with $q = \Lambda + \Lambda'$ and $q = -\Lambda - \Lambda'$ will be zero because of the requirement that $|q| \le k$. Therefore, Eq. (5.77) reduces to

$$\mu_{ba,p} = \frac{\sqrt{(2J'+1)(2J+1)}}{2} \begin{pmatrix} J' & 1 & J \\ -M' & p & M \end{pmatrix} (-1)^{M'-\Lambda'} \begin{pmatrix} J' & 1 & J \\ -\Lambda' & \Delta\Lambda & \Lambda \end{pmatrix} \times \left[\tilde{\mu}_{\Delta\Lambda} - (-1)^{2J'+2J+1+2\Lambda'} \tilde{\mu}_{-\Delta\Lambda}\right]. \tag{5.78}$$

Since $2J$, $2J'$, and $2\Lambda'$ are all even numbers, Eq. (5.78) reduces to

$$\mu_{ba,p} = \frac{\sqrt{(2J'+1)(2J+1)}}{2} \begin{pmatrix} J' & 1 & J \\ -M' & p & M \end{pmatrix} (-1)^{M'-\Lambda'} \begin{pmatrix} J' & 1 & J \\ -\Lambda' & \Delta\Lambda & \Lambda \end{pmatrix} [\tilde{\mu}_{\Delta\Lambda} + \tilde{\mu}_{-\Delta\Lambda}]. \tag{5.79}$$

Assuming that $L - \Lambda = L' - \Lambda'$, then $\tilde{\mu}_{\Delta\Lambda} = \tilde{\mu}_{-\Delta\Lambda}$, and Eq. (5.79) reduces to

$$\mu_{ba,p} = \sqrt{(2J'+1)(2J+1)} \begin{pmatrix} J' & 1 & J \\ -M' & p & M \end{pmatrix} (-1)^{M'-\Lambda'} \begin{pmatrix} J' & 1 & J \\ -\Lambda' & \Delta\Lambda & \Lambda \end{pmatrix} \tilde{\mu}_{\Delta\Lambda}. \tag{5.80}$$

The Hönl–London factor becomes

$$S_{J''}^{\Delta J} = S_{J'}^{\Delta J} = 3 \sum_{M'} \sum_{M} \frac{\mu_{ba,0} \mu_{ba,0}^{*}}{\tilde{\mu}_{BA,\Delta\Lambda}^{2}} = (2J'+1)(2J+1) \begin{pmatrix} J & J' & 1 \\ \Lambda & -\Lambda' & \Delta\Lambda \end{pmatrix}^2. \tag{5.81}$$

The Hönl–London factors are given below for different values of $\Delta\Lambda$ and ΔJ.

$\Delta\Lambda = \Lambda' - \Lambda = -1$

$\Delta J = J' - J = -1$

$$S_{J''}^{\Delta J} = (2J'+1)(2J+1) \begin{pmatrix} J & J-1 & 1 \\ \Lambda & -(\Lambda-1) & -1 \end{pmatrix}^2 = (2J'+1)(2J+1) \begin{pmatrix} J & J-1 & 1 \\ -\Lambda & \Lambda-1 & 1 \end{pmatrix}^2 = (2J-1)(2J+1) \frac{(J+\Lambda-1)(J+\Lambda)}{(2J-1)2J(2J+1)} = \frac{(J+\Lambda-1)(J+\Lambda)}{2J}. \tag{5.82}$$

$\Delta J = 0$

$$S_{J''}^{\Delta J} = (2J+1)^2 \begin{pmatrix} J & J & 1 \\ -\Lambda & \Lambda-1 & 1 \end{pmatrix}^2 = (2J+1)^2 \frac{2(J+\Lambda)(J+1-\Lambda)}{(2J+2)(2J+1)2J} = \frac{(2J+1)(J+\Lambda)(J+1-\Lambda)}{2J(J+1)}. \tag{5.83}$$

$\Delta J = +1$

$$S_{J''}^{\Delta J} = (2J+3)(2J+1)\begin{pmatrix} J & J+1 & 1 \\ -\Lambda & \Lambda-1 & 1 \end{pmatrix}^2 = (2J+3)(2J+1)\frac{(J-2+\Lambda)(J-1+\Lambda)}{(2J+3)(2J+2)(2J+1)}$$
$$= \frac{(J+2-\Lambda)(J-1+\Lambda)}{2(J+1)}. \tag{5.84}$$

$\Delta\Lambda = \Lambda' - \Lambda = 0$

$\Delta J = J' - J = -1$

$$S_{J''}^{\Delta J} = (2J-1)(2J+1)\begin{pmatrix} J & J-1 & 1 \\ \Lambda & -\Lambda & 0 \end{pmatrix}^2 = (2J-1)(2J+1)\frac{2(J+\Lambda)(J-\Lambda)}{(2J+1)2J(2J-1)}$$
$$= \frac{(J+\Lambda)(J-\Lambda)}{J}. \tag{5.85}$$

$\Delta J = 0$

$$S_{J''}^{\Delta J} = (2J+1)^2\begin{pmatrix} J & J & 1 \\ \Lambda & -\Lambda & 0 \end{pmatrix}^2 = (2J+1)^2(2J+1)\frac{\Lambda^2}{(2J+1)J(J+1)}$$
$$= \frac{(2J+1)\Lambda^2}{J(J+1)}. \tag{5.86}$$

$\Delta J = +1$

$$S_{J''}^{\Delta J} = (2J+3)(2J+1)\begin{pmatrix} J & J+1 & 1 \\ \Lambda & -\Lambda & 0 \end{pmatrix}^2 = (2J+3)(2J+1)\frac{2(J+1-\Lambda)(J+1+\Lambda)}{(2J+3)(2J+2)(2J+1)}$$
$$= \frac{(J+1-\Lambda)(J+1+\Lambda)}{(J+1)}. \tag{5.87}$$

$\Delta\Lambda = \Lambda' - \Lambda = +1$

$\Delta J = J' - J = -1$

$$S_{J''}^{\Delta J} = (2J-1)(2J+1)\begin{pmatrix} J & J-1 & 1 \\ \Lambda & -\Lambda-1 & 1 \end{pmatrix}^2 = (2J-1)(2J+1)\frac{(J-1-\Lambda)(J-\Lambda)}{(2J+1)2J(2J-1)}$$
$$= \frac{(J-1-\Lambda)(J-\Lambda)}{2J}. \tag{5.88}$$

$\Delta J = 0$

$$S_{J''}^{\Delta J} = (2J+1)^2\begin{pmatrix} J & J & 1 \\ \Lambda & -\Lambda-1 & 1 \end{pmatrix}^2 = (2J+1)^2\frac{2(J-\Lambda)(J+1+\Lambda)}{(2J+2)(2J+1)2J}$$
$$= \frac{(2J+1)(J-\Lambda)(J+1+\Lambda)}{2J(J+1)}. \tag{5.89}$$

Table 5.1 Rotational line strengths for singlet transitions. Table reproduced from similar tables in Kovacs (1969) and Zare (1988)

	$S_{J''}^{\Delta J}$		
	$\Delta\Lambda = +1$	$\Delta\Lambda = 0$	$\Delta\Lambda = -1$
$P(J)$	$\dfrac{(J-1-\Lambda)(J-\Lambda)}{2J}$	$\dfrac{(J+\Lambda)(J-\Lambda)}{J}$	$\dfrac{(J-1+\Lambda)(J+\Lambda)}{2J}$
$Q(J)$	$\dfrac{(2J+1)(J-\Lambda)(J+1+\Lambda)}{2J(J+1)}$	$\dfrac{(2J+1)\Lambda^2}{J(J+1)}$	$\dfrac{(J+1-\Lambda)(J+\Lambda)(2J+1)}{2J(J+1)}$
$R(J)$	$\dfrac{(J+1+\Lambda)(J+2+\Lambda)}{2(J+1)}$	$\dfrac{(J+1-\Lambda)(J+1+\Lambda)}{J+1}$	$\dfrac{(J+2-\Lambda)(J+1-\Lambda)}{2(J+1)}$

$\Delta\Lambda = \Lambda' - \Lambda \quad J = J_A = J'' \quad \Lambda = \Lambda_A \quad \Lambda' = \Lambda_B$

$\Delta J = +1$

$$S_{J''}^{\Delta J} = (2J+3)(2J+1)\begin{pmatrix} J & J+1 & 1 \\ \Lambda & -\Lambda-1 & 1 \end{pmatrix}^2 = (2J+3)(2J+1)\frac{(J+1+\Lambda)(J+2+\Lambda)}{(2J+3)(2J+2)(2J+1)}$$
$$= \frac{(J+1+\Lambda)(J+2+\Lambda)}{2(J+1)}. \tag{5.90}$$

Table 5.1 is reproduced from similar tables in Kovacs (1969) and Zare (1988) for singlet transitions.

5.3 Doublet Electronic Transitions

The calculation of radiative transition rates for transitions between rotational levels contained within doublet electronic levels is discussed in this section. A different formulation for the tabulation of rotational line strengths is introduced that enables the straightforward inclusion of higher-order correction factors for effects such as centrifugal distortion and Lambda-doubling in the calculation of the radiative transition rates. The formulae for the rotational line strength factors are developed using Hund's case (a) basis states, and the correction factors for spin–orbit splitting, centrifugal distortion, spin–rotation interactions, and Lambda-doubling are included in the Hund's case (a) coefficients for the state wavefunctions. The tables of rotational line strength factors also feature the inclusion of molecule-fixed vibrational band intensities $\tilde{\mu}$ that are corrected for rotational distortion. The molecule-fixed vibrational band intensities are calculated for both the $\Omega = \Lambda - \frac{1}{2}$ and $\Omega = \Lambda + \frac{1}{2}$ case (a) states and are associated with the appropriate case (a) coefficients in the tables of rotational line strength factors. The formulation of the rotational line strength factors in terms of the Hund's case (a) coefficients leads to increased physical insight, makes it much easier to incorporate spectroscopic data at desired levels of approximation, and allows investigation of the effects of different spectroscopic parameters on the values of the rotational line strengths. Calculation of the radiative transition rates is illustrated for the hydroxyl radical (OH) and nitric oxide (NO).

5.3.1 $F_1(J) \rightarrow F_1(J')$ Transitions

First consider transitions from positive-parity F_1 lower quantum states (state a in level A) to negative-parity F_1 upper quantum states (state b in level B). A general expression for these transitions will be developed, which can then be modified to account for transitions from negative-parity lower states to positive-parity upper states and for transitions involving F_2 states. For the lower F_1 state, the wavefunctions are given by

$$\begin{aligned}|F_1; \mathrm{v}; JM; +\rangle = {} & \frac{a_J}{\sqrt{2}}\left[\left|\Lambda; \frac{1}{2} - \frac{1}{2}; \mathrm{v}; J\, \Omega_- M\right\rangle + (-1)^{\rho}\left|-\Lambda; \frac{1}{2}\, \frac{1}{2}; \mathrm{v}; J - \Omega_- M\right\rangle\right] \\ & + \frac{b_J}{\sqrt{2}}\left[\left[\left|\Lambda; \frac{1}{2} \frac{1}{2}; \mathrm{v}; J\, \Omega_+ M\right\rangle + (-1)^{\rho}\left|-\Lambda; \frac{1}{2} - \frac{1}{2}; \mathrm{v}; J - \Omega_+ M\right\rangle\right]\right],\end{aligned} \tag{5.91}$$

where $\rho = J - \frac{1}{2} + s$ and

$$\Omega_{\pm} = \Lambda \pm \frac{1}{2}. \tag{5.92}$$

For the upper state the wavefunctions are given by

$$\begin{aligned}|F_1; \mathrm{v}'; J'M'; -\rangle = {} & \frac{a_J'}{\sqrt{2}}\left[\left|\Lambda'; \frac{1}{2} - \frac{1}{2}; \mathrm{v}'; J'\, \Omega_-' M'\right\rangle - (-1)^{\rho'}\left|-\Lambda'; \frac{1}{2} \frac{1}{2}; \mathrm{v}'; J' - \Omega_-' M'\right\rangle\right] \\ & + \frac{b_J'}{\sqrt{2}}\left[\left[\left|\Lambda'; \frac{1}{2} \frac{1}{2}; \mathrm{v}'; J'\, \Omega_+' M'\right\rangle - (-1)^{\rho'}\left|-\Lambda'; \frac{1}{2} - \frac{1}{2}; \mathrm{v}'; J' - \Omega_+' M'\right\rangle\right]\right],\end{aligned} \tag{5.93}$$

where $\rho' = J' - \frac{1}{2} + s'$ and

$$\Omega_{\pm}' = \Lambda' \pm \frac{1}{2}. \tag{5.94}$$

In Eq. (5.91) we have made the substitutions

$$a_J = S_{\Lambda-1/2, N+1/2} \tag{5.95}$$

and

$$b_J = S_{\Lambda+1/2, N+1/2}. \tag{5.96}$$

This is standard notation for $^2\Pi$ electronic levels (Zare, 1988). For the remainder of the doublet transition section we will also use the basis wavefunction coefficients c_J and d_J for the F_2 levels as given by

$$c_J = S_{\Lambda-1/2,N-1/2} \tag{5.97}$$

and

$$d_J = S_{\Lambda+1/2,N-1/2}. \tag{5.98}$$

The component p of the electric dipole moment matrix element in the space-fixed reference frame is given by (Brown & Carrington, 2003)

$$\begin{aligned}
\mu_{ba,p} &= \left\langle F_1;\mathrm{v}';J'M';-\left|\sum_q \hat{\mu}_q D^{1*}_{pq}(\omega)\right|F_1;\mathrm{v};JM;+\right\rangle \\
&= \frac{1}{2}\left\{a_J a'_J\left\langle \Lambda';-\frac{1}{2};\mathrm{v}';J'\Omega'_-M'\left|\sum_q \hat{\mu}_q D^{1*}_{pq}(\omega)\right|\Lambda;-\frac{1}{2};\mathrm{v};J\Omega_-M\right\rangle\right. \\
&\quad - a_J b'_J(-1)^{p'}\left\langle -\Lambda';-\frac{1}{2};\mathrm{v}';J'-\Omega'_+M'\left|\sum_q \hat{\mu}_q D^{1*}_{pq}(\omega)\right|\Lambda;-\frac{1}{2};\mathrm{v};J\Omega_-M\right\rangle \\
&\quad - a_J a'_J(-1)^{p+p'}\left\langle -\Lambda';\frac{1}{2};\mathrm{v}';J'-\Omega'_-M'\left|\sum_q \hat{\mu}_q D^{1*}_{pq}(\omega)\right|-\Lambda;\frac{1}{2};\mathrm{v};J-\Omega_-M\right\rangle \\
&\quad + a_J b'_J(-1)^{p}\left\langle \Lambda';\frac{1}{2};\mathrm{v}';J'\Omega'_+M'\left|\sum_q \hat{\mu}_q D^{1*}_{pq}(\omega)\right|-\Lambda;\frac{1}{2};\mathrm{v};J-\Omega_-M\right\rangle \\
&\quad + b_J b'_J\left\langle \Lambda';\frac{1}{2};\mathrm{v}';J'\Omega'_+M'\left|\sum_q \hat{\mu}_q D^{1*}_{pq}(\omega)\right|\Lambda;\frac{1}{2};\mathrm{v};J\Omega_+M\right\rangle \\
&\quad - b_J a'_J(-1)^{p'}\left\langle -\Lambda';\frac{1}{2};\mathrm{v}';J'-\Omega'_-M'\left|\sum_q \hat{\mu}_q D^{1*}_{pq}(\omega)\right|\Lambda;\frac{1}{2};\mathrm{v};J\Omega_+M\right\rangle \\
&\quad - b_J b'_J(-1)^{p+p'}\left\langle -\Lambda';-\frac{1}{2};\mathrm{v}';J'-\Omega'_+M'\left|\sum_q \hat{\mu}_q D^{1*}_{pq}(\omega)\right|-\Lambda;-\frac{1}{2};\mathrm{v};J-\Omega_+M\right\rangle \\
&\quad \left. + b_J a'_J(-1)^{p}\left\langle \Lambda';-\frac{1}{2};\mathrm{v}';J'\Omega'_-M'\left|\sum_q \hat{\mu}_q D^{1*}_{pq}(\omega)\right|-\Lambda;-\frac{1}{2};\mathrm{v};J-\Omega_+M\right\rangle\right\}.
\end{aligned} \tag{5.99}$$

In Eq. (5.99), terms with $\Sigma \neq \Sigma'$ are equal to zero and are thus dropped from the equation. The terms that involve integration over angular variables and terms that involve integration over the coordinate r along the internuclear axis can be separated, and Eq. (5.99) can be written as

$$\mu_{ba,p}(\mathrm{v}'J'M';\mathrm{v}JM) = \langle F_1;\mathrm{v}'J'M';-|\sum_q \hat{\mu}_q D^{1*}_{pq}(\omega)|F_1;\mathrm{v}JM;+\rangle$$

$$= \frac{1}{2}\sum_q a_J a'_J \left\langle \Lambda';-\frac{1}{2};\mathrm{v}'J'\Omega'_- \middle| \hat{\mu}_q \middle| \Lambda;-\frac{1}{2};\mathrm{v}J\Omega_- \right\rangle \langle J'\Omega'_- M'|D^{1*}_{pq}(\omega)|J\Omega_- M\rangle$$

$$-a_J b'_J(-1)^{p'} \left\langle -\Lambda';-\frac{1}{2};\mathrm{v}'J'-\Omega'_+ \middle| \hat{\mu}_q \middle| \Lambda;\mathrm{v};-\frac{1}{2}J\Omega_- \right\rangle \langle J'-\Omega'_+ M'|D^{1*}_{pq}(\omega)|J,\Omega_- M\rangle$$

$$-a_J a'_J(-1)^{p+p'} \left\langle -\Lambda';\frac{1}{2};\mathrm{v}'J'-\Omega'_- \middle| \hat{\mu}_q \middle| -\Lambda;\frac{1}{2};\mathrm{v}J-\Omega_- \right\rangle \langle J'-\Omega'_- M'|D^{1*}_{pq}(\omega)|J-\Omega_- M\rangle$$

$$+a_J b'_J(-1)^{p} \left\langle \Lambda';\frac{1}{2};\mathrm{v}'J'\Omega'_+ \middle| \hat{\mu}_q \middle| -\Lambda;\frac{1}{2};\mathrm{v}J-\Omega_- \right\rangle \langle J'\Omega'_+ M'|D^{1*}_{pq}(\omega)|J-\Omega_- M\rangle$$

$$+b_J b'_J \left\langle \Lambda';\frac{1}{2};\mathrm{v}'J'\Omega'_+ \middle| \hat{\mu}_q \middle| \Lambda;\frac{1}{2};\mathrm{v}J\Omega_+ \right\rangle \langle J'\Omega'_+ M'|D^{1*}_{pq}(\omega)|J\Omega_+ M\rangle$$

$$-b_J a'_J(-1)^{p'} \left\langle -\Lambda';\frac{1}{2};\mathrm{v}'J'-\Omega'_- \middle| \hat{\mu}_q \middle| \Lambda;\frac{1}{2};\mathrm{v}J\Omega_+ \right\rangle \langle J'-\Omega'_- M'|D^{1*}_{pq}(\omega)|J\Omega_+ M\rangle$$

$$-b_J b'_J(-1)^{p+p'} \left\langle -\Lambda';-\frac{1}{2};\mathrm{v}'J'-\Omega'_+ \middle| \hat{\mu}_q \middle| -\Lambda;-\frac{1}{2};\mathrm{v}J-\Omega_+ \right\rangle \langle J'-\Omega'_+ M'|D^{1*}_{pq}(\omega)|J-\Omega_+ M\rangle$$

$$+b_J a'_J(-1)^{p} \left\langle \Lambda';-\frac{1}{2};\mathrm{v}'J'\Omega'_- \middle| \hat{\mu}_q \middle| -\Lambda;-\frac{1}{2};\mathrm{v}J-\Omega_+ \right\rangle \langle J'\Omega'_- M'|D^{1*}_{pq}(\omega)|J-\Omega_+ M\rangle \bigg\}. \tag{5.100}$$

The molecule-fixed electric dipole moment matrix elements are given by

$$\langle \pm\Lambda'\,\mathrm{v}'\,J' \pm \Omega'_k|\hat{\mu}_q|\pm\Lambda\,\mathrm{v}\,J \pm \Omega_n\rangle \tag{5.101}$$

or

$$\langle \pm\Lambda'\,\mathrm{v}'\,J' \pm \Omega'_k|\hat{\mu}_q|\mp\Lambda\,\mathrm{v}\,J \mp \Omega_n\rangle, \tag{5.102}$$

where the symbols k and n in Eqs. (5.101) and (5.102) each represent either a "plus" sign or a "minus" sign. Although the terms J and Ω appear in these expressions, the evaluation of these terms does not involve integration over angular variables. Rather, the terms J and Ω appear because of the centrifugal distortion of the vibrational wavefunction, which becomes more significant as the rotational quantum number J increases. We will assume that the matrix elements in Eqs. (5.101) and (5.102)will be the same regardless of the signs on Λ and Ω, i.e.,

$$\langle \Lambda'\,\mathrm{v}'\,J'\,\Omega'_\pm|\hat{\mu}_q|\Lambda\,\mathrm{v}\,J\,\Omega_\pm\rangle = \langle -\Lambda'\,\mathrm{v}'\,J' - \Omega'_\pm|\hat{\mu}_q|-\Lambda\,\mathrm{v}\,J - \Omega_\pm\rangle. \tag{5.103}$$

$$\langle \Lambda'\,\mathrm{v}'\,J'\,\Omega'_+|\hat{\mu}_q|-\Lambda\,\mathrm{v}\,J - \Omega_-\rangle = \langle -\Lambda'\,\mathrm{v}'\,J' - \Omega'_+|\hat{\mu}_q|\Lambda\,\mathrm{v}\,J\,\Omega_-\rangle. \tag{5.104}$$

$$\langle \Lambda'\,\mathrm{v}'\,J'\,\Omega'_-|\hat{\mu}_q|-\Lambda\,\mathrm{v}\,J - \Omega_+\rangle = \langle -\Lambda'\,\mathrm{v}'\,J' - \Omega'_-|\hat{\mu}_q|\Lambda\,\mathrm{v}\,J\,\Omega_+\rangle. \tag{5.105}$$

The matrix element in Eq. (5.101) can be written as

$$\begin{aligned}\tilde{\mu}_{qkn}(\mathrm{v}',J',\Omega'_k,\mathrm{v}J\Omega_n) &= \langle \Lambda'\mathrm{v}'J'\Omega'_k|\hat{\mu}_q|\Lambda\mathrm{v}J\Omega_n\rangle \\ &= \int_0^\infty \psi_{\mathrm{v}'J'\Omega'_k}(r)\,R_{e,\,q}(r)\,\psi_{\mathrm{v}J\Omega_n}(r)dr.\end{aligned} \tag{5.106}$$

The vibration-rotation functions $\psi_{\mathrm{v}'J'\Omega'}(r)$ and $\psi_{\mathrm{v}J\Omega}(r)$ are calculated using the programs RKR1 (Le Roy, 2017a) and LEVEL (Le Roy, 2017b) developed by R. J. Le Roy. The program RKR1 is used to determine the turning points of the potential function for the diatomic molecules of interest. The Dunham coefficients for the molecules such as OH and NO are the input parameters needed for the RKR1 calculations. The potential function turning points were then used as input for the program LEVEL. The program LEVEL was developed to solve the radial SWE

$$-\frac{\hbar^2}{2\mu}\frac{d^2\psi_{\mathrm{v},J}(r)}{dr^2}+\left\{V(r)+\frac{\left[J(J+1)-\Lambda^2\right]\hbar^2}{2\mu r^2}\right\}\psi_{\mathrm{v},J}(r)=E_{\mathrm{v},J}(r)\psi_{\mathrm{v},J}(r). \tag{5.107}$$

using RKR methods. In Eq. (5.107), $V(r)$ is the rotationless vibrational potential, and $\left[J(J+1)-\Lambda^2\right]\hbar^2/2\mu r^2$ is the potential due to centrifugal forces induced by the rotation of the molecule; the symbol μ in Eq. (5.107) is the reduced mass (kg). The results of the LEVEL calculation cannot directly be applied to doublet intermediate states or to case (a) wavefunctions because Λ is always assumed to be an integer in the LEVEL code. We use an interpolation scheme to calculate the wavefunctions for the case (a) states with the centrifugal potential term given by $\left[J(J+1)-\Omega^2\right]\hbar^2/2\mu r^2$, where Ω is half-integral; this will be discussed below in calculations for OH and NO.

Combining Eqs. (5.100), (5.106), and (5.36), we obtain

$$\begin{aligned}\mu_{ba,p} = {} & \frac{\sqrt{(2J+1)(2J'+1)}}{2}\begin{pmatrix} J & J' & 1 \\ M & -M' & p\end{pmatrix}(-1)^{M'+\frac{1}{2}} \\ & \times\sum_q\left\{a_J d'_J\tilde{\mu}_{q--}\left[(-1)^{-\Lambda'}\begin{pmatrix} J & J' & 1 \\ \Lambda-\frac{1}{2} & -\Lambda'+\frac{1}{2} & \Delta\Lambda\end{pmatrix}-(-1)^{\Lambda'-1+p+p'}\begin{pmatrix} J & J' & 1 \\ -\Lambda+\frac{1}{2} & \Lambda'-\frac{1}{2} & -\Delta\Lambda\end{pmatrix}\right]\right. \\ & +b_J b'_J\tilde{\mu}_{q++}\left[(-1)^{-\Lambda'-1}\begin{pmatrix} J & J' & 1 \\ \Lambda+\frac{1}{2} & -\Lambda'-\frac{1}{2} & \Delta\Lambda\end{pmatrix}-(-1)^{\Lambda'+p+p'}\begin{pmatrix} J & J' & 1 \\ -\Lambda-\frac{1}{2} & \Lambda'+\frac{1}{2} & -\Delta\Lambda\end{pmatrix}\right] \\ & +a_J b'_J\tilde{\mu}_{q+-}\left[-(-1)^{\Lambda'+p'}\begin{pmatrix} J & J' & 1 \\ \Lambda-\frac{1}{2} & \Lambda'+\frac{1}{2} & -\Sigma\Lambda\end{pmatrix}+(-1)^{-\Lambda'-1+p}\begin{pmatrix} J & J' & 1 \\ -\Lambda+\frac{1}{2} & -\Lambda'-\frac{1}{2} & \Sigma\Lambda\end{pmatrix}\right] \\ & \left.+b_J d'_J\tilde{\mu}_{q-+}\left[-(-1)^{\Lambda'-1+p'}\begin{pmatrix} J & J' & 1 \\ \Lambda+\frac{1}{2} & \Lambda'-\frac{1}{2} & -\Sigma\Lambda\end{pmatrix}+(-1)^{-\Lambda'+p}\begin{pmatrix} J & J' & 1 \\ -\Lambda-\frac{1}{2} & -\Lambda'+\frac{1}{2} & \Sigma\Lambda\end{pmatrix}\right]\right\},\end{aligned} \tag{5.108}$$

where $\Delta\Lambda = \Lambda' - \Lambda$ and $\Sigma\Lambda = \Lambda + \Lambda'$. The expression in Eq. (5.108) can be adapted for analysis of (1) $F_1 \rightarrow F_2$ transitions by substituting c'_J and d'_J for a'_J and b'_J, respectively; of (2) $F_2 \rightarrow F_1$ transitions by substituting c_J and d_J for a_J and b_J, respectively; and of (3) $F_2 \rightarrow F_2$ transitions by substituting c_J, c'_J, d_J, and d'_J for a_J, a'_J, b_J, and b'_J, respectively.

The spontaneous emission rate coefficient $A_{J'J''}$ (s^{-1}) for transitions from the upper level J' to the lower level J is given by (Bernath, 2016; Zare, 1988)

$$A_{J'J} = \frac{16\pi^3 \nu^3_{J'J''}}{3\varepsilon_0 c^3 (2J'+1)} \sum_p \sum_M \sum_{M'} \mu_{ba,p}\mu^*_{ba,p} = \frac{16\pi^3 \tilde{\nu}^3_{J'J''}}{3\varepsilon_0 (2J'+1)} \left[3 \sum_M \sum_{M'} \mu_{ba,p=0}\mu^*_{ba,p=0} \right]. \tag{5.109}$$

In Eq. (5.109), $\nu_{J'J''}$ is the transition frequency in Hz, $\tilde{\nu}_{J'J''}$ is the transition frequency in wavenumbers (cm^{-1}), and due to the isotropy of space

$$\sum_M \sum_{M'} \mu_{ba,p=0}\mu^*_{ba,p=0} = \sum_M \sum_{M'} \mu_{ba,p=+1}\mu^*_{ba,p=+1} = \sum_M \sum_{M'} \mu_{ba,p=-1}\mu^*_{ba,p=-1}. \tag{5.110}$$

5.3.2 Radiative Transitions between Doublet Electronic Levels with $\Delta\Lambda = 0$

For $\Delta\Lambda = 0$ ($\Lambda = \Lambda'$), Eq. (5.108) reduces to

$$\begin{aligned}
\mu_{ba,p} = {} & \frac{\sqrt{(2J+1)(2J'+1)}}{2} \begin{pmatrix} J & J' & 1 \\ M & -M' & p \end{pmatrix} (-1)^{M'+\frac{1}{2}} \\
& \times \sum_q \left\{ a_J d'_J \tilde{\mu}_{0--} (-1)^{-\Lambda} \begin{pmatrix} J & J' & 1 \\ \Lambda-\frac{1}{2} & -\Lambda+\frac{1}{2} & \Delta\Lambda \end{pmatrix} \left[1-(-1)^{2\Lambda-1+J-\frac{1}{2}+s+J'-\frac{1}{2}+s'+J+J'+1}\right] \right. \\
& + b_J b'_J \tilde{\mu}_{0++} (-1)^{-\Lambda-1} \begin{pmatrix} J & J' & 1 \\ \Lambda+\frac{1}{2} & -\Lambda-\frac{1}{2} & \Delta\Lambda \end{pmatrix} \left[1-(-1)^{2\Lambda+1+J-\frac{1}{2}+s+J'-\frac{1}{2}+s'+J+J'+1}\right] \\
& + a_J b'_J \begin{pmatrix} J & J' & 1 \\ \Lambda-\frac{1}{2} & \Lambda+\frac{1}{2} & -2\Lambda \end{pmatrix} \left[-(-1)^{\Lambda+J'-\frac{1}{2}+s'} \tilde{\mu}_{-2\Lambda+-} + (-1)^{-\Lambda-1+J-\frac{1}{2}+s+J+J'+1} \tilde{\mu}_{2\Lambda+-}\right] \\
& \left. + b_J a'_J \tilde{\mu}_{q-+} \begin{pmatrix} J & J' & 1 \\ \Lambda+\frac{1}{2} & \Lambda'-\frac{1}{2} & -2\Lambda \end{pmatrix} \left[-(-1)^{\Lambda'-1+J'-\frac{1}{2}+s'} \tilde{\mu}_{-2\Lambda-+} + (-1)^{-\Lambda'+J-\frac{1}{2}+s+J+J'+1} \tilde{\mu}_{2\Lambda-+}\right] \right\},
\end{aligned} \tag{5.111}$$

where $\Lambda = \Lambda'$. Collecting terms in Eq. (5.111) and simplifying, we obtain

$$\begin{aligned}
\mu_{ba,p} = {} & \frac{\sqrt{(2J+1)(2J'+1)}}{2} \begin{pmatrix} J & J' & 1 \\ M & -M' & p \end{pmatrix} (-1)^{M'+\frac{1}{2}} \\
& \times \left\{ a_J d'_J (-1)^{-\Lambda} \begin{pmatrix} J & J' & 1 \\ \Lambda-\frac{1}{2} & -\Lambda+\frac{1}{2} & 0 \end{pmatrix} \tilde{\mu}_{0--} \left[1-(-1)^{2\Lambda-1+2J+2J'+s+s'}\right] \right. \\
& + b_J b'_J (-1)^{-\Lambda-1} \begin{pmatrix} J & J' & 1 \\ \Lambda+\frac{1}{2} & -\Lambda-\frac{1}{2} & 0 \end{pmatrix} \tilde{\mu}_{0++} \left[1-(-1)^{2\Lambda+1+2J+2J'+s+s'}\right] \\
& + (-1)^{\Lambda+J'-\frac{1}{2}+s'} a_J b'_J \begin{pmatrix} J & J' & 1 \\ \Lambda-\frac{1}{2} & \Lambda+\frac{1}{2} & -2\Lambda \end{pmatrix} \left[-\tilde{\mu}_{-2\Lambda+-} + (-1)^{-2\Lambda+2J+2+s-s'} \tilde{\mu}_{2\Lambda+-}\right] \\
& \left. + (-1)^{\Lambda+J'-\frac{3}{2}+s'} b_J a'_J \begin{pmatrix} J & J' & 1 \\ \Lambda+\frac{1}{2} & \Lambda-\frac{1}{2} & -2\Lambda \end{pmatrix} \left[-\tilde{\mu}_{-2\Lambda-+} + (-1)^{-2\Lambda+2J+2+s-s'} \tilde{\mu}_{2\Lambda-+}\right] \right\}.
\end{aligned} \tag{5.112}$$

In Eq. (5.112), 2Λ is always an even number, and $2J$ and $2J'$ are always odd numbers, and rewriting Eq. (5.112), we obtain

$$\begin{aligned}\mu_{ba,p} &= \frac{\sqrt{(2J+1)(2J'+1)}}{2}\begin{pmatrix} J & J' & 1 \\ M & -M' & p \end{pmatrix}(-1)^{M'+\frac{1}{2}} \\ &\times\left\{a_J a'_J(-1)^{-\Lambda}\begin{pmatrix} J & J' & 1 \\ \Lambda-\frac{1}{2} & -\Lambda+\frac{1}{2} & 0 \end{pmatrix}\tilde{\mu}_{0--}\left[1+(-1)^{s+s'}\right]\right. \\ &+b_J b'_J(-1)^{-\Lambda-1}\begin{pmatrix} J & J' & 1 \\ \Lambda+\frac{1}{2} & -\Lambda-\frac{1}{2} & 0 \end{pmatrix}\tilde{\mu}_{0++}\left[1+(-1)^{s+s'}\right] \\ &+(-1)^{\Lambda+J'-\frac{1}{2}+s'}a_J b'_J\begin{pmatrix} J & J' & 1 \\ \Lambda-\frac{1}{2} & \Lambda+\frac{1}{2} & -2\Lambda \end{pmatrix}\left[-\tilde{\mu}_{-2\Lambda+-}-(-1)^{s-s'}\tilde{\mu}_{2\Lambda+-}\right] \\ &\left.+(-1)^{\Lambda+J'-\frac{3}{2}+s'}b_J a'_J\begin{pmatrix} J & J' & 1 \\ \Lambda+\frac{1}{2} & \Lambda-\frac{1}{2} & -2\Lambda \end{pmatrix}\left[-\tilde{\mu}_{-2\Lambda-+}-(-1)^{s-s'}\tilde{\mu}_{2\Lambda-+}\right]\right\}.\end{aligned} \tag{5.113}$$

The *3j* symbols with $q = 0$ can be evaluated using formulae given in Appendix A3. The *3j* symbols are given by

$J' = J + 1$

$$\begin{pmatrix} J & J+1 & 1 \\ \Lambda-\frac{1}{2} & -\Lambda+\frac{1}{2} & 0 \end{pmatrix} = (-1)^{J-\Lambda-\frac{1}{2}}\sqrt{\frac{(2J+2\Lambda+1)(2J-2\Lambda+3)}{4(2J+3)(J+1)(2J+1)}}. \tag{5.114}$$

$$\begin{pmatrix} J & J+1 & 1 \\ \Lambda+\frac{1}{2} & -\Lambda-\frac{1}{2} & 0 \end{pmatrix} = (-1)^{J-\Lambda+\frac{1}{2}}\sqrt{\frac{(2J+2\Lambda+3)(2J-2\Lambda+1)}{4(2J+3)(J+1)(2J+1)}}. \tag{5.115}$$

$J' = J$

$$\begin{pmatrix} J & J & 1 \\ \Lambda-\frac{1}{2} & -\Lambda+\frac{1}{2} & 0 \end{pmatrix} = (-1)^{J-\Lambda+\frac{1}{2}}\frac{\Lambda-\frac{1}{2}}{\sqrt{(2J+1)J(J+1)}}. \tag{5.116}$$

$$\begin{pmatrix} J & J & 1 \\ \Lambda+\frac{1}{2} & -\Lambda-\frac{1}{2} & 0 \end{pmatrix} = (-1)^{J-\Lambda-\frac{1}{2}}\frac{\Lambda+\frac{1}{2}}{\sqrt{(2J+1)J(J+1)}}. \tag{5.117}$$

$J' = J - 1$

$$\begin{pmatrix} J & J-1 & 1 \\ \Lambda-\frac{1}{2} & -\Lambda+\frac{1}{2} & 0 \end{pmatrix} = (-1)^{J-\Lambda-\frac{1}{2}}\sqrt{\frac{(2J+2\Lambda-1)(2J-2\Lambda+1)}{4(2J+1)J(2J-1)}}. \tag{5.118}$$

$$\begin{pmatrix} J & J-1 & 1 \\ \Lambda+\frac{1}{2} & -\Lambda-\frac{1}{2} & 0 \end{pmatrix} = (-1)^{J-\Lambda+\frac{1}{2}}\sqrt{\frac{(2J+2\Lambda+1)(2J-2\Lambda-1)}{4(2J+1)J(2J-1)}}. \tag{5.119}$$

Radiative Transitions between Doublet Electronic Levels with $\Delta\Lambda = 0$, $\Lambda = 0$ ($^2\Sigma$–$^2\Sigma$ Transitions)

For $\Lambda = \Lambda' = 0$, again assuming that the lower level has positive parity and the upper level has negative parity, the Hund's case (a) wavefunction coefficients become $a_J = [1+(-1)^\rho]/4$, $a'_J = [1-(-1)^{\rho'}]/4$, $b_J = [1+(-1)^\rho]/4$, $b'_J = [1-(-1)^{\rho'}]/4$, and Eq. (5.113) reduces to

$$\mu_{ba,p} = \frac{\sqrt{(2J+1)(2J'+1)}}{32}\begin{pmatrix} J & J' & 1 \\ M & -M' & p \end{pmatrix}(-1)^{M'+\frac{1}{2}}\tilde{\mu}_0\begin{pmatrix} J & J' & 1 \\ -\frac{1}{2} & \frac{1}{2} & 0 \end{pmatrix}$$
$$\times\Big\{\left[1+(-1)^\rho\right]\left[1+(-1)^{\rho'}\right]\left[-(-1)^{J+J'+1}+1\right]\left[1+(-1)^{s+s'}\right]$$
$$+(-1)^{\rho'}\left[1+(-1)^\rho\right]\left[1+(-1)^{\rho'}\right]\left[(-1)^{J+J'+1}-1\right]\left[1+(-1)^{s-s'}\right]\Big\}. \tag{5.120}$$

It is immediately evident that when $s \neq s'$, the dipole moment matrix element will be zero; in other words, single-photon transitions coupling levels in $^2\Sigma^+$ $(s=0)$ and $^2\Sigma^-$ $(s=1)$ levels are not allowed. Furthermore, there is no difference in the magnitude of Ω_+ and Ω_-, so $\tilde{\mu}_{0--} = \tilde{\mu}_{0++} = \tilde{\mu}_{0-+} = \tilde{\mu}_{0+-} = \tilde{\mu}_0$. Assuming that $s = s'$ is zero, Eq. (5.120) can be written as

$$\mu_{ba,p} = -\frac{\sqrt{(2J+1)(2J'+1)}}{16}\begin{pmatrix} J & J' & 1 \\ M & -M' & p \end{pmatrix}(-1)^{M'+\frac{1}{2}}\tilde{\mu}_0\begin{pmatrix} J & J' & 1 \\ -\frac{1}{2} & \frac{1}{2} & 0 \end{pmatrix}$$
$$\times\left[1-(-1)^{J+J'+1}\right]\left[1+(-1)^\rho\right]\left[1-(-1)^{\rho'}\right]\left[1-(-1)^{\rho'}\right]. \tag{5.121}$$

The dipole moment matrix element will be zero for odd ρ and even ρ', consistent with our assumption of a positive-parity lower level and negative-parity upper level for the $F_1 \to F_1$ transition. The dipole moment matrix element will also be zero when $J = J'$ because $2J$ is always an odd number. For even ρ, odd ρ', and $J' = J \pm 1$, we obtain

$$\mu_{ba,p} = -\sqrt{(2J+1)(2J'+1)}\begin{pmatrix} J & J' & 1 \\ M & -M' & p \end{pmatrix}(-1)^{M'+\frac{1}{2}}\tilde{\mu}_0\begin{pmatrix} J & J' & 1 \\ -\frac{1}{2} & \frac{1}{2} & 0 \end{pmatrix}. \tag{5.122}$$

Taking the modulus squared of the dipole moment, summing over all M and M', and summing over all p, we obtain

$$\sum_p\sum_M\sum_{M'}\mu_{ba,p}\mu^*_{ba,p} = (2J+1)(2J'+1)\begin{pmatrix} J & J' & 1 \\ -\frac{1}{2} & \frac{1}{2} & 0 \end{pmatrix}^2\tilde{\mu}_0^2. \tag{5.123}$$

The Einstein coefficient for spontaneous emission is obtained by substituting Eq. (5.123) into Eq. (5.109). The results will be exactly the same if it is assumed that the lower level is a negative-parity level and the upper level is a positive-parity level.

For $F_2 \to F_2$ transitions, the dipole moment matrix elements can be calculated by replacing a_J, a'_J, b_J, and b'_J by c_J, c'_J, d_J, and d'_J, respectively, in Eq. (5.113). For $F_2 \to F_2$ transitions from a positive-parity lower level to a negative-parity upper level, $c_J = -[1-(-1)^p]/4$, $c'_J = -[1+(-1)^{p'}]/4$, $d_J = [1-(-1)^p]/4$, $d'_J = [1+(-1)^{p'}]/4$, and Eq. (5.113) becomes

$$\mu_{ba,p} = \frac{\sqrt{(2J+1)(2J'+1)}}{32}\begin{pmatrix} J & J' & 1 \\ M & -M' & p \end{pmatrix}(-1)^{M'+\frac{1}{2}}\tilde{\mu}_0\begin{pmatrix} J & J' & 1 \\ -\frac{1}{2} & \frac{1}{2} & 0 \end{pmatrix}$$
$$\times\Big\{\Big[(-1)^{J+J'+1}-1\Big]\Big[1-(-1)^p\Big]\Big[1+(-1)^{p'}\Big]\Big[1+(-1)^{s+s'}\Big]$$
$$-(-1)^{p'}\Big[-(-1)^{J+J'+1}+1\Big]\Big[1-(-1)^p\Big]\Big[1+(-1)^{p'}\Big]\Big[1+(-1)^{s-s'}\Big]\Big\}. \tag{5.124}$$

For $F_2 \to F_2$ transitions, P_2 and R_2 transitions are allowed, but the dipole moment matrix elements are zero for Q_2 transitions. Equation (5.124) is identical to Eq. (5.120), and so is the final expression for the dipole moment matrix elements,

$$\sum_p\sum_M\sum_{M'}\mu_{ba,p}\mu^*_{ba,p} = (2J+1)(2J'+1)\begin{pmatrix} J & J' & 1 \\ -\frac{1}{2} & \frac{1}{2} & 0 \end{pmatrix}^2\tilde{\mu}_0^2. \tag{5.125}$$

For $F_1 \to F_2$ transitions, the dipole moment matrix elements can be calculated by replacing a'_J and b'_J by c'_J and d'_J, respectively, in Eq. (5.113). Equation (5.113) becomes

$$\mu_{ba,p} = \frac{\sqrt{(2J+1)(2J'+1)}}{32}\begin{pmatrix} J & J' & 1 \\ M & -M' & p \end{pmatrix}(-1)^{M'+\frac{1}{2}}\tilde{\mu}_0\begin{pmatrix} J & J' & 1 \\ -\frac{1}{2} & \frac{1}{2} & 0 \end{pmatrix}$$
$$\times\Big\{\Big[1+(-1)^p\Big]\Big[1+(-1)^{p'}\Big]\Big[1+(-1)^{J+J'+1}\Big]\Big[1+(-1)^{s+s'}\Big]$$
$$-(-1)^{p'}\Big[1+(-1)^p\Big]\Big[1+(-1)^{p'}\Big]\Big[1+(-1)^{J+J'+1}\Big]\Big[1+(-1)^{s-s'}\Big]\Big\}. \tag{5.126}$$

Equation (5.126) is different in that now the dipole moment matrix elements will be nonzero only when $J = J'$. In particular, the only nonzero matrix elements for $F_1 \to F_2$ transitions are for ${}^RQ_{21}$ transitions. Similarly, the only nonzero matrix elements for $F_2 \to F_1$ transitions are for ${}^PQ_{12}$ transitions.

Referring to Eq. (5.109), we can define the line strength $\tilde{S}_{J''J'}$ as

$$\tilde{S}_{JJ'} = A_{J'J}\frac{3\varepsilon_0(2J'+1)}{16\pi^3\,\tilde{\nu}^3_{J'J}} = 3\sum_M\sum_{M'}\mu_{ba,p}\mu^*_{ba,p}. \tag{5.127}$$

Substituting Eqs. (5.114), (5.116), and (5.118) into Eq. (5.123) we obtain the results listed in Table 5.2 for the line strength $\tilde{S}_{JJ'}$. For vibrational bands where vibration-

Table 5.2 Line strength expressions for vibration-rotation transitions between a ground electronic level with $\Lambda = 0$ and an excited electronic level with $\Lambda' = 0$

$^2\Sigma^\pm \to (^2\Sigma^\pm)'$, $\Lambda = \Lambda'$	
$J = J''$	$\tilde{S}_{JJ'}$
$P_1(J)$	$\tilde{\mu}_0^2(2J+1)(2J-1)/4J$
$Q_1(J)$	0
$R_1(J)$	$\tilde{\mu}_0^2(2J+1)(2J+3)/4(J+1)$
$^QP_{21}(J)$	0
$^RQ_{21}(J)$	$\tilde{\mu}_0^2(2J+1)/4J(J+1)$
$^SR_{21}(J)$	0
$^OP_{12}(J)$	0
$^PQ_{12}(J)$	$\tilde{\mu}_0^2(2J+1)/4J(J+1)$
$^QR_{12}(J)$	0
$P_2(J)$	$\tilde{\mu}_0^2(2J+1)(2J-1)/4J$
$Q_2(J)$	0
$R_2(J)$	$\tilde{\mu}_0^2(2J+1)(2J+3)/4(J+1)$

rotation interactions are negligible, we can define the Hönl–London factor. The Hönl–London factor is given by (Bernath, 2016)

$$S_J^{\Delta J} = 3\sum_{M'}\sum_{M}\frac{\mu_{ba.p=0}\mu^*_{ba,p=0}}{\tilde{\mu}_0^2} = \frac{\tilde{S}_{JJ'}}{\tilde{\mu}_0^2} = (2J+1)(2J'+1)\begin{pmatrix} J & J' & 1 \\ \frac{1}{2} & -\frac{1}{2} & 0 \end{pmatrix}^2. \tag{5.128}$$

The Hönl–London factors are useful when the band intensity $\tilde{\mu}_0$ is independent of J, i.e., when vibration-rotation interactions are insignificant (Herman & Wallis, 1955). The allowed transitions for the CN $B^2\Sigma^+(\mathrm{v}') - X^2\Sigma^+(\mathrm{v})$ system are shown in Figure 5.1.

Radiative Transitions between Doublet Electronic Levels with $\Delta\Lambda = 0$, $\Lambda > 0$

For $\Lambda \geq 1$, the $3j$ symbols with $q = \pm 2\Lambda$ will be zero because the magnitude of q is greater than $k = 1$. Equation (5.113) therefore reduces to

$$\mu_p = \sqrt{(2J+1)(2J'+1)}\begin{pmatrix} J & J' & 1 \\ M & -M' & p \end{pmatrix}(-1)^{M'+\frac{1}{2}+\Lambda}$$
$$\times\left[-a_J a'_J\tilde{\mu}_{0--}\begin{pmatrix} J & J' & 1 \\ \Lambda-\frac{1}{2} & -\Lambda+\frac{1}{2} & 0 \end{pmatrix} + b_J b'_J\tilde{\mu}_{0++}\begin{pmatrix} J & J' & 1 \\ \Lambda+\frac{1}{2} & -\Lambda-\frac{1}{2} & 0 \end{pmatrix}\right]. \tag{5.129}$$

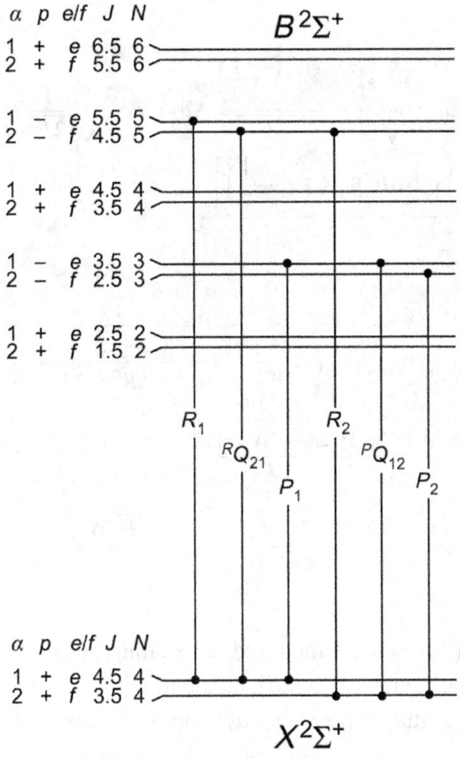

Figure 5.1 Allowed rotational transitions for $^2\Sigma^+(\mathrm{v}') - {}^2\Sigma^+(\mathrm{v})$ electronic systems. This particular diagram pertains to the CN molecule.

The square of the electric dipole moment matrix elements is given by

$$\mu_p\mu_p^* = (2J+1)(2J'+1)\begin{pmatrix} J & J' & 1 \\ M & -M' & p \end{pmatrix}^2$$
$$\times\left[-a_J a'_J \tilde{\mu}_{0--}\begin{pmatrix} J & J' & 1 \\ \Lambda-\frac{1}{2} & -\Lambda+\frac{1}{2} & 0 \end{pmatrix} + b_J b'_J \tilde{\mu}_{0++}\begin{pmatrix} J & J' & 1 \\ \Lambda+\frac{1}{2} & -\Lambda-\frac{1}{2} & 0 \end{pmatrix}\right]^2. \tag{5.130}$$

For $R_1(J)$, $Q_1(J)$, and $P_1(J)$ main-branch $F_1(J) \to F_1(J')$ transitions, the normalized line strengths are given by

$P_1(J): \quad J' = J - 1$

$$\tilde{S}_{J''J'} = (2J+1)(2J-1)\left[-a_J a'_J \tilde{\mu}_{0--}(-1)^{J-\Lambda-\frac{1}{2}}\sqrt{\frac{(2J+2\Lambda-1)(2J-2\Lambda+1)}{4(2J+1)J(2J-1)}}\right.$$
$$\left.+ b_J b'_J \tilde{\mu}_{0++}(-1)^{J-\Lambda+\frac{1}{2}}\sqrt{\frac{(2J+2\Lambda+1)(2J-2\Lambda-1)}{4(2J+1)J(2J-1)}}\right]^2$$
$$= \frac{\left[a_J a'_J \tilde{\mu}_{0--}\sqrt{(2J+2\Lambda-1)(2J-2\Lambda+1)} + b_J b'_J \tilde{\mu}_{0++}\sqrt{(2J+2\Lambda+1)(2J-2\Lambda-1)}\right]^2}{4J}. \tag{5.131}$$

$Q_1(J): \quad J' = J$

$$\tilde{S}_{J''J'} = (2J+1)(2J+1)\left[-a_J a'_J \tilde{\mu}_{0--}\frac{(-1)^{J-\Lambda+\frac{1}{2}}\left(\Lambda - \frac{1}{2}\right)}{\sqrt{(2J+1)J(J+1)}} + b_J b'_J \tilde{\mu}_{0++}\frac{(-1)^{J-\Lambda-\frac{1}{2}}\left(\Lambda+\frac{1}{2}\right)}{\sqrt{(2J+1)J(J+1)}}\right]^2$$
$$= \frac{(2J+1)\left[a_J a'_J \tilde{\mu}_{0--}\left(\Lambda - \frac{1}{2}\right)\right] + b_J b'_J \tilde{\mu}_{0++}\left(\Lambda + \frac{1}{2}\right)\Big]^2}{J(J+1)}. \tag{5.132}$$

$R_1(J): \quad J' = J+1$

$$\tilde{S}_{J''J'} = (2J+1)(2J+3)\left[-a_J a'_J \tilde{\mu}_{0--}(-1)^{J-\Lambda-\frac{1}{2}}\sqrt{\frac{(2J+2\Lambda+1)(2J-2\Lambda+3)}{4(2J+3)(J+1)(2J+1)}}\right.$$
$$\left. + b_J b'_J \tilde{\mu}_{0++}(-1)^{J-\Lambda+\frac{1}{2}}\sqrt{\frac{(2J+2\Lambda+3)(2J-2\Lambda+1)}{4(2J+3)(J+1)(2J+1)}}\right]^2$$
$$= \frac{\left[a_J a'_J \tilde{\mu}_{0--}\sqrt{(2J+2\Lambda+1)(2J-2\Lambda+3)} + b_J b'_J \tilde{\mu}_{0++}\sqrt{(2J+2\Lambda+3)(2J-2\Lambda+1)}\right]^2}{4(J+1)}. \tag{5.133}$$

As discussed above, Eq. (5.130) can be modified in a straightforward manner to calculate the square of the dipole moment matrix elements for $F_1 \to F_2$, $F_2 \to F_1$, and $F_2 \to F_2$ transitions by substituting for a_J, b_J, a'_J, and b'_J. The results are listed in Table 5.3.

5.3.3 Radiative Transitions between Doublet Electronic Levels with $\Delta\Lambda = +1$

Analyzing Eq. (5.108) for $\Delta\Lambda = \Lambda' - \Lambda = +1,\ \Lambda' = \Lambda + 1$, we obtain

$$\mu_p = \frac{\sqrt{(2J+1)(2J'+1)}}{2}\begin{pmatrix} J & J' & 1 \\ M & -M' & p \end{pmatrix}(-1)^{M'+\frac{1}{2}}$$
$$\times\left\{a_J a'_J\left[(-1)^{-\Lambda-1}\begin{pmatrix} J & J' & 1 \\ \Lambda-\frac{1}{2} & -\Lambda-\frac{1}{2} & 1 \end{pmatrix}\tilde{\mu}_{1--} - (-1)^{\Lambda+\rho+\rho'}\begin{pmatrix} J & J' & 1 \\ -\Lambda+\frac{1}{2} & \Lambda+\frac{1}{2} & -1 \end{pmatrix}\tilde{\mu}_{-1--}\right]\right.$$
$$+ b_J b'_J\left[(-1)^{-\Lambda}\begin{pmatrix} J & J' & 1 \\ \Lambda+\frac{1}{2} & -\Lambda-\frac{3}{2} & 1 \end{pmatrix}\tilde{\mu}_{1++} - (-1)^{\Lambda+1+\rho+\rho'}\begin{pmatrix} J & J' & 1 \\ -\Lambda-\frac{1}{2} & \Lambda+\frac{3}{2} & -1 \end{pmatrix}\tilde{\mu}_{-1++}\right]$$
$$+ a_J b'_J\left[-(-1)^{\Lambda+1+\rho'}\begin{pmatrix} J & J' & 1 \\ \Lambda-\frac{1}{2} & \Lambda+\frac{3}{2} & -(2\Lambda+1) \end{pmatrix}\tilde{\mu}_{-(2\Lambda+1)+-}\right.$$
$$\left. +(-1)^{-\Lambda+\rho}\begin{pmatrix} J & J' & 1 \\ -\Lambda+\frac{1}{2} & -\Lambda-\frac{3}{2} & 2\Lambda+1 \end{pmatrix}\tilde{\mu}_{(2\Lambda+1)+-}\right]$$
$$+ b_J a'_J\left[-(-1)^{\Lambda+\rho'}\begin{pmatrix} J & J' & 1 \\ \Lambda+\frac{1}{2} & \Lambda+\frac{1}{2} & -(2\Lambda+1) \end{pmatrix}\tilde{\mu}_{-(2\Lambda+1)-+}\right.$$
$$\left.\left. +(-1)^{-\Lambda-1+\rho}\begin{pmatrix} J & J' & 1 \\ -\Lambda-\frac{1}{2} & -\Lambda-\frac{1}{2} & 2\Lambda+1 \end{pmatrix}\tilde{\mu}_{(2\Lambda+1)-+}\right]\right\}. \tag{5.134}$$

Table 5.3 Line strength expressions for vibration-rotation transitions between a ground electronic level with $\Lambda > 0$ and an excited electronic level with $\Lambda' > 0$, and for $\Lambda = \Lambda'$

$^2X \rightarrow {}^2X'$, $\Lambda' = \Lambda > 1$
$\tilde{S}_{J''J'}$

$P_1(J)$	$(1/4J)\left[a_J a'_J \tilde{\mu}_{0--}\sqrt{(2J+2\Lambda-1)(2J-2\Lambda+1)}\right.$ $\left.+\, b_J b'_J \tilde{\mu}_{0++}\sqrt{(2J+2\Lambda+1)(2J-2\Lambda-1)}\right]^2$
$Q_1(J)$	$[(2J+1)/J(J+1)]\left[a_J a'_J \tilde{\mu}_{0--}\left(\Lambda-\tfrac{1}{2}\right)+b_J b'_J \tilde{\mu}_{0++}\left(\Lambda+\tfrac{1}{2}\right)\right]^2$
$R_1(J)$	$[1/4(J+1)]\left[a_J a'_J \tilde{\mu}_{0--}\sqrt{(2J+2\Lambda+1)(2J-2\Lambda+3)}\right.$ $\left.+\, b_J b'_J \tilde{\mu}_{0++}\sqrt{(2J+2\Lambda+3)(2J-2\Lambda+1)}\right]^2$
$^QP_{21}(J)$	$(1/4J)\left[a_J c'_J \tilde{\mu}_{0--}\sqrt{(2J+2\Lambda-1)(2J-2\Lambda+1)}\right.$ $\left.+\, b_J d'_J \tilde{\mu}_{0++}\sqrt{(2J+2\Lambda+1)(2J-2\Lambda-1)}\right]^2$
$^RQ_{21}(J)$	$[(2J+1)/J(J+1)]\left[a_J c'_J \tilde{\mu}_{0--}\left(\Lambda-\tfrac{1}{2}\right)+b_J d'_J \tilde{\mu}_{0++}\left(\Lambda+\tfrac{1}{2}\right)\right]^2$
$^SR_{21}(J)$	$[1/4(J+1)]\left[a_J c'_J \tilde{\mu}_{0--}\sqrt{(2J+2\Lambda+1)(2J-2\Lambda+3)}\right.$ $\left.+\, b_J d'_J \tilde{\mu}_{0++}\sqrt{(2J+2\Lambda+3)(2J-2\Lambda+1)}\right]^2$
$^OP_{12}(J)$	$(1/4J)\left[c_J a'_J \tilde{\mu}_{0--}\sqrt{(2J+2\Lambda-1)(2J-2\Lambda+1)}\right.$ $\left.+\, d_J b'_J \tilde{\mu}_{0++}\sqrt{(2J+2\Lambda+1)(2J-2\Lambda-1)}\right]^2$
$^PQ_{12}(J)$	$[(2J+1)/J(J+1)]\left[c_J a'_J \tilde{\mu}_{0--}\left(\Lambda-\tfrac{1}{2}\right)+d_J b'_J \tilde{\mu}_{0++}\left(\Lambda+\tfrac{1}{2}\right)\right]^2$
$^QR_{12}(J)$	$[1/4(J+1)]\left[c_J a'_J \tilde{\mu}_{0--}\sqrt{(2J+2\Lambda+1)(2J-2\Lambda+3)}\right.$ $\left.+\, d_J b'_J \tilde{\mu}_{0++}\sqrt{(2J+2\Lambda+3)(2J-2\Lambda+1)}\right]^2$
$P_2(J)$	$(1/4J)\left[c_J c'_J \tilde{\mu}_{0--}\sqrt{(2J+2\Lambda-1)(2J-2\Lambda+1)}\right.$ $\left.+\, d_J d'_J \tilde{\mu}_{0++}\sqrt{(2J+2\Lambda+1)(2J-2\Lambda-1)}\right]^2$
$Q_2(J)$	$[(2J+1)/J(J+1)]\left[c_J c'_J \tilde{\mu}_{0--}\left(\Lambda-\tfrac{1}{2}\right)+d_J d'_J \tilde{\mu}_{0++}\left(\Lambda+\tfrac{1}{2}\right)\right]^2$
$R_2(J)$	$[1/4(J+1)]\left[c_J c'_J \tilde{\mu}_{0--}\sqrt{(2J+2\Lambda+1)(2J-2\Lambda+3)}\right.$ $\left.+\, d_J d'_J \tilde{\mu}_{0++}\sqrt{(2J+2\Lambda+3)(2J-2\Lambda+1)}\right]^2$

Substituting for $\rho = J - \frac{1}{2} + s$ and $\rho' = J' - \frac{1}{2} + s'$, we obtain

$$\begin{aligned}
\mu_p = {} & \frac{\sqrt{(2J+1)(2J'+1)}}{2} \begin{pmatrix} J & J' & 1 \\ M & -M' & p \end{pmatrix} (-1)^{M'+\frac{1}{2}-\Lambda} \\
& \times \left\{ a_J a'_J \begin{pmatrix} J & J' & 1 \\ \Lambda - \frac{1}{2} & -\Lambda - \frac{1}{2} & 1 \end{pmatrix} \left[-\tilde{\mu}_{1--} - (-1)^{2\Lambda+2J+2J'+s+s'} \tilde{\mu}_{-1--} \right] \right. \\
& + b_J b'_J \begin{pmatrix} J & J' & 1 \\ \Lambda + \frac{1}{2} & -\Lambda - \frac{3}{2} & 1 \end{pmatrix} \left[+\tilde{\mu}_{1++} - (-1)^{2\Lambda+1+2J+2J'+s+s'} \tilde{\mu}_{-1++} \right] \\
& + a_J b'_J (-1)^{J-\frac{1}{2}+s} \begin{pmatrix} J & J' & 1 \\ -\Lambda + \frac{1}{2} & -\Lambda - \frac{3}{2} & 2\Lambda + 1 \end{pmatrix}_J \left[-(-1)^{2\Lambda+2J'+2-s+s'} \tilde{\mu}_{-(2\Lambda+1)+-} + \tilde{\mu}_{(2\Lambda+1)+-} \right] \\
& \left. + b_J a'_J (-1)^{J-\frac{1}{2}+s} \begin{pmatrix} J & J' & 1 \\ -\Lambda - \frac{1}{2} & -\Lambda - \frac{1}{2} & 2\Lambda + 1 \end{pmatrix} \left[-(-1)^{2\Lambda+2J'+1-s+s'} \tilde{\mu}_{-(2\Lambda+1)-+} - \tilde{\mu}_{(2\Lambda+1)-+} \right] \right\}.
\end{aligned} \tag{5.135}$$

In Eq. (5.135), $s' = 0$ because $\Lambda' \geq 1$, 2Λ is always an even number, and $2J$ and $2J'$ are odd numbers. Equation (5.135) therefore reduces to

$$\begin{aligned}
\mu_p = {} & \frac{\sqrt{(2J+1)(2J'+1)}}{2} \begin{pmatrix} J & J' & 1 \\ M & -M' & p \end{pmatrix} (-1)^{M'+\frac{1}{2}-\Lambda} \\
& \times \left\{ a_J a'_J \begin{pmatrix} J & J' & 1 \\ \Lambda - \frac{1}{2} & -\Lambda - \frac{1}{2} & 1 \end{pmatrix} \left[-\tilde{\mu}_{+1--} - (-1)^{s} \tilde{\mu}_{-1--} \right] \right. \\
& + b_J b'_J \begin{pmatrix} J & J' & 1 \\ \Lambda + \frac{1}{2} & -\Lambda - \frac{3}{2} & 1 \end{pmatrix} \left[\tilde{\mu}_{+1++} + (-1)^{s} \tilde{\mu}_{-1++} \right] \\
& + a_J b'_J (-1)^{\rho} \begin{pmatrix} J & J' & 1 \\ -\Lambda + \frac{1}{2} & -\Lambda - \frac{3}{2} & 2\Lambda + 1 \end{pmatrix}_J \left[(-1)^{-s} \tilde{\mu}_{-(2\Lambda+1)+-} + \tilde{\mu}_{+(2\Lambda+1)+-} \right] \\
& \left. + b_J a'_J (-1)^{\rho} \begin{pmatrix} J & J' & 1 \\ -\Lambda - \frac{1}{2} & -\Lambda - \frac{1}{2} & 2\Lambda + 1 \end{pmatrix} \left[-(-1)^{-s} \tilde{\mu}_{-(2\Lambda+1)-+} - \tilde{\mu}_{+(2\Lambda+1)-+} \right] \right\}.
\end{aligned} \tag{5.136}$$

The $3j$ symbols with $q = +1$ are given by

$\underline{J' = J - 1}$

$$\begin{pmatrix} J & J-1 & 1 \\ \Lambda - \frac{1}{2} & -\Lambda - \frac{1}{2} & 1 \end{pmatrix} = (-1)^{J-\Lambda+\frac{1}{2}} \sqrt{\frac{(2J-2\Lambda-1)(2J-2\Lambda+1)}{8J(2J+1)(2J-1)}}. \tag{5.137}$$

$$\begin{pmatrix} J & J-1 & 1 \\ \Lambda + \frac{1}{2} & -\Lambda - \frac{3}{2} & 1 \end{pmatrix} = (-1)^{J-\Lambda-\frac{1}{2}} \sqrt{\frac{(2J-2\Lambda-3)(2J-2\Lambda-1)}{8J(2J+1)(2J-1)}}. \tag{5.138}$$

$J' = J$

$$\begin{pmatrix} J & J & 1 \\ \Lambda - \frac{1}{2} & -\Lambda - \frac{1}{2} & 1 \end{pmatrix} = (-1)^{J-\Lambda+\frac{1}{2}} \sqrt{\frac{(2J-2\Lambda-1)(2J+2\Lambda+1)}{8J(J+1)(2J+1)}}. \tag{5.139}$$

$$\begin{pmatrix} J & J & 1 \\ \Lambda + \frac{1}{2} & -\Lambda - \frac{3}{2} & 1 \end{pmatrix} = (-1)^{J-\Lambda-\frac{1}{2}} \sqrt{\frac{(2J-2\Lambda-1)(2J+2\Lambda+3)}{8J(J+1)(2J+1)}}. \tag{5.140}$$

$J' = J + 1$

$$\begin{pmatrix} J & J+1 & 1 \\ \Lambda - \frac{1}{2} & -\Lambda - \frac{1}{2} & 1 \end{pmatrix} = (-1)^{3J-\Lambda+\frac{3}{2}} \sqrt{\frac{(2J+2\Lambda+1)(2J+2\Lambda+3)}{8(2J+3)(J+1)(2J+1)}}. \tag{5.141}$$

$$\begin{pmatrix} J & J+1 & 1 \\ \Lambda + \frac{1}{2} & -\Lambda - \frac{3}{2} & 1 \end{pmatrix} = (-1)^{3J-\Lambda+\frac{1}{2}} \sqrt{\frac{(2J+2\Lambda+3)(2J+2\Lambda+5)}{8(2J+3)(J+1)(2J+1)}}. \tag{5.142}$$

Radiative Transitions between Doublet Electronic Levels with $\Delta\Lambda = +1$, $\Lambda = 0$

First consider the case where $\Lambda = 0$. At this point we will not specify whether the lower state is an F_1 state or an F_2 state, just that it has positive parity. The molecule-fixed electric dipole moment matrix element $\tilde{\mu}_{\pm 1kn} = \tilde{\mu}_{\pm 1k}$ is given by

$$\begin{aligned} \tilde{\mu}_{\pm 1k} &= \langle v' J' \Omega'_k | \langle \Lambda'; S' \Sigma' | \hat{\mu}_x \pm i\hat{\mu}_y | \Lambda; S\Sigma \rangle | v J \Omega \rangle \\ &= \langle v' J' \Omega'_k | \left\langle 1; \frac{1}{2}\frac{1}{2} \middle| \hat{\mu}_x \pm i\hat{\mu}_y \middle| 0^{\pm}; \frac{1}{2}\frac{1}{2} \right\rangle | v J \Omega \rangle. \end{aligned} \tag{5.143}$$

Because $\Lambda = 0$, we have again dropped the subscript specifying whether we are dealing with Ω_+ or Ω_- for the lower level. In the molecule-fixed reference frame (x, y, z), the coordinate z is coincident with the internuclear axis. Applying the symmetry operation σ_v, representing reflection of coordinates across a plane that contains the internuclear axis, will leave the value of the dipole moment matrix element unchanged because the dipole moment matrix element is calculated by performing an integration over all coordinates (Hougen, 1970). From Hougen (1970) we know that

$$\sigma_v \left| 0^{\pm}; \frac{1}{2}\frac{1}{2} \right\rangle = \pm(-1)^{S-\Sigma} \left| 0^{\pm}; \frac{1}{2} - \frac{1}{2} \right\rangle = \pm \left| 0^{\pm}; \frac{1}{2} - \frac{1}{2} \right\rangle. \tag{5.144}$$

$$\sigma_v \left(\hat{\mu}_x \pm i\hat{\mu}_y \right) = \hat{\mu}_x \mp i\hat{\mu}_y. \tag{5.145}$$

For the $^2\Pi$ electronic level, Hougen (1970) states that the result of the symmetry operation is given by

$$\sigma_v\left\langle 1;\frac{1}{2}\frac{1}{2}\right| = (-1)^{L-\Lambda}(-1)^{S-\Sigma}\left\langle -1;\frac{1}{2}-\frac{1}{2}\right|. \tag{5.146}$$

The quantum number L is not well defined. Choosing $L = 1$ we obtain

$$\sigma_v\left\langle 1;\frac{1}{2}\frac{1}{2}\right| = \left\langle -1;\frac{1}{2}-\frac{1}{2}\right|. \tag{5.147}$$

Therefore

$$\begin{aligned}\tilde{\mu}_{+1kn} &= \sigma_v(\tilde{\mu}_{\pm 1kn}) = \sigma_v\left(\langle \mathrm{v}' J' \Omega_k' | \left\langle 1;\frac{1}{2}\frac{1}{2}\right| \hat{\mu}_x \pm i\hat{\mu}_y \left| 0^{\pm};\frac{1}{2}\frac{1}{2}\right\rangle |\mathrm{v} J \Omega_n\rangle \right) \\ &= \pm \langle \mathrm{v}' J' \Omega_k' | \left\langle -1;\frac{1}{2}\frac{1}{2}\right| \hat{\mu}_x - i\hat{\mu}_y \left| 0^{\pm};\frac{1}{2}\frac{1}{2}\right\rangle |\mathrm{v} J \Omega_n\rangle = \pm \tilde{\mu}_{-1kn}\end{aligned}. \tag{5.148}$$

When the initial state is in a $^2\Sigma^+$ electronic level, then $s = 0$ and $\tilde{\mu}_{+1k} = \tilde{\mu}_{-1k} = \tilde{\mu}_{1k}$ (the subscript n is dropped because the lower level is a $^2\Sigma$ electronic level). When the initial state is in a $^2\Sigma^-$ electronic level, then $s = 1$ and $\tilde{\mu}_{+1k} = -\tilde{\mu}_{-1k} = \tilde{\mu}_{1k}$. In both cases Eq. (5.136) reduces to

$$\begin{aligned}\mu_{ba,p} &= \sqrt{(2J+1)(2J+1)}\begin{pmatrix} J & J' & 1 \\ M & -M' & p \end{pmatrix}(-1)^{M'+\frac{1}{2}} \\ &\times \left\{ \left[-a_J a_J' \begin{pmatrix} J & J' & 1 \\ -\frac{1}{2} & -\frac{1}{2} & 1 \end{pmatrix}\tilde{\mu}_{1-} + b_J b_J' \begin{pmatrix} J & J' & 1 \\ \frac{1}{2} & -\frac{3}{2} & 1 \end{pmatrix}\tilde{\mu}_{1+} \right] \right. \\ &\left. + (-1)^{\rho}\left[+a_J b_J' \begin{pmatrix} J & J' & 1 \\ +\frac{1}{2} & -\frac{3}{2} & 1 \end{pmatrix}\tilde{\mu}_{1+} - b_J a_J' \begin{pmatrix} J & J' & 1 \\ -\frac{1}{2} & -\frac{1}{2} & 1 \end{pmatrix}\tilde{\mu}_{1-} \right] \right\}.\end{aligned} \tag{5.149}$$

For $\Lambda = 0$, if the lower state is an F_1 state with positive parity, then $a_J = [1 + (-1)^{\rho}]/4$ and $b_J = [1 + (-1)^{\rho}]/4$. Substituting for a_J and b_J, we can rewrite Eq. (5.149) as

$$\begin{aligned}\mu_{ba,p} &= -\frac{\sqrt{(2J+1)(2J+1)}}{4}\begin{pmatrix} J & J' & 1 \\ M & -M' & p \end{pmatrix}(-1)^{M'+\frac{1}{2}}[1 + (-1)^{\rho}]^2 \\ &\times \left[a_J' \begin{pmatrix} J & J' & 1 \\ -\frac{1}{2} & -\frac{1}{2} & 1 \end{pmatrix}\tilde{\mu}_{1-} - b_J' \begin{pmatrix} J & J' & 1 \\ \frac{1}{2} & -\frac{3}{2} & 1 \end{pmatrix}\tilde{\mu}_{1+} \right].\end{aligned} \tag{5.150}$$

For the lower electronic level, ρ is even for positive-parity states, and Eq. (5.150) reduces to

$$\begin{aligned}\mu_{ba,p} &= -\sqrt{(2J+1)(2J+1)}\begin{pmatrix} J & J' & 1 \\ M & -M' & p \end{pmatrix}(-1)^{M'+\frac{1}{2}} \\ &\times \left[a_J' \begin{pmatrix} J & J' & 1 \\ -\frac{1}{2} & -\frac{1}{2} & 1 \end{pmatrix}\tilde{\mu}_{1-} - b_J' \begin{pmatrix} J & J' & 1 \\ \frac{1}{2} & -\frac{3}{2} & 1 \end{pmatrix}\tilde{\mu}_{1+} \right].\end{aligned} \tag{5.151}$$

When ρ is odd, then $\mu_p = 0$, reflecting the fact that there are no positive-parity states with odd ρ for $\Sigma^{\pm}$ electronic levels.

For the case where $J' = J - 1$, we can substitute Eqs. (5.137) and (5.138) into Eq. (5.151) to obtain for $P_1(J)$ transitions:

$\underline{J' = J - 1}$

$$\mu_{ba,p} = -\sqrt{(2J+1)(2J-1)}\begin{pmatrix} J & J' & 1 \\ M & -M' & p \end{pmatrix}(-1)^{M'+\frac{1}{2}}(-1)^{J+\frac{1}{2}}$$
$$\times\left[a'_J\sqrt{\frac{(2J-1)(2J+1)}{8J(2J+1)(2J-1)}}\tilde{\mu}_{1-}+b'_J\sqrt{\frac{(2J-3)(2J-1)}{8J(2J+1)(2J-1)}}\tilde{\mu}_{1+}\right]. \tag{5.152}$$

$$\tilde{S}_{J''J'} = 3\sum_M\sum_{M'}\mu_{ba,p}\mu^*_{ba,p} = \frac{(2J-1)}{8J}\left[a'_J\sqrt{(2J+1)}\tilde{\mu}_{1-}+b'_J\sqrt{(2J-3)}\tilde{\mu}_{1+}\right]^2. \tag{5.153}$$

When the lower level is a positive-parity F_2 level then $c_J = -[1-(-1)^{\rho}]/4$ and $d_J = [1-(-1)^{\rho}]/4$; positive-parity F_2 levels have odd ρ. An analysis for $F_2 \to F_1$ transitions similar to that for $F_1 \to F_2$ results in the following expression for ${}^{O}P_{12}(J)$ transitions:

$$\tilde{S}_{J''J'} = 3\sum_M\sum_{M'}\mu_{ba,p}\mu^*_{ba,p} = \frac{(2J-1)}{8J}\left[-a'_J\sqrt{(2J+1)}\tilde{\mu}_{1-}+b'_J\sqrt{(2J-3)}\tilde{\mu}_{1+}\right]^2. \tag{5.154}$$

The results are shown in Table 5.4 for when similar reasoning is applied to the other 10 transitions for Σ^+ or Σ^- lower electronic levels.

Radiative Transitions between Doublet Electronic Levels with $\Delta\Lambda = +1$, $\Lambda > 0$

Now consider the case where $\Lambda \geq 1$. In this case $s = s' = 0$ and $2\Lambda + 1 \geq 3$. The $3j$ symbols in Eq. (5.136) with $q = \pm(2\Lambda+1)$ will be zero because $|q| > k = 1$, and Eq. (5.136) becomes

$$\mu_{ba,p} = -\frac{\sqrt{(2J+1)(2J'+1)}}{2}\begin{pmatrix} J & J' & 1 \\ M & -M' & p \end{pmatrix}(-1)^{M'+\frac{1}{2}+\Lambda}$$
$$\times\left\{a_Ja'_J\begin{pmatrix} J & J' & 1 \\ \Lambda-\frac{1}{2} & -\Lambda-\frac{1}{2} & 1 \end{pmatrix}[\tilde{\mu}_{+1--}+\tilde{\mu}_{-1--}]\right. \tag{5.155}$$
$$\left.-b_Jb'_J\begin{pmatrix} J & J' & 1 \\ \Lambda+\frac{1}{2} & -\Lambda-\frac{3}{2} & 1 \end{pmatrix}[\tilde{\mu}_{+1++}+\tilde{\mu}_{-1++}]\right\}.$$

Table 5.4 Line strength expressions for vibration-rotation transitions between a ground electronic level with $\Lambda = 0$ and an excited electronic level with $\Lambda' = 1$

	$^2\Sigma^{\pm}-{}^2\Pi$
	$\tilde{S}_{J''J'}$
$P_1(J)$	$[(2J-1)/8J]\left[a'_J\tilde{\mu}_{1-}\sqrt{2J+1}+b'_J\tilde{\mu}_{1+}\sqrt{2J-3}\right]^2$
$Q_1(J)$	$[(2J+1)/8J(J+1)]\left[a'_J\tilde{\mu}_{1-}(2J+1)+b'_J\tilde{\mu}_{1+}\sqrt{(2J-1)(2J+3)}\right]^2$
$R_1(J)$	$[(2J+3)/8(J+1)]\left[a'_J\tilde{\mu}_{1-}\sqrt{2J+1}+b'_J\tilde{\mu}_{1+}\sqrt{2J+5}\right]^2$
$^{Q}P_{21}(J)$	$[(2J-1)/8J]\left[c'_J\tilde{\mu}_{1-}\sqrt{2J+1}+d'_J\tilde{\mu}_{1+}\sqrt{2J-3}\right]^2$
$^{R}Q_{21}(J)$	$[(2J+1)/8J(J+1)]\left[c'_J\tilde{\mu}_{1-}(2J+1)+d'_J\tilde{\mu}_{1+}\sqrt{(2J-1)(2J+3)}\right]^2$
$^{S}R_{21}(J)$	$[(2J+3)/8(J+1)]\left[c'_J\tilde{\mu}_{1-}\sqrt{2J+1}+d'_J\tilde{\mu}_{1+}\sqrt{2J+5}\right]^2$
$^{O}P_{12}(J)$	$[(2J-1)/8J]\left[-a'_J\tilde{\mu}_{1-}\sqrt{2J+1}+b'_J\tilde{\mu}_{1+}\sqrt{2J-3}\right]^2$
$^{P}Q_{12}(J)$	$[(2J+1)/8J(J+1)]\left[-a'_J\tilde{\mu}_{1-}(2J+1)+b'_J\tilde{\mu}_{1+}\sqrt{(2J-1)(2J+3)}\right]^2$
$^{Q}R_{12}(J)$	$[(2J+3)/8(J+1)]\left[-a'_J\tilde{\mu}_{1-}\sqrt{2J+1}+b'_J\tilde{\mu}_{1+}\sqrt{2J+5}\right]^2$
$P_2(J)$	$[(2J-1)/8J]\left[-c'_J\tilde{\mu}_{1-}\sqrt{2J+1}+d'_J\tilde{\mu}_{1+}\sqrt{2J-3}\right]^2$
$Q_2(J)$	$[(2J+1)/8J(J+1)]\left[-c'_J\tilde{\mu}_{1-}\sqrt{(2J+1)(2J+1)}+d'_J\tilde{\mu}_{1++}\sqrt{(2J-1)(2J+3)}\right]^2$
$R_2(J)$	$[(2J+3)/8(J+1)]\left[-c'_J\tilde{\mu}_{1-}\sqrt{2J+1}+d'_J\tilde{\mu}_{1+}\sqrt{2J+5}\right]^2$

Defining

$$\tilde{\mu}_{1--} = \frac{\tilde{\mu}_{+1--}+\tilde{\mu}_{-1--}}{2} \tag{5.156}$$

and

$$\tilde{\mu}_{1++} = \frac{\tilde{\mu}_{+1++}+\tilde{\mu}_{-1++}}{2}, \tag{5.157}$$

we obtain

$$\mu_{ba,p} = \sqrt{(2J+1)(2J'+1)}\begin{pmatrix} J & J' & 1 \\ M & -M' & p \end{pmatrix}(-1)^{M'+\frac{1}{2}+\Lambda}$$
$$\times\left[a_J a'_J\begin{pmatrix} J & J' & 1 \\ \Lambda-\frac{1}{2} & -\Lambda-\frac{1}{2} & 1 \end{pmatrix}\tilde{\mu}_{1--} - b_J b'_J\begin{pmatrix} J & J' & 1 \\ \Lambda+\frac{1}{2} & -\Lambda-\frac{3}{2} & 1 \end{pmatrix}\tilde{\mu}_{1++}\right]. \tag{5.158}$$

The normalized line strength is given by

$$\tilde{S}_{J''J'} = 3\sum_{M}\sum_{M}\mu_{ba,p}\mu^{*}_{ba,p}$$
$$= (2J+1)(2J'+1)\left[a_J a'_J\begin{pmatrix} J & J' & 1 \\ \Lambda-\frac{1}{2} & -\Lambda-\frac{1}{2} & 1\end{pmatrix}\tilde{\mu}_{1--} - b_J b'_J\begin{pmatrix} J & J' & 1 \\ \Lambda+\frac{1}{2} & -\Lambda-\frac{3}{2} & 1\end{pmatrix}\tilde{\mu}_{1++}\right]^2. \tag{5.159}$$

Substituting in Eq. (5.159) for the $3j$ symbols using Eqs. (5.137)–(5.142) and substituting for $F_1 \rightarrow F_2$, $F_2 \rightarrow F_1$, and $F_2 \rightarrow F_2$ transitions by substituting for a_J, b_J, a'_J, and b'_J as appropriate with c_J, d_J, c'_J, and d'_J, we obtain the results shown in Table 5.5.

Table 5.5 Line strength expressions for vibration-rotation transitions between a ground electronic level with $\Lambda > 0$ and an excited electronic level with $\Lambda' = \Lambda + 1$

$^2X_A \rightarrow {}^2X_B,\ \Lambda' = \Lambda + 1$	$\tilde{S}_{J''J'}$
$P_1(J)$	$[(2J-2\Lambda-1)/8J]\left[a_J a'_J\tilde{\mu}_{1--}\sqrt{2J-2\Lambda+1} + b_J b'_J\tilde{\mu}_{1++}\sqrt{2J-2\Lambda-3}\right]^2$
$Q_1(J)$	$[(2J+1)/8J(J+1)]\left[a_J a'_J\tilde{\mu}_{1--}\sqrt{(2J-2\Lambda+1)(2J+2\Lambda+1)} + b_J b'_J\tilde{\mu}_{1++}\sqrt{(2J-2\Lambda-1)(2J+2\Lambda+3)}\right]^2$
$R_1(J)$	$[(2J+2\Lambda+3)/8(J+1)]\left[a_J a'_J\tilde{\mu}_{1--}\sqrt{2J+2\Lambda+1} + b_J b'_J\tilde{\mu}_{1++}\sqrt{2J+2\Lambda+5}\right]^2$
${}^QP_{21}(J)$	$[(2J-2\Lambda-1)/8J]\left[a_J c'_J\tilde{\mu}_{1--}\sqrt{2J-2\Lambda+1} + b_J d'_J\tilde{\mu}_{1++}\sqrt{2J-2\Lambda-3}\right]^2$
${}^RQ_{21}(J)$	$[(2J+1)/8J(J+1)]\left[a_J c'_J\tilde{\mu}_{1--}\sqrt{(2J-2\Lambda+1)(2J+2\Lambda+1)} + b_J d'_J\tilde{\mu}_{1++}\sqrt{(2J-2\Lambda-1)(2J+2\Lambda+3)}\right]^2$
${}^SR_{21}(J)$	$[(2J+2\Lambda+3)/8(J+1)]\left[a_J c'_J\tilde{\mu}_{1--}\sqrt{2J+2\Lambda+1} + b_J d'_J\tilde{\mu}_{1++}\sqrt{2J+2\Lambda+5}\right]^2$
${}^OP_{12}(J)$	$[(2J-2\Lambda-1)/8J]\left[c_J a'_J\tilde{\mu}_{1--}\sqrt{2J-2\Lambda+1} + d_J b'_J\tilde{\mu}_{1++}\sqrt{2J-2\Lambda-3}\right]^2$
${}^PQ_{12}(J)$	$[(2J+1)/8J(J+1)]\left[c_J a'_J\tilde{\mu}_{1--}\sqrt{(2J-2\Lambda+1)(2J+2\Lambda+1)} + d_J b'_J\tilde{\mu}_{1++}\sqrt{(2J-2\Lambda-1)(2J+2\Lambda+3)}\right]^2$
${}^QR_{12}(J)$	$[(2J+2\Lambda+3)/8(J+1)]\left[c_J a'_J\tilde{\mu}_{1--}\sqrt{2J+2\Lambda+1} + d_J b'_J\tilde{\mu}_{1++}\sqrt{2J+2\Lambda+5}\right]^2$
$P_2(J)$	$[(2J-2\Lambda-1)/8J]\left[c_J c'_J\tilde{\mu}_{1--}\sqrt{2J-2\Lambda+1} + d_J d'_J\tilde{\mu}_{1++}\sqrt{2J-2\Lambda-3}\right]^2$
$Q_2(J)$	$[(2J+1)/8J(J+1)]\left[c_J c'_J\tilde{\mu}_{1--}\sqrt{(2J-2\Lambda+1)(2J+2\Lambda+1)} + d_J d'_J\tilde{\mu}_{1++}\sqrt{(2J-2\Lambda-1)(2J+2\Lambda+3)}\right]^2$
$R_2(J)$	$[(2J+2\Lambda+3)/8(J+1)]\left[c_J c'_J\tilde{\mu}_{1--}\sqrt{2J+2\Lambda+1} + d_J d'_J\tilde{\mu}_{1++}\sqrt{2J+2\Lambda+5}\right]^2$

5.3.4 Radiative Transitions between Doublet Electronic Levels with $\Delta\Lambda = -1$

Substituting $\Delta\Lambda = \Lambda' - \Lambda = -1, \Lambda' = \Lambda - 1$ in Eq. (5.108), we obtain

$$\begin{aligned}
\mu_{ba,p} = {} & \frac{\sqrt{(2J+1)(2J'+1)}}{2}\begin{pmatrix} J & J' & 1 \\ M & -M' & p \end{pmatrix}(-1)^{M'+\frac{1}{2}} \\
& \times \sum_q \left\{ a_J a'_J \tilde{\mu}_{q--}\left[(-1)^{-\Lambda+1}\begin{pmatrix} J & J' & 1 \\ \Lambda-\frac{1}{2} & -\Lambda+\frac{3}{2} & -1 \end{pmatrix} - (-1)^{\Lambda-2+\rho+\rho'}\begin{pmatrix} J & J' & 1 \\ -\Lambda+\frac{1}{2} & \Lambda-\frac{3}{2} & 1 \end{pmatrix}\right]\right. \\
& + b_J b'_J \tilde{\mu}_{q++}\left[(-1)^{-\Lambda}\begin{pmatrix} J & J' & 1 \\ \Lambda+\frac{1}{2} & -\Lambda+\frac{1}{2} & -1 \end{pmatrix} - (-1)^{\Lambda-1+\rho+\rho'}\begin{pmatrix} J & J' & 1 \\ -\Lambda-\frac{1}{2} & \Lambda-\frac{1}{2} & 1 \end{pmatrix}\right] \\
& + a_J b'_J \tilde{\mu}_{q+-}\left[-(-1)^{\Lambda-1+\rho'}\begin{pmatrix} J & J' & 1 \\ \Lambda-\frac{1}{2} & \Lambda-\frac{1}{2} & -(2\Lambda-1) \end{pmatrix} + (-1)^{-\Lambda+\rho}\begin{pmatrix} J & J' & 1 \\ -\Lambda+\frac{1}{2} & -\Lambda+\frac{1}{2} & 2\Lambda-1 \end{pmatrix}\right] \\
& \left. + b_J a'_J \tilde{\mu}_{q-+}\left[-(-1)^{\Lambda-2+\rho'}\begin{pmatrix} J & J' & 1 \\ \Lambda+\frac{1}{2} & \Lambda-\frac{3}{2} & -(2\Lambda-1) \end{pmatrix} + (-1)^{-\Lambda+1+\rho}\begin{pmatrix} J & J' & 1 \\ -\Lambda-\frac{1}{2} & -\Lambda+\frac{3}{2} & 2\Lambda-1 \end{pmatrix}\right]\right\}.
\end{aligned} \tag{5.160}$$

Substituting for ρ and ρ', we obtain

$$\begin{aligned}
\mu_{ba,p} = {} & \frac{\sqrt{(2J+1)(2J'+1)}}{2}\begin{pmatrix} J & J' & 1 \\ M & -M' & p \end{pmatrix}(-1)^{M'+\frac{1}{2}} \\
& \times \left\{ a_J a'_J \begin{pmatrix} J & J' & 1 \\ -\Lambda+\frac{1}{2} & \Lambda-\frac{3}{2} & 1 \end{pmatrix}\left[(-1)^{-\Lambda+J+J'}\tilde{\mu}_{-1--} - (-1)^{\Lambda+J+J'-1++s+s'}\tilde{\mu}_{+1--}\right]\right. \\
& + b_J b'_J \begin{pmatrix} J & J' & 1 \\ -\Lambda-\frac{1}{2} & \Lambda-\frac{1}{2} & 1 \end{pmatrix}\left[(-1)^{-\Lambda+J+J'+1}\tilde{\mu}_{-1++} - (-1)^{\Lambda+J+J'+s+s'}\tilde{\mu}_{+1++}\right] \\
& + a_J b'_J (-1)^{J-\frac{1}{2}+s}\begin{pmatrix} J & J' & 1 \\ -\Lambda+\frac{1}{2} & -\Lambda+\frac{1}{2} & 2\Lambda-1 \end{pmatrix}\left[-(-1)^{\Lambda+2J'+s'-s}\tilde{\mu}_{-(2\Lambda-1)+-} + (-1)^{-\Lambda}\tilde{\mu}_{+(2\Lambda-1)+-}\right] \\
& \left. + b_J a'_J (-1)^{J-\frac{1}{2}+s}\begin{pmatrix} J & J' & 1 \\ -\Lambda-\frac{1}{2} & -\Lambda+\frac{3}{2} & 2\Lambda-1 \end{pmatrix}\left[-(-1)^{\Lambda-1+2J'+s'-s}\tilde{\mu}_{-(2\Lambda-1)-+} + (-1)^{-\Lambda+1}\tilde{\mu}_{+(2\Lambda-1)-+}\right]\right\}.
\end{aligned} \tag{5.161}$$

Collecting terms and simplifying, we obtain

$$\begin{aligned}
\mu_{ba,p} = {} & \frac{\sqrt{(2J+1)(2J'+1)}}{2}\begin{pmatrix} J & J' & 1 \\ M & -M' & p \end{pmatrix}(-1)^{M'+\frac{1}{2}-\Lambda} \\
& \times \left\{ a_J a'_J \begin{pmatrix} J & J' & 1 \\ -\Lambda+\frac{1}{2} & \Lambda-\frac{3}{2} & 1 \end{pmatrix}\left[\tilde{\mu}_{-1--} + (-1)^{2\Lambda+s+s'}\tilde{\mu}_{+1--}\right](-1)^{J+J'}\right. \\
& + b_J b'_J \begin{pmatrix} J & J' & 1 \\ -\Lambda-\frac{1}{2} & \Lambda-\frac{1}{2} & 1 \end{pmatrix}\left[-\tilde{\mu}_{-1++} - (-1)^{2\Lambda+s+s'}\tilde{\mu}_{+1++}\right](-1)^{J+J'} \\
& + a_J b'_J (-1)^{\rho}\begin{pmatrix} J & J' & 1 \\ -\Lambda+\frac{1}{2} & -\Lambda+\frac{1}{2} & 2\Lambda-1 \end{pmatrix}\left[-(-1)^{2\Lambda+2J'+s'-s}\tilde{\mu}_{-(2\Lambda-1)+-} + \tilde{\mu}_{+(2\Lambda-1)+-}\right] \\
& \left. + b_J a'_J (-1)^{\rho}\begin{pmatrix} J & J' & 1 \\ -\Lambda-\frac{1}{2} & -\Lambda+\frac{3}{2} & 2\Lambda-1 \end{pmatrix}\left[(-1)^{2\Lambda+2J'+s'-s}\tilde{\mu}_{-(2\Lambda-1)-+} - \tilde{\mu}_{+(2\Lambda-1)-+}\right]\right\}.
\end{aligned} \tag{5.162}$$

For the lower level, $\Lambda \geq 1$ so $s = 0$. Because 2Λ is always an even number and $2J$ is always an odd number, Eq. (5.162) can be further simplified to obtain

$$\mu_{ba,p} = \frac{\sqrt{(2J+1)(2J'+1)}}{2}\begin{pmatrix} J & J' & 1 \\ M & -M' & p \end{pmatrix}(-1)^{M'+\frac{1}{2}+\Lambda}$$
$$\times\Bigg\{ a_J a'_J \begin{pmatrix} J & J' & 1 \\ -\Lambda+\frac{1}{2} & \Lambda-\frac{3}{2} & 1 \end{pmatrix}\left[\tilde{\mu}_{-1--}+(-1)^{s'}\tilde{\mu}_{+1--}\right](-1)^{J+J'}$$
$$- b_J b'_J \begin{pmatrix} J & J' & 1 \\ -\Lambda-\frac{1}{2} & \Lambda-\frac{1}{2} & 1 \end{pmatrix}\left[\tilde{\mu}_{-1++}+(-1)^{s'}\tilde{\mu}_{+1++}\right](-1)^{J+J'}$$
$$+ a_J b'_J (-1)^{\rho+s'} \begin{pmatrix} J & J' & 1 \\ -\Lambda+\frac{1}{2} & -\Lambda+\frac{1}{2} & 2\Lambda-1 \end{pmatrix}\left[\tilde{\mu}_{-(2\Lambda-1)+-}+(-1)^{s'}\tilde{\mu}_{+(2\Lambda-1)+-}\right]$$
$$- b_J a'_J (-1)^{\rho+s'} \begin{pmatrix} J & J' & 1 \\ -\Lambda-\frac{1}{2} & -\Lambda+\frac{3}{2} & 2\Lambda-1 \end{pmatrix}\left[\tilde{\mu}_{-(2\Lambda-1)-+}+(-1)^{s'}\tilde{\mu}_{+(2\Lambda-1)-+}\right]\Bigg\}. \tag{5.163}$$

The *3j* symbols are given by

$\underline{J' = J-1}$

$$\begin{pmatrix} J & J-1 & 1 \\ -\Lambda+\frac{1}{2} & \Lambda-\frac{3}{2} & 1 \end{pmatrix} = (-1)^{J+\Lambda-\frac{1}{2}}\sqrt{\frac{(2J+2\Lambda-3)(2J+2\Lambda-1)}{8J(2J+1)(2J-1)}}. \tag{5.164}$$

$$\begin{pmatrix} J & J-1 & 1 \\ -\Lambda-\frac{1}{2} & \Lambda-\frac{1}{2} & 1 \end{pmatrix} = (-1)^{J+\Lambda+\frac{1}{2}}\sqrt{\frac{(2J+2\Lambda-1)(2J+2\Lambda+1)}{8J(2J+1)(2J-1)}}. \tag{5.165}$$

$\underline{J' = J}$

$$\begin{pmatrix} J & J & 1 \\ -\Lambda+\frac{1}{2} & \Lambda-\frac{3}{2} & 1 \end{pmatrix} = (-1)^{J+\Lambda-\frac{1}{2}}\sqrt{\frac{(2J+2\Lambda-1)(2J-2\Lambda+3)}{8J(J+1)(2J+1)}}. \tag{5.166}$$

$$\begin{pmatrix} J & J & 1 \\ -\Lambda-\frac{1}{2} & \Lambda-\frac{1}{2} & 1 \end{pmatrix} = (-1)^{J+\Lambda+\frac{1}{2}}\sqrt{\frac{(2J+2\Lambda+1)(2J-2\Lambda+1)}{8J(J+1)(2J+1)}}. \tag{5.167}$$

$\underline{J' = J+1}$

$$\begin{pmatrix} J & J+1 & 1 \\ -\Lambda+\frac{1}{2} & \Lambda-\frac{3}{2} & 1 \end{pmatrix} = (-1)^{3J+\Lambda+\frac{1}{2}}\sqrt{\frac{(2J-2\Lambda+3)(2J-2\Lambda+5)}{8(2J+3)(J+1)(2J+1)}}. \tag{5.168}$$

$$\begin{pmatrix} J & J+1 & 1 \\ -\Lambda-\frac{1}{2} & \Lambda-\frac{1}{2} & 1 \end{pmatrix} = (-1)^{3J+\Lambda-\frac{1}{2}}\sqrt{\frac{(2J-2\Lambda+1)(2J-2\Lambda+3)}{8(2J+3)(J+1)(2J+1)}}. \tag{5.169}$$

Radiative Transitions between Doublet Electronic Levels with $\Delta\Lambda = -1$, $\Lambda = 1$

Again, in deriving Eq. (5.163) we are considering transitions from positive-parity F_1 lower quantum states to negative-parity upper quantum states. For the negative-parity upper quantum states, $a'_J = b'_J = [1 - (-1)^{\rho'}]/4$ for F_1 states and $-c'_J = d'_J = [1 - (-1)^{\rho'}]/4$ for F_2 states. For the case of $^2\Pi\ F_1 \rightarrow {}^2\Sigma^\pm\ F_1$ transitions, Eq. (5.163) becomes

$$\mu_{ba,p} = \frac{\sqrt{(2J+1)(2J'+1)}}{8}\begin{pmatrix} J & J' & 1 \\ M & -M' & p \end{pmatrix}(-1)^{M'-\frac{1}{2}}\left[1-(-1)^{\rho'}\right]\left[(-1)^{J+J'}+(-1)^{\rho+s'}\right]$$
$$\times\left\{-a_J\begin{pmatrix} J & J' & 1 \\ -\frac{1}{2} & -\frac{1}{2} & 1 \end{pmatrix}\left[\tilde{\mu}_{-1-}+(-1)^{s'}\tilde{\mu}_{+1-}\right]+b_J\begin{pmatrix} J & J' & 1 \\ -\frac{3}{2} & \frac{1}{2} & 1 \end{pmatrix}\left[\tilde{\mu}_{-1+}+(-1)^{s'}\tilde{\mu}_{+1+}\right]\right\}. \tag{5.170}$$

When the final state is in a $^2\Sigma^+$ electronic level, then $s' = 0$ and $\tilde{\mu}_{+1n} = \tilde{\mu}_{-1n} = \tilde{\mu}_{1n}$. When the final state is in a $^2\Sigma^-$ electronic level, then $s' = 1$ and $\tilde{\mu}_{+1n} = -\tilde{\mu}_{-1n} = \tilde{\mu}_{1n}$. In both cases, Eq. (5.170) becomes

$$\mu_p = \frac{\sqrt{(2J+1)(2J'+1)}}{4}\begin{pmatrix} J & J' & 1 \\ M & -M' & p \end{pmatrix}(-1)^{M'-\frac{1}{2}}\left[1-(-1)^{\rho'}\right]\left[(-1)^{J+J'}+(-1)^{\rho+s'}\right]$$
$$\times\left\{-a_J\begin{pmatrix} J & J' & 1 \\ -\frac{1}{2} & -\frac{1}{2} & 1 \end{pmatrix}\tilde{\mu}_{1-}+b_J\begin{pmatrix} J & J' & 1 \\ -\frac{3}{2} & \frac{1}{2} & 1 \end{pmatrix}\tilde{\mu}_{1+}\right\}. \tag{5.171}$$

In Eq. (5.171), the dipole moment matrix element will be zero for even ρ', consistent with our assumption of a negative-parity upper level. Given that we have a negative-parity upper level, only transitions with $J' = J \pm 1$ or $J' = J$, not both, will be allowed from a given positive-parity lower level J. Assuming that the lower level is a negative-parity level and the upper level is a positive-parity level, a similar analysis results in

$$\mu_p = \frac{\sqrt{(2J+1)(2J'+1)}}{4}\begin{pmatrix} J & J' & 1 \\ M & -M' & p \end{pmatrix}(-1)^{M'-\frac{1}{2}}\left[1+(-1)^{\rho'}\right]\left[(-1)^{J+J'}-(-1)^{\rho+s'}\right]$$
$$\times\left\{-a_J\begin{pmatrix} J & J' & 1 \\ -\frac{1}{2} & -\frac{1}{2} & 1 \end{pmatrix}\tilde{\mu}_{1-}+b_J\begin{pmatrix} J & J' & 1 \\ -\frac{3}{2} & \frac{1}{2} & 1 \end{pmatrix}\tilde{\mu}_{1+}\right\}. \tag{5.172}$$

So if a particular transition is not allowed when the lower level is a positive-parity level, that transition will be allowed from a negative-parity lower level, and vice versa. Assuming that the transition is allowed, Eqs. (5.171) and (5.172) both reduce to

$$\mu_p = \sqrt{(2J+1)(2J'+1)}\begin{pmatrix} J & J' & 1 \\ M & -M' & p \end{pmatrix}(-1)^{M'-\frac{1}{2}}$$
$$\times\left\{-a_J\begin{pmatrix} J & J' & 1 \\ -\frac{1}{2} & -\frac{1}{2} & 1 \end{pmatrix}\tilde{\mu}_{1-}+b_J\begin{pmatrix} J & J' & 1 \\ -\frac{3}{2} & \frac{1}{2} & 1 \end{pmatrix}\tilde{\mu}_{1+}\right\}, \tag{5.173}$$

Table 5.6 Line strength expressions for vibration-rotation transitions between a ground electronic level with $\Lambda = 1$ and an excited electronic level with $\Lambda' = 0$

	$^2\Pi - {}^2\Sigma^{\pm}$
	$\tilde{S}_{J''J'}$
$P_1(J)$	$[(2J-1)/8J]\left[a_J\tilde{\mu}_{1-}\sqrt{2J-3}+b_J\tilde{\mu}_{1+}\sqrt{2J+1}\right]^2$
$Q_1(J)$	$[(2J+1)/8J(J+1)]\left[a_J\tilde{\mu}_{1-}(2J+1)+b_J\tilde{\mu}_{1+}\sqrt{(2J+3)(2J-1)}\right]^2$
$R_1(J)$	$[(2J+3)/8(J+1)]\left[a_J\tilde{\mu}_{1-}\sqrt{2J+5}+b_J\tilde{\mu}_{1+}\sqrt{2J+1}\right]^2$
${}^QP_{21}(J)$	$[(2J-1)/8J]\left[a_J\tilde{\mu}_{1-}\sqrt{2J-3}-b_J\tilde{\mu}_{1+}\sqrt{2J+1}\right]^2$
${}^RQ_{21}(J)$	$[(2J+1)/8J(J+1)]\left[a_J\tilde{\mu}_{1-}(2J+1)-b_J\tilde{\mu}_{1+}\sqrt{(2J+3)(2J-1)}\right]^2$
${}^SR_{21}(J)$	$[(2J+3)/8(J+1)]\left[a_J\tilde{\mu}_{1-}\sqrt{2J+5}-b_J\tilde{\mu}_{1+}\sqrt{2J+1}\right]^2$
${}^OP_{12}(J)$	$[(2J-1)/8J]\left[c_J\tilde{\mu}_{1-}\sqrt{2J-3}+d_J\tilde{\mu}_{1+}\sqrt{2J+1}\right]^2$
${}^PQ_{12}(J)$	$[(2J+1)/8J(J+1)]\left[c_J\tilde{\mu}_{1-}(2J+1)+d_J\tilde{\mu}_{1+}\sqrt{(2J+3)(2J-1)}\right]^2$
${}^QR_{12}(J)$	$[(2J+3)/8(J+1)]\left[c_J\tilde{\mu}_{1-}\sqrt{2J+5}+d_J\tilde{\mu}_{1+}\sqrt{2J+1}\right]^2$
$P_2(J)$	$[(2J-1)/8J]\left[c_J\tilde{\mu}_{1-}\sqrt{2J-3}-d_J\tilde{\mu}_{1+}\sqrt{2J+1}\right]^2$
$Q_2(J)$	$[(2J+1)/8J(J+1)]\left[c_J\tilde{\mu}_{1-}(2J+1)-d_J\tilde{\mu}_{1+}\sqrt{(2J+3)(2J-1)}\right]^2$
$R_2(J)$	$[(2J+3)/8(J+1)]\left[c_J\tilde{\mu}_{1-}\sqrt{2J+5}-d_J\tilde{\mu}_{1+}\sqrt{2J+1}\right]^2$

and the line strengths are given by

$$\tilde{S}_{J''J'} = 3\sum_{M}\sum_{M'}\mu_p\mu_p^* = \left[-a_J\begin{pmatrix} J & J' & 1 \\ -\frac{1}{2} & -\frac{1}{2} & 1 \end{pmatrix}\tilde{\mu}_{1-} + b_J\begin{pmatrix} J & J' & 1 \\ -\frac{3}{2} & \frac{1}{2} & 1 \end{pmatrix}\tilde{\mu}_{1+}\right]^2. \tag{5.174}$$

The line strengths for the different allowed transitions are listed in Table 5.6.

Radiative Transitions between Doublet Electronic Levels with $\Delta\Lambda = -1$, $\Lambda > 0$

For $\Delta\Lambda = -1$ with $\Lambda \geq 2$, $s = s' = 0$, and the $3j$ symbols with $q = 2\Lambda - 1$ will be zero because $q > 1$. With these modifications to Eq. (5.163), the dipole moment matrix element is given by

$$\mu_p = \sqrt{(2J+1)(2J'+1)}\begin{pmatrix} J & J' & 1 \\ M & -M' & p \end{pmatrix}(-1)^{M'+\frac{1}{2}+\Lambda+J+J'}$$
$$\times\left[-a_J a'_J\begin{pmatrix} J & J' & 1 \\ -\Lambda+\frac{1}{2} & \Lambda-\frac{3}{2} & 1 \end{pmatrix}\tilde{\mu}_{1--} + b_J b'_J\begin{pmatrix} J & J' & 1 \\ -\Lambda-\frac{1}{2} & \Lambda-\frac{1}{2} & 1 \end{pmatrix}\tilde{\mu}_{1++}\right], \tag{5.175}$$

Table 5.7 Line strength expressions for vibration-rotation transitions between a ground electronic level with $\Lambda > 1$ and an excited electronic level with $\Lambda' = \Lambda - 1$

$^2X \rightarrow {}^2X'$, $\Lambda' = \Lambda - 1$	
	$\tilde{S}_{J''J'}$
$P_1(J)$	$[(2J+2\Lambda-1)/8J]\left[a_J a'_J \tilde{\mu}_{1--}\sqrt{2J+2\Lambda-3}+b_J b'_J \tilde{\mu}_{1++}\sqrt{2J+2\Lambda+1}\right]^2$
$Q_1(J)$	$[(2J+1)/8J(J+1)]\left[a_J a'_J \tilde{\mu}_{1--}\sqrt{(2J+2\Lambda-1)(2J-2\Lambda+3)}\right.$ $\left.+b_J b'_J \tilde{\mu}_{1++}\sqrt{(2J+2\Lambda+1)(2J-2\Lambda+1)}\right]^2$
$R_1(J)$	$[(2J-2\Lambda+3)/8(J+1)]\left[a_J a'_J \tilde{\mu}_{1--}\sqrt{2J-2\Lambda+5}+b_J b'_J \tilde{\mu}_{1++}\sqrt{2J-2\Lambda+1}\right]^2$
$^QP_{21}(J)$	$[(2J+2\Lambda-1)/8J]\left[a_J c'_J \tilde{\mu}_{1--}\sqrt{2J+2\Lambda-3}+b_J d'_J \tilde{\mu}_{1++}\sqrt{2J+2\Lambda+1}\right]^2$
$^RQ_{21}(J)$	$[(2J+1)/8J(J+1)]\left[a_J c'_J \tilde{\mu}_{1--}\sqrt{(2J+2\Lambda-1)(2J-2\Lambda+3)}\right.$ $\left.+b_J d'_J \tilde{\mu}_{1++}\sqrt{(2J+2\Lambda+1)(2J-2\Lambda+1)}\right]^2/8J(J+1)$
$^SR_{21}(J)$	$[(2J-2\Lambda+3)/8(J+1)]\left[a_J c'_J \tilde{\mu}_{1--}\sqrt{2J-2\Lambda+5}+b_J d'_J \tilde{\mu}_{1++}\sqrt{2J-2\Lambda+1}\right]^2$
$^OP_{12}(J)$	$[(2J+2\Lambda-1)/8J]\left[c_J a'_J \tilde{\mu}_{1--}\sqrt{2J+2\Lambda-3}+d_J b'_J \tilde{\mu}_{1++}\sqrt{2J+2\Lambda+1}\right]^2$
$^PQ_{12}(J)$	$[(2J+1)/8J(J+1)]\left[c_J a'_J \tilde{\mu}_{1--}\sqrt{(2J+2\Lambda-1)(2J-2\Lambda+3)}\right.$ $\left.+d_J b'_J \tilde{\mu}_{1++}\sqrt{(2J+2\Lambda+1)(2J-2\Lambda+1)}\right]^2/8J(J+1)$
$^QR_{12}(J)$	$[(2J-2\Lambda+3)/8(J+1)]\left[c_J a'_J \tilde{\mu}_{1--}\sqrt{2J-2\Lambda+5}+d_J b'_J \tilde{\mu}_{1++}\sqrt{2J-2\Lambda+1}\right]^2$
$P_2(J)$	$[(2J+2\Lambda-1)/8J]\left[c_J c'_J \tilde{\mu}_{1--}\sqrt{2J+2\Lambda-3}+d_J d'_J \tilde{\mu}_{1++}\sqrt{2J+2\Lambda+1}\right]^2$
$Q_2(J)$	$[(2J+1)/8J(J+1)]\left[c_J c'_J \tilde{\mu}_{1--}\sqrt{(2J+2\Lambda-1)(2J-2\Lambda+3)}\right.$ $\left.+d_J d'_J \tilde{\mu}_{1++}\sqrt{(2J+2\Lambda+1)(2J-2\Lambda+1)}\right]^2$
$R_2(J)$	$[(2J-2\Lambda+3)/8(J+1)]\left[c_J c'_J \tilde{\mu}_{1--}\sqrt{2J-2\Lambda+5}+d_J d'_J \tilde{\mu}_{1++}\sqrt{2J-2\Lambda+1}\right]^2$

where $\tilde{\mu}_{1--}$ and $\tilde{\mu}_{1++}$ are given Eqs. (5.156) and (5.157), respectively. The normalized line strengths for the 12 allowed transitions are found by substituting Eqs. (5.164)–(5.169) into Eq. (5.175) and substituting for $a_J, a'_J, b_J,$ and b'_J with $c_J, c'_J, d_J,$ and d'_J as appropriate; we obtain the results shown in Table 5.7.

5.4 Calculation of Radiative Transition Rates for the $X^2\Pi(\mathrm{v})$ – $A^2\Sigma^+(\mathrm{v}')$ Bands of OH and NO

In this section radiative transitions for NO and OH, two of the most important species for gas-phase diagnostics in reacting flows, are discussed in detail. These two molecules have very different values for $Y_{\mathrm{v}} = A_{\mathrm{v}}/B_{\mathrm{v}}$, the ratio of spin–orbit splitting to the rotation constant, and thus the dependence of the line strength of the satellite transitions on J is very different. The effect of the radial dependence of the electronic transition moment $R_{e,q}(r)$ on the relative vibrational band strengths for these two species is also of great interest.

5.4.1 Einstein Coefficients for the $X^2\Pi(\mathrm{v}) - A^2\Sigma^+(\mathrm{v}')$ Vibrational Bands of OH and NO

The calculation of radiative transition rates using the results of Section 5.3 are illustrated in this section for the $X^2\Pi(\mathrm{v}) - A^2\Sigma^+(\mathrm{v}')$ bands of OH and the $X^2\Pi(\mathrm{v}) - A^2\Sigma^+(\mathrm{v}')$ bands of NO. Both OH and NO are important molecules in atmospheric chemistry, combustion, and other reacting flow applications, and accurate determination of radiative transition rates is critical for quantitative application of techniques such as absorption spectroscopy and laser-induced fluorescence spectroscopy. Furthermore, the effects of vibration-rotation interaction are much more significant for OH than for NO, and the two molecules have very different values of $Y_\mathrm{v} = A_\mathrm{v}/B_\mathrm{v}$, which has a significant effect on the relative intensities of satellite and main-branch transitions. The results of the calculations in Section 5.4 are compared with recent calculations of OH transition strengths by Yousefi et al. (2018) and with LIFBASE (Luque & Crosley, 1999a) for NO. For both OH and NO, the calculations were performed for values for the vibrational quantum numbers for both the ground and excited electronic states ranging from 0 to 2, corresponding to the vibrational bands of most interest for combustion diagnostics.

For OH, the Dunham coefficients from table 3 of Brooke et al. (2016) for the $X^2\Pi$ level and from table I in Luque and Crosley (1998) for the $A^2\Sigma^+$ electronic level were used in the RKR calculations. For NO, the Dunham coefficients from eqs. (1)–(3) and table IV of Engleman and Rouse (1971) were used for both the $X^2\Pi$ and $A^2\Sigma^+$ electronic levels in the RKR calculations. The RKR calculations to determine the turning points of the potential energy function $V(r)$ were performed using the code RKR1 described by Le Roy (2017a). The turning points for the potential energy function in turn were used as input to the LEVEL code (Le Roy, 2017b) to calculate the vibration-rotation wave function $\psi_{\mathrm{v}J\Lambda}(r)$. As noted in Section 5.3.1, the LEVEL code assumes integral values of J and Λ. Therefore, we performed a Lagrangian interpolation (Ketter & Prawel, 1969) on the LEVEL wavefunctions $\psi_{\mathrm{v}K_i\Lambda}(r)$ to obtain the case (a) wavefunction $\psi_{\mathrm{v}J\Omega}(r)$ using the centrifugal potential as the interpolation variable,

$$\psi_{\mathrm{v}J\Omega}(r) = L_1\psi_{\mathrm{v}K_1\Lambda}(r) + L_2\psi_{\mathrm{v}K_2\Lambda}(r) + L_3\psi_{\mathrm{v}K_3\Lambda}(r), \tag{5.176}$$

where

$$L_1 = \left\{\frac{[J(J+1)-\Omega^2]-[K_2(K_2+1)-\Lambda^2]}{[K_1(K_1+1)-\Lambda^2]-[K_2(K_2+1)-\Lambda^2]}\right\}\left\{\frac{[J(J+1)-\Omega^2]-[K_3(K_3+1)-\Lambda^2]}{[K_1(K_1+1)-\Lambda^2]-[K_3(K_3+1)-\Lambda^2]}\right\} \tag{5.177}$$

$$L_2 = \left\{ \frac{[J(J+1) - \Omega^2] - [K_1(K_1+1) - \Lambda^2]}{[K_2(K_2+1) - \Lambda^2] - [K_1(K_1+1) - \Lambda^2]} \right\} \left\{ \frac{[J(J+1) - \Omega^2] - [K_3(K_3+1) - \Lambda^2]}{[K_2(K_2+1) - \Lambda^2] - [K_3(K_3+1) - \Lambda^2]} \right\} \tag{5.178}$$

$$L_3 = \left\{ \frac{[J(J+1) - \Omega^2] - [K_1(K_1+1) - \Lambda^2]}{[K_3(K_3+1) - \Lambda^2] - [K_1(K_1+1) - \Lambda^2]} \right\} \left\{ \frac{[J(J+1) - \Omega^2] - [K_2(K_2+1) - \Lambda^2]}{[K_3(K_3+1) - \Lambda^2] - [K_2(K_2+1) - \Lambda^2]} \right\} \tag{5.179}$$

and

$$K_1 = J - \Omega + \Lambda - 1. \tag{5.180}$$

$$K_2 = J - \Omega + \Lambda. \tag{5.181}$$

$$K_3 = J - \Omega + \Lambda + 1. \tag{5.182}$$

The OH $X^2\Pi$ vibration-rotation wavefunctions for $\mathrm{v} = 0, J = 4.5, 20.5,$ and $\Omega = 1/2, 3/2$ are shown in Figure 5.2a, and the NO $X^2\Pi$ vibration-rotation wavefunctions for $\mathrm{v} = 0, J = 4.5, 60.5,$ and $\Omega = 1/2, 3/2$ are shown in Figure 5.2b. As shown in Figure 5.2a for OH, there is a significant shift in the wavefunction to higher values of the internuclear separation r as J increases from 4.5 to 20.5 due to the increase in the centrifugal potential $J(J+1) - \Omega^2$. The difference between the wavefunctions $\psi_{\mathrm{v}J3/2}$ and $\psi_{\mathrm{v}J1/2}$ is also plotted in Figure 5.2, and again the wavefunction $\psi_{\mathrm{v}J1/2}$ is shifted to higher values of r compared to $\psi_{\mathrm{v}J3/2}$, again consistent with the higher value of the centrifugal potential for $\psi_{\mathrm{v}J1/2}$ compared to $\psi_{\mathrm{v}J3/2}$. However, the differences between the wavefunctions $\psi_{\mathrm{v}J1/2}$ and $\psi_{\mathrm{v}J3/2}$ are very small, as is evident from the overlap of the wavefunctions in Figure 5.2a. For NO, there is again a

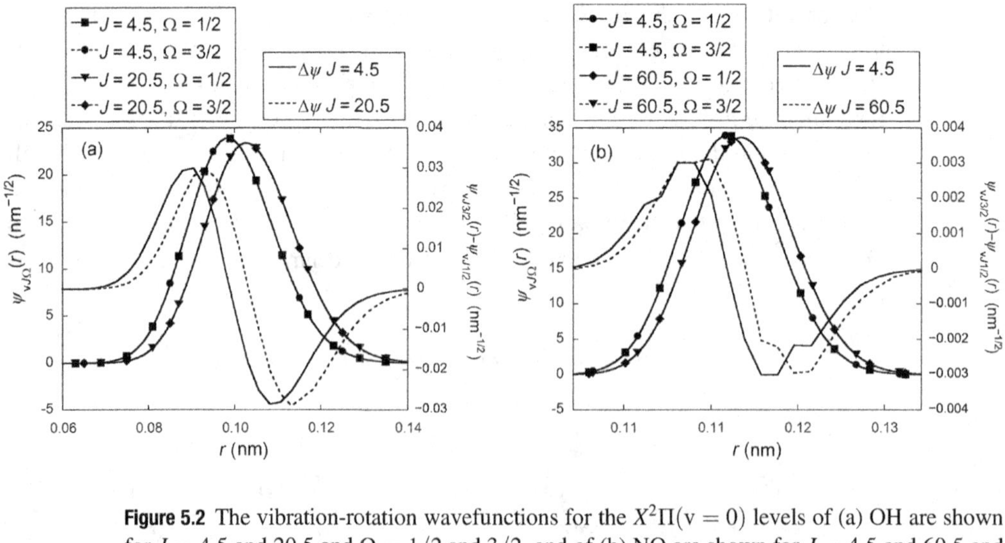

Figure 5.2 The vibration-rotation wavefunctions for the $X^2\Pi(\mathrm{v} = 0)$ levels of (a) OH are shown for $J = 4.5$ and 20.5 and $\Omega = 1/2$ and $3/2$, and of (b) NO are shown for $J = 4.5$ and 60.5 and $\Omega = 1/2$ and $3/2$.

shift in the wavefunctions to higher values of r as the rotational quantum number increases. The differences between the wavefunctions $\psi_{vJ1/2}$ and $\psi_{vJ3/2}$ are an order of magnitude smaller than for OH, as is shown in Figure 5.2b; noise due to round-off errors is clearly evident in the curves showing the wavefunction differences in Figure 5.2b.

The molecule-fixed electric dipole moment matrix element $\tilde{\mu}_{qkn}(v', J', \Omega'_k, vJ\Omega_n)$ is calculated using Eq. (5.106). The electronic transition moment $R_{e,q=1}(r)$ for OH was determined by Yousefi et al. (2018) for OH using the quantum chemistry program MOLPRO (Werner et al., 2012); the data are listed in the supplementary data for Yousefi et al. (2018); we performed a polynomial fit to this data and obtained the following expression for the electronic transition moment $R_{e,q=1}(r)$ (D):

$$R_{e,q=1}(r) = 0.015334 + 34.431\,r - 601.26\,r^2 + 3747.2\,r^3 - 10,068\,r^4 + 9933.4\,r^5, \tag{5.183}$$

where 1 D = 3.335640952 × 10^{-30} Cm (Bernath, 2016), and the units of r are nm in Eq. (5.183). The polynomial fit coefficients for the square of the electronic transition moment for NO, in units of ea_0, were given by Settersten et al. (2009). Converting the expression in Settersten et al. (2009) to an expression with units of D for $R_{e,q=1}(r)$ and units of *nm* for r, we obtain

$$R_{e,q=1}(r) = 78.700 - 2620.00\,r + 32,919.1\,r^2 - 184,068\,r^3 + 385,960\,r^4. \tag{5.184}$$

The method that we describe here to calculate the molecule-fixed band intensity matrix elements is an alternate approach to the approach discussed in detail by Brooke et al. (2016). Brooke et al. (2016) performed a transformation of the Hund's case (b) band intensity matrix element to Hund's case (a). We tried to use this transformation but there were nonphysical oscillations in the Hund's case (a) TDM matrix elements at low J due to the $3j$ symbols in the expression, and we decided to use a Lagrangian interpolation of the LEVEL vibration-rotation wavefunctions using the centrifugal potential as an interpolation parameter.

The wavefunctions for the (0,0), (0,1), (1,0), and (1,1) bands of OH and NO are shown in Figures 5.3 and 5.4, respectively. In Figures 5.3 and 5.4, the electronic transition moment $R_{e,q=1}(r)$ is plotted along with the upper (ψ_e), and lower (ψ_g) wavefunctions are plotted for the (0,0) bands; for the (0,1), (1,0), and (1,1) bands, the product $\psi_e R_e \psi_g$ is plotted instead of the electronic transition moment.

In Tables 5.8 and 5.9 we list the values that we calculate for $A_{v'v''}$ for OH and NO, respectively, using the formula

$$A_{v'v''} = \frac{16\pi^3\,\tilde{\nu}^3_{v'v''}}{3h\varepsilon_0}\left[\frac{2-\Lambda_{0,\Lambda''+\Lambda'}}{2-\Lambda_{0,\Lambda'}}\right]\left[\frac{\tilde{\mu}_{1+}(v', v'', J' = J'' = 1.5) + \tilde{\mu}_{1-}(v', v'', J' = J'' = 0.5)}{2}\right]^2, \tag{5.185}$$

where Eq. (5.185) is a slightly modified version of similar equations in Luque and Crosley (1998) and Bernath (2016); the molecule-fixed electric dipole moment matrix

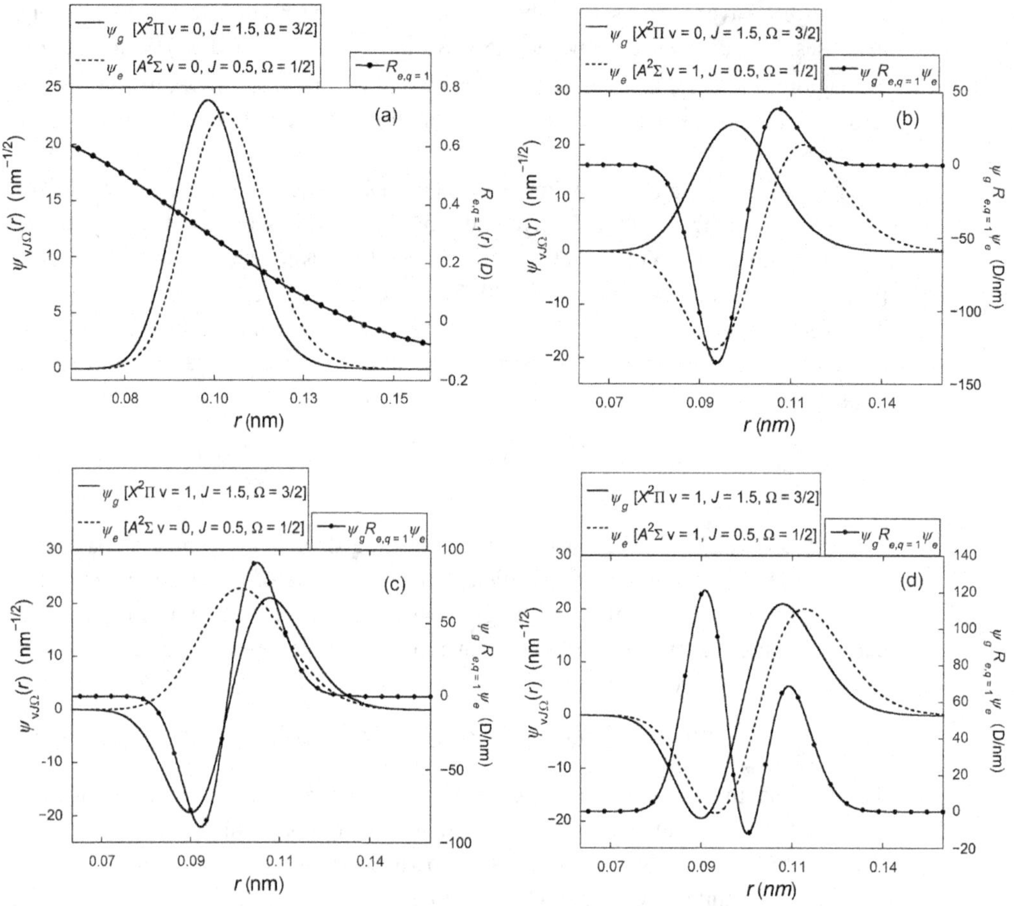

Figure 5.3 Vibration-rotation wavefunctions and $R_{e,q=1}(r)$ for (a) the OH (0,0) band. Vibration-rotation wavefunctions and the products $\psi_e(r)R_{e,q=1}(r)\psi_g(r)$ for the OH (b) (1,0), (c) (0,1), and (d) (1,1) bands.

elements for the lowest J transitions are averaged in Eq. (5.185). For strong transitions, the calculated values of $A_{v'v''}$ are in good agreement with values listed by Yousefi et al. (2018) – in most cases, 1 or 2% higher. For weak transitions – the (0,1), (1,2), and (0,2) bands – our calculated values are lower by 20–40%. The reason for the differences is apparent from Figure 5.3c. The plot of the integrand $\psi_e R_{e,q=1} \psi_g$ has large and nearly equal negative and positive lobes, so the value of the integral is dependent on the difference between two large numbers. For the calculations of the OH Einstein coefficients for the vibration-rotation transitions discussed below, the values of $A_{v'v''}$ are corrected to agree with the results of Yousefi et al. (2018). The potential function for $X^2\Pi$ electronic level depicted in figure 1 in Yousefi et al. (2018) has many more turning points in the repulsive part of the potential than we were able to extract from our RKR1 calculations. Therefore, our results for the weak (0,1), (1,2),

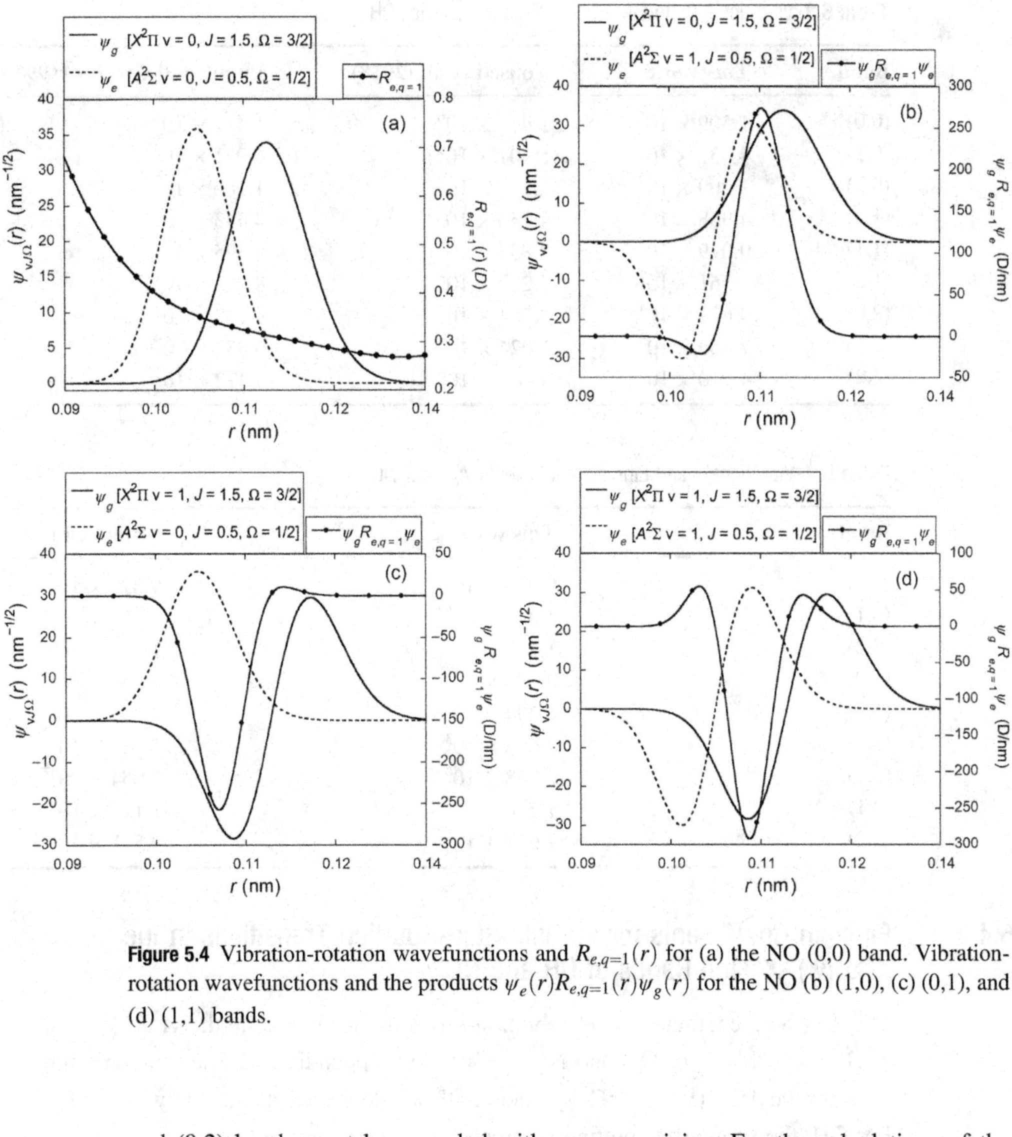

Figure 5.4 Vibration-rotation wavefunctions and $R_{e,q=1}(r)$ for (a) the NO (0,0) band. Vibration-rotation wavefunctions and the products $\psi_e(r)R_{e,q=1}(r)\psi_g(r)$ for the NO (b) (1,0), (c) (0,1), and (d) (1,1) bands.

and (0,2) bands must be regarded with some suspicion. For the calculations of the Einstein coefficients for vibration-rotation transitions in OH, the molecule-fixed electric dipole moment functions were scaled so that $A_{v'v''}$ for each band was equal to the value listed in Yousefi et al. (2018).

For the NO molecule, however, our calculations of $A_{v'v''}$ agree to better than 1% with the values listed in Settersten et al. (2009) for all of the bands. It is evident from Figure 5.4 that the situation depicted in Figure 5.3c for the weak OH (0,1) band does not occur for any of the NO bands that are depicted. This is partly due to the weak dependence on $R_{e,q=1}(r)$ in the regions of significant wavefunction overlap and also because the wavefunctions for the $A^2\Sigma^+$ level are shifted significantly to lower values of r compared to the wavefunctions for the $X^2\Pi$ level.

Table 5.8 Vibrational band Einstein coefficients $A_{v'v''}$ for OH

Band	This work	Yousefi et al. (2018)	Luque and Crosley (1998)
(0,0)	1.509×10^6	1.467×10^6	1.451×10^6
(0,1)	4.451×10^3	5.901×10^3	6.921×10^3
(0,2)	1.009×10^2	1.252×10^2	1.986×10^2
(1,0)	4.958×10^5	4.750×10^5	4.643×10^5
(1,1)	9.079×10^5	8.835×10^5	8.595×10^5
(1,2)	4.361×10^3	7.284×10^3	8.207×10^3
(2,0)	9.627×10^4	9.704×10^4	9.202×10^4
(2,1)	7.166×10^5	7.024×10^5	6.852×10^5
(2,2)	4.910×10^5	4.700×10^5	4.472×10^5

Table 5.9 Vibrational band Einstein coefficients $A_{v'v''}$ for NO

Band	This work	Settersten et al.
(0,0)	9.946×10^5	9.877×10^5
(0,1)	1.469×10^6	1.459×10^6
(0,2)	1.233×10^6	1.225×10^6
(1,0)	2.173×10^6	2.158×10^6
(1,1)	6.294×10^5	6.254×10^5
(1,2)	5.603×10^3	5.532×10^3
(2,0)	2.068×10^6	2.054×10^6
(2,1)	1.116×10^5	1.111×10^5
(2,2)	9.647×10^5	9.571×10^5

5.4.2 Einstein Coefficients for the Vibration-Rotation Transitions in the $A^2\Sigma^+(v')$–$X^2\Pi(v)$ Bands of OH and NO

The Einstein coefficients for vibration-rotation transitions in the $A^2\Sigma^+(v' = 0) - X^2\Pi(v = 0)$ bands of OH and NO are listed in Appendix A7. The rotational transitions for the $A^2\Sigma^+(v') - X^2\Pi(v)$ band of OH are depicted schematically in Figure 5.5. For OH, the spin–orbit splitting parameter A_v is negative, so that the $\Omega = 3/2$ levels are lower in energy than the $\Omega = 1/2$ levels. For the NO molecule, parameter A_v is positive, so that the $\Omega = 1/2$ levels are lower in energy than the $\Omega = 3/2$ levels. However, in both cases, the F_1 levels are lower in energy, so that these levels are associated with $\Omega = 3/2$ for OH and $\Omega = 1/2$ for NO.

The OH calculations were performed using the spectroscopic parameters listed in Stark et al. (1994). The parameters H_{11}, H_{12}, and H_{22} were calculated using the following equations from Chapter 3:

$$a_J = d_J = \sqrt{\frac{1}{2}\left(1 - \frac{\beta}{X}\right)}, \tag{5.186}$$

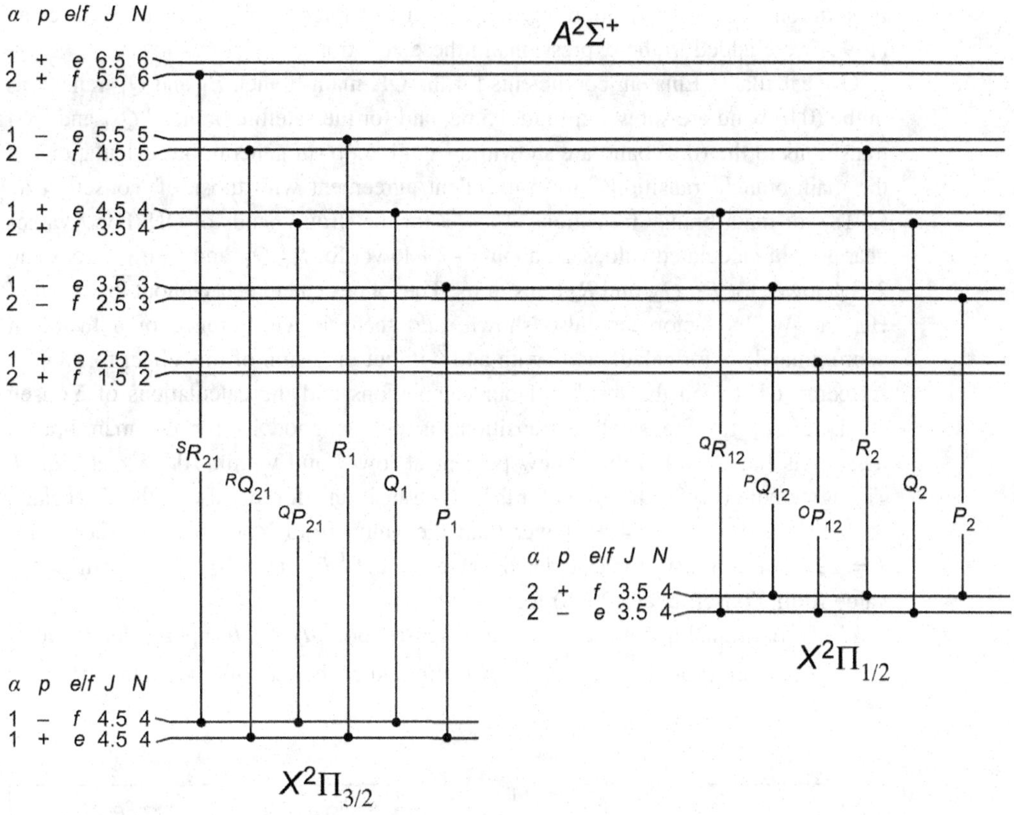

Figure 5.5 Schematic diagram of the rotational transitions in the OH $A^2\Sigma^+(v') - X^2\Pi(v)$ system.

$$b_J = -c_J = \sqrt{\frac{1}{2}\left(1 + \frac{\beta}{X}\right)}, \tag{5.187}$$

where the parameters α, β, and X are given by

$$\alpha = J(J+1) + \frac{1}{4} - \Lambda^2 = \left(J + \frac{1}{2}\right)^2 - \Lambda^2. \tag{5.188}$$

$$\beta = \frac{1}{2}(H_{11} - H_{22}) = \Lambda\left[B_v - \frac{1}{2}A_v - \frac{1}{2}\gamma_v - \alpha\left(\frac{1}{2}A_{Dv} + 2D_v\right)\right]. \tag{5.189}$$

$$X = \left[\left(\frac{H_{11} - H_{22}}{2}\right)^2 + H_{12}^2\right]^{1/2}$$
$$= \sqrt{\left(J + \frac{1}{2}\right)^2\left[B_v - \frac{1}{2}\gamma_v - 2D_v\alpha\right]^2 + \Lambda^2(A_v + \alpha A_{Dv})\left[-B_v + \frac{1}{2}\gamma_v + 2D_v\alpha + \frac{1}{4}(A_v + \alpha A_{Dv})\right]}. \tag{5.190}$$

In addition, terms for Λ-doubling and higher-order terms discussed in Stark et al. (1994) were added to the expressions in these equations.

Our calculated Einstein coefficients for the OH main-branch P_1 and Q_1 transitions in the (0,0) band are shown in Figure 5.6a, and for the satellite branch ${}^{R}Q_{21}$ and ${}^{S}R_{21}$ transitions in the (0,0) band are shown in Figure 5.6b. In general, our calculations of the main-branch transitions are in excellent agreement with those of Yousefi et al. (2018) for main-branch transitions at low J for the strong bands of OH. For J-values near 30, our calculated values are about 6–7% lower for P_1, Q_1, and R_1 transitions and 2–3% lower for P_2, Q_2, and R_2 transitions. The values calculated without inclusion of Herman–Wallis factors are also shown, and there is a difference of a factor of approximately 3 for calculations with and without inclusion of this effect for high J. Agreement between the results of our calculations and the calculations of Yousefi et al. (2018) for the satellite transitions is not as good as for the main-branch transitions but is still within a few percent at low J and within 10–15% at high J. The agreement is typically worst for the ${}^{S}R_{21}$ transitions; for the case of the (0,0) band, our calculated value is 40% lower than the value from Yousefi et al. (2018) for J = 29.5. In contrast, our calculated value for the ${}^{O}P_{12}$ is only 5% lower than the value from Yousefi et al. (2018) for J = 29.5.

As the rotational quantum number J increases, both a_J and b_J approach a value of $1/\sqrt{2}$. In the limit of very high J, the line strength of the main-branch rotational lines

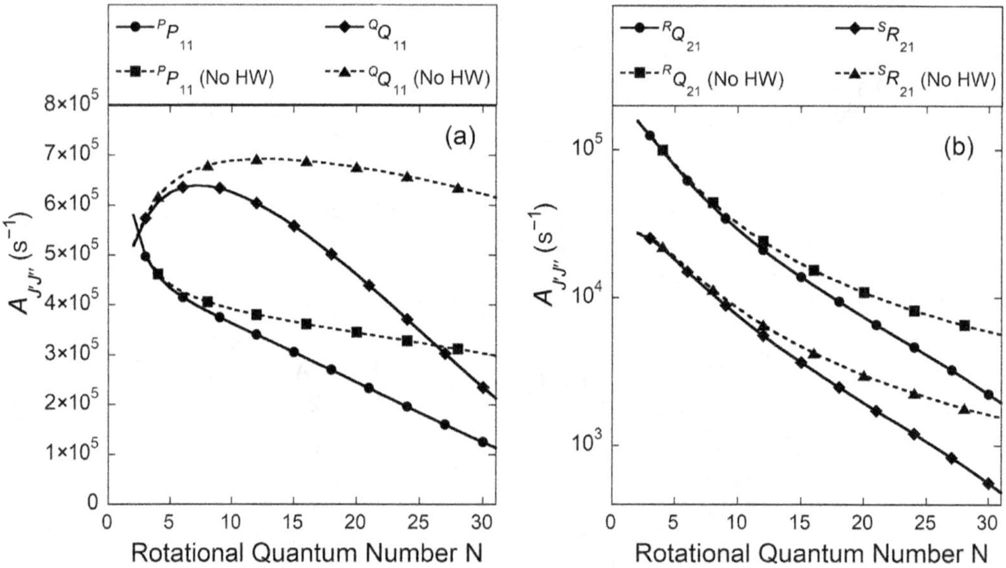

Figure 5.6 Calculated Einstein coefficients for rotational transitions in the OH $A^2\Sigma^+$ (v′ = 0) – $X^2\Pi$ (v″ = 0) band with and without the inclusion of Herman–Wallis effects. The results for the main-branch transitions ${}^{P}P_{11}$ and ${}^{Q}Q_{11}$ are shown in panel (a), and the results for the satellite branch transitions ${}^{S}R_{21}$ and ${}^{R}Q_{21}$ are shown in panel (b).

P_1, Q_1, and R_1 for $^2\Pi - {}^2\Sigma^\pm$ transitions is proportional to $a_J + b_J$. We can use the expansion $\sqrt{1+x} \cong 1 + \frac{1}{2}x$ for small x to obtain the following expression:

$$a_J + b_J = \frac{1}{\sqrt{2}}\sqrt{1 - \frac{\beta}{X}} + \frac{1}{\sqrt{2}}\sqrt{1 + \frac{\beta}{X}} \cong \frac{1}{\sqrt{2}}\left(1 - \frac{\beta}{2X}\right) + \frac{1}{\sqrt{2}}\left(1 + \frac{\beta}{2X}\right) = \frac{2}{\sqrt{2}}. \tag{5.191}$$

The line strength of main-branch rotational transitions is insensitive to the values of spectroscopic constants in the limit of high J. Conversely, the line strength of the main-branch rotational lines $^QP_{21}(J)$, $^RQ_{21}(J)$, and $^SR_{21}(J)$ for $^2\Pi - {}^2\Sigma^\pm$ transitions is proportional to $a_J - b_J$ in the limit of high J. Using Eqs. (3.189)–(3.185), we obtain

$$a_J - b_J = \frac{1}{\sqrt{2}}\sqrt{1 - \frac{\beta}{X}} - \frac{1}{\sqrt{2}}\sqrt{1 + \frac{\beta}{X}} \cong \frac{1}{\sqrt{2}}\left(1 - \frac{\beta}{2X}\right) + \frac{1}{\sqrt{2}}\left(1 + \frac{\beta}{2X}\right) = \frac{1}{\sqrt{2}}\frac{\beta}{X}$$
$$= \frac{\Lambda\left[B_v - \frac{1}{2}A_v - \frac{1}{2}\gamma_v - \alpha\left(\frac{1}{2}A_{Dv} + 2D_v\right)\right]/\sqrt{2}}{\sqrt{\left(J+\frac{1}{2}\right)^2\left[B_v - \frac{1}{2}\gamma_v - 2D_v\alpha\right]^2 + \Lambda^2(A_v + \alpha A_{Dv})\left[-B_v + \frac{1}{2}\gamma_v + 2D_v\alpha + \frac{1}{4}(A_v + \alpha A_{Dv})\right]}}. \tag{5.192}$$

Therefore, the decrease in the line strengths of the satellite branch rotational transitions with increasing J is very dependent on the value of the spectroscopic constants. Neglecting centrifugal distortion effects and spin–rotation coupling in Eq. (5.192), we obtain

$$a_J - b_J \cong \frac{\Lambda[2 - Y_v]}{\sqrt{8\left(J + \frac{1}{2}\right)^2 + 2\Lambda^2 Y_v[Y_v - 4]}}. \tag{5.193}$$

For the $X^2\Pi(\mathrm{v}=0)$ vibrational bands of OH and NO, $Y_v = -7.4959$ and 72.596, respectively. Consequently, the line strength of satellite transitions compared to main-branch transitions falls much more rapidly with increasing J for OH than for NO.

Our calculated Einstein coefficients for the NO main-branch P_1 and Q_1 transitions in the (0,0) band are shown in Figure 5.7a, and for the satellite branch $^RQ_{21}$ and $^SR_{21}$ transitions in the (0,0) band are shown in Figure 5.7b. The NO calculations were performed using the spectroscopic parameters listed in Amiot et al. (1978) for the lower $\mathrm{v}'' = 0, 1, 2$ levels, in Paul (1997) for the $\mathrm{v}' = 0, 1$ levels, and in Engleman and Rouse (1971) for the $\mathrm{v}' = 2$ level. In general, our calculations of Einstein coefficients for the main-branch transitions are in excellent agreement, to within 1–2%, with those of LIFBASE (Luque & Crosley, 1999a) for main-branch transitions at J values up to 80 for nearly all bands of NO. The values calculated without inclusion of Herman–Wallis factors are also shown, and there is a difference of a factor of approximately 10–20% for calculations with and without inclusion of this effect for high J. Agreement between the results of our calculations and the values listed in LIFBASE (Luque & Crosley, 1999a) for the satellite transitions is not as good as for the main-branch

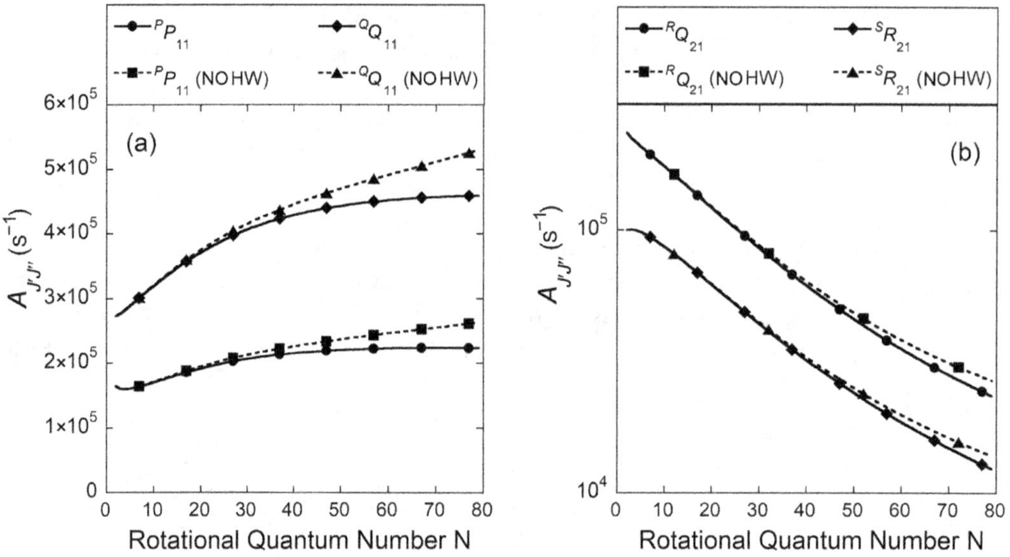

Figure 5.7 Calculated Einstein coefficients for rotational transitions in the NO $A^2\Sigma^+$ ($v' = 0$) – $X^2\Pi$ ($v'' = 0$) band with and without the inclusion of Herman–Wallis effects. The results for the main-branch transitions $^PP_{11}$ and $^QQ_{11}$ are shown in panel (a), and the results for the satellite branch transitions $^SR_{21}$ and $^RQ_{21}$ are shown in panel (b).

transitions but is still within a few percent at low J and within 5–15% at high J. The comparison for these transitions is shown in Figure 5.7b, along with our calculations without the inclusion of the Herman–Wallis effect. Our calculated values of the Einstein coefficients for these transitions are lower than the values listed in LIFBASE (Luque & Crosley, 1999a); we assume that this is due to our incorporation of terms such as spin–rotation coupling and centrifugal distortion in the expressions for the line strengths.

5.5 Triplet Electronic Transitions

The calculation of radiative transition rates for transitions between rotational levels contained within triplet electronic levels is discussed in this section. As was the case for doublet transitions, a Hund's case (a) basis set is used in the calculation of rotational line strengths, and the Herman–Wallis effect is taken into account. The rotational line strengths for the $X^3\Sigma^-\,(v=0) - A^3\Pi\,(v=0)$ vibrational band are calculated to illustrate the calculation methodology for triplet transitions. Unlike the case for singlet and doublet transitions, tables for Hönl–London factors or rotational line strengths are not developed. The formulae for rotational line strengths would be quite complicated following the evaluation of the appropriate $3j$ symbols, so the rotational line strength expressions include the terms $\begin{pmatrix} J & J' & 1 \\ \Omega & -\Omega' & \Omega'-\Omega \end{pmatrix}$. These terms can be evaluated using the appropriate equations in Appendix A3.

5.5.1 Radiative Transitions between Triplet Electronic Levels with $\Lambda = 0$, $\Lambda' = 0$

The allowed rotational transitions for the $^3\Sigma^- - {}^3\Sigma^-$ system are shown in Figure 5.8. For $^3\Sigma^-$ electronic levels, the basis wavefunctions are given by

$$\left|^3\Sigma^-_{1e};\mathrm{v};JM\right\rangle = \frac{1}{\sqrt{2}}[|0^-;11;\mathrm{v};J1M\rangle + |0^-;1-1;\mathrm{v};J-1M\rangle]. \tag{5.194}$$

$$\left|^3\Sigma^-_{0e};\mathrm{v};JM\right\rangle = |0^-;10;\mathrm{v};J0M\rangle. \tag{5.195}$$

$$\left|^3\Sigma^-_{1f};\mathrm{v};JM\right\rangle = \frac{1}{\sqrt{2}}[|0^-;11;\mathrm{v};J1M\rangle - |0^-;1-1;\mathrm{v};J-1M\rangle]. \tag{5.196}$$

For $^3\Sigma^-$ electronic levels, the state wavefunctions are given by

$$|F_1;\mathrm{v};JM;e\rangle = S_{1e,N+1}\left|^3\Sigma^-_{1e};\mathrm{v};JM\right\rangle + S_{0e,N+1}\left|^3\Sigma^-_{0e};\mathrm{v};JM\right\rangle. \tag{5.197}$$

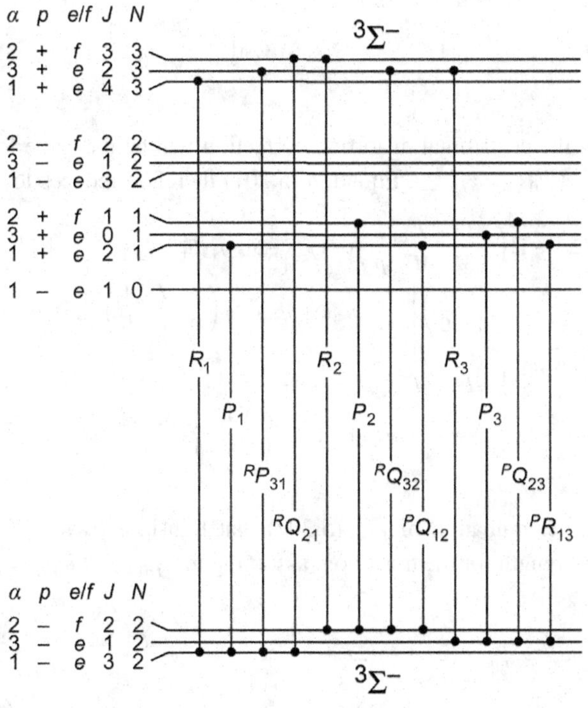

Figure 5.8 Allowed rotational transitions for the $^3\Sigma^- - {}^3\Sigma^-$ system.

$$|F_2; \mathrm{v}; JM; f\rangle = S_{1f,N}\left|{}^3\Sigma^-_{1f}; \mathrm{v}; JM\right\rangle = \left|{}^3\Sigma^-_{1f}; \mathrm{v}; JM\right\rangle. \tag{5.198}$$

$$|F_3; \mathrm{v}; JM; e\rangle = S_{1e,N-1}\left|{}^3\Sigma^-_{1e}; \mathrm{v}; JM\right\rangle + S_{0e,N-1}\left|{}^3\Sigma^-_{0e}; \mathrm{v}; JM\right\rangle. \tag{5.199}$$

$F_1(^3\Sigma^-) \rightarrow F_1(^3\Sigma^-)$ Transitions

The component p of the electric dipole moment matrix element in the space-fixed reference frame is given by

$$\begin{aligned}
\mu_{ba,p} &= \langle F_1; \mathrm{v}'; J'M'; e|\sum_q \hat{\mu}_q D^{1*}_{pq}(\omega)|F_1; \mathrm{v}; JM; e\rangle \\
&= \sum_q \Big\{\frac{1}{2} S_{1e,N+1} S'_{1e,N+1} \langle 0^-; 11; \mathrm{v}'|\hat{\mu}_q|0^-; 11; \mathrm{v}\rangle \langle J' 1M'|D^{1*}_{pq}(\omega)|J 1M\rangle \\
&\quad + \frac{1}{2} S_{1e,N+1} S'_{1e,N+1} \langle 0^-; 1-1; \mathrm{v}'|\hat{\mu}_q|0^-; 1-1; \mathrm{v}\rangle \langle J' -1M'|D^{1*}_{pq}(\omega)|J -1M\rangle \\
&\quad + S_{0e,N+1} S'_{0e,N+1} \langle 0^-; 10; \mathrm{v}'|\hat{\mu}_q|0^-; 10; \mathrm{v}\rangle \langle J' 0M'|D^{1*}_{pq}(\omega)|J 0M\rangle \Big\}.
\end{aligned} \tag{5.200}$$

The rotation- and angle-dependent terms in the expression in Eq. (5.200) can be analyzed using Eq. (5.36) to obtain

$$\begin{aligned}
\mu_{ba,p} = \sqrt{(2J+1)(2J'+1)} \begin{pmatrix} J & J' & 1 \\ -M & M' & p \end{pmatrix} (-1)^M \Big\{ \frac{1}{2} S_{1e,N+1} S'_{1e,N+1} (-1)^{-1} \Big[\tilde{\mu}_{0++} \begin{pmatrix} J & J' & 1 \\ 1 & -1 & 0 \end{pmatrix} \\
+ \tilde{\mu}_{0--} (-1)^1 \begin{pmatrix} J & J' & 1 \\ -1 & 1 & 0 \end{pmatrix} \Big] + S_{0e,N+1} S'_{0e,N+1} \tilde{\mu}_{000} \begin{pmatrix} J & J' & 1 \\ 0 & 0 & 0 \end{pmatrix} \Big\}.
\end{aligned} \tag{5.201}$$

The magnitude of the centrifugal distortion correction will be the same for $\tilde{\mu}_{0++}$ and $\tilde{\mu}_{0--}$, so we can set $\tilde{\mu}_{0++} = \tilde{\mu}_{0--}$. Equation (5.201) therefore reduces to

$$\begin{aligned}
\mu_{ba,p} &= \sqrt{(2J+1)(2J'+1)} \begin{pmatrix} J & J' & 1 \\ -M & M' & p \end{pmatrix} (-1)^M \Big\{ \frac{1}{2} S_{1e,N+1} S'_{1e,N+1} \tilde{\mu}_{0++} \Big[(-1)^{-1} \begin{pmatrix} J & J' & 1 \\ 1 & -1 & 0 \end{pmatrix} \\
&\quad + (-1)^{1+J+J'+1} \begin{pmatrix} J & J' & 1 \\ -1 & 1 & 0 \end{pmatrix} \Big] + S_{0e,N+1} S'_{0e,N+1} \tilde{\mu}_{000} \begin{pmatrix} J & J' & 1 \\ 0 & 0 & 0 \end{pmatrix} \Big\} \\
&= \sqrt{(2J+1)(2J'+1)} \begin{pmatrix} J & J' & 1 \\ -M & M' & p \end{pmatrix} (-1)^M \Big\{ -\frac{1}{2} S_{1e,N+1} S'_{1e,N+1} \tilde{\mu}_{0++} \begin{pmatrix} J & J' & 1 \\ 1 & -1 & 0 \end{pmatrix} \\
&\quad \times \Big[1 + (-1)^{J+J'+1} \Big] + S_{0e,N+1} S'_{0e,N+1} \tilde{\mu}_{000} \begin{pmatrix} J & J' & 1 \\ 0 & 0 & 0 \end{pmatrix} \Big\}.
\end{aligned} \tag{5.202}$$

It is obvious from examination of Eq. (5.202) that transitions with $J = J'$ are not allowed. The line strength for the transition, assuming that $\tilde{\mu}_{0++} = \tilde{\mu}_{000} = \tilde{\mu}_0$, is given by

$$\begin{aligned}
\tilde{S}_{J''J'} &= \sum_p \sum_M \sum_M \mu_{ba,p} \mu^*_{ba,p} = (2J+1)(2J'+1) \tilde{\mu}_0^2 \\
&\quad \times \Big\{ -\frac{1}{2} S_{1e,N+1} S'_{1e,N+1} \begin{pmatrix} J & J' & 1 \\ 1 & -1 & 0 \end{pmatrix} \Big[1 + (-1)^{J+J'+1} \Big] + S_{0e,N+1} S'_{0e,N+1} \begin{pmatrix} J & J' & 1 \\ 0 & 0 & 0 \end{pmatrix} \Big\}^2.
\end{aligned} \tag{5.203}$$

$F_2(^3\Sigma^-) \rightarrow F_2(^3\Sigma^-)$ Transitions

For these transitions the component p of the electric dipole moment matrix element in the space-fixed reference frame is given by

$$\begin{aligned}
\mu_{ba,p} &= \langle F_2; \mathrm{v}'; J'M'; f| \sum_q \hat{\mu}_q D_{pq}^{1*}(\omega)|F_2; \mathrm{v}; JM; f\rangle \\
&\sum_q \left\{ \frac{1}{2}\langle 0^-; 11; \mathrm{v}'|\hat{\mu}_q|0^-; 11; \mathrm{v}\rangle\langle J'1M'|D_{pq}^{1*}(\omega)|J1M\rangle \right. \\
&\left. +\frac{1}{2}\langle 0^-; 1-1; \mathrm{v}'|\hat{\mu}_q|0^-; 1-1; \mathrm{v}\rangle\langle J'-1M'|D_{pq}^{1*}(\omega)|J-1M\rangle \right\} \\
&= \frac{\sqrt{(2J+1)(2J'+1)}}{2}\begin{pmatrix} J & J' & 1 \\ -M & M' & p \end{pmatrix}(-1)^M\left\{(-1)^{-1}\left[\tilde{\mu}_{0++}\begin{pmatrix} J & J' & 1 \\ 1 & -1 & 0 \end{pmatrix}\right.\right. \\
&\left.\left. +\tilde{\mu}_{0--}(-1)^1\begin{pmatrix} J & J' & 1 \\ -1 & 1 & 0 \end{pmatrix}\right]\right\}.
\end{aligned} \tag{5.204}$$

The magnitude of the centrifugal distortion correction will be the same for $\tilde{\mu}_{0++}$ and $\tilde{\mu}_{0--}$, so we can again set $\tilde{\mu}_{0++} = \tilde{\mu}_{0--}$. The line strength for the transition is given by

$$\tilde{S}_{J''J'} = \sum_p \sum_M \sum_M \mu_{ba,p}\mu_{ba,p}^* = (2J+1)(2J'+1)\tilde{\mu}_{0++}^2\begin{pmatrix} J & J' & 1 \\ 1 & -1 & 0 \end{pmatrix}^2\left[1+(-1)^{J+J'+1}\right]^2. \tag{5.205}$$

$F_3(^3\Sigma^-) \rightarrow F_3(^3\Sigma^-)$ Transitions

The component p of the electric dipole moment matrix element in the space-fixed reference frame is given by

$$\begin{aligned}
\mu_{ab,p} &= \langle F_3; \mathrm{v}'; J'M'; e| \sum_q \hat{\mu}_q D_{pq}^{1*}(\omega)|F_3; \mathrm{v}; JM; e\rangle \\
&= \sum_q \left\{ \frac{1}{2}S_{1e,N-1}S'_{1e,N-1}\langle 0^-; 1-1; \mathrm{v}'|\hat{\mu}_q|0^-; 1-1; \mathrm{v}\rangle\langle J'-1M'|D_{pq}^{1*}(\omega)|J-1M\rangle \right. \\
&+\frac{1}{2}S_{1e,N-1}S'_{1e,N-1}\langle 0^-; 11; \mathrm{v}'|\hat{\mu}_q|0^-; 11; \mathrm{v}\rangle\langle J'1M'|D_{pq}^{1*}(\omega)|J1M\rangle \\
&\left. +S_{0e,N-1}S'_{0e,N-1}\langle 0^-; 10; \mathrm{v}'|\hat{\mu}_q|0^-; 10; \mathrm{v}\rangle\langle J'0M'|D_{pq}^{1*}(\omega)|J0M\rangle \right\}.
\end{aligned} \tag{5.206}$$

The rotation- and angle-dependent terms in the expression in Eq. (5.200) can be analyzed using Eq. (5.36) to obtain

$$\begin{aligned}
\mu_{ba,p} &= \sqrt{(2J+1)(2J'+1)}\begin{pmatrix} J & J' & 1 \\ -M & M' & p \end{pmatrix}(-1)^M\left\{\frac{1}{2}S_{1e,N-1}S'_{1e,N-1}(-1)^{-1}\left[\tilde{\mu}_{0++}\begin{pmatrix} J & J' & 1 \\ 1 & -1 & 0 \end{pmatrix}\right.\right. \\
&\left.\left. +\tilde{\mu}_{0--}(-1)^1\begin{pmatrix} J & J' & 1 \\ -1 & 1 & 0 \end{pmatrix}\right] + S_{0e,N-1}S'_{0e,N-1}\tilde{\mu}_{000}\begin{pmatrix} J & J' & 1 \\ 0 & 0 & 0 \end{pmatrix}\right\}.
\end{aligned} \tag{5.207}$$

The magnitude of the centrifugal distortion correction will be the same for $\tilde{\mu}_{0++}$ and $\tilde{\mu}_{0--}$, so we can set $\tilde{\mu}_{0++} = \tilde{\mu}_{0--}$. Equation (5.201) therefore reduces to

$$\begin{aligned}\mu_{ba,p} &= \sqrt{(2J+1)(2J'+1)}\begin{pmatrix} J & J' & 1 \\ -M & M' & p \end{pmatrix}(-1)^M\left\{\frac{1}{2}S_{1e,N-1}S'_{1e,N-1}\tilde{\mu}_{0++}\left[(-1)^{-1}\begin{pmatrix} J & J' & 1 \\ 1 & -1 & 0 \end{pmatrix}\right.\right.\\ &\left.+(-1)^{1+J+J'+1}\begin{pmatrix} J & J' & 1 \\ -1 & 1 & 0 \end{pmatrix}\right] \left.+S_{0e,N-1}S'_{0e,N-1}\tilde{\mu}_{000}\begin{pmatrix} J & J' & 1 \\ 0 & 0 & 0 \end{pmatrix}\right\}\\ &= \sqrt{(2J+1)(2J'+1)}\begin{pmatrix} J & J' & 1 \\ -M & M' & p \end{pmatrix}(-1)^M\left\{-\frac{1}{2}S_{1e,N-1}S'_{1e,N-1}\tilde{\mu}_{0++}\begin{pmatrix} J & J' & 1 \\ 1 & -1 & 0 \end{pmatrix}\right.\\ &\left.\times\left[1+(-1)^{J+J'+1}\right]+S_{0e,N-1}S'_{0e,N-1}\tilde{\mu}_{000}\begin{pmatrix} J & J' & 1 \\ 0 & 0 & 0 \end{pmatrix}\right\}.\end{aligned} \tag{5.208}$$

It is again obvious from examination of Eq. (5.208) that transitions with $J = J'$ are not allowed. The line strength for the transition, assuming that $\tilde{\mu}_{0++} = \tilde{\mu}_{000} = \tilde{\mu}_0$, is given by

$$\begin{aligned}\tilde{S}_{J''J'} &= \sum_p\sum_M\sum_M \mu_{ba,p}\mu^*_{ba,p} = (2J+1)(2J'+1)\tilde{\mu}_0^2\\ &\times\left\{-\frac{1}{2}S_{1e,N-1}S'_{1e,N-1}\begin{pmatrix} J & J' & 1 \\ 1 & -1 & 0 \end{pmatrix}\left[1+(-1)^{J+J'+1}\right]+S_{0e,N-1}S'_{0e,N-1}\begin{pmatrix} J & J' & 1 \\ 0 & 0 & 0 \end{pmatrix}\right\}^2.\end{aligned} \tag{5.209}$$

$F_1(^3\Sigma^-) \rightarrow F_3(^3\Sigma^-)$ and $F_3(^3\Sigma^-) \rightarrow F_1(^3\Sigma^-)$ Transitions

The component p of the electric dipole moment matrix element in the space-fixed reference frame for $F_1 \rightarrow F_3$ transitions is given by

$$\begin{aligned}\mu_{ba,p} &= \langle F_3; v'; J'M'; e|\sum_q \hat{\mu}_q D^{1*}_{pq}(\omega)|F_1; v; JM; e\rangle\\ &= \sum_q\left\{\frac{1}{2}S_{1e,N+1}S'_{1e,N-1}\langle 0^-;11;v'|\hat{\mu}_q|0^-;11;v\rangle\langle J'1M'|D^{1*}_{pq}(\omega)|J1M\rangle\right.\\ &+\frac{1}{2}S_{1e,N+1}S'_{1e,N-1}\langle 0^-;1-1;v'|\hat{\mu}_q|0^-;1-1;v\rangle\langle J'-1M'|D^{1*}_{pq}(\omega)|J-1M\rangle\\ &\left.+S_{0e,N+1}S'_{0e,N-1}\langle 0^-;10;v'|\hat{\mu}_q|0^-;10;v\rangle\langle J'0M'|D^{1*}_{pq}(\omega)|J0M\rangle\right\}.\end{aligned} \tag{5.210}$$

Following the same procedure we used for the $F_1 \rightarrow F_1$ and $F_3 \rightarrow F_3$ transitions, the matrix element becomes

$$\begin{aligned}\mu_{ba,p} &= \sqrt{(2J+1)(2J'+1)}\begin{pmatrix} J & J' & 1 \\ -M & M' & p \end{pmatrix}(-1)^M\left\{-\frac{1}{2}S_{1e,N+1}S'_{1e,N-1}\tilde{\mu}_{0++}\begin{pmatrix} J & J' & 1 \\ 1 & -1 & 0 \end{pmatrix}\right.\\ &\left.\times\left[1+(-1)^{J+J'+1}\right]+S_{0e,N+1}S'_{0e,N-1}\tilde{\mu}_{000}\begin{pmatrix} J & J' & 1 \\ 0 & 0 & 0 \end{pmatrix}\right\}.\end{aligned} \tag{5.211}$$

The line strength for $F_1 \rightarrow F_3$ transitions, assuming that $\tilde{\mu}_{0++} = \tilde{\mu}_{000} = \tilde{\mu}_0$, is given by

$$\tilde{S}_{J''J'} = \sum_p \sum_M \sum_M \mu_{ba,p}\mu^*_{ba,p} = (2J+1)(2J'+1)\tilde{\mu}_0^2$$
$$\times \left\{ -\frac{1}{2} S_{1e,N+1} S'_{1e,N-1} \begin{pmatrix} J & J' & 1 \\ 1 & -1 & 0 \end{pmatrix} \left[1 + (-1)^{J+J'+1} \right] + S_{0e,N+1} S'_{0e,N-1} \begin{pmatrix} J & J' & 1 \\ 0 & 0 & 0 \end{pmatrix} \right\}^2. \tag{5.212}$$

The line strength for $F_3 \rightarrow F_1$ transitions, assuming that $\tilde{\mu}_{0++} = \tilde{\mu}_{000} = \tilde{\mu}_0$, is given by

$$\tilde{S}_{J''J'} = \sum_p \sum_M \sum_M \mu_{ba,p}\mu^*_{ba,p} = (2J+1)(2J'+1)\tilde{\mu}_0^2$$
$$\times \left\{ -\frac{1}{2} S_{1e,N-1} S'_{1e,N+1} \begin{pmatrix} J & J' & 1 \\ 1 & -1 & 0 \end{pmatrix} \left[1 + (-1)^{J+J'+1} \right] + S_{0e,N-1} S'_{0e,N+1} \begin{pmatrix} J & J' & 1 \\ 0 & 0 & 0 \end{pmatrix} \right\}^2. \tag{5.213}$$

$F_1(^3\Sigma^-) \rightarrow F_2(^3\Sigma^-)$, $F_2(^3\Sigma^-) \rightarrow F_1(^3\Sigma^-)$, $F_2(^3\Sigma^-) \rightarrow F_3(^3\Sigma^-)$, and $F_3(^3\Sigma^-) \rightarrow F_2(^3\Sigma^-)$ Transitions

For $F_1 \rightarrow F_2$ transitions, the component p of the electric dipole moment matrix element in the space-fixed reference frame is given by

$$\mu_{ba,p} = \langle F_2; \mathrm{v}'; J'M'; f | \sum_q \hat{\mu}_q D^{1*}_{pq}(\omega) | F_1; \mathrm{v}; JM; e \rangle$$

$$\sum_q \left\{ \frac{1}{\sqrt{2}} S_{1e,N+1} \langle 0^-; 11; \mathrm{v}' | \hat{\mu}_q | 0^-; 11; \mathrm{v} \rangle \langle J' 1M' | D^{1*}_{pq}(\omega) | J 1M \rangle \right.$$
$$\left. - \frac{1}{\sqrt{2}} S_{1e,N+1} \langle 0^-; 1-1; \mathrm{v}' | \hat{\mu}_q | 0^-; 1-1; \mathrm{v} \rangle \langle J' - 1M' | D^{1*}_{pq}(\omega) | J - 1M \rangle \right\}$$
$$= \frac{\sqrt{(2J+1)(2J'+1)}}{\sqrt{2}} \begin{pmatrix} J & J' & 1 \\ -M & M' & p \end{pmatrix} (-1)^M S_{1e,N+1} \left\{ (-1)^{-1} \left[\tilde{\mu}_{0++} \begin{pmatrix} J & J' & 1 \\ 1 & -1 & 0 \end{pmatrix} \right. \right.$$
$$\left. \left. - \tilde{\mu}_{0--} (-1)^1 \begin{pmatrix} J & J' & 1 \\ -1 & 1 & 0 \end{pmatrix} \right] \right\}. \tag{5.214}$$

The line strength for $F_1 \rightarrow F_2$ transitions is given by

$$\tilde{S}_{J''J'} = \sum_p \sum_M \sum_M \mu_{ba,p}\mu^*_{ba,p} = \frac{(2J+1)(2J'+1)}{2} \tilde{\mu}^2_{0++} S_{1e,N+1} \begin{pmatrix} J & J' & 1 \\ 1 & -1 & 0 \end{pmatrix}^2 \left[1 - (-1)^{J+J'+1} \right]^2. \tag{5.215}$$

For $F_1 \rightarrow F_2$ transitions, $J' = J$ transitions are allowed but $J' = J \pm 1$ transitions are not allowed. The line strengths for $F_2 \rightarrow F_1$, $F_3 \rightarrow F_2$, and $F_2 \rightarrow F_3$ transitions are given by Eqs. (5.216), (5.217), and (5.218), respectively.

$$\tilde{S}_{J''J'} = \frac{(2J+1)(2J'+1)}{2}\tilde{\mu}^2_{0++}S'_{1e,N+1}\begin{pmatrix} J & J' & 1 \\ 1 & -1 & 0 \end{pmatrix}^2 \left[1-(-1)^{J+J'+1}\right]^2. \quad (5.216)$$

$$\tilde{S}_{J''J'} = \frac{(2J+1)(2J'+1)}{2}\tilde{\mu}^2_{0++}S_{1e,N-1}\begin{pmatrix} J & J' & 1 \\ 1 & -1 & 0 \end{pmatrix}^2 \left[1-(-1)^{J+J'+1}\right]^2. \quad (5.217)$$

$$\tilde{S}_{J''J'} = \frac{(2J+1)(2J'+1)}{2}\tilde{\mu}^2_{0++}S'_{1e,N-1}\begin{pmatrix} J & J' & 1 \\ 1 & -1 & 0 \end{pmatrix}^2 \left[1-(-1)^{J+J'+1}\right]^2. \quad (5.218)$$

5.5.2 Radiative Transitions between Triplet Electronic Levels with $\Lambda = 0$, $\Lambda' = 1$

For ${}^3\Sigma^-$ electronic levels, the basis wavefunctions and state wavefunctions are given by Eqs. (5.194)–(5.199). The wavefunctions for the vibration-rotation levels for the ${}^3\Pi$ electronic levels with $\Lambda = 1$ are again assumed to be intermediate between case (a) and case (b). The line strength expressions are developed below under the assumptions that the ${}^3\Sigma$ level is the lower level in the transition and the ${}^3\Pi$ level is the upper level. However, line strengths for the case where the ${}^3\Pi$ level is the lower level in the transition and the ${}^3\Sigma$ level is the upper level can be obtain by switching the primed and unprimed parameters because of the symmetry of the line-strength equations.

The wavefunctions for the ${}^3\Pi$ level are given by

$$|F_1; \text{v}; JM; \pm\rangle = S_{0,N+1}\left|{}^3\Pi_0; \Lambda = 1; \text{v}; JM; \pm\right\rangle + S_{1,N+1}\left|{}^3\Pi_1; 1; \text{v}; JM; \pm\right\rangle + S_{2,N+1}\left|{}^3\Pi_2; 1; \text{v}; JM; \pm\right\rangle, \quad (5.219)$$

$$|F_2; \text{v}; JM; \pm\rangle = S_{0,N}\left|{}^3\Pi_0; 1; \text{v}; JM; \pm\right\rangle + S_{1,N}\left|{}^3\Pi_1; 1; \text{v}; JM; \pm\right\rangle + S_{2,N}\left|{}^3\Pi_2; 1; \text{v}; JM; \pm\right\rangle, \quad (5.220)$$

$$|F_3; \text{v}; JM; \pm\rangle = S_{0,N-1}\left|{}^3\Pi_0; 1; \text{v}; JM; \pm\right\rangle + S_{1,N-1}\left|{}^3\Pi_1; 1; \text{v}; JM; \pm\right\rangle + S_{2,N-1}\left|{}^3\Pi_2; 1; \text{v}; JM; \pm\right\rangle, \quad (5.221)$$

where

$$\left|{}^3\Pi_0; \Lambda; \text{v}; JM; \pm\right\rangle = \frac{1}{\sqrt{2}}\left[|\Lambda = 1; S = 1\,\Sigma = -1; \text{v}; J\ \Omega_- = \Lambda - 1\ M\rangle \pm (-1)^{\rho}|-1; 1\,1; \text{v}; J - \Omega_-\,M\rangle\right]. \quad (5.222)$$

$$\left|{}^3\Pi_1; 1; \text{v}; JM; \pm\right\rangle = \frac{1}{\sqrt{2}}\left[|1; 1\,0; \text{v}; J\Omega_0 M\rangle \pm (-1)^{\rho}|-1; 1\,0; \text{v}; J - \Omega_0\,M\rangle\right]. \quad (5.223)$$

$$\left|{}^3\Pi_2; 1; \text{v}; JM; \pm\right\rangle = \frac{1}{\sqrt{2}}\left[|1; 1\,1; \text{v}; J\Omega_+ M\rangle \pm (-1)^{\rho}|-1; 1 - 1; \text{v}; J - \Omega_+ M\rangle\right]. \quad (5.224)$$

$F_1(^3\Sigma^-) \rightarrow F_1(^3\Pi)$ and $F_3(^3\Sigma^-) \rightarrow F_3(^3\Pi)$ Transitions

The component p of the electric dipole moment matrix element in the space-fixed reference frame for $F_1(^3\Sigma^-) \rightarrow F_1(^3\Pi)$ transitions is given by

$$
\begin{aligned}
\mu_{ba,p} &= \langle F_1; v'; J'M'; \pm | \sum_q \hat{\mu}_q D^{1*}_{pq}(\omega) | F_1; v; JM; e\rangle \\
&= \sum_q \left\{ \frac{1}{2} S_{1e,N+1} S'_{0,N+1} \langle 1; 1-1; v' | \hat{\mu}_q | 0; 1-1; v\rangle \langle J' 0M' | D^{1*}_{pq}(\omega) | J-1M\rangle \right. \\
&\pm \frac{1}{2} S_{1e,N+1} S'_{0,N+1} (-1)^{\rho'} \langle -1; 11; v' | \hat{\mu}_q | 0; 11; v\rangle \langle J' 0M' | D^{1*}_{pq}(\omega) | J 1M\rangle \\
&+ \frac{1}{\sqrt{2}} S_{0e,N+1} S'_{1,N+1} \langle 1; 10; v' | \hat{\mu}_q | 0; 10; v\rangle \langle J' 1M' | D^{1*}_{pq}(\omega) | J 0M\rangle \\
&\pm \frac{1}{\sqrt{2}} S_{0e,N+1} S'_{1,N+1} (-1)^{\rho'} \langle -1; 10; v' | \hat{\mu}_q | 0; 10; v\rangle \langle J' -1M' | D^{1*}_{pq}(\omega) | J 0M\rangle \\
&+ \frac{1}{2} S_{1e,N+1} S'_{2,N+1} \langle 1; 11; v' | \hat{\mu}_q | 0; 11; v\rangle \langle J' 2M' | D^{1*}_{pq}(\omega) | J 1M\rangle \\
&\left. \pm \frac{1}{2} S_{1e,N+1} S'_{2,N+1} (-1)^{\rho'} \langle -1; 1-1; v' | \hat{\mu}_q | 0; 1-1; v\rangle \langle J' -2M' | D^{1*}_{pq}(\omega) | J-1M\rangle \right\}.
\end{aligned}
\tag{5.225}
$$

The rotation- and angle-dependent terms in the expression in Eq. (5.225) can be analyzed using Eq. (5.36),

$$
\begin{aligned}
\mu_{ba,p} &= \frac{\sqrt{(2J+1)(2J'+1)}}{2} \begin{pmatrix} J & J' & 1 \\ -M & M' & p \end{pmatrix} (-1)^{M'} \\
&\times \left\{ S_{1e,N+1} S'_{0,N+1} \left[\tilde{\mu}_{1,01} \begin{pmatrix} J & J' & 1 \\ -1 & 0 & 1 \end{pmatrix} \pm \tilde{\mu}_{-1,01} (-1)^{\rho'} \begin{pmatrix} J & J & 1 \\ 1 & 0 & -1 \end{pmatrix} \right] \right. \\
&+ \sqrt{2} S_{0e,N+1} S'_{1,N+1} \left[\tilde{\mu}_{1,10} (-1)^{-1} \begin{pmatrix} J & J' & 1 \\ 0 & -1 & 1 \end{pmatrix} \pm \tilde{\mu}_{-1,10} (-1)^{\rho'+1} \begin{pmatrix} J & J' & 1 \\ 0 & 1 & -1 \end{pmatrix} \right] \\
&\left. + S_{1e,N+1} S'_{2,N+1} \left[\tilde{\mu}_{1,21} (-1)^{-2} \begin{pmatrix} J & J' & 1 \\ 1 & -2 & 1 \end{pmatrix} \pm \tilde{\mu}_{-1,21} (-1)^{\rho'+2} \begin{pmatrix} J & J' & 1 \\ -1 & 2 & -1 \end{pmatrix} \right] \right\}.
\end{aligned}
\tag{5.226}
$$

At this point we will neglect the dependence of the electric dipole moment matrix elements on the values of Ω and Ω', and set $\tilde{\mu}_1 = \tilde{\mu}_{1,01} = \tilde{\mu}_{1,10} = \tilde{\mu}_{1,21}$ and $\tilde{\mu}_{-1} = \tilde{\mu}_{-1,01} = \tilde{\mu}_{-1,10} = \tilde{\mu}_{-1,21}$. This expression can also be simplified using Eq. (A3.2) from Appendix A3 and noting that $\rho' = J' - 1$:

$$
\begin{aligned}
\mu_{ba,p} &= \frac{\sqrt{(2J'+1)(2J+1)}}{2} \begin{pmatrix} J & J' & 1 \\ -M & M' & p \end{pmatrix} (-1)^{M'} \\
&= \left\{ S_{1e,N+1} S'_{0,N+1} \begin{pmatrix} J & J' & 1 \\ -1 & 0 & 1 \end{pmatrix} \left[\tilde{\mu}_1 \pm \tilde{\mu}_{-1} (-1)^{J+2J'} \right] \right. \\
&+ \sqrt{2} S_{0e,N+1} S'_{1,N+1} \begin{pmatrix} J & J' & 1 \\ 0 & -1 & 1 \end{pmatrix} \left[\tilde{\mu}_1 (-1)^{-1} \pm \tilde{\mu}_{-1} (-1)^{J+2J'+1} \right] \\
&\left. + S_{1e,N+1} S_{2,N+1} \begin{pmatrix} J & J' & 1 \\ 1 & -2 & 1 \end{pmatrix} \left[\tilde{\mu}_1 \pm \tilde{\mu}_{-1} (-1)^{J+2J'} \right] \right\}.
\end{aligned}
\tag{5.227}
$$

Since J' is an integer, $2J'$ is always an even number, and Eq. (5.227) reduces to

$$\mu_{ba,p} = \frac{\sqrt{(2J+1)(2J'+1)}}{2}\begin{pmatrix} J & J' & 1 \\ -M & M' & p \end{pmatrix}(-1)^{M'}\left[\tilde{\mu}_1 \pm \tilde{\mu}_{-1}(-1)^{J}\right]$$
$$\times\left\{S_{1e,N+1}S'_{0,N+1}\begin{pmatrix} J & J' & 1 \\ -1 & 0 & 1 \end{pmatrix} - \sqrt{2}S_{0e,N+1}S'_{1,N+1}\begin{pmatrix} J & J' & 1 \\ 0 & -1 & 1 \end{pmatrix}\right.$$
$$\left. + S_{1e,N+1}S'_{2,N+1}\begin{pmatrix} J & J' & 1 \\ 1 & -2 & 1 \end{pmatrix}\right\}. \tag{5.228}$$

For the ${}^3\Sigma^-$ electronic level, the symmetry properties imply that $\tilde{\mu}_{-1} = -\tilde{\mu}_1$. Substituting in Eq. (5.228), we obtain

$$\mu_{ba,p} = \frac{\sqrt{(2J+1)(2J'+1)}}{2}\begin{pmatrix} J & J' & 1 \\ -M & M' & p \end{pmatrix}(-1)^{M}\tilde{\mu}_1\left[1 \mp (-1)^{J}\right]$$
$$\times\left\{S_{1e,N+1}S'_{0,N+1}\begin{pmatrix} J & J' & 1 \\ -1 & 0 & 1 \end{pmatrix} - \sqrt{2}S_{0e,N+1}S'_{1,N+1}\begin{pmatrix} J & J' & 1 \\ 0 & -1 & 1 \end{pmatrix}\right.$$
$$\left. + S_{1e,N+1}S'_{2,N+1}\begin{pmatrix} J & J' & 1 \\ 1 & -2 & 1 \end{pmatrix}\right\}. \tag{5.229}$$

The line strength is given by

$$\tilde{S}_{J''J'} = \sum_{p}\sum_{M}\sum_{M}\mu_{ba,p}\mu^{*}_{ba,p} = \frac{(2J+1)(2J'+1)}{4}\tilde{\mu}_1^2\left[1 \mp (-1)^{J}\right]^2$$
$$\times\left\{S_{1e,N+1}S'_{0,N+1}\begin{pmatrix} J & J' & 1 \\ -1 & 0 & 1 \end{pmatrix} - \sqrt{2}S_{0e,N+1}S'_{1,N+1}\begin{pmatrix} J & J' & 1 \\ 0 & -1 & 1 \end{pmatrix}\right.$$
$$\left. + S_{1e,N+1}S'_{2,N+1}\begin{pmatrix} J & J' & 1 \\ 1 & -2 & 1 \end{pmatrix}\right\}^2. \tag{5.230}$$

Examining Eq. (5.229), it is clear that $F_1({}^3\Sigma^-)$ rotational levels with positive parity, which have even J, will have allowed radiative transitions only with ${}^3\Pi(F_1)$ levels with negative parity (lower symbol in the $\pm$ or $\mp$ signs in Eqs. (5.228) and (5.229)). Following similar reasoning, $F_1({}^3\Sigma^-)$rotational levels with negative parity, which have odd J, will have allowed radiative transitions only with ${}^3\Pi(F_1)$ levels with positive parity. Following a very similar analysis to that outlined in Eqs. (5.225)–(5.229), the line strength for $F_3({}^3\Sigma^-) \rightarrow F_3({}^3\Pi)$ transitions is given by

$$\tilde{S}_{J''J'} = \frac{(2J+1)(2J'+1)}{4}\tilde{\mu}_1^2\left[1 \mp (-1)^{J}\right]^2$$
$$\times\left\{S_{1e,N-1}S'_{0,N-1}\begin{pmatrix} J & J' & 1 \\ -1 & 0 & 1 \end{pmatrix} - \sqrt{2}S_{0e,N-1}S'_{1,N-1}\begin{pmatrix} J & J' & 1 \\ 0 & -1 & 1 \end{pmatrix}\right.$$
$$\left. + S_{1e,N-1}S'_{2,N-1}\begin{pmatrix} J & J' & 1 \\ 1 & -2 & 1 \end{pmatrix}\right\}^2. \tag{5.231}$$

$F_2(^3\Sigma^-) \rightarrow F_2(^3\Pi)$ Transitions

For these transitions, the component p of the electric dipole moment matrix element in the space-fixed reference frame for $F_2(^3\Sigma^-) \rightarrow F_2(^3\Pi)$ transitions is given by

$$\begin{aligned}\mu_{ba,p} &= \langle F_2; \mathrm{v}'; J'M'; \pm | \sum_q \hat{\mu}_q D^{1*}_{pq}(\omega) | F_2; \mathrm{v}; JM; f\rangle \\ &= \frac{1}{2}\sum_q \Big\{ -S'_{0,N}\langle 1; 1-1; \mathrm{v}'|\hat{\mu}_q|0; 1-1; \mathrm{v}\rangle\langle J'0M'|D^{1*}_{pq}(\omega)|J-1M\rangle \\ &\quad \pm S'_{0,N}(-1)^{\rho'}\langle -1; 11; \mathrm{v}'|\hat{\mu}_q|0; 11; \mathrm{v}\rangle\langle J'0M'|D^{1*}_{pq}(\omega)|J1M\rangle \\ &\quad + S'_{2,N}\langle 1; 11; \mathrm{v}'|\hat{\mu}_q|0; 11; \mathrm{v}\rangle\langle J'2M'|D^{1*}_{pq}(\omega)|J1M\rangle \\ &\quad \mp S'_{2,N}(-1)^{\rho'}\langle -1; 1-1; \mathrm{v}'|\hat{\mu}_q|0; 1-1; \mathrm{v}\rangle\langle J'-2M'|D^{1*}_{pq}(\omega)|J-1M\rangle \Big\}. \end{aligned} \tag{5.232}$$

Using Eq. (5.36), we obtain

$$\begin{aligned}\mu_{ba,p} &= \frac{\sqrt{(2J'+1)(2J+1)}}{2}\begin{pmatrix} J & J' & 1 \\ -M & M' & p \end{pmatrix}(-1)^{M'} \\ &= \Big\{ -S'_{0,N}\Big[\tilde{\mu}_{-1}\begin{pmatrix} J & J' & 1 \\ 1 & 0 & -1 \end{pmatrix} \mp \tilde{\mu}_1(-1)^{\rho'}\begin{pmatrix} J & J' & 1 \\ -1 & 0 & 1 \end{pmatrix}\Big] \\ &\quad + S'_{2,N}\Big[\tilde{\mu}_{-1}(-1)^2\begin{pmatrix} J & J' & 1 \\ -1 & 2 & -1 \end{pmatrix} \mp \tilde{\mu}_1(-1)^{\rho'-2}\begin{pmatrix} J & J' & 1 \\ 1 & -2 & 1 \end{pmatrix}\Big]\Big\}. \end{aligned} \tag{5.233}$$

Substituting $\rho' = J' - 1$ and using Eq. (A3.2) from Appendix A3, we obtain

$$\begin{aligned}\mu_{ba,p} &= -\frac{\sqrt{(2J+1)(2J'+1)}}{2}\begin{pmatrix} J & J' & 1 \\ -M & M' & p \end{pmatrix}(-1)^{M'+J'}\left[\tilde{\mu}_{-1}(-1)^J \mp \tilde{\mu}_1\right] \\ &\quad \times \Big\{ -S'_{-0,N}\begin{pmatrix} J & J' & 1 \\ -1 & 0 & 1 \end{pmatrix} + S'_{2,N}\begin{pmatrix} J & J' & 1 \\ 1 & -2 & 1 \end{pmatrix}\Big\}. \end{aligned} \tag{5.234}$$

For the $^3\Sigma^-$ electronic level, the symmetry properties again imply that $\tilde{\mu}_{-1} = -\tilde{\mu}_1$. Substituting in Eq. (5.234), we obtain

$$\begin{aligned}\mu_{ba,p} &= \frac{\sqrt{(2J'+1)(2J+1)}}{2}\begin{pmatrix} J & J' & 1 \\ -M & M' & p \end{pmatrix}(-1)^{M+J'}\tilde{\mu}_1\left[(-1)^J \pm 1\right] \\ &\quad \times \Big\{ -S'_{0,N}\begin{pmatrix} J & J' & 1 \\ -1 & 0 & 1 \end{pmatrix} + S'_{2,N}\begin{pmatrix} J & J' & 1 \\ 1 & -2 & 1 \end{pmatrix}\Big\}. \end{aligned} \tag{5.235}$$

The line strength is given by

$$\tilde{S}_{J''J'} = \frac{(2J+1)(2J'+1)}{4}\tilde{\mu}_1^2\left[(-1)^J \pm 1\right]^2\Big\{ -S'_{0,N}\begin{pmatrix} J & J' & 1 \\ -1 & 0 & 1 \end{pmatrix} + S'_{2,N}\begin{pmatrix} J & J' & 1 \\ 1 & -2 & 1 \end{pmatrix}\Big\}^2. \tag{5.236}$$

Examining Eq. (5.235), it is clear that $^3\Sigma^-(F_2)$ rotational levels with positive parity, which have odd J, will have allowed radiative transitions only with $^3\Pi(F_2)$ levels with negative parity (lower symbol in the $\pm$ or $\mp$ signs in Eqs. (5.232)–(5.235)). Following similar reasoning, $^3\Sigma^-(F_1)$ rotational levels with negative parity, which have even J, will have allowed radiative transitions only with $^3\Pi(F_1)$ levels with positive parity.

$F_1(^3\Sigma^-) \rightarrow F_2(^3\Pi)$ and $F_2(^3\Sigma^-) \rightarrow F_1(^3\Pi)$ Transitions

The component p of the electric dipole moment matrix element in the space-fixed reference frame for $F_1(^3\Sigma^-) \rightarrow F_2(^3\Pi)$ transitions is given by

$$\begin{aligned}
\mu_{ba,p} &= \langle F_2; \mathrm{v}'; J'M'; \pm | \sum_q \hat{\mu}_q D^{1*}_{pq}(\omega) | F_1; \mathrm{v}; JM; e\rangle \\
&= \sum_q \left\{ \frac{1}{2} S_{1e,N+1} S'_{0,N} \langle 1; 1-1; \mathrm{v}' | \hat{\mu}_q | 0; 1-1; \mathrm{v}\rangle \langle J'0M' | D^{1*}_{pq}(\omega) | J-1M\rangle \right. \\
&\pm \frac{1}{2} S_{1e,N+1} S'_{0,N} (-1)^{p'} \langle -1; 11; \mathrm{v}' | \hat{\mu}_q | 0; 11; \mathrm{v}\rangle \langle J'0M' | D^{1*}_{pq}(\omega) | J1M\rangle \\
&+ \frac{1}{\sqrt{2}} S_{0e,N+1} S'_{1,N} \langle 1; 10; \mathrm{v}' | \hat{\mu}_q | 0; 10; \mathrm{v}\rangle \langle J'1M' | D^{1*}_{pq}(\omega) | J0M\rangle \\
&\pm \frac{1}{\sqrt{2}} S_{0e,N+1} S'_{1,N} (-1)^{p'} \langle -1; 10; \mathrm{v}' | \hat{\mu}_q | 0; 10; \mathrm{v}\rangle \langle J'-1M' | D^{1*}_{pq}(\omega) | J0M\rangle \\
&+ \frac{1}{2} S_{1e,N+1} S'_{2,N} \langle 1; 11; \mathrm{v}' | \hat{\mu}_q | 0; 11; \mathrm{v}\rangle \langle J'2M' | D^{1*}_{pq}(\omega) | J1M\rangle \\
&\left. \pm \frac{1}{2} S_{1e,N+1} S'_{2,N} (-1)^{p'} \langle -1; 1-1; \mathrm{v}' | \hat{\mu}_q | 0; 1-1; \mathrm{v}\rangle \langle J'-2M' | D^{1*}_{pq}(\omega) | J-1M\rangle \right\}.
\end{aligned} \tag{5.237}$$

Following an analysis very similar to that outlined above for $F_1(^3\Sigma^-) \rightarrow F_1(^3\Pi)$ transitions, we obtain

$$\begin{aligned}
\tilde{S}_{J''J'} &= \frac{(2J+1)(2J'+1)}{4} \tilde{\mu}_1^2 \left[(-1)^J \mp 1\right]^2 \\
&\times \left\{ S_{1e,N+1} S'_{0,N} \begin{pmatrix} J & J' & 1 \\ -1 & 0 & 1 \end{pmatrix} - \sqrt{2} S_{0e,N+1} S'_{1,N} \begin{pmatrix} J & J' & 1 \\ 0 & -1 & 1 \end{pmatrix} \right. \\
&\left. + S_{1e,N+1} S'_{2,N} \begin{pmatrix} J & J' & 1 \\ 1 & -2 & 1 \end{pmatrix} \right\}^2.
\end{aligned} \tag{5.238}$$

Following a very similar analysis, the expression for $F_2(^3\Sigma^-) \rightarrow F_1(^3\Pi)$ transitions is

$$\begin{aligned}
\tilde{S}_{J''J'} &= \frac{(2J+1)(2J'+1)}{4} \tilde{\mu}_1^2 \left[(-1)^J \mp 1\right]^2 \\
&\times \left\{ S_{1e,N} S'_{0,N+1} \begin{pmatrix} J & J' & 1 \\ -1 & 0 & 1 \end{pmatrix} - \sqrt{2} S_{0e,N} S'_{1,N+1} \begin{pmatrix} J & J' & 1 \\ 0 & -1 & 1 \end{pmatrix} \right. \\
&\left. + S_{1e,N} S'_{2,N+1} \begin{pmatrix} J & J' & 1 \\ 1 & -2 & 1 \end{pmatrix} \right\}^2.
\end{aligned} \tag{5.239}$$

$F_1(^3\Sigma^-) \to F_3(^3\Pi)$ and $F_3(^3\Sigma^-) \to F_1(^3\Pi)$ Transitions

The component p of the electric dipole moment matrix element in the space-fixed reference frame is given by

$$
\begin{aligned}
\mu_{ba,p} &= \langle F_1;\mathrm{v}';J'M';\pm|\sum_q \hat{\mu}_q D^{1*}_{pq}(\omega)|F_3;\mathrm{v};JM;e\rangle \\
&= \sum_q \Bigg\{ \frac{1}{2} S_{1e,N-1}S'_{0,N+1}\langle 1;1-1;\mathrm{v}'|\hat{\mu}_q|0;1-1;\mathrm{v}\rangle\langle J'0M'|D^{1*}_{pq}(\omega)|J-1M\rangle \\
&\pm \frac{1}{2} S_{1e,N-1}S'_{0,N+1}(-1)^{p'}\langle -1;11;\mathrm{v}'|\hat{\mu}_q|0;11;\mathrm{v}\rangle\langle J'0M'|D^{1*}_{pq}(\omega)|J1M\rangle \\
&+ \frac{1}{\sqrt{2}} S_{0e,N-1}S'_{1,N+1}\langle 1;10;\mathrm{v}'|\hat{\mu}_q|0;10;\mathrm{v}\rangle\langle J'1M'|D^{1*}_{pq}(\omega)|J0M\rangle \\
&\pm \frac{1}{\sqrt{2}} S_{0e,N-1}S'_{1,N+1}(-1)^{p'}\langle -1;10;\mathrm{v}'|\hat{\mu}_q|0;10;\mathrm{v}\rangle\langle J'-1M'|D^{1*}_{pq}(\omega)|J0M\rangle \\
&+ \frac{1}{2} S_{1e,N-1}S'_{2,N+1}\langle 1;11;\mathrm{v}'|\hat{\mu}_q|0;11;\mathrm{v}\rangle\langle J'2M'|D^{1*}_{pq}(\omega)|J1M\rangle \\
&\pm \frac{1}{2} S_{1e,N-1}S'_{2,N+1}(-1)^{p'}\langle -1;1-1;\mathrm{v}'|\hat{\mu}_q|0;1-1;\mathrm{v}\rangle\langle J'-2M'|D^{1*}_{pq}(\omega)|J-1M\rangle \Bigg\}.
\end{aligned}
\tag{5.240}
$$

For $F_3(^3\Sigma^-) \to F_1(^3\Pi)$ transitions, we obtain,

$$
\begin{aligned}
\tilde{S}_{J''J'} &= \frac{(2J+1)(2J'+1)}{4}\tilde{\mu}_1^2\left[(-1)^J \mp 1\right]^2 \\
&\times \left\{ S_{1e,N-1}S'_{0,N+1}\begin{pmatrix} J & J' & 1 \\ -1 & 0 & 1 \end{pmatrix} - \sqrt{2}S_{0e,N-1}S'_{1,N+1}\begin{pmatrix} J & J' & 1 \\ 0 & -1 & 1 \end{pmatrix} \right. \\
&\left. + S_{1e,N-1}S'_{2,N+1}\begin{pmatrix} J & J' & 1 \\ 1 & -2 & 1 \end{pmatrix} \right\}^2.
\end{aligned}
\tag{5.241}
$$

For $F_1(^3\Sigma^-) \to F_3(^3\Pi)$ transitions, we obtain

$$
\begin{aligned}
\tilde{S}_{J''J'} &= \frac{(2J+1)(2J'+1)}{4}\tilde{\mu}_1^2\left[(-1)^J \mp 1\right]^2 \\
&\times \left\{ S_{1e,N+1}S'_{0,N-1}\begin{pmatrix} J & J' & 1 \\ -1 & 0 & 1 \end{pmatrix} - \sqrt{2}S_{0e,N+1}S'_{1,N-1}\begin{pmatrix} J & J' & 1 \\ 0 & -1 & 1 \end{pmatrix} \right. \\
&\left. + S_{1e,N+1}S'_{2,N-1}\begin{pmatrix} J & J' & 1 \\ 1 & -2 & 1 \end{pmatrix} \right\}^2.
\end{aligned}
\tag{5.242}
$$

$F_2(^3\Sigma^-) \to F_1(^3\Pi)$ and $F_2(^3\Sigma^-) \to F_3(^3\Pi)$ Transitions

For $F_2(^3\Sigma^-) \to F_1(^3\Pi)$ transitions, the component p of the electric dipole moment matrix element in the space-fixed reference frame is given by

$$
\begin{aligned}
\mu_{ba,p} &= \langle F_1; \mathrm{v}'; J'M'; \pm | \sum_q \hat{\mu}_q D^{1*}_{pq}(\omega) | F_2; \mathrm{v}; JM; f \rangle \\
&= \frac{1}{2} \sum_q \Big\{ -S'_{-1,N+1} \langle 1; 1-1; \mathrm{v}' | \hat{\mu}_q | 0; 1-1; \mathrm{v} \rangle \langle J' 0 M' | D^{1*}_{pq}(\omega) | J -1 M \rangle \\
&\quad \pm S'_{-1,N+1} (-1)^{p'} \langle -1; 11; \mathrm{v}' | \hat{\mu}_q | 0; 11; \mathrm{v} \rangle \langle J' 0 M' | D^{1*}_{pq}(\omega) | J 1 M \rangle \\
&\quad + S'_{1,N+1} \langle 1; 11; \mathrm{v}' | \hat{\mu}_q | 0; 11; \mathrm{v} \rangle \langle J' 2 M' | D^{1*}_{pq}(\omega) | J 1 M \rangle \\
&\quad \mp S'_{1,N+1} (-1)^{p'} \langle -1; 1-1; \mathrm{v}' | \hat{\mu}_q | 0; 1-1; \mathrm{v} \rangle \langle J' -2 M' | D^{1*}_{pq}(\omega) | J -1 M \rangle \Big\}.
\end{aligned}
\tag{5.243}
$$

The final expression for the line strength for $F_2(^3\Sigma^-) \rightarrow F_1(^3\Pi)$ transitions is given by

$$
\tilde{S}_{J''J'} = \frac{(2J+1)(2J'+1)}{4} \tilde{\mu}_1^2 \left[(-1)^J \pm 1\right]^2 \left\{ -S'_{-1,N+1} \begin{pmatrix} J & J' & 1 \\ -1 & 0 & 1 \end{pmatrix} + S'_{1,N+1} \begin{pmatrix} J & J' & 1 \\ 1 & -2 & 1 \end{pmatrix} \right\}^2. \tag{5.244}
$$

The final expression for the line strength for $F_2(^3\Sigma^-) \rightarrow F_3(^3\Pi)$ transitions is given by

$$
\tilde{S}_{J''J'} = \frac{(2J+1)(2J'+1)}{4} \tilde{\mu}_1^2 \left[(-1)^J \pm 1\right]^2 \left\{ -S'_{-1,N-1} \begin{pmatrix} J & J' & 1 \\ -1 & 0 & 1 \end{pmatrix} + S'_{1,N-1} \begin{pmatrix} J & J' & 1 \\ 1 & -2 & 1 \end{pmatrix} \right\}^2. \tag{5.245}
$$

5.5.3 Radiative Transitions between Triplet Electronic Levels with $\Lambda > 0$, $\Lambda' > 0$

First consider transitions from positive-parity F_1 lower quantum states to negative-parity F_1 upper quantum states. A general expression for these transitions will be developed, which can then be modified to account for transitions from negative-parity lower states to positive-parity upper states and for transitions involving F_2 and F_3 states. The wavefunctions for the vibration-rotation levels for ^3X electronic levels with $\Lambda > 0$ are again assumed to be intermediate between case (a) and case (b). The wavefunctions are given by

$$
\begin{aligned}
|F_1; \mathrm{v}; JM; \pm\rangle &= S_{\Lambda-1,N+1} \left|^3X_{\Lambda-1}; \Lambda; \mathrm{v}; JM; \pm\right\rangle + S_{\Lambda,N+1} \left|^3X_{\Lambda}; \Lambda; \mathrm{v}; JM; \pm\right\rangle \\
&\quad + S_{\Lambda+1,N+1} \left|^3X_{\Lambda+1}; \Lambda; \mathrm{v}; JM; \pm\right\rangle,
\end{aligned}
\tag{5.246}
$$

$$
\begin{aligned}
|F_2; \mathrm{v}; JM; \pm\rangle &= S_{\Lambda-1,N} \left|^3X_{\Lambda-1}; \Lambda; \mathrm{v}; JM; \pm\right\rangle + S_{\Lambda,N} \left|^3X_{\Lambda}; \Lambda; \mathrm{v}; JM; \pm\right\rangle \\
&\quad + S_{\Lambda+1,N} \left|^3X_{\Lambda+1}; \Lambda; \mathrm{v}; JM; \pm\right\rangle,
\end{aligned}
\tag{5.247}
$$

$$
\begin{aligned}
|F_3; \mathrm{v}; JM; \pm\rangle &= S_{\Lambda-1,N-1} \left|^3X_{\Lambda-1}; \Lambda; \mathrm{v}; JM; \pm\right\rangle + S_{\Lambda,N-1} \left|^3X_{\Lambda}; \Lambda; \mathrm{v}; JM; \pm\right\rangle \\
&\quad + S_{\Lambda+1,N-1} \left|^3X_{\Lambda+1}; \Lambda; \mathrm{v}; JM; \pm\right\rangle,
\end{aligned}
\tag{5.248}
$$

where

$$|^3X_{\Lambda-1};\Lambda;\mathrm{v};JM;\pm\rangle = \frac{1}{\sqrt{2}}[|\Lambda;S=1\,\Sigma=-1;\mathrm{v};J\,\Omega_-=\Lambda-1\,M\rangle \\ \pm(-1)^{\rho}|-\Lambda;1\,1;\mathrm{v};J-\Omega_1\,M\rangle], \tag{5.249}$$

$$|^3X_{\Lambda};\Lambda;\mathrm{v};JM;\pm\rangle = \frac{1}{\sqrt{2}}[|\Lambda;1\,0;\mathrm{v};J\Omega_0M\rangle \pm(-1)^{\rho}|-\Lambda;1\,0;\mathrm{v};J-\Omega_0\,M\rangle], \tag{5.250}$$

$$|^3X_{\Lambda+1};\Lambda;\mathrm{v};JM;\pm\rangle = \frac{1}{\sqrt{2}}[|\Lambda;1\,1;\mathrm{v};J\Omega_+M\rangle \pm(-1)^{\rho}|-\Lambda;1-1;\mathrm{v};J-\Omega_+M\rangle], \tag{5.251}$$

where $\rho = J - S + s = J - 1$, $\Omega_0 = \Lambda$, and $\Omega_\pm = \Lambda \pm 1$.

The component p of the electric dipole moment matrix element in the space-fixed reference frame is given by

$$\begin{aligned}
\mu_{ba,p} &= \langle F_1;\mathrm{v}'J'M';\mp|\sum_q \hat{\mu}_q D^{1*}_{pq}(\omega)|F_1;\mathrm{v}JM;\pm\rangle \\
&= \frac{1}{2}\sum_q \Big\{ S_{\Lambda-1,N+1}S'_{\Lambda'-1,N+1}\langle\Lambda';1-1;\mathrm{v}'|\hat{\mu}_q|\Lambda;1-1;\mathrm{v}\rangle\langle J'\Omega'_-M'|D^{1*}_{pq}(\omega)|J\Omega_-M\rangle \\
&+ S_{\Lambda,N+1}S'_{\Lambda',N+1}\langle\Lambda';10;\mathrm{v}'|\hat{\mu}_q|\Lambda;10;\mathrm{v}\rangle\langle J'\Omega'_0M'|D^{1*}_{pq}(\omega)|J\Omega_0M\rangle \\
&+ S_{\Lambda+1,N+1}S'_{\Lambda'+1,N+1}\langle\Lambda';11;\mathrm{v}'|\hat{\mu}_q|\Lambda;11;\mathrm{v}\rangle\langle J'\Omega'_+M'|D^{1*}_{pq}(\omega)|J\Omega_+M\rangle \\
&- S_{\Lambda-1,N+1}S'_{\Lambda'-1,N+1}(-1)^{\rho+\rho'}\langle-\Lambda';11;\mathrm{v}'|\hat{\mu}_q|-\Lambda;11;\mathrm{v}\rangle\langle J'-\Omega'_-M'|D^{1*}_{pq}(\omega)|J-\Omega_-M\rangle \\
&- S_{\Lambda,N+1}S'_{\Lambda',N+1}(-1)^{\rho+\rho'}\langle-\Lambda';10;\mathrm{v}'|\hat{\mu}_q|-\Lambda;10;\mathrm{v}\rangle\langle J'-\Omega'_0M'|D^{1*}_{pq}(\omega)|J-\Omega_0M\rangle \\
&- S_{\Lambda+1,N+1}S'_{\Lambda'+1,N+1}(-1)^{\rho+\rho'}\langle-\Lambda';1-1;\mathrm{v}'|\hat{\mu}_q|-\Lambda;1-1;\mathrm{v}\rangle\langle J'-\Omega'_+M'|D^{1*}_{pq}(\omega)|J-\Omega_+M\rangle \\
&\pm S_{\Lambda+1,N+1}S'_{\Lambda'-1,N+1}(-1)^{\rho}\langle\Lambda';1-1;\mathrm{v}'|\hat{\mu}_q|-\Lambda;1-1;\mathrm{v}\rangle\langle J'\Omega'_-M'|D^{1*}_{pq}(\omega)|J-\Omega_+M\rangle \\
&\mp S_{\Lambda+1,N+1}S'_{\Lambda'-1,N+1}(-1)^{\rho'}\langle-\Lambda';11;\mathrm{v}'|\hat{\mu}_q|\Lambda;11;\mathrm{v}\rangle\langle J'-\Omega'_-M'|D^{1*}_{pq}(\omega)|J\Omega_+M\rangle \\
&\pm S_{\Lambda-1,N+1}S'_{\Lambda'+1,N+1}(-1)^{\rho}\langle\Lambda';11;\mathrm{v}'|\hat{\mu}_q|-\Lambda;11;\mathrm{v}\rangle\langle J'\Omega'_+M'|D^{1*}_{pq}(\omega)|J-\Omega_-M\rangle \\
&\mp S_{\Lambda-1,N+1}S'_{\Lambda'+1,N+1}(-1)^{\rho'}\langle-\Lambda';1-1;\mathrm{v}'|\hat{\mu}_q|\Lambda;1-1;\mathrm{v}\rangle\langle J'-\Omega'_+M'|D^{1*}_{pq}(\omega)|J\Omega_-M\rangle \Big\}.
\end{aligned} \tag{5.252}$$

Again using Eq. (5.36), we obtain

$$\begin{aligned}
\mu_{ba,p} &= \frac{\sqrt{(2J+1)(2J'+1)}}{2}\begin{pmatrix} J & J' & 1 \\ M & -M' & p \end{pmatrix}(-1)^{M'} \\
&\times\sum_q \Big\{ S_{-1,N+1}S'_{-1,N+1}\tilde{\mu}_{q--}\left[(-1)^{-\Lambda'+1}\begin{pmatrix} J & J' & 1 \\ \Lambda-1 & -\Lambda'+1 & \Delta\Lambda \end{pmatrix} - (-1)^{\Lambda'-1+\rho+\rho'}\begin{pmatrix} J & J' & 1 \\ -\Lambda+1 & \Lambda'-1 & -\Delta\Lambda \end{pmatrix}\right] \\
&+ S_{0,N+1}S'_{0,N+1}\tilde{\mu}_{q00}\left[(-1)^{-\Lambda'}\begin{pmatrix} J & J' & 1 \\ \Lambda & -\Lambda' & \Delta\Lambda \end{pmatrix} - (-1)^{\Lambda'+\rho+\rho'}\begin{pmatrix} J & J' & 1 \\ -\Lambda & \Lambda' & -\Delta\Lambda \end{pmatrix}\right] \\
&+ S_{1,N+1}S'_{1,N+1}\tilde{\mu}_{q++}\left[(-1)^{-\Lambda'-1}\begin{pmatrix} J & J' & 1 \\ \Lambda+1 & -\Lambda'-1 & \Delta\Lambda \end{pmatrix} - (-1)^{\Lambda'+1+\rho+\rho'}\begin{pmatrix} J & J' & 1 \\ -\Lambda-1 & \Lambda'+1 & -\Delta\Lambda \end{pmatrix}\right] \\
&\pm S_{-1,N+1}S'_{1,N+1}\tilde{\mu}_{q+-}\left[(-1)^{\Lambda'+1+\rho'}\begin{pmatrix} J & J' & 1 \\ \Lambda-1 & \Lambda'+1 & -\Sigma\Lambda \end{pmatrix} - (-1)^{-\Lambda'-1+\rho}\begin{pmatrix} J & J' & 1 \\ -\Lambda+1 & -\Lambda'-1 & \Sigma\Lambda \end{pmatrix}\right] \\
&\mp S_{1,N+1}S'_{-1,N+1}\tilde{\mu}_{q-+}\left[(-1)^{\Lambda'-1+\rho'}\begin{pmatrix} J & J' & 1 \\ \Lambda+1 & \Lambda'-1 & -\Sigma\Lambda \end{pmatrix} - (-1)^{-\Lambda'+1+\rho}\begin{pmatrix} J & J' & 1 \\ -\Lambda-1 & -\Lambda'+1 & \Sigma\Lambda \end{pmatrix}\right]\Big\},
\end{aligned} \tag{5.253}$$

where $\Delta\Lambda = \Lambda' - \Lambda$ and $\Sigma\Lambda = \Lambda + \Lambda'$. For the cases we are considering here $\Lambda \geq 1$ and $\Lambda' \geq 1$, so Eq. (5.253) reduces to

$$\mu_{ba,p} = \frac{\sqrt{(2J+1)(2J'+1)}}{2}\begin{pmatrix} J & J' & 1 \\ M & -M' & p \end{pmatrix}(-1)^{M'}$$
$$\times\left\{S_{-1,N+1}S'_{-1,N+1}\begin{pmatrix} J & J' & 1 \\ \Lambda-1 & -\Lambda'+1 & \Delta\Lambda \end{pmatrix}\left[(-1)^{-\Lambda'+1}\tilde{\mu}_{\Delta\Lambda--} - (-1)^{\Lambda'+J+J'+\rho+\rho'}\tilde{\mu}_{-\Delta\Lambda--}\right]\right.$$
$$+ S_{0,N+1}S'_{0,N+1}\begin{pmatrix} J & J' & 1 \\ \Lambda & -\Lambda' & \Delta\Lambda \end{pmatrix}\left[(-1)^{-\Lambda'}\tilde{\mu}_{\Delta\Lambda 00} + (-1)^{\Lambda'+J+J'+\rho+\rho'}\tilde{\mu}_{-\Delta\Lambda 00}\right]$$
$$\left. + S_{1,N+1}S'_{1,N+1}\begin{pmatrix} J & J' & 1 \\ \Lambda+1 & -\Lambda'-1 & \Delta\Lambda \end{pmatrix}\left[(-1)^{-\Lambda'-1}\tilde{\mu}_{\Delta\Lambda++} - (-1)^{\Lambda'+J+J'+\rho+\rho'}\tilde{\mu}_{-\Delta\Lambda++}\right]\right\}. \tag{5.254}$$

Substituting for $\rho = J - 1$ and $\rho' = J' - 1$ and simplifying, we obtain

$$\mu_{ba,p} = \frac{\sqrt{(2J+1)(2J'+1)}}{2}\begin{pmatrix} J & J' & 1 \\ M & -M' & p \end{pmatrix}(-1)^{M'}(-1)^{-\Lambda'}$$
$$\times\left\{S_{-1,N+1}S'_{-1,N+1}\begin{pmatrix} J & J' & 1 \\ \Lambda-1 & -\Lambda'+1 & \Delta\Lambda \end{pmatrix}\left[-\tilde{\mu}_{\Delta\Lambda--} - (-1)^{2J+2J'-2}\tilde{\mu}_{-\Delta\Lambda--}\right]\right.$$
$$+ S_{0,N+1}S'_{0,N+1}\begin{pmatrix} J & J' & 1 \\ \Lambda & -\Lambda' & \Delta\Lambda \end{pmatrix}\left[\tilde{\mu}_{\Delta\Lambda 00} + (-1)^{2J+2J'-2}\tilde{\mu}_{-\Delta\Lambda 00}\right]$$
$$\left. + S_{1,N+1}S'_{1,N+1}\begin{pmatrix} J & J' & 1 \\ \Lambda+1 & -\Lambda'-1 & \Delta\Lambda \end{pmatrix}\left[-\tilde{\mu}_{\Delta\Lambda++} - (-1)^{2J+2J'-2}\tilde{\mu}_{-\Delta\Lambda++}\right]\right\}. \tag{5.255}$$

For triplet states, J and J' are always integers, and Eq. (5.255) reduces to

$$\mu_{ba,p} = \frac{\sqrt{(2J+1)(2J'+1)}}{2}\begin{pmatrix} J & J' & 1 \\ M & -M' & p \end{pmatrix}(-1)^{M'}(-1)^{-\Lambda'}$$
$$\times\left\{-S_{-1,N+1}S'_{-1,N+1}\begin{pmatrix} J & J' & 1 \\ \Lambda-1 & -\Lambda'+1 & \Delta\Lambda \end{pmatrix}\left[\tilde{\mu}_{\Delta\Lambda--} + \tilde{\mu}_{-\Delta\Lambda--}\right]\right.$$
$$+ S_{0,N+1}S'_{0,N+1}\begin{pmatrix} J & J' & 1 \\ \Lambda & -\Lambda' & \Delta\Lambda \end{pmatrix}\left[\tilde{\mu}_{\Delta\Lambda 00} + \tilde{\mu}_{-\Delta\Lambda 00}\right]$$
$$\left. - S_{1,N+1}S'_{1,N+1}\begin{pmatrix} J & J' & 1 \\ \Lambda+1 & -\Lambda'-1 & \Delta\Lambda \end{pmatrix}\left[\tilde{\mu}_{\Delta\Lambda++} + \tilde{\mu}_{-\Delta\Lambda++}\right]\right\}. \tag{5.256}$$

Radiative Transitions between Triplet Electronic Levels with ΔΛ = 0

For $\Delta\Lambda = 0$ $(\Lambda = \Lambda' \geq 1)$ we obtain

$$\mu_{ba,p} = \sqrt{(2J+1)(2J'+1)}\begin{pmatrix} J & J' & 1 \\ M & -M' & p \end{pmatrix}(-1)^{M'}(-1)^{-\Lambda'}$$
$$\times\left\{-S_{-1,N+1}S'_{-1,N+1}\begin{pmatrix} J & J' & 1 \\ \Lambda-1 & -\Lambda+1 & 0 \end{pmatrix}\tilde{\mu}_{0--} + S_{0,N+1}S'_{0,N+1}\begin{pmatrix} J & J' & 1 \\ \Lambda & -\Lambda & 0 \end{pmatrix}\tilde{\mu}_{000}\right.$$
$$\left. - S_{1,N+1}S'_{1,N+1}\begin{pmatrix} J & J' & 1 \\ \Lambda+1 & -\Lambda-1 & 0 \end{pmatrix}\tilde{\mu}_{0++}\right\}, \tag{5.257}$$

where $\Lambda = \Lambda'$. The *3j* symbols with $q = 0$, evaluated using formulae from Appendix A3, are given by

$\underline{J' = J+1}$

$$\begin{pmatrix} J & J' & 1 \\ \Lambda-1 & -\Lambda+1 & 0 \end{pmatrix} = \begin{pmatrix} J & J+1 & 1 \\ \Lambda-1 & -\Lambda+1 & 0 \end{pmatrix} = (-1)^{3J+\Lambda}\sqrt{\frac{(J-\Lambda+2)(J+\Lambda)}{(2J+3)(J+1)(2J+1)}}$$
$$= (-1)^{J+\Lambda}\sqrt{\frac{(J-\Lambda+2)(J+\Lambda)}{(2J+3)(J+1)(2J+1)}}. \quad (5.258)$$

$$\begin{pmatrix} J & J+1 & 1 \\ \Lambda & -\Lambda & 0 \end{pmatrix} = (-1)^{3J+\Lambda-1}\sqrt{\frac{(J-\Lambda+1)(J+\Lambda+1)}{(2J+3)(J+1)(2J+1)}}$$
$$= (-1)^{J+\Lambda-1}\sqrt{\frac{(J-\Lambda+1)(J+\Lambda+1)}{(2J+3)(J+1)(2J+1)}}. \quad (5.259)$$

$$\begin{pmatrix} J & J+1 & 1 \\ \Lambda+1 & -\Lambda-1 & 0 \end{pmatrix} = (-1)^{3J+\Lambda}\sqrt{\frac{(J-\Lambda)(J+\Lambda+2)}{(2J+3)(J+1)(2J+1)}}$$
$$= (-1)^{J+\Lambda}\sqrt{\frac{(J-\Lambda)(J+\Lambda+2)}{(2J+3)(J+1)(2J+1)}}. \quad (5.260)$$

$\underline{J = J'}$

$$\begin{pmatrix} J & J' & 1 \\ \Lambda-1 & -\Lambda+1 & 0 \end{pmatrix} = \begin{pmatrix} J & J & 1 \\ \Lambda-1 & -\Lambda+1 & 0 \end{pmatrix} = (-1)^{J-\Lambda+1}\frac{\Lambda-1}{\sqrt{(2J+1)J(J+1)}}. \quad (5.261)$$

$$\begin{pmatrix} J & J & 1 \\ \Lambda & -\Lambda & 0 \end{pmatrix} = (-1)^{J-\Lambda}\frac{\Lambda}{\sqrt{(2J+1)J(J+1)}}. \quad (5.262)$$

$$\begin{pmatrix} J & J & 1 \\ \Lambda+1 & -\Lambda-1 & 0 \end{pmatrix} = (-1)^{J-\Lambda-1}\frac{\Lambda+1}{\sqrt{(2J+1)J(J+1)}}. \quad (5.263)$$

$\underline{J' = J-1}$

$$\begin{pmatrix} J & J-1 & 1 \\ \Lambda-1 & -\Lambda+1 & 0 \end{pmatrix} = (-1)^{J-\Lambda+1}\sqrt{\frac{(J+\Lambda-1)(J-\Lambda+1)}{(2J+1)J(2J-1)}}. \quad (5.264)$$

$$\begin{pmatrix} J & J' & 1 \\ \Lambda & -\Lambda & 0 \end{pmatrix} = \begin{pmatrix} J & J-1 & 1 \\ \Lambda & -\Lambda & 0 \end{pmatrix} = (-1)^{J-\Lambda}\sqrt{\frac{(J+\Lambda)(J-\Lambda)}{(2J+1)J(2J-1)}}. \quad (5.265)$$

$$\begin{pmatrix} J & J-1 & 1 \\ \Lambda+1 & -\Lambda-1 & 0 \end{pmatrix} = (-1)^{J-\Lambda-1}\sqrt{\frac{(J+\Lambda+1)(J-\Lambda-1)}{(2J+1)J(2J-1)}}. \quad (5.266)$$

Radiative Transitions between Triplet Electronic Levels with $\Delta\Lambda = +1$
For $\Delta\Lambda = +1$ ($\Lambda' = \Lambda + 1$), we obtain

$$\begin{aligned}\mu_p = {} & \sqrt{(2J+1)(2J'+1)}\begin{pmatrix} J & J' & 1 \\ M & -M' & p \end{pmatrix}(-1)^{M'}(-1)^{-\Lambda-1} \\ & \times \Bigg\{ -S_{-1,N+1}S'_{-1,N+1}\begin{pmatrix} J & J' & 1 \\ \Lambda-1 & -\Lambda & 1 \end{pmatrix}\tilde{\mu}_{0--} \\ & + S_{0,N+1}S'_{0,N+1}\begin{pmatrix} J & J' & 1 \\ \Lambda & -\Lambda-1 & 1 \end{pmatrix}\tilde{\mu}_{000} \\ & - S_{1,N+1}S'_{1,N+1}\begin{pmatrix} J & J' & 1 \\ \Lambda+1 & -\Lambda-2 & 1 \end{pmatrix}\tilde{\mu}_{0++} \Bigg\},\end{aligned} \tag{5.267}$$

where $\Lambda = \Lambda'$. The *3j* symbols with $q = 1$ are again evaluated using formulae given Appendix A3:

$\underline{J' = J+1}$

$$\begin{aligned}\begin{pmatrix} J & J' & 1 \\ \Lambda-1 & -\Lambda & 1 \end{pmatrix} = \begin{pmatrix} J & J+1 & 1 \\ \Lambda-1 & -\Lambda & 1 \end{pmatrix} &= (-1)^{3J+\Lambda-1}\sqrt{\frac{(J+\Lambda)(J+\Lambda+1)}{2(2J+3)(J+1)(2J+1)}} \\ &= (-1)^{J+\Lambda-1}\sqrt{\frac{(J+\Lambda)(J+\Lambda+1)}{2(2J+3)(J+1)(2J+1)}}.\end{aligned} \tag{5.268}$$

$$\begin{aligned}\begin{pmatrix} J & J+1 & 1 \\ \Lambda & -\Lambda-1 & 1 \end{pmatrix} &= (-1)^{3J+\Lambda}\sqrt{\frac{(J+\Lambda+1)(J+\Lambda+2)}{2(2J+3)(J+1)(2J+1)}} \\ &= (-1)^{J+\Lambda}\sqrt{\frac{(J+\Lambda+1)(J+\Lambda+2)}{2(2J+3)(J+1)(2J+1)}}.\end{aligned} \tag{5.269}$$

$$\begin{aligned}\begin{pmatrix} J & J+1 & 1 \\ \Lambda+1 & -\Lambda-2 & 1 \end{pmatrix} &= (-1)^{3J+\Lambda+1}\sqrt{\frac{(J+\Lambda+2)(J+\Lambda+3)}{2(2J+3)(J+1)(2J+1)}} \\ &= (-1)^{J+\Lambda+1}\sqrt{\frac{(J+\Lambda+2)(J+\Lambda+3)}{2(2J+3)(J+1)(2J+1)}}.\end{aligned} \tag{5.270}$$

$\underline{J = J'}$

$$\begin{pmatrix} J & J' & 1 \\ \Lambda-1 & -\Lambda & 1 \end{pmatrix} = \begin{pmatrix} J & J & 1 \\ \Lambda-1 & -\Lambda & 1 \end{pmatrix} = (-1)^{J-\Lambda+1}\sqrt{\frac{(J-\Lambda+1)(J+\Lambda)}{2(J+1)(2J+1)J}}. \tag{5.271}$$

$$\begin{pmatrix} J & J & 1 \\ \Lambda & -\Lambda-1 & 1 \end{pmatrix} = (-1)^{J-\Lambda}\sqrt{\frac{(J-\Lambda)(J+\Lambda+1)}{2(J+1)(2J+1)J}}. \tag{5.272}$$

$$\begin{pmatrix} J & J & 1 \\ \Lambda+1 & -\Lambda-2 & 0 \end{pmatrix} = (-1)^{J-\Lambda-1}\sqrt{\frac{(J-\Lambda-1)(J+\Lambda+2)}{2(J+1)(2J+1)J}}. \tag{5.273}$$

$J' = J - 1$

$$\begin{pmatrix} J & J' & 1 \\ \Lambda-1 & -\Lambda & 1 \end{pmatrix} = \begin{pmatrix} J & J-1 & 1 \\ \Lambda-1 & -\Lambda & 1 \end{pmatrix} = (-1)^{J-\Lambda+1}\sqrt{\frac{(J-\Lambda)(J-\Lambda+1)}{2(2J+1)J(2J-1)}}. \tag{5.274}$$

$$\begin{pmatrix} J & J-1 & 1 \\ \Lambda & -\Lambda-1 & 1 \end{pmatrix} = (-1)^{J-\Lambda}\sqrt{\frac{(J-\Lambda-1)(J-\Lambda)}{2(2J+1)J(2J-1)}}. \tag{5.275}$$

$$\begin{pmatrix} J & J-1 & 1 \\ \Lambda+1 & -\Lambda-2 & 1 \end{pmatrix} = (-1)^{J-\Lambda-1}\sqrt{\frac{(J-\Lambda-2)(J-\Lambda-1)}{2(2J+1)J(2J-1)}}. \tag{5.276}$$

Radiative Transitions between Triplet Electronic Levels with $\Delta\Lambda = -1$

For $\Delta\Lambda = -1$ ($\Lambda' = \Lambda - 1$), we obtain

$$\begin{aligned} \mu_p = {} & \sqrt{(2J+1)(2J'+1)}\begin{pmatrix} J & J' & 1 \\ M & -M' & p \end{pmatrix}(-1)^{M'}(-1)^{-\Lambda-1} \\ & \times\Bigg\{-S_{-1,N+1}S'_{-1,N+1}\begin{pmatrix} J & J' & 1 \\ \Lambda-1 & -\Lambda+2 & -1 \end{pmatrix}\tilde{\mu}_{0--} \\ & + S_{0,N+1}S'_{0,N+1}\begin{pmatrix} J & J' & 1 \\ \Lambda & -\Lambda+1 & -1 \end{pmatrix}\tilde{\mu}_{000} \\ & - S_{1,N+1}S'_{1,N+1}\begin{pmatrix} J & J' & 1 \\ \Lambda+1 & -\Lambda & -1 \end{pmatrix}\tilde{\mu}_{0++}\Bigg\}, \end{aligned} \tag{5.277}$$

where $\Lambda = \Lambda'$. The *3j* symbols with $q = -1$ are given by

$J' = J + 1$

$$\begin{aligned} \begin{pmatrix} J & J' & 1 \\ \Lambda-1 & -\Lambda+2 & -1 \end{pmatrix} &= \begin{pmatrix} J & J+1 & 1 \\ \Lambda-1 & -\Lambda+2 & -1 \end{pmatrix} = (-1)^{2J+1}\begin{pmatrix} J & J+1 & 1 \\ -\Lambda+1 & \Lambda-2 & 1 \end{pmatrix} \\ &= (-1)^{5J+1-(\Lambda-1)}\sqrt{\frac{(J-\Lambda+2)(J-\Lambda+3)}{2(2J+3)(J+1)(2J+1)}} = (-1)^{J-\Lambda}\sqrt{\frac{(J-\Lambda+2)(J-\Lambda+3)}{2(2J+3)(J+1)(2J+1)}}. \end{aligned} \tag{5.278}$$

$$\begin{aligned} \begin{pmatrix} J & J' & 1 \\ \Lambda & -\Lambda+1 & -1 \end{pmatrix} &= \begin{pmatrix} J & J+1 & 1 \\ \Lambda & -\Lambda+1 & -1 \end{pmatrix} = (-1)^{2J+1}\begin{pmatrix} J & J+1 & 1 \\ -\Lambda & \Lambda-1 & 1 \end{pmatrix} \\ &= (-1)^{5J+1-\Lambda}\sqrt{\frac{(J-\Lambda+1)(J-\Lambda+2)}{2(2J+3)(J+1)(2J+1)}} = (-1)^{J-\Lambda+1}\sqrt{\frac{(J-\Lambda+1)(J-\Lambda+2)}{2(2J+3)(J+1)(2J+1)}}. \end{aligned} \tag{5.279}$$

$$\begin{pmatrix} J & J+1 & 1 \\ \Lambda+1 & -\Lambda & -1 \end{pmatrix} = (-1)^{2J+1}\begin{pmatrix} J & J+1 & 1 \\ -\Lambda-1 & \Lambda & 1 \end{pmatrix}$$
$$= (-1)^{5J+1-\Lambda-1}\sqrt{\frac{(J-\Lambda)(J-\Lambda+1)}{2(2J+3)(J+1)(2J+1)}} = (-1)^{J-\Lambda}\sqrt{\frac{(J-\Lambda)(J-\Lambda+1)}{2(2J+3)(J+1)(2J+1)}}. \tag{5.280}$$

$\underline{J = J'}$

$$\begin{pmatrix} J & J' & 1 \\ \Lambda-1 & -\Lambda+2 & -1 \end{pmatrix} = \begin{pmatrix} J & J & 1 \\ \Lambda-1 & -\Lambda+2 & -1 \end{pmatrix} = (-1)^{2J+1}\begin{pmatrix} J & J & 1 \\ -\Lambda+1 & \Lambda-2 & 1 \end{pmatrix}$$
$$= (-1)^{3J+\Lambda}\sqrt{\frac{(J+\Lambda-1)(J-\Lambda+2)}{2(J+1)(2J+1)J}} = (-1)^{J+\Lambda}\sqrt{\frac{(J+\Lambda-1)(J-\Lambda+2)}{2(J+1)(2J+1)J}}. \tag{5.281}$$

$$\begin{pmatrix} J & J & 1 \\ \Lambda & -\Lambda+1 & -1 \end{pmatrix} = (-1)^{2J+1}\begin{pmatrix} J & J & 1 \\ -\Lambda & \Lambda-1 & 1 \end{pmatrix}$$
$$= (-1)^{3J+\Lambda+1}\sqrt{\frac{(J+\Lambda)(J-\Lambda+1)}{2(J+1)(2J+1)J}} = (-1)^{J+\Lambda+1}\sqrt{\frac{(J+\Lambda-1)(J-\Lambda+2)}{2(J+1)(2J+1)J}}. \tag{5.282}$$

$$\begin{pmatrix} J & J & 1 \\ \Lambda+1 & -\Lambda & -1 \end{pmatrix} = (-1)^{2J+1}\begin{pmatrix} J & J & 1 \\ -\Lambda-1 & \Lambda & 1 \end{pmatrix}$$
$$= (-1)^{3J+\Lambda+2}\sqrt{\frac{(J+\Lambda+1)(J-\Lambda)}{2(J+1)(2J+1)J}} = (-1)^{J+\Lambda}\sqrt{\frac{(J+\Lambda+1)(J-\Lambda)}{2(J+1)(2J+1)J}}. \tag{5.283}$$

$\underline{J' = J-1}$

$$\begin{pmatrix} J & J' & 1 \\ \Lambda-1 & -\Lambda+2 & -1 \end{pmatrix} = \begin{pmatrix} J & J-1 & 1 \\ \Lambda-1 & -\Lambda+2 & -1 \end{pmatrix} = (-1)^{2J+1}\begin{pmatrix} J & J-1 & 1 \\ -\Lambda+1 & \Lambda-2 & 1 \end{pmatrix}$$
$$= (-1)^{3J+\Lambda}\sqrt{\frac{(J+\Lambda-2)(J+\Lambda-1)}{2(2J+1)J(2J-1)}} = (-1)^{J+\Lambda}\sqrt{\frac{(J+\Lambda-2)(J+\Lambda-1)}{2(2J+1)J(2J-1)}}. \tag{5.284}$$

$$\begin{pmatrix} J & J-1 & 1 \\ \Lambda & -\Lambda+1 & -1 \end{pmatrix} = (-1)^{2J+1}\begin{pmatrix} J & J-1 & 1 \\ -\Lambda & \Lambda-1 & 1 \end{pmatrix}$$
$$= (-1)^{3J+\Lambda+1}\sqrt{\frac{(J+\Lambda-1)(J+\Lambda)}{2(2J+1)J(2J-1)}} = (-1)^{J+\Lambda+1}\sqrt{\frac{(J+\Lambda-1)(J+\Lambda)}{2(2J+1)J(2J-1)}}. \tag{5.285}$$

$$\begin{pmatrix} J & J-1 & 1 \\ \Lambda+1 & -\Lambda & -1 \end{pmatrix} = (-1)^{2J+1}\begin{pmatrix} J & J-1 & 1 \\ -\Lambda-1 & \Lambda & 1 \end{pmatrix}$$
$$= (-1)^{3J+\Lambda+2}\sqrt{\frac{(J+\Lambda)(J+\Lambda+1)}{2(2J+1)J(2J-1)}} = (-1)^{J+\Lambda}\sqrt{\frac{(J+\Lambda)(J+\Lambda+1)}{2(2J+1)J(2J-1)}}. \tag{5.286}$$

5.6 Calculation of Radiative Transitions Rates for the $X^3\Sigma^-(\mathrm{v}) - A^3\Pi^+(\mathrm{v}')$ Bands of NH

The calculation of radiative transition rates using the results of Section 5.5 are illustrated in this section for the $X^3\Sigma^-(\mathrm{v}) - A^3\Pi(\mathrm{v}')$ bands of NH. The rotational transitions for this band are depicted schematically in Figure 5.9. The molecule NH is important in reaction chemistry, especially for nitrogen-containing fuels such as ammonia.

5.6.1 Einstein Coefficients for the $X^3\Sigma^-(\mathrm{v}) - A^3\Pi^+(\mathrm{v}')$ Vibrational Bands of NH

The calculation of the vibrational-band Einstein coefficients for the $X^3\Sigma^-(\mathrm{v}) - A^3\Pi(\mathrm{v}')$ bands is discussed in this section. The results of these calculations are compared with recent calculations of NH transition strengths by Fernando et al. (2018); the LIFBASE database (Luque & Crosley, 1999a) does not consider NH transitions. The calculations were performed for values for the vibrational quantum numbers for both the ground and excited electronic states ranging from 0 to 2, corresponding to the vibrational bands of most interest for combustion diagnostics.

The Dunham coefficients for the $X^3\Sigma^-$ and $A^3\Pi$ electronic levels of NH are listed in table 1 of Fernando et al. (2018). The RKR calculations to determine the turning points of the potential energy function $V(r)$ for these electronic levels were performed using the code RKR1 described by Le Roy (2017a). The turning points for the potential energy function in turn were used as input to the LEVEL code (Le Roy, 2017b) to calculate the vibration-rotation wave function $\psi_{\mathrm{v}J\Lambda}(r)$. As noted in Section 5.3.1, the LEVEL code assumes integral values of J and Λ, but for these triplet electronic levels this was not an issue, and the level wavefunctions $\psi_{\mathrm{v}J\Lambda}(r)$ were

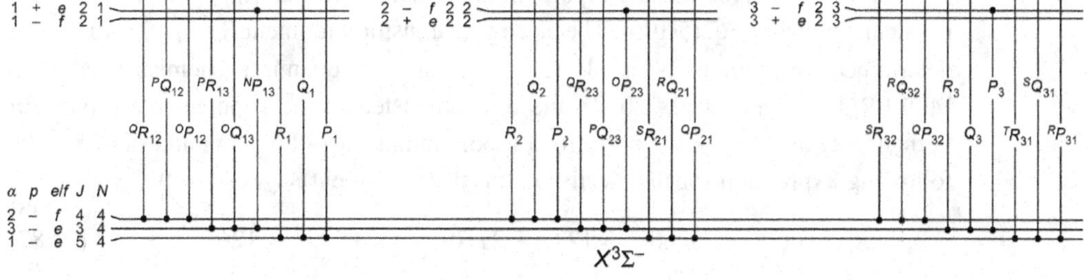

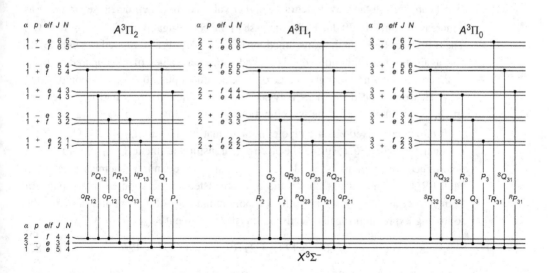

Figure 5.9 Rotational transitions in the $X^3\Sigma^-(\mathrm{v}) - A^3\Pi(\mathrm{v}')$ system of NH.

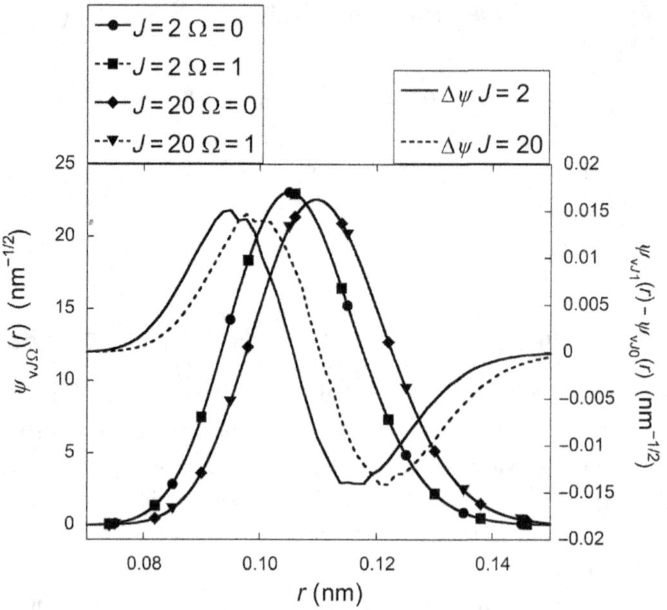

Figure 5.10 The vibration-rotation wavefunctions for the $X^3\Sigma^-(\mathrm{v}=0)$ levels of NH are shown for $J=2$ and 20 and $\Omega=0$ and 1. The difference between the wavefunctions is also plotted.

calculated directly without the need to perform the Lagrangian interpolation used for doublet electronic levels.

The NH $A^3\Pi$ vibration-rotation wavefunctions for $\mathrm{v}=0, J=2$, and $\mathrm{v}=0, J=20$ for values of $\Omega=0$ and 1 are shown in Figure 5.10. As shown in this figure, there is a significant shift in the wavefunction to higher values of the internuclear separation r as J increases from 2 to 20 due to the increase in the centrifugal potential $J(J+1)-\Omega^2$. The difference between the wavefunctions $\psi_{\mathrm{v}J0}$ and $\psi_{\mathrm{v}J1}$ is also plotted in Figure 5.10, and again the wavefunction $\psi_{\mathrm{v}J0}$ is shifted to higher values of r compared to $\psi_{\mathrm{v}J1}$, again consistent with the higher value of the centrifugal potential for $\psi_{\mathrm{v}J0}$ compared to $\psi_{\mathrm{v}J1}$. However, the differences between the wavefunctions $\psi_{\mathrm{v}J0}$ and $\psi_{\mathrm{v}J1}$ are very small, as was the case for the wavefunctions $\psi_{\mathrm{v}J1/2}$ and $\psi_{\mathrm{v}J3/2}$ for OH and NO.

The molecule-fixed electric dipole moment matrix element $\tilde{\mu}_{qkn}(\mathrm{v}',J',\Omega'_k,\mathrm{v}J\Omega_n)$ is calculated using Eq. (5.106). The electronic transition moment $R_{e,q=1}(r)$ for NH was determined by Fernando et al. (2018) using the quantum chemistry program MOLPRO (Werner et al., 2012); the data are listed in the supplementary data for Fernando et al. (2018); we performed a polynomial fit to this data and obtained the following expression for the electronic transition moment $R_{e,q=1}(r)$ (D):

$$R_{e,q=1}(r) = -1.260 + 3.122\,r + 21.79\,r^2 + 233.2\,r^3 - 1267\,r^4. \qquad (5.287)$$

where $1\ D = 3.335640952 \times 10^{-30}\ Cm$ (Bernath, 2016) and the units of r are nm in Eq. (5.287).

The wavefunctions for the (0,0), (0,1), (1,0), and (1,1) bands of NH are shown in Figure 5.11. In Figure 5.11a, the electronic transition moment $R_{e,q=1}(r)$ (subscript q = 1 is dropped) is plotted along with the upper (ψ_e) and lower (ψ_g) wavefunctions for the (0,0) bands; for the (0,1), (1,0), and (1,1) bands, the product $\psi_e R_{e,q=1} \psi_g$ is plotted instead of the electronic transition moment. The values of the equilibrium internuclear separation r_e and the vibrational frequency ω_e are nearly the same for the ground $X^3\Sigma^-$ and excited $A^3\Pi$ electronic levels. Because of this, the wavefunctions for the $\mathrm{v} = 0, J = 0, \Omega = 0$ and $\mathrm{v} = 1, J = 0, \Omega = 0$ levels are nearly identical for the ground $X^3\Sigma^-$ and excited $A^3\Pi$ electronic levels, and the wavefunctions for levels with different values of the vibrational quantum numbers are very close to orthogonal. Consequently, the Einstein coefficients for $X^3\Sigma^- - A^3\Pi$ vibrational bands with $\Delta\mathrm{v} \neq 0$ will be very small compared to vibrational bands with $\Delta\mathrm{v} = 0$.

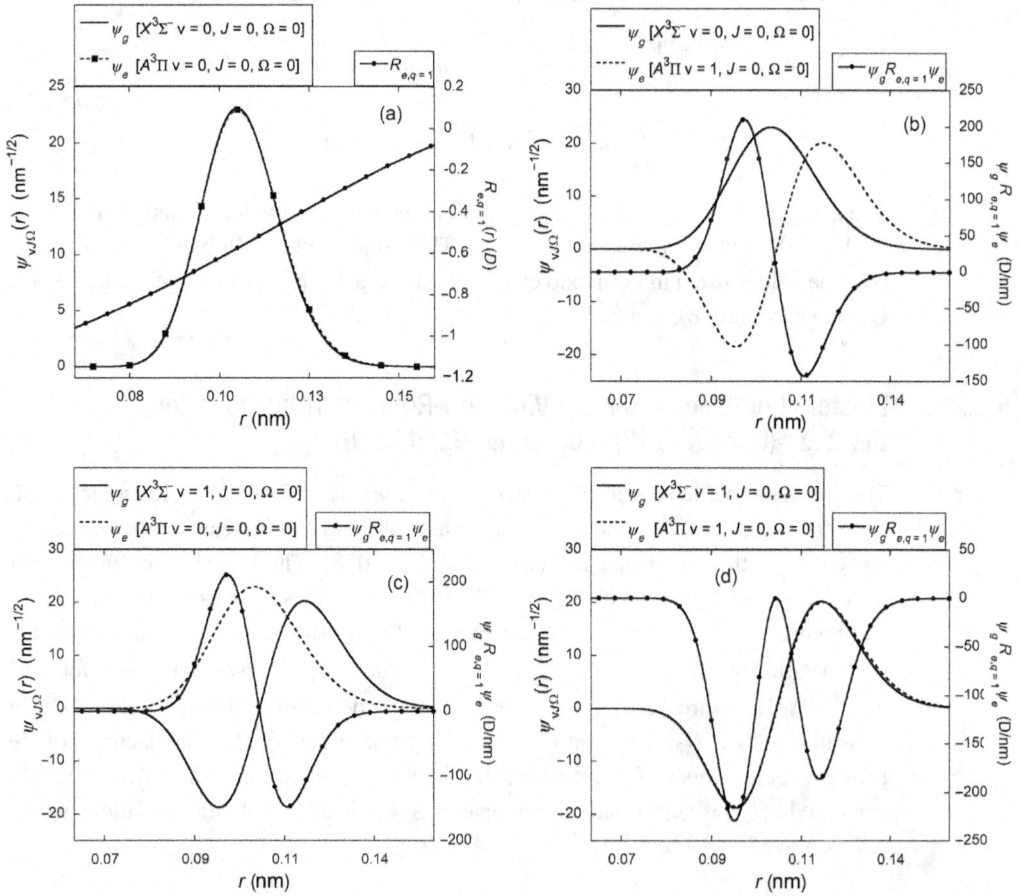

Figure 5.11 Vibration-rotation wavefunctions and $R_{e,q=1}(r)$ for (a) the NH (0,0) band. Vibration-rotation wavefunctions and the products $\psi_e(r) R_{e,q=1}(r) \psi_g(r)$ for the NH (b) (1,0), (c) (0,1), and (d) (1,1) bands.

Table 5.10 Vibrational band Einstein coefficients $A_{\mathrm{v}'\mathrm{v}''}$ for NH

Band	This work	Fernando et al. (2018)	Owono et al. (2008)
(0,0)	2.68×10^6	2.54×10^6	2.60×10^6
(0,1)	2.76×10^4		
(0,2)	5.02×10^2		
(1,0)	6.23×10^4	5.87×10^4	5.78×10^4
(1,1)	2.30×10^6	2.18×10^6	2.26×10^6
(1,2)	4.74×10^4		
(2,0)	4.14×10^2	3×10^2	1×10^3
(2,1)	1.45×10^5	1.39×10^5	1.16×10^5
(2,2)	1.92×10^6	4.700×10^5	1.93×10^6

In Table 5.10 we list the values that we calculate for $A_{\mathrm{v}'\mathrm{v}''}$ using the formula

$$\begin{aligned} A_{\mathrm{v}'\mathrm{v}''} &= \frac{16\pi^3\,\tilde{\nu}^3_{\mathrm{v}'\mathrm{v}''}}{3h\varepsilon_0}\left[\frac{2-\Lambda_{0,\Lambda''+\Lambda'}}{2-\Lambda_{0,\Lambda'}}\right]\tilde{\mu}^2_{1,01}(\mathrm{v}',\mathrm{v}'',J'=1,J''=0) \\ &= \frac{16\pi^3\,\tilde{\nu}^3_{\mathrm{v}'\mathrm{v}''}}{3h\varepsilon_0}\tilde{\mu}^2_{1,\,01}(\mathrm{v}',\mathrm{v}'',J'=1,J''=0). \end{aligned} \tag{5.288}$$

The calculated values of $A_{\mathrm{v}'\mathrm{v}''}$ are in good agreement with values listed by Fernando et al. (2018) and by Owono et al. (2008). The value for the (0,0) band is 5.5% higher than the values listed in Fernando et al. (2018) and 3.1% higher than the value listed in Owono et al. (2008).

5.6.2 Einstein Coefficients for the Vibration-Rotation Transitions for the $X^3\Sigma^-(0) - A^3\Pi\ (0)$ Vibrational Band of NH

The Einstein coefficients for the (0,0) band of the NH $A^3\Pi - X^3\Sigma^-$ electronic transition are listed in Appendix A7; the listed Einstein coefficients are in excellent agreement with those listed in Fernando et al. (2018). The Einstein coefficients for the main-branch transitions ${}^QQ_{11}$ and ${}^RR_{11}$ ($\Delta N = \Delta J$) are shown in Figure 5.12a, with and without the inclusion of Herman–Wallis effects. The Herman–Wallis effects are very significant for the NH $A^3\Pi - X^3\Sigma^-$ transition, as was the case for OH $A^2\Sigma^+ - X^2\Pi$ electronic transition. The Einstein coefficients for the satellite branch transitions ${}^QP_{21}$, ${}^RQ_{21}$, ${}^RP_{31}$, and ${}^SQ_{31}$ are shown in Figure 5.12b. The decrease of the Einstein coefficients with increasing J for the satellite transitions where $\Delta N = \Delta J \pm 1$ (${}^QP_{21}$ and ${}^RQ_{21}$) is very rapid but not nearly as drastic as for satellite transitions where $\Delta N = \Delta J \pm 2$ (${}^RP_{31}$ and ${}^SQ_{31}$).

5.7 Forbidden Transitions

Weak radiative transitions are also observed for electronic systems in which it would appear that the transitions should not occur; these are so-called forbidden transitions.

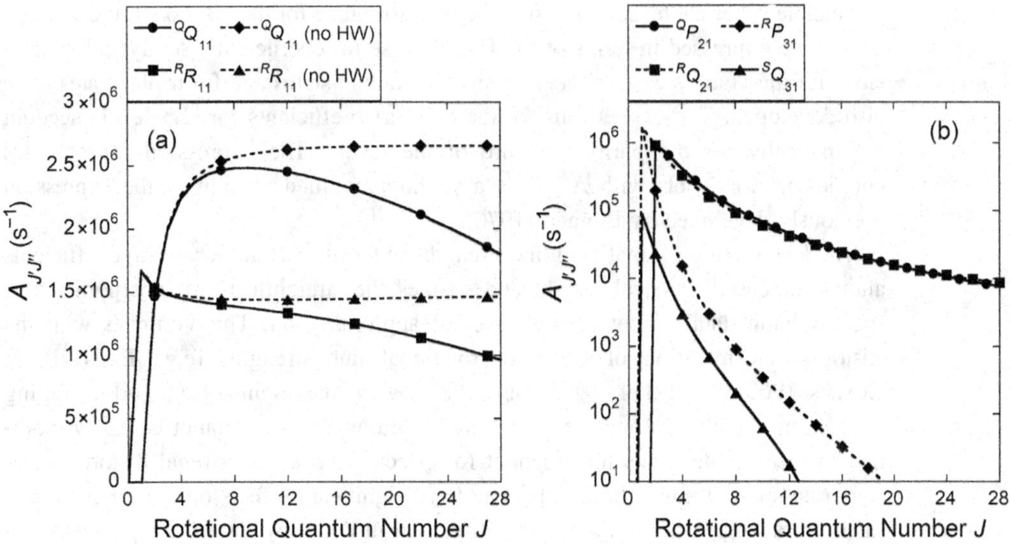

Figure 5.12 Calculated Einstein coefficients for rotational transitions in the NH $A^3\Pi$ (v′ = 0) – $X^3\Sigma^-$ (v″ = 0) band. The results for the main-branch transitions $^{Q}Q_{11}$ and $^{R}R_{11}$ are shown in panel (a) with and without the inclusion of Herman–Wallis effects. The results for the satellite branch transitions $^{Q}P_{21}$, $^{R}Q_{21}$, $^{R}P_{31}$, and $^{S}Q_{31}$ are shown in panel (b).

For example, in his discussion of radiative transitions for triplet electronic levels, Kovacs (1969) includes a table of rotational line strengths for $^3\Sigma - {}^3\Delta$ transitions. However, these transitions with $\Delta\Lambda = \pm 2$ should not occur according to the results presented in Section 5.5. However, due to the spin–orbit interaction, the $^3\Delta$ level can take on the character of electronic levels with different values of S and Λ. The spin–orbit interaction term couples electronic levels with $\Delta S = 0, \pm 1$, $\Delta\Lambda = 0, \pm 1$, and $\Delta\Omega = 0$ (Kovacs, 1960). Thus the perturbed wave function $\psi(^3\Delta_\Omega)$ would contain contributions from unperturbed wavefunctions $\psi_0(^{2S+1}\Lambda_\Omega)$for singlet electronic levels and for electronic levels with different values of Λ. As an example, the wavefunction for the $^3\Delta_2$ can be written (Kovacs, 1960)

$$\psi(^3\Delta_{+2}) = a_0\psi_0(^3\Delta_{+2}) + a_1\psi_0(^3\Pi_{+2}) + a_0\psi_0(^3\Phi_{+2}) + a_0\psi_0(^1\Delta_{+2}). \quad (5.289)$$

Kovacs (1960) outlines the calculation of rotational line strengths for forbidden $^1\Sigma - {}^3\Delta$, $^3\Sigma - {}^3\Delta$, and $^1\Pi - {}^3\Delta$ electronic transitions. The calculation of intensities for the $^1\Sigma - {}^3\Pi_r$ transition of SiO and for the $a^3\Pi - X^1\Sigma$ Cameron system of CO are discussed by James (1963) and James (1971), respectively.

5.8 Summary and Extension to Other Molecules

A general framework for the calculation of radiative transition rates in diatomic molecules was outlined in this chapter. The rotational line strengths for the vibration-rotation transitions were developed using Hund's case (a) basis states and irreducible spherical tensor analysis. The final formulae for the rotational line strength factors are expressed in terms of the Hund's case (a) coefficients. Analytical expressions for the Hund's case (a) coefficients for doublet states and for triplet states were also developed. The expressions for the case (a) coefficients for $^2\Sigma^{\pm}$ levels account automatically for the parity structure of the levels. The expression for case (a) coefficients for levels with $\Lambda \geq 1$ is a slightly modified version of the expression previously developed by Bennett (1970).

The formulation of rotational line strengths in terms of Hund's case (a) coefficients allows increased physical insight and enables the straightforward incorporation of spectroscopic data to any desired level of approximation. This contrasts with the customary formulation of tables of rotational line strengths (e.g., Earls, 1935; Kovacs, 1969; Reisel et al., 1992; Zare, 1988), where the dominant spin–orbit splitting interaction is included, and the final complicated expressions cannot be easily modified to incorporate terms that account for effects such as centrifugal distortion and spin–rotation splitting. These terms can have significant effects on the intensities of satellite transitions at high J, as was discussed in detail by Paul (1997) for satellite transitions in the $A^2\Sigma^+ - X^2\Pi$ system of NO.

This formulation also allows the incorporation of Herman–Wallis factors in transition line strengths in a straightforward manner through the calculation of molecule-fixed transition dipole moment matrix elements for given values of J and for the different absolute values of Ω – either 1/2 or 3/2 for doublet levels, and either 0, 1, or 2 for triplet levels. The vibration-rotation wavefunctions $\psi_{vJ\Omega}$ needed for the calculations of the molecule-fixed TDM matrix elements are determined by interpolating the wavefunctions $\psi_{vJ\Omega}$ that are the output of the program LEVEL (Le Roy, 2017b). For doublet states, interpolation is necessary because it is assumed that the parameter Λ is an integer in the program LEVEL, and the interpolation variable used is the centrifugal potential term $J(J+1) - \Omega^2$. The molecule-fixed electronic TDM matrix elements $\tilde{\mu}_{q++}, \tilde{\mu}_{q+-}, \tilde{\mu}_{q-+}$, and $\tilde{\mu}_{q--}$ (where + denotes $\Omega = 3/2$ and – denotes $\Omega = 1/2$) are associated with $\Omega_+ \to \Omega_+$, $\Omega_+ \to \Omega_-, \Omega_- \to \Omega_+$, and $\Omega_- \to \Omega_-$ transitions. The contribution of each of these factors to the overall line strength of the transition is determined by the value of the Hund's case (a) coefficients. For example, for low-J $F_1 \to F_1$ rotational transitions in the OH $X^2\Pi - A^2\Sigma^+$ electronic transition system, the line strength will be determined primarily by the $\tilde{\mu}_{1++}$ because $a_J \ll b_J$ and $a'_J \ll b'_J$ for these transitions. In the absence of significant vibration-rotation interactions, the expressions for the rotational line strength factors can be modified easily to obtain expressions for Hönl–London factors.

The Einstein coefficients for the OH and NO $X^2\Pi - A^2\Sigma^+$ rotational transitions and for the NH $X^3\Sigma^- - A^2\Pi$ are calculated to illustrate the general methods outlined in this chapter. For the OH and NO $X^2\Pi - A^2\Sigma^+$rotational transitions, the rotational line strength formulae in Table 5.6 are used. The species OH, NO, and NH were selected because of the importance of performing quantitative measurements of these species in, for example, studies of combustion chemistry. The results of the OH calculations were compared with recently published tables of Einstein coefficients (Yousefi et al.,

2018) and with tables extracted from the widely used spectroscopic database LIFBASE (Luque & Crosley, 1999a). The results of the NO calculations were compared with LIFBASE (Luque & Crosley, 1999a). The results of the NH calculations were compared with recently published tables of Einstein coefficients (Fernando et.al., 2018).

The methods outlined in this chapter can be used to calculate radiative transition rates for numerous diatomic species, both neutral molecules and ions. The accuracy of the calculated values will depend on the availability of accurate spectroscopic constants and on the use of quantum chemical codes for the calculations of vibrational wavefunctions as a function of v and J, and for calculations of the electronic transition dipole moment as a function of the internuclear separation. The methods outlined can be used to develop computer programs for the calculation of theoretical spectra for numerous diatomic species. Recent examples of papers describing such programs include Billoux et al. (2014), with tables of radiative transition rates for OH, CH, CH^+, CO, and CO^+, and Qin et al. (2017), with tables of radiative transition rates for $N_2, N_2^+, NO, O_2, CO, CO^+, CN, C_2$, and H_2.

In summary, our new formulation of rotational line strength factors (1) provides increased physical insight, (2) makes it straightforward to incorporate spectroscopic parameters in the calculation to any desired level of approximation, and (3) makes it straightforward to incorporate vibration-rotation interactions (the Herman–Wallis effect). The methods illustrated here can be extended to other single-photon transitions, such as transitions between quadruplet electronic levels; to two-photon transitions (Kulatilaka & Lucht, 2017); and to Raman transitions, as will be discussed in Chapter 7.

6 Absorption and Emission Spectroscopy

6.1 Introduction

Laser absorption spectroscopy is widely used for sensitive and quantitative detection of trace species. In this chapter the density matrix approach is used to introduce laser absorption spectroscopy. Spectroscopic quantities that characterize the absorption process are defined, and the relationships among these quantities are discussed. Broadening processes for spectral line shapes are also discussed, and the Doppler, Voigt, and Galatry profiles are introduced. The chapter concludes with an analysis of the emission properties of transitions incorporating the use of unit dyads.

6.2 Density Matrix Analysis for the Interaction of a Two-State Resonance with Monochromatic Laser Radiation

The time-dependent density matrix equations for a multi-state system irradiated by laser radiation are given by

$$\frac{\partial \rho_{bb}(\mathbf{r},t)}{\partial t} = -\frac{i}{\hbar}\sum_{c}\left[V_{bc}(\mathbf{r},t)\rho_{cb}(\mathbf{r},t) - \rho_{bc}(\mathbf{r},t)V_{cb}(\mathbf{r},t)\right] - \Gamma_b\rho_{bb}(\mathbf{r},t) + \sum_{c}\Gamma_{cb}\rho_{cc}(\mathbf{r},t). \tag{6.1}$$

$$\frac{\partial \rho_{ba}(\mathbf{r},t)}{\partial t} = -\rho_{ba}(\mathbf{r},t)(i\omega_{ba}+\gamma_{ba}) - \frac{i}{\hbar}\sum_{c}\left[V_{bc}(\mathbf{r},t)\rho_{ca}(\mathbf{r},t) - \rho_{bc}(\mathbf{r},t)V_{ca}(\mathbf{r},t)\right], \tag{6.2}$$

where the diagonal matrix element ρ_{bb} is the occupation probability for the excited state and is proportional to the population of state b, and the off-diagonal matrix element ρ_{ba} describes the coherence between state b and the lower state a. In Eqs. (6.1) and (6.2), ω_{ba} is the resonance frequency (s^{-1}) between states b and a, and phenomenological population transfer and coherence dephasing rate coefficients are introduced. The rate coefficient for population transfer from state c to state b is Γ_{cb} (s^{-1}), and the rate coefficient for the collisional dephasing of coherence between state b and state a is γ_{ba} (s^{-1}). The interaction term V_{bc} (J) is given by

$$V_{bc} = -\boldsymbol{\mu}_{bc} \bullet \boldsymbol{E}_L(\boldsymbol{r}, t), \tag{6.3}$$

where $\boldsymbol{\mu}_{bc} = \langle \psi_b | \boldsymbol{\mu} | \psi_c \rangle$ is the electric dipole matrix element (Cm) and $\boldsymbol{E}_L(\boldsymbol{r}, t)$ is the laser electric field (J/Cm). The laser field is given by

$$\boldsymbol{E}_L(\boldsymbol{r}, t) = \frac{1}{2} \hat{\boldsymbol{e}}_L A_L(\boldsymbol{r}, t) \exp[+i(\boldsymbol{k}_L \bullet \boldsymbol{r} - \omega_L t)] + c.c. \tag{6.4}$$

where $\hat{\boldsymbol{e}}_L$ is the complex unit vector that describes the polarization state of the field $(\hat{\boldsymbol{e}}_L \bullet \hat{\boldsymbol{e}}_L^* = 1)$, $\boldsymbol{r}$ is the position vector, and $A_L(\boldsymbol{r}, t)$ (J/Cm) is the slowly varying amplitude function for the electric field of the laser.

6.2.1 The Two-State System

To gain some physical insight into the radiative interaction terms and the effect of the phenomenological constants, consider the case of an isolated two-state resonance interacting with monochromatic laser radiation. The two-state resonance $b \leftrightarrow a$ is depicted in Figure 6.1. The two states b and a are included in levels B and A, respectively. For this case, Eqs. (6.1) and (6.2) reduce to

$$\frac{\partial \rho_{bb}(\boldsymbol{r}, t)}{\partial t} = -\frac{i}{\hbar}[V_{ba}(\boldsymbol{r}, t)\rho_{ab}(\boldsymbol{r}, t) - \rho_{ba}(\boldsymbol{r}, t)V_{ab}(\boldsymbol{r}, t)] - \Gamma_{ba}\rho_{bb}(\boldsymbol{r}, t) + \Gamma_{ab}\rho_{aa}(\boldsymbol{r}, t). \tag{6.5}$$

$$\frac{\partial \rho_{ba}(\boldsymbol{r}, t)}{\partial t} = -\rho_{ba}(i\omega_{ba} + \gamma_{ba}) - \frac{i}{\hbar}[V_{ba}(\boldsymbol{r}, t)\rho_{aa}(\boldsymbol{r}, t) - \rho_{bb}(\boldsymbol{r}, t)V_{ba}(\boldsymbol{r}, t)]. \tag{6.6}$$

It is assumed that the off-diagonal matrix element ρ_{ba} will oscillate at a frequency close to that of the driving laser field. The off-diagonal matrix element can be written as the product of a slowly varying amplitude function and an exponential term oscillating at the laser frequency, $\rho_{ba} = \sigma_{ba} \exp(-i\omega_L t)$. Substituting this expression and Eq. (6.4) into Eqs. (6.5) and (6.6), setting $\boldsymbol{r} = 0$, noting that $\rho_{ab} = \rho_{ba}^*$ and $\boldsymbol{\mu}_{ab} = \boldsymbol{\mu}_{ba}^*$, and rearranging, we obtain

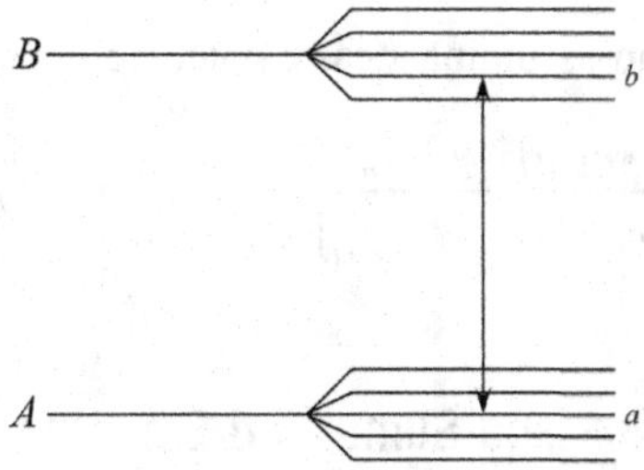

Figure 6.1 Energy level diagram for the absorption transition from level A to level B. The Zeeman states a and b are included in levels A and B, respectively. The energy level splitting of the states a and b would occur only in the presence of a magnetic field.

$$\frac{\partial\rho_{bb}(t)}{\partial t}=\dot{\rho}_{bb}=-\frac{i}{2\hbar}\left\{-\boldsymbol{\mu}_{ba}\bullet\left[\hat{\boldsymbol{e}}_L A_L+\hat{\boldsymbol{e}}_L^* A_L^*\exp\left(+2i\omega_L t\right)\right]\sigma_{ba}^*\right.$$
$$\left.+\sigma_{ba}\boldsymbol{\mu}_{ba}^*\bullet\left[\hat{\boldsymbol{e}}_L^* A_L^*+\hat{\boldsymbol{e}}_L A_L\exp\left(-2i\omega_L t\right)\right]\right\}-\Gamma_{ba}\rho_{bb}+\Gamma_{ab}\rho_{aa}. \tag{6.7}$$

$$\dot{\sigma}_{ba}=-\sigma_{ba}\left[i(\omega_{ba}-\omega_L)+\gamma_{ba}\right]+\frac{i}{2\hbar}(\rho_{aa}-\rho_{bb})\vec{\mu}_{ba}\bullet\left[\hat{e}_L A_L+\hat{e}_L^* A_L^*\exp(+2i\omega_L t)\right]. \tag{6.8}$$

Except for femtosecond (or now attosecond) laser pulses, the level populations and the amplitude function σ_{ba} change on timescales much longer than the timescale associated with the optical field. Thus, the contributions of the terms containing $\exp(\pm 2i\omega_L t)$ average to zero over timescales on the order of a few optical cycles. These terms are therefore neglected; this is the *rotating wave approximation.* Rewriting Eqs. (6.7) and (6.8), we obtain

$$\dot{\rho}_{bb}=-\frac{i}{2\hbar}\left[-\boldsymbol{\mu}_{ba}\bullet\hat{\boldsymbol{e}}_L A_L\sigma_{ba}^*+\sigma_{ba}\boldsymbol{\mu}_{ba}^*\bullet\hat{\boldsymbol{e}}_L^* A_L^*\right]-\Gamma_{ba}\rho_{bb}+\Gamma_{ab}\rho_{aa}. \tag{6.9}$$

$$\dot{\sigma}_{ba}=-\sigma_{ba}\left[i\left(\omega_{ba}-\omega_L\right)+\gamma_{ba}\right]+\frac{i}{2\hbar}(\rho_{aa}-\rho_{bb})\boldsymbol{\mu}_{ba}\bullet\hat{\boldsymbol{e}}_L A_L. \tag{6.10}$$

Solving Eq. (6.10) for σ_{ba} in the steady state limit, we obtain

$$\sigma_{ba}=-\frac{(\rho_{aa}-\rho_{bb})\boldsymbol{\mu}_{ba}\bullet\hat{\boldsymbol{e}}_L A_L}{2\hbar\left[(\omega_L-\omega_{ba})+i\gamma_{ba}\right]}. \tag{6.11}$$

Note from Eq. (6.11) that using $\rho_{ba}=\sigma_{ba}\exp(+i\omega_L t)$ rather than $\rho_{ba}=\sigma_{ba}\exp(-i\omega_L t)$ would have resulted in a term $\left[(-\omega_L-\omega_{ba})+i\gamma_{ba}\right]$ in the denominator, and σ_{ba} would not exhibit resonant behavior. This result can also be written in terms of the Rabi frequency Ω (s^{-1}),

$$\sigma_{ba}=-\frac{(\rho_{aa}-\rho_{bb})\,\Omega}{2\left[(\omega_L-\omega_{ba})+i\gamma_{ba}\right]}, \tag{6.12}$$

where

$$\Omega=\frac{\boldsymbol{\mu}_{ba}\bullet\hat{\boldsymbol{e}}_L A_L}{\hbar}. \tag{6.13}$$

Substituting Eq. (6.11) into Eq. (6.9) and assuming steady state, we obtain

$$0=\frac{\left|\boldsymbol{\mu}_{ba}\bullet\hat{\boldsymbol{e}}_L A_L\right|^2\gamma_{ba}\rho_{aa}}{2\hbar^2\left[(\omega_L-\omega_{ba})^2+\gamma_{ba}^2\right]}-\frac{\left|\boldsymbol{\mu}_{ba}\bullet\hat{\boldsymbol{e}}_L A_L\right|^2\gamma_{ba}\rho_{bb}}{2\hbar^2\left[(\omega_L-\omega_{ba})^2+\gamma_{ba}^2\right]}-\Gamma_{ba}\rho_{bb}+\Gamma_{ab}\rho_{aa}. \tag{6.14}$$

6.2.2 Calculation of the Induced Absorption and Stimulated Emission Rates

Examination of Eq. (6.14) leads to the interpretation of the first term as the induced absorption rate and the second term as the stimulated emission rate for the resonant system formed by states *b* and *a*. In the analysis used to obtain Eqs. (6.11) and (6.14), we have assumed steady state. This is frequently a questionable assumption, especially when pulsed lasers are used in an experiment. We have implicitly assumed monochromatic laser radiation in the above analysis. For the case of monochromatic laser radiation, the phase of the laser field can be chosen such that the amplitude function A_L is a real number without loss of generality. This is not the case for multi-axial-mode laser radiation. In addition, we have considered only the interaction between states b and a, assuming that they form an isolated system. The effects of the level degeneracy will be considered in the next section.

The induced absorption and stimulated emission rate coefficients for a homogeneously broadened absorption interacting with a monochromatic, steady-state laser field are given by

$$W_{a\to b} = W_{b\to a} = \frac{|\boldsymbol{\mu}_{ba} \bullet \hat{\mathbf{e}}_L A_L|^2 \gamma_{ba}}{2\hbar^2 \left[(\omega_L - \omega_{ba})^2 + \gamma_{ba}^2\right]} = \frac{\pi g_H(\omega_L) I_L}{\hbar^2 c \varepsilon_0} |\boldsymbol{\mu}_{ba} \bullet \hat{\mathbf{e}}_L|^2, \tag{6.15}$$

where $W_{a\to b}$ (s^{-1}) is the induced absorption coefficient, and $W_{b\to a}$ (s^{-1}) is the stimulated emission coefficient. The normalized line shape function for the resonance is given by

$$g_H(\omega_L - \omega_{ba}) = \frac{\gamma_{ba}}{\pi\left[\gamma_{ba}^2 + (\omega_L - \omega_{ba})^2\right]} = \frac{2\Delta\omega_H}{\pi\left[\Delta\omega_H^2 + 4(\omega_L - \omega_{ba})^2\right]}, \tag{6.16}$$

where $\Delta\omega_H = 2\gamma_{ba}$ (s^{-1}) is the full-width-at-half-maximum (FWHM) of the resonance line. At this point we are considering broadening due only to collisions and to spontaneous emission, and these processes are assumed to be the same for all atoms or molecules in the medium at position r and time t; this is referred to as homogenous broadening, and the line shape is Lorentzian. The laser intensity $I_L(\mathbf{r}, t)$ (W/m^2) is given by

$$I_L(\mathbf{r}, t) = \frac{1}{2} c \varepsilon_0 |A_L(\mathbf{r}, t)|^2. \tag{6.17}$$

Note that the relationship between the laser field intensity and electric field amplitude must be consistent with the definition of the electric field from Eq. (6.4).

6.3 The Absorption Coefficient

The calculation of various parameters of importance for characterizing the electric dipole interaction of monochromatic laser fields with atomic and molecular resonances is now discussed. The analysis is presented for the general case of an anisotropic medium, and then the general result is specialized to the case of an isotropic medium.

As the laser beam propagates through the medium, the intensity will change because of the induced absorption and stimulated emission processes. The derivative

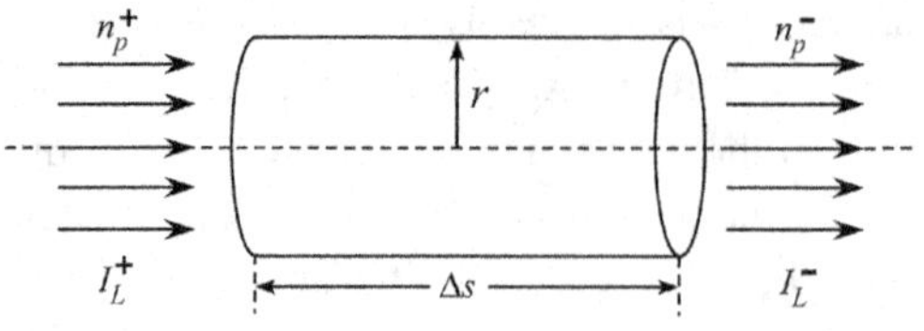

Figure 6.2 Schematic diagram for the absorption of a laser beam with initial intensity I_L^+ in a cylindrical volume with radius r and length Δs.

of intensity with respect to distance for a monochromatic laser beam of frequency ω_L and intensity I_L is given by

$$\frac{dI_L}{ds} = -\alpha_{ba}(\omega_L) I_L, \tag{6.18}$$

where the absorption coefficient $\alpha_{ba}(\omega_L)$ (m^{-1}) will be positive when the number density n_a (m^{-3}) of the ground state is greater than the number density n_b of the excited state, and negative for $n_a < n_b$ (the two-state system will then have positive gain). Note that Eq. (6.18) is rigorously correct only when the laser spectral width is much smaller than the resonance line width. Consider the cylindrical column of collimated laser radiation shown in Figure 6.2. The net number of photons absorbed in the volume element $\pi r^2 \Delta s$ is given by

$$\frac{dN_p}{dt} = -(n_a - n_b) W_{a\to b} \pi r^2 \Delta s. \tag{6.19}$$

The number of photons that pass through the left edge of the volume element per unit time is given by $n_p^+ c\pi r^2$, and the number that pass through the right edge of the volume element per unit time is given by $n_p^- c\pi r^2$, where n_p^+ (m^{-3}) and n_p^- are the photon number densities at the left and right edges, respectively, of the cylindrical volume element. The difference between these two numbers is the net number of photons absorbed in the volume element per unit time. Using the relation $I_L = \hbar\omega_L c n_p$, we obtain

$$\frac{dN_p}{dt} = n_p^+ c\pi r^2 - n_p^- c\pi r^2 = \frac{(I_L^+ - I_L^-)}{\hbar\omega_L} \pi r^2. \tag{6.20}$$

Equating Eqs. (6.19) and (6.20), we obtain

$$\frac{(I_L^+ - I_L^-)}{\Delta s} = -(n_a - n_b) W_{a\to b} \hbar\omega_L, \quad \lim \Delta s \to 0 \Rightarrow \frac{dI_L}{ds} = -(n_a - n_b) W_{a\to b} \hbar\omega_L. \tag{6.21}$$

From Eqs. (6.15), (6.18), and (6.21), we obtain the following expression for the absorption coefficient:

$$\alpha_{ba}(\omega_L - \omega_{ba}) = \frac{2\pi^2 g_H(\omega_L - \omega_{ba})}{\varepsilon_0 \hbar \lambda_L} (n_a - n_b) |\boldsymbol{\mu}_{ba} \bullet \hat{\mathbf{e}}_L|^2. \tag{6.22}$$

The laser will in general interact with many different pairs of coupled states. The absorption coefficient for the irradiance of a laser beam propagating through a medium is given by

$$\alpha_{BA}(\omega_L - \omega_{BA}) = \frac{2\pi^2 g_H(\omega_L - \omega_{BA})}{\varepsilon_0 \hbar \lambda_L} \sum_b \sum_a \left[(n_a - n_b) |\boldsymbol{\mu}_{ba} \bullet \hat{\boldsymbol{e}}_L|^2 \right]. \tag{6.23}$$

We have assumed that the dephasing rates γ_{ba} for the coherences for any two coupled states b and a in the levels B and A are equal and that $\omega_{ba} = \omega_{BA}$ is the same for all transitions ba.

6.3.1 Interaction of Linearly Polarized Light with Resonances between Energy Levels

Consider a linearly polarized laser beam propagating along the y-axis through a medium. Assume that the laser beam is polarized in the z-direction, $\hat{\boldsymbol{e}}_L = \hat{z}$. Then

$$\boldsymbol{E}_L(y,t) = \frac{1}{2} A_L(y,t)\, \hat{z} \left\{ \exp[+i(k_L y - \omega_L t)] + \exp[-i(k_L y - \omega_L t)] \right\}, \tag{6.24}$$

where we can assume without loss of generality that the slowly varying amplitude function $A_L(y,t)$ is real. Assume that the laser is interacting with a resonance transition with $J_b = J_B = J' = J_a = J_A = J$ and that the laser intensity is low enough that the populations of the states of the resonance are not perturbed significantly by the laser interaction. From examination of Eqs. (A4.4)–(A4.6), it can be seen that the z-polarized laser field will interact only with those states b and a for which $M_b = M' = M_a = M$. Substituting Eq. (A4.5) into Eq. (6.23), we obtain

$$\alpha_{BA}(\omega_L - \omega_{BA}) = \frac{2\pi^2 g_H(\omega_L - \omega_{BA})}{\varepsilon_0 \hbar \lambda_L} \sum_b \sum_a \left[(n_a - n_b) \langle \alpha J \| \boldsymbol{\mu} \| \alpha' J' \rangle^2 M^2 \right]. \tag{6.25}$$

The reduced matrix element $\langle \alpha J \| \boldsymbol{\mu} \| \alpha' J' \rangle$ is independent of the states b and a and can be moved outside the summation symbol. Assuming that the population distributions for the upper and lower Zeeman states are initially isotropic, then

$$n_a = \frac{n_A}{2J+1} \tag{6.26}$$

and

$$n_b = \frac{n_B}{2J'+1} \tag{6.27}$$

for all states b and a. Substituting Eqs. (6.26) and (6.27) into Eq. (6.25), we obtain

$$\alpha_{BA}(\omega_L) = \alpha_{J'J}(\omega_L) = \frac{2\pi^2 g_H(\omega_L)}{\varepsilon_0 \hbar \lambda_L} \left(\frac{n_J}{2J+1} - \frac{n_{J'}}{2J'+1} \right) \langle \alpha J \| \boldsymbol{\mu} \| \alpha' J' \rangle^2 \sum_{M'=-J}^{+J} M^2. \tag{6.28}$$

The summation over the upper Zeeman states is given by (Condon & Shortley, 1951)

$$\sum_{M'=-J}^{+J} M^2 = \frac{1}{3} J(J+1)(2J+1). \tag{6.29}$$

From Eq. (4.194), we obtain

$$\langle \alpha J \| \boldsymbol{\mu} \| \alpha' J' \rangle^2 = \frac{A_{J'J}(2J'+1)3\pi\varepsilon_0 \hbar c^3}{\omega_{J'J}^3} = \frac{A_{J'J}(2J'+1)3\varepsilon_0 \hbar \lambda_{J'J}^3}{8\pi^2}. \tag{6.30}$$

Substituting Eqs. (6.29) and (6.30) into Eq. (6.28) and rearranging, we obtain

$$\alpha_{J'J}(\omega_L - \omega_{J'J}) = \frac{\lambda_{J'J}^2 A_{J'J}\, g_H(\omega_L - \omega_{J'J})}{4} \left(n_J \frac{2J'+1}{2J+1} - n_{J'} \right). \tag{6.31}$$

As will be discussed in Section 6.5, there are other broadening effects, such as Doppler broadening, that may be important, so in the general case Eq. (6.31) can be written as

$$\alpha_{J'J}(\omega_L - \omega_{J'J}) = \frac{\lambda_{J'J}^2 A_{J'J}\, g(\omega_L - \omega_{J'J})}{4} \left(n_J \frac{2J'+1}{2J+1} - n_{J'} \right), \tag{6.32}$$

where $g(\omega_L - \omega_{J'J})$ is a normalized line shape function of unspecified form at this point. In writing Eqs. (6.30)–(6.32) we have assumed that $\lambda_L \cong \lambda_{ba} = \lambda_{BA} = \lambda_{J'J}$. The line strength $K_{J'J}$ ($\mathrm{m}^{-1}\mathrm{s}^{-1}$) is defined by

$$K_{J'J} = \frac{\lambda_{J'J}^2 A_{J'J}}{4} \left(n_J \frac{2J'+1}{2J+1} - n_{J'} \right). \tag{6.33}$$

For propagation of a monochromatic beam of light in a medium, Eq. (6.18) can be rewritten as

$$\frac{dI_L}{ds} = -K_{J'J}\, g(\omega_L - \omega_{J'J}) I_L. \tag{6.34}$$

6.3.2 Interaction of Circularly Polarized Light with Resonances between Energy Levels

Consider a circularly polarized laser beam propagating along the z-axis through the medium. The polarization vector for the beam is given by $\hat{\boldsymbol{e}}_L = \hat{\boldsymbol{e}}_{+1} = (\hat{\boldsymbol{x}} + i\hat{\boldsymbol{y}})/\sqrt{2}$. Then

$$\boldsymbol{E}_L(z,t) = \frac{A_L(z,t)}{2\sqrt{2}} \{ (\hat{\boldsymbol{x}} + i\hat{\boldsymbol{y}}) \exp[+i(k_L z - \omega_L t)] + (\hat{\boldsymbol{x}} - i\hat{\boldsymbol{y}}) \exp[-i(k_L z - \omega_L t)] \}. \tag{6.35}$$

Assume that the laser is interacting with a resonance transition with $J' = J + 1$. From an examination of Eqs. (4.183)–(4.184) it can be seen that the product $\boldsymbol{\mu}_{ba} \bullet \hat{\boldsymbol{e}}_L^*$ will be nonzero only for those states b and a for which $M' = M + 1$. Substituting Eq. (A3.19) into Eq. (6.23), we obtain

$$\begin{aligned}\alpha_{J'J}(\omega_L - \omega_{J'J}) &= \frac{2\pi^2 g(\omega_L - \omega_{J'J})\langle \alpha J \| \boldsymbol{\mu}_1 \| \alpha' J' \rangle^2}{\varepsilon_0 \hbar \lambda_L (2J+3)(2J+2)(2J+1)} \\ &\quad \times \sum_b \sum_a [(n_a - n_b)(J - M + 1)(J - M + 2)] \\ &= \frac{2\pi^2 g(\omega_L - \omega_{J'J})\langle \alpha J \| \boldsymbol{\mu}_1 \| \alpha' J' \rangle^2}{\varepsilon_0 \hbar \lambda_L (2J+3)(2J+2)(2J+1)} \left(\frac{n_J}{2J+1} - \frac{n_{J'}}{2J'+1} \right) \\ &\quad \times \sum_{M=-J}^{M=+J} (J + M + 1)(J + M + 2).\end{aligned} \tag{6.36}$$

We have assumed once again that the populations are not significantly perturbed by the laser interaction. The summation over the lower Zeeman states is given by

$$\begin{aligned}\sum_{M=-J}^{M=+J} (J + M + 1)(J + M + 2) &= \sum_{M=-J}^{M=+J} (J^2 + 2JM + M^2 + 3J + 3M + 2) \\ &= (J^2 + 3J + 2)(2J + 1) + \frac{1}{3} J(J + 1)(2J + 1) \\ &= (J + 2)(J + 1)(2J + 1) + \frac{1}{3} J(J + 1)(2J + 1) \\ &= \frac{2}{3}(2J + 3)(J + 1)(2J + 1),\end{aligned} \tag{6.37}$$

where Eq. (6.29) has been used to simplify the expression. Substituting Eqs. (6.30) and (6.37) into Eq. (6.36) and simplifying, we obtain

$$\alpha_{J'J}(\omega_L - \omega_{J'J}) = \frac{\lambda_{J'J}^2 A_{J'J}\, g_H(\omega_L - \omega_{J'J})}{4} \left(n_J \frac{2J'+1}{2J+1} - n_{J'} \right). \tag{6.38}$$

As expected, we obtain the same result as in Eq. (6.31) for linearly polarized laser radiation. The transition cross section $\sigma_{J'J}(\mathrm{m}^2)$ is given by the absorption coefficient divided by the number density of $J'J$ oscillators in the medium,

$$\sigma_{J'J}(\omega_L - \omega_{J'J}) = \frac{\alpha_{J'J}(\omega_L - \omega_{J'J})}{n_J + n_{J'}} = \frac{\lambda_{J'J}^2 A_{J'J}\, g_H(\omega_L - \omega_{J'J})}{4(n_J + n_{J'})} \left(n_J \frac{2J'+1}{2J+1} - n_{J'} \right). \tag{6.39}$$

If all the oscillators are in the ground state, then $n = n_A$, and Eq. (6.39) reduces to

$$\sigma_{J'J}(\omega_L - \omega_{J'J}) = \frac{\lambda_{J'J}^2 A_{J'J}\, g_H(\omega_L - \omega_{J'J})}{4} \frac{2J'+1}{2J+1} = \frac{\lambda_{J'J}^2 A_{J'J}\, g_H(\omega_L - \omega_{J'J}) g_{J'}}{4 g_J}, \tag{6.40}$$

where g_J and $g_{J'}$ are the degeneracies of the ground and excited levels, respectively.

6.3.3 The Oscillator Strength

The oscillator strength for a transition between two levels is the ratio of the spontaneous emission coefficient and the decay rate for a classical electric dipole oscillator (Hilborn, 1982; Siegman, 1986). The decay rate for the classical electric dipole oscillator is the maximum possible decay rate. Therefore, the strongest transitions have oscillator strengths on the order of one. Strong electric dipole transitions in some atoms will have oscillator strengths on the order of one, but oscillator strengths for molecular transitions are almost always much less than one. The oscillator strengths for emission, $f_{J'J}$, and absorption, $f_{JJ'}$, are dimensionless and are given by

$$f_{J'J} = \frac{g_J}{g_{J'}} f_{JJ'} = \frac{2\pi\varepsilon_0 m_e c^3 A_{J'J}}{\omega_{J'J}^2 e^2}, \tag{6.41}$$

where m_e (kg) is the electron mass and e (C) is the electron charge.

6.4 Medium Polarization and Susceptibility

The macroscopic medium polarization and the complex susceptibility are discussed in great detail by Siegman (1986). These concepts are quite useful in the analysis of laser absorption and laser gain processes. The polarization $\boldsymbol{P}(\omega_L, t)$ of the medium (C/m^2) at $\boldsymbol{r} = 0$ due to the resonance between the states b and a is given by (Siegman, 1986)

$$\boldsymbol{P}(\omega_L, t) = \varepsilon_0 \tilde{\chi}(\omega_L) \boldsymbol{E}(t) = \frac{1}{2} \boldsymbol{P}_0(\omega_L, t) \exp(-i\omega_L t) + \frac{1}{2} \boldsymbol{P}_0^*(\omega_L, t) \exp(+i\omega_L t), \tag{6.42}$$

where

$$\boldsymbol{P}_0(\omega_L, t) = \varepsilon_0 \tilde{\chi}(\omega_L) \boldsymbol{A}_L(t) = \varepsilon_0 \tilde{\chi}(\omega_L) A_L(t) \hat{\boldsymbol{e}}_L. \tag{6.43}$$

Note that if the amplitude of the monochromatic laser field is constant in time, then the amplitude of the polarization $\boldsymbol{P}_0(\omega_L, t)$ will be a function of ω_L only. The complex susceptibility tensor $\tilde{\chi}(\omega_L)$ is given by

$$\begin{aligned}
\tilde{\chi}(\omega_L) &= -\frac{n(\rho_{aa} - \rho_{bb})}{\hbar\varepsilon_0[(\omega_L - \omega_{ba}) + i\gamma_{ba}]} \boldsymbol{\mu}_{ba} \boldsymbol{\mu}_{ab} \\
&= -\frac{(n_a - n_b)}{\hbar\varepsilon_0[(\omega_L - \omega_{ba}) + i\gamma_{ba}]} \begin{bmatrix} \mu_{ba,x} \\ \mu_{ba,y} \\ \mu_{ba,z} \end{bmatrix} \times \begin{bmatrix} \mu_{ba,x}^* & \mu_{ba,y}^* & \mu_{ba,z}^* \end{bmatrix} \\
&= -\frac{(n_a - n_b)}{\hbar\varepsilon_0[(\omega_L - \omega_{ba}) + i\gamma_{ba}]} \begin{bmatrix} \mu_x\mu_x^* & \mu_x\mu_y^* & \mu_x\mu_z^* \\ \mu_y\mu_x^* & \mu_y\mu_y^* & \mu_y\mu_z^* \\ \mu_z\mu_x^* & \mu_z\mu_y^* & \mu_z\mu_z^* \end{bmatrix}.
\end{aligned} \tag{6.44}$$

Substituting Eq. (6.44) into Eq. (6.43), the amplitude of the medium polarization is given by

$$\boldsymbol{P}_0(\omega_L, t) = \begin{bmatrix} P_{0,x} \\ P_{0,y} \\ P_{0,z} \end{bmatrix} = -\frac{(n_a - n_b)}{\hbar[(\omega_L - \omega_{ba}) + i\gamma_{ba}]} \begin{bmatrix} \mu_x\mu_x^* & \mu_x\mu_y^* & \mu_x\mu_z^* \\ \mu_y\mu_x^* & \mu_y\mu_y^* & \mu_y\mu_z^* \\ \mu_z\mu_x^* & \mu_z\mu_y^* & \mu_z\mu_z^* \end{bmatrix} \times \begin{bmatrix} A_{L,x} \\ A_{L,y} \\ A_{L,z} \end{bmatrix}. \tag{6.45}$$

6.4.1 Linear Susceptibility for Linearly Polarized Light

As an example, consider the interaction of a laser field that is linearly polarized in the z-direction interacting with a resonance transition for which $J_B = J' = J_A + 1 = J + 1$. For this laser field, $A_{L,x} = A_{L,y} = 0$, $A_{L,z} = A_L$, and $\hat{\boldsymbol{e}}_L = \hat{\boldsymbol{z}}$. For coupled states b and a for which $M' = M \pm 1$, the matrix element μ_z is zero, and the polarization contribution from these states will also be zero. For coupled states b and a for which $M = M'$, the matrix elements μ_x and μ_y are zero, and the polarization contribution from these states is thus given by

$$\begin{aligned} \boldsymbol{P}_{0,ba} &= -\frac{(n_a - n_b)\mu_z\mu_z^*}{\hbar[(\omega_L - \omega_{ba}) + i\gamma_{ba}]} A_L\hat{\boldsymbol{z}} \\ &= -\frac{(n_a - n_b)\langle \alpha J \| \boldsymbol{\mu}_1 \| \alpha' J' \rangle^2 (J + M + 1)(J - M + 1)}{(2J + 3)(J + 1)(2J + 1)\hbar[(\omega_L - \omega_{J'J}) + i\gamma_{ba}]} A_L\hat{\boldsymbol{z}} \\ &= -\frac{3\varepsilon_0 A_{J'J}\lambda_{J'J}^3 (n_a - n_b)\left(J^2 - M^2 + 2J + 1\right)}{8\pi^2 (J + 1)(2J + 1)[(\omega_L - \omega_{J'J}) + i\gamma_{ba}]} A_L\hat{\boldsymbol{z}}, \end{aligned} \tag{6.46}$$

where Eqs. (6.30) and (A3.10) have been used. Summing over all coupled states b and a, again using Eqs. (6.26), (6.27), and (6.29) to simplify the result, and assuming that $\gamma_{ba} = \gamma_{BA} = \gamma_{J'J}$ is the same for all state transitions $b \leftrightarrow a$, we obtain the medium polarization due to the interaction of the laser with the resonance:

$$\begin{aligned} \boldsymbol{P}_{0,J'J} &= -\frac{3\varepsilon_0 A_{J'J}\lambda_{J'J}^3 \left(\frac{n_J}{2J + 1} - \frac{n_{J'}}{2J' + 1}\right)\left[J^2 + 2J + 1 - \frac{1}{3}J^2 - \frac{1}{3}J\right]}{8\pi^2 (J + 1)[(\omega_L - \omega_{J'J}) + i\gamma_{J'J}]} A_L\hat{\boldsymbol{z}} \\ &= -\frac{\varepsilon_0 A_{J'J}\lambda_{J'J}^3 \left(\frac{n_J}{2J + 1} - \frac{n_{J'}}{2J' + 1}\right)[2J^2 + 5J + 3]}{8\pi^2 (J + 1)[(\omega_L - \omega_{J'J}) + i\gamma_{J'J}]} A_L\hat{\boldsymbol{z}} \\ &= -\frac{\varepsilon_0 A_{J'J}\lambda_{J'J}^3 \left(n_J \frac{2J + 3}{2J + 1} - n_{J'}\right)}{8\pi^2 [(\omega_L - \omega_{J'J}) + i\gamma_{J'J}]} A_L\hat{\boldsymbol{z}}. \end{aligned} \tag{6.47}$$

6.4.2 Linear Susceptibility for Circularly Polarized Light

Now consider the interaction of a circularly polarized laser field propagating in the y-direction with a resonance transition for which $J = J'$. For this laser field,

$\hat{\boldsymbol{e}}_L = (\hat{\boldsymbol{x}} - i\hat{\boldsymbol{z}})/\sqrt{2}$, or equivalently $A_{L,x} = A_L/\sqrt{2}$, $A_{L,y} = 0$, and $A_{L,z} = -iA_L/\sqrt{2}$. Substituting into Eq. (6.45) and rearranging, we obtain

$$\boldsymbol{P}_{0,ba} = -\frac{(n_a - n_b)}{\hbar[(\omega_L - \omega_{J'J}) + i\gamma_{ba}]}\left[(\mu_x\mu_x^* A_{Lx} + \mu_x\mu_z^* A_{Lz})\hat{\boldsymbol{x}} + (\mu_z\mu_x^* A_{Lx} + \mu_z\mu_z^* A_{Lz})\hat{\boldsymbol{z}}\right]. \tag{6.48}$$

In the basis set of quantum states that we have selected by our choice of the z-axis in this problem, the x- and z-components of the laser field will interact with different pairs of coupled Zeeman states. For coupled states b and a for which $M' = M \pm 1$, the matrix element μ_z is zero. For coupled states b and a for which $M' = M$, the matrix elements μ_x and μ_y are zero. From Eqs. (A3.8)–(A3.10), the components μ_x and μ_z will be given by

$$\mu_x = \langle \alpha J \| \boldsymbol{\mu}_1 \| \alpha' J' \rangle \sqrt{\frac{(J - M + 1)(J + M)}{(2J + 1)(2J)(J + 1)}}\frac{1}{\sqrt{2}} \quad \text{for } M' = M - 1. \tag{6.49}$$

$$\mu_x = \langle \alpha J \| \boldsymbol{\mu}_1 \| \alpha' J' \rangle \sqrt{\frac{(J + M + 1)(J - M)}{(2J + 1)(2J)(J + 1)}}\frac{1}{\sqrt{2}} \quad \text{for } M' = M + 1. \tag{6.50}$$

$$\mu_z = \langle \alpha J \| \boldsymbol{\mu}_1 \| \alpha' J' \rangle \frac{M}{\sqrt{(2J + 1)J(J + 1)}} \quad \text{for } M' = M. \tag{6.51}$$

Since no state transitions $b \leftrightarrow a$ are coupled by both x- and z-polarized laser radiation, the cross terms $\mu_x\mu_z^*$ and $\mu_z\mu_x^*$ will be zero. Using Eqs. (6.49)–(6.51) and summing Eq. (6.48) over all states a and b, we obtain

$$\boldsymbol{P}_{0,J'J} = -\frac{\left(\frac{n_J}{2J+1} - \frac{n_{J'}}{2J+1}\right)\langle \alpha J \| \boldsymbol{\mu}_1 \| \alpha' J' \rangle^2}{\hbar[(\omega_L - \omega_{J'J}) + i\gamma_{J'J}]J(2J + 1)(J + 1)} \times \left\{\left[\sum_M \frac{(J + M + 1)(J - M)}{4} + \sum_M \frac{(J - M + 1)(J + M)}{4}\right] A_{Lx}\hat{\boldsymbol{x}} + \sum_M M^2 A_{Lz}\hat{\boldsymbol{z}}\right\}, \tag{6.52}$$

where again we have assumed that the population in both levels J' and J is equally distributed over the Zeeman states b and a, respectively, and that $\gamma_{ba} = \gamma_{BA} = \gamma_{J'J}$ for all state transitions $b \leftrightarrow a$. In Eq. (6.52) the summations over M for the x-component of the polarization are given by

$$\begin{aligned}\sum_M \frac{(J + M + 1)(J - M)}{4} &= \frac{1}{4}\sum_M (J^2 + J - M^2 - M) \\ &= \frac{1}{4}\left[(2J + 1)J(J + 1) - \frac{1}{3}J(J + 1)(2J + 1)\right] = \frac{J(2J + 1)(J + 1)}{6}.\end{aligned} \tag{6.53}$$

$$\sum_M \frac{(J - M + 1)(J + M)}{4} = \frac{1}{4}\sum_M (J^2 + J - M^2 + M_e) = \frac{J(2J + 1)(J + 1)}{6}. \tag{6.54}$$

Substituting Eqs. (6.53), (6.54), and (6.29) into Eq. (6.52) and simplifying, we obtain

$$\boldsymbol{P}_{0,J'J} = -\frac{\left(\frac{n_J}{2J+1} - \frac{n_{J'}}{2J+1}\right)\langle \alpha J \| \boldsymbol{\mu}_1 \| \alpha' J' \rangle^2}{3\hbar[(\omega_L - \omega_{J'J}) + i\gamma_{J'J}]}[A_{Lx}\hat{\boldsymbol{x}} + A_{Lz}\hat{\boldsymbol{z}}]. \tag{6.55}$$

Using Eq. (4.203), we obtain

$$\boldsymbol{P}_{0,J'J} = -\frac{\varepsilon_0 A_{J'J}\lambda^3_{J'J}(n_J - n_{J'})}{8\pi^2[(\omega_L - \omega_{J'J}) + i\gamma_{J'J}]}[A_{Lx}\hat{\boldsymbol{x}} + A_{Lz}\hat{\boldsymbol{z}}]. \tag{6.56}$$

Note from Eqs. (6.47) and (6.56) that the ratio $\boldsymbol{P}_{0,BA}/\varepsilon_0 \boldsymbol{A}_L$ is the same for both cases that we have considered, which suggests, as expected, that the susceptibility may be represented by a scalar quantity for an isotropic medium,

$$\chi_{BA}(\omega_L) = -\frac{A_{BA}\lambda^3_{BA}\left[n_A\frac{2J'+1}{2J+1} - n_B\right]}{8\pi^2[(\omega_L - \omega_{BA}) + i\gamma_{BA}]}. \tag{6.57}$$

The expression that we obtain in Eq. (6.57) is the same as that given by Siegman (1986) for an isotropic medium, for which the factor 3* = 1. Note that this expression for the scalar susceptibility will not be correct when the medium is not isotropic. This can occur in the gas phase if the resonance *BA* is pumped strongly by polarized laser radiation, for example.

The same results as in Eqs. (6.46) and (6.52) can be obtained by calculating the polarization as the trace of the density matrix product (Boyd, 2008):

$$\begin{aligned}\boldsymbol{P}(\omega_L,t) &= nTr(\boldsymbol{\mu}\rho) = n[\boldsymbol{\mu}_{ab}\rho_{ba}(\omega_L,t) + \boldsymbol{\mu}_{ba}\rho_{ab}(\omega_L,t)]\\ &= n[\boldsymbol{\mu}_{ab}\sigma_{ba}(\omega_L,t)\exp(-i\omega_L t) + \boldsymbol{\mu}_{ba}\sigma_{ab}(\omega_L,t)\exp(+i\omega_L t)]\\ &= \frac{1}{2}\boldsymbol{P}_{0,ba}(\omega_L,t)\exp(-i\omega_L t) + \frac{1}{2}\boldsymbol{P}^*_{0,ba}(\omega_L,t)\exp(+i\omega_L t).\end{aligned} \tag{6.58}$$

From Eqs. (6.11) and (6.58), it can be seen that

$$\boldsymbol{P}_{0,ba} = 2n\boldsymbol{\mu}_{ab}\sigma_{ba} = 2n\boldsymbol{\mu}^*_{ba}\sigma_{ab} = -\frac{n(\rho_{aa} - \rho_{bb})\boldsymbol{\mu}^*_{ba}(\boldsymbol{\mu}_{ba} \bullet \hat{\boldsymbol{e}}_L)A_L}{\hbar[(\omega_L - \omega_{ba}) + i\gamma_{ba}]}. \tag{6.59}$$

For the case of z-polarized laser radiation interacting with a resonance transition for which $J' = J + 1$, the polarization amplitude due to coupled states b and a for which $M' = M$ is given by

$$\boldsymbol{P}_{0,ba} = -\frac{(n_a - n_b)\langle \alpha J \| \boldsymbol{\mu}_1 \| \alpha' J' \rangle^2 (J + M + 1)(J - M + 1)}{(2J+3)(J+1)(2J+1)\hbar[(\omega_L - \omega_{J'J}) + i\gamma_{J'J}]}A_L\hat{\boldsymbol{z}}. \tag{6.60}$$

The polarization amplitude expressions are identical in Eqs. (6.46) and (6.60).

6.5 Spectral Line Shape Models for Absorption Spectroscopy

The phenomenon of homogeneous broadening of resonance transitions due to collisions and spontaneous emission has been introduced in discussions of the classical electron oscillator and in treatments of resonance transitions using perturbation theory and the density matrix. However, for gas-phase species, Doppler broadening due to the motion of the molecules or atoms is an important line-broadening process. The convolution of the Lorentzian lineshape resulting from homogenous broadening and the inhomogeneous Doppler lineshape is the Voigt line shape. Furthermore, collisions can result not only in line broadening but in line narrowing and shifting under certain circumstances, leading to more complicated line shapes such as the Galatry line shape.

6.5.1 Homogeneous Broadening

Thus far, discussion of line-broadening processes has been limited to the introduction of the phenomenological rate coefficients for population transfer (Γ_{ba} and Γ_{ab}) and coherence dephasing (γ_{ba}). These phenomena were introduced in Chapter 2 in the discussion of the classical electron oscillator, and these effects result in homogeneous broadening of gas-phase resonances; the term homogeneous broadening is used to indicate that the broadening process is the same for each atom or molecule in the assembly. The macroscopic coherence between states b and a can decay by pure dephasing collisions, but it will also be destroyed by population transfer out of either state b or state a. For the two-state system,

$$\Delta\omega_H = 2\gamma_{ba} = 2\gamma_{ab} = \Gamma_{ba} + \Gamma_{ab} + Q_{dp}, \tag{6.61}$$

where it is assumed in writing Eq. (6.61) that population transfer occurs only between states b and a. In general, of course, states b and a will be coupled to numerous other states and levels by population transfer collisions or by spontaneous emission. The more general expression for the homogeneous line width is

$$\Delta\omega_H = 2\gamma_{ba} = 2\gamma_{ab} = \sum_m \Gamma_{bm} + \sum_m \Gamma_{am} + Q_{dp}, \tag{6.62}$$

where Γ_{bm} and Γ_{am} (s^{-1}) are the rate coefficients for population transfer from states b and a, respectively, to state m. Population transfer may result from collisions, spontaneous emission, or by effects such as intersystem crossing in more complicated polyatomic molecules

6.5.2 Inhomogeneous Broadening

Inhomogeneous broadening occurs when the central resonant frequency is shifted, either due to the motion of the atoms or molecules in a gas or due to variations in the local environment for atoms in a crystal. For gas-phase atoms and molecules in translational equilibrium, the velocity distribution is described by the Maxwell–Boltzmann distribution,

$$g(\mathrm{v}_z) = \left(\frac{1}{2\pi\sigma_{\mathrm{v}}^2}\right)\exp\left(-\frac{\mathrm{v}_z^2}{2\sigma_{\mathrm{v}}^2}\right), \tag{6.63}$$

where $\sigma_{\mathrm{v}}^2 = k_B T/m_a$. Similar expressions apply for the x- and y-components of velocity. If the laser radiation is propagating in the +z direction, the frequency of the laser will appear to be shifted in the rest frame attached to the moving atom or molecule. The frequency shift is given by

$$\omega_L' = \left(1 - \frac{\mathrm{v}_z}{c}\right)\omega_L. \tag{6.64}$$

In the laboratory frame, it appears that the resonant frequency $\omega_0' = \omega_{BA}'$ of the molecule will be shifted in the opposite direction,

$$\omega_0' = \left(1 + \frac{\mathrm{v}_z}{c}\right)\omega_0 = \omega_0 + \omega_0\frac{\mathrm{v}_z}{c}, \tag{6.65}$$

where $\omega_0 = \omega_{BA,0}$ is the resonant frequency for molecules with $\mathrm{v}_z = 0$. Using the relation $g(\mathrm{v}_z)d\mathrm{v}_z = g_D(\omega_0' - \omega_0)$, we obtain

$$g_D(\omega_0' - \omega_0) = \sqrt{\frac{\ln 2}{\pi}}\frac{2}{\Delta\omega_D}\exp\left[-4\ln 2\left(\frac{\omega_0' - \omega_0}{\Delta\omega_D}\right)^2\right], \tag{6.66}$$

where $\Delta\omega_D$ is the FWHM of the Doppler line shape,

$$\Delta\omega_D = \omega_0\sqrt{\frac{8\ln 2k_BT}{m_ac^2}} = \omega_0\sqrt{\frac{\pi}{\ln 2}}\frac{\bar{\mathrm{v}}}{c}. \tag{6.67}$$

In Eq. (6.67), $\bar{\mathrm{v}}$ is the average molecular velocity.

6.5.3 The Voigt Line Shape

The Voigt line shape is the result of the convolution of the Lorentzian and Doppler line shapes. For a gas, for a given velocity v_z, the central frequency will be shifted, but it can be assumed that the broadening of the radiative transition line shape due to collisions and spontaneous emission will be the same for all molecules regardless of the value of v_z. With this assumption, the absorption coefficient for an assembly of gas molecules can be calculated using the convolution integral

$$\alpha_{J'J}(\omega_L - \omega_0) = \int_{-\infty}^{+\infty} \alpha_{J'J}(\omega_L - \omega_0')g_D(\omega_0' - \omega_0)d\omega_0', \tag{6.68}$$

where $\omega_0 = \omega_{J'J}$ in the laboratory frame of reference. Substituting for $\alpha_{J'J}(\omega_L - \omega_0')$ using Eq. (6.31), we obtain

$$\alpha_{J'J}(\omega_L - \omega_0) = \frac{\lambda_{J'J}^2 A_{J'J}}{4}\left[n_J\left(\frac{2J'+1}{2J+1}\right) - n_{J'}\right]\int_{-\infty}^{+\infty} g_H(\omega_L - \omega_0')g_D(\omega_0' - \omega_0)d\omega_0'. \tag{6.69}$$

The convolution integral in Eq. (6.69) is the Voigt integral,

$$g_V(\omega_L - \omega_0) = \int_{-\infty}^{+\infty} g_H(\omega_L - \omega_0')g_D(\omega_0' - \omega_0)d\omega_0'$$

$$= \int_{-\infty}^{\infty}\left[\frac{2\Delta\omega_H}{\pi(\Delta\omega_H)^2 + 4\pi(\omega_L - \omega_0')^2}\right]\left\{\sqrt{\frac{\ln 2}{\pi}}\frac{2}{\Delta\omega_D}\exp\left[-4\ln 2\left(\frac{\omega_0' - \omega_0}{\Delta\omega_D}\right)^2\right]\right\}d\omega_0', \tag{6.70}$$

and the absorption coefficient is given by

$$\alpha_{J'J}(\omega_L - \omega_0) = \frac{\lambda_{J'J}^2 A_{J'J}}{4}\left[n_J\left(\frac{2J'+1}{2J+1}\right) - n_{J'}\right]g_V(\omega_L - \omega_0) = K_{J'J}g_V(\omega_L - \omega_0). \tag{6.71}$$

At this point we define the dimensionless variables

$$y = 2\sqrt{\ln 2}\left(\frac{\omega_0' - \omega_0}{\Delta\omega_D}\right) \tag{6.72}$$

$$a = \sqrt{\ln 2}\left(\frac{\Delta\omega_H}{\Delta\omega_D}\right) \tag{6.73}$$

and

$$x = 2\sqrt{\ln 2}\left(\frac{\omega_L - \omega_0}{\Delta\omega_D}\right). \tag{6.74}$$

Substituting Eqs. (6.72)–(6.74) into Eq. (6.70) and rearranging, we obtain

$$g_V(\omega_L - \omega_0) = \frac{2}{\Delta\omega_D}\sqrt{\frac{\ln 2}{\pi}}\frac{a}{\pi}\int_{-\infty}^{\infty}\frac{\exp(-y^2)dy}{a^2 + (x-y)^2} = \frac{2}{\Delta\omega_D}\sqrt{\frac{\ln 2}{\pi}}V(a,x), \tag{6.75}$$

where $V(a,x)$ is the Voigt function,

$$V(a,x) = \frac{a}{\pi}\int_{-\infty}^{\infty}\frac{\exp(-y^2)dy}{a^2 + (x-y)^2}. \tag{6.76}$$

Appendix A9 contains tables of the Voigt function $V(a,x)$.

6.5.4 The Effect of Velocity-Changing Collisions

The Voigt line shape results from the assumption that the Lorentzian line shape is the same for every velocity group, that is, that the cross section for phase-changing and state-changing collisions does not depend on the speed of the species that is absorbing (or emitting). The effect of velocity-changing collisions also needs to be considered in a more complete line shape theory. The line shape can be significantly narrowed by velocity-changing collisions under certain conditions of pressure and temperature, especially for light species such as HF; the line shape of this species has been the subject of numerous experimental studies (Chou et al., 1999a, 1999b; Pine, 1980). The physical origin of the effect of velocity-changing collisions is discussed by Rautian and Sobel'man (1967), who consider an atom or molecule with an excited resonance with central frequency ω_0 and velocity v_z in the direction of an observer. In the frame of reference of the observer, the apparent central resonant frequency for the molecule will be given by Eq. (6.65),

$$\omega_0' = \left(1 + \frac{v_z}{c}\right)\omega_0 = \omega_0 + \omega_0 \frac{v_z}{c}. \tag{6.65}$$

Integration of Eq. (6.65) over the assumed Maxwell–Boltzmann velocity distribution results in the Doppler profile. However, as pointed out by Rautian and Sobel'man (1967), this is strictly true only if the molecule maintains the same velocity v_z for an infinite time. If the velocity is changed by collisions after a time τ, then the emission or absorption for the resonance will be characterized by a frequency spread of $\Delta\omega_{vcc} \cong \tau^{-1} = \overline{v}\pi d^2 n$, where d is the hard-sphere diameter of the molecule and n is the number density of the gas assembly. The Doppler profile will be accurate for the transition only when the frequency spread due to velocity changing collisions is much less than the Doppler width,

$$\Delta\omega_{vcc} = \overline{v}\pi d^2 n = \frac{\overline{v}\sqrt{2}}{\ell} \ll \Delta\omega_D = \omega_0\sqrt{\frac{\pi}{\ln 2}}\frac{\overline{v}}{c} = \frac{2\pi c}{\lambda_0}\sqrt{\frac{\pi}{\ln 2}}\frac{\overline{v}}{c}. \tag{6.77}$$

Ignoring numerical terms of the order of unity, the Doppler profile will be accurate when the mean free path ℓ (m) is much greater that the resonance central wavelength λ_0,

$$\ell \gg \lambda_0. \tag{6.78}$$

Depending on the details of the collisional process, pressure-induced shifts in the central frequency of the resonance are also observed.

Two line-shape models that incorporate the effects of velocity-changing collisions are the Galatry or soft-collision model (Galatry, 1961) and the hard-collision model due to Rautian and Sobel'man (1967) and Gersten and Foley (1968). In the simplest forms of these models, it is assumed that the velocity-changing collisions and the phase-changing collisions are uncorrelated.

The normalized line shape functions for Voigt, Galatry, and Rautian broadening were developed by Herbert (1974). These line-shape models are further discussed and used to model collisional narrowing effects in diode laser absorption measurements by Hanson and co-workers (Chou et al., 1999a; Chou et al., 1999b; Ouyang & Varghese, 1989; Varghese & Hanson, 1984). The line profiles $K(x, y, z, \ldots)$ are normalized such that

$$\int_{-\infty}^{+\infty} K(x, a, z, \ldots)dx = \sqrt{\pi}. \tag{6.79}$$

The normalized, dimensionless parameters x and a are given by Eqs. (6.74) and (6.73), respectively. The dimensionless parameter z is associated with velocity-changing collisions and with the correlations between velocity-changing collisions and phase-changing collisions. The more general line shape expressions discussed in this section also allow for pressure-induced shifts in the central resonant frequency. For a pressure shift in the central resonant frequency $\Delta\omega_S$, the dimensionless parameters s and x' are defined as

$$s = \frac{\omega_0' - \omega_0}{\Delta\omega_D'} \tag{6.80}$$

$$x' = x - s \tag{6.81}$$

and

$$\Delta\omega_D' = \frac{\Delta\omega_D}{2\sqrt{\ln 2}}. \tag{6.82}$$

The Galatry profile (Galatry, 1961) was derived with the assumptions that velocity-changing collisions and phase-changing collisions are separate events and that the velocity of the absorbing (or emitting) molecule after a velocity-changing collision is nearly the same as before the collision; this is the so-called soft-collision model. With these assumptions, the Galatry profile is given by

$$G(x', a, z) = \frac{1}{\sqrt{\pi}} \operatorname{Re} \int_0^{\infty} \left[(-ix' - a)t + \frac{1 - zt - \exp(-zt)}{2z^2} \right] dt, \tag{6.83}$$

where

$$z = \frac{\beta}{\Delta\omega_D'} \tag{6.84}$$

and β is the rate of velocity-changing collisions. The soft-collision model will be most appropriate for the radiative transitions of a heavy species colliding with much lighter species; the velocity change for the heavy species will be much less than for the light species as a result of the collision. Therefore, the velocity of the heavy, radiating species after the collision will be strongly correlated with the velocity before the collision. Considering the effects of correlated phase-changing and velocity-changing collisions, the line profile becomes

$$G(x', a, z, \zeta, s) = G(x', a, Z), \tag{6.85}$$

where

$$Z = z\left(1 - \frac{a}{\zeta}\right) - \frac{isz}{\zeta} \tag{6.86}$$

and

$$\zeta = \frac{\Omega}{\Delta\omega_D'}. \tag{6.87}$$

In Eq. (6.87), Ω is the total collision frequency. Note that the shift parameter s appears in the complex parameter Z, and therefore the correlated line shape can exhibit asymmetry.

In the hard-collision model it is assumed that the velocity of the absorbing molecule is completely randomized after a velocity-changing collision. Under the assumption that the final velocity distribution is a Maxwell–Boltzmann distribution, then the line shape is given by

$$R(x', a, z) = \mathrm{Re}\left[\frac{W(x', a+z)}{1 - \sqrt{\pi} z W(x', a+z)}\right], \tag{6.88}$$

where again

$$z = \frac{\beta}{\Delta\omega'_D} \tag{6.89}$$

and β is the rate of velocity-changing collisions for the hard-collision model. Similarly, the expression for correlated phase-changing and velocity-changing collisions in the hard-collision limit results in the following line shape function:

$$R(x', a, \zeta, s) = \mathrm{Re}\left[\frac{W(x', a+\zeta)}{1 - \sqrt{\pi}(\zeta - a - is) W(x', a+\zeta)}\right], \tag{6.90}$$

where again

$$\zeta = \frac{\Omega}{\Delta\omega'_D} \tag{6.91}$$

and again Ω is the total collision frequency.

6.6 Intensity, Polarization, and Angular Distribution of Spontaneous Emission from Excited States

Consider a group of oscillators in a small volume element located at the center of the coordinate system shown in Figure 6.3. The electric field amplitude at a distance r from a single oscillator due to spontaneous emission from upper state b to ground state a is given by (Condon & Shortley, 1951; Shore & Menzel, 1968)

$$\boldsymbol{A}_{SP} = \frac{k_{ba}^2}{2\pi\varepsilon_0 r}(\hat{\boldsymbol{x}}\hat{\boldsymbol{x}} + \hat{\boldsymbol{y}}\hat{\boldsymbol{y}} + \hat{\boldsymbol{z}}\hat{\boldsymbol{z}} - \hat{\boldsymbol{r}}\hat{\boldsymbol{r}}) \bullet \boldsymbol{\mu}_{ba}, \tag{6.92}$$

where $\hat{\boldsymbol{x}}\hat{\boldsymbol{x}}$, $\hat{\boldsymbol{y}}\hat{\boldsymbol{y}}$, $\hat{\boldsymbol{z}}\hat{\boldsymbol{z}}$, and $\hat{\boldsymbol{r}}\hat{\boldsymbol{r}}$ are unit dyads. The unit vector $\hat{\boldsymbol{r}}$ is given by

$$\hat{\boldsymbol{r}} = \sin\theta\cos\varphi\hat{\boldsymbol{x}} + \sin\theta\sin\varphi\hat{\boldsymbol{y}} + \cos\theta\hat{\boldsymbol{z}}. \tag{6.93}$$

For an isotropic distribution of upper states (i.e., equal populations in each upper Zeeman state b), spontaneous emission will be unpolarized and isotropic. Using Eqs. (A4.4)–(A4.6) and (6.92), the electric field amplitude for a transition where $J = J'$ will be given by

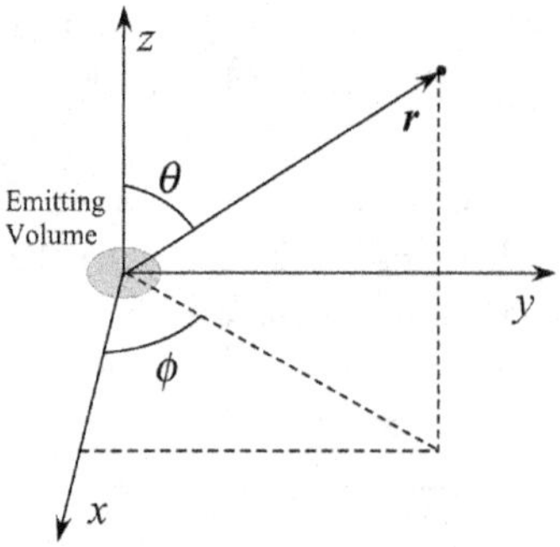

Figure 6.3 Coordinate system for the analysis of spontaneous emission from a volume element at the origin.

$$A_{SP0} = \frac{k^2_{J'J}\langle \alpha J\|\boldsymbol{\mu}\|\alpha' J'\rangle}{2\pi\varepsilon_0 r}\frac{M'}{\sqrt{(2J'+1)J'(J'+1)}} \qquad \Delta M = M - M' = 0. \tag{6.94}$$

$$A_{SP+1} = \frac{k^2_{J'J}\langle \alpha J\|\boldsymbol{\mu}\|\alpha' J'\rangle}{2\pi\varepsilon_0 r}\frac{1}{2}\sqrt{\frac{(J'-M')(J'+M'+1)}{(2J'+2)(2J'+1)(2J')}} \times \left[(1-\sin^2\theta\cos^2\varphi)\hat{\boldsymbol{x}} + i(1-\sin^2\theta\sin^2\varphi)\hat{\boldsymbol{y}}\right] \qquad \Delta M = +1. \tag{6.95}$$

$$A_{SP-1} = \frac{k^2_{J'J}\langle \alpha J\|\boldsymbol{\mu}\|\alpha' J'\rangle}{2\pi\varepsilon_0 r}\frac{1}{2}\sqrt{\frac{(J'+M')(J'-M'+1)}{(2J'+2)(2J'+1)(2J')}} \times \left[(1-\sin^2\theta\cos^2\varphi)\hat{\boldsymbol{x}} - i(1-\sin^2\theta\sin^2\varphi)\hat{\boldsymbol{y}}\right] \qquad \Delta M = -1. \tag{6.96}$$

Consider the case where spontaneous emission is observed at different angles φ in the x–y plane. The amplitudes of the spontaneous emission due to a single oscillator will be

$$A_{SP0} = \frac{k^2_{J'J}\langle \alpha J\|\boldsymbol{\mu}\|\alpha' J'\rangle}{2\pi\varepsilon_0 r}\frac{M'}{\sqrt{(2J'+1)J'(J'+1)}} \qquad \Delta M = 0. \tag{6.97}$$

$$A_{SP+1} = \frac{k^2_{J'J}\langle \alpha J\|\boldsymbol{\mu}\|\alpha' J\rangle}{4\pi\varepsilon_0 r}\sqrt{\frac{(J'-M')(J'+M'+1)}{(2J'+2)(2J'+1)(2J')}}(\sin^2\varphi\hat{\boldsymbol{x}} + i\cos^2\varphi\hat{\boldsymbol{y}}) \quad \Delta M = +1. \tag{6.98}$$

$$A_{SP-1} = \frac{k^2_{J'J}\langle \alpha J\|\boldsymbol{\mu}\|\alpha' J'\rangle}{4\pi\varepsilon_0 r}\sqrt{\frac{(J'+M')(J'-M'+1)}{(2J'+2)(2J'+1)(2J')}}(\sin^2\varphi\hat{\boldsymbol{x}} - i\cos^2\varphi\hat{\boldsymbol{y}}) \quad \Delta M = -1. \tag{6.99}$$

The relation $\cos^2\varphi + \sin^2\varphi = 1$ has been used to simplify the expressions in Eqs. (6.98) and (6.99). The intensity of the spontaneous emission is given by the sum of the intensities from each of the transitions $b \to a$ because of the incoherent nature of the spontaneous emission process. The intensity of light emitted from each transition will be proportional to the number density of state b, again because of the incoherent nature of the emission. The sums of the intensities for the transitions $b \rightarrow a$ in the x–y plane are thus given by

$$I_{SP0} = \frac{1}{2} c\varepsilon_0 \mathbf{A}_{SP0} \bullet \mathbf{A}^*_{SP0} = \frac{ck^4_{J'J} \langle \alpha J \| \boldsymbol{\mu} \| \alpha' J' \rangle^2}{8\pi^2 \varepsilon_0 r^2} \sum_{M'=-J'}^{M'=+J'} n_b \frac{(M')^2}{J'(J'+1)(2J'+1)} = \frac{A_{J'J} \hbar \omega_{J'J}}{8\pi r^2} n_{J'}, \tag{6.100}$$

$$I_{SP+1} = \frac{ck^4_{J'J} \langle \alpha J \| \boldsymbol{\mu} \| \alpha' J' \rangle^2}{32\pi^2 \varepsilon_0 r^2} \sum_{M'=-J'}^{M'=+J'} n_b \left[\frac{J'(J'+1) - (M')^2 - M'}{(2J'+2)(2J'+1)(2J')} \right] = \frac{A_{J'J} \hbar \omega_{J'J}}{16\pi r^2} n_{J'}, \tag{6.101}$$

$$I_{SP-1} = \frac{ck^4_{J'J} \langle \alpha J \| \boldsymbol{\mu} \| \alpha' J' \rangle^2}{32\pi^2 \varepsilon_0 r^2} \sum_{M'=-J'}^{M'=+J'} n_b \left[\frac{J'(J'+1) - (M')^2 + M'}{(2J'+2)(2J'+1)(2J')} \right] = \frac{A_{J'J} \hbar \omega_{J'J}}{16\pi r^2} n_{J'}, \tag{6.102}$$

where we have assumed that n_b is the same for each Zeeman state and that the frequencies of all the transitions $b \rightarrow a$ are the same, as will be the case in the absence of an external magnetic field. Therefore $k_{ba} = k_{BA}$ and $\omega_{ba} = \omega_{BA}$. In the x–y plane, the intensity of spontaneous emission is independent of angle φ for each of the three different types of transitions ($\Delta M = 0, \pm 1$). This will not be true in general, but the sum of the intensities for the different ΔM transitions will be independent of both φ and θ so long as the upper Zeeman states b are populated equally. Therefore, the total intensity of spontaneous emission will be the same everywhere on a sphere of radius r surrounding a collection of isotropic oscillators,

$$I_{SP} = I_{SP0} + I_{SP+1} + I_{SP-1} = \frac{A_{J'J} \hbar \omega_{J'J}}{4\pi r^2} n_{J'}, \tag{6.103}$$

and the rate at which energy is radiated into all solid angles by the oscillators will be $A_{J'J} \hbar \omega_{J'J} n_{J'}$.

Consider the character of the spontaneous emission in the x–y plane as viewed by an observer looking back at the origin along the $+y$ axis. In this case $\varphi = \pi/2$, and spontaneous emission from $\Delta M = 0$ will be polarized in the z-direction. Spontaneous emission from $\Delta M = \pm 1$ transitions will be polarized in the x-direction. The intensities of the z-polarized and x-polarized spontaneous emission will be the same, and the spontaneous emission is thus unpolarized. This will be the case regardless of the angles φ and θ, although the analysis is more complicated for arbitrary angles.

6.7 Example Problem: Absorption Spectroscopy

A 0.2 m-long gas cell contains nitric oxide (NO) at a partial pressure of 0.1 kPa and a temperature of 700 K. A monochromatic laser beam is directed through the cell, and its frequency is tuned near the ${}^{O}P_{12}(2)$ transition ($N = 2, J = 1.5$; $N' = 0$, $J' = 0.5$) in the ($\text{v} = \text{v}' = 0$) vibrational band of the $A^2\Sigma^+ - X^2\Pi$ electronic transition. The cell also contains helium buffer gas at a pressure of 50 kPa, and the collisional broadening of the resonance is due almost entirely to the helium buffer gas. At a He pressure of

50 kPa, the homogeneous linewidth of the NO resonance is 0.165 cm^{-1}, and it can be assumed that the broadening is due to collisions with helium alone (neglect NO–NO collisions). Assume that a Voigt profile accurately describes the transition line shape. For a monochromatic laser beam tuned to line center, what is the fractional transmission through the 0.2 m–long cell? What is the fractional transmission for $\tilde{\nu}_L - \tilde{\nu}_0 = 0.1\ \text{cm}^{-1}$?

Solution: From Table 3.6, the upper level (F_{1f}) energy is 44,198.943 cm^{-1}, and from Table 3.3 the lower level (F_{2f}) energy level is 124.914 cm^{-1}. The central resonance frequency is therefore $\tilde{\nu}_{J'J} = \tilde{\nu}_0 = 44,074\ \text{cm}^{-1}$, and $A_{J'J} = 4.7886 \times 10^5\ \text{s}^{-1}$ (Appendix A7) for this transition.

We need to find the line strength $K_{J'J}$ for the ${}^{O}P_{12}(2)$ transition:

$$K_{J'J} = \frac{\left(\frac{g_{J'}}{g_J} n_J - n_{J'}\right)\lambda_{J'J}^2 A_{J'J}}{4} = \frac{n_J \lambda_{J'J}^2 A_{J'J} g_{J'}}{4 g_J}, \tag{6.104}$$

where in writing Eq. (6.103) we have assumed that the population of the upper level of the transition is negligible. The wavelength of the transition is given by

$$\lambda_{J'J} = \frac{1}{\tilde{\nu}_{J'J}} = \frac{1}{44,074\ \text{cm}^{-1}} = 226.89\ \text{nm}. \tag{6.105}$$

To find the number density n_J of the lower level of the transition, we first need to find the number density of NO. From the ideal gas law,

$$n_{NO} = \frac{p_{NO}}{k_B T} = \frac{0.10 \times 10^3\ \text{J/m}^3}{\left(1.381 \times 10^{-23}\ \text{J/K}\right)(700\ \text{K})} = 1.035 \times 10^{22}\ \text{m}^{-3}. \tag{6.106}$$

For calculating the population fraction in the lower level of the transition, we assume that Boltzmann statistics apply and that NO is a rigid rotator and harmonic oscillator (RRHO); this is not the most accurate method of calculating the population fraction, but the alternative method is to develop a full computer model of the NO energy levels and sum over all levels to compute the partition function. For the RRHO calculation, the formulae for the partition functions and spectroscopic constants are taken from Laurendeau (2005). The partition functions and energy level expressions are also discussed by Hanson et al. (2016). The population fraction is given by

$$\frac{n_J}{n_{NO}} = \frac{g_1 \exp\left[-\left(\frac{\varepsilon_J}{hc}\right)\left(\frac{hc}{k_B}\right)/T\right]}{Z_{elec} Z_{vib} Z_{rot}}. \tag{6.107}$$

For the lower level,

$$g_J = 2J + 1 = 4, \quad \frac{\varepsilon_J}{hc} = F(J) + G(\text{v}) = 124.914\ \text{cm}^{-1}.$$

The RRHO partition functions and the spectroscopic constants for NO are given by (Laurendeau, 2005):

$$B_e = 1.704\ \text{cm}^{-1}, \quad \omega_e = 1,904\ \text{cm}^{-1}, \quad g_{0,elec} = 4.$$

$$Z_{elec} = g_{0,elec} = 4, \quad \theta_{vib} = \frac{hc\omega_e}{k_B} = 2,740\ \text{K}, \quad Z_{vib} = \frac{1}{1-\exp(-\theta_{vib}/T)} = 1.020.$$

$$\theta_{rot} = B_0\left(\frac{hc}{k_B}\right) = 2.44\ \text{K}, \quad \frac{hc}{k_B} = 1.439\ \text{K/cm}^{-1}, \quad Z_{rot} = \frac{T}{\theta_{rot}} = 287.$$

Substituting into Eq. (6.107), we obtain

$$\frac{n_J}{n_{\text{NO}}} = \frac{4\exp[-(124.914\ \text{cm}^{-1})(1.439\ \text{K/cm}^{-1})/(700\ \text{K})]}{(4)(1.020)(287)} = 2.64\times10^{-3}, \quad (6.108)$$

and the lower-level number density is given by

$$n_J = \left(2.64\times10^{-3}\right)\left(1.035\times10^{22}\ \text{m}^{-3}\right) = 2.73\times10^{19}\ \text{m}^{-3}. \quad (6.109)$$

Now we can calculate $K_{J'J}$:

$$\begin{aligned} K_{J'J} &= \frac{g_{J'} n_J \lambda_{J'J}^2 A_{J'J}}{4 g_J} \\ &= \frac{2\left(2.73\times10^{19}\ \text{m}^{-3}\right)\left(226.89\times10^{-9}\ \text{m}\right)^2\left(4.7886\times10^{5}\ \text{s}^{-1}\right)}{4(4)} \\ &= 8.41\times10^{10}\ \text{m}^{-1}\ \text{s}^{-1}. \end{aligned} \quad (6.110)$$

Now we need to find the Doppler width. From Eq. (6.67),

$$\Delta\tilde{\nu}_D = \tilde{\nu}_0\sqrt{\frac{8\ln 2 k_B T}{m_{\text{NO}} c^2}}, \quad m_{\text{NO}} = (30\ \text{amu})\left(1.661\times10^{-27}\frac{\text{kg}}{\text{amu}}\right) = 4.98\times10^{-26}\ \text{kg}.$$

Therefore

$$\begin{aligned} \Delta\tilde{\nu}_D &= \left(44,074\ \text{cm}^{-1}\right)\sqrt{\frac{8\ln 2\left(1.381\times10^{-23}\ \text{J/K}\right)(700\ \text{K})}{\left(4.98\times10^{-26}\ \text{kg}\right)\left(2.998\times10^{8}\ \text{m/s}\right)^2}} \\ &= \left(44,074\ \text{cm}^{-1}\right)\left(3.46\times10^{-6}\right) = 0.153\ \text{cm}^{-1}. \end{aligned} \quad (6.111)$$

The collisional width of $\Delta\tilde{\nu}_C = 0.165\ \text{cm}^{-1}$ is given in the problem statement. Therefore the Voigt parameter a is given by

$$a = \sqrt{\ln 2}\frac{\Delta\tilde{\nu}_C}{\Delta\tilde{\nu}_D} = \sqrt{\ln 2}\frac{0.165}{0.153} = 0.90. \quad (6.112)$$

Voigt line shape factors $V(a,x)$ are listed in Appendix A9. For $x = 0$, $V(a,x) = V(0.90,0) = 0.457$. The value of the Voigt line shape function at line center is given by

$$g_V(\omega) = \sqrt{\frac{\ln 2}{\pi}} \frac{2}{\Delta\omega_D} V(a, x).$$
$$\Delta\omega_D = \Delta\tilde{\nu}_D(2\pi c) = \left(0.153\ \mathrm{cm}^{-1}\right)(2\pi)\left(2.998 \times 10^{10}\ \mathrm{cm/s}\right) = 2.88 \times 10^{10}\ \mathrm{s}^{-1}.$$
$$g_V(\omega) = \sqrt{\frac{\ln 2}{\pi}} \left(\frac{2}{2.88 \times 10^{10}\ \mathrm{s}^{-1}}\right)(0.456) = 1.49 \times 10^{-11}\ \mathrm{s}. \tag{6.113}$$

The absorption at line center is given by

$$\begin{aligned} A(\omega) &= 1 - \exp\left[-K_{J'J}\, g_V(\omega) L\right] \\ &= 1 - \exp\left[-\left(8.41 \times 10^{10}\ \mathrm{m}^{-1}\mathrm{s}^{-1}\right)\left(1.49 \times 10^{-11}\ \mathrm{s}\right)(0.2\ \mathrm{m})\right] = 0.222. \end{aligned} \tag{6.114}$$

When the laser is detuned from line center by 0.1 cm^{-1},

$$\tilde{\nu}_L - \tilde{\nu}_0 = 0.1\ \mathrm{cm}^{-1}, \quad x = \frac{2\sqrt{\ln 2}(\tilde{\nu}_L - \tilde{\nu}_0)}{\Delta\tilde{\nu}_D} = \frac{2\sqrt{\ln 2}(0.1\ \mathrm{cm}^{-1})}{(0.153\ \mathrm{cm}^{-1})} = 1.56. \tag{6.115}$$

From Appendix A9, we find that $V(a, x) = V(0.90, 1.56) = 0.2012$. The value of the Voigt function and the absorption for this detuning are given by

$$g_V(\omega) = \sqrt{\frac{\ln 2}{\pi}} \frac{2}{2.88 \times 10^{10}\ \mathrm{s}^{-1}}(0.2012) = 6.56 \times 10^{-12}\ \mathrm{s} \tag{6.116}$$

and

$$A(\omega) = 1 - \exp\left[-\left(8.41 \times 10^{10}\ \mathrm{m}^{-1}\mathrm{s}^{-1}\right)\left(6.56 \times 10^{-12}\ \mathrm{s}\right)(0.2\ \mathrm{m})\right] = 0.104. \tag{6.117}$$

7 Raman Spectroscopy

7.1 Introduction

Raman scattering spectroscopy is widely used in analytical chemistry, for structural analysis of materials and molecules and, most importantly for our purposes, as a gas-phase diagnostic technique. Raman scattering is a two-photon scattering process, and the mathematical treatment of Raman scattering is very similar to the mathematical treatment of two-photon absorption. Many of the molecules of interest for quantitative gas-phase spectroscopy are diatomic molecules with nondegenerate $^1\Sigma$ ground electronic levels, including N_2, CO, and H_2. In this chapter we will also discuss Raman scattering from O_2, which has a nondegenerate $^3\Sigma_g^-$ ground electronic level, to illustrate methods for dealing with the nondegenerate case. Raman spectroscopy is a very useful technique for detecting and analyzing the structure of polyatomic molecules, but that is beyond the scope of this book. Raman scattering for polyatomic molecules is discussed in detail in a number of excellent texts including Long (2002).

7.2 Interaction of Laser Radiation with Raman Transitions

The geometry for the Raman scattering process that will be analyzed in this chapter is shown in Figure 7.1. This is by far the most common Raman scattering geometry used for gas-phase diagnostics. The effects of the finite solid angle subtended by the collection lens for the geometry shown in Figure 7.1 and scattering geometries other than that shown in Figure 7.1 are analyzed by Long (2002). The polarization of a molecule due to Z-polarized incident light in the geometry shown in Figure 7.1 is given by

$$\mu_X = \alpha_{XZ} E_Z \qquad \mu_Y = \alpha_{YZ} E_Z \qquad \mu_Z = \alpha_{ZZ} E_Z, \tag{7.1}$$

where μ_X, μ_Y, and μ_Z are the induced molecular dipole moments amplitude (Cm), E_Z is the amplitude of the incident electric field $(\mathrm{J C^{-1} m^{-1}})$, and α_{XZ}, α_{YZ}, and α_{ZZ} are

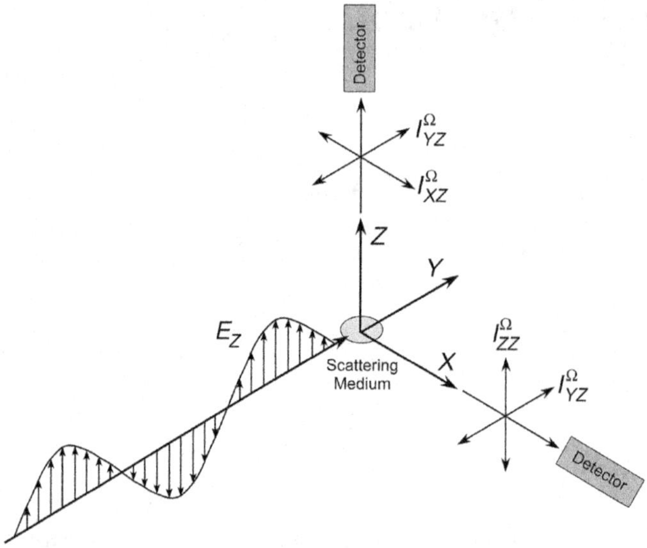

Figure 7.1 Typical geometry for laser excitation and detection of Raman scattering signals for gas-phase diagnostics.

components of the Raman polarizability tensor $\left(C^2 m^2 J^{-1}\right)$. The polarized molecule radiates Z-polarized radiation along the direction of the X-axis with an intensity I^{Ω}_{ZZ} $\left(W\, sr^{-1}\right)$ given by

$$I^{\Omega}_{ZZ} = \frac{\left(\omega_L + \omega_{ab}\right)^4 N_a}{32c^3\pi^2\varepsilon_0}\left|\alpha_{ZZ}(a,b)\right|^2 E_Z^2 = \frac{\left(\omega_L + \omega_{ab}\right)^4 N_a}{16c^4\pi^2\varepsilon_0^2}\left|\alpha_{ZZ}(a,b)\right|^2 I_L, \tag{7.2}$$

where a and b are the initial and final quantum states of the Raman transition, ω_L is the angular frequency of the incident laser field (s^{-1}), c is the speed of light ($m\,s^{-1}$), and ε_0 is the dielectric permittivity ($C^2 J^{-1}\, m^{-1}$). The angular frequency ω_{ab} is given by

$$\omega_{ab} = \left(\varepsilon_a - \varepsilon_b\right)/\hbar, \tag{7.3}$$

where ε_a and ε_b are the energies of quantum states a and b, respectively, and $\hbar$ is Planck's constant (J s). The polarized molecule radiates Y-polarized radiation along the direction of the X-axis with an intensity I^{Ω}_{YZ} given by

$$I^{\Omega}_{YZ} = \frac{\left(\omega_L + \omega_{ab}\right)^4 N_a}{32c^3\pi^2\varepsilon_0}\left|\alpha_{YZ}(a,b)\right|^2 E_Z^2 = \frac{\left(\omega_L + \omega_{ab}\right)^4 N_a}{16c^4\pi^2\varepsilon_0^2}\left|\alpha_{YZ}(a,b)\right|^2 I_L. \tag{7.4}$$

The components of the Raman polarizability tensor are given by

$$\alpha_{ZZ}(a,b) = \frac{1}{\hbar}\sum_c \left\{\frac{\mu_{Zbc}\mu_{Zca}}{\omega_{ca} - \omega_L - i\Gamma_c} + \frac{\mu_{Zbc}\mu_{Zca}}{\omega_{cb} + \omega_L + i\Gamma_c}\right\}. \tag{7.5}$$

$$\alpha_{YZ}(a,b) = \frac{1}{\hbar}\sum_{c}\left\{\frac{\mu_{Ybc}\mu_{Zca}}{\omega_{ca}-\omega_L-i\Gamma_c}+\frac{\mu_{Zbc}\mu_{Yca}}{\omega_{cb}+\omega_L+i\Gamma_c}\right\}. \tag{7.6}$$

In Eqs. (7.5) and (7.6), the electric dipole moment component μ_{Zbc} (Cm) is given by

$$\mu_{Zbc} = \langle T_B J_B M_b|\hat{\mu}_Z|T_C J_C M_c\rangle = \mu_R(T_B J_B, T_C J_C)(-1)^{J_B-M_b}\begin{pmatrix} J_B & 1 & J_C \\ -M_b & 0 & M_c \end{pmatrix}, \tag{7.7}$$

where T_b indicates all quantum numbers other than the rotational quantum number $J_B = J_b$ and orientation quantum number M_b for state b (and similarly for state a). The electric dipole moment operator $\hat{\mu}_Z$ is given by

$$\hat{\mu}_Z = \sum_k q_k \mathbf{Z}_k = \hat{\mathbf{Z}}\sum_k q_k Z_k, \tag{7.8}$$

where the summation is over all particles with charge q_k (C) and Z-coordinate Z_k (m) in the molecule. The reduced matrix element $\mu_R(T_B J_B, T_C J_C)$ is given by

$$\mu_R(T_B J_B, T_C J_C) = \pm\sqrt{\frac{A_{J_B J_C}(2J_B+1)3\pi\varepsilon_0\hbar c^3}{\omega^3_{J_B J_C}}}. \tag{7.9}$$

In Eq. (7.9), the negative sign applies for $J_C = J_B + 1$ and the positive sign applies for $J_C = J_B$ or $J_C = J_B - 1$. Tables of values of the $3j$ symbols $\begin{pmatrix} J_B & 1 & J_C \\ -M_b & q & M_c \end{pmatrix}$, where $q = -1,\ 0,\ +1$, can be found in, among many possible references, Edmonds (1960), Rose (1957), Weissbluth (1978), and Zare (1988), and in Appendix A3. In Eq. (7.6), the electric dipole moment component μ_{Ybc} is given by

$$\begin{aligned}\mu_{Ybc} &= \langle T_B J_B M_b|\hat{\mu}_Y|T_C J_C M_c\rangle \\ &= i\mu_R(T_B J_B, T_C J_C)(-1)^{J_B-M_b+1}\left\{\begin{pmatrix} J_B & 1 & J_C \\ -M_b & -1 & M_c \end{pmatrix} - \begin{pmatrix} J_B & 1 & J_C \\ -M_b & +1 & M_c \end{pmatrix}\right\}.\end{aligned} \tag{7.10}$$

If there are N_a molecules in the initial state then the total intensity I^{Ω}_{ZZ} will be the incoherent sum of the scattering from each of the molecules; Raman scattering from one molecule has a random phase with respect to scattering from other molecules in the probe volume. Also, the molecules in the probe volume will, for freely rotating molecules, have a random orientation, corresponding to an isotropic distribution over the orientation quantum numbers $M_a = -J_A, -J_A+1, \ldots, J_A-1, J_A$. For the ensemble of molecules N_A, an average over the different molecular orientations in the initial level must be performed to calculate the scattered intensity. For freely rotating molecules, this corresponds to a summation over the possible values of the initial states a (with projection number M_a) in energy level A, divided by the number of quantum states, $2J_A + 1$, in level A. For each state a, a summation over all possible final quantum states b (with projection number M_b) in energy level B must be performed. The final relations for the scattered intensities are given by

$$I_{ZZ}^{\Omega} = \frac{(\omega_L + \omega_{AB})^4 N_A}{32 c^3 \pi^2 \varepsilon_0} \left\langle |\alpha_{ZZ}(A,B)|^2 \right\rangle E_Z^2 = \frac{(\omega_L + \omega_{AB})^4 N_A}{16 c^4 \pi^2 \varepsilon_0^2} \left\langle |\alpha_{ZZ}(A,B)|^2 \right\rangle I_L, \quad (7.11)$$

$$I_{YZ}^{\Omega} = \frac{(\omega_L + \omega_{AB})^4 N_A}{32 c^3 \pi^2 \varepsilon_0} \left\langle |\alpha_{YZ}(A,B)|^2 \right\rangle E_Z^2 = \frac{(\omega_L + \omega_{AB})^4 N_A}{16 c^4 \pi^2 \varepsilon_0^2} \left\langle |\alpha_{YZ}(A,B)|^2 \right\rangle I_L, \quad (7.12)$$

where

$$\begin{aligned} \left\langle |\alpha_{ZZ}(A,B)|^2 \right\rangle &= \frac{1}{(2J_A+1)\hbar^2} \sum_{a,b} \left| \sum_c \left\{ \frac{\mu_{Zbc}\mu_{Zca}}{\omega_{ca} - \omega_L - i\Gamma_c} + \frac{\mu_{Zbc}\mu_{Zca}}{\omega_{cb} + \omega_L + i\Gamma_c} \right\} \right|^2 \\ &= \left(\frac{\partial \sigma}{\partial \Omega} \right)_{ZZ} \frac{16 \pi^2 \varepsilon_0^2 c^4}{(\omega_L - \omega_{BA})^4} \end{aligned} \quad (7.13)$$

$$\begin{aligned} \left\langle |\alpha_{YZ}(A,B)|^2 \right\rangle &= \frac{1}{(2J_A+1)\hbar^2} \sum_{a,b} \left| \sum_c \left\{ \frac{\mu_{Ybc}\mu_{Zca}}{\omega_{ca} - \omega_L - i\Gamma_c} + \frac{\mu_{Zbc}\mu_{Yca}}{\omega_{cb} + \omega_L + i\Gamma_c} \right\} \right|^2 \\ &= \left(\frac{\partial \sigma}{\partial \Omega} \right)_{YZ} \frac{16 \pi^2 \varepsilon_0^2 c^4}{(\omega_L - \omega_{BA})^4}. \end{aligned} \quad (7.14)$$

or for the general case

$$\begin{aligned} \left\langle |\alpha_{ij}(A,B)|^2 \right\rangle &= \frac{1}{(2J_A+1)\hbar^2} \sum_{a,b} \left| \sum_c \left\{ \frac{\mu_{ibc}\mu_{jca}}{\omega_{ca} - \omega_L - i\Gamma_c} + \frac{\mu_{jbc}\mu_{ica}}{\omega_{cb} + \omega_L + i\Gamma_c} \right\} \right|^2 \\ &= \left(\frac{\partial \sigma}{\partial \Omega} \right)_{ij} \frac{16 \pi^2 \varepsilon_0^2 c^4}{(\omega_L - \omega_{BA})^4}. \end{aligned} \quad (7.15)$$

7.3 Placzek Polarizability Theory

7.3.1 The Polarizability and Polarizability Operator Tensors

Although Eqs. (7.13) and (7.14) are accurate expressions, in most cases they are not very useful expressions. The reason for this is can be illustrated by referring to publications such as Lofthus and Krupenie (1977), an exhaustive review of the spectrum of N_2. The ground electronic level of N_2 is a nondegenerate $^1\Sigma_g^+$ level and is coupled by single-photon-allowed transitions to numerous $^1\Sigma_u^+$ and $^1\Pi_u$ excited electronic levels and by forbidden transitions to $^1\Sigma_g^+$, $^1\Sigma_u^-$, $^1\Pi_g$, and $^1\Delta_u$ excited electronic levels. For diatomic molecules like nitrogen, the summations over all intermediate states c indicated in Eqs. (7.13) and (7.14) is not possible to perform with any accuracy because the number of intermediate states is very large and the single-photon dipole moment matrix elements are not very well known, especially for intermediate states in highly excited electronic energy levels. Placzek polarizability theory is used to simplify the expression shown in Eqs. (7.13) and (7.14) so that the properties of the Raman scattering signal can be calculated in detail in terms of an experimentally determined Raman transition strength. As will be described later in the chapter, it is also possible to compute the Raman polarizability of molecules such as nitrogen using quantum chemistry methods that do not require detailed knowledge of the excited electronic levels.

The first step in the simplification of Eqs. (7.13) and (7.14) is to use the Born–Oppenheimer approximation so that the molecular wavefunction can be written as the product of the electronic, vibrational, and rotational wave functions, and the energy of the state can be written as the sum of the energies for the individual modes,

$$|\eta_c \text{v}_c J_c K_c M_c\rangle = |\eta_c\rangle|\text{v}_c\rangle|J_c K_c M_c\rangle \tag{7.16}$$

$$\varepsilon_c = \varepsilon_{ec} + \varepsilon_{\text{v}c} + \varepsilon_{Jc} \tag{7.17}$$

Similar relations also apply for states a and b. Using the notation of Eq. (7.10), we can rewrite Eq. (7.15) as

$$\left\langle \left|\alpha_{ij}(A,B)\right|^2 \right\rangle = \frac{1}{(2J_A+1)\hbar^2} \sum_{a,b} \left| \sum_c \left\{ \frac{\langle \eta_b \text{v}_b R_b | \hat{\mu}_i | \eta_c \text{v}_c R_c\rangle \langle \eta_c \text{v}_c R_c | \hat{\mu}_j | \eta_a \text{v}_a R_a\rangle}{\omega_{ca} - \omega_L - i\Gamma_c} + \frac{\langle \eta_b \text{v}_b R_b | \hat{\mu}_j | \eta_c \text{v}_c R_c\rangle \langle \eta_c \text{v}_c R_c | \hat{\mu}_i | \eta_a \text{v}_a R_a\rangle}{\omega_{cb} + \omega_L + i\Gamma_c} \right\} \right|^2, \tag{7.18}$$

where the symbol R_i is meant to imply the angular momentum quantum number set $J_i K_i M_i$ for the symmetric top wavefunction. Applying Eqs. (7.16) and (7.17) to Eq. (7.18), we obtain

$$\left\langle \left|\alpha_{ij}(A,B)\right|^2 \right\rangle = \frac{1}{(2J_A+1)\hbar^2} \sum_{a,b} \left| \sum_c \left\{ \frac{\langle R_b | \langle \text{v}_b | \langle \eta_b | \hat{\mu}_i | \eta_c\rangle | \text{v}_c\rangle | R_c\rangle \langle R_c | \langle \text{v}_c | \langle \eta_c | \hat{\mu}_j | \eta_a\rangle | \text{v}_a\rangle | R_a\rangle}{\omega_{eca} + \omega_{\text{v}ca} + \omega_{Jca} - \omega_L - i\Gamma_c} + \frac{\langle R_b | \langle \text{v}_b | \langle \eta_b | \hat{\mu}_j | \eta_c\rangle | \text{v}_c\rangle | R_c\rangle \langle R_c | \langle \text{v}_c | \langle \eta_c | \hat{\mu}_i | \eta_a\rangle | \text{v}_a\rangle | R_a\rangle}{\omega_{ecb} + \omega_{\text{v}cb} + \omega_{Jcb} + \omega_L + i\Gamma_c} \right\} \right|^2, \tag{7.19}$$

where

$$\omega_{ca} = \omega_{eca} + \omega_{\text{v}ca} + \omega_{Jca} = \hbar(\varepsilon_{ec} - \varepsilon_{ea}) + \hbar(\varepsilon_{\text{v}c} - \varepsilon_{\text{v}a}) + \hbar(\varepsilon_{Jc} - \varepsilon_{Ja}), \tag{7.20}$$

and similar relations apply for ω_{cb}. Now consider the evaluation of Eq. (7.20) for the case where both states a and b are in the ground electronic level ($\eta_a = \eta_b = \eta$), and the summation over states c excludes states in the ground electronic level. Further, we can assume that the frequency difference between the excited electronic level and the ground electronic level is much greater than any differences in frequency due to vibrational and rotational energy level differences,

$$\omega_{eca} \gg \omega_{\text{v}ca} + \omega_{Jca} \qquad \omega_{eca} = \omega_{ecb} \gg \omega_{\text{v}cb} + \omega_{Jcb}. \tag{7.21}$$

Under conditions where the laser frequency is far below electronic resonance, we can neglect the frequency differences due to vibrational and rotational energy level differences and also the population decay terms, so we can rewrite Eq. (7.19) as

$$\begin{aligned}\left\langle\left|\alpha_{ij}(A,B)\right|^2\right\rangle &= \frac{1}{(2J_A+1)\hbar^2}\sum_{a,b}\left|\sum_c\left\{\frac{\langle R_b|\langle v_b|\langle\eta|\hat{\mu}_i|\eta_c\rangle|v_c\rangle|R_c\rangle\langle R_c|\langle v_c|\langle\eta_c|\hat{\mu}_j|\eta\rangle|v_a\rangle|R_a\rangle}{\omega_{eca}-\omega_L}\right.\right.\\ &\left.\left.+\frac{\langle R_b|\langle v_b|\langle\eta|\hat{\mu}_j|\eta_c\rangle|v_c\rangle|R_c\rangle\langle R_c|\langle v_c|\langle\eta_c|\hat{\mu}_i|\eta\rangle|v_a\rangle|R_a\rangle}{\omega_{ecb}+\omega_L}\right\}\right|^2.\end{aligned} \tag{7.22}$$

We can now sum over all the vibrational and rotational levels in the excited electronic levels T_c and apply the closure relations

$$\sum_{R_c}|R_c\rangle\langle R_c| = \sum_{J_cK_cM_c}|J_cK_cM_c\rangle\langle J_cK_cM_c| = 1 \tag{7.23}$$

and

$$\sum_{v_c}|v_c\rangle\langle v_c| = 1. \tag{7.24}$$

After applying the closure relations to Eq. (7.22), we obtain

$$\begin{aligned}\left\langle\left|\alpha_{ij}(A,B)\right|^2\right\rangle &= \frac{1}{(2J_A+1)\hbar^2}\sum_{a,b}\left|\sum_{T_c}\left\{\frac{\langle R_b|\langle v_b|\langle\eta|\hat{\mu}_i|\eta_c\rangle\langle\eta_c|\hat{\mu}_j|\eta\rangle|v_a\rangle|R_a\rangle}{\omega_{eCA}-\omega_L}\right.\right.\\ &\left.\left.+\frac{\langle R_b|\langle v_b|\langle\eta|\hat{\mu}_j|\eta_c\rangle\langle\eta_c|\hat{\mu}_i|\eta\rangle|v_a\rangle|R_a\rangle}{\omega_{eCA}+\omega_L}\right\}\right|^2.\end{aligned} \tag{7.25}$$

At this point the polarizability operator $\hat{\alpha}_{ij}$ is defined:

$$\hat{\alpha}_{ij} = \frac{1}{\hbar}\sum_{T_c}\left[\frac{\hat{\mu}_i|\eta_c\rangle\langle\eta_c|\hat{\mu}_j}{\omega_{eCA}-\omega_L}+\frac{\hat{\mu}_j|\eta_c\rangle\langle\eta_c|\hat{\mu}_i}{\omega_{eCA}+\omega_L}\right], \tag{7.26}$$

and the polarizability for a transition from state a to state b can be written as

$$\alpha_{ij}(a,b) = \langle J_BK_BM_b|\langle v_B|\langle\eta|\hat{\alpha}_{ij}|\eta\rangle|v_A\rangle|J_AK_AM_a\rangle. \tag{7.27}$$

In Eq. (7.27), lower-case subscripts on J, K, and v have been replaced by upper-case subscripts because these values will be the same for each quantum state a or b in the energy levels A or B, respectively.

7.3.2 Cartesian and Irreducible Spherical Tensor Representations

The polarizability can also be expressed in terms of a space-fixed irreducible spherical coordinate system rather than a space-fixed Cartesian coordinate system,

$$\alpha_p^k(a,b) = \langle J_BK_BM_b|\langle v_B|\langle\eta|\hat{\alpha}_p^k|\eta\rangle|v_A\rangle|J_AK_AM_a\rangle. \tag{7.28}$$

The second-rank irreducible spherical tensor components p of ranks $k = 0, k = 1$, and $k = 2$ can be expressed in terms of Cartesian polarizability tensor components. The component index $p = -k, -k+1, \ldots, k-1, k$, that is, each of the irreducible spherical tensor components has $2k + 1$ terms. In terms of the Cartesian tensor components, the irreducible spherical tensor components are given by

$$\alpha_0^0 = -\frac{1}{\sqrt{3}}(\alpha_{XX} + \alpha_{YY} + \alpha_{ZZ}). \tag{7.29}$$

$$\alpha_{-1}^1 = -\frac{1}{2}(\alpha_{XZ} - \alpha_{ZX}) + \frac{i}{2}(\alpha_{YZ} - \alpha_{ZY}). \tag{7.30}$$

$$\alpha_0^1 = \frac{i}{\sqrt{2}}(\alpha_{XY} - \alpha_{YX}). \tag{7.31}$$

$$\alpha_1^1 = -\frac{1}{2}(\alpha_{XZ} - \alpha_{ZX}) - \frac{i}{2}(\alpha_{YZ} - \alpha_{ZY}). \tag{7.32}$$

$$\alpha_{-2}^2 = \frac{1}{2}(\alpha_{XX} - \alpha_{YY}) - \frac{i}{2}(\alpha_{XY} + \alpha_{YX}). \tag{7.33}$$

$$\alpha_{-1}^2 = \frac{1}{2}(\alpha_{XZ} + \alpha_{ZX}) - \frac{i}{2}(\alpha_{YZ} + \alpha_{ZY}). \tag{7.34}$$

$$\alpha_0^2 = -\frac{1}{\sqrt{6}}(\alpha_{XX} + \alpha_{YY}) + \frac{2}{\sqrt{6}}\alpha_{ZZ}. \tag{7.35}$$

$$\alpha_1^2 = -\frac{1}{2}(\alpha_{XZ} + \alpha_{ZX}) - \frac{i}{2}(\alpha_{YZ} + \alpha_{ZY}). \tag{7.36}$$

$$\alpha_2^2 = \frac{1}{2}(\alpha_{XX} - \alpha_{YY}) + \frac{i}{2}(\alpha_{XY} + \alpha_{YX}). \tag{7.37}$$

In Eqs. (7.29)–(7.37), the uppercase Cartesian coordinates X, Y, Z indicate a space-fixed frame of reference. The lowercase Cartesian coordinates x, y, z indicate a molecule-fixed frame of reference. We can also write the Cartesian components of the polarizability tensor in terms of the irreducible spherical tensor components:

$$\alpha_{XX} = -\frac{1}{\sqrt{3}}\alpha_0^0 + \frac{1}{2}\alpha_2^2 - \frac{1}{\sqrt{6}}\alpha_0^2 + \frac{1}{2}\alpha_{-2}^2. \tag{7.38}$$

$$\alpha_{YY} = -\frac{1}{\sqrt{3}}\alpha_0^0 - \frac{1}{2}\alpha_2^2 - \frac{1}{\sqrt{6}}\alpha_0^2 - \frac{1}{2}\alpha_{-2}^2. \tag{7.39}$$

$$\alpha_{ZZ} = -\frac{1}{\sqrt{3}}\alpha_0^0 + \frac{2}{\sqrt{6}}\alpha_0^2. \tag{7.40}$$

$$\alpha_{XY} = -\frac{i}{\sqrt{2}}\alpha_0^1 - \frac{i}{2}\alpha_2^2 + \frac{i}{2}\alpha_{-2}^2. \tag{7.41}$$

$$\alpha_{YX} = \frac{i}{\sqrt{2}}\alpha_0^1 - \frac{i}{2}\alpha_2^2 + \frac{i}{2}\alpha_{-2}^2. \tag{7.42}$$

$$\alpha_{XZ} = -\frac{1}{2}\alpha_1^1 - \frac{1}{2}\alpha_{-1}^1 - \frac{1}{2}\alpha_1^2 + \frac{1}{2}\alpha_{-1}^2. \tag{7.43}$$

$$\alpha_{ZX} = \frac{1}{2}\alpha_1^1 + \frac{1}{2}\alpha_{-1}^1 - \frac{1}{2}\alpha_1^2 + \frac{1}{2}\alpha_{-1}^2. \tag{7.44}$$

$$\alpha_{YZ} = \frac{i}{2}\alpha_1^1 - \frac{i}{2}\alpha_{-1}^1 + \frac{i}{2}\alpha_1^2 + \frac{i}{2}\alpha_{-1}^2. \tag{7.45}$$

$$\alpha_{ZY} = -\frac{i}{2}\alpha_1^1 + \frac{i}{2}\alpha_{-1}^1 + \frac{i}{2}\alpha_1^2 + \frac{i}{2}\alpha_{-1}^2. \tag{7.46}$$

7.3.3 Relation between Space-Fixed and Molecule-Fixed Tensor Components

One of the major advantages of using the irreducible spherical tensor components is that the spherical tensor components of rank k transform under rotations in the same manner as spherical harmonics of rank k; this is in fact the definition of an irreducible spherical tensor component. Thus, there is no mixing of different irreducible spherical tensor components as a result of the rotation operation. Consequently, we can write the space-fixed irreducible spherical tensor component of the polarizability operator $\hat{\alpha}_p^k$ in terms of the molecule-fixed irreducible spherical tensor operator component $\hat{\alpha}_q^k$:

$$\hat{\alpha}_p^k = \sum_q \hat{\alpha}_q^k D_{pq}^{k*}(\omega), \tag{7.47}$$

where $D_{pq}^{k*}(\omega)$ is a rotation matrix element, and the molecule-fixed coordinate system is attained via a rotation of the space-fixed coordinate system through the Euler angles $\omega = (\phi, \theta, \chi)$. The space-fixed polarizability irreducible spherical tensor component α_p^k can also be expressed in terms of the molecule-fixed irreducible tensor component α_q^k,

$$\alpha_p^k = \sum_q \alpha_q^k D_{pq}^{k*}(\omega). \tag{7.48}$$

A useful expression for the orientation average of the product of two space-fixed polarizability irreducible spherical tensor components will now be developed. The product of two space-fixed tensor components α_p^k and $\alpha_{p'}^{k'*}$ is thus given by

$$\alpha_p^k \alpha_{p'}^{k'*} = \sum_{q'}\sum_q \alpha_q^k \alpha_{q'}^{k'*} D_{pq}^{k*}(\omega) D_{p'q'}^{k'}(\omega). \tag{7.49}$$

The orientation average of the product is found by integrating over all possible orientations of the molecule,

$$\left\langle \alpha_p^k \alpha_{p'}^{k'*} \right\rangle = \frac{\sum_{q'}\sum_q \alpha_q^k \alpha_{q'}^{k'*} \int_\omega D_{pq}^{k*}(\omega) D_{p'q'}^{k'}(\omega) d\omega}{\int_\omega d\omega}. \tag{7.50}$$

But from Chapter 3 we know that

$$\int_{\omega} D^{k*}_{pq}(\omega) D^{k'}_{p'q'}(\omega) d\omega = \frac{8\pi^2}{2k+1} \delta_{kk'} \delta_{pp'} \delta_{qq'}. \tag{7.51}$$

Therefore, we can write

$$\left\langle \alpha^{k}_{p} \alpha^{k'*}_{p'} \right\rangle = \frac{\delta_{kk'} \delta_{pp'}}{2k+1} \sum_{q} \alpha^{k}_{q} \alpha^{k*}_{q}. \tag{7.52}$$

$$\left\langle \alpha^{k}_{p} \alpha^{k*}_{p} \right\rangle = \left| \alpha^{k}_{p} \right|^2 = \frac{1}{2k+1} \sum_{q} \left| \alpha^{k}_{q} \right|^2. \tag{7.53}$$

It is important to note that the orientation average of the square of the tensor component p for a given rank k will be the same for each component p; thus, for example,

$$\left\langle \alpha^{2}_{2} \alpha^{2*}_{2} \right\rangle = \left| \alpha^{2}_{2} \right|^2 = \left\langle \alpha^{2}_{1} \alpha^{2*}_{1} \right\rangle = \left| \alpha^{2}_{1} \right|^2 = \cdots \left\langle \alpha^{2}_{-2} \alpha^{2*}_{-2} \right\rangle = \left| \alpha^{2}_{-2} \right|^2 = \frac{1}{5} \sum_{q} \left| \alpha^{2}_{q} \right|^2. \tag{7.54}$$

If we now sum over the components of α^{k}_{p}, we obtain

$$\sum_{p} \left| \alpha^{k}_{p} \right|^2 = \sum_{q} \left| \alpha^{k}_{q} \right|^2. \tag{7.55}$$

The summation $\sum_{p} \left| \alpha^{k}_{p} \right|^2$ is obviously independent of the coordinate system.

We can also determine the orientation-averaged values of the squares of the Cartesian tensor components in terms of the space-fixed spherical tensor components. For example,[1]

$$\begin{aligned} \left\langle \alpha^{2}_{ZZ} \right\rangle &= \left\langle \left(-\frac{1}{\sqrt{3}} \alpha^{0}_{p=0} + \frac{2}{\sqrt{6}} \alpha^{2}_{p=0} \right) \left(-\frac{1}{\sqrt{3}} \alpha^{0}_{p=0} + \frac{2}{\sqrt{6}} \alpha^{2}_{p=0} \right)^{*} \right\rangle \\ &= \left\langle \frac{1}{3} \left| \alpha^{0}_{p=0} \right|^2 + \frac{2}{3} \left| \alpha^{2}_{p=0} \right|^2 - \frac{2}{3\sqrt{2}} \alpha^{0}_{p=0} \alpha^{2*}_{p=0} - \frac{2}{3\sqrt{2}} \alpha^{0*}_{p=0} \alpha^{2*}_{p=0} \right\rangle \end{aligned} \tag{7.56}$$

The orientation average of the last two terms will be zero according to Eq. (7.52). Therefore

$$\left\langle \alpha^{2}_{ZZ} \right\rangle = \frac{1}{3} \left| \alpha^{0}_{p=0} \right|^2 + \frac{2}{3} \left| \alpha^{2}_{p=0} \right|. \tag{7.57}$$

[1] As a reminder, in Eq. (7.56) the symbol p, followed by the numerical value of p, is included in the subscript of α to make it clear that this is a spherical tensor component in the space-fixed coordinate system. The symbol q in the subscript, followed by the numerical value of q, will be used to indicate a spherical tensor component in the molecule-fixed reference system.

Similarly we can show that

$$\begin{aligned}\langle\alpha_{ZX}^2\rangle &= \left\langle\left(\frac{1}{2}\alpha_{p=1}^1+\frac{1}{2}\alpha_{p=-1}^1-\frac{1}{2}\alpha_{p=1}^2+\frac{1}{2}\alpha_{p=-1}^2\right)\left(\frac{1}{2}\alpha_{p=1}^1+\frac{1}{2}\alpha_{p=-1}^1-\frac{1}{2}\alpha_{p=1}^2+\frac{1}{2}\alpha_{p=-1}^2\right)^*\right\rangle\\ &=\frac{1}{4}\left\langle\left|\alpha_{p=1}^1\right|^2\right\rangle+\frac{1}{4}\left\langle\left|\alpha_{p=-1}^1\right|^2\right\rangle+\frac{1}{4}\left\langle\left|\alpha_{p=1}^2\right|^2\right\rangle+\frac{1}{4}\left\langle\left|\alpha_{p=-1}^2\right|^2\right\rangle\\ &=\frac{1}{2}\left\langle\left|\alpha_{p=1}^1\right|^2\right\rangle+\frac{1}{2}\left\langle\left|\alpha_{p=1}^2\right|^2\right\rangle.\end{aligned} \tag{7.58}$$

$$\begin{aligned}\langle\alpha_{YZ}^2\rangle &= \left\langle\left(\frac{i}{2}\alpha_{p=1}^1-\frac{i}{2}\alpha_{p=-1}^1+\frac{i}{2}\alpha_{p=1}^2+\frac{i}{2}\alpha_{p=-1}^2\right)\left(-\frac{i}{2}\alpha_{p=1}^1+\frac{i}{2}\alpha_{p=-1}^1-\frac{i}{2}\alpha_{p=1}^2-\frac{i}{2}\alpha_{p=-1}^2\right)^*\right\rangle\\ &=\frac{1}{4}\left\langle\left|\alpha_{p=1}^1\right|^2\right\rangle+\frac{1}{4}\left\langle\left|\alpha_{p=-1}^1\right|^2\right\rangle+\frac{1}{4}\left\langle\left|\alpha_{p=1}^2\right|^2\right\rangle+\frac{1}{4}\left\langle\left|\alpha_{p=-1}^2\right|^2\right\rangle\\ &=\frac{1}{2}\left\langle\left|\alpha_{p=1}^1\right|^2\right\rangle+\frac{1}{2}\left\langle\left|\alpha_{p=1}^2\right|^2\right\rangle=\langle\alpha_{ZX}^2\rangle.\end{aligned} \tag{7.59}$$

7.3.4 The Polarizability and Derived Polarizability Tensors

Expressions for the space-fixed polarizability irreducible tensor components α_p^k using the molecule-fixed irreducible tensor operator components $\hat{\alpha}_q^k$ will now be developed. The molecule-fixed polarizability operator components are functions only of the normal vibrational coordinates Q of the molecule and are independent of molecular orientation. Consequently, the molecule-fixed polarizability does not depend on the quantum numbers associated with the angular momentum of the molecule, and the resulting expressions for α_p^k can be simplified considerably. Substituting Eq. (7.47) into Eq. (7.28), we obtain

$$\alpha_p^k(a,b)=\langle J_BK_BM_b|\langle \mathrm{v}_B|\langle\eta|\sum_q\hat{\alpha}_q^kD_{pq}^{k^*}(\omega)|\eta\rangle|\mathrm{v}_A\rangle|J_AK_AM_a\rangle. \tag{7.60}$$

The molecule-fixed polarizability components do not depend on the orientation of the molecule or the angular coordinates ω. Thus, we can write

$$\alpha_p^k(a,b)=\sum_q\langle \mathrm{v}_B|\langle\eta|\hat{\alpha}_q^k|\eta\rangle|\mathrm{v}_A\rangle\langle J_BK_BM_b|D_{pq}^{k^*}(\omega)|J_AK_AM_a\rangle. \tag{7.61}$$

The rotational wave functions for a symmetric top molecule can be expressed as rotational matrix elements as discussed in Chapter 3. Thus, we can write

$$\begin{aligned}&\langle J_BK_BM_b|D_{pq}^{k^*}(\omega)|J_AK_AM_a\rangle\\ &=\frac{\sqrt{(2J_B+1)(2J_A+1)}}{8\pi^2}\int_\omega D_{M_bK_B}^{J_B}(\omega)D_{pq}^{k^*}(\omega)D_{M_aK_A}^{J_A*}(\omega)\,d\omega.\end{aligned} \tag{7.62}$$

But we know that

$$D_{pq}^{k*}(\omega) = (-1)^{p-q} D_{-p,-q}^{k}(\omega) \tag{7.63}$$

and

$$D_{M_a K_A}^{J_A *}(\omega) = (-1)^{M_a - K_A} D_{-M_a,-K_A}^{J_A}(\omega). \tag{7.64}$$

Rewriting the integral in Eq. (7.62) using Eqs. (7.63) and (7.64), we obtain

$$\begin{aligned} &\int_\omega D_{M_b K_B}^{J_B}(\omega) D_{pq}^{k*}(\omega) D_{M_a K_A}^{J_A *}(\omega)\, d\omega \\ &= (-1)^{p-q}(-1)^{M_a - K_A} \int_\omega D_{M_b K_B}^{J_B}(\omega) D_{-p,-q}^{k}(\omega) D_{-M_a,-K_A}^{J_A}(\omega)\, d\omega. \end{aligned} \tag{7.65}$$

The integral term can now be evaluated using

$$\begin{aligned} &\int_\omega D_{M_b K_B}^{J_B}(\omega) D_{-p,-q}^{k}(\omega) D_{-M_a,-K_A}^{J_A}(\omega) d\omega = 8\pi^2 \begin{pmatrix} J_B & k & J_A \\ M_b & -p & -M_a \end{pmatrix} \begin{pmatrix} J_B & k & J_A \\ K_B & -q & -K_A \end{pmatrix} \\ &= (-1)^{2(J_B + k + J_A)} 8\pi^2 \begin{pmatrix} J_B & k & J_A \\ -M_b & p & M_a \end{pmatrix} \begin{pmatrix} J_B & k & J_A \\ -K_B & q & K_A \end{pmatrix} \\ &= 8\pi^2 \begin{pmatrix} J_B & k & J_A \\ -M_b & p & M_a \end{pmatrix} \begin{pmatrix} J_B & k & J_A \\ -K_B & q & K_A \end{pmatrix}. \end{aligned} \tag{7.66}$$

The *3j* symbols are subject to the restriction that they are nonzero only when $|J_B - J_A| \leq k$; Raman transitions will therefore differ from the single-photon transitions discussed earlier because the rank of the polarizability $k = 2$. Substituting Eqs. (7.65) and (7.66) into Eq. (7.62), we obtain

$$\begin{aligned} &\langle J_B K_B M_b | D_{pq}^{k*}(\omega) | J_A K_A M_b \rangle \\ &= (-1)^{p-q}(-1)^{M_a - K_A} \sqrt{(2J_B + 1)(2J_A + 1)} \begin{pmatrix} J_B & k & J_A \\ -M_b & p & M_a \end{pmatrix} \begin{pmatrix} J_B & k & J_A \\ -K_B & q & K_B \end{pmatrix}. \end{aligned} \tag{7.67}$$

For diatomic and linear polyatomic molecules with a $^1\Sigma$ ground electronic level, $K_A = \Omega_A = K_B = \Omega_B = 0$; this also implies that J_A, J_B, M_a, and M_b are integers. The value of a *3j* symbol will be zero unless the sum of the arguments in the lower row is equal to zero; consequently, the *3j* symbol involving K_A and K_B in Eq. (7.67) will be zero unless $q = 0$. Consequently, for $K_A = K_B = 0$, $J = J_A$, $M = M_a$, $J' = J_B$, and $M' = M_b$, Eq. (5.15) reduces to

$$\begin{aligned} &\langle J' 0 M' | D_{pq}^{k*}(\omega) | J 0 M \rangle \\ &= (-1)^{-M'} \sqrt{(2J' + 1)(2J + 1)} \begin{pmatrix} J' & k & J \\ -M' & p & M \end{pmatrix} \begin{pmatrix} J' & k & J \\ 0 & 0 & 0 \end{pmatrix}. \end{aligned} \tag{7.68}$$

In Eq. (7.68) we have also used the relation that the first *3j* symbol is zero unless $p + M = -M'$. Substituting Eq. (7.68) into Eq. (7.61), we obtain

$$\alpha_p^k(a,b) = \sum_q \langle \mathrm{v}'|\langle\eta|\hat{\alpha}_q^k|\eta\rangle|\mathrm{v}\rangle\langle J'\,0M'|D_{pq}^{k*}(\omega)|J\,0M\rangle$$
$$= (-1)^{-M'}\langle \mathrm{v}'|\langle\eta|\hat{\alpha}_{q=0}^k|\eta\rangle|\mathrm{v}\rangle\sqrt{(2J'+1)(2J+1)}\begin{pmatrix} J' & k & J \\ -M' & p & M\end{pmatrix}\begin{pmatrix} J' & k & J \\ 0 & 0 & 0\end{pmatrix}, \tag{7.69}$$

where $\mathrm{v} = \mathrm{v}_A$, $\mathrm{v}' = \mathrm{v}_B$, and the subscript q in the polarizability term $\hat{\alpha}_q^k$ is a reminder that it is evaluated in a molecule-fixed reference frame. The orientation-averaged tensor element $\left\langle |\alpha_p^k(A,B)|^2 \right\rangle$ is found by summing $|\alpha_p^k(a,b)|^2$ over all possible values of M and M' and then dividing by the number of quantum states, $2J+1$, in level A:

$$\left\langle \left|\alpha_p^k(A,B)\right|^2 \right\rangle = (2J'+1)\langle \mathrm{v}'|\langle\eta|\hat{\alpha}_{q=0}^k|\eta\rangle|\mathrm{v}\rangle^2\begin{pmatrix} J' & k & J \\ 0 & 0 & 0\end{pmatrix}^2\sum_M\sum_{M'}\begin{pmatrix} J' & k & J \\ -M' & p & M\end{pmatrix}^2$$
$$= \frac{(2J'+1)}{2k+1}\langle \mathrm{v}'|\langle\eta|\hat{\alpha}_{q=0}^k|\eta\rangle|\mathrm{v}\rangle^2\begin{pmatrix} J' & k & J \\ 0 & 0 & 0\end{pmatrix}^2. \tag{7.70}$$

The molecule-fixed polarizability tensor operators can be expanded in a Taylor series about the equilibrium value of zero for the mass-weighted normal coordinate Q,

$$\hat{\alpha}_q^k = \left(\hat{\alpha}_q^k\right)_0 + \left(\frac{\partial\hat{\alpha}_q^k}{\partial Q}\right)_0 Q + \frac{1}{2}\left(\frac{\partial^2\hat{\alpha}_q^k}{\partial Q^2}\right)_0 Q^2 + \cdots. \tag{7.71}$$

For a diatomic molecule, $Q = (r - r_e)$, where r is the internuclear separation, and r_e is the equilibrium internuclear separation. Keeping only the first and second terms on the right-hand side and substituting them into the vibrational polarizability matrix element expression, we obtain

$$\langle \mathrm{v}'|\langle\eta|\hat{\alpha}_q^k|\eta\rangle|\mathrm{v}\rangle = \langle \mathrm{v}'|\langle\eta|\left(\hat{\alpha}_{q=0}^k\right)_0|\eta\rangle|\mathrm{v}\rangle + \langle \mathrm{v}'|\langle\eta|\left(\hat{\alpha}_{q=0}^k/\partial Q\right)_0 Q|\eta\rangle|\mathrm{v}\rangle$$
$$= \langle\eta|\left(\hat{\alpha}_{q=0}^k\right)_0|\eta\rangle\langle \mathrm{v}'\,|\,\mathrm{v}\rangle + \langle\eta|\left(\hat{\alpha}_{q=0}^k/\partial Q_0\right)_0|\eta\rangle\langle \mathrm{v}'|Q|\mathrm{v}\rangle. \tag{7.72}$$

If $\mathrm{v}' = \mathrm{v}$, then

$$\langle \mathrm{v}'|\langle\eta|\hat{\alpha}_{q=0}^k|\eta\rangle|\mathrm{v}\rangle = \langle\eta|\left(\hat{\alpha}_{q=0}^k\right)_0|\eta\rangle\langle \mathrm{v}\,|\,\mathrm{v}\rangle = \langle\eta|\left(\hat{\alpha}_{q=0}^k\right)_0|\eta\rangle \tag{7.73}$$

because $\langle \mathrm{v}\,|\,\mathrm{v}\rangle = 1$ and

$$\langle \mathrm{v}|Q|\mathrm{v}\rangle = \langle Q\rangle = 0 \tag{7.74}$$

If $\mathrm{v}' = \mathrm{v} + 1$, then

$$\langle \mathrm{v}'|\langle\eta|\hat{\alpha}_{q=0}^{\,k}|\eta\rangle|\mathrm{v}\rangle = \langle\eta|\left(\partial\hat{\alpha}_{q=0}^k/\partial Q\right)_0|\eta\rangle\langle \mathrm{v}'|Q|\mathrm{v}\rangle$$
$$= \langle\eta|\left(\partial\hat{\alpha}_{q=0}^k/\partial Q\right)_0|\eta\rangle(\mathrm{v}+1)^{1/2}\left(\frac{\hbar}{2\mu_R\,\omega_R}\right)^{1/2}, \tag{7.75}$$

where μ_R is the reduced mass of the molecule, and ω_R is the angular frequency of the vibrational mode. This term will be used to calculate polarizabilities for Stokes vibrational Raman scattering. If $\mathrm{v}' = \mathrm{v} - 1$ then

$$\langle \mathrm{v}' | \langle \eta | \hat{\alpha}_q^k | \eta \rangle | \mathrm{v} \rangle = \langle \eta | \left(\partial \hat{\alpha}_{q=0}^k / \partial Q \right)_0 | \eta \rangle \mathrm{v}^{1/2} \left(\frac{\hbar}{2\mu_R \omega_R} \right)^{1/2}. \tag{7.76}$$

This term will be used to calculate polarizabilities for anti-Stokes vibrational Raman scattering.

The polarizability tensor components can be written as the sum of the equilibrium and derived polarizability tensor components,

$$\begin{aligned} \alpha_p^k(a,b) &= (\alpha_0)_p^k(a,b) + (\alpha')_p^k(a,b) \\ &= (-1)^{-M'} \langle \eta | \left(\hat{\alpha}_{q=0}^k \right)_0 | \eta \rangle \langle \mathrm{v}' \,|\, \mathrm{v} \rangle \sqrt{(2J'+1)(2J+1)} \begin{pmatrix} J' & k & J \\ -M' & p & M \end{pmatrix} \begin{pmatrix} J' & k & J \\ 0 & 0 & 0 \end{pmatrix} \\ &\quad + (-1)^{-M'} \langle \eta | \left(\partial \hat{\alpha}_{q=0}^k / \partial Q \right)_0 | \eta \rangle \langle \mathrm{v}' | Q | \mathrm{v} \rangle \sqrt{(2J'+1)(2J+1)} \\ &\quad \begin{pmatrix} J' & k & J \\ -M' & p & M \end{pmatrix} \begin{pmatrix} J' & k & J \\ 0 & 0 & 0 \end{pmatrix}. \end{aligned} \tag{7.77}$$

For $\mathrm{v}' = \mathrm{v}$, Eq. (7.77) reduces to

$$\begin{aligned} \alpha_p^k(a,b) &= (\alpha_0)_p^k(a,b) \\ &= (-1)^{-M_b} \langle \eta | \left(\hat{\alpha}_{q0}^k \right)_0 | \eta \rangle \sqrt{(2J'+1)(2J+1)} \begin{pmatrix} J' & k & J \\ -M' & p & M \end{pmatrix} \begin{pmatrix} J' & k & J \\ 0 & 0 & 0 \end{pmatrix}. \end{aligned} \tag{7.78}$$

At this point the isotropic and anisotropic invariants a_0^2 and γ_0^2 are defined:

$$a_0^2 = \frac{1}{3} \left| \langle \eta | \left(\hat{\alpha}_{q=0}^0 \right)_0 | \eta \rangle \right|^2 = \frac{1}{9} \left| \langle \eta | \alpha_{xx} + \alpha_{yy} + \alpha_{zz} | \eta \rangle \right|^2 = \frac{1}{9} \left| \langle \eta | 2\alpha_{xx} + \alpha_{zz} | \eta \rangle \right|^2, \tag{7.79}$$

$$\gamma_0^2 = \frac{3}{2} \left| \langle \eta | \left(\hat{\alpha}_{q=0}^2 \right)_0 | \eta \rangle \right|^2 = \frac{3}{2} \left| \langle \eta | -\frac{1}{\sqrt{6}} (\alpha_{xx} + \alpha_{yy}) + \frac{2}{\sqrt{6}} \alpha_{zz} | \eta \rangle \right|^2 = \left| \langle \eta | \alpha_{zz} - \alpha_{xx} | \eta \rangle \right|^2, \tag{7.80}$$

where the molecule-fixed reference frame (MFRF) forms of Eqs. (7.29) and (7.35) have been substituted into Eqs. (7.79) and (7.80), respectively. The antisymmetric anisotropy term $(\alpha_0)_{q=0}^1$ will be negligible except in the case of resonance Raman (Long, 2002). The orientation-averaged squares of the Cartesian tensor components $\alpha_{ZZ}(A,B)$ and $\alpha_{ZY}(A,B)$ are found using Eqs. (7.57) and (7.58):

$$\left\langle |\alpha_{ZZ}(A,B)|^2 \right\rangle = \frac{1}{3}\left\langle \left|\alpha^0_{p=0}(A,B)\right|^2 \right\rangle + \frac{2}{3}\left\langle \left|\alpha^2_{p=0}(A,B)\right|^2 \right\rangle$$
$$= (2J'+1)\begin{pmatrix} J' & 0 & J \\ 0 & 0 & 0 \end{pmatrix}^2 \left(\frac{1}{3}\left|\langle\eta|\left(\hat{\alpha}^0_{q=0}\right)_0|\eta\rangle\right|^2\right)$$
$$+\frac{(2J'+1)}{5}\begin{pmatrix} J' & 2 & J \\ 0 & 0 & 0 \end{pmatrix}^2 \left(\frac{2}{3}\left|\langle\eta|\left(\hat{\alpha}^2_{q=0}\right)_0|\eta\rangle\right|^2\right) \quad (7.81)$$
$$= (a_0^2)\delta_{JJ'} + (2J'+1)\begin{pmatrix} J' & 2 & J \\ 0 & 0 & 0 \end{pmatrix}^2 \left(\frac{4}{45}\gamma_0^2\right).$$

$$\left\langle |\alpha_{ZY}(A,B)|^2 \right\rangle = \frac{1}{4}\left\langle \left|\alpha^2_{p=1}(A,B)\right|^2 \right\rangle + \frac{1}{4}\left\langle \left|\alpha^2_{p=-1}(A,B)\right|^2 \right\rangle = \frac{1}{2}\left\langle \left|\alpha^2_{p=1}(A,B)\right|^2 \right\rangle$$
$$= \frac{(2J'+1)}{10}\begin{pmatrix} J' & 2 & J \\ 0 & 0 & 0 \end{pmatrix}^2 \left|\langle\eta|\left(\hat{\alpha}^2_{q=0}\right)_0|\eta\rangle\right|^2 = \frac{(2J'+1)}{15}\begin{pmatrix} J' & 2 & J \\ 0 & 0 & 0 \end{pmatrix}^2 \gamma_0^2. \quad (7.82)$$

The $3j$ symbol $\begin{pmatrix} J' & k & J \\ 0 & 0 & 0 \end{pmatrix}$ will be nonzero only when $J + J' + k$ is an even integer and $k \geq |J - J'|$. The $3j$ symbol $\begin{pmatrix} J' & k & J \\ 0 & 0 & 0 \end{pmatrix}$ is evaluated using the Eq. (A3.5) formula from Appendix A3 (also from Edmonds, 1960),

$$\begin{pmatrix} J' & k & J \\ 0 & 0 & 0 \end{pmatrix}$$
$$= (-1)^\beta \sqrt{\frac{(J'+k-J)!(J+J'-k)!(k+J-J')!}{(2\beta+1)!}}\frac{\beta!}{(\beta-J)!(\beta-k)!(\beta-J')!}, \quad (7.83)$$

where

$$\beta = \frac{J'+k+J}{2}. \quad (7.84)$$

The Placzek–Teller coefficient is defined by the formula

$$b^{(k)}_{J'J} = (2J'+1)\begin{pmatrix} J' & k & J \\ 0 & 0 & 0 \end{pmatrix}^2. \quad (7.85)$$

Substituting Eq. (7.85) into Eqs. (7.81) and (7.82), we obtain

$$\left\langle |\alpha_{ZZ}(A,B)|^2 \right\rangle = \left(a_0^2\delta_{J'J} + \frac{4}{45}b^{(2)}_{J'J}\gamma_0^2\right), \quad (7.86)$$

$$\left\langle |\alpha_{YZ}(A,B)|^2 \right\rangle = \left(\frac{1}{15}b^{(2)}_{J'J}\gamma_0^2\right), \quad (7.87)$$

where

$$\begin{pmatrix} J' & 0 & J \\ 0 & 0 & 0 \end{pmatrix}^2 = \frac{1}{2J'+1}\delta_{J'J}. \tag{7.88}$$

The Placzek–Teller coefficients $b^{(2)}_{J'J}$ are evaluated using Eq. (A3.5) in Appendix A3 and are listed below:

$$b^{(2)}_{J+2,J} = \frac{3(J+2)(J+1)}{2(2J+3)(2J+1)}. \tag{7.89}$$

$$b^{(2)}_{J,J} = \frac{J(J+1)}{(2J+3)(2J-1)}. \tag{7.90}$$

$$b^{(2)}_{J-2,J} = \frac{3J(J-1)}{2(2J+1)(2J-1)}. \tag{7.91}$$

The transitions with $\Delta J = 0$ are at zero frequency; this is essentially Rayleigh scattering. The pure rotational Raman O- and S-branch-transitions for a $^1\Sigma$ molecule are depicted in Figure 7.2. The O- and S-branch-transitions have $\Delta J = J' - J = -2$ and $+2$, respectively.

The derivation of the orientation-averaged squares of the polarizability tensor components for transitions with $\mathrm{v}' = \mathrm{v} \pm 1$ is very similar to the analysis for $\mathrm{v}' = \mathrm{v}$. The invariants a' and γ' are defined by the relations

$$(a')^2 = \frac{1}{3}\left|(\alpha')^0_{q=0}\right|^2 = \frac{1}{9}\left|\alpha'_{zz} + 2\alpha'_{xx}\right|^2 = \frac{1}{9}\left|\alpha'_{\parallel} + 2\alpha'_{\perp}\right|^2. \tag{7.92}$$

$$(\gamma')^2 = \frac{3}{2}\left|(\alpha')^2_{q=0}\right|^2 = \left|\alpha'_{zz} - \alpha'_{xx}\right|^2 = \left|\alpha'_{\parallel} - \alpha'_{\perp}\right|^2. \tag{7.93}$$

The result of the analysis is for $\mathrm{v}' = \mathrm{v} + 1$ is

$$\left\langle |\alpha_{ZZ}(A,B)|^2 \right\rangle = \frac{\hbar(\mathrm{v}+1)}{2\mu_R\omega_R}\left[(a')^2\delta_{J'J} + \frac{4}{45}b^{(2)}_{J'J}(\gamma')^2\right]. \tag{7.94}$$

$$\left\langle |\alpha_{YZ}(A,B)|^2 \right\rangle = \frac{\hbar(\mathrm{v}+1)}{2\mu_R\omega_R}\left[\frac{1}{15}b^{(2)}_{J'J}(\gamma')^2\right]. \tag{7.95}$$

The result of the analysis is for $\mathrm{v}' = \mathrm{v} - 1$ is

$$\left\langle |\alpha_{ZZ}(A,B)|^2 \right\rangle = \frac{\hbar\mathrm{v}}{2\mu_R\omega_R}\left[(a')^2\delta_{J'J} + \frac{4}{45}b^{(2)}_{J'J}(\gamma')^2\right]. \tag{7.96}$$

$$\left\langle |\alpha_{YZ}(A,B)|^2 \right\rangle = \frac{\hbar\mathrm{v}}{2\mu_R\omega_R}\left[\frac{1}{15}b^{(2)}_{J'J}(\gamma')^2\right]. \tag{7.97}$$

The allowed Stokes vibrational Raman transitions for a molecule with a $^1\Sigma$ electronic level are depicted schematically in Figure 7.3. The intensity of the Raman

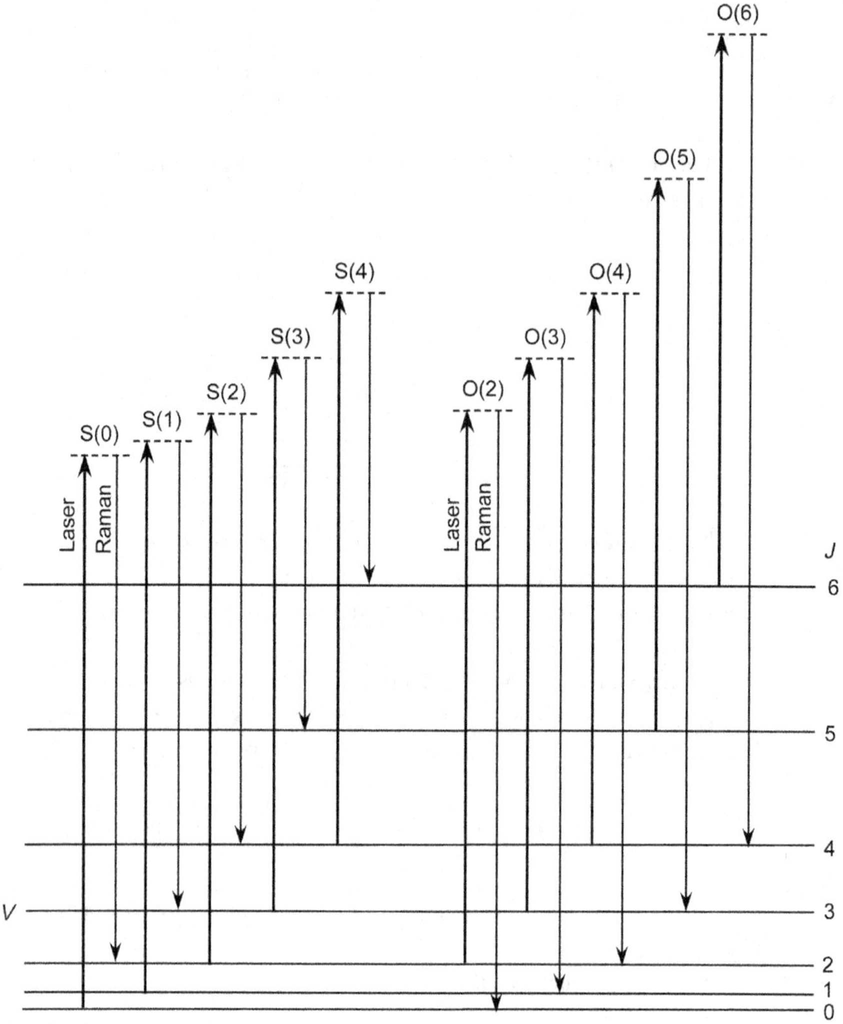

Figure 7.2 Structure of pure rotational Raman transitions for molecule with a $^1\Sigma$ ground electronic level.

scattered radiation is calculated by substituting the orientation-averaged squares of the polarizability tensor elements into Eqs. (7.2) and (7.4):

$$I_{ZZ}^{\Omega} = \frac{(\omega_L + \omega_{AB})^4 N_A}{32c^3\pi^2\varepsilon_0}\left\langle |\alpha_{ZZ}(A,B)|^2 \right\rangle E_Z^2 = \frac{(\omega_L + \omega_{AB})^4 N_A}{16c^4\pi^2\varepsilon_0^2}\left\langle |\alpha_{ZZ}(A,B)|^2 \right\rangle I_L, \quad (7.98)$$

$$I_{YZ}^{\Omega} = \frac{(\omega_L + \omega_{AB})^4 N_A}{32c^3\pi^2\varepsilon_0}\left\langle |\alpha_{YZ}(A,B)|^2 \right\rangle E_Z^2 = \frac{(\omega_L + \omega_{AB})^4 N_A}{16c^4\pi^2\varepsilon_0^2}\left\langle |\alpha_{YZ}(A,B)|^2 \right\rangle I_L, \quad (7.99)$$

where $\omega_{AB} = (\varepsilon_A - \varepsilon_B)/\hbar$. For Stokes scattering, $\omega_{AB} < 0$, and for anti-Stokes scattering, $\omega_{AB} > 0$. Equations (7.98) and (7.99) are valid under the assumptions of

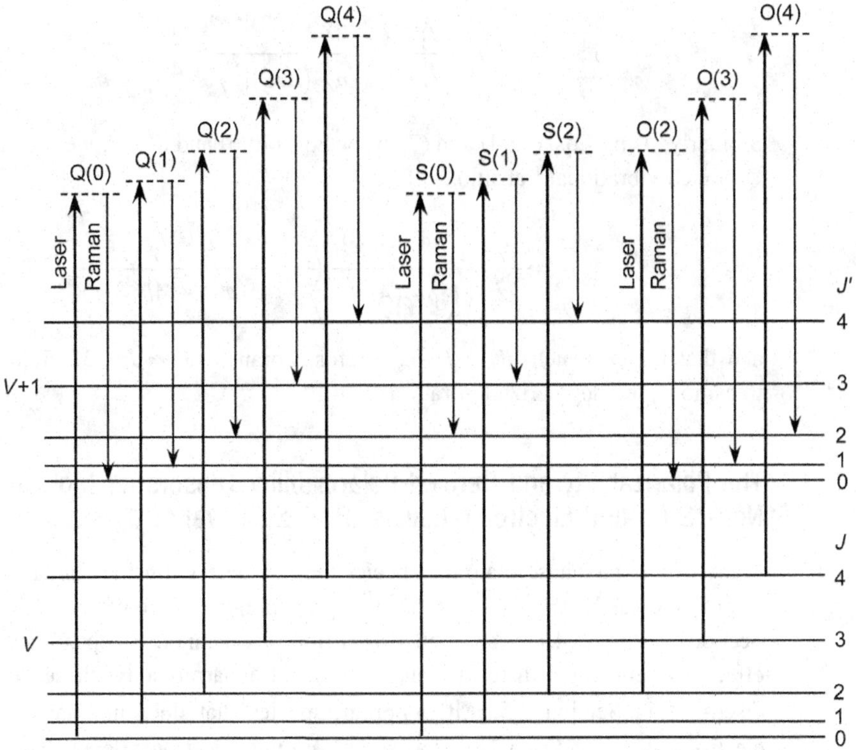

Figure 7.3 Structure of vibrational Stokes Raman transitions for molecule with a $^1\Sigma$ ground electronic level.

uniform irradiation and scattered light collection efficiency for the molecular assembly, and of incoherent scattered radiation. For Raman scattering, the scattered radiation will be incoherent because the final state for the process is different than the initial state. For Rayleigh scattering, the initial and final states are the same, and scattered Rayleigh light is coherent under some conditions.

The depolarization ratio ρ for a given transition is defined by

$$\rho = \frac{I_{YZ}^{\Omega}}{I_{ZZ}^{\Omega}} = \frac{\left\langle |\alpha_{YZ}(A,B)|^2 \right\rangle}{\left\langle |\alpha_{ZZ}(A,B)|^2 \right\rangle}. \tag{7.100}$$

The depolarization ratio can be expressed in terms of the tensor invariants. For Rayleigh scattering, $J' = J$, and using Eqs. (7.86) and (7.87) we can show that the depolarization ratio is given by

$$\rho = \frac{I_{YZ}^{\Omega}}{I_{ZZ}^{\Omega}} = \frac{\left\langle |\alpha_{YZ}(A,B)|^2 \right\rangle}{\left\langle |\alpha_{ZZ}(A,B)|^2 \right\rangle} = \frac{3b_{JJ}^{(2)}\gamma_0^2}{45a_0^2 + 4b_{JJ}^{(2)}\gamma_0^2}. \tag{7.101}$$

For O-branch ($J' = J - 2$) or for S-branch ($J' = J + 2$) pure rotational transitions, the depolarization ratio is given by

$$\rho = \frac{I_{YZ}^{\Omega}}{I_{ZZ}^{\Omega}} = \frac{\left\langle |\alpha_{YZ}(A,B)|^2 \right\rangle}{\left\langle |\alpha_{ZZ}(A,B)|^2 \right\rangle} = \frac{3}{4}. \tag{7.102}$$

Similarly, using Eqs. (7.94) and (7.95) or Eqs. (7.96) and (7.97), we can show that for Q-branch vibrational transitions, $J' = J$,

$$\rho = \frac{I_{YZ}^{\Omega}}{I_{ZZ}^{\Omega}} = \frac{\left\langle |\alpha_{YZ}(A,B)|^2 \right\rangle}{\left\langle |\alpha_{ZZ}(A,B)|^2 \right\rangle} = \frac{3b_{JJ}^{(2)}(\gamma')^2}{45(a')^2 + 4b_{JJ}^{(2)}(\gamma')^2} \tag{7.103}$$

and that for O-branch $(J' = J - 2)$ or for S-branch $(J' = J + 2)$ vibration-rotation transitions, the depolarization ratio is $3/4$.

7.3.5 The Polarizability and Derived Polarizability Tensors for Molecules with Non-$^1\Sigma$ Ground Electronic Levels: The $^3\Sigma_g^-$ Level of O_2

Many of the gas-phase diatomic species that are detected using Raman scattering or CARS, such as N_2, H_2, and CO, have $^1\Sigma$ ground electronic levels, and the analysis of Sections 7.3.1–7.3.4 can be applied in a rigorous manner, except to account for the effects of centrifugal distortion and vibrational anharmonicity; these effects will be discussed in Section 7.5. An important species that does not have a $^1\Sigma$ ground electronic level is O_2, which has a $^3\Sigma^-$ ground electronic level. The space-fixed polarizability components for molecules with non-$^1\Sigma$ ground electronic levels can be calculated by assuming that the wavefunctions for the rotational levels can be expressed in terms of Hund's case (a) basis wavefunctions. The Raman polarizabilities are then calculated using procedures similar to those discussed in Chapter 5 for single-photon transitions. The structure of the $^3\Sigma^-$ electronic level of NH was discussed in detail in Chapter 3. For $^3\Sigma^-$ electronic levels, the basis wavefunctions are given by

$$\left|{}^3\Sigma_{1e}^-; \mathrm{v}; JM\right\rangle = \frac{1}{\sqrt{2}}[|0^-; 11; \mathrm{v}; J1M\rangle + |0^-; 1-1; \mathrm{v}; J-1M\rangle]. \tag{7.104}$$

$$\left|{}^3\Sigma_{0e}^-; \mathrm{v}; JM\right\rangle = |0^-; 10; \mathrm{v}; J0M\rangle. \tag{7.105}$$

$$\left|{}^3\Sigma_{1f}^-; \mathrm{v}; JM\right\rangle = \frac{1}{\sqrt{2}}[|0^-; 11; \mathrm{v}; J1M\rangle - |0^-; 1-1; \mathrm{v}; J-1M\rangle]. \tag{7.106}$$

The vibration-rotation state wavefunctions for $^3\Sigma^-$ electronic levels are given by

$$|F_1; \mathrm{v}; JM; e\rangle = S_{1e,N+1}\left|{}^3\Sigma_{1e}^-; \mathrm{v}; JM\right\rangle + S_{0e,N+1}\left|{}^3\Sigma_{0e}^-; \mathrm{v}; JM\right\rangle. \tag{7.107}$$

$$|F_2; \mathrm{v}; JM; f\rangle = S_{1f,N}\left|{}^3\Sigma_{1f}^-; \mathrm{v}; JM\right\rangle = \left|{}^3\Sigma_{1f}^-; \mathrm{v}; JM\right\rangle. \tag{7.108}$$

$$|F_3; \mathrm{v}; JM; e\rangle = S_{1e,N-1}\left|{}^3\Sigma_{1e}^-; \mathrm{v}; JM\right\rangle + S_{0e,N-1}\left|{}^3\Sigma_{0e}^-; \mathrm{v}; JM\right\rangle. \tag{7.109}$$

The pure rotational Raman Stokes transitions for the ground $^3\Sigma_g^-$ level of O_2 are depicted in Figure 7.4. Expressions for the Raman transition strengths for the main

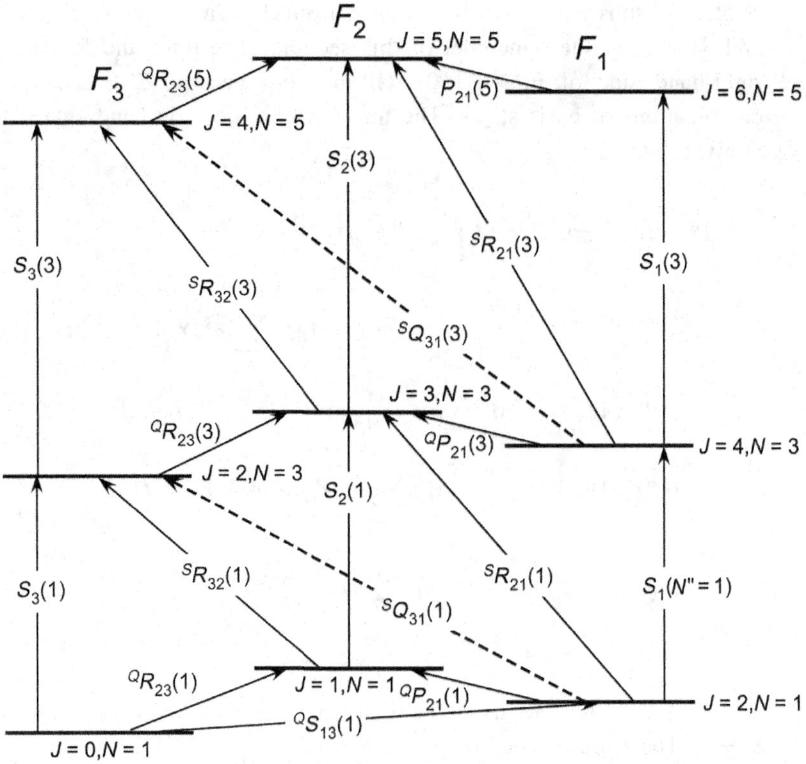

Figure 7.4 Energy levels and allowed pure rotational Raman transitions $^{\Delta N}\Delta J(N'')$ for the ground $^3\Sigma_g^-$ electronic level of O_2. The $^SQ_{31}$ transitions are depicted with dotted lines. Although these transitions are nominally allowed, the Placzek Teller coefficients for $^SQ_{31}$ transitions are zero. The energy level splitting is between the levels depicted is only approximate.

branch transitions $F_1 \to F_1$, $F_2 \to F_2$, and $F_3 \to F_3$, and for the satellite transitions $F_1 \leftrightarrow F_2$, $F_2 \leftrightarrow F_3$, and $F_1 \leftrightarrow F_3$, are derived below.

$F_1 \to F_1$ Transitions

Consider transitions between positive parity F_1 vibration-rotation levels and positive parity F_1 levels. For $F_1 \to F_1$ transitions, the polarizability is given by

$$
\begin{aligned}
\alpha_p^k(a,b) &= \langle F_1; \mathrm{v}'; J'M'; e| \sum_q \hat{\alpha}_q^k D_{pq}^{k*}(\omega) |F_1; \mathrm{v}; JM; e\rangle \\
&= S'_{1,N'+1} S_{1,N+1} \langle \Sigma_{1e}^-; \mathrm{v}'; J'M'| \sum_q \hat{\alpha}_q^k D_{pq}^{k*}(\omega) |\Sigma_{1e}^-; \mathrm{v}; JM\rangle \\
&\quad + S'_{0,N'+1} S_{0,N+1} \langle \Sigma_{0e}^-; \mathrm{v}'; J'M'| \sum_q \hat{\alpha}_q^k D_{pq}^{k*}(\omega) |\Sigma_{0e}^-; \mathrm{v}; JM\rangle \\
&\quad + S'_{0,N'+1} S_{1,N+1} \langle \Sigma_{0e}^-; \mathrm{v}'; J'M'| \sum_q \hat{\alpha}_q^k D_{pq}^{k*}(\omega) |\Sigma_{1e}^-; \mathrm{v}; JM\rangle \\
&\quad + S'_{1,N'+1} S_{0,N+1} \langle \Sigma_{1e}^-; \mathrm{v}; JM| \sum_q \hat{\alpha}_q^k D_{pq}^{k*}(\omega) |\Sigma_{0e}^-; \mathrm{v}'; J'M'\rangle.
\end{aligned}
\tag{7.110}
$$

(Note: the subscript e will be dropped from the wavefunction coefficients $S_{1e,N\pm 1}$ and $S_{0e,N\pm 1}$ for the remainder of this section.) The third and fourth terms on the right-hand side of Eq. (7.110) will be zero because $\Sigma' \neq \Sigma$ for any possible combinations of basis states. The first term on the right-hand side of Eq. (7.110) is given by

$$
\begin{aligned}
&S'_{1,N'+1}S_{1,N+1}\langle\Sigma^-_{1e};\mathrm{v}';J'M'|\sum_q \hat{\alpha}^k_q D^{k*}_{pq}(\omega)|\Sigma^-_{1e};\mathrm{v};JM\rangle \\
&= \frac{S'_{1,N'+1}S_{1,N+1}}{2}\Bigg\{\langle 0^-;1-1;\mathrm{v}';J'-1M'|\sum_q \hat{\alpha}^k_q D^{k*}_{pq}(\omega)|0^-;1-1;\mathrm{v};J-1M\rangle \\
&+\langle 0^-;11;\mathrm{v}';J'1M'|\sum_q \hat{\alpha}^k_q D^{k*}_{pq}(\omega)|0^-;1-1;\mathrm{v};J-1M\rangle \\
&+\langle 0^-;1-1;\mathrm{v}';J'-1M'|\sum_q \hat{\alpha}^k_q D^{k*}_{pq}(\omega)|0^-;11;\mathrm{v};J1M\rangle \\
&+\langle 0^-;11;\mathrm{v}';J'1M'|\sum_q \hat{\alpha}^k_q D^{k*}_{pq}(\omega)|\eta;0^-;11;\mathrm{v};J1M\rangle\Bigg\}.
\end{aligned}
\tag{7.111}
$$

The second and third terms on the right-hand side of Eq. (7.111) will be zero because $\Sigma' \neq \Sigma$. Therefore,

$$
\begin{aligned}
&S'_{1,N'+1}S_{1,N+1}\langle\Sigma^-_{1e};\mathrm{v}';J'M'|\sum_q \hat{\alpha}^k_q D^{k*}_{pq}(\omega)|\Sigma^-_{1e};\mathrm{v};JM\rangle \\
&= \frac{S'_{1,N'+1}S_{1,N+1}}{2}\Bigg\{\sum_q \langle\eta;\mathrm{v}'|\hat{\alpha}^k_q|\eta;\mathrm{v}\rangle\langle J'-1M'|D^{k*}_{pq}(\omega)|J-1M\rangle \\
&+\sum_q \langle\eta;\mathrm{v}'|\hat{\alpha}^k_q|\eta;\mathrm{v}\rangle\langle J'1M'|D^{k*}_{pq}(\omega)|J1M\rangle\Bigg\} \\
&= \frac{S'_{1,N'+1}S_{1,N+1}}{2}\sqrt{(2J'+1)(2J+1)}\begin{pmatrix} J' & k & J \\ -M' & p & M\end{pmatrix} \\
&\times \left\langle\eta;\mathrm{v}'\left|\hat{\alpha}^k_q\right|\eta;\mathrm{v}\right\rangle\left[(-1)^{M'-1}\begin{pmatrix} J' & k & J \\ 1 & 0 & -1\end{pmatrix}+(-1)^{M'+1}\begin{pmatrix} J' & k & J \\ -1 & 0 & 1\end{pmatrix}\right] \\
&= \frac{S'_{1,N'+1}S_{1,N+1}}{2}\sqrt{(2J'+1)(2J+1)}\begin{pmatrix} J' & k & J \\ -M' & p & M\end{pmatrix}(-1)^{M'} \\
&\times \left\langle\eta;\mathrm{v}'\left|\hat{\alpha}^k_{q=0}\right|\eta;\mathrm{v}\right\rangle\left[-(-1)^{J+J'+k}\begin{pmatrix} J & J' & k \\ 1 & -1 & 0\end{pmatrix}-\begin{pmatrix} J & J' & k \\ 1 & -1 & 0\end{pmatrix}\right].
\end{aligned}
\tag{7.112}
$$

The second term on the right-hand side of Eq. (7.110) is given by

$$\begin{aligned} & S'_{0,N'+1}S_{0,N+1}\langle \Sigma^-_{0e}; \mathrm{v}'; JM'|\sum_q \hat{\alpha}^k_q D^{k*}_{pq}(\omega)|\Sigma^-_{0e}; \mathrm{v}; JM\rangle \\ & = S'_{0,N'+1}S_{0,N+1}\langle \eta; \mathrm{v}'|\hat{\alpha}^k_0|\eta; \mathrm{v}\rangle \langle J'\,0M'|D^{k*}_{pq}(\omega)|J\,0M\rangle \\ & = S'_{0,N'+1}S_{0,N+1}\sqrt{(2J'+1)(2J+1)}\begin{pmatrix} J' & k & J \\ -M' & p & M \end{pmatrix}(-1)^{M'} \\ & \times \langle \eta; \mathrm{v}'|\hat{\alpha}^k_{q=0}|\eta; \mathrm{v}\rangle \begin{pmatrix} J & J' & k \\ 0 & 0 & 0 \end{pmatrix}. \end{aligned} \tag{7.113}$$

Combining Eqs. (7.112) and (7.113), we obtain

$$\begin{aligned} \alpha^k_p(a,b) = {} & \langle \eta; \mathrm{v}'|\hat{\alpha}^{\,k}_{q=0}|\eta; \mathrm{v}\rangle \sqrt{(2J'+1)(2J+1)}\begin{pmatrix} J' & k & J \\ -M' & p & M \end{pmatrix}(-1)^{M'} \\ & \times \left\{ -\frac{S'_{1,N'+1}S_{1,N+1}}{2}\left[(-1)^{J+J'+k}\begin{pmatrix} J & J' & k \\ 1 & -1 & 0 \end{pmatrix} + \begin{pmatrix} J & J' & k \\ 1 & -1 & 0 \end{pmatrix}\right] \right. \\ & \left. + S'_{0,N'+1}S_{0,N+1}\begin{pmatrix} J & J' & k \\ 0 & 0 & 0 \end{pmatrix}\right\}. \end{aligned} \tag{7.114}$$

Evaluating Eq. (7.114) for $k=2$, it is obvious that $\alpha^2_p(a,b)=0$ unless $J+J'+k$ is an even number. Assuming that $J+J'+k$ is an even number, we obtain

$$\begin{aligned} \alpha^2_p(a,b) = {} & \langle \eta; \mathrm{v}'|\hat{\alpha}^{\,2}_{q=0}|\eta; \mathrm{v}\rangle \sqrt{(2J'+1)(2J+1)}\begin{pmatrix} J' & 2 & J \\ -M' & p & M \end{pmatrix}(-1)^{M'} \\ & \times \left\{ -\frac{S'_{1,N'+1}S_{1,N+1}}{2}\begin{pmatrix} J & J' & 2 \\ 1 & -1 & 0 \end{pmatrix} + S'_{0,N'+1}S_{0,N+1}\begin{pmatrix} J & J' & 2 \\ 0 & 0 & 0 \end{pmatrix}\right\}. \end{aligned} \tag{7.115}$$

For $F_1 \to F_1$, the only possible pure rotational transitions are ${}^sS_1(J)$ transitions for which $\Delta N = N' - N = +2$ and $\Delta J = J' - J = +2$ and ${}^oO_1(J)$ transitions for which $\Delta N = N' - N = -2$ and $\Delta J = J' - J = -2$. For vibrational transitions, $\Delta N = N' - N = 0, \pm 2$ and $\Delta J = J' - J = 0, \pm 2$. Setting $p=0$, taking the square of Eq. (7.115), summing over all M and M', and dividing by the number of quantum states $2J+1$ in the initial level A, we obtain

$$\begin{aligned} \left\langle \left|\alpha^2_{p=0}(A,B)\right|^2 \right\rangle = {} & \langle \eta; \mathrm{v}'|\hat{\alpha}^2_{q=0}|\eta; \mathrm{v}\rangle^2 \frac{(2J'+1)}{5} \\ & \times \left\{ S'^2_{1,N'+1}S^2_{1,N+1}\begin{pmatrix} J' & J & 2 \\ -1 & 1 & 0 \end{pmatrix}^2 + S'^2_{0,N'+1}S^2_{0,N+1}\begin{pmatrix} J' & J & 2 \\ 0 & 0 & 0 \end{pmatrix}^2 \right. \\ & \left. - 2S'_{1,N'+1}S_{1,N+1}S'_{0,N'+1}S_{0,N+1}\begin{pmatrix} J' & J & 2 \\ -1 & 1 & 0 \end{pmatrix}\begin{pmatrix} J' & J & 2 \\ 0 & 0 & 0 \end{pmatrix}^2 \right\} \\ = {} & \frac{\langle \eta; \mathrm{v}'|\hat{\alpha}^2_{q=0}|\eta; \mathrm{v}\rangle^2}{5} b^{(2)}_{N'+1,N+1}, \end{aligned} \tag{7.116}$$

where

$$b^{(2)}_{N'+1,N+1} = (2J'+1)\left\{S'^2_{1,N'+1}S^2_{1,N+1}\begin{pmatrix} J' & J & 2 \\ -1 & 1 & 0 \end{pmatrix}^2 + S'^2_{0,N'+1}S^2_{0,N+1}\begin{pmatrix} J' & J & 2 \\ 0 & 0 & 0 \end{pmatrix}^2 \right.$$
$$\left. - 2S'_{1,N'+1}S_{1,N+1}S'_{0,N'+1}S_{0,N+1}\begin{pmatrix} J' & J & 2 \\ -1 & 1 & 0 \end{pmatrix}\begin{pmatrix} J' & J & 2 \\ 0 & 0 & 0 \end{pmatrix}^2\right\}. \tag{7.117}$$

Evaluating Eq. (7.114) for $k = 0$ and $p = 0$, we obtain

$$\left\langle \left|\alpha^0_{p=0}(A,B)\right|^2 \right\rangle = \langle \eta; \mathrm{v}'|\hat{\alpha}^0_{q=0}|\eta; \mathrm{v}\rangle^2 (2J'+1)\left\{ S'^2_{1,N'+1}S^2_{1,N+1}\begin{pmatrix} J' & J & 0 \\ -1 & 1 & 0 \end{pmatrix}^2 \right.$$
$$- 2S'_{1,N'+1}S_{1,N+1}S'_{0,N'+1}S_{0,N+1}\begin{pmatrix} J' & J & 0 \\ -1 & 1 & 0 \end{pmatrix}\begin{pmatrix} J' & J & 0 \\ 0 & 0 & 0 \end{pmatrix}$$
$$\left. + S'^2_{0,N'+1}S^2_{0,N+1}\begin{pmatrix} J' & J & 0 \\ 0 & 0 & 0 \end{pmatrix}^2 \right\}$$
$$= \langle \eta; \mathrm{v}'|\hat{\alpha}^0_{q=0}|\eta; \mathrm{v}\rangle^2 b^{(0)}_{N'+1,N+1}, \tag{7.118}$$

where

$$b^{(0)}_{N'+1,N+1} = (2J'+1)\left\{ S'^2_{1,N'+1}S^2_{1,N+1}\begin{pmatrix} J' & J & 0 \\ -1 & 1 & 0 \end{pmatrix}^2 \right.$$
$$- 2S'_{1,N'+1}S_{1,N+1}S'_{0,N'+1}S_{0,N+1}\begin{pmatrix} J' & J & 0 \\ -1 & 1 & 0 \end{pmatrix}\begin{pmatrix} J' & J & 0 \\ 0 & 0 & 0 \end{pmatrix}$$
$$\left. + S'^2_{0,N'+1}S^2_{0,N+1}\begin{pmatrix} J' & J & 0 \\ 0 & 0 & 0 \end{pmatrix}^2 \right\}. \tag{7.119}$$

For pure rotational transitions with $\Delta J = \pm 2$, $\mathrm{v}' = \mathrm{v}$, we obtain

$$\left\langle |\alpha_{ZZ}(A,B)|^2 \right\rangle = \frac{2}{3}\left\langle \left|\alpha^2_{p=0}(A,B)\right|^2 \right\rangle = \frac{4}{15}\gamma_0^2 b^{(2)}_{N'+1,N+1} \tag{7.120}$$

and

$$\left\langle |\alpha_{YZ}(A,B)|^2 \right\rangle = \frac{3}{4}\left\langle |\alpha_{ZZ}(A,B)|^2 \right\rangle, \tag{7.121}$$

where

$$\langle \eta; \mathrm{v}'|\hat{\alpha}^2_{q=0}|\eta; \mathrm{v}\rangle^2 = \langle \eta|\hat{\alpha}^0_{q=0}|\eta\rangle^2 \langle \mathrm{v}\,|\,\mathrm{v}\rangle^2 = \frac{2}{3}\delta_0^2. \tag{7.122}$$

For vibrational transitions with $\Delta J = 0, \pm 2, \mathrm{v}' = \mathrm{v} + 1$, we obtain

$$\begin{aligned}\left\langle |\alpha_{ZZ}(A,B)|^2 \right\rangle &= \frac{1}{3}\left\langle \left|(\alpha')^0_{p=0}(A,B)\right|^2 \right\rangle + \frac{2}{3}\left\langle \left|(\alpha')^2_{p=0}(A,B)\right|^2 \right\rangle \\ &= \frac{\hbar(\mathrm{v}+1)}{2\,\mu_R\,\omega_{vib}}\left[(a')^2 b^{(0)}_{N'+1,N+1} + \frac{4}{45}(\gamma')^2 b^{(2)}_{N'+1,N+1}\right]\end{aligned} \tag{7.123}$$

and

$$\begin{aligned}\left\langle |\alpha_{YZ}(A,B)|^2 \right\rangle &= \frac{1}{2}\left\langle \left|(\alpha')^2_{p=1}(A,B)\right|^2 \right\rangle \\ &= \frac{\hbar(\mathrm{v}+1)}{2\,\mu_R\,\omega_{vib}}\left[\frac{1}{15}(\gamma')^2 b^{(2)}_{N'+1,N+1}\right],\end{aligned} \tag{7.124}$$

where

$$\langle \eta;\mathrm{v}'|(\hat{\alpha}')^0_{q=0}|\eta;\mathrm{v}\rangle^2 = \langle \eta|(\hat{\alpha}')^0_{q=0}|\eta\rangle^2 \langle \mathrm{v}+1\,|\,\mathrm{v}\rangle^2 = \frac{\hbar(\mathrm{v}+1)}{2\,\mu_R\,\omega_{vib}}\left[3(a')^2\right] \tag{7.125}$$

and

$$\langle \eta;\mathrm{v}'|(\hat{\alpha}')^2_{q=0}|\eta;\mathrm{v}\rangle^2 = \langle \eta|(\hat{\alpha}')^0_{q=0}|\eta\rangle^2 \langle \mathrm{v}+1\,|\,\mathrm{v}\rangle^2 = \frac{\hbar(\mathrm{v}+1)}{2\,\mu_R\,\omega_{vib}}\left[\frac{2}{3}(\gamma')^2\right]. \tag{7.126}$$

$F_3 \rightarrow F_3$ Transitions

For $F_3 \rightarrow F_3$ transitions, the analysis is very similar to that for $F_1 \rightarrow F_1$ transitions. The orientation-averaged squares of the polarizabilities in irreducible coordinates are given by

$$\begin{aligned}\left\langle \left|\alpha^2_{p=0}(A,B)\right|^2 \right\rangle &= \langle \eta;\mathrm{v}'|\hat{\alpha}^2_{q=0}|\eta;\mathrm{v}\rangle^2 \frac{(2J'+1)}{5}\Bigg\{ S'^2_{1,N'-1}S^2_{1,N-1}\begin{pmatrix} J' & J & 2 \\ -1 & 1 & 0\end{pmatrix}^2 \\ &\quad - 2S'_{1,N'-1}S_{1,N-1}S'_{0,N'-1}S_{0,N-1}\begin{pmatrix} J' & J & 2 \\ -1 & 1 & 0\end{pmatrix}\begin{pmatrix} J' & J & 2 \\ 0 & 0 & 0\end{pmatrix} \\ &\quad + S'^2_{0,N'-1}S^2_{0,N-1}\begin{pmatrix} J' & J & 2 \\ 0 & 0 & 0\end{pmatrix}^2\Bigg\} \\ &= \frac{\langle \eta;\mathrm{v}'|\hat{\alpha}^2_{q=0}|\eta;\mathrm{v}\rangle^2}{5} b^{(2)}_{N'-1,N-1},\end{aligned} \tag{7.127}$$

where

$$\begin{aligned}b^{(2)}_{N'-1,N-1} &= (2J'+1)\Bigg\{ S'^2_{1,N'-1}S^2_{1,N-1}\begin{pmatrix} J' & J & 2 \\ -1 & 1 & 0\end{pmatrix}^2 \\ &\quad - 2S'_{1,N'-1}S_{1,N-1}S'_{0,N'-1}S_{0,N-1}\begin{pmatrix} J' & J & 2 \\ -1 & 1 & 0\end{pmatrix}\begin{pmatrix} J' & J & 2 \\ 0 & 0 & 0\end{pmatrix} \\ &\quad + S'^2_{0,N'-1}S^2_{0,N-1}\begin{pmatrix} J' & J & 2 \\ 0 & 0 & 0\end{pmatrix}^2\Bigg\}\end{aligned} \tag{7.128}$$

and

$$\left\langle\left|\alpha^0_{p=0}(A,B)\right|^2\right\rangle = \langle\eta;\mathrm{v}'|\hat{\alpha}^0_{q=0}|\eta;\mathrm{v}\rangle^2(2J'+1)\left\{S'^2_{1,N'-1}S^2_{1,N-1}\begin{pmatrix} J' & J & 0 \\ -1 & 1 & 0 \end{pmatrix}^2\right.$$
$$-2S'_{1,N'-1}S_{1,N-1}S'_{0,N'-1}S_{0,N-1}\begin{pmatrix} J' & J & 0 \\ -1 & 1 & 0 \end{pmatrix}\begin{pmatrix} J' & J & 0 \\ 0 & 0 & 0 \end{pmatrix}$$
$$\left.+S'^2_{0,N'-1}S^2_{0,N-1}\begin{pmatrix} J' & J & 0 \\ 0 & 0 & 0 \end{pmatrix}^2\right\}$$
$$= \langle\eta;\mathrm{v}'|\hat{\alpha}^0_{q=0}|\eta;\mathrm{v}\rangle^2 b^{(0)}_{N'-1,N-1}, \tag{7.129}$$

where

$$b^{(0)}_{N'-1,N-1} = (2J'+1)\left\{S'^2_{1,N'-1}S^2_{1,N-1}\begin{pmatrix} J' & J & 0 \\ -1 & 1 & 0 \end{pmatrix}^2\right.$$
$$-2S'_{1,N'-1}S_{1,N-1}S'_{0,N'-1}S_{0,N-1}\begin{pmatrix} J' & J & 0 \\ -1 & 1 & 0 \end{pmatrix}\begin{pmatrix} J' & J & 0 \\ 0 & 0 & 0 \end{pmatrix} \tag{7.130}$$
$$\left.+S'^2_{0,N'-1}S^2_{0,N-1}\begin{pmatrix} J' & J & 0 \\ 0 & 0 & 0 \end{pmatrix}^2\right\}.$$

The expressions for $\left\langle|\alpha_{ZZ}(A,B)|^2\right\rangle$ and $\left\langle|\alpha_{YZ}(A,B)|^2\right\rangle$ for $F_3 \to F_3$ transitions can be obtained by substituting $b^{(2)}_{N'-1,N-1}$ and $b^{(0)}_{N'-1,N-1}$ for $b^{(2)}_{N'+1,N+1}$ and $b^{(0)}_{N'+1,N+1}$, respectively, in Eqs. (7.120), (7.123), and (7.124).

$F_2 \to F_2$ Transitions

For $F_2 \to F_2$ transitions, the polarizability is given by

$$\alpha^k_p(a,b) = \langle F_2;\mathrm{v}';J'M'|\sum_q \hat{\alpha}^k_q D^{k*}_{pq}(\omega)|F_2;\mathrm{v};JM\rangle$$
$$= \left\langle\Sigma^-_{1f};\mathrm{v}';J'M'\middle|\sum_q \hat{\alpha}^k_q D^{k*}_{pq}(\omega)\middle|\Sigma^-_{1f};\mathrm{v};JM\right\rangle$$
$$= \frac{1}{2}\langle 0^-;11;\mathrm{v}';J'1M'|\sum_q \hat{\alpha}^k_q D^{k*}_{pq}(\omega)|0^-;11;\mathrm{v};J1M\rangle$$
$$+\frac{1}{2}\langle 0^-;1-1;\mathrm{v}';J'-1M'|\sum_q \hat{\alpha}^k_q D^{k*}_{pq}(\omega)|0^-;1-1;\mathrm{v};J-1M\rangle$$
$$= \frac{1}{2}\langle\eta;\mathrm{v}'|\hat{\alpha}^k_{q=0}|\eta;\mathrm{v}\rangle\sqrt{(2J'+1)(2J+1)}\begin{pmatrix} J' & k & J \\ -M' & p & M \end{pmatrix}(-1)^{M'}$$
$$\times\left[(-1)^{J+J'+k-1}\begin{pmatrix} J & J' & k \\ 1 & -1 & 0 \end{pmatrix} - \begin{pmatrix} J & J' & k \\ 1 & -1 & 0 \end{pmatrix}\right]. \tag{7.131}$$

For $F_2 \to F_2$ transitions, $J + J'$ is always an even number, and therefore the polarizability is zero for $k = 1$. Setting $k = 2$ and $p = 0$ in Eq. (7.131), squaring $\alpha^2_{p=0}(a,b)$, summing over all M and M', and dividing by $2J + 1$, we obtain

$$\left\langle \left|\alpha^2_{p=0}(A,B)\right|^2\right\rangle = \frac{\langle \eta;\mathrm{v}'|\hat{\alpha}^2_{q=0}|\eta;\mathrm{v}\rangle^2(2J'+1)}{5}\begin{pmatrix} J & J' & 2 \\ 1 & -1 & 0\end{pmatrix}^2 = \frac{\langle \eta;\mathrm{v}'|\hat{\alpha}^2_{q=0}|\eta;\mathrm{v}\rangle^2}{5} b^{(2)}_{N',N}, \tag{7.132}$$

where

$$b^{(2)}_{N',N} = (2J'+1)\begin{pmatrix} J & J' & 2 \\ 1 & -1 & 0\end{pmatrix}^2. \tag{7.133}$$

Setting $k=0$ and $p=0$ in Eq. (7.131), squaring $\alpha^0_{p=0}(a,b)$, summing over all M and M', and dividing by $2J+1$, we obtain

$$\left\langle \left|\alpha^0_{p=0}(A,B)\right|^2\right\rangle = \langle \eta;\mathrm{v}'|\hat{\alpha}^0_{q=0}|\eta;\mathrm{v}\rangle^2(2J'+1)\begin{pmatrix} J & J' & 0 \\ 1 & -1 & 0\end{pmatrix}^2 = \langle \eta;\mathrm{v}'|\hat{\alpha}^0_{q=0}|\eta;\mathrm{v}\rangle^2 b^{(0)}_{N',N}, \tag{7.134}$$

where

$$b^{(0)}_{N',N} = (2J'+1)\begin{pmatrix} J & J' & 0 \\ 1 & -1 & 0\end{pmatrix}^2. \tag{7.135}$$

The expressions for $\left\langle |\alpha_{ZZ}(A,B)|^2\right\rangle$ and $\left\langle |\alpha_{YZ}(A,B)|^2\right\rangle$ for $F_2 \to F_2$ transitions can be obtained by substituting $b^{(2)}_{N',N}$ and $b^{(0)}_{N',N}$ for $b^{(2)}_{N'+1,N+1}$ and $b^{(0)}_{N'+1,N+1}$, respectively, in Eqs. (7.120), (7.123), and (7.124).

$F_1 \leftrightarrow F_2$ and $F_3 \leftrightarrow F_2$ Transitions

The polarizability is given by

$$\begin{aligned}
\alpha^k_p(a,b) &= \langle F_2;\mathrm{v}';J'M'|\sum_q \hat{\alpha}^k_q D^{k*}_{pq}(\omega)|F_1;\mathrm{v};JM\rangle \\
&= -\frac{S_{1,N+1}}{2}\langle 0^-;1-1;\mathrm{v}';J'-1M'|\sum_q \hat{\alpha}^k_q D^{k*}_{pq}(\omega)|0^-;1-1;\mathrm{v};J-1M\rangle \\
&\quad + \frac{S_{1,N+1}}{2}\langle 0^-;11;\mathrm{v}';J'1M'|\sum_q \hat{\alpha}^k_q D^{k*}_{pq}(\omega)|0^-;11;\mathrm{v};J1M\rangle \\
&= \frac{S_{1,N+1}}{2}\langle \eta;\mathrm{v}'|\hat{\alpha}^k_{q=0}|\eta;\mathrm{v}\rangle\sqrt{(2J'+1)(2J+1)}\begin{pmatrix} J' & k & J \\ -M' & p & M\end{pmatrix} \\
&\quad \times\left[-(-1)^{J+J'+k-1}\begin{pmatrix} J & J' & k \\ 1 & -1 & 0\end{pmatrix} - \begin{pmatrix} J & J' & k \\ 1 & -1 & 0\end{pmatrix}\right].
\end{aligned} \tag{7.136}$$

For $F_1 \to F_2$ transitions, $J+J'$ is always odd, so the polarizability will be zero for $k=1$. For $k=2$ and $k=0$, we obtain

$$\alpha^2_p(a,b) = \langle \eta;\mathrm{v}'|\hat{\alpha}^2_{q=0}|\eta;\mathrm{v}\rangle S_{1,N+1}\sqrt{(2J'+1)(2J+1)}\begin{pmatrix} J' & 2 & J \\ -M' & p & M\end{pmatrix}\begin{pmatrix} J' & J & 2 \\ 1 & -1 & 0\end{pmatrix} \tag{7.137}$$

and

$$\alpha_p^0(a,b) = \langle \eta; \mathrm{v}' | \hat{\alpha}_{q=0}^0 | \eta; \mathrm{v} \rangle S_{1,N+1} \sqrt{(2J'+1)(2J+1)} \begin{pmatrix} J' & 0 & J \\ -M' & p & M \end{pmatrix} \begin{pmatrix} J' & J & 0 \\ 1 & -1 & 0 \end{pmatrix}. \tag{7.138}$$

However, the second $3j$ symbol in Eq. (7.138) will be zero because $|J' - J| \geq 1 > k = 0$. Setting $p = 0$ in Eq. (7.137), squaring $\alpha_{p=0}^2(a,b)$, and then summing over all M and M', we obtain

$$\begin{aligned} \left|\alpha_{p=0}^2(A,B)\right|^2 &= \frac{\langle \eta; \mathrm{v}' | \hat{\alpha}_{q=0}^2 | \eta; \mathrm{v} \rangle^2}{5} (2J'+1) \begin{pmatrix} J' & J & 2 \\ 1 & -1 & 0 \end{pmatrix}^2 S_{1,N+1}^2 \\ &= \frac{\langle \eta; \mathrm{v}' | \hat{\alpha}_{q=0}^2 | \eta; \mathrm{v} \rangle^2}{5} b_{N',N+1}^{(2)}, \end{aligned} \tag{7.139}$$

where

$$b_{N',N+1}^{(2)} = (2J'+1) \begin{pmatrix} J' & J & 2 \\ 1 & -1 & 0 \end{pmatrix}^2 S_{1,N+1}^2. \tag{7.140}$$

Following the same procedures as for the $F_1 \to F_1$ transitions, we obtain for pure rotational transitions with $\Delta J = \pm 1, \mathrm{v}' = \mathrm{v}$,

$$\left\langle |\alpha_{ZZ}(A,B)|^2 \right\rangle = \frac{2}{3} \left\langle \left|\alpha_{p=0}^2(A,B)\right|^2 \right\rangle = \frac{4}{15} \gamma_0^2 b_{N,N+1}^{(2)} \tag{7.141}$$

and

$$\left\langle |\alpha_{YZ}(A,B)|^2 \right\rangle = \frac{3}{4} \left\langle |\alpha_{ZZ}(A,B)|^2 \right\rangle \tag{7.142}$$

and for vibrational transitions with $\Delta J = \pm 1, \mathrm{v}' = \mathrm{v} + 1$.

$$\begin{aligned} \left\langle |\alpha_{ZZ}(A,B)|^2 \right\rangle &= \frac{1}{3} \left\langle \left|(\alpha')_{p=0}^0(A,B)\right|^2 \right\rangle + \frac{2}{3} \left\langle \left|(\alpha')_{p=0}^2(A,B)\right|^2 \right\rangle \\ &= \frac{\hbar(\mathrm{v}+1)}{2\mu_R \omega_{vib}} \left[\frac{4}{45} (\gamma')^2 b_{N',N+1}^{(2)} \right] \end{aligned} \tag{7.143}$$

and

$$\left\langle |\alpha_{YZ}(A,B)|^2 \right\rangle = \frac{1}{2} \left\langle \left|(\alpha')_{p=0}^2(A,B)\right|^2 \right\rangle = \frac{\hbar(\mathrm{v}+1)}{2\mu_R \omega_{vib}} \left[\frac{1}{15} (\gamma')^2 b_{N,N+1}^{(2)} \right] \tag{7.144}$$

For $F_3 \to F_2$ transitions, Eqs. (7.139)–(7.144) will apply except that $S_{1,N-1}$ and $b_{N',N-1}^{(2)}$ will be substituted for $S_{1,N+1}$ and $b_{N'+1,N}^{(2)}$, respectively.

For pure rotational $F_2 \to F_1$ transitions, we obtain

$$\left\langle |\alpha_{ZZ}(A,B)|^2 \right\rangle = \frac{2}{3} \left\langle \left|\alpha_{p=0}^2(A,B)\right|^2 \right\rangle = \frac{4}{15} \gamma_0^2 b_{N'+1,N}^{(2)} \tag{7.145}$$

and

$$\left\langle|\alpha_{YZ}(A,B)|^2\right\rangle = \frac{3}{4}\left\langle|\alpha_{ZZ}(A,B)|^2\right\rangle, \tag{7.146}$$

where

$$b^{(2)}_{N'+1,N} = (2J'+1)\begin{pmatrix} J' & J & 2 \\ 1 & -1 & 0 \end{pmatrix}^2 S'^2_{1,N'+1}. \tag{7.147}$$

For vibrational $F_2 \to F_1$ transitions, we obtain

$$\left\langle|\alpha_{ZZ}(A,B)|^2\right\rangle = \frac{\hbar(\mathrm{v}+1)}{2\mu_R\omega_{vib}}\left[\frac{4}{45}(\gamma')^2 b^{(2)}_{N'+1,N}\right] \tag{7.148}$$

and

$$\left\langle|\alpha_{YZ}(A,B)|^2\right\rangle = \frac{\hbar(\mathrm{v}+1)}{2\mu_R\omega_{vib}}\left[\frac{1}{15}(\gamma')^2 b^{(2)}_{N'+1,N}\right]. \tag{7.149}$$

$F_1 \leftrightarrow F_3$ Transitions

For $F_1 \to F_3$ transitions, the polarizability is given by

$$\begin{aligned}\alpha^k_p(a,b) &= \langle F_3;\mathrm{v}';J'M';e|\sum_q \hat{\alpha}^k_q D^{k*}_{pq}(\omega)|F_1;\mathrm{v};JM;e\rangle \\ &= S'_{1,N'-1}S_{1,N+1}\langle\Sigma^-_{1e};\mathrm{v}';J'M'|\sum_q \hat{\alpha}^k_q D^{k*}_{pq}(\omega)|\Sigma^-_{1e};\mathrm{v};JM\rangle \\ &\quad + S'_{0,N'-1}S_{0,N+1}\langle\Sigma^-_{0e};\mathrm{v}';J'M'|\sum_q \hat{\alpha}^k_q D^{k*}_{pq}(\omega)|\Sigma^-_{0e};\mathrm{v};JM\rangle.\end{aligned} \tag{7.150}$$

The first term on the right-hand side of Eq. (7.150) is given by

$$\begin{aligned}&S'_{1,N'-1}S_{1,N+1}\langle\Sigma^-_{1e};\mathrm{v}';J'M'|\sum_q \hat{\alpha}^k_q D^{k*}_{pq}(\omega)|\Sigma^-_{1e};\mathrm{v};JM\rangle \\ &= \frac{S'_{1,N'-1}S_{1,N+1}}{2}\Bigg\{\langle 0^-;1-1;\mathrm{v}';J'-1M'|\sum_q \hat{\alpha}^k_q D^{k*}_{pq}(\omega)|0^-;1-1;\mathrm{v};J-1M\rangle \\ &\quad + \langle 0^-;11;\mathrm{v}';J'1M'|\sum_q \hat{\alpha}^k_q D^{k*}_{pq}(\omega)|0^-;1-1;\mathrm{v};J-1M\rangle \\ &\quad + \langle 0^-;1-1;\mathrm{v}';J'-1M'|\sum_q \hat{\alpha}^k_q D^{k*}_{pq}(\omega)|0^-;11;\mathrm{v};J1M\rangle \\ &\quad + \langle 0^-;11;\mathrm{v}';J'1M'|\sum_q \hat{\alpha}^k_q D^{k*}_{pq}(\omega)|0^-;11;\mathrm{v};J1M\rangle\Bigg\}.\end{aligned} \tag{7.151}$$

The second and third terms on the right-hand side of Eq. (7.151) will be zero because $\Sigma_a \neq \Sigma_b$. Therefore,

$$
\begin{aligned}
&S'_{1,N'-1}S_{1,N+1}\left\langle \Sigma^-_{1e};\mathrm{v}';J'M'\right|\sum_q \hat{\alpha}^k_q D^{k*}_{pq}(\omega)\left|\Sigma^-_{1e};\mathrm{v};JM\right\rangle \\
&= \frac{S'_{1,N'-1}S_{1,N+1}}{2}\left\{\sum_q \langle\eta;\mathrm{v}'|\hat{\alpha}^k_q|\eta;\mathrm{v}\rangle\langle J'-1M'|D^{k*}_{pq}(\omega)|\mathrm{v};J-1M\rangle\right. \\
&\quad \left. +\sum_q \langle\eta;\mathrm{v}'|\hat{\alpha}^k_q|\eta;\mathrm{v}\rangle\langle J'1M'|D^{k*}_{pq}(\omega)|\mathrm{v};J1M\rangle\right\} \\
&= \frac{S'_{1,N'-1}S_{1,N+1}}{2}\sqrt{(2J'+1)(2J+1)}\begin{pmatrix} J' & k & J \\ -M' & p & M\end{pmatrix} \\
&\quad \times \langle\eta;\mathrm{v}'|\hat{\alpha}^k_q|\eta;\mathrm{v}\rangle\left[(-1)^{M'-1}\begin{pmatrix} J' & k & J \\ 1 & 0 & -1\end{pmatrix}+(-1)^{M'+1}\begin{pmatrix} J' & k & J \\ -1 & 0 & 1\end{pmatrix}\right] \\
&= \frac{S'_{1,N'-1}S_{1,N+1}}{2}\sqrt{(2J'+1)(2J+1)}\begin{pmatrix} J' & k & J \\ -M' & p & M\end{pmatrix}(-1)^{M'}(-1)^{J+J'+k} \\
&\quad \times \langle\eta;\mathrm{v}_B|\hat{\alpha}^k_{q=0}|\eta;\mathrm{v}_A\rangle\left[-\begin{pmatrix} J' & J & k \\ 1 & -1 & 0\end{pmatrix}-(-1)^{J+J'+k}\begin{pmatrix} J' & J & k \\ 1 & -1 & 0\end{pmatrix}\right].
\end{aligned}
\tag{7.152}
$$

The second term on the right-hand side of Eq. (7.150) is given by

$$
\begin{aligned}
&S'_{0,N'-1}S_{0,N+1}\left\langle \Sigma^-_{0e};\mathrm{v}';J'M'\right|\sum_q \hat{\alpha}^k_q D^{k*}_{pq}(\omega)\left|\Sigma^-_{0e};\mathrm{v};JM\right\rangle \\
&= S'_{0,N'-1}S_{0,N+1}\langle\eta;\mathrm{v}'|\hat{\alpha}^k_{q=0}|\eta;\mathrm{v}\rangle\langle J'0M'|D^{k*}_{pq}(\omega)|J0M\rangle \\
&= S'_{0,N'-1}S_{0,N+1}\sqrt{(2J'+1)(2J+1)}\begin{pmatrix} J' & k & J \\ -M' & p & M\end{pmatrix}(-1)^{M'}(-1)^{J+J'+k} \\
&\quad \times \langle\eta;\mathrm{v}'|\hat{\alpha}^k_{q=0}|\eta;\mathrm{v}\rangle\begin{pmatrix} J' & J & k \\ 0 & 0 & 0\end{pmatrix}.
\end{aligned}
\tag{7.153}
$$

Combining Eqs. (7.152) and (7.153), we obtain

$$
\begin{aligned}
\alpha^k_p(a,b) &= \langle\eta;\mathrm{v}'|\hat{\alpha}^k_{q=0}|\eta;\mathrm{v}\rangle\sqrt{(2J'+1)(2J+1)}\begin{pmatrix} J' & k & J \\ -M' & p & M\end{pmatrix} \\
&\quad \times (-1)^{M'}(-1)^{J+J'+k}\left\{-\frac{S'_{1,N'-1}S_{1,N+1}}{2}\left[1+(-1)^{J+J'+k}\right]\begin{pmatrix} J' & J & k \\ 1 & -1 & 0\end{pmatrix}\right. \\
&\quad \left. +S'_{0,N'-1}S_{0,N+1}\begin{pmatrix} J' & J & k \\ 0 & 0 & 0\end{pmatrix}\right\}.
\end{aligned}
\tag{7.154}
$$

Possible $F_1 \to F_3$ pure rotational Stokes transitions include ${}^{Q}O_{31}(J)$ transitions for which $\Delta N = N' - N = 0$ and $\Delta J = J' - J = -2$, ${}^{Q}S_{13}(J)$ transitions for which $\Delta N = N' - N = 0$ and $\Delta J = J' - J = -2$, ${}^{S}Q_{31}(J)$ transitions for which $\Delta N = N' - N = +2$ and $\Delta J = J' - J = 0$, and ${}^{U}S_{31}(J)$ transitions for which $\Delta N = N' - N = +4$ and $\Delta J = J' - J = +2$. Evaluating Eq. (7.154) for $k = 2$, and realizing that $J + J' + 2$ is always an even number, we obtain

$$\alpha_p^2(a,b) = \langle \eta; \mathrm{v}'|\hat{\alpha}^2_{q=0}|\eta; \mathrm{v}\rangle \sqrt{(2J'+1)(2J+1)} \begin{pmatrix} J' & 2 & J \\ -M' & p & M \end{pmatrix} (-1)^{M'} \times \left\{ -S'_{1,N'-1} S_{1,N+1} \begin{pmatrix} J' & J & 2 \\ 1 & -1 & 0 \end{pmatrix} + S'_{0,N'-1} S_{0,N+1} \begin{pmatrix} J' & J & 2 \\ 0 & 0 & 0 \end{pmatrix} \right\}. \tag{7.155}$$

Setting $p = 0$ in Eq. (7.155), squaring $\alpha^2_{p=0}(a,b)$, summing over all M and M', and dividing by $2J+1$, we obtain

$$\left\langle \left| \alpha^2_{p=0}(A,B) \right|^2 \right\rangle = \frac{\langle \eta; \mathrm{v}'|\hat{\alpha}^2_{q=0}|\eta; \mathrm{v}\rangle^2 (2J'+1)}{5} \left\{ S'^2_{1,N'-1} S^2_{1,N+1} \begin{pmatrix} J' & J & 2 \\ 1 & -1 & 0 \end{pmatrix}^2 - 2S'_{1,N'-1} S_{1,N+1} S'_{0,N'-1} S_{0,N+1} \begin{pmatrix} J' & J & 2 \\ 1 & -1 & 0 \end{pmatrix} \begin{pmatrix} J' & J & 2 \\ 0 & 0 & 0 \end{pmatrix} + S'^2_{0,N'-1} S^2_{0,N+1} \begin{pmatrix} J' & J & 2 \\ 0 & 0 & 0 \end{pmatrix}^2 \right\} = \frac{\langle \eta; \mathrm{v}'|\hat{\alpha}^2_{q=0}|\eta; \mathrm{v}\rangle^2}{5} b^{(2)}_{N'-1,N+1}, \tag{7.156}$$

where

$$b^{(2)}_{N'-1,N+1} = (2J'+1) \left\{ S'^2_{1,N'-1} S^2_{1,N+1} \begin{pmatrix} J' & J & 2 \\ 1 & -1 & 0 \end{pmatrix}^2 - 2S'_{1,N'-1} S_{1,N+1} S'_{0,N'-1} S_{0,N+1} \begin{pmatrix} J' & J & 2 \\ 1 & -1 & 0 \end{pmatrix} \begin{pmatrix} J' & J & 2 \\ 0 & 0 & 0 \end{pmatrix} + S'^2_{0,N'-1} S^2_{0,N+1} \begin{pmatrix} J' & J & 2 \\ 0 & 0 & 0 \end{pmatrix}^2 \right\}. \tag{7.157}$$

Evaluating Eq. (7.154) for $k = 1$, and realizing that $J + J' + 1$ is always an odd number, we obtain $\alpha_p^1(a,b) = 0$. Evaluating Eq. (7.154) for $k = 0$, and realizing that $J + J'$ is always an even number, we obtain

$$\alpha_p^0(a,b) = \langle \eta; \mathrm{v}'|\hat{\alpha}^0_{q=0}|\eta; \mathrm{v}\rangle \sqrt{(2J'+1)(2J+1)} \begin{pmatrix} J' & 0 & J \\ -M' & p & M \end{pmatrix} (-1)^{M'} \times \left\{ -S'_{1,N'-1} S_{1,N+1} \begin{pmatrix} J' & J & 0 \\ 1 & -1 & 0 \end{pmatrix} + S'_{0,N'-1} S_{0,N+1} \begin{pmatrix} J' & J & 0 \\ 0 & 0 & 0 \end{pmatrix} \right\}. \tag{7.158}$$

Setting $p = 0$ in Eq.(7.158), squaring $\alpha^0_{p=0}(a,b)$, summing over all M and M', and dividing by $2J+1$, we obtain

$$\left\langle \left| \alpha^0_{p=0}(A,B) \right|^2 \right\rangle = \langle \eta; \mathrm{v}'|\hat{\alpha}^0_{q=0}|\eta; \mathrm{v}\rangle^2 (2J'+1) \left\{ S'^2_{1,N'-1} S^2_{1,N+1} \begin{pmatrix} J' & J & 0 \\ 1 & -1 & 0 \end{pmatrix}^2 - 2S'_{1,N'-1} S_{1,N+1} S'_{0,N'-1} S_{0,N+1} \begin{pmatrix} J' & J & 0 \\ 1 & -1 & 0 \end{pmatrix} \begin{pmatrix} J' & J & 0 \\ 0 & 0 & 0 \end{pmatrix} + S'^2_{0,N'-1} S^2_{0,N+1} \begin{pmatrix} J' & J & 0 \\ 0 & 0 & 0 \end{pmatrix}^2 \right\} = \langle \eta; \mathrm{v}'|\hat{\alpha}^0_{q=0}|\eta; \mathrm{v}\rangle^2 b^{(0)}_{N'-1,N+1}, \tag{7.159}$$

where

$$b^{(0)}_{N'-1,N+1} = (2J'+1)\left\{ S'^2_{1,N'-1} S^2_{1,N+1} \begin{pmatrix} J' & J & 0 \\ 1 & -1 & 0 \end{pmatrix}^2 \right.$$
$$- 2S'_{1,N'-1} S_{1,N+1} S'_{0,N'-1} S_{0,N+1} \begin{pmatrix} J' & J & 0 \\ 1 & -1 & 0 \end{pmatrix} \begin{pmatrix} J' & J & 0 \\ 0 & 0 & 0 \end{pmatrix} \tag{7.160}$$
$$\left. + S'^2_{0,N'-1} S^2_{0,N+1} \begin{pmatrix} J' & J & 0 \\ 0 & 0 & 0 \end{pmatrix}^2 \right\}.$$

Following the same procedures as for the $F_1 \to F_1$ transitions, we obtain for pure rotational transitions with $\Delta J = 0, \ \pm 2, \ \mathrm{v}' = \mathrm{v}$,

$$\left\langle |\alpha_{ZZ}(A,B)|^2 \right\rangle = \frac{2}{3} \left\langle \left| \alpha^2_{p=0}(A,B) \right|^2 \right\rangle = \frac{4}{15} \gamma_0^2 b^{(2)}_{N'+1,N+1} \tag{7.161}$$

and

$$\left\langle |\alpha_{YZ}(A,B)|^2 \right\rangle = \frac{3}{4} \left\langle |\alpha_{ZZ}(A,B)|^2 \right\rangle \tag{7.162}$$

and for vibrational transitions with $\Delta J = 0, \ \pm 2, \ \mathrm{v}' = \mathrm{v} + 1$.

$$\begin{aligned} \left\langle |\alpha_{ZZ}(A,B)|^2 \right\rangle &= \frac{1}{3} \left\langle \left| (\alpha')^2_{p=0}(A,B) \right|^2 \right\rangle + \frac{2}{3} \left\langle \left| (\alpha')^2_{p=0}(A,B) \right|^2 \right\rangle \\ &= \frac{\hbar(\mathrm{v}+1)}{2\mu_R \omega_{vib}} \left[(a')^2 b^{(0)}_{N'-1,N+1} + \frac{4}{45} (\gamma')^2 b^{(2)}_{N'-1,N+1} \right] \end{aligned} \tag{7.163}$$

and

$$\begin{aligned} \left\langle |\alpha_{YZ}(A,B)|^2 \right\rangle &= \frac{1}{2} \left\langle \left| (\alpha')^2_{p=1}(A,B) \right|^2 \right\rangle = \frac{1}{2} \left\langle \left| (\alpha')^2_{p=0}(A,B) \right|^2 \right\rangle \\ &= \frac{\hbar(\mathrm{v}+1)}{2\mu_R \omega_{vib}} \left[\frac{1}{15} (\gamma')^2 b^{(2)}_{N'-1,N+1} \right] \end{aligned} \tag{7.164}$$

For $F_3 \to F_1$ transitions, the Placzek–Teller coefficients $b^{(0)}_{N'-1,N+1}$ and $b^{(2)}_{N'-1,N+1}$ are replaced by $b^{(0)}_{N'+1,N-1}$ and $b^{(2)}_{N'+1,N-1}$, respectively, in Eqs. (7.161)–(7.164), where

$$b^{(2)}_{N'+1,N-1} = (2J'+1)\left\{ S^2_{1,N-1} S'^2_{1,N'+1} \begin{pmatrix} J' & J & 2 \\ 1 & -1 & 0 \end{pmatrix}^2 \right.$$
$$- 2S_{1,N-1} S'_{1,N'+1} S_{0,N-1} S'_{1,N'+1} \begin{pmatrix} J' & J & 2 \\ 1 & -1 & 0 \end{pmatrix} \begin{pmatrix} J' & J & 2 \\ 0 & 0 & 0 \end{pmatrix} \tag{7.165}$$
$$\left. + S^2_{0,N-1} S'^2_{0,N'+1} \begin{pmatrix} J' & J & 2 \\ 0 & 0 & 0 \end{pmatrix}^2 \right\}$$

and

$$b^{(0)}_{N'+1,N-1} = (2J'+1)\left\{ S^2_{1,N'-1} S'^2_{1,N+1} \begin{pmatrix} J' & J & 0 \\ 1 & -1 & 0 \end{pmatrix}^2 \right.$$
$$- 2S_{1,N-1} S'_{1,N'+1} S_{0,N-1} S'_{0,N'+1} \begin{pmatrix} J' & J & 0 \\ 1 & -1 & 0 \end{pmatrix} \begin{pmatrix} J' & J & 0 \\ 0 & 0 & 0 \end{pmatrix} \tag{7.166}$$
$$\left. + S^2_{0,N-1} S'^2_{0,N'+1} \begin{pmatrix} J' & J & 0 \\ 0 & 0 & 0 \end{pmatrix}^2 \right\}.$$

The pure rotational Raman spectrum of O_2 was measured with a spectral resolution of approximately 0.4 cm^{-1} by Renschler et al. (1969) and also by Bérard et al. (1983). The data from Figure 2 in Renschler et al. (1969) were digitized and are shown in Figure 7.5 along with a theoretical spectrum calculated using the results of this section. The spectrum was recorded for both the Stokes and anti-Stokes spectral regions around the very strong Rayleigh line at zero Raman shift. The anti-Stokes intensities were calculated by defining level A as the upper, initial level for the transition in Eqs. (7.11) and (7.12) and in the expressions in Section 7.3 for $\left\langle |\alpha_{ZZ}(A,B)|^2 \right\rangle$ and $\left\langle |\alpha_{YZ}(A,B)|^2 \right\rangle$.

As shown in Figure 7.5, there are allowed Stokes transitions with $\Delta J = \pm 1$ and ± 2. The N-dependence of the Placzek–Teller coefficients for some of these transitions is shown in Figure 7.6. The Placzek–Teller coefficient $b^{(2)}_{N'+1,N+1}$ for the $\Delta N = \Delta J = 2$ main-branch transition ${}^sS_{11}(N)$ is shown in Figure 7.6. The value of the Placzek–Teller coefficient for ${}^sS_{11}(N)$ is nearly independent of N and reaches a limiting value of 3/8 at high N; the behavior of the Placzek–Teller coefficients for the

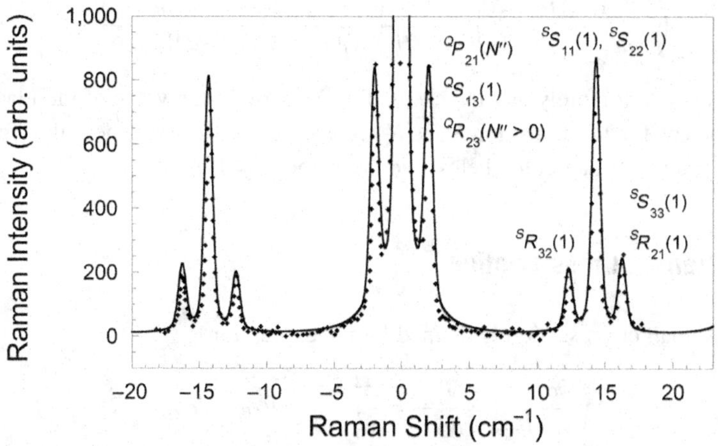

Figure 7.5 Experimental (solid diamonds) and theoretical (solid line) pure rotational Raman spectra for the ground $^3\Sigma_g^-$ electronic level of O_2. The spectrum was acquired and published by Renschler et al. (1969) as Figure 2 in that paper (© Elsevier, data used with permission). The spectrum shown in Figure 2 of Renschler et al. (1969) was digitized, and a sloping background in the experimental spectrum was corrected for the experimental data shown in this figure.

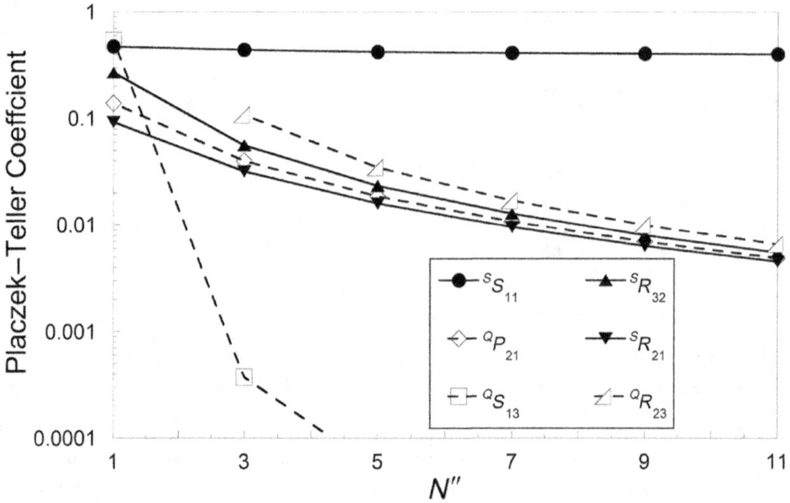

Figure 7.6 Calculated values of the Placzek–Teller coefficients for both main branch ($\Delta N = \Delta J$) and satellite ($\Delta N \neq \Delta J$) pure rotational Raman transitions in the ground $^3\Sigma_g^-$ electronic level of O_2.

main branch transitions $^sS_{22}(N)$ and $^sS_{33}(N)$ are similar. Conversely, for the satellite transitions $^QP_{21}(N)$, $^QR_{23}(N)$, $^SR_{21}(N)$, and $^SR_{23}(N)$ with $\Delta J = \pm 1$ and $\Delta J \neq \Delta N$, the value of the Placzek–Teller coefficient decreases rapidly with increasing N. Examining the appropriate equations for $3j$ symbols in Appendix A3, we find that the value of the square of the appropriate $3j$ symbol,

$$\begin{pmatrix} j+1 & j & 2 \\ m & -m & 0 \end{pmatrix}^2 = \frac{4m^2[6(j+m+1)(j-m+1)]}{(2j+4)(2j+3)(2j+2)(2j+1)2j}, \tag{7.167}$$

decreases very rapidly as j increases. The decrease in the value of the Placzek–Teller coefficient for the transition $^QS_{13}(N)$ is even more drastic: The value for $N = 3$ is three orders of magnitude below the value for $N = 1$.

7.4 The Raman Cross-Section

The Raman cross-section is defined by the expression

$$\left(\frac{\partial\sigma}{\partial\Omega}\right)_{ZZ} = \frac{I_{ZZ}^{\Omega}}{N_A I_L} = \frac{\pi^2}{\varepsilon_0^2}(\tilde{\nu}_0 + \tilde{\nu}_{AB})^4\left\langle |\alpha_{ZZ}(A,B)|^2 \right\rangle. \tag{7.168}$$

$$\left(\frac{\partial\sigma}{\partial\Omega}\right)_{YZ} = \frac{I_{YZ}^{\Omega}}{N_A I_L} = \frac{\pi^2}{\varepsilon_0^2}(\tilde{\nu}_0 + \tilde{\nu}_{AB})^4\left\langle |\alpha_{YZ}(A,B)|^2 \right\rangle. \tag{7.169}$$

The results of experimental measurements of Raman transition strengths are most often reported in terms of Raman cross-sections because I_{ZZ}^{Ω} and I_{YZ}^{Ω} are measured directly, and N_A and I_L are usually controllable parameters in these experiments. Schrötter and Klöckner (1979) have compiled an extensive list of Raman cross-sections for numerous gas-phase and liquid-phase species. In general, they list Raman cross-sections from experiments where both Y- and Z-polarized scattering was detected. Typically in these experiments, Q-branch Stokes Raman scattering was collected from an entire vibrational band for the species. The appropriate cross-section for these experiments is given by

$$\left(\frac{\partial\sigma}{\partial\Omega}\right) = \left(\frac{\partial\sigma}{\partial\Omega}\right)_{ZZ} + \left(\frac{\partial\sigma}{\partial\Omega}\right)_{YZ} = \frac{\pi^2}{\varepsilon_0^2}(\tilde{\nu}_0 - \tilde{\nu}_R)^4\left[\left\langle \alpha_{ZZ}|^2\right\rangle + \left\langle |\alpha_{YZ}|^2\right\rangle\right]. \tag{7.170}$$

$$\left\langle |\alpha_{ZZ}(A,B)|^2\right\rangle = \frac{\hbar(\mathrm{v}+1)}{4\pi\mu_R c\tilde{\nu}_R}\left[(a')^2\delta_{J'J} + \frac{4}{45}b_{J'J}^{(2)}(\gamma')^2\right]. \tag{7.171}$$

$$\left\langle |\alpha_{YZ}(A,B)|^2\right\rangle = \frac{\hbar(\mathrm{v}+1)}{4\pi\mu_R c\tilde{\nu}_R}\left[\frac{1}{15}b_{J'J}^{(2)}(\gamma')^2\right]. \tag{7.172}$$

The Raman cross-section for a single Q-branch transition $Q(J)$can be expressed in terms of the tensor invariants $(a')^2$ and $(\gamma')^2$:

$$\left(\frac{\partial\sigma}{\partial\Omega}\right) = \left(\frac{\partial\sigma}{\partial\Omega}\right)_{ZZ} + \left(\frac{\partial\sigma}{\partial\Omega}\right)_{YZ} = \frac{\hbar\pi(\tilde{\nu}_0 - \tilde{\nu}_R)^4(\mathrm{v}+1)}{4\varepsilon_0^2\mu_R c\,\tilde{\nu}_R}\left[(a')^2 + \frac{7}{45}b_{JJ}^{(2)}(\gamma')^2\right]. \tag{7.173}$$

For a given band $\mathrm{v} \to \mathrm{v}+1$, the invariants $(a')^2$ and $(\gamma')^2$ are nearly independent of rotational quantum number J except for Herman–Wallis effects due to centrifugal distortion, which will be discussed in detail in Section 7.5. For comparison with band-averaged measurements, we can assume that $b_{JJ} = 0.25$, the high-J asymptote for the Placzek–Teller coefficient. The band-averaged cross-section is thus given by

$$\left(\frac{\partial\sigma}{\partial\Omega}\right) = \left(\frac{\partial\sigma}{\partial\Omega}\right)_{ZZ} + \left(\frac{\partial\sigma}{\partial\Omega}\right)_{YZ} = \frac{\hbar\pi(\tilde{\nu}_0 - \tilde{\nu}_R)^4(\mathrm{v}+1)}{4\varepsilon_0^2\mu_R c\,\tilde{\nu}_R}\left[(a')^2 + \frac{7}{180}(\gamma')^2\right]. \tag{7.174}$$

We can obtain an expression for $(\gamma')^2$ in terms of ρ and $(a')^2$ from Eq. (7.103):

$$(\gamma')^2 = \frac{45\rho}{(3-4\rho)b_{JJ}^{(2)}}(a')^2 = \frac{180\rho}{(3-4\rho)}(a')^2. \tag{7.175}$$

Substituting Eq. (7.175) into Eq. (7.174) and rearranging, we obtain

$$\begin{aligned}(a')^2 &= \left(\frac{\partial\sigma}{\partial\Omega}\right)\frac{4\varepsilon_0^2\mu_R c\,\tilde{\nu}_R}{\hbar\pi(\tilde{\nu}_0 - \tilde{\nu}_R)^4(\mathrm{v}+1)\{1+[7\rho/(3-4\rho)]\}}\\ &= \left(\frac{\partial\sigma}{\partial\Omega}\right)_{zz}\frac{4\varepsilon_0^2\mu_R c\,\tilde{\nu}_R}{\hbar\pi(\tilde{\nu}_0 - \tilde{\nu}_R)^4(\mathrm{v}+1)\{1+[4\rho/(3-4\rho)]\}}.\end{aligned} \tag{7.176}$$

The Raman frequencies, cross sections, depolarization ratios, invariants, and polarizabilities are listed for four important gas phase species in Table 7.1. This table will

Table 7.1 Raman cross-sections, invariants, depolarization ratios, and polarizabilities for Q-branch transitions in N_2, O_2, H_2, and CO. The values of Σ_{ZZ} were taken from Lapp (1980) who analyzed literature values of experimentally determined Raman cross-sections listed in Schrötter and Klöckner (1979). The listed depolarization ratios apply to high-J Q-branch transitions for which $b_{JJ} = 0.25$

Species	Σ_{ZZ}*	$\tilde{\nu}_R$ $\left(\text{cm}^{-1}\right)$	$\left(\frac{\partial\sigma}{\partial\Omega}\right)_{ZZ}$ $\left(\frac{\text{m}^2}{\text{sr}}\right)$ $\lambda_L = 500$ nm	ρ	$(a')^2$ $\left(\frac{\text{C}^4\,\text{m}^2}{\text{J}^2}\right)$	$(\gamma')^2$ $\left(\frac{\text{C}^4\,\text{m}^2}{\text{J}^2}\right)$	$\langle a_{ZZ}^2\rangle_{10}$ $\left(\frac{\text{C}^4\,\text{m}^2}{\text{J}^2}\right)$
N_2	1.00	2331	4.92×10^{-35}	0.022	3.79×10^{-60}	5.15×10^{-60}	4.01×10^{-83}
O_2	1.04	1555	6.08×10^{-35}	0.047	2.90×10^{-60}	8.73×10^{-60}	4.17×10^{-83}
H_2	3.86	4156	12.3×10^{-35}	0.012	1.89×10^{-60}	1.38×10^{-60}	15.5×10^{-83}
CO	0.93	2143	4.77×10^{-35}	0.04	3.69×10^{-60}	9.35×10^{-60}	3.73×10^{-83}

$$^*\Sigma_{ZZ,j} = \frac{(\partial\sigma/\partial\Omega)_{ZZ,j}}{(\partial\sigma/\partial\Omega)_{ZZ,N_2}} \frac{\left(\tilde{\nu}_L - 2,331\ \text{cm}^{-1}\right)^4}{\left(\tilde{\nu}_L - \tilde{\nu}_{R,j}\right)^4}\left[1 - \exp\left(-\frac{hc\tilde{\nu}_{R,j}}{k_B T}\right)\right]$$

provide the reader with a means to check the units and conversion factors associated with the calculation of important quantities for Raman spectral analysis.

7.5 Effects of Vibration-Rotation Interactions and Vibrational Anharmonicity on Raman Tensor Invariants

The effect of centrifugal distortion on Raman intensities has been the subject of numerous studies since an early paper by James and Klemperer (1959). Taking into account centrifugal distortion, the calculation of vibrational Raman intensities for molecules with $^1\Sigma$ ground electronic levels is performed by writing Eq. (7.70) in a slightly modified form:

$$\left\langle \left(\alpha_p^k\right)^2 \right\rangle = \frac{(2J'+1)}{2k+1} \langle \psi_{\text{v}'J'}(r) | \langle \psi_{elec}(r) | \hat{\alpha}_q^k | \psi_{elec}(r) \rangle | \psi_{\text{v}J}(r) \rangle^2 \begin{pmatrix} J' & k & J \\ 0 & 0 & 0 \end{pmatrix}^2. \tag{7.177}$$

In Eq. (7.177), the vibrational and rotational quantum numbers for the initial and final levels are v, J and v$'$, J', respectively, and $\psi_{\text{v}J}(r)$ and $\psi_{\text{v}'J'}(r)$ are the vibration-rotation wavefunctions for the initial and final levels, respectively, $\psi_{elec}(r)$ is the electronic wavefunction, and r is the internuclear separation. The electronic integral is solved to calculate the molecule-fixed reference frame (MFRF) polarizability at a specified value of the internuclear separation r,

$$\alpha_q^k(r) = \langle \psi_{elec}(r) | \hat{\alpha}_q^k(r) | \psi_{elec}(r) \rangle. \tag{7.178}$$

As is the case with our earlier analysis of Eq. (7.70), the polarizability can then be expanded in a Taylor series around the equilibrium internuclear distance $r = r_e$:

$$\alpha_q^k(Q) = \alpha_q^k(0) + \left(\frac{\partial \alpha_q^k}{\partial Q}\right)_{Q=0} Q + \frac{1}{2}\left(\frac{\partial^2 \hat{\alpha}_q^k}{\partial Q^2}\right)_{Q=0} Q^2 + \cdots, \tag{7.179}$$

where $Q = r - r_e$. If (1) vibration-rotation interactions are neglected so that $\left(\partial \alpha_q^k / \partial Q\right)_{Q=0}$ is not a function of Q, (2) terms in Eq. (7.179) that are quadratic and higher are neglected, and (3) the molecule is assumed to be a harmonic oscillator, the polarizability tensor elements that lead to vibrational Raman scattering are given by

$$\left\langle \alpha_{ZZ}^2 \right\rangle = \frac{1}{3}\left(\alpha_{p=0}^0\right)^2 + \frac{2}{3}\left(\alpha_{p=0}^2\right)^2 = \frac{\hbar(\mathrm{v}+1)}{2\mu_R \omega_R}\left[(a')^2 \delta_{JJ'} + \frac{4}{45}(2J'+1)\begin{pmatrix} J' & k & J \\ 0 & 0 & 0 \end{pmatrix}^2 (\gamma')^2\right] \tag{7.180}$$

and

$$\left\langle \alpha_{YZ}^2 \right\rangle = \frac{1}{2}\left(\alpha_{p=1}^2\right)^2 = \frac{\hbar(\mathrm{v}+1)}{2\mu_R \omega_R}\left[\frac{1}{15}(2J'+1)\begin{pmatrix} J' & k & J \\ 0 & 0 & 0 \end{pmatrix}^2 (\gamma')^2\right], \tag{7.181}$$

where

$$(a')^2 = \frac{1}{3}\left|\left(\partial \alpha_{q=0}^0 / \partial Q\right)_{Q=0}\right|^2 = \frac{1}{9}\left|\left[\partial(\alpha_{zz} + 2\alpha_{xx})/\partial Q\right]_{Q=0}\right|^2 = \frac{1}{9}\left|(\alpha'_{zz} + 2\alpha'_{xx})\right|^2. \tag{7.182}$$

$$(\gamma')^2 = \frac{3}{2}\left|\left(\partial \alpha_{q=0}^2 / \partial Q\right)_{Q=0}\right|^2 = \left|\left[\partial(\alpha'_{zz} - \alpha'_{xx})/\partial Q\right]_{Q=0}\right|^2. \tag{7.183}$$

In general, the MFRF isotropic and anisotropic polarizability tensor components $\alpha_{q=0}^0$ and $\alpha_{q=0}^2$, respectively, are the only components that contribute significantly to Raman intensities. In Eqs. (7.182) and (7.183), $\alpha'_{zz} = \alpha'_{\parallel}$ and $\alpha'_{xx} = \alpha'_{yy} = \alpha'_{\perp}$ are the polarizability derivatives parallel to and perpendicular to the internuclear axis in the MFRF, respectively.

In computer programs such as CARSFT (Palmer, 1989) from Sandia, assumptions (2) and (3) above are used, and the vibration-rotation interaction is modeled using the results from James and Klemperer (1959). There have been a number of papers devoted to the calculation of the Herman–Wallis factors in Raman spectra in the intervening years (Bohlin et al., 2011; Marrocco, 2009, 2010, 2012; Marrocco et al., 2012; Tipping & Ogilvie, 1984; Utsav & Varghese, 2013). Following Marocco (2009), we define the Herman–Wallis correction factor for $^1\Sigma$ Raman transitions from an initial level v, J to a final level v', J' as

$$F_{\mathrm{v},J}^{\mathrm{v}',J'} = \frac{\langle \mathrm{v}, J|\hat{\alpha}|\mathrm{v}', J'\rangle^2}{\langle \mathrm{v}, J_0|\hat{\alpha}|\mathrm{v}', J_0'\rangle^2}, \tag{7.184}$$

where $\hat{\alpha}$ is a polarizability operator, and $J_0 = J_0' = 0$ will equal to 0 for Q-branch transitions, $J_0 = 0$, $J_0' = 2$ for S-branch transitions, and $J_0 = 2$, $J_0' = 0$ for O-branch transitions. Restricting our discussion to Q-branch transitions for the moment, and again following the analysis of Marrocco (2009), we define vibration-level-dependent values of the Raman invariants $\left(a'_{\mathrm{vv}'}\right)^2$ and $\left(\gamma'_{\mathrm{vv}'}\right)^2$:

$$\left(a'_{\mathrm{vv}'}\right)^2 = \frac{\left\langle \psi_{\mathrm{v},0}(r) \middle| \left[\alpha'_{zz}(r) + 2\alpha'_{xx}(r)\right] \middle| \psi_{\mathrm{v}',0}(r) \right\rangle^2}{9}. \tag{7.185}$$

$$\left(\gamma'_{\mathrm{vv}'}\right)^2 = \left\langle \psi_{\mathrm{v},0}(r) \middle| \left[\alpha'_{zz}(r) - \alpha'_{xx}(r)\right] \middle| \psi_{\mathrm{v}',0}(r) \right\rangle^2. \tag{7.186}$$

James and Klemperer (1959) performed an analysis of vibrational and pure rotational transitions assuming a potential function given by the harmonic oscillator potential "plus the first three terms of the expansion of the centrifugal potential $\left[\hbar^2 J(J+1)\right]/2\mu r^2$." They also neglected all terms except for the first two terms in the Taylor series expansion of the polarizability given in Eq. (7.179). The result of their analysis for vibrational Q-branch transitions is given by

$$\left(F_{\mathrm{v},J}^{\mathrm{v}',J'}\right)_{JK} = 1 - 3\eta^2 J(J+1)/2, \tag{7.187}$$

where $\eta = 2B_e/\omega_e$, and Eq. (7.187) applies for both the mean polarizability $(a')^2$ and the anisotropy $(\gamma')^2$. More complicated expressions were derived by James and Klemperer for vibrational O- and S-branch transitions and for pure rotational transitions. An improved theory that began to incorporate the effect of vibrational anharmonicity was developed by Bouanich and Brodbeck (1976) and Luthe et al. (1986). The Herman–Wallis factor of Luthe et al. (1986) for vibrational Q-branch transitions is given by

$$\left(F_{\mathrm{v},J}^{\mathrm{v}',J'}\right)_{LBY} = \left[1 - 3\eta^2(a_1+1)J(J+1)/4\right]^2, \tag{7.188}$$

where $a_1 = -\left(\alpha_e\omega_e/6B_e^2\right) - 1$. For this analysis, any variation in the polarizability beyond the first two terms in the Taylor series expansion of Eq. (7.179) is again neglected. Higher order terms in the Taylor series expansion were considered in the analysis of Tipping and Ogilvie (1984) and Tipping and Bouanic (2001), resulting in the following expression for the Herman–Wallis factor:

$$\left(F_{\mathrm{v},J}^{\mathrm{v}',J'}\right)_{TB} = 1 - \left[3(a_1+1)/2 - 4p_2/p_1\right]\eta^2 J(J+1), \tag{7.189}$$

where p_2 and p_1 are coefficients in the following expansions for the mean and anisotropic polarizabilities:

$$\alpha = \frac{\left(\alpha_{\parallel} + 2\alpha_{\perp}\right)}{3} = \frac{(ea_0)^2}{E_h}\left(p_{0\alpha} + p_{1\alpha}x + p_{2\alpha}x^2 + p_{3\alpha}x^3 + \cdots\right), \tag{7.190}$$

$$\gamma = \frac{\left(\alpha_{\parallel} - \alpha_{\perp}\right)}{3} = \frac{(ea_0)^2}{E_h}\left(p_{0\gamma} + p_{1\gamma}x + p_{2\gamma}x^2 + p_{3\gamma}x^3 + \cdots\right), \tag{7.191}$$

where $x = (r - r_e)/r_e$, and the Hartree energy $E_h = 4.3597482 \times 10^{-18}$ J. Substitution of Eqs. (7.190) or (7.191) into Eq. (7.189) results in the calculation of

Table 7.2 Coefficients for the calculation of the mean polarizability α and the anisotropy γ using Eqs. (7.190) and (7.191), respectively. The coefficients given in Maroulis (2003) and Buldakov et al. (2003) were adapted to conform to the format of Eqs. (7.190) and (7.191). The values for α and γ listed in Rychlewski (1980) and Wolniewicz (1993) were curve-fitted to find the coefficients listed in the table

Coefficient	N_2 Maroulis (2003)	O_2 Buldakov et al. (2003)	H_2 Rychlewski (1980) Wolniewicz (1993)
$p_{0\alpha}$	11.848	11.9	5.2211
$p_{1\alpha}$	12.811	12.6	6.4116
$p_{2\alpha}$	3.9551	4.05	1.2733
$p_{3\alpha}$	−7.3295	−5.17	−1.7623
$p_{4\alpha}$	0.011108		0.11292
$p_{0\gamma}$	4.6032	4.79	1.8680
$p_{1\gamma}$	14.583	15.0	5.3991
$p_{2\gamma}$	8.3216	8.77	3.1588
$p_{3\gamma}$	−5.0946	−6.07	−1.9129
$p_{4\gamma}$	3.6084		−0.30493

Herman–Wallis factors for the mean and anisotropic polarizabilities, respectively. A table of coefficients for the molecules H_2, N_2, and O_2 is given in Table 7.2.

Advances in calculation methods for quantum chemistry since that time now enable a different approach for the calculation of these effects in Raman and CARS spectroscopy. The use of the software programs RKR1 (Le Roy, 2017a) and LEVEL (Le Roy, 2017b), developed by Robert Le Roy to calculate transition strengths in the ultraviolet spectra of OH and NO, was discussed in Chapter 5. We use these same programs to calculate the wavefunctions $\psi_{v,J}(r)$ in the ground electronic $^1\Sigma^+$ level of N_2 for $v = 0$ to $v = 4$ and for $J = 0$ to $J = 80$. This allows us to use the results of ab initio quantum chemistry calculations of the polarizability terms $\alpha(r)$ and $\gamma(r)$ (Maroulis, 2003) to directly calculate Raman polarizability components using Eq. (7.177), tensor invariants using

$$\left(a'_{vv'}\right)^2 = \left\langle \psi_{v,0}(r) \middle| \left[\alpha_{\parallel}(r) + 2\alpha_{\perp}(r)\right]/3 \middle| \psi_{v',0}(r) \right\rangle^2 = \left\langle \psi_{v,0}(r) \middle| \alpha(r) \middle| \psi_{v',0}(r) \right\rangle^2, \quad (7.192)$$

$$\left(\gamma'_{vv'}\right)^2 = \left\langle \psi_{v,0}(r) \middle| \left[\alpha_{\parallel}(r) - \alpha_{\perp}(r)\right] \middle| \psi_{v',0}(r) \right\rangle^2 = \left\langle \psi_{v,0}(r) \middle| \gamma(r) \middle| \psi_{v',0}(r) \right\rangle^2, \quad (7.193)$$

and Herman–Wallis correction factors using Eq. (7.184).

The results of the calculation of the polarizability tensor element $\langle \alpha_{ZZ}^2 \rangle$ for the fundamental band ($v'' = 0 \rightarrow v' = 1$) and the second hot band ($v'' = 2 \rightarrow v' = 3$) are shown in Figure 7.7a and b. There are several interesting aspects of the plots shown in Figure 7.7. First, for the fundamental band, the effect of the vibration-rotation interaction is that the Raman polarizability increases slowly with increasing rotational quantum number J compared to the value calculated with the assumption that the molecule is a rigid rotator and harmonic oscillator (RRHO). For $J = 50$, the

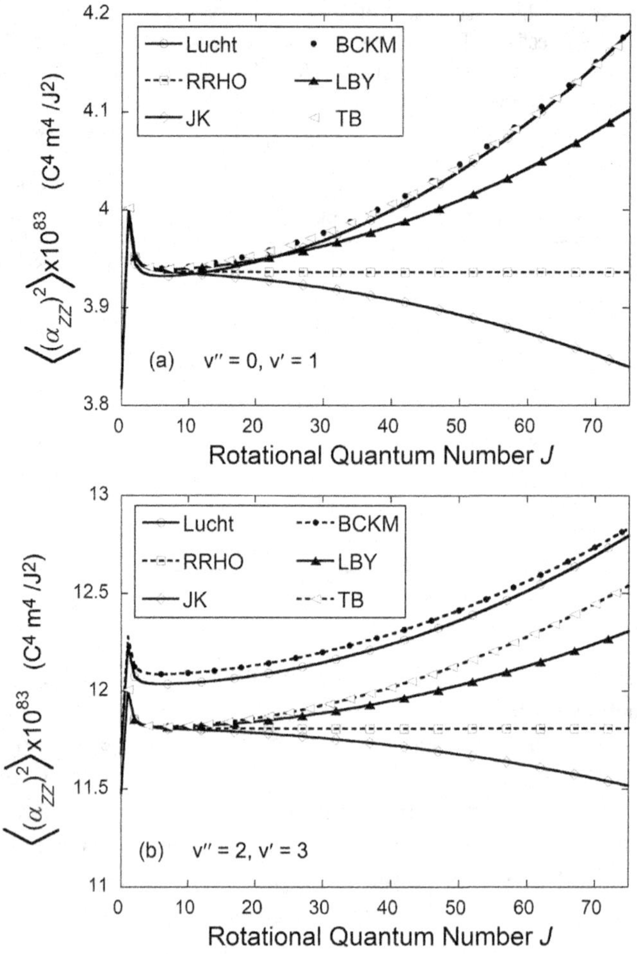

Figure 7.7 The values of the polarizability tensor element $\langle \alpha_{ZZ}^2 \rangle$ for the (a) fundamental band and (b) the second hot band of N_2 for the case where both the rigid-rotator harmonic oscillator case (RRHO) and where vibration-rotation interactions and vibration anharmonicity are considered. The RRHO results are multiplied by the Herman–Wallis factors listed in Eqs. (7.187), (7.188), and (7.189) for the curves labeled JK, LBY, and TB, respectively. The curve labeled BCKM is from Buldakov et al. (2003) who use the Herman–Wallis correction factor from Tipping and Bouanic (2001) but also include the effects of vibrational anharmonicity and higher-order polarizability terms in their analysis.

polarizability tensor element $\langle \alpha_{ZZ}^2 \rangle$ is about 2.6% greater than the RRHO value due to the vibration-rotation interaction. The reason for this increase can be understood by examining Figure 7.8. The derivative of the isotropic polarizability with respect to r increases with increasing r. At higher J, the peak of the vibration-rotation wavefunction shifts to higher values of r due to centrifugal stretching, and consequently the integral $\langle \psi_{v'J'}(r) | [d(\alpha_{zz} + 2\alpha_{xx})/dr](r - r_e) | \psi_{vJ}(r) \rangle$ also increases

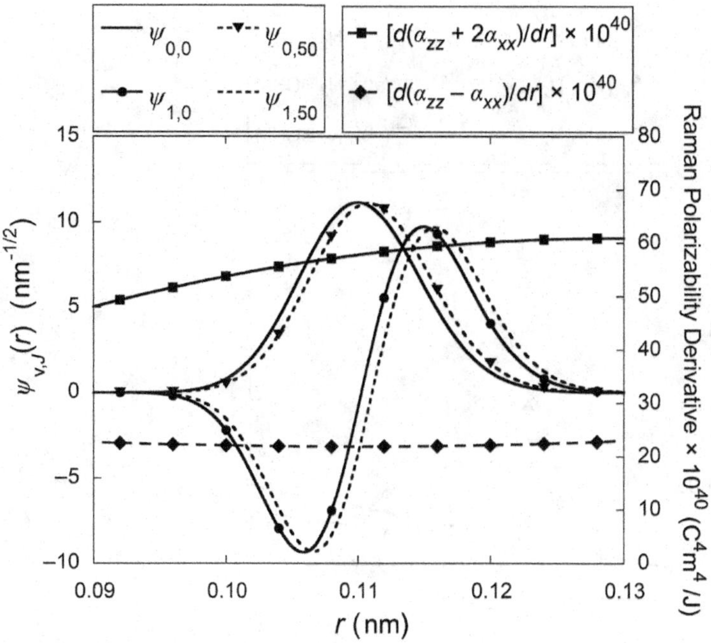

Figure 7.8 The N_2 vibration rotation wavefunctions for $(\mathrm{v}=0, J=0)$, $(\mathrm{v}=0, J=50)$, $(\mathrm{v}=1, J=0)$, and $(\mathrm{v}=1, J=50)$ are plotted along with the isotropic and anisotropic polarizability derivatives.

slightly. Neglecting the effects of the vibration-rotation interaction in the spectral fitting process will result in fitted temperatures that are higher than the true temperature.

For the second hot band, there is a significant deviation between the two values even for low J. This is due to the anharmonic nature of the vibrational wave function and the inclusion of higher-order terms in the polarizability for the integral $\langle\psi_{\mathrm{v}'0}(r)|\alpha(r)|\psi_{\mathrm{v}0}(r)\rangle$. Again, the peak values of the vibration-rotation wavefunctions will shift to higher r as the vibrational quantum number increases. Therefore, the polarizability tensor element $\langle\alpha_{ZZ}^2\rangle$ increases slightly faster than the $\mathrm{v}+1$ dependence that would be characteristic of a harmonic oscillator. The dependence of the tensor invariants $\left(a'_{\mathrm{vv}'}\right)^2$ and $\left(\gamma'_{\mathrm{vv}'}\right)^2$ is shown in Figure 7.9, where the ratios $R_a(\mathrm{v},\mathrm{v}+1)=\left(a'_{\mathrm{vv}'}\right)^2/\left(a'_{01}\right)^2$ and $R_\gamma(\mathrm{v},\mathrm{v}+1)=\left(\gamma'_{\mathrm{vv}'}\right)^2/\left(\gamma'_{01}\right)^2$ are plotted as functions of initial vibrational quantum number for both N_2 and H_2. The results of Buldakov et al. (2003) are in excellent agreement with our calculations for N_2. The vibrational-level dependence of the anisotropic invariant $\left(\gamma'_{\mathrm{vv}'}\right)^2$ for H_2 is especially significant. Again, neglecting the effects of the vibration anharmonicity and the higher-order terms in the polarizability for the spectral fitting process will result in fitted temperatures that are higher than the true temperature.

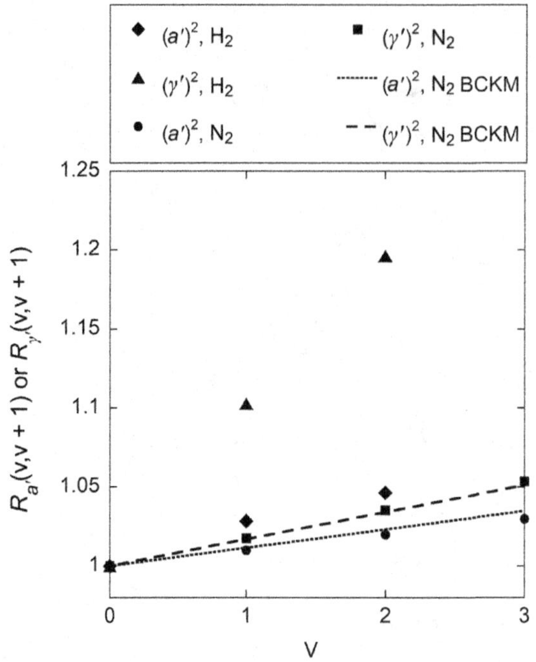

Figure 7.9 Vibrational-level dependence of the ratios $R_a(\mathrm{v},\mathrm{v}+1)=\left(a'_{\mathrm{vv'}}\right)^2/\left(a'_{01}\right)^2$ and $R_\gamma(\mathrm{v},\mathrm{v}+1)=\left(\gamma'_{\mathrm{vv'}}\right)^2/\left(\gamma'_{01}\right)^2$ for N_2 and H_2. The curves labeled BCKM are calculated from formulae listed in Buldakov et al. (2003).

7.6 Raman Scattering Spectroscopy Example Problem

A gas cell contains 50% H_2 and 50% N_2 by volume at a pressure of 200 kPa and a temperature of 500 K. A laser beam with a pulse energy of 1.0 J, a pulse length of 1 μsec, and a wavelength of 400 nm is focused to a diameter of 200 μm in the cell. Raman scattering is collected by a 100-mm-diameter, 100-mm-focal-length lens placed 200 mm from the 1-mm-long probe volume. The laser is linearly polarized in the *Z*-direction, and the collection lens is placed along the *X*-axis as depicted in Figure 7.1. For this example problem, Herman–Wallis effects will be neglected. Calculate the number of photons collected by the lens in a single laser pulse due to Raman scattering from:

(a) the nitrogen Q-branch, $\mathrm{v}=0\rightarrow\mathrm{v}=1$, both polarizations detected;
(b) the Q(1) line of the $\mathrm{v}=0\rightarrow\mathrm{v}=1$ transition of hydrogen, I^{Ω}_{ZZ} detected;
(c) the Q(6) line of the $\mathrm{v}=0\rightarrow\mathrm{v}=1$ transition of nitrogen, both polarizations detected.

Solution: (a) We will need to calculate the scattered intensity I^{Ω}_{ZZ} (W/sr) from Eq. (7.11),

$$I_{ZZ}^{\Omega} = \frac{(\omega_L + \omega_{AB})^4 N_A}{16 c^4 \pi^2 \varepsilon_0^2} \left\langle |\alpha_{ZZ}(A,B)|^2 \right\rangle I_L = \frac{\pi^2}{\varepsilon_0^2} (\tilde{\nu}_L - \tilde{\nu}_R)^4 \left\langle |\alpha_{ZZ}(A,B)|^2 \right\rangle I_L N_A, \tag{7.194}$$

where $\tilde{\nu}_{AB} = -\tilde{\nu}_R$ for the Stokes scattering process that we are considering. The orientation-averaged tensor element is given by

$$\left\langle |\alpha_{ZZ}(A,B)|^2 \right\rangle = \frac{(\mathrm{v}+1)\hbar}{2\pi\mu_R c\tilde{\nu}_R} \left[(a')^2 + \frac{4}{45} b_{J,J} (\gamma')^2 \right] \quad \mathrm{v} \rightarrow \mathrm{v}+1,\ \Delta J = 0. \tag{7.195}$$

For nitrogen, from Table 7.1, we find

$$(a')^2 = 3.79 \times 10^{-60} \frac{\mathrm{C^4 m^2}}{\mathrm{J^2}}, \quad (\gamma')^2 = 5.15 \times 10^{-60} \frac{\mathrm{C^4 m^2}}{\mathrm{J^2}}.$$

At high J, the Placzek coefficient $b_{J,J}$ approaches an asymptotic limit of 1/4. We will use this value to calculate the ZZ tensor component. The ZZ tensor component is thus given by

$$\begin{aligned} \left\langle |\alpha_{ZZ}(A,B)|^2 \right\rangle &= \frac{(0+1)(1.054 \times 10^{-34}\ \mathrm{J\,s})}{4\pi(1.169 \times 10^{-26}\ \mathrm{kg})(2.998 \times 10^8\ \mathrm{m/s})(2330 \times 10^2\ \mathrm{m^{-1}})} \\ &\quad \times \left[\left(3.79 \times 10^{-60} \frac{\mathrm{C^4 m^2}}{\mathrm{J^2}} \right) + \frac{4}{45} \left(\frac{1}{4} \right) \left(5.15 \times 10^{-60} \frac{\mathrm{C^4 m^2}}{\mathrm{J^2}} \right) \right] \\ &= 4.01 \times 10^{-83} \frac{\mathrm{C^4 m^4}}{\mathrm{J^2}}. \end{aligned} \tag{7.196}$$

The laser irradiance is given by

$$I_L = \frac{E_L}{\Delta t_L A_L} = \frac{(1.0\ J)}{(1.0 \times 10^{-6}\ \mathrm{s}) \left(\frac{\pi}{4}\right) (200 \times 10^{-6}\ \mathrm{m})^2} = 3.18 \times 10^{13} \frac{\mathrm{J}}{\mathrm{s-m^2}}. \tag{7.197}$$

The number of molecules that contribute to the Q-branch signal is approximately equal to the number of nitrogen molecules in the probe volume. At 500 K, virtually all of the N_2 molecules are in the $\mathrm{v} = 0$ level, and we assume that we collect Raman scattered light from all Q-branch Raman transitions regardless of J. Therefore,

$$\begin{aligned} N_A = n_{N_2} V_c &= \frac{p_{N_2}}{k_B T} \frac{\pi}{4} D_c^2 \ell_c \\ &= \frac{(10^5\ \mathrm{J/m^3})}{(1.3806 \times 10^{-23}\ \mathrm{J/K})(500\ \mathrm{K})} \left(\frac{\pi}{4} \right) (200 \times 10^{-6}\ \mathrm{m})^2 (10^{-3}\ \mathrm{m}) \\ &= 4.55 \times 10^{14}. \end{aligned} \tag{7.198}$$

The frequency of the Raman Stokes photons in units of $\mathrm{m^{-1}}$ is given by

$$(\tilde{\nu}_L - \tilde{\nu}_R)^4 = (2.500 \times 10^6 - 2.330 \times 10^5\ \mathrm{m^{-1}})^4 = 2.64 \times 10^{25}\ \mathrm{m^{-4}}. \tag{7.199}$$

The intensity in W/sr of the Z-polarized Raman scattered light emitted from the probe volume for the geometry shown in Figure 7.1 is given by

$$I_{ZZ}^{\Omega} = \frac{\pi^2 \left(2.64\times 10^{25}\ \mathrm{m^{-4}}\right)\left(4.03\times 10^{-83}\ \frac{\mathrm{C^4\,m^4}}{\mathrm{J^2}}\right)}{\left(8.854\times 10^{-12}\ \frac{\mathrm{C^2}}{\mathrm{Jm}}\right)^2}\left(3.18\times 10^{13}\ \frac{\mathrm{J}}{\mathrm{s\,m^2}}\right)\left(4.55\times 10^{14}\right)$$
$$= 1.94\times 10^{-6}\ \mathrm{W/sr}. \quad (7.200)$$

The intensity in Watts/steradian of the Raman scattered light emitted for both polarizations is

$$I^{\Omega} = I_{ZZ}^{\Omega}(1+\rho) = \left(1.94\times 10^{-6}\ \frac{\mathrm{J}}{\mathrm{s}}\right)(1.022) = 1.98\times 10^{-6}\ \frac{\mathrm{J}}{\mathrm{s}}. \quad (7.201)$$

The solid angle of the collection optics is given by

$$\Omega_c = \frac{A_{lens}}{\left(\ell_{pv\text{-}lens}\right)^2} = \frac{(\pi/4)D_{lens}^2}{\left(\ell_{pv\text{-}lens}\right)^2} = \frac{(\pi/4)(100)^2}{(200)^2} = 0.196. \quad (7.202)$$

The number of photons collected is given by

$$N_p = \frac{I^{\Omega}\Omega_c \Delta t_L}{hc(\tilde{\nu}_L - \tilde{\nu}_R)} = \frac{\left(1.98\times 10^{-6}\ \frac{\mathrm{J}}{\mathrm{s}}\right)(0.196)\left(10^{-6}\ \mathrm{s}\right)}{\left(6.626\times 10^{-34}\ \mathrm{J-s}\right)\left(2.998\times 10^{8}\ \frac{\mathrm{m}}{\mathrm{s}}\right)\left(2.267\times 10^{6}\ \mathrm{m^{-1}}\right)}$$
$$= 862{,}000 \text{ photons}. \quad (7.203)$$

(b) For the Q(1) line of hydrogen, we detect only Z-polarized Raman scattering,

$$N_p = \frac{I_{ZZ}^{\Omega}\Omega_c \Delta t_L}{hc(\tilde{\nu}_L - \tilde{\nu}_R)}. \quad (7.204)$$

For part (b), the solid angle of the collection optics and the laser pulse length are the same as for part (a). The scattered intensity scattered intensity I_{ZZ}^{Ω} is given by Eq. (7.194). The laser irradiance is the same as for part (a). The *ZZ* tensor element is given by Eq. (7.195). For hydrogen, from Table 7.1, we find

$$(a')^2 = 1.89\times 10^{-60}\frac{\mathrm{C^4\,m^2}}{\mathrm{J^2}} \quad (\gamma')^2 = 1.38\times 10^{-60}\frac{\mathrm{C^4\,m^2}}{\mathrm{J^2}} \quad \rho = 0.012.$$

The Placzek coefficient is given by Eq. (7.90)

$$b_{J,J} = \frac{J(J+1)}{(2J-1)(2J+3)} = \frac{2}{5}. \quad (7.205)$$

The *ZZ* tensor component is given by

$$\left\langle |\alpha_{ZZ}(A,B)|^2 \right\rangle = \frac{(0+1)(6.626\times10^{-34}\ \mathrm{J-s})}{8\pi^2(0.835\times10^{-27}\ \mathrm{kg})\left(2.998\times10^8\ \frac{\mathrm{m}}{\mathrm{s}}\right)(4156\times10^2\ \mathrm{m}^{-1})}$$
$$\times\left[\left(1.89\times10^{-60}\ \frac{\mathrm{C}^4\,\mathrm{m}^2}{\mathrm{J}^2}\right)+\frac{4}{45}\left(\frac{2}{5}\right)\left(1.38\times10^{-60}\ \frac{\mathrm{C}^4\,\mathrm{m}^2}{\mathrm{J}^2}\right)\right]$$
$$= 1.55\times10^{-82}\ \frac{\mathrm{C}^4\,\mathrm{m}^4}{\mathrm{J}^2}. \tag{7.206}$$

The number of hydrogen molecules in the probe volume is equal to the number of nitrogen molecules in the probe volume, 4.55×10^{14}. We need to find the number of hydrogen molecules in the probe volume in the $J=1, \mathrm{v}=0$ level. At a temperature of 500 K, all the hydrogen molecules will be in the ground vibrational and electronic levels. The fraction of H_2 molecules in a rotational level J is given by (Hanson, 2016; Laurendeau, 2005)

$$\frac{N_J}{N_{H_2}} = \frac{NSSW(2J+1)\exp[-hcF(J)/k_BT]}{Z_{rot}Z_{nuc}}. \tag{7.207}$$

For odd J levels in H_2, the nuclear spin statistical weight is $NSSW=3$. For even J levels in H_2, $NSSW=1$. The spectroscopic parameters needed for the calculation of the lower level energy and the Raman transition frequency are (Laurendeau, 2005)

$$\omega_e = 4{,}401.21\ \mathrm{cm}^{-1},\quad \omega_e x_e = 121.336\ \mathrm{cm}^{-1},\quad B_e = 60.853\ \mathrm{cm}^{-1},\quad \alpha_e = 3.062\ \mathrm{cm}^{-1},$$
$$D_e = 4.71\times10^{-2}\ \mathrm{cm}^{-1}$$
$$G(\mathrm{v}) = \omega_e\left(\mathrm{v}+\frac{1}{2}\right) - \omega_e x_e\left(\mathrm{v}+\frac{1}{2}\right)\quad G(0)=2{,}170.3\ \mathrm{cm}^{-1}\quad G(1)=6{,}328.8\ \mathrm{cm}^{-1}$$
$$B_\mathrm{v} = B_e - \alpha_e\left(\mathrm{v}+\frac{1}{2}\right)\quad B_0 = 59.32\ \mathrm{cm}^{-1}\quad B_1 = 56.26\ \mathrm{cm}^{-1}$$
$$F_\mathrm{v}(J) = B_\mathrm{v}J(J+1) - D_eJ^2(J+1)^2\quad F_0(1) = 118.5\ \mathrm{cm}^{-1}\quad F_1(1) = 112.3\ \mathrm{cm}^{-1}.$$

The rotational partition function is given by

$$Z_{rot} = \frac{1}{\sigma}\frac{T}{\theta_{rot}} = \frac{1}{\sigma}\frac{T}{(hc/k_B)B_0} = \frac{1}{2}\frac{500\ \mathrm{K}}{(1.439\ \mathrm{K/cm^{-1}})(59.32\ \mathrm{cm}^{-1})} = 2.93, \tag{7.208}$$

where the parameter $\sigma=2$ for homonuclear diatomic molecules, and $\sigma=1$ for heteronuclear diatomic molecules. The nuclear partition function is given by

$$Z_{nuc} = (2I_H+1)^2 = 4. \tag{7.209}$$

The population fraction for the $J=1, \mathrm{v}=0$ level is given by

$$\frac{N_{J=1}}{N_{H_2}} = \frac{3(3)\exp[-(1.439\ \mathrm{K\,cm})(118.5\ \mathrm{cm}^{-1})/(500\ \mathrm{K})]}{(2.93)(4)} = 0.546. \tag{7.210}$$

$$N_{J=1} = (0.546)\left(4.55 \times 10^{14}\right) = 2.48 \times 10^{14}. \tag{7.211}$$

The Raman frequency is given by

$$\begin{aligned} \tilde{\nu}_R &= F_1(1) + G(1) - F_0(1) + G(0) = 4,152.3 \text{ cm}^{-1} = 415,230 \text{ m}^{-1} \\ \omega_j &= 2\pi c\tilde{\nu}_R = 7.8217 \times 10^{14} \text{ s}^{-1}. \end{aligned} \tag{7.212}$$

$$\left(\tilde{\nu}_L - \tilde{\nu}_R\right)^4 = \left(2.500 \times 10^6 - 4.1523 \times 10^5 \text{ m}^{-1}\right)^4 = 1.889 \times 10^{25} \text{ m}^{-4}. \tag{7.213}$$

The $Z-$polarized Raman scattered intensity is given by

$$\begin{aligned} I_{ZZ}^{\Omega} &= \frac{\pi^2\left(1.889 \times 10^{25} \text{ m}^{-4}\right)\left(1.55 \times 10^{-82} \dfrac{\text{C}^4 \text{ m}^4}{\text{J}^2}\right)}{\left(8.8542 \times 10^{-12} \dfrac{\text{C}^2}{\text{Jm}}\right)^2}\left(3.18 \times 10^{13} \frac{\text{J}}{\text{sm}^2}\right)\left(2.48 \times 10^{14}\right) \\ &= 2.91 \times 10^{-6} \text{ W/sr}. \end{aligned} \tag{7.214}$$

The number of photons collected for the Z-polarized Raman scattering from the Q(1) transition is

$$\begin{aligned} N_p &= \frac{I_{ZZ}^{\Omega}\Omega_c\Delta t_L}{hc(\tilde{\nu}_L - \tilde{\nu}_R)} = \frac{\left(2.91 \times 10^{-6} \dfrac{\text{J}}{\text{s}}\right)(0.196)\left(10^{-6} \text{ s}\right)}{\left(6.626 \times 10^{-34} \text{ J} - \text{s}\right)\left(2.998 \times 10^8 \dfrac{\text{m}}{\text{s}}\right)\left(2.084 \times 10^6 \text{ m}^{-1}\right)} \\ &= 1.38 \times 10^6 \text{ photons}. \end{aligned} \tag{7.215}$$

(c) For the O(6) line of nitrogen, we detect both Z-polarized and Y-polarized Raman scattering. We will calculate the number of photons in the Z-polarized and Y-polarized Raman scattering signals separately. For part (c), the following equations apply:

$$N_{pZ} = \frac{I_{ZZ}^{\Omega}\Omega_c\Delta t_L}{hc(\tilde{\nu}_L - \tilde{\nu}_R)}, \quad N_{pY} = \frac{I_{YZ}^{\Omega}\Omega_c\Delta t_L}{hc(\tilde{\nu}_L - \tilde{\nu}_R)} = N_{pZ}\frac{I_{YZ}^{\Omega}}{I_{ZZ}^{\Omega}} \tag{7.216}$$

and

$$I_{ZZ}^{\Omega} = \frac{\pi^2}{\varepsilon_0^2}(\tilde{\nu}_L - \tilde{\nu}_R)^4\left\langle|\alpha_{ZZ}(A,B)|^2\right\rangle I_L N_A, \quad I_{YZ}^{\Omega} = \frac{\pi^2}{\varepsilon_0^2}(\tilde{\nu}_L - \tilde{\nu}_R)^4\left\langle|\alpha_{YZ}(A,B)|^2\right\rangle I_L N_A. \tag{7.217}$$

For part (c), the calculation of the Raman scattering from the O(6) line of N_2, the solid angle of the collection optics, and the laser parameters are the same as for part (a). Now let's calculate the frequency of the Raman transition. For the O(6) line,

$$J_i = J'' = 6 \quad J_f = J' = 4 \quad \text{v}_i = \text{v}'' = 0 \quad \text{v}_f = \text{v}' = 1.$$

For nitrogen (Laurendeau, 2005),

$$\omega_e = 2357.6\ \text{cm}^{-1} \quad \omega_e x_e = 14.06\ \text{cm}^{-1} \quad B_e = 1.998\ \text{cm}^{-1} \quad \alpha_e = 0.0179\ \text{cm}^{-1}$$
$$D_e = 5.76 \times 10^{-6}\ \text{cm}^{-1}$$
$$G(\text{v}) = \omega_e\left(\text{v} + \frac{1}{2}\right) - \omega_e x_e\left(\text{v} + \frac{1}{2}\right) \quad G(0) = 1{,}175.3\ \text{cm}^{-1} \quad G(1) = 3{,}504.8\ \text{cm}^{-1}$$
$$B_\text{v} = B_e - \alpha_e\left(\text{v} + \frac{1}{2}\right) \Rightarrow \quad B_0 = 1.989\ \text{cm}^{-1} \quad B_1 = 1.971\ \text{cm}^{-1}$$
$$F_\text{v}(J) = B_\text{v} J(J+1) - D_e J^2 (J+1)^2 \Rightarrow \quad F_0(6) = 83.5\ \text{cm}^{-1} \quad F_1(4) = 39.4\ \text{cm}^{-1}.$$

For the O(6) line,

$$\frac{\varepsilon_A}{hc} = G(0) + F_0(6) = 1{,}258.8\ \text{cm}^{-1} \qquad \frac{\varepsilon_B}{hc} = G(1) + F_1(4) = 3{,}544.2\ \text{cm}^{-1}$$
$$\tilde{\nu}_R = \frac{\varepsilon_B}{hc} - \frac{\varepsilon_A}{hc} = 2{,}285.4\ \text{cm}^{-1} = 2.2854 \times 10^5\ \text{m}^{-1}$$
$$\tilde{\nu}_L - \tilde{\nu}_R = 2.5 \times 10^6 - 2.2854 \times 10^5\ \text{m}^{-1} = 2.2715 \times 10^6\ \text{m}^{-1}.$$

The *ZZ* and *YZ* tensor elements are given by

$$\left\langle |\alpha_{ZZ}(A,B)|^2 \right\rangle = \frac{(\text{v}+1)\hbar}{4\pi\mu_R c\tilde{\nu}_R}\left[\frac{4}{45} b_{J-2,J} (\gamma')^2\right], \quad \text{v} = 0 \rightarrow \text{v}' = 1,\ J = 6 \rightarrow J' = 4. \tag{7.218}$$

$$\left\langle |\alpha_{YZ}(A,B)|^2 \right\rangle = \frac{(\text{v}+1)\hbar}{4\pi\mu_R c\,\tilde{\nu}_R}\left[\frac{1}{15} b_{J-2,J} (\gamma')^2\right], \quad \text{v} = 0 \rightarrow \text{v}' = 1,\ J = 6 \rightarrow J' = 4. \tag{7.219}$$

The Placzek coefficient is given by Eq. (7.91),

$$b_{J-2,J} = \frac{3J(J-1)}{2(2J+1)(2J-1)} = \frac{3(6)(5)}{2(13)(11)} = 0.315 \quad J \rightarrow J-2. \tag{7.220}$$

For nitrogen, from Table 7.1, we find

$$(\gamma')^2 = 5.15 \times 10^{-60}\ \frac{\text{C}^4\,\text{m}^2}{\text{J}^2}.$$

The *ZZ* tensor component is thus given by

$$\left\langle |\alpha_{ZZ}(A,B)|^2 \right\rangle = \frac{(0+1)\left(1.055 \times 10^{-34}\ \text{J} - \text{s}\right)}{4\pi\left(1.169 \times 10^{-26}\ \text{kg}\right)\left(2.998 \times 10^8\ \frac{\text{m}}{\text{s}}\right)\left(2.2854 \times 10^5\ \text{m}^{-1}\right)} \times \left[\frac{4}{45}(0.315)\left(5.15 \times 10^{-60}\ \frac{\text{C}^4\,\text{m}^2}{\text{J}^2}\right)\right] = 1.51 \times 10^{-84}\ \frac{\text{C}^4\,\text{m}^4}{\text{J}^2}. \tag{7.221}$$

The YZ tensor component is given by

$$\left\langle|\alpha_{YZ}(A,B)|^2\right\rangle = \frac{(0+1)(1.055\times10^{-34}\ \mathrm{J-s})}{4\pi(1.169\times10^{-26}\ \mathrm{kg})\left(2.998\times10^8\ \frac{\mathrm{m}}{\mathrm{s}}\right)(2.2854\times10^5\ \mathrm{m}^{-1})} \times\left[\frac{1}{15}(0.315)\left(5.15\times10^{-60}\ \frac{\mathrm{C}^4\,\mathrm{m}^2}{\mathrm{J}^2}\right)\right] = 1.14\times10^{-84}\ \frac{\mathrm{C}^4\,\mathrm{m}^4}{\mathrm{J}^2}. \tag{7.222}$$

Now we need to find the number of nitrogen molecules in the probe volume in the $\mathrm{v}=0$, $J=6$ level. Again, we will assume that virtually all the molecules are in the $\mathrm{v}=0$ ground vibrational level. The number of nitrogen molecules in the probe volume is equal to 4.55×10^{14} from part (a). The fraction of N_2 molecules in $J=6$ is given by

$$\frac{N_{J=6}}{N_{N_2}} = \frac{NSSW(2J+1)\exp[-hcF_0(6)/k_BT]}{Z_{rot}Z_{nuc}}. \tag{7.223}$$

$$Z_{rot} = \frac{1}{\sigma}\frac{T}{\theta_{rot}} = \frac{1}{\sigma}\frac{T}{(hc/k_B)B_0} = \frac{1}{2}\frac{500\ \mathrm{K}}{(1.439\ \mathrm{cm\,K})(1.989\ \mathrm{cm}^{-1})} = 87.3. \tag{7.224}$$

$$Z_{nuc} = (2I_N+1)^2 = [2(1)+1]^2 = 9. \tag{7.225}$$

For odd J and even J levels in N_2, $NSSW = 3$ and $NSSW = 6$, respectively. Therefore

$$\frac{N_{J=6}}{N_{N_2}} = \frac{6(13)\exp[-(1.439\ \mathrm{cm\,K})(83.5\ \mathrm{cm}^{-1})/(500\ \mathrm{K})]}{(87.3)(9)} = 0.0781. \tag{7.226}$$

and

$$N_A = (0.0781)(4.55\times10^{14}) = 3.55\times10^{13}. \tag{7.227}$$

The Z-polarized Raman scattered intensity is given by

$$I_{ZZ}^{\Omega} = \frac{\pi^2(2.271\times10^6\ \mathrm{m}^{-1})^4\left(1.51\times10^{-84}\ \frac{\mathrm{C}^4\,\mathrm{m}^4}{\mathrm{J}^2}\right)}{\left(8.8542\times10^{-12}\ \frac{\mathrm{C}^2}{\mathrm{J\,m}}\right)^2}\left(3.18\times10^{13}\ \frac{\mathrm{W}}{\mathrm{m}^2}\right)(3.55\times10^{13}) = 5.72\times10^{-9}\ \mathrm{W/sr}. \tag{7.228}$$

The Y–polarized Raman scattered intensity is given by

$$I_{YZ}^{\Omega} = \frac{\pi^2(2.271\times10^6\ \mathrm{m}^{-1})^4\left(1.14\times10^{-84}\ \frac{\mathrm{C}^4\,\mathrm{m}^4}{\mathrm{J}^2}\right)}{\left(8.8542\times10^{-12}\ \frac{\mathrm{C}^2}{\mathrm{J\,m}}\right)^2}\left(3.18\times10^{13}\ \frac{\mathrm{W}}{\mathrm{m}^2}\right)(3.55\times10^{13}) = 4.30\times10^{-9}\ \frac{\mathrm{J}}{\mathrm{s}}. \tag{7.229}$$

The number of photons collected for the Z-polarized Raman scattering from the O(6) transition is

$$
\begin{aligned}
N_{pZ} &= \frac{I_{ZZ}^{\Omega}\Omega_c \Delta t_L}{hc(\tilde{\nu}_L - \tilde{\nu}_R)} \\
&= \frac{\left(5.72 \times 10^{-9}\ \frac{\text{J}}{\text{s}}\right)(0.196)\left(10^{-6}\ \text{s}\right)}{\left(6.626 \times 10^{-34}\ \text{J} - \text{s}\right)\left(2.998 \times 10^{8}\ \frac{\text{m}}{\text{s}}\right)\left(2.27146 \times 10^{6}\ \text{m}^{-1}\right)} \qquad (7.230) \\
&= 2,480 \text{ photons.}
\end{aligned}
$$

The number of photons collected for the Y-polarized Raman scattering from the O(6) transition is

$$
N_{pY} = \frac{I_{IZ}^{\Omega}\Omega_c \Delta t_L}{hc(\tilde{\nu}_L - \tilde{\nu}_R)} = 1,860 \text{ photons} = 0.75 N_{pZ}. \qquad (7.231)
$$

The total number of photons collected for both polarizations for the O(6) transition is 4,340.

8 Coherent Anti-Stokes Raman Scattering (CARS) Spectroscopy

8.1 Introduction

Coherent anti-Stokes Raman scattering (CARS) spectroscopy is a technique that has been widely applied for temperature measurements in combustion and for microscopic imaging of cell structures. CARS spectroscopy is discussed in detail in this chapter as an example of a nonlinear optical technique. The concept of nonlinear susceptibility is introduced, and the derivation of the susceptibility tensor appropriate for CARS spectroscopy is described in detail. A key aspect of this derivation is the incorporation of the electric dipole transition matrix elements for the Raman scattering process into the susceptibility tensor.

8.2 Perturbative Solution of the Density Matrix Equations

The development of the expressions for the susceptibility tensor $\chi^{(n)}(-\omega_\sigma;\omega_1,\omega_2,\ldots\omega_n)$ is described succinctly in Boyd (2008). We will discuss the development of the linear susceptibility and the nonlinear second-order susceptibility in detail, and then develop the expression for the CARS susceptibility from the general expression for the third-order nonlinear susceptibility.

8.2.1 The Linear Susceptibility

The interaction of laser radiation with atomic or molecular resonances is described in general by the time-dependent density matrix equation,

$$\begin{aligned}\dot{\rho}_{ab} &= -i\omega_{ab}\rho_{ab} - \frac{i}{\hbar}\sum_c (V_{ac}\rho_{cb} - \rho_{ac}V_{cb}) - \gamma_{ab}\left(\rho_{ab} - \rho_{ab}^{eq}\right) \\ &= -i\omega_{ab}\rho_{ab} - \frac{i}{\hbar}[V,\rho]_{ab} - \gamma_{ab}\left(\rho_{ab} - \rho_{ab}^{eq}\right).\end{aligned} \tag{8.1}$$

The density matrix element is expanded as a power series in the perturbation parameter λ,

$$\rho_{ab} = \rho_{ab}^{(0)} + \lambda\rho_{ab}^{(1)} + \lambda^2\rho_{ab}^{(2)} + \lambda^3\rho_{ab}^{(3)} + \cdots. \tag{8.2}$$

Substituting Eq. (8.2) into Eq. (8.1) and solving for terms corresponding to the same value of the perturbation parameter λ, we obtain

$$\dot{\rho}_{ab}^{(0)} = -i\omega_{ab}\rho_{ab}^{(0)} - \gamma_{ab}\left(\rho_{ab}^{(0)} - \rho_{ab}^{eq}\right). \tag{8.3}$$

$$\dot{\rho}_{ab}^{(1)} = -(i\omega_{ab} + \gamma_{ab})\rho_{ab}^{(1)} - \frac{i}{\hbar}\left[\hat{V},\hat{\rho}^{(0)}\right]_{ab}. \tag{8.4}$$

$$\dot{\rho}_{ab}^{(2)} = -(i\omega_{ab} + \gamma_{ab})\rho_{ab}^{(2)} - \frac{i}{\hbar}\left[\hat{V},\hat{\rho}^{(1)}\right]_{ab}. \tag{8.5}$$

$$\dot{\rho}_{ab}^{(3)} = -(i\omega_{ab} + \gamma_{ab})\rho_{ab}^{(3)} - \frac{i}{\hbar}\left[\hat{V},\hat{\rho}^{(2)}\right]_{ab}. \tag{8.6}$$

The zero-order terms describe the condition of the atomic or molecular assembly in the absence of laser irradiation. The off-diagonal coherence terms will be zero in the absence of laser irradiation,

$$\rho_{ab}^{(0)} = 0 \quad \mathit{if} \quad a \neq b. \tag{8.7}$$

The diagonal terms are equal to the equilibrium state population fractions,

$$\rho_{aa}^{(0)} = \rho_{aa}^{eq}, \quad \rho_{bb}^{(0)} = \rho_{bb}^{eq}. \tag{8.8}$$

Using these values of the zero-order density matrix elements, Eq. (8.4) can be solved by direct integration. The first-order density matrix elements are expressed in terms of a slowly varying amplitude function $\sigma_{ab}^{(1)}$,

$$\rho_{ab}^{(1)} = \sigma_{ab}^{(1)} \exp[-(i\omega_{ab} + \gamma_{ab})t]. \tag{8.9}$$

Substituting Eq. (8.9) into Eq. (8.4) and simplifying, we obtain

$$\dot{\sigma}_{ab}^{(1)} = -\frac{i}{\hbar}\left[V(t),\rho^{(0)}\right]_{ab} \exp[(i\omega_{ab} + \gamma_{ab})t]. \tag{8.10}$$

Integrating Eq. (8.10), we obtain

$$\sigma_{ab}^{(1)} = \int_{-\infty}^{t} -\frac{i}{\hbar}\left[V(t_1),\rho^{(0)}\right]_{ab} \exp[(i\omega_{ab} + \gamma_{ab})t_1]dt_1. \tag{8.11}$$

Substituting for the slowly varying amplitude using Eq. (8.9), we obtain

$$\begin{aligned} \rho_{ab}^{(1)} &= \int_{-\infty}^{t} -\frac{i}{\hbar}\left[V(t_1),\rho^{(0)}\right]_{ab} \exp\left[(i\omega_{ab} + \gamma_{ab})(t_1 - t)\right]dt_1 \\ &= \exp\left[-(i\omega_{ab} + \gamma_{ab})t\right]\int_{-\infty}^{t} -\frac{i}{\hbar}\left[V(t_1),\rho^{(0)}\right]_{ab} \exp\left[(i\omega_{ab} + \gamma_{ab})t_1\right]dt_1. \end{aligned} \tag{8.12}$$

We will assume that the zero-order off-diagonal density matrix terms are zero,

$$\rho_{ab}^{(0)} = 0 \quad \mathit{if} \quad a \neq b. \tag{8.13}$$

In other words, the coherence terms will be zero in the absence of laser irradiation. The laser field is written as the sum of frequency components for fields of different frequencies ω_p, which are given by

$$\boldsymbol{E}(t) = \frac{1}{2}\sum_p \boldsymbol{E}(\omega_p)\exp(-i\omega_p t). \tag{8.14}$$

For the first-order polarization term, the term $\left[\hat{V}(t_1), \hat{\rho}^{(0)}\right]_{ab}$ is given by

$$\left[\hat{V}, \hat{\rho}^{(0)}\right]_{ab} = \sum_c \left(V_{ac}\rho_{cb}^{(0)} - \rho_{ac}^{(0)} V_{cb}\right) = \sum_c \left[-\boldsymbol{\mu}_{ac} \bullet \boldsymbol{E}(t_1)\rho_{cb}^{(0)} + \rho_{ac}^{(0)}\boldsymbol{\mu}_{cb} \bullet \boldsymbol{E}(t_1)\right]. \tag{8.15}$$

Because the zero-order off-diagonal terms are zero, this reduces to

$$\left[\hat{V}, \hat{\rho}^{(0)}\right]_{ab} = \left(\rho_{aa}^{(0)} - \rho_{bb}^{(0)}\right)\boldsymbol{\mu}_{ab} \bullet \boldsymbol{E}(t_1). \tag{8.16}$$

Substituting Eqs. (8.14) and (8.16) into Eq. (8.12) and simplifying, we obtain

$$\begin{aligned} \rho_{ab}^{(1)} &= \frac{i}{\hbar}\left(\rho_{bb}^{(0)} - \rho_{aa}^{(0)}\right)\sum_p \boldsymbol{\mu}_{ab} \bullet \boldsymbol{E}(\omega_p)\exp\left[-(i\omega_{ab} + \gamma_{ab})t\right] \\ &\quad \times \int_{-\infty}^{t} \exp\left\{\left[i(\omega_{ab} - \omega_p) + \gamma_{ab}\right]t_1\right\} dt_1 \\ &= \frac{i}{\hbar}\left(\rho_{bb}^{(0)} - \rho_{aa}^{(0)}\right)\sum_p \boldsymbol{\mu}_{ab} \bullet \boldsymbol{E}(\omega_p)\exp\left[-(i\omega_{ab} + \gamma_{ab})t\right] \\ &\quad \times \left[\frac{\exp\left\{\left[i(\omega_{ab} - \omega_p) + \gamma_{ab}\right]t_1\right\}}{i(\omega_{ab} - \omega_p) + \gamma_{ab}}\right]_{-\infty}^{t}. \end{aligned} \tag{8.17}$$

Evaluating the direct integral and simplifying, we obtain

$$\rho_{ab}^{(1)} = \frac{\left(\rho_{bb}^{(0)} - \rho_{aa}^{(0)}\right)}{\hbar}\sum_p \frac{\boldsymbol{\mu}_{ab} \bullet \boldsymbol{E}(\omega_p)\exp(-i\omega_p t)}{(\omega_{ab} - \omega_p) - i\gamma_{ab}}. \tag{8.18}$$

The expectation value of the dipole moment is given by

$$\begin{aligned} \langle \boldsymbol{\mu}(t) \rangle &= \mathrm{Tr}\left[\rho^{(1)}(t)\boldsymbol{\mu}\right] = \sum_a \sum_b \rho_{ab}^{(1)}(t)\boldsymbol{\mu}_{ba} \\ &= \sum_a \sum_b \frac{\left(\rho_{bb}^{(0)} - \rho_{aa}^{(0)}\right)}{2\hbar}\sum_p \frac{\boldsymbol{\mu}_{ba}\boldsymbol{\mu}_{ab} \bullet \boldsymbol{E}(\omega_p)\exp(-i\omega_p t)}{(\omega_{ab} - \omega_p) - i\gamma_{ab}}. \end{aligned} \tag{8.19}$$

We can write the time-dependent dipole moment in terms of its frequency components,

$$\langle \boldsymbol{\mu}(t) \rangle = \frac{1}{2}\sum_p \langle \boldsymbol{\mu}(\omega_p) \rangle \exp(-i\omega_p t). \tag{8.20}$$

The medium polarization amplitude at frequency ω_p is given by

$$\boldsymbol{P}^{(1)}(\omega_p) = n\left\langle \boldsymbol{\mu}^{(1)}(\omega_p)\right\rangle = \varepsilon_0 \boldsymbol{\chi}^{(1)}(\omega_p) \bullet \boldsymbol{E}(\omega_p), \tag{8.21}$$

where n is the total number density (m^{-3}) of molecules or atoms in the medium. Equation (8.21) serves to define the first-order, linear susceptibility. Combining Eqs. (8.19), (8.20), and (8.21), we obtain

$$\chi^{(1)}(\omega_p) = \frac{n}{\varepsilon_0 \hbar} \sum_a \sum_b \left(\rho_{bb}^{(0)} - \rho_{aa}^{(0)}\right) \frac{\boldsymbol{\mu}_{ba} \boldsymbol{\mu}_{ab}}{(\omega_{ab} - \omega_p) - i\gamma_{ab}}. \tag{8.22}$$

In Cartesian components we can write

$$P_i(\omega_p) = n\langle \mu_i(\omega_p) \rangle = \varepsilon_0 \chi_{ij}^{(1)}(\omega_p) E_j(\omega_p). \tag{8.23}$$

The Cartesian components of the first-order susceptibility are given by

$$\chi_{ij}^{(1)}(\omega_p) = \frac{n}{\varepsilon_0 \hbar} \sum_a \sum_b \left(\rho_{bb}^{(0)} - \rho_{aa}^{(0)}\right) \frac{\mu_{ba,i} \mu_{ab,j}}{(\omega_{ab} - \omega_p) - i\gamma_{ab}}. \tag{8.24}$$

Equation (8.24) can be written in another form as

$$\begin{aligned} \chi_{ij}^{(1)}(\omega_p) &= \frac{n}{\varepsilon_0 \hbar} \sum_a \sum_b \rho_{aa}^{(0)} \left[\frac{\mu_{ab,i} \mu_{ba,j}}{(\omega_{ba} - \omega_p) - i\gamma_{ba}} - \frac{\mu_{ba,i} \mu_{ab,j}}{(\omega_{ab} - \omega_p) - i\gamma_{ab}} \right] \\ &= \frac{n}{\varepsilon_0 \hbar} \sum_a \sum_b \rho_{aa}^{(0)} \left[\frac{\mu_{ab,i} \mu_{ba,j}}{(\omega_{ba} - \omega_p) - i\gamma_{ba}} + \frac{\mu_{ba,i} \mu_{ab,j}}{(\omega_{ba} + \omega_p) + i\gamma_{ab}} \right]. \end{aligned} \tag{8.25}$$

8.2.2 The Second-Order Nonlinear Susceptibility

Starting from Eq. (8.5) and following the same procedure as for the first-order term, the second-order term in the perturbative expansion of the density matrix is given by

$$\rho_{ab}^{(2)} = -\exp\left[-(i\omega_{ab} + \gamma_{ab})t\right] \int_{-\infty}^{t} \frac{i}{\hbar} \left[V(t_1), \rho^{(1)}\right]_{ab} \exp\left[(i\omega_{ab} + \gamma_{ab})t_1\right] dt_1. \tag{8.26}$$

The commutator is given by

$$\left[V(t_1), \rho^{(1)}\right]_{ab} = -\sum_c \left(\boldsymbol{\mu}_{ac} \rho_{cb}^{(1)} - \rho_{ac}^{(1)} \boldsymbol{\mu}_{cb}\right) \bullet \boldsymbol{E}(t). \tag{8.27}$$

From the solution for the first-order density matrix terms, we obtain

$$\rho_{ac}^{(1)} = \frac{\left(\rho_{cc}^{(0)} - \rho_{aa}^{(0)}\right)}{2\hbar} \sum_p \frac{\boldsymbol{\mu}_{ac} \bullet \boldsymbol{E}(\omega_p) \exp(-i\omega_p t)}{(\omega_{ac} - \omega_p) - i\gamma_{ac}} \tag{8.28}$$

and

$$\rho_{cb}^{(1)} = \frac{\left(\rho_{bb}^{(0)} - \rho_{cc}^{(0)}\right)}{2\hbar} \sum_p \frac{\boldsymbol{\mu}_{cb} \bullet \boldsymbol{E}(\omega_p) \exp(-i\omega_p t)}{(\omega_{cb} - \omega_p) - i\gamma_{cb}}. \tag{8.29}$$

In Eq. (8.27) we write the electric field as

$$\boldsymbol{E}(t) = \frac{1}{2}\sum_{q}\boldsymbol{E}(\omega_q)\exp(-i\omega_q t). \tag{8.30}$$

Substituting Eqs. (8.27)–(8.30) into Eq. (8.26), simplifying, and performing the integration, we obtain

$$\begin{aligned}\rho_{ab}^{(2)} = \sum_{c}\sum_{p}\sum_{q} &\exp[-i(\omega_p+\omega_q)t] \\ \times &\left\{\frac{\left(\rho_{bb}^{(0)}-\rho_{cc}^{(0)}\right)}{4\hbar^2}\frac{[\boldsymbol{\mu}_{ac}\bullet\boldsymbol{E}(\omega_q)][\boldsymbol{\mu}_{cb}\bullet\boldsymbol{E}(\omega_p)]}{[(\omega_{ab}-\omega_p-\omega_q)-i\gamma_{ab}][(\omega_{cb}-\omega_p)-i\gamma_{cb}]}\right. \\ &\left.-\frac{\left(\rho_{cc}^{(0)}-\rho_{aa}^{(0)}\right)}{4\hbar^2}\frac{[\boldsymbol{\mu}_{ac}\bullet\boldsymbol{E}(\omega_p)][\boldsymbol{\mu}_{cb}\bullet\boldsymbol{E}(\omega_q)]}{[(\omega_{ab}-\omega_p-\omega_q)-i\gamma_{ab}][(\omega_{ac}-\omega_p)-i\gamma_{ac}]}\right\}.\end{aligned} \tag{8.31}$$

The expectation value of the dipole moment due to the second-order density matrix terms is given by

$$\begin{aligned}\left\langle\boldsymbol{\mu}^{(2)}(t)\right\rangle = \mathrm{Tr}\left[\rho^{(2)}(t)\boldsymbol{\mu}\right] = \sum_{a}\sum_{b}\rho_{ab}^{(2)}(t)\boldsymbol{\mu}_{ba} = \sum_{a,b,c}\sum_{p,q}&\exp[-i(\omega_p+\omega_q)t] \\ \times &\left\{\frac{\left(\rho_{bb}^{(0)}-\rho_{cc}^{(0)}\right)}{4\hbar^2}\frac{\boldsymbol{\mu}_{ba}[\boldsymbol{\mu}_{ac}\bullet\boldsymbol{E}(\omega_q)][\boldsymbol{\mu}_{cb}\bullet\boldsymbol{E}(\omega_p)]}{[(\omega_{ab}-\omega_p-\omega_q)-i\gamma_{ab}][(\omega_{cb}-\omega_p)-i\gamma_{cb}]}\right. \\ &\left.-\frac{\left(\rho_{cc}^{(0)}-\rho_{aa}^{(0)}\right)}{4\hbar^2}\frac{\boldsymbol{\mu}_{ba}[\boldsymbol{\mu}_{ac}\bullet\boldsymbol{E}(\omega_p)][\boldsymbol{\mu}_{cb}\bullet\boldsymbol{E}(\omega_q)]}{[(\omega_{ab}-\omega_p-\omega_q)-i\gamma_{ab}][(\omega_{ac}-\omega_p)-i\gamma_{ac}]}\right\}.\end{aligned} \tag{8.32}$$

The second-order nonlinear polarization induced in the medium at frequency $\omega_1+\omega_2$ by excitation by laser fields with frequencies ω_1 and ω_2 is given by

$$\begin{aligned}\left\langle\boldsymbol{\mu}^{(2)}(t)\right\rangle &= \frac{1}{2}\left\langle\boldsymbol{\mu}^{(2)}(\omega_1+\omega_2)\right\rangle\exp[-i(\omega_1+\omega_2)t] \\ &+\frac{1}{2}\left\langle\boldsymbol{\mu}^{(2)}(-\omega_1-\omega_2)\right\rangle\exp[+i(\omega_1+\omega_2)t].\end{aligned} \tag{8.33}$$

The second-order polarization amplitude at frequency $\omega_1+\omega_2$ is given by

$$\begin{aligned}\boldsymbol{P}^{(2)}(\omega_1+\omega_2;\omega_1,\omega_2) &= N\left\langle\boldsymbol{\mu}^{(2)}(\omega_1+\omega_2)\right\rangle \\ &= \varepsilon_0\boldsymbol{\chi}^{(2)}(\omega_1+\omega_2;\omega_1,\omega_2)\bullet\bullet\boldsymbol{E}(\omega_1)\boldsymbol{E}(\omega_2).\end{aligned} \tag{8.34}$$

In component form, the second-order polarization is given by

$$P_i^{(2)}(\omega_1+\omega_2;\omega_1,\omega_2) = \varepsilon_0\sum_{j}\sum_{k}\chi_{ijk}^{(2)}(\omega_1+\omega_2;\omega_1,\omega_2)E_j(\omega_1)E_k(\omega_2), \tag{8.35}$$

and the second-order nonlinear susceptibility is given by

$$\chi_{ijk}^{(2)}(\omega_1+\omega_2;\omega_1,\omega_2)=\frac{n}{2\varepsilon_0\hbar^2}\sum_{a,b,c}\left\{\frac{\left(\rho_{bb}^{(0)}-\rho_{cc}^{(0)}\right)\mu_{ba,i}\mu_{ac,j}\mu_{cb,k}}{[(\omega_{ab}-\omega_1-\omega_2)-i\gamma_{ab}][(\omega_{cb}-\omega_2)-i\gamma_{cb}]}\right.$$
$$\left.-\frac{\left(\rho_{cc}^{(0)}-\rho_{aa}^{(0)}\right)\mu_{ba,i}\mu_{ac,k}\mu_{cb,j}}{[(\omega_{ab}-\omega_1-\omega_2)-i\gamma_{ab}][(\omega_{ac}-\omega_2)-i\gamma_{ac}]}\right\}. \tag{8.36}$$

The expression for the susceptibility in Eq. (8.36) is not invariant when the laser field frequency–polarization pairs (ω_1,j) and (ω_2,k) are permuted; this is referred to as intrinsic permutation symmetry. A suitable expression is obtained by adding two terms in which these terms are permuted and then dividing the resultant expression by a factor of 2. The result is

$$\chi_{ijk}^{(2)}(\omega_p+\omega_q,\omega_p,\omega_q)=\frac{n}{4\varepsilon_0\hbar^2}\sum_{a,b,c}\left\{\frac{\left(\rho_{bb}^{(0)}-\rho_{cc}^{(0)}\right)\mu_{ba,i}\mu_{ac,j}\mu_{cb,k}}{[(\omega_{ab}-\omega_2-\omega_1)-i\gamma_{ab}][(\omega_{cb}-\omega_2)-i\gamma_{cb}]}\right.$$
$$+\frac{\left(\rho_{bb}^{(0)}-\rho_{cc}^{(0)}\right)\mu_{ba,i}\mu_{ac,k}\mu_{cb,j}}{[(\omega_{ab}-\omega_1-\omega_2)-i\gamma_{ab}][(\omega_{cb}-\omega_1)-i\gamma_{cb}]}$$
$$-\frac{\left(\rho_{cc}^{(0)}-\rho_{aa}^{(0)}\right)\mu_{ba,i}\mu_{ac,k}\mu_{cb,j}}{[(\omega_{ab}-\omega_2-\omega_1)-i\gamma_{ab}][(\omega_{ac}-\omega_2)-i\gamma_{ac}]}$$
$$\left.-\frac{\left(\rho_{cc}^{(0)}-\rho_{aa}^{(0)}\right)\mu_{ba,i}\mu_{ac,j}\mu_{cb,k}}{[(\omega_{ab}-\omega_1-\omega_2)-i\gamma_{ab}][(\omega_{ac}-\omega_1)-i\gamma_{ac}]}\right\}. \tag{8.37}$$

8.3 The Third-Order Nonlinear Susceptibility for CARS

Starting from Eq. (8.6) and following the same procedure as for the first-order and second-order terms, the third-order term in the perturbative expansion of the density matrix is given by

$$\rho_{ab}^{(3)}=-\exp[-(i\omega_{ab}+\gamma_{ab})t]\int_{-\infty}^{t}\frac{i}{\hbar}\left[V(t_1),\rho^{(2)}\right]_{ab}\exp[(i\omega_{ab}+\gamma_{ab})t_1]dt_1. \tag{8.38}$$

The commutator is given by

$$\left[V(t_1),\rho^{(2)}\right]_{ab}=-\sum_{c}\left(\boldsymbol{\mu}_{ac}\rho_{cb}^{(2)}-\rho_{ac}^{(2)}\boldsymbol{\mu}_{cb}\right)\bullet\boldsymbol{E}(t). \tag{8.39}$$

In Eq. (8.39) we write the electric field as

$$\boldsymbol{E}(t)=\frac{1}{2}\sum_{r}\boldsymbol{E}(\omega_r)\exp(-i\omega_r t). \tag{8.40}$$

The development of the general expression for the third-order susceptibility is much more complicated than for the second-order susceptibility because of the complexity of the second-order density matrix terms $\rho_{cb}^{(2)}$ and $\rho_{ac}^{(2)}$. The basic steps of the derivation are outlined in Boyd (2008) and Butcher and Cotter (1990), and the third-order susceptibility in its most general form is given by

$$\chi_{ijkm}^{(3)}(-\omega_4;\omega_1,\omega_2,\omega_3) = \frac{n}{\varepsilon_0}\frac{1}{3!\hbar^3}\mathbf{S}_{op}\sum_{abcd}\rho_{aa}^{(0)}$$
$$\times\left\{\frac{\mu_{ad,i}\mu_{db,j}\mu_{bc,k}\mu_{ca,m}}{[(\Omega_{da}-\omega_4)-i\gamma_{da}][(\Omega_{ba}-\omega_3-\omega_2)-i\gamma_{ba}][(\Omega_{ca}-\omega_3)-i\gamma_{ca}]}\right.$$
$$+\frac{\mu_{ad,m}\mu_{db,i}\mu_{bc,j}\mu_{ca,k}}{[(\Omega_{bd}-\omega_4)-i\gamma_{bd}][(\Omega_{cd}-\omega_3-\omega_2)-i\gamma_{cd}][(\Omega_{da}+\omega_3)+i\gamma_{da}]}$$
$$+\frac{\mu_{ad,k}\mu_{db,i}\mu_{bc,j}\mu_{ca,m}}{[(\Omega_{bd}-\omega_4)-i\gamma_{bd}][(\Omega_{dc}+\omega_3+\omega_2)+i\gamma_{dc}][(\Omega_{ca}-\omega_3)-i\gamma_{ca}]}$$
$$+\frac{\mu_{ad,m}\mu_{db,k}\mu_{bc,i}\mu_{ca,j}}{[(\Omega_{cb}-\omega_4)-i\gamma_{cb}][(\Omega_{ba}+\omega_3+\omega_2)+i\gamma_{ba}][(\Omega_{da}+\omega_3)+i\gamma_{da}]}$$
$$+\frac{\mu_{ad,j}\mu_{db,i}\mu_{bc,k}\mu_{ca,m}}{[(\Omega_{db}+\omega_4)+i\gamma_{db}][(\Omega_{ba}-\omega_3-\omega_2)-i\gamma_{ba}][(\Omega_{ca}-\omega_3)-i\gamma_{ca}]}$$
$$+\frac{\mu_{ad,m}\mu_{db,j}\mu_{bc,i}\mu_{ca,k}}{[(\Omega_{bc}+\omega_4)+i\gamma_{bc}][(\Omega_{cd}-\omega_3-\omega_2)-i\gamma_{cd}][(\Omega_{da}+\omega_3)+i\gamma_{da}]}$$
$$+\frac{\mu_{ad,k}\mu_{db,j}\mu_{bc,i}\mu_{ca,m}}{[(\Omega_{bc}+\omega_4)+i\gamma_{bc}][(\Omega_{dc}+\omega_3+\omega_2)+i\gamma_{dc}][(\Omega_{ca}-\omega_3)-i\gamma_{ca}]}$$
$$\left.+\frac{\mu_{ad,m}\mu_{db,k}\mu_{bc,j}\mu_{ca,i}}{[(\Omega_{ca}+\omega_4)+i\gamma_{ca}][(\Omega_{ba}+\omega_3+\omega_2)+i\gamma_{ba}][(\Omega_{da}+\omega_3)+i\gamma_{da}]}\right\}, \tag{8.41}$$

where the permutation operator $\mathbf{S}_{op}$ indicates that the term in brackets be summed over the 3! possible permutations of the frequency–polarization pairs (ω_1, j), (ω_2, k) and (ω_3, m).

Consider now the CARS process depicted in Figure 8.1, where all laser frequencies are far from electronic resonance. In this case the frequency of the four-wave mixing signal will be given by $\omega_4 = \omega_1 - \omega_2 + \omega_3$, and the susceptibility expression becomes

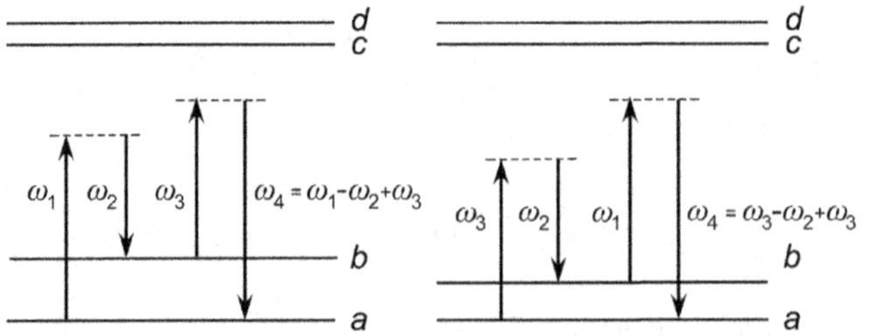

Figure 8.1 Energy level diagrams for the CARS process. For the diagram on the left, ω_1, ω_2, ω_3 and ω_4 are the frequencies of the pump, Stokes, probe, and CARS signal beams, respectively. For the diagram on the right, ω_3, ω_2, ω_1 and ω_4 are the frequencies of the pump, Stokes, probe, and CARS signal beams, respectively.

$$
\begin{aligned}
\chi^{(3)}_{ijkm}(-\omega_4;\omega_1,\,-\omega_2,\omega_3) &= \frac{n}{\varepsilon_0}\frac{1}{3!\hbar^3}\mathbf{S}_{op}\sum_{abcd}\rho^{(0)}_{ca} \\
&\times\left\{\frac{\mu_{ad,i}\mu_{db,j}\mu_{bc,k}\mu_{ca,m}}{[(\Omega_{da}-\omega_4)-i\gamma_{da}][(\Omega_{ba}-\omega_3+\omega_2)-i\gamma_{ba}][(\Omega_{ca}-\omega_3)-i\gamma_{ca}]}\right. \\
&+\frac{\mu_{ad,m}\mu_{db,i}\mu_{bc,j}\mu_{ca,k}}{[(\Omega_{bd}-\omega_4)-i\gamma_{bd}][(\Omega_{cd}-\omega_3+\omega_2)-i\gamma_{cd}][(\Omega_{da}+\omega_3)+i\gamma_{da}]} \\
&+\frac{\mu_{ad,k}\mu_{db,i}\mu_{bc,j}\mu_{ca,m}}{[(\Omega_{bd}-\omega_4)-i\gamma_{bd}][(\Omega_{dc}+\omega_3-\omega_2)+i\gamma_{dc}][(\Omega_{ca}-\omega_3)-i\gamma_{ca}]} \\
&+\frac{\mu_{ad,m}\mu_{db,k}\mu_{bc,i}\mu_{ca,j}}{[(\Omega_{cb}-\omega_4)-i\gamma_{cb}][(\Omega_{ba}+\omega_3-\omega_2)+i\gamma_{ba}][(\Omega_{da}+\omega_3)+i\gamma_{da}]} \\
&+\frac{\mu_{ad,j}\mu_{db,i}\mu_{bc,k}\mu_{ca,m}}{[(\Omega_{db}+\omega_4)+i\gamma_{db}][(\Omega_{ba}-\omega_3+\omega_2)-i\gamma_{ba}][(\Omega_{ca}-\omega_3)-i\gamma_{ca}]} \\
&+\frac{\mu_{ad,m}\mu_{db,j}\mu_{bc,i}\mu_{ca,k}}{[(\Omega_{bc}+\omega_4)+i\gamma_{bc}][(\Omega_{cd}-\omega_3+\omega_2)-i\gamma_{cd}][(\Omega_{da}+\omega_3)+i\gamma_{da}]} \\
&+\frac{\mu_{ad,k}\mu_{db,j}\mu_{bc,i}\mu_{ca,m}}{[(\Omega_{bc}+\omega_4)+i\gamma_{bc}][(\Omega_{dc}+\omega_3-\omega_2)+i\gamma_{dc}][(\Omega_{ca}-\omega_3)-i\gamma_{ca}]} \\
&\left.+\frac{\mu_{ad,m}\mu_{db,k}\mu_{bc,j}\mu_{ca,i}}{[(\Omega_{ca}+\omega_4)+i\gamma_{ca}][(\Omega_{ba}+\omega_3-\omega_2)+i\gamma_{ba}][(\Omega_{da}+\omega_3)+i\gamma_{da}]}\right\}.
\end{aligned}
$$

The CARS signal will be enhanced by Raman resonance when $\Omega_{ba} \cong \omega_1 - \omega_2$ or when $\Omega_{ba} \cong \omega_3 - \omega_2$. For the moment let us assume that the species under consideration only has Raman resonances when $\Omega_{ba} \cong \omega_3 - \omega_2 > 0$. Keeping only terms that are near resonance in the denominator, we obtain

$$
\begin{aligned}
\chi^{(3)}_{ijkm}(-\omega_4;\omega_1,\,-\omega_2,\omega_3) &= \frac{n}{\varepsilon_0}\frac{1}{3!\hbar^3}\mathbf{S}_{op}\sum_{abcd}\rho^{(0)}_{aa} \\
&\times\left\{\frac{\mu_{ad,i}\mu_{db,j}\mu_{bc,k}\mu_{ca,m}}{[(\Omega_{da}-\omega_4)-i\gamma_{da}][(\Omega_{ba}-\omega_3+\omega_2)-i\gamma_{ba}][(\Omega_{ca}-\omega_3)-i\gamma_{ca}]}\right. \\
&\left.+\frac{\mu_{ad,j}\mu_{db,i}\mu_{bc,k}\mu_{ca,m}}{[(\Omega_{db}+\omega_4)+i\gamma_{db}][(\Omega_{ba}-\omega_3+\omega_2)-i\gamma_{ba}][(\Omega_{ca}-\omega_3)-i\gamma_{ca}]}\right\}.
\end{aligned}
\tag{8.43}
$$

Now performing the pair-permutation operation $(-\omega_2, k) \Leftrightarrow (\omega_3, m)$, we obtain two additional Raman-resonant terms:

$$
\begin{aligned}
\chi^{(3)}_{ijkm}(-\omega_4;\omega_1,\,-\omega_2,\omega_3) &= \frac{n}{\varepsilon_0}\frac{1}{3!\hbar^3}\mathbf{S}_{\rm op}\sum_{abcd}\rho^{(0)}_{aa} \\
&\times\left\{\frac{\mu_{ad,i}\mu_{db,j}\mu_{bc,k}\mu_{ca,m}}{[(\Omega_{da}-\omega_4)-i\gamma_{da}][(\Omega_{ba}-\omega_3+\omega_2)-i\gamma_{ba}][(\Omega_{ca}-\omega_3)-i\gamma_{ca}]}\right. \\
&+\frac{\mu_{ad,i}\mu_{db,j}\mu_{bc,m}\mu_{ca,k}}{[(\Omega_{da}-\omega_4)-i\gamma_{da}][(\Omega_{ba}+\omega_2-\omega_3)-i\gamma_{ba}][(\Omega_{ca}+\omega_2)-i\gamma_{ca}]} \\
&+\frac{\mu_{ad,j}\mu_{db,i}\mu_{bc,k}\mu_{ca,m}}{[(\Omega_{db}+\omega_4)+i\gamma_{db}][(\Omega_{ba}-\omega_3+\omega_2)-i\gamma_{ba}][(\Omega_{ca}-\omega_3)-i\gamma_{ca}]} \\
&\left.+\frac{\mu_{ad,j}\mu_{db,i}\mu_{bc,m}\mu_{ca,k}}{[(\Omega_{db}+\omega_4)+i\gamma_{db}][(\Omega_{ba}+\omega_2-\omega_3)-i\gamma_{ba}][(\Omega_{ca}+\omega_2)-i\gamma_{ca}]}\right\}.
\end{aligned}
\tag{8.44}
$$

Performing the pair-permutation operation $(\omega_1, i) \Leftrightarrow (\omega_3, m)$ and keeping only terms with near-resonant denominators, we obtain

$$\chi^{(3)}_{ijkm}(-\omega_4; \omega_1, -\omega_2, \omega_3) = \frac{n}{\varepsilon_0} \frac{1}{3!\hbar^3} \sum_{abcd} \rho^{(0)}_{aa}$$
$$\times \left\{ \frac{\mu_{ad,i}\mu_{db,j}\mu_{bc,k}\mu_{ca,m}}{[(\Omega_{da} - \omega_4) - i\gamma_{da}][(\Omega_{ba} - \omega_3 + \omega_2) - i\gamma_{ba}][(\Omega_{ca} - \omega_3) - i\gamma_{ca}]} \right.$$
$$+ \frac{\mu_{ad,i}\mu_{db,j}\mu_{bc,m}\mu_{ca,k}}{[(\Omega_{da} - \omega_4) - i\gamma_{da}][(\Omega_{ba} + \omega_2 - \omega_3) - i\gamma_{ba}][(\Omega_{ca} + \omega_2) - i\gamma_{ca}]}$$
$$+ \frac{\mu_{ad,j}\mu_{db,i}\mu_{bc,k}\mu_{ca,m}}{[(\Omega_{db} + \omega_4) + i\gamma_{db}][(\Omega_{ba} - \omega_3 + \omega_2) - i\gamma_{ba}][(\Omega_{ca} - \omega_3) - i\gamma_{ca}]}$$
$$+ \frac{\mu_{ad,j}\mu_{db,i}\mu_{bc,m}\mu_{ca,k}}{[(\Omega_{db} + \omega_4) + i\gamma_{db}][(\Omega_{ba} + \omega_2 - \omega_3) - i\gamma_{ba}][(\Omega_{ca} + \omega_2) - i\gamma_{ca}]} \tag{8.45}$$
$$+ \frac{\mu_{ad,i}\mu_{db,m}\mu_{bc,k}\mu_{ca,j}}{[(\Omega_{da} - \omega_4) - i\gamma_{da}][(\Omega_{ba} - \omega_1 + \omega_2) - i\gamma_{ba}][(\Omega_{ca} - \omega_1) - i\gamma_{ca}]}$$
$$+ \frac{\mu_{ad,i}\mu_{db,m}\mu_{bc,j}\mu_{ca,k}}{[(\Omega_{da} - \omega_4) - i\gamma_{da}][(\Omega_{ba} + \omega_2 - \omega_1) - i\gamma_{ba}][(\Omega_{ca} + \omega_2) - i\gamma_{ca}]}$$
$$+ \frac{\mu_{ad,m}\mu_{db,i}\mu_{bc,k}\mu_{ca,j}}{[(\Omega_{db} + \omega_4) + i\gamma_{db}][(\Omega_{ba} - \omega_1 + \omega_2) - i\gamma_{ba}][(\Omega_{ca} - \omega_1) - i\gamma_{ca}]}$$
$$\left. + \frac{\mu_{ad,m}\mu_{db,i}\mu_{bc,j}\mu_{ca,k}}{[(\Omega_{db} + \omega_4) + i\gamma_{db}][(\Omega_{ba} + \omega_2 - \omega_1) - i\gamma_{ba}][(\Omega_{ca} + \omega_2) - i\gamma_{ca}]} \right\}.$$

Assuming now that only $\omega_1 - \omega_2$ is close to a Raman resonance and $\omega_3 - \omega_2$ is far from resonance, we obtain

$$\chi^{(3)}_{ijkm}(-\omega_4; \omega_1, -\omega_2, \omega_3) = \frac{n}{\varepsilon_0} \frac{1}{3!\hbar^3} \sum_{abcd} \rho^{(0)}_{aa}$$
$$\times \left\{ \frac{\mu_{ad,i}\mu_{db,m}\mu_{bc,k}\mu_{ca,j}}{[(\Omega_{da} - \omega_4) - i\gamma_{da}][(\Omega_{ba} - \omega_1 + \omega_2) - i\gamma_{ba}][(\Omega_{ca} - \omega_1) - i\gamma_{ca}]} \right.$$
$$+ \frac{\mu_{ad,i}\mu_{db,m}\mu_{bc,j}\mu_{ca,k}}{[(\Omega_{da} - \omega_4) - i\gamma_{da}][(\Omega_{ba} + \omega_2 - \omega_1) - i\gamma_{ba}][(\Omega_{ca} + \omega_2) - i\gamma_{ca}]} \tag{8.46}$$
$$+ \frac{\mu_{ad,m}\mu_{db,i}\mu_{bc,k}\mu_{ca,j}}{[(\Omega_{db} + \omega_4) + i\gamma_{db}][(\Omega_{ba} - \omega_1 + \omega_2) - i\gamma_{ba}][(\Omega_{ca} - \omega_1) - i\gamma_{ca}]}$$
$$\left. + \frac{\mu_{ad,m}\mu_{db,i}\mu_{bc,j}\mu_{ca,k}}{[(\Omega_{db} + \omega_4) + i\gamma_{db}][(\Omega_{ba} + \omega_2 - \omega_1) - i\gamma_{ba}][(\Omega_{ca} + \omega_2) - i\gamma_{ca}]} \right\}.$$

Equation (8.46) can be rewritten as

$$\chi^{(3)}_{ijkm}(-\omega_4; \omega_1, -\omega_2, \omega_3) = \frac{n}{\varepsilon_0} \frac{1}{3!\hbar^3} \sum_{abcd} \frac{\rho^{(0)}_{aa}}{[(\Omega_{ba} - \omega_1 + \omega_2) - i\gamma_{ba}]}$$
$$\times \left\{ \frac{\mu_{ad,i}\mu_{db,m}\mu_{bc,k}\mu_{ca,j}}{(\Omega_{da} - \omega_4)(\Omega_{ca} - \omega_1)} + \frac{\mu_{ad,i}\mu_{db,m}\mu_{bc,j}\mu_{ca,k}}{(\Omega_{da} - \omega_4)(\Omega_{ca} + \omega_2)} \right. \tag{8.47}$$
$$\left. + \frac{\mu_{ad,m}\mu_{db,i}\mu_{bc,k}\mu_{ca,j}}{(\Omega_{db} + \omega_4)(\Omega_{ca} - \omega_1)} + \frac{\mu_{ad,m}\mu_{db,i}\mu_{bc,j}\mu_{ca,k}}{(\Omega_{db} + \omega_4)(\Omega_{ca} + \omega_2)} \right\}.$$

Using the relation $(\Omega_{ca} + \omega_2) \cong (\Omega_{cb} + \omega_1)$, , we can rewrite this expression as

$$\chi^{(3)}_{ijkm}(-\omega_4; \omega_1, -\omega_2, \omega_3) = \frac{n}{\varepsilon_0} \frac{1}{3!\hbar^3} \sum_{abcd} \frac{\rho^{(0)}_{aa}}{[(\Omega_{ba} - \omega_1 + \omega_2) - i\gamma_{ba}]} \times \left[\frac{\mu_{ad,i}\mu_{db,m}}{(\Omega_{da} - \omega_4)} + \frac{\mu_{ad,m}\mu_{db,i}}{(\Omega_{db} + \omega_4)} \right] \left[\frac{\mu_{bc,k}\mu_{ca,j}}{(\Omega_{ca} - \omega_1)} + \frac{\mu_{bc,j}\mu_{ca,k}}{(\Omega_{cb} + \omega_1)} \right]. \tag{8.48}$$

In performing the summation in Eq. (8.48) it is understood that $\varepsilon_b > \varepsilon_a$. Returning again to Eq. (8.42), it is clear that resonant terms can result from the fourth and eighth terms in the square bracket if $\Omega_{ba} \cong \omega_2 - \omega_1 < 0$, in other words if $\varepsilon_a > \varepsilon_b$. These resulting four terms are given by

$$\begin{aligned} \chi^{(3)}_{ijkm}(-\omega_4; \omega_1, -\omega_2, \omega_3) &= \frac{n}{\varepsilon_0} \frac{1}{3!\hbar^3} \sum_{abcd} \rho^{(0)}_{aa} \\ &\times \left\{ \frac{\mu_{ad,j}\mu_{db,k}\mu_{bc,i}\mu_{ca,m}}{[(\Omega_{cb} - \omega_4) - i\gamma_{cb}][(\Omega_{ba} + \omega_1 - \omega_2) + i\gamma_{ba}][(\Omega_{da} + \omega_1) + i\gamma_{da}]} \right. \\ &+ \frac{\mu_{ad,k}\mu_{db,j}\mu_{bc,i}\mu_{ca,m}}{[(\Omega_{cb} - \omega_4) - i\gamma_{cb}][(\Omega_{ba} - \omega_2 + \omega_1) + i\gamma_{ba}][(\Omega_{da} - \omega_2) + i\gamma_{da}]} \\ &+ \frac{\mu_{ad,j}\mu_{db,k}\mu_{bc,m}\mu_{ca,i}}{[(\Omega_{ca} + \omega_4) + i\gamma_{ca}][(\Omega_{ba} + \omega_1 - \omega_2) + i\gamma_{ba}][(\Omega_{da} + \omega_1) + i\gamma_{da}]} \\ &+ \left. \frac{\mu_{ad,k}\mu_{db,j}\mu_{bc,m}\mu_{ca,i}}{[(\Omega_{ca} + \omega_4) + i\gamma_{ca}][(\Omega_{ba} - \omega_2 + \omega_1) + i\gamma_{ba}][(\Omega_{da} - \omega_2) + i\gamma_{da}]} \right\}. \end{aligned} \tag{8.49}$$

Equation (8.49) can be written as

$$\begin{aligned} \chi^{(3)}_{ijkm}(-\omega_4; \omega_1, -\omega_2, \omega_3) &= \frac{n}{\varepsilon_0} \frac{1}{3!\hbar^3} \sum_{abcd} \frac{\rho^{(0)}_{aa}}{[(\Omega_{ba} + \omega_1 - \omega_2) + i\gamma_{ba}]} \\ &\times \left\{ \frac{\mu_{ad,j}\mu_{db,k}\mu_{bc,i}\mu_{ca,m}}{(\Omega_{cb} - \omega_4)(\Omega_{da} + \omega_1)} + \frac{\mu_{ad,k}\mu_{db,j}\mu_{bc,i}\mu_{ca,m}}{(\Omega_{cb} - \omega_4)(\Omega_{da} - \omega_2)} \right. \\ &+ \left. \frac{\mu_{ad,j}\mu_{db,k}\mu_{bc,m}\mu_{ca,i}}{(\Omega_{ca} + \omega_4)(\Omega_{da} + \omega_1)} + \frac{\mu_{ad,k}\mu_{db,j}\mu_{bc,m}\mu_{ca,i}}{(\Omega_{ca} + \omega_4)(\Omega_{da} - \omega_2)} \right\}. \end{aligned} \tag{8.50}$$

However, at this point it is somewhat confusing as to how to combine Eqs. (8.47) and (8.50) to obtain a complete expression. We can rewrite Eq. (8.50) by permuting the state indices a and b, resulting in

$$\begin{aligned} \chi^{(3)}_{ijkm}(-\omega_4; \omega_1, -\omega_2, \omega_3) &= \frac{n}{\varepsilon_0} \frac{1}{3!\hbar^3} \sum_{abcd} \frac{\rho^{(0)}_{bb}}{[(\Omega_{ab} + \omega_1 - \omega_2) + i\gamma_{ab}]} \\ &\times \left\{ \frac{\mu_{bd,j}\mu_{da,k}\mu_{ac,i}\mu_{cb,m}}{(\Omega_{ca} - \omega_4)(\Omega_{db} + \omega_1)} + \frac{\mu_{bd,k}\mu_{da,j}\mu_{ac,i}\mu_{cb,m}}{(\Omega_{ca} - \omega_4)(\Omega_{db} - \omega_2)} \right. \\ &+ \left. \frac{\mu_{bd,j}\mu_{da,k}\mu_{ac,m}\mu_{cb,i}}{(\Omega_{cb} + \omega_4)(\Omega_{db} + \omega_1)} + \frac{\mu_{bd,k}\mu_{da,j}\mu_{ac,m}\mu_{cb,i}}{(\Omega_{cb} + \omega_4)(\Omega_{db} - \omega_2)} \right\}, \end{aligned} \tag{8.51}$$

where it is understood that for a particular state a, the summation will only be performed for states b where $\varepsilon_b > \varepsilon_a$. Using the relations $\Omega_{ab} = -\Omega_{ba}$, $\gamma_{ab} = \gamma_{ba}$, and $(\Omega_{db} - \omega_2) \cong (\Omega_{da} - \omega_1)$ and then permuting the indices c and d, we obtain

$$\chi^{(3)}_{ijkm}(-\omega_4;\omega_1,-\omega_2,\omega_3) = \frac{n}{\varepsilon_0}\frac{1}{3!\hbar^3}\sum_{abcd}\frac{-\rho^{(0)}_{bb}}{[(\Omega_{ba}-\omega_1+\omega_2)-i\gamma_{ba}]}$$
$$\times\left\{\frac{\mu_{bc,j}\mu_{ca,k}\mu_{ad,i}\mu_{db,m}}{(\Omega_{da}-\omega_4)(\Omega_{cb}+\omega_1)}+\frac{\mu_{bc,k}\mu_{ca,j}\mu_{ad,i}\mu_{db,m}}{(\Omega_{da}-\omega_4)(\Omega_{ca}-\omega_1)}\right. \tag{8.52}$$
$$\left.+\frac{\mu_{bc,j}\mu_{ca,k}\mu_{ad,m}\mu_{db,i}}{(\Omega_{db}+\omega_4)(\Omega_{cb}+\omega_1)}+\frac{\mu_{bc,k}\mu_{ca,j}\mu_{ad,m}\mu_{db,i}}{(\Omega_{db}+\omega_4)(\Omega_{ca}-\omega_1)}\right\}.$$

Equation (8.52) can now be rewritten as

$$\chi^{(3)}_{ijkm}(-\omega_4;\omega_1,-\omega_2,\omega_3) = \frac{n}{\varepsilon_0}\frac{1}{3!\hbar^3}\sum_{abcd}\frac{-\rho^{(0)}_{bb}}{[(\Omega_{ba}-\omega_1+\omega_2)-i\gamma_{ba}]}$$
$$\times\left[\frac{\mu_{ad,i}\mu_{db,m}}{(\Omega_{da}-\omega_4)}+\frac{\mu_{ad,m}\mu_{db,i}}{(\Omega_{db}+\omega_4)}\right]\left[\frac{\mu_{bc,k}\mu_{ca,j}}{(\Omega_{ca}-\omega_1)}+\frac{\mu_{bc,j}\mu_{ca,k}}{(\Omega_{cb}+\omega_1)}\right]. \tag{8.53}$$

Adding Eqs. (8.48) and (8.53) and dividing by a factor of 2 to account for the permutation of states a and b, we obtain

$$\chi^{(3)}_{ijkm}(-\omega_4;\omega_1,-\omega_2,\omega_3) = \frac{n}{2\varepsilon_0}\frac{1}{3!\hbar^3}\sum_{abcd}\frac{\rho^{(0)}_{aa}-\rho^{(0)}_{bb}}{[(\Omega_{ba}-\omega_1+\omega_2)-i\gamma_{ba}]}$$
$$\times\left[\frac{\mu_{ad,i}\mu_{db,m}}{(\Omega_{da}-\omega_4)}+\frac{\mu_{ad,m}\mu_{db,i}}{(\Omega_{db}+\omega_4)}\right]\left[\frac{\mu_{bc,k}\mu_{ca,j}}{(\Omega_{ca}-\omega_1)}+\frac{\mu_{bc,j}\mu_{ca,k}}{(\Omega_{cb}+\omega_1)}\right]. \tag{8.54}$$

But the terms in the square brackets are the polarizabilities that were introduced in Chapter 7. By analogy with Eqs. (7.5) and (7.6), we can write

$$\alpha_{im}(a,b) = \frac{1}{\hbar}\sum_d\left[\frac{\mu_{ad,i}\mu_{db,m}}{(\Omega_{da}-\omega_4)}+\frac{\mu_{ad,m}\mu_{db,i}}{(\Omega_{db}+\omega_4)}\right]. \tag{8.55}$$

$$\alpha_{kj}(a,b) = \frac{1}{\hbar}\sum_c\left[\frac{\mu_{bc,k}\mu_{ca,j}}{(\Omega_{ca}-\omega_1)}+\frac{\mu_{bc,k}\mu_{ca,j}}{(\Omega_{cb}+\omega_1)}\right]. \tag{8.56}$$

Substituting Eqs. (8.56) and (8.55) into Eq. (8.54), we obtain

$$\chi^{(3)}_{imkj}(-\omega_4;\omega_3,-\omega_2,\omega_1) = \frac{n}{2\varepsilon_0}\frac{1}{3!\hbar}\sum_{ab}\frac{\left[\rho^{(0)}_{aa}-\rho^{(0)}_{bb}\right]\alpha_{im}(a,b)\alpha_{kj}(a,b)}{[\Omega_{ba}-(\omega_1-\omega_2)-i\gamma_{ba}]}. \tag{8.57}$$

Assuming that the linewidth is the same for all the state-to-state transitions for the transition from level A to level B, $\gamma_{ba}=\gamma_{BA}$, and the zero-order Zeeman state populations are given by $n\rho^{(0)}_{aa}=n^{(0)}_A/(2J_A+1)$ and $n\rho^{(0)}_{bb}=n^{(0)}_B/(2J_B+1)$, we obtain

$$\chi^{(3)}_{imkj}(-\omega_4;\omega_3,-\omega_2,\omega_1) = \frac{\left[\frac{n^{(0)}_A}{(2J_A+1)}-\frac{n^{(0)}_B}{(2J_B+1)}\right]\langle\alpha_{im}(A,B)\alpha_{kj}(A,B)\rangle}{12\varepsilon_0\hbar[\Omega_{BA}-(\omega_1-\omega_2)-i\gamma_{BA}]}, \tag{8.58}$$

where

$$\left\langle \alpha_{im}(A,B)\alpha_{kj}(A,B)\right\rangle = \frac{1}{\hbar^2}\sum_{abcd}\left[\frac{\mu_{ad,i}\mu_{db,m}}{(\Omega_{da}-\omega_4)}+\frac{\mu_{ad,m}\mu_{db,i}}{(\Omega_{db}+\omega_4)}\right]\left[\frac{\mu_{bc,k}\mu_{ca,j}}{(\Omega_{ca}-\omega_1)}+\frac{\mu_{bc,k}\mu_{ca,j}}{(\Omega_{cb}+\omega_1)}\right]. \tag{8.59}$$

In some cases, both $\omega_1 - \omega_2$ and $\omega_3 - \omega_2$ may be near Raman resonance. Evaluating Eq. (8.42) for the case where both terms may be near Raman resonance and summing over all possible levels A and B , we obtain

$$\chi^{(3)}_{imkj}(-\omega_4;\omega_3,-\omega_2,\omega_1) = \frac{1}{12\varepsilon_0\hbar}\sum_{AB}\left[\frac{n_A^{(0)}}{(2J_A+1)}-\frac{n_B^{(0)}}{(2J_B+1)}\right]\left\{\frac{\left\langle \alpha_{im}(A,B)\alpha_{kj}(A,B)\right\rangle}{[\Omega_{BA}-(\omega_1-\omega_2)-i\gamma_{BA}]}\right.$$
$$\left.+\frac{\left\langle \alpha_{ij}(A,B)\alpha_{km}(A,B)\right\rangle}{[\Omega_{BA}-(\omega_3-\omega_2)-i\gamma_{BA}]}\right\}, \tag{8.60}$$

where

$$\left\langle \alpha_{ij}(A,B)\alpha_{km}(A,B)\right\rangle = \frac{1}{\hbar^2}\sum_{abcd}\left[\frac{\mu_{ad,i}\mu_{db,j}}{(\Omega_{da}-\omega_4)}+\frac{\mu_{ad,j}\mu_{db,i}}{(\Omega_{db}+\omega_4)}\right]\left[\frac{\mu_{bc,k}\mu_{ca,m}}{(\Omega_{ca}-\omega_3)}+\frac{\mu_{bc,k}\mu_{ca,m}}{(\Omega_{cb}+\omega_3)}\right]. \tag{8.61}$$

8.4 Orientation Averaging for the CARS Polarizability Products

Calculation of $P_i^{(3)}(t)$ requires an orientation average over the product of the polarizabilities, and the value of the orientation average will depend on the polarization components j,k,m of the input laser beams. Consider first the case where $i = m \neq j = k$. In this case we must calculate the orientation averages, such as $\langle \alpha_{XX}\alpha_{YY}\rangle$. Expressing the Cartesian polarization derivative tensor elements in terms of space-fixed spherical tensor components, we obtain

$$\begin{aligned}\langle \alpha_{XX}\alpha_{YY}\rangle &= \left\langle\left(-\frac{1}{\sqrt{3}}\alpha^0_{p0}+\frac{1}{2}\alpha^2_{p2}-\frac{1}{\sqrt{6}}\alpha^2_{p0}+\frac{1}{2}\alpha^2_{p-2}\right)\right.\\ &\quad\times\left.\left(-\frac{1}{\sqrt{3}}\alpha^0_{p0}-\frac{1}{2}\alpha^2_{p2}-\frac{1}{\sqrt{6}}\alpha^2_{p0}-\frac{1}{2}\alpha^2_{p-2}\right)\right\rangle\\ &= \left\langle\left(-\frac{1}{\sqrt{3}}\alpha^0_{p0}+\frac{1}{2}\alpha^2_{p2}-\frac{1}{\sqrt{6}}\alpha^2_{p0}+\frac{1}{2}\alpha^2_{p-2}\right)\left(-\frac{1}{\sqrt{3}}\alpha^{0*}_{p0}-\frac{1}{2}\alpha^{2*}_{p-2}-\frac{1}{\sqrt{6}}\alpha^{2*}_{p0}-\frac{1}{2}\alpha^{2*}_{p2}\right)\right\rangle,\end{aligned} \tag{8.62}$$

where the relations $\alpha^2_2 = \left(\alpha^2_{-2}\right)^*$, $\alpha^2_{-2} = \left(\alpha^2_2\right)^*$, $\alpha^0_0 = \left(\alpha^0_0\right)^*$, and $\alpha^2_0 = \left(\alpha^2_0\right)^*$ have been used in Eq. (8.62). The orientation average for the terms in Eq. (8.62) is given by

$$\left\langle \alpha^k_p\alpha^{k'*}_{p'}\right\rangle = \frac{\delta_{kk'}\delta_{pp'}}{2k+1}\sum_q \alpha^k_q\alpha^{k*}_q, \tag{8.63}$$

so that the orientation average for the cross terms ($k \neq k'$ or $p \neq p'$) in the product in Eq. (8.62) will be zero. Analyzing Eq. (8.62) using Eq. (8.63), we obtain

$$\begin{aligned}\langle \alpha_{XX}\alpha_{YY}\rangle &= \frac{1}{3}\left\langle \alpha_{p0}^{0}\alpha_{p0}^{0\,*}\right\rangle - \frac{1}{4}\left\langle \alpha_{p2}^{2}\alpha_{p2}^{2\,*}\right\rangle + \frac{1}{6}\left\langle \alpha_{p0}^{2}\alpha_{p0}^{2*}\right\rangle - \frac{1}{4}\left\langle \alpha_{p-2}^{2}\alpha_{p-2}^{2\,*}\right\rangle \\ &= \frac{1}{3}\left\langle \alpha_{p0}^{0}\alpha_{p0}^{0*}\right\rangle - \frac{1}{3}\left\langle \alpha_{p0}^{2}\alpha_{p0}^{2\,*}\right\rangle = \frac{1}{3}\left|\alpha_{p0}^{0}\right|^{2} - \frac{1}{3}\left|\alpha_{p0}^{2}\right|^{2},\end{aligned} \tag{8.64}$$

where we have used the relation, implied by Eq. (8.63),

$$\left\langle \alpha_{p0}^{2}\alpha_{p0}^{2*}\right\rangle = \left\langle \alpha_{p2}^{2}\alpha_{p2}^{2\,*}\right\rangle = \left\langle \alpha_{p-2}^{2}\alpha_{p-2}^{2\,*}\right\rangle. \tag{8.65}$$

That is, all orientation-averaged products of the space-fixed tensor components of the same rank are equal.

Next consider the case where $i = k \neq j = m$. In this case we must calculate the orientation average of terms such as $\langle \alpha_{XY}\alpha_{XY}\rangle$. Expressing the Cartesian polarization derivative tensor elements in their spherical tensor form ,we obtain

$$\begin{aligned}\langle \alpha_{XY}\alpha_{XY}\rangle &= \left\langle\left(-\frac{i}{\sqrt{2}}\alpha_{p0}^{1} - \frac{i}{2}\alpha_{p2}^{2} + \frac{i}{2}\alpha_{p-2}^{2}\right)\left(-\frac{i}{\sqrt{2}}\alpha_{p0}^{1} - \frac{i}{2}\alpha_{p2}^{2} + \frac{i}{2}\alpha_{p-2}^{2}\right)\right\rangle \\ &= \left\langle\left(-\frac{i}{\sqrt{2}}\alpha_{p0}^{1} - \frac{i}{2}\alpha_{p2}^{2} + \frac{i}{2}\alpha_{p-2}^{2}\right)\left(+\frac{i}{\sqrt{2}}\alpha_{p0}^{1*} - \frac{i}{2}\alpha_{p-2}^{2\,*} + \frac{i}{2}\alpha_{p2}^{2\,*}\right)\right\rangle \\ &= \frac{1}{2}\left\langle \alpha_{p0}^{1}\alpha_{p0}^{1*}\right\rangle + \frac{1}{4}\left\langle \alpha_{p2}^{2}\alpha_{p2}^{2*}\right\rangle + \frac{1}{4}\left\langle \alpha_{p-2}^{2}\alpha_{p-2}^{2\,*}\right\rangle = \frac{1}{2}\left|\alpha_{p0}^{1}\right|^{2} + \frac{1}{2}\left|\alpha_{p2}^{2}\right|^{2},\end{aligned} \tag{8.66}$$

where the relations $\alpha_{p2}^{2} = \alpha_{p-2}^{2*}$, $\alpha_{p-2}^{2} = \alpha_{p2}^{2*}$, and $\alpha_{p0}^{1} = \alpha_{p0}^{1*}$ have been used in Eq. (8.66). For the case $i = j \neq k = m$, we obtain

$$\begin{aligned}\langle \alpha_{XY}\alpha_{YX}\rangle &= \left\langle\left(-\frac{i}{\sqrt{2}}\alpha_{p0}^{1} - \frac{i}{2}\alpha_{p2}^{2} + \frac{i}{2}\alpha_{p-2}^{2}\right)\left(+\frac{i}{\sqrt{2}}\alpha_{p0}^{1} - \frac{i}{2}\alpha_{p2}^{2} + \frac{i}{2}\alpha_{p-2}^{2}\right)\right\rangle \\ &= \left\langle\left(-\frac{i}{\sqrt{2}}\alpha_{p0}^{1} - \frac{i}{2}\alpha_{p2}^{2} + \frac{i}{2}\alpha_{p-2}^{2}\right)\left(-\frac{i}{\sqrt{2}}\alpha_{p0}^{1*} - \frac{i}{2}\alpha_{p-2}^{2\,*} + \frac{i}{2}\alpha_{p2}^{2\,*}\right)\right\rangle \\ &= -\frac{1}{2}\left\langle \alpha_{p0}^{1}\alpha_{p0}^{1*}\right\rangle + \frac{1}{4}\left\langle \alpha_{p2}^{2}\alpha_{p2}^{2\,*}\right\rangle + \frac{1}{4}\left\langle \alpha_{p-2}^{2}\alpha_{p-2}^{2\,*}\right\rangle = -\frac{1}{2}\left|\alpha_{p0}^{1}\right|^{2} + \frac{1}{2}\left|\alpha_{p2}^{2}\right|^{2}.\end{aligned} \tag{8.67}$$

Terms such as $\langle \alpha_{XX}\alpha_{XY}\rangle$ with odd numbers of like coordinates will be zero for isotropic media. This is illustrated below for $\langle \alpha_{XX}\alpha_{XY}\rangle$:

$$\begin{aligned}\langle \alpha_{XX}\alpha_{XY}\rangle &= \left\langle\left(-\frac{1}{\sqrt{3}}\alpha_{p0}^{0} + \frac{1}{2}\alpha_{p2}^{2} - \frac{1}{\sqrt{6}}\alpha_{p0}^{2} + \frac{1}{2}\alpha_{p-2}^{2}\right)\left(-\frac{i}{\sqrt{2}}\alpha_{p0}^{1} - \frac{i}{2}\alpha_{p2}^{2} + \frac{i}{2}\alpha_{p-2}^{2}\right)\right\rangle \\ &= \left\langle\left(-\frac{1}{\sqrt{3}}\alpha_{p0}^{0} + \frac{1}{2}\alpha_{p2}^{2} - \frac{1}{\sqrt{6}}\alpha_{p0}^{2} + \frac{1}{2}\alpha_{p-2}^{2}\right)\left(+\frac{i}{\sqrt{2}}\alpha_{p0}^{1*} - \frac{i}{2}\alpha_{p-2}^{2\,*} + \frac{i}{2}\alpha_{p2}^{2\,*}\right)\right\rangle \\ &= \frac{i}{2}\left\langle \alpha_{2}^{2}\alpha_{2}^{2*}\right\rangle - \frac{i}{2}\left\langle \alpha_{-2}^{2}\alpha_{-2}^{2\,*}\right\rangle = 0.\end{aligned} \tag{8.68}$$

8.4.1 Pure Rotational CARS

Consider first the case where $v_a = v_A = v = v_b = v_B = v'$ for molecules with $^1\Sigma$ electronic ground states. In Chapter 7 the following expression for the space-fixed spherical tensor components in terms of the molecule-fixed spherical tensor components was derived:

$$\left\langle \alpha_p^k(A,B)\alpha_p^{k*}(A,B)\right\rangle = \left|\alpha_p^k(A,B)\right|^2 = \frac{(2J+1)(2J'+1)}{(2k+1)}\left\langle \eta v'\left|\hat{\alpha}_{q0}^k\right|\eta v\right\rangle^2 \begin{pmatrix} J' & k & J \\ 0 & 0 & 0 \end{pmatrix}^2, \tag{8.69}$$

where $J = J_A$ and $J' = J_B$. Substituting Eq. (8.69) into Eq. (8.64), we obtain

$$\begin{aligned} \langle \alpha_{XX}\alpha_{YY}\rangle_{AB} &= \frac{1}{3}(2J+1)(2J'+1)\left\langle \eta v'\left|\hat{\alpha}_{q0}^0\right|\eta v\right\rangle^2 \begin{pmatrix} J' & 0 & J \\ 0 & 0 & 0 \end{pmatrix}^2 \\ &- \frac{1}{3}\frac{(2J+1)(2J'+1)}{5}\left\langle \eta v'\left|\hat{\alpha}_{q0}^2\right|\eta v\right\rangle^2 \begin{pmatrix} J' & 2 & J \\ 0 & 0 & 0 \end{pmatrix}^2. \end{aligned} \tag{8.70}$$

Substituting for $\left\langle \eta v'\left|\hat{\alpha}_{q0}^0\right|\eta v\right\rangle^2 = 3a_0^2$ and $\left\langle \eta v'\left|\hat{\alpha}_{q0}^2\right|\eta v\right\rangle^2 = 2\gamma_0^2/3$ using Eqs. (7.79) and (7.80), we obtain

$$\begin{aligned} \langle \alpha_{XX}\alpha_{YY}\rangle_{AB} &= a_0^2(2J+1)(2J'+1)\begin{pmatrix} J' & 0 & J \\ 0 & 0 & 0 \end{pmatrix}^2 \\ &- \frac{2}{45}\gamma_0^2(2J+1)(2J'+1)\begin{pmatrix} J' & 2 & J \\ 0 & 0 & 0 \end{pmatrix}^2. \end{aligned} \tag{8.71}$$

The $3j$ symbol $\begin{pmatrix} J' & k & J \\ 0 & 0 & 0 \end{pmatrix}$ will be nonzero only when $J + J' + k$ is an even integer and $k \geq |J - J'|$. Consequently, $J' = J$ for the first term in Eq. (8.71), and $J' - J = 0, \pm 2$ for the second term. Earlier, we restricted our analysis such that $\Omega_{ba} \geq 0$, so for the second term $J' - J = 0, +2$. Evaluating the $3j$ symbols using the appropriate equations from Appendix A3, we obtain

$$\langle \alpha_{XX}\alpha_{YY}\rangle_{AB} = \left[a_0^2(2J+1) - \frac{2}{45}b_{J'J}^{(2)}\gamma_0^2(2J+1)\right]\delta_{JJ'} - \frac{2}{45}b_{J+2,J}^{(2)}\gamma_0^2(2J+1). \tag{8.72}$$

The first term on the RHS of Eq. (8.72) is connected with coherent Rayleigh scattering, and the second term on the RHS is connected with pure rotational CARS. Ignoring the coherent Rayleigh scattering term, we obtain

$$\langle \alpha_{XX}\alpha_{YY}\rangle_{AB} = -\frac{2}{45}b_{J+2,J}^{(2)}\gamma_0^2(2J+1). \tag{8.73}$$

Now substituting Eq. (8.69) into Eq. (8.66) and performing a similar analysis, we obtain

$$
\begin{aligned}
\langle \alpha_{XY}\alpha_{XY}\rangle_{AB} &= +\frac{1}{2}\frac{(2J+1)(2J'+1)}{3}\left\langle \eta \mathrm{v}'\left|\hat{\alpha}^{1}_{q0}\right|\eta \mathrm{v}\right\rangle^2 \begin{pmatrix} J' & 1 & J \\ 0 & 0 & 0 \end{pmatrix}^2 \\
&+\frac{1}{2}\frac{(2J+1)(2J'+1)}{5}\left\langle \eta \mathrm{v}'\left|\hat{\alpha}^{2}_{q0}\right|\eta \mathrm{v}\right\rangle^2 \begin{pmatrix} J' & 2 & J \\ 0 & 0 & 0 \end{pmatrix}^2 .
\end{aligned} \tag{8.74}
$$

The molecule fixed antisymmetric anisotropy terms $\left\langle \eta \mathrm{v}'\left|\hat{\alpha}^{1}_{q0}\right|\eta \mathrm{v}\right\rangle$ can be written as

$$
\left\langle \eta \mathrm{v}\left|\hat{\alpha}^{1}_{q0}\right|\eta \mathrm{v}'\right\rangle = \left\langle \eta \mathrm{v}'\left|i(\hat{\alpha}_{xy}-\hat{\alpha}_{yx})/\sqrt{2}\right|\eta \mathrm{v}\right\rangle . \tag{8.75}
$$

These terms will be negligible except in the case of resonance Raman (Long, 2002). Setting $\left\langle \eta \mathrm{v}'\left|\hat{\alpha}^{1}_{q0}\right|\eta \mathrm{v}\right\rangle$ to zero in Eq. (8.74), we obtain

$$
\begin{aligned}
\langle \alpha_{XY}\alpha_{XY}\rangle_{AB} &= \frac{1}{10}\left\langle \eta \mathrm{v}'\left|\hat{\alpha}^{2}_{q0}\right|\eta \mathrm{v}\right\rangle^2 (2J+1)(2J'+1)\begin{pmatrix} J' & 2 & J \\ 0 & 0 & 0 \end{pmatrix}^2 \\
&= \frac{1}{15}\gamma_0^2(2J+1)(2J'+1)\begin{pmatrix} J' & 2 & J \\ 0 & 0 & 0 \end{pmatrix}^2 = \frac{1}{15}b^{(2)}_{J'J}\gamma_0^2(2J+1) \\
&= \frac{1}{15}b^{(2)}_{J+2,J}\gamma_0^2(2J+1).
\end{aligned} \tag{8.76}
$$

A similar analysis for $\langle \alpha_{XY}\alpha_{YX}\rangle_{AB}$ results in

$$
\langle \alpha_{XY}\alpha_{YX}\rangle_{AB} = \langle \alpha_{XY}\alpha_{XY}\rangle_{AB} = \frac{1}{15}b^{(2)}_{J+2,J}\gamma_0^2(2J+1). \tag{8.77}
$$

8.4.2 Vibrational CARS

Now consider the case where $\mathrm{v}_B = \mathrm{v}' = \mathrm{v}_A + 1 = \mathrm{v}+1$ for molecules with $^1\Sigma$ electronic ground states. Substituting Eq. (8.69) into Eq. (8.64), we obtain

$$
\begin{aligned}
\langle \alpha_{XX}\alpha_{YY}\rangle_{AB} &= \frac{1}{3}\left\langle \eta\left|\left(\partial\hat{\alpha}^{0}_{q0}/\partial Q\right)_0\right|\eta\right\rangle^2 \langle \mathrm{v}'|Q|\mathrm{v}\rangle^2(2J'+1)(2J+1)\begin{pmatrix} J' & 0 & J \\ 0 & 0 & 0 \end{pmatrix}^2 \\
&-\frac{1}{3}\left\langle \eta\left|\left(\partial\hat{\alpha}^{2}_{q0}/\partial Q\right)_0\right|\eta\right\rangle^2 \langle \mathrm{v}'|Q|\mathrm{v}\rangle^2(2J'+1)(2J+1)\begin{pmatrix} J' & 2 & J \\ 0 & 0 & 0 \end{pmatrix}^2 .
\end{aligned} \tag{8.78}
$$

For vibrational Raman scattering, the invariants $(a')^2$ and $(\gamma')^2$ are defined as follows:

$$
\left\langle \eta\left|\left(\partial\hat{\alpha}^{0}_{q=0}/\partial Q\right)_0\right|\eta\right\rangle^2 = 3(a')^2. \tag{8.79}
$$

$$
\left\langle \eta\left|\left(\partial\hat{\alpha}^{2}_{q0}/\partial Q\right)_0\right|\eta\right\rangle^2 = 2(\gamma')^2/3. \tag{8.80}
$$

Performing an analysis similar to that for the pure rotational case, we obtain

$$
\langle \alpha_{XX}\alpha_{YY}\rangle_{AB} = \frac{\hbar(\mathrm{v}+1)}{2\mu_R\omega_R}\left[(a')^2(2J+1)\delta_{J'J} - \frac{2}{45}b^{(2)}_{J'J}(\gamma')^2(2J+1)\right]. \tag{8.81}
$$

Because for vibrational CARS the condition $\mathrm{v}' = \mathrm{v} + 1$ guarantees that $\varepsilon_B > \varepsilon_A$, the rotational quantum number J' can take on the values of $J - 2$, J, or $J + 2$, corresponding to the vibrational O-, Q-, and S-branches, respectively. The other polarizability terms are given by

$$\langle \alpha_{XY}\alpha_{XY}\rangle_{AB} = \langle \alpha_{XY}\alpha_{YX}\rangle_{AB} = \frac{\hbar(\mathrm{v}+1)}{2\mu_R\omega_R}\left[\frac{1}{15}b^{(2)}_{J'J}(\gamma')^2(2J+1)\right]. \tag{8.82}$$

8.5 General Expression for the CARS Susceptibility

In an isotropic medium there are 21 nonzero components of the third-order susceptibility tensor:

$$\chi^{(3)}_{XXYY} = \chi^{(3)}_{YYXX} = \chi^{(3)}_{XXZZ} = \chi^{(3)}_{ZZXX} = \chi^{(3)}_{YYZZ} = \chi^{(3)}_{ZZYY} = \chi^{(3)}_{1122}. \tag{8.83}$$

$$\chi^{(3)}_{XYYX} = \chi^{(3)}_{YXXY} = \chi^{(3)}_{XZZX} = \chi^{(3)}_{ZXXZ} = \chi^{(3)}_{YZZY} = \chi^{(3)}_{ZYYZ} = \chi^{(3)}_{1221}. \tag{8.84}$$

$$\chi^{(3)}_{XYXY} = \chi^{(3)}_{YXYX} = \chi^{(3)}_{XZXZ} = \chi^{(3)}_{ZXZX} = \chi^{(3)}_{YZYZ} = \chi^{(3)}_{ZYZY} = \chi^{(3)}_{1212}. \tag{8.85}$$

$$\chi^{(3)}_{XXXX} = \chi^{(3)}_{YYYY} = \chi^{(3)}_{ZZZZ} = \chi^{(3)}_{1111} = \chi^{(3)}_{1122} + \chi^{(3)}_{1221} + \chi^{(3)}_{1212}. \tag{8.86}$$

The general expression for the components $\chi^{(3)}_{1122}, \chi^{(3)}_{1221}$, and $\chi^{(3)}_{1221}$ of the CARS susceptibility can be written as

$$\chi^{(3)}_{1122}(-\omega_4;\omega_3,-\omega_2,\omega_1) = \frac{1}{12\varepsilon_0\hbar}\sum_{AB}\left[\frac{n_A^{(0)}}{(2J+1)} - \frac{n_B^{(0)}}{(2J'+1)}\right]\left\{\frac{\langle\alpha_{XX}(A,B)\alpha_{YY}(A,B)\rangle}{[\Omega_{BA}-(\omega_1-\omega_2)-i\gamma_{BA}]} + \frac{\langle\alpha_{XY}(A,B)\alpha_{YX}(A,B)\rangle}{[\Omega_{BA}-(\omega_3-\omega_2)-i\gamma_{BA}]}\right\} = \frac{1}{24}\sum_{AB}[2a_{AB}(\omega_1-\omega_2)+b_{AB}(\omega_3-\omega_2)]. \tag{8.87}$$

$$\chi^{(3)}_{1221}(-\omega_4;\omega_3,-\omega_2,\omega_1) = \frac{1}{12\varepsilon_0\hbar}\sum_{AB}\left[\frac{n_A^{(0)}}{(2J+1)} - \frac{n_B^{(0)}}{(2J'+1)}\right]\left\{\frac{\langle\alpha_{XY}(A,B)\alpha_{YX}(A,B)\rangle}{[\Omega_{BA}-(\omega_1-\omega_2)-i\gamma_{BA}]} + \frac{\langle\alpha_{XX}(A,B)\alpha_{YY}(A,B)\rangle}{[\Omega_{BA}-(\omega_3-\omega_2)-i\gamma_{BA}]}\right\} = \frac{1}{24}\sum_{AB}[b_{AB}(\omega_1-\omega_2)+2a_{AB}(\omega_3-\omega_2)]. \tag{8.88}$$

$$\chi^{(3)}_{1212}(-\omega_4;\omega_3,-\omega_2,\omega_1) = \frac{1}{12\varepsilon_0\hbar}\sum_{AB}\left[\frac{n_A^{(0)}}{(2J+1)} - \frac{n_B^{(0)}}{(2J'+1)}\right]\left\{\frac{\langle\alpha_{XY}(A,B)\alpha_{XY}(A,B)\rangle}{[\Omega_{BA}-(\omega_1-\omega_2)-i\gamma_{BA}]} + \frac{\langle\alpha_{XY}(A,B)\alpha_{YX}(A,B)\rangle}{[\Omega_{BA}-(\omega_3-\omega_2)-i\gamma_{BA}]}\right\} = \frac{1}{24}\sum_{AB}[b_{AB}(\omega_1-\omega_2)+b_{AB}(\omega_3-\omega_2)]. \tag{8.89}$$

For $^1\Sigma$ electronic levels and for pure rotational CARS,

$$a_{AB}(\omega_n-\omega_2)=\frac{\left[n_A^{(0)}-n_B^{(0)}\frac{(2J+1)}{(2J'+1)}\right]}{\varepsilon_0\hbar[\Omega_{BA}-(\omega_n-\omega_2)-i\gamma_{BA}]}\left(-\frac{2}{45}b_{J+2,J}^{(2)}\gamma_0^2\right)\quad n=1,3. \tag{8.90}$$

$$b_{AB}(\omega_n-\omega_2)=\frac{\left[n_A^{(0)}-n_B^{(0)}\frac{(2J+1)}{(2J'+1)}\right]}{\varepsilon_0\hbar[\Omega_{BA}-(\omega_n-\omega_2)-i\gamma_{BA}]}\left[\frac{2}{15}b_{J+2,J}^{(2)}\gamma_0^2\right]\quad n=1,3. \tag{8.91}$$

For $^1\Sigma$ electronic levels and for vibrational CARS with $\mathrm{v}'=\mathrm{v}+1$,

$$a_{AB}(\omega_n-\omega_2)=\frac{\left[n_A^{(0)}-n_B^{(0)}\frac{(2J+1)}{(2J'+1)}\right](\mathrm{v}+1)}{2\mu_R\omega_R\varepsilon_0[\Omega_{BA}-(\omega_n-\omega_2)-i\gamma_{BA}]}\times\left[(a')^2\delta_{J'J}-\frac{2}{45}b_{J'J}^{(2)}(\gamma')^2\right]\quad n=1,3. \tag{8.92}$$

$$b_{AB}(\omega_n-\omega_2)=\frac{\left[n_A^{(0)}-n_B^{(0)}\frac{(2J+1)}{(2J'+1)}\right](\mathrm{v}+1)}{2\mu_R\omega_R\varepsilon_0[\Omega_{BA}-(\omega_n-\omega_2)-i\gamma_{BA}]}\left[\frac{2}{15}b_{J'J}^{(2)}(\gamma')^2\right]\quad n=1,3. \tag{8.93}$$

The nonresonant response of the molecule to the input laser beams differs from the Raman response in that the response is instantaneous and requires the simultaneous presence of all three input beams. Including the nonresonant response $\sigma=8\chi_{nr}/3$, the general expressions for the susceptibility elements can be written as

$$\chi_{1122}^{(3)}(-\omega_4;\omega_3,-\omega_2,\omega_1)=\frac{1}{24}\left\{\sigma+\sum_{AB}[2a_{AB}(\omega_1-\omega_2)+b_{AB}(\omega_3-\omega_2)]\right\}. \tag{8.94}$$

$$\chi_{1221}^{(3)}(-\omega_4;\omega_3,-\omega_2,\omega_1)=\frac{1}{24}\left\{\sigma+\sum_{AB}[b_{AB}(\omega_1-\omega_2)+2a_{AB}(\omega_3-\omega_2)]\right\}. \tag{8.95}$$

$$\chi_{1212}^{(3)}(-\omega_4;\omega_3,-\omega_2,\omega_1)=\frac{1}{24}\left\{\sigma+\sum_{AB}[b_{AB}(\omega_1-\omega_2)+b_{AB}(\omega_3-\omega_2)]\right\}. \tag{8.96}$$

Values of the nonresonant susceptibility χ_{nr}/n for selected gases are given in Table 8.1 reproduced from Eckbreth (1996).

Table 8.1 Values of the nonresonant susceptibility for selected gases (Eckbreth, 1996)

Species	χ_{nr}/n ($m^5\,C^2/J^2$)	Species	χ_{nr}/n ($m^5\,C^2/J^2$)
Ar	2.74×10^{-51}	NO	6.69×10^{-51}
N_2	2.19×10^{-51}	H_2O	4.82×10^{-51}
O_2	2.95×10^{-51}	C_2H_2	14.1×10^{-51}
CO	3.24×10^{-51}	C_2H_4	19.3×10^{-51}
CO_2	2.37×10^{-51}	C_2H_6	11.1×10^{-51}
H_2	1.67×10^{-51}	C_3H_8	25.2×10^{-51}
CH_4	6.90×10^{-51}	C_4H_{10}	31.5×10^{-51}

8.6 CARS Signal Strengths

In component form, the third-order polarization is given by

$$P_i^{(3)}(\omega_4;\omega_3,-\omega_2,\omega_1) = \varepsilon_0 \chi_{imkj}^{(3)}(\omega_4;\omega_3,-\omega_2,\omega_1)E_m(\omega_3)E_k^*(-\omega_2)E_j(\omega_1), \quad (8.97)$$

where a sum over repeated indices is implied on the RHS of Eq. (8.97). The wave equation in a polarizable medium is given by (Eckbreth, 1996)

$$\nabla^2 \boldsymbol{E} - \mu_0\varepsilon_0 \frac{\partial^2 \boldsymbol{E}}{\partial t^2} = \nabla^2 \boldsymbol{E} - \frac{1}{c_0^2}\frac{\partial^2 \boldsymbol{E}}{\partial t^2} = \mu_0 \frac{\partial^2 \boldsymbol{P}}{\partial t^2}. \quad (8.98)$$

The input electric fields and the resulting polarization field can be written as

$$\boldsymbol{E}(\boldsymbol{r},t) = \frac{1}{2}\sum_q \boldsymbol{E}(\omega_q)\exp\left[i(\boldsymbol{k}_q \bullet \boldsymbol{r} - \omega_q t)\right] = \frac{1}{2}\sum_q \boldsymbol{E}(\omega_q,\boldsymbol{r})\exp(-i\omega_q t). \quad (8.99)$$

$$\boldsymbol{P}(\boldsymbol{r},t) = \frac{1}{2}\sum_q \boldsymbol{P}(\omega_q)\exp\left[i(\boldsymbol{k}_q \bullet \boldsymbol{r} - \omega_q t)\right] = \frac{1}{2}\sum_q \boldsymbol{P}(\omega_q,\boldsymbol{r})\exp(-i\omega_q t). \quad (8.100)$$

Remember that in the expressions above, the frequencies can be both positive and negative, and that the field amplitudes are related by

$$\boldsymbol{E}(-\omega_q,\boldsymbol{r}) = \boldsymbol{E}^*(\omega_q,\boldsymbol{r}) \quad (8.101)$$

and

$$\boldsymbol{P}(-\omega_q,\boldsymbol{r}) = \boldsymbol{P}^*(\omega_q,\boldsymbol{r}). \quad (8.102)$$

Considering a sum over both positive and negative frequencies ω_q in Eqs. (8.99) and (8.100), it can be seen that both the electric field and the polarization field terms are real. Substituting Eqs. (8.99) and (8.100) into Eq. (8.98), we obtain

$$\nabla^2 \boldsymbol{E}(\omega_q,\boldsymbol{r}) + \frac{\omega_q^2}{c_0^2}\boldsymbol{E}(\omega_q,\boldsymbol{r}) = -\mu_0\omega_q^2\boldsymbol{P}(\omega_q,\boldsymbol{r}) \quad (8.103)$$

for each frequency ω_q. In an isotropic medium, the polarization is given by

$$\boldsymbol{P}(\omega_q,\boldsymbol{r}) = \boldsymbol{P}^{(1)}(\omega_q,\boldsymbol{r}) + \boldsymbol{P}^{(3)}(\omega_q,\boldsymbol{r}) = \varepsilon_0\chi^{(1)}\boldsymbol{E}(\omega_q,\boldsymbol{r}) + \boldsymbol{P}^{(3)}(\omega_q,\boldsymbol{r}). \quad (8.104)$$

Substituting Eq. (8.104) into Eq. (8.103), we obtain

$$\begin{aligned}&\nabla^2 \boldsymbol{E}(\omega_q,\boldsymbol{r}) + \frac{\omega_q^2}{c_0^2}\left(1+\chi^{(1)}\right)\boldsymbol{E}(\omega_q,\boldsymbol{r})\\ &= \nabla^2 \boldsymbol{E}(\omega_q,\boldsymbol{r}) + \frac{\omega_q^2}{c^2}n_{q,refr}^2\boldsymbol{E}(\omega_q,\boldsymbol{r}) = -\mu_0\omega_q^2\boldsymbol{P}^{(3)}(\omega_q,\boldsymbol{r}),\end{aligned} \quad (8.105)$$

where

$$k_q = n_{q,refr}\frac{\omega_q}{c} = n_{q,refr}\frac{2\pi}{\lambda_q} \cong \frac{2\pi}{\lambda_q}. \quad (8.106)$$

For gas-phase media far from single-photon resonances, we will assume that the refractive index $n_{q,refr}$ at frequency ω_q is equal to one.

Now assume that the laser beams are propagating as plane waves at a small angle with respect to the z-axis and that the direction of CARS signal generation is precisely along the z-axis. Substituting Eq. (8.97) into Eq. (8.105) and solving for the ith component of the electric field at the CARS signal frequency ω_4, we obtain

$$\begin{aligned}&\nabla^2 E_i(\omega_4, \boldsymbol{r}) + k_4^2 E_i(\omega_4, \boldsymbol{r})\\&= -\frac{\omega_4^2}{c_0^2}\chi^{(3)}_{imkj} E_m(\omega_3) E_k^*(\omega_2) E_j(\omega_1) \exp[i(\boldsymbol{k}_3 - \boldsymbol{k}_2 + \boldsymbol{k}_1) \bullet \boldsymbol{r}].\end{aligned} \tag{8.107}$$

Now assume that the CARS signal is propagating such that $\boldsymbol{k}_4$ is parallel to the z-axis. Under these conditions, Eq. (8.107) can now be written as

$$\begin{aligned}\nabla^2 E_i(\omega_4, \boldsymbol{r}) + k_4^2 E_i(\omega_3, \boldsymbol{r}) &= \frac{\partial^2 [E_i(\omega_4) \exp{(ik_4 z)}]}{\partial z^2} + k_4^2 E_i(\omega_3) \exp{(ik_4 z)}\\&= \frac{\partial^2 E_i(\omega_4)}{\partial z^2} \exp{(ik_4 z)} + 2ik_4 \frac{\partial E_i(\omega_4)}{\partial z} \exp{(ik_4 z)}\\&\quad - k_4^2 E_i(\omega_3) \exp{(ik_4 z)} + k_4^2 E_i(\omega_3) \exp{(ik_4 z)}\\&= \frac{\partial^2 E_i(\omega_4)}{\partial z^2} \exp{(ik_4 z)} + 2ik_4 \frac{\partial E_i(\omega_4)}{\partial z} \exp{(ik_4 z)}.\end{aligned} \tag{8.108}$$

Substituting Eq. (8.108) into Eq. (8.107), we obtain

$$\begin{aligned}&\frac{\partial^2 E_i(\omega_4)}{\partial z^2} \exp(ik_4 z) + 2ik_4 \frac{\partial E_i(\omega_4)}{\partial z} \exp(ik_4 z)\\&= -k_4^2 \chi^{(3)}_{imkj} E_m(\omega_3) E_k^*(\omega_2) E_j(\omega_1) \exp[i(\boldsymbol{k}_3 - \boldsymbol{k}_2 + \boldsymbol{k}_1) \bullet \boldsymbol{r}].\end{aligned} \tag{8.109}$$

This can be rewritten as

$$\begin{aligned}&\frac{\partial^2 E_i(\omega_4)}{\partial z^2} + 2ik_4 \frac{\partial E_i(\omega_4)}{\partial z}\\&= -k_4^2 \chi^{(3)}_{imkj} E_m(\omega_3) E_k^*(\omega_2) E_j(\omega_1) \exp[i(\boldsymbol{k}_1 - \boldsymbol{k}_2 + \boldsymbol{k}_3 - \boldsymbol{k}_4) \bullet \boldsymbol{r}].\end{aligned} \tag{8.110}$$

The CARS signal field amplitude will build up to appreciable levels only under phase-matched conditions, or when

$$|\boldsymbol{k}_3 - \boldsymbol{k}_2 + \boldsymbol{k}_1 - \boldsymbol{k}_4| L_{CARS} \ll 1, \tag{8.111}$$

where L_{CARS} is the length of the probe volume where the CARS beams overlap. Under these conditions the variation of the CARS signal amplitude over distances on the order of a wavelength will be small, and the second derivative term in Eq. (8.110) can be neglected. Consequently Eq. (8.110) can be rewritten as

$$i\frac{\partial E_i(\omega_4)}{\partial z} = -k_4^2 \chi^{(3)}_{imkj} E_m(\omega_3) E_k^*(\omega_2) E_j(\omega_1) \exp[i\Delta k z], \tag{8.112}$$

where $\Delta k z = (\boldsymbol{k}_1 - \boldsymbol{k}_2 + \boldsymbol{k}_3 - \boldsymbol{k}_4) \bullet \boldsymbol{r}$. Assuming that the input laser field amplitudes are constant over the length of the CARS probe volume and then integrating Eq. (8.112) over the length of the CARS probe volume, we obtain

$$E_i(\omega_4) = i\frac{k_4}{2}\chi^{(3)}_{imkj}E_m(\omega_3)E^*_k(\omega_2)E_j(\omega_1)\left[\frac{\exp(i\Delta k L_{CARS}) - 1}{i\Delta k}\right]. \tag{8.113}$$

But under conditions of near-perfect phase-matching,

$$\exp(i\Delta k L_{CARS}) \cong 1 + i\Delta k L_{CARS}. \tag{8.114}$$

Substituting Eq. (8.114) into Eq. (8.113), we obtain

$$E_i(\omega_4) = i\frac{k_4}{2}\chi^{(3)}_{imkj}E_m(\omega_3)E^*_k(\omega_2)E_j(\omega_1)L_{CARS}. \tag{8.115}$$

The formula for the CARS signal electric field can be written in a compact form as

$$\begin{aligned}\boldsymbol{E}(\omega_4) = i\frac{\omega_4}{2c}&\left[\chi^{(3)}_{1122}\hat{\boldsymbol{e}}_3(\hat{\boldsymbol{e}}_2 \bullet \hat{\boldsymbol{e}}_1) + \chi^{(3)}_{1212}\hat{\boldsymbol{e}}_2(\hat{\boldsymbol{e}}_3 \bullet \hat{\boldsymbol{e}}_1) + \chi^{(3)}_{1221}\hat{\boldsymbol{e}}_1(\hat{\boldsymbol{e}}_3 \bullet \hat{\boldsymbol{e}}_2)\right] \\ &\times E(\omega_1)E^*(\omega_2)E(\omega_3)L_{CARS},\end{aligned} \tag{8.116}$$

where

$$E(\omega_q) = \frac{\boldsymbol{E}(\omega_q)}{\hat{\boldsymbol{e}}_q}. \tag{8.117}$$

The CARS signal irradiance I_4 $(\mathrm{W/m^2})$ is given by

$$\begin{aligned}I_4(\omega_4) &= \frac{1}{2}c_0\varepsilon_0|E(\omega_4)|^2 = \frac{\varepsilon_0\omega_4^2}{8c_0}\left|\chi^{(3)}\right|^2|E(\omega_1)|^2|E(\omega_2)|^2|E(\omega_3)|^2L^2_{CARS} \\ &= \frac{\omega_4^2}{\varepsilon_0^2c_0^4}\left|\chi^{(3)}\right|^2 I_1(\omega_1)I_2(\omega_2)I_3(\omega_3)L^2_{CARS},\end{aligned} \tag{8.118}$$

where

$$\begin{aligned}\left|\chi^{(3)}\right|^2 = &\left[\chi^{(3)}_{1122}\hat{\boldsymbol{e}}_3(\hat{\boldsymbol{e}}_2 \bullet \hat{\boldsymbol{e}}_1) + \chi^{(3)}_{1212}\hat{\boldsymbol{e}}_2(\hat{\boldsymbol{e}}_3 \bullet \hat{\boldsymbol{e}}_1) + \chi^{(3)}_{1221}\hat{\boldsymbol{e}}_1(\hat{\boldsymbol{e}}_3 \bullet \hat{\boldsymbol{e}}_2)\right] \\ &\bullet \left[\chi^{(3)}_{1122}\hat{\boldsymbol{e}}_3(\hat{\boldsymbol{e}}_2 \bullet \hat{\boldsymbol{e}}_1) + \chi^{(3)}_{1212}\hat{\boldsymbol{e}}_2(\hat{\boldsymbol{e}}_3 \bullet \hat{\boldsymbol{e}}_1) + \chi^{(3)}_{1221}\hat{\boldsymbol{e}}_1(\hat{\boldsymbol{e}}_3 \bullet \hat{\boldsymbol{e}}_2)\right]^*.\end{aligned} \tag{8.119}$$

As discussed by Eckbreth (1996), the CARS susceptibility in Eq. 8.119 should be multiplied a factor of $6/n!$, where n is the number of CARS beams with the same frequency and polarization.

8.7 CARS Spectral Modeling

The modeling of CARS spectra acquired in the regime where the laser pulse lengths are much longer than the mean time between collisions in the medium is discussed in this section. The modeling of CARS spectra acquired with narrowband lasers is discussed first. These spectra are referred to as scanning CARS spectra because either the Stokes or pump laser frequency must be varied to generate a CARS spectrum, and spectral acquisition times can be very long. For practical applications in areas such as combustion diagnostics and biomedical imaging, broadband Stokes sources are typically employed, so that the CARS spectra can be acquired on single laser shots. The modeling of broadband or multiplex CARS spectra is discussed in detail in this section. The modeling of femtosecond CARS spectra, where the laser pulse lengths are in general much shorter than characteristic collision times, is deferred to later sections.

In deriving Eq. (8.118) it is assumed that the spectral widths of the lasers are much narrower than the spectral widths of the Raman resonances. This is rarely the case in actual CARS experiments, and the spectral profiles of the laser sources must be taken into account to extract accurate temperature and species information from the CARS signal. Also, the laser amplitudes and irradiances in Eq. (8.118) will be functions of time for pulsed laser radiation. Experimentally, the scanning CARS signal is detected with a photodiode or a photomultiplier, and the signal is typically integrated over the duration of the laser pulses. For broadband CARS, the CARS signal is dispersed using a spectrometer and is then detected using a multichannel array detector.

8.7.1 Spectral Modeling of the Electric Fields

The notation for the frequency decomposition of the electric field that we first introduced in Eq. (8.14) is widely used in the CARS literature, but it is somewhat imprecise and is really appropriate only for truly monochromatic laser radiation. We will discuss the electric fields in more detail in this section. The electric field for a laser with frequency ω_p can be written in the form

$$\boldsymbol{E}_p(t) = \frac{1}{2}\boldsymbol{A}_p(t)\exp\left(-i\omega_{p0}\,t\right) + \frac{1}{2}\boldsymbol{A}_p^*(t)\exp\left(+i\omega_{p0}t\right). \tag{8.120}$$

For continuous-wave (cw) lasers emitting monochromatic laser radiation, the amplitudes $\boldsymbol{A}(t)$ and $\boldsymbol{A}^*(t)$ will be equal and independent of time. The electric field $\boldsymbol{E}_p(t)$ can also be written as a Fourier transform,

$$\boldsymbol{E}_p(t) = \frac{1}{2}\int_{-\infty}^{+\infty} \tilde{\boldsymbol{E}}_p(\omega)\exp(-i\omega t)\,d\omega, \tag{8.121}$$

where

$$\tilde{\boldsymbol{E}}_p(\omega_p) = \boldsymbol{A}_p(\omega_p - \omega_{p0}) + \boldsymbol{A}_p^*(-\omega_p - \omega_{p0}) \tag{8.122}$$

and

$$\boldsymbol{A}_p(t) = \int_{-\infty}^{+\infty} \boldsymbol{A}_p(\omega_p - \omega_{p0})\exp\left[-i(\omega_p - \omega_{p0})t\right]d(\omega_p - \omega_{p0}). \tag{8.123}$$

Again, in the limit of cw, monochromatic laser radiation,

$$A_p(\omega_p - \omega_{p0}) = E(\omega_p)\delta(\omega_p - \omega_{p0}). \tag{8.124}$$

Substituting Eq. (8.124) into Eq. (8.123), we obtain

$$A_p(t) = E_p(\omega_p). \tag{8.125}$$

Equation (8.125) makes the connection back to Eq. (8.14) in Section 8.2.

The use of pulsed laser sources with spectral widths comparable to or greater than the Raman spectral widths significantly complicates the analysis of CARS spectra. For pulsed laser sources we assume that the envelope function $A_p(\omega_p - \omega_{p0})$ is still sharply peaked around zero frequency $(\omega_p \cong \omega_{p0})$ but is not a delta function. The envelope function $A_p(t)$ will be peaked around $t = t_{p0}$, where t_{p0} is the time at which the intensity of the laser pulse peaks, but will have some finite temporal width τ_p. For stationary fields, defined by the relation $\langle E_p(t) \bullet E_p(t+\tau)\rangle = f(\tau)$, the spectral intensity of the laser pulse will be given by (Yuratich, 1979)

$$S_p(\omega_p) = \frac{1}{2}c_0\varepsilon_0 \int_{-\infty}^{+\infty} \left\langle E_p^*(\omega_p) \bullet E_p(\omega_p + \omega)\right\rangle d\omega. \tag{8.126}$$

The spectral intensity can be written as

$$S_p(\omega_p) = I_p(t)g_p(\omega_p - \omega_{p0}). \tag{8.127}$$

Note that the spectral intensity is a function of time, but we assume that the normalized spectral function $g_p(\omega_p - \omega_{p0})$ is constant throughout the laser pulse. For monochromatic laser radiation, $g_p(\omega_p - \omega_{p0}) = \delta(\omega_p - \omega_{p0})$ and

$$S_p(\omega_p) = \frac{1}{2}c_0\varepsilon_0 \left|E_p(\omega_p)\right|^2 = \frac{1}{2}c_0\varepsilon_0 \left|A_p\right|^2. \tag{8.128}$$

8.7.2 Scanning CARS: Monochromatic Input Laser Beams

Consider first the case where the pump, Stokes, and probe beams are all monochromatic. In this case

$$I(\omega_1, t) = I_1(t)g_1(\omega_1 - \omega_{10}) = I_1(t)\delta(\omega_1 - \omega_{10}). \tag{8.129}$$

$$I(\omega_2, t) = I_2(t)g_2(\omega_2 - \omega_{20}) = I_2(t)\delta(\omega_2 - \omega_{20}). \tag{8.130}$$

$$I(\omega_3, t) = I_3(t)g_3(\omega_3 - \omega_{30}) = I_3(t)\delta(\omega_3 - \omega_{30}). \tag{8.131}$$

This situation is depicted in Figure 8.2. In Figure 8.2 it is assumed that the pump and probe beams are obtained from the same laser, so that $\omega_{10} = \omega_{30}$. Initially, the Stokes laser frequency is resulting in the generation of a CARS signal at a frequency $\omega_4 = \omega_{10} - \omega_{20} + \omega_{30} = 2\omega_{10} - \omega_{20}$. For scanned CARS with monochromatic lasers, the Stokes laser is typically tuned to different frequencies ω'_{20} to produce the CARS spectrum

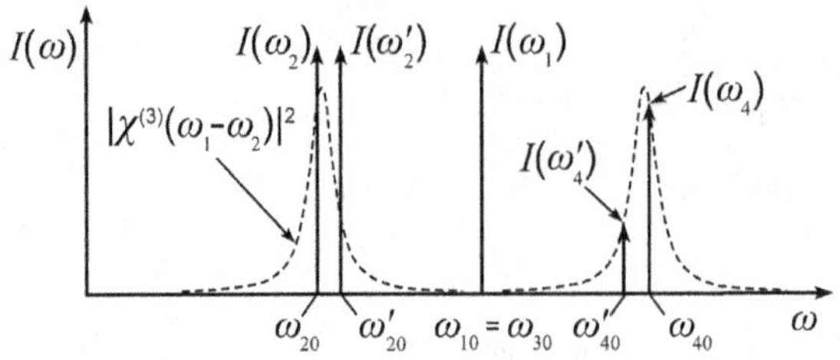

Figure 8.2 Schematic diagram of scanning CARS signal generation for case where the pump, Stokes, and probe beams are all monochromatic, and $\omega_{10} = \omega_{30}$.

$$I(\omega_4, t) = I_4(t)\delta(\omega_{10} - \omega_{20} + \omega_{30} - \omega_4) \propto \left|\chi^{(3)}(\omega_{10} - \omega_{20})\right|^2 I_1(t)I_2(t)I_3(t). \tag{8.132}$$

For gas-phase measurements, the CARS signal would be generated using pulsed laser radiation. The pump and probe beams would typically be generated from an injection-seeded, Q-switched Nd:YAG[1] laser. The tunable Stokes beam would be generated either from a pulse-dye-amplified ring dye laser or a pulse-dye-amplified optical parametric generator. The CARS signal would be detected with a photodiode or photomultiplier, and typically the signal would be integrated over the laser pulse and then recorded on a shot-by-shot basis as the Stokes laser was scanned across the CARS resonances. In this case the recorded CARS spectrum is given by

$$S_4(\omega_{40}) = \int_0^{\infty}\int_{-\infty}^{+\infty} I(\omega_4, t)\, d\omega\, dt \propto \int_0^{\infty} \left|\chi^{(3)}(\omega_{10} - \omega_{20})\right|^2 I_1(t)I_2(\omega_{20}, t)I_3(t)\, dt, \tag{8.133}$$

where $\omega_{40} = \omega_{10} - \omega_{20} + \omega_{30}$. The frequency dependence is explicitly noted for the Stokes beam because there may be substantial variation in the pulse energy and the magnitude of the irradiance I_2 as ω_{20} is scanned.

8.7.3 Scanning CARS: Monochromatic Pump and Probe Beams; Narrowband Stokes Beam

Now consider the case where the pump and probe beams are monochromatic, but the Stokes beam has a spectral width that is significant compared to the CARS line width. In this case, depicted in Figure 8.3, the spectral line shape function for the Stokes beam is not a delta function. The CARS spectral irradiance is given by

$$\begin{aligned} I_{4\omega}(\omega_4, t) &\propto \left|\chi^{(3)}(\omega_{10} - \omega_2)\right|^2 I_1(t)I_{2\omega}(\omega_2, t)I_3(t)\delta(\omega_4 - \omega_{10} - \omega_{30} + \omega_2) \\ &= I_1(t)I_2(t)I_3(t)\int_{-\infty}^{+\infty} \left|\chi^{(3)}(\omega_{10} - \omega_2)\right|^2 g_2(\omega_2 - \omega_{20})\delta(\omega_4 - \omega_{10} - \omega_{30} + \omega_2)\, d\omega_2. \end{aligned} \tag{8.134}$$

[1] Neodymium-doped yttrium-aluminum-garnet

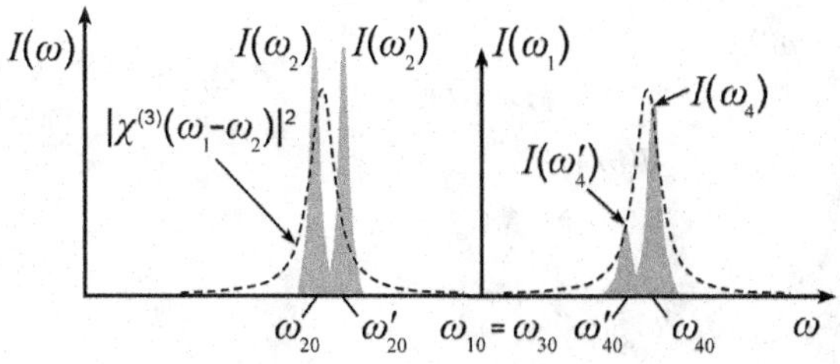

Figure 8.3 Schematic diagram of scanning CARS signal generation for case where the pump and probe beams are monochromatic, the narrowband Stokes beam has a spectral width that is significant compared to the Raman linewidth, and $\omega_{10} = \omega_{30}$.

The CARS signal is found by integrating Eq. (8.134) over the laser pulse lengths and the spectrum of the Stokes laser:

$$S_4(\omega_{40}) \propto \int_0^{\infty} I_1(t)I_2(t)I_3(t)\left[\int_{-\infty}^{+\infty} \left|\chi^{(3)}(\omega_{10}-\omega_2)\right|^2 g_2(\omega_2-\omega_{20}) \times \delta(\omega_4-\omega_{10}-\omega_{30}+\omega_2)d\omega_2\right]dt. \tag{8.135}$$

As would be the case for monochromatic lasers, the Stokes laser would be tuned to different frequencies ω_{20}' to produce the CARS spectrum, and the time-integrated CARS signal would be recorded on a shot-by-shot basis as the Stokes laser was scanned over the CARS resonances. Numerical modeling of the spectrum would be required to account for the spectral width of the Stokes laser; in particular, a convolution integral over the Stokes spectral width must be performed.

8.7.4 Scanning CARS: Narrowband Pump, Probe, and Stokes Beams

Now consider the case where the pump, Stokes, and probe beams all have spectral widths that are significant compared to the CARS line width. This was typically the case in the early 1980s before the advent of commercially available injection-seeded Nd:YAG lasers. For this case, shown in Figure 8.4, the CARS signal is given by

$$S_4(\omega_{40}) \propto \int_0^{\infty} I_1(t)I_2(t)I_3(t)\left[\int_{-\infty}^{+\infty}\int_{-\infty}^{+\infty}\int_{-\infty}^{+\infty} \left|\chi^{(3)}(\omega_1-\omega_2)+\chi^{(3)}(\omega_3-\omega_2)\right|^2 \times g_1(\omega_1-\omega_{10})g_2(\omega_2-\omega_{20})g_3(\omega_3-\omega_{30})\delta(\omega_4-\omega_1-\omega_3+\omega_2)d\omega_1 d\omega_2 d\omega_3\right]dt. \tag{8.136}$$

Just as discussed in the previous two sections, the Stokes laser would be tuned to different frequencies ω_{20}' to produce the CARS spectrum, and the time-integrated CARS signal would be recorded on a shot-by-shot basis as the Stokes laser was scanned over the CARS resonances. Numerical modeling of the spectrum would again be required to account for the spectral width of the Stokes laser. However, now a triple convolution over the spectral widths of the pump, Stokes, and probe lasers must be performed.

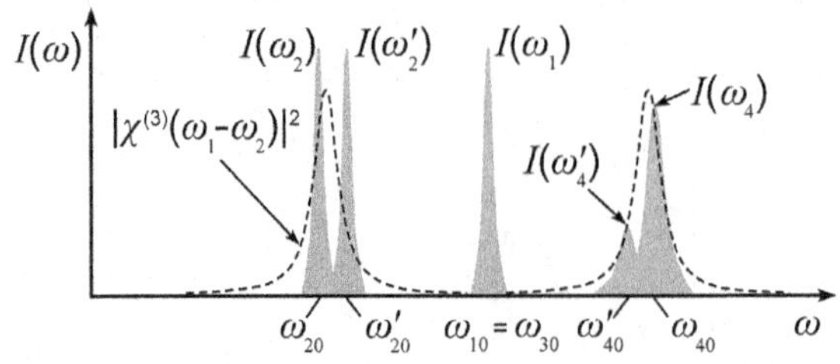

Figure 8.4 Schematic diagram of scanning CARS signal generation for case where the pump, Stokes, and probe beams are narrowband with spectral widths that are significant compared to the Raman linewidth, and $\omega_{10} = \omega_{30}$.

8.7.5 Broadband CARS: Monochromatic Pump and Probe Beams and Broadband Stokes Beam

In combustion, most CARS experiments employ a broadband Stokes laser so that the CARS spectra can be acquired on single laser shots. Because of the widespread commercial availability of injection-seeded, Q-switched Nd:YAG lasers, we will consider the case where the pump beam is monochromatic. The generation of the broadband CARS spectrum is depicted schematically in Figure 8.5. For this case the CARS signal is given by

$$\begin{aligned} S_{4\omega}(\omega_4) &\propto \left[\int_0^{\infty} I_1(t)I_2(t)I_3(t)dt\right] \\ &\times \left[\int_{-\infty}^{+\infty} \left|\chi^{(3)}(\omega_{10}-\omega_2)+\chi^{(3)}(\omega_{30}-\omega_2)\right|^2 g_2(\omega_2-\omega_{20})\delta(\omega_4-\omega_{10}-\omega_{30}+\omega_2)d\omega_2\right] \\ &= \left[\int_0^{\infty} I_1(t)I_2(t)I_3(t)dt\right] \left|\chi^{(3)}(\omega_4-\omega_{30})+\chi^{(3)}(\omega_4-\omega_{10})\right|^2 g_2(\omega_4-\omega_{40}), \end{aligned} \tag{8.137}$$

where

$$\omega_{40} = \omega_{10} + \omega_{30} - \omega_{20}. \tag{8.138}$$

The Stokes laser in broadband CARS has a typical spectral width of several hundred cm^{-1} and exhibits significant spectral fluctuations from laser shot to laser shot. The effect of the Stokes laser spectrum on the generated CARS signal is usually accounted for by measuring the nonresonant four-wave mixing spectrum from a gas such as argon. The nonresonant susceptibility is independent of frequency, so Eq. (8.137) yields

$$S_{4\omega,nr}(\omega_4) \propto \chi_{nr}^2 g_2(\omega_4 - \omega_{40}). \tag{8.139}$$

The broadband CARS signal is detected using a multichannel array detector, usually an unintensified CCD[1] camera (Rakestraw et al., 1989), although many detectors

[1] Charge-coupled device

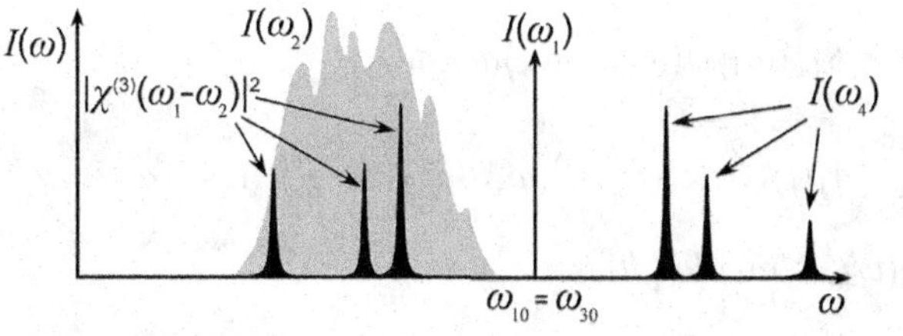

Figure 8.5 Schematic diagram of broadband CARS signal generation for case where the pump and probe beams are monochromatic, the Stokes beam is broadband, and $\omega_{10} = \omega_{30}$.

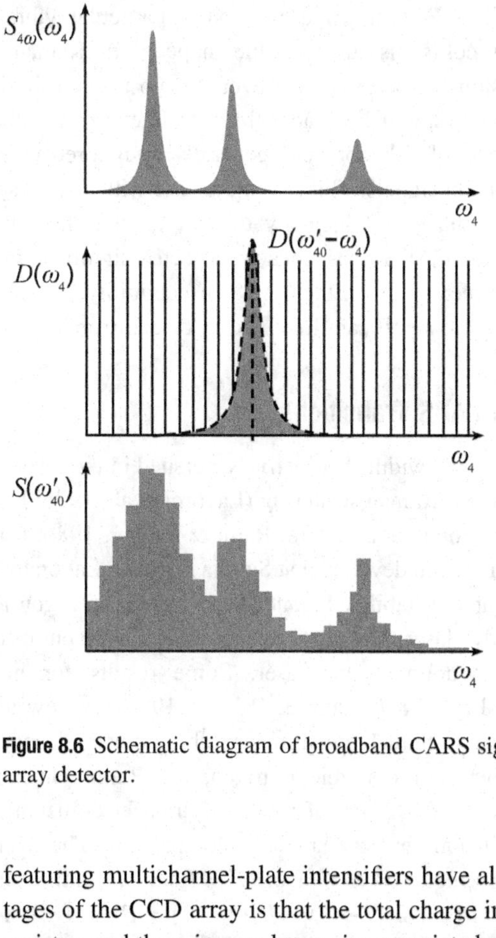

Figure 8.6 Schematic diagram of broadband CARS signal detection on a multichannel array detector.

featuring multichannel-plate intensifiers have also been employed. One of the advantages of the CCD array is that the total charge in a column can be binned in the serial register, and there is very low noise associated with the charge transfer. The detection process is indicated in Figure 8.6, where each vertical box represents a column of pixels, and each pixel column is associated with a central frequency ω'_{40}. The signal for the pixel centered at ω'_{40} is found by convolving the broadband CARS spectral intensity with the instrument function $D(\omega'_{40} - \omega_4)$ and integrating in time over the laser pulse,

$$S_4(\omega'_{40}) = \int_0^\infty \int_{-\infty}^{+\infty} S_{4\omega}(\omega_4) D(\omega'_{40} - \omega_4) d\omega_4 dt$$
$$\propto \int_0^\infty \int_{-\infty}^{+\infty} I_1(t) I_2(t) I_3(t) \left| \chi^{(3)}(\omega_4 - \omega_{30}) + \chi^{(3)}(\omega_4 - \omega_{10}) \right|^2 g_2(\omega_4 - \omega_{40})$$
$$\times D(\omega'_{40} - \omega_4) d\omega_4 dt. \tag{8.140}$$

8.8 Effects of Collisions on CARS Lineshapes and Intensities

The intensity of spontaneous Raman scattering from a particular vibration-rotation transition is unaffected by collisions, although the shape of the Raman spectral line will be affected by collisional processes. However, in most cases the instrumental width of the detection system is much broader than the linewidths of the individual Raman lines, and the effects of collisions can generally be ignored when analyzing gas-phase spontaneous Raman signals. This is not the case with CARS spectroscopy. Collisions affect both the shape and intensity of individual Raman transitions in CARS spectroscopy, and the CARS lineshapes are affected significantly by interference effects due to the complex nature of the CARS susceptibility. In addition, at high pressure, collisions can lead to significant line mixing or collisional narrowing.

8.8.1 Broadening of Isolated CARS Transitions

The measurement of Raman linewidths was actively pursued in the 1980s using cw or quasi-cw stimulated or inverse Raman scattering (Lavorel et al., 1986; Lempert et al., 1984; Owyoung, 1978; Owyoung et al., 1978; Rahn & Palmer, 1986; Rosasco et al., 1989). The inverse Raman system developed at Sandia National Laboratories by Larry Rahn and co-workers featured a stabilized, continuous-wave (cw) argon-ion laser and a pulse-dye amplified cw dye laser. The pulse-dye amplifiers were pumped by a 22-ns injection-seeded, frequency-doubled dye laser. Some results for nitrogen self-broadening are shown in Figure 8.7 (Rahn & Palmer, 1986). At low temperatures, the linewidths of the Raman Q-branch transitions exhibit a drastic dependence on the rotational quantum number. For diatomic nitrogen, the Raman linewidth at low pressure is determined by the frequency of rotational transfer collisions; the rate of vibrational dephasing collisions is very small in comparison. The average kinetic energy of the nitrogen molecules is $3k_BT/2hc$ or $1.439T\,\text{cm}^{-1}$, and the rotational level spacing is approximately $B(J+2)(J+3) - BJ(J+1) = B(4J+6)$; for diatomic nitrogen, $B \cong 2\,\text{cm}^{-1}$. At low temperature, the colliding nitrogen molecules do not have enough kinetic energy to cause transitions between high-J rotational levels. At high temperature, the J-dependence of the Raman linewidths is drastically reduced due to the higher average kinetic energy of the colliding molecules. The data shown in Figure 8.7 is analyzed using a modified-exponential-gap (MEG) model with the upward rotational transfer rate coefficient from level i to level j, with $J_j > J_i$ given by

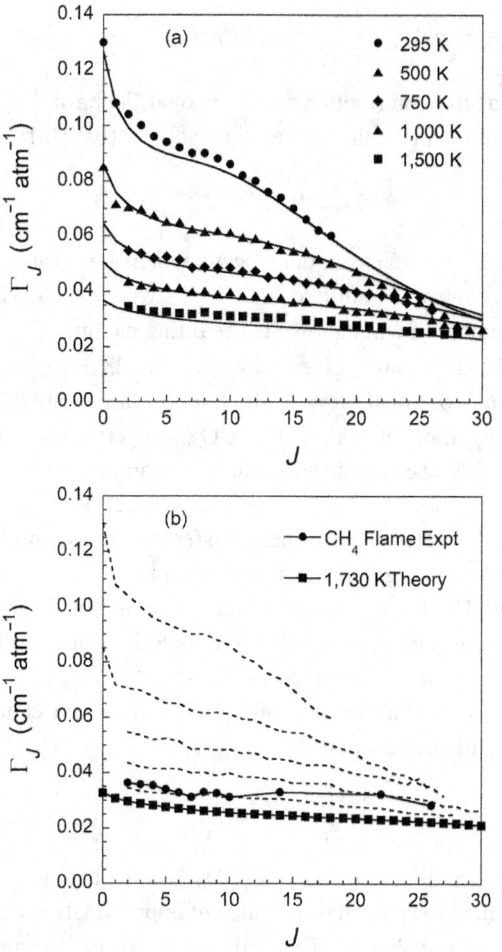

Figure 8.7 (a) Nitrogen self-broadening linewidths for Q-branch Raman transitions at different temperatures. The symbols represent experimental data, and the solid lines are calculated using the MEG law, Eq. (8.141), with the coefficients $\alpha = 0.023\ \mathrm{cm}^{-1}\ \mathrm{atm}^{-1}$, $\beta = 1.67$, $\delta = 1.26$, and $n = 1.346$. (b) Nitrogen collisional broadening linewidths for Q-branch Raman transitions from the postflame region of a lean methane/air flame with a measured temperature of 1,730 K (solid circles), and the predicted Raman linewidths (solid squares) from the MEG law with the same coefficients as for the self-broadening calculations shown in (a). The dashed lines in (b) are lines drawn through the experimental data from (a). Figures adapted with permission from Rahn and Palmer (1986), © The Optical Society.

$$\gamma_{ji} = \alpha p \left(\frac{T_0}{T}\right)^n \left(\frac{1 + 1.5E_i/k_BT\gamma}{1 + 1.5E_i/k_BT}\right)^2 \exp\left(-\beta \Delta E_{ji}/k_BT\right), \tag{8.141}$$

where p is the pressure in atm, the reference temperature $T_0 = 295$ K, E_i is the N_2 rotational term energy in J, and $\Delta E_{ji} = E_j - E_i > 0$. The downward rotational transition rate coefficient is given by

$$\gamma_{ij} = \gamma_{ji}\left(\frac{2J_i + 1}{2J_j + 1}\right)\exp\left(\Delta E_{ji}/k_B T\right). \tag{8.142}$$

For N_2, it is assumed that the contribution of vibrational dephasing to the Raman linewidth is negligible, and therefore the Raman linewidth Γ_j (FWHM) is given by

$$\Gamma_j = 2\sum_k \gamma_{kj} = 2\gamma_j. \tag{8.143}$$

Note that for N_2, rotational transfer can only occur when the rotational quantum numbers J_i and J_i are both even or both odd; transfer between levels with even and odd rotational quantum numbers are not allowed. The fitting parameters as determined from the experimental data are given by $\alpha = 0.023 \pm 0.003\ \text{cm}^{-1}\ \text{atm}^{-1}$, $\beta = 1.67 \pm 0.15$, $\delta = 1.26 \pm 0.06$, and $n = 1.346 \pm 0.06$. Similar studies have been done by other groups for N_2 (Lavorel et al., 1986), NO (Lempert et al., 1984), and CO (Rosasco et al., 1989). These species all have similar rotational level spacing with rotational constants close to 2 cm^{-1}. The Raman linewidths at atmospheric pressure are determined almost exclusively by rotational transfer collisions, and for the most part the effects of velocity-changing collisions can be neglected.

This is not the case for H_2 due to its low molecular weight and large rotational constant. The Raman linewidths of H_2 were measured by Rahn et al. (1991). Motional narrowing and elastic vibrational dephasing collisions are both very important for the determination of Raman linewidths for near-atmospheric pressure conditions. The Raman linewidth for H_2 is given by (Rahn et al., 1991)

$$\Gamma_j = \frac{4\pi D_0 \nu_R^2}{c^2 \rho} + 2\rho\gamma_j. \tag{8.144}$$

The dependence of the optical diffusion coefficient D_0 (cm^2 amagat s^{-1}) is discussed in more detail in Rahn et al. (1991) and has the value of approximately 1.35–1.42 cm^2 amagat s^{-1} for the Q(1) line at 295 K. The collisional-broadening coefficient has contributions due to both elastic vibrational dephasing and inelastic rotational transfer collisions:

$$\gamma_j = \gamma_{j,vdeph} + \gamma_{j,rtrans}. \tag{8.145}$$

The modeling of the elastic vibrational dephasing and inelastic rotational transfer collisions is discussed in detail in the paper.

8.8.2 Collisional Narrowing

For CARS spectroscopy of molecules performed near atmospheric pressure, the Raman broadening of the Raman transitions can be treated as due to dephasing collisions, and the susceptibility expressions derived in Section 8.5 can be applied with the substitution of the appropriate values for the Raman linewidth Γ_j for each transition. However, for Q-branch transitions in molecules such as N_2, O_2, and CO, the Raman polarization is not actually dephased by the rotational transfer collisions that dominate the broadening process. The phase of the Raman polarization is determined by the internuclear spacing, and these collisions are not energetic enough to change the internuclear spacing. As stated by Dion and May (1973), "Vibrationally inelastic collisions or collisions that modify the phase of the vibrational motion, broaden or shift vibrational Raman lines." These are referred to as vibrational dephasing collisions, but for the molecules listed above, these are in most cases negligible compared to rotationally inelastic collisions.

The treatment of collisional narrowing (also referred to as line mixing) requires a more sophisticated analysis of collisions that the phenomenological dephasing rate coefficients used in Eq. (8.1). The steps in this analysis are summarized by Alekseyev (1968), Hall et al. (1980), and Hall (1983). The result of the analysis in Hall et al. (1980) is the following expression for the CARS susceptibility:

$$\chi_3 = \sum_t \frac{N\alpha_t}{\hbar} \sum_s \alpha_s \Delta\rho_s^0 [G]_{ts}^{-1}. \tag{8.146}$$

Neglecting pressure shifts and vibrational dephasing collisions, the elements G_{st} of the G matrix are given by (Koszykowski et al., 1985)

$$G_{st} = \left[-(\omega_p - \omega_S - \omega_t) - i\frac{\Gamma_t}{2}\right]\delta_{st} + i\gamma_{st}(1 - \delta_{ts}) \tag{8.147}$$

and

$$\Gamma_t = 2\sum_{s \neq t} \gamma_{st}. \tag{8.148}$$

To illustrate the phenomenon of collisional narrowing, we will first consider a two-level system with transitions 1 and 2 having Raman frequencies of ω_1 and ω_2, respectively. The analysis of this two-level system is outlined by Alekseyev et al. (1968). The G matrix for this two-level system is given by

$$[G] = \begin{bmatrix} -(\Delta\omega - \omega_1) - i\dfrac{\Gamma_1}{2} & i\gamma_{12} \\ i\gamma_{21} & -(\Delta\omega - \omega_2) - i\dfrac{\Gamma_2}{2} \end{bmatrix}, \tag{8.149}$$

where $\gamma_{12} = \Gamma_2/2$, $\gamma_{21} = \Gamma_1/2$, and $\Delta\omega = \omega_p - \omega_S$. The inverse of a 2×2 matrix is given by (Schneider & Barker, 1973)

$$[A] = \begin{bmatrix} a & b \\ c & d \end{bmatrix} \Rightarrow [A]^{-1} = \frac{1}{ad - bc}\begin{bmatrix} d & -b \\ -c & a \end{bmatrix}. \tag{8.150}$$

Applying Eq. (8.150) to Eq. (8.149), we obtain

$$[G]^{-1} = \frac{\begin{bmatrix} -(\Delta\omega - \omega_2) - i\dfrac{\Gamma_2}{2} & -i\gamma_{12} \\ -i\gamma_{21} & -(\Delta\omega - \omega_1) - i\dfrac{\Gamma_1}{2} \end{bmatrix}}{\left[(\Delta\omega - \omega_2) + i\dfrac{\Gamma_2}{2}\right]\left[(\Delta\omega - \omega_1) + i\dfrac{\Gamma_1}{2}\right] + \gamma_{12}\gamma_{21}}. \tag{8.151}$$

Substituting the results of Eq. (8.151) into Eq. (8.146), we obtain

$$\chi_3 = \frac{iN\alpha_1}{\hbar}\left(\alpha_1\Delta\rho_1^0[G]_{11}^{-1} + \alpha_2\Delta\rho_2^0[G]_{12}^{-1}\right) + \frac{iN\alpha_2}{\hbar}\left(\alpha_2\Delta\rho_2^0[G]_{22}^{-1} + \alpha_1\Delta\rho_1^0[G]_{21}^{-1}\right)$$

$$= \frac{iN\alpha_1\left\{\alpha_1\Delta\rho_1^0\left[-(\Delta\omega-\omega_2) - i\frac{\Gamma_2}{2}\right] - i\alpha_2\Delta\rho_2^0\gamma_{12}\right\}}{\hbar\left\{(\Delta\omega-\omega_1)(\Delta\omega-\omega_2) - \frac{\Gamma_1\Gamma_2}{4} + \gamma_{12}\gamma_{21} + i\left[(\Delta\omega-\omega_1)\frac{\Gamma_2}{2} + (\Delta\omega-\omega_2)\frac{\Gamma_1}{2}\right]\right\}}$$

$$+ \frac{iN\alpha_2\left\{\alpha_2\Delta\rho_2^0\left[-(\Delta\omega-\omega_1) - i\frac{\Gamma_1}{2}\right] - \alpha_1\Delta\rho_1^0\gamma_{21}\right\}}{\hbar\left\{(\Delta\omega-\omega_1)(\Delta\omega-\omega_2) - \frac{\Gamma_1\Gamma_2}{4} + \gamma_{12}\gamma_{21} + i\left[(\Delta\omega-\omega_1)\frac{\Gamma_2}{2} + (\Delta\omega-\omega_2)\frac{\Gamma_1}{2}\right]\right\}}. \tag{8.152}$$

Collecting real and imaginary terms and simplifying using the relations $\Gamma_2 = 2\gamma_{12}$ and $\Gamma_1 = 2\gamma_{21}$, we obtain

$$\chi_3 = N\frac{-i\alpha_1^2\Delta\rho_1^0(\Delta\omega-\omega_2) + \gamma_{12}\left[\alpha_1^2\Delta\rho_1^0 + \alpha_1\alpha_2\Delta\rho_2^0\right]}{\hbar\{(\Delta\omega-\omega_1)(\Delta\omega-\omega_2) + i[(\Delta\omega-\omega_1)\gamma_{12} + (\Delta\omega-\omega_2)\gamma_{21}]\}}$$
$$+ N\frac{-i\alpha_2^2\Delta\rho_2^0(\Delta\omega-\omega_1) + \gamma_{21}\left[\alpha_2^2\Delta\rho_2^0 + \alpha_2\alpha_1\Delta\rho_1^0\right]}{\hbar\{(\Delta\omega-\omega_1)(\Delta\omega-\omega_2) + i[(\Delta\omega-\omega_1)\gamma_{12} + (\Delta\omega-\omega_2)\gamma_{21}]\}}. \tag{8.153}$$

Now let us assume that $\alpha_1 = \alpha_2 = \alpha$ and that $\Delta\rho_1^0 = \Delta\rho_2^0 = \Delta\rho^0$. Under the condition that $\Delta\rho_1^0 = \Delta\rho_2^0$, detailed balancing implies that $\gamma_{12} = \gamma_{21} = \gamma$ and $\Gamma_1 = \Gamma_2$. Substituting into Eq. (8.153) and simplifying, we obtain

$$\chi_3 = 2N\frac{\alpha^2\Delta\rho^0\{i[(\Delta\omega-\omega_2) + (\Delta\omega-\omega_1)] + \gamma\}}{\hbar\{(\Delta\omega-\omega_1)(\Delta\omega-\omega_2) + i\gamma[(\Delta\omega-\omega_1) + (\Delta\omega-\omega_2)]\}}. \tag{8.154}$$

Now defining the detuning frequencies $\Omega = \Delta\omega - (\omega_2 - \omega_1)/2$ and $\delta = (\omega_2 - \omega_1)/2$, we can rewrite Eq. (8.154) as

$$\chi_3 = 2N\frac{\alpha^2\Delta\rho^0\{i[(\Omega-\delta) + (\Omega+\delta)] + \gamma\}}{\hbar\{(\Omega+\delta)(\Omega-\delta) + i\gamma[(\Omega+\delta) + (\Omega-\delta)]\}}. \tag{8.155}$$

Taking the square of Eq. (8.155), we obtain

$$\chi_3^2 = \frac{4\mathrm{N}^2(\Delta\rho^0)^2\alpha^4}{\hbar^2}\,\frac{4\Omega^2 + \gamma^2}{(\Omega^2-\delta^2)^2 + 4\Omega^2\gamma^2}. \tag{8.156}$$

Now consider the resonance behavior of Eq. (8.156) for $\Omega = 0, \pm\delta$. For $\Omega = \pm\delta$, we obtain

$$\chi_3^2 = \frac{4\mathrm{N}^2(\Delta\rho^0)^2\alpha^4}{\hbar^2}\,\frac{4\delta^2+\gamma^2}{4\delta^2\gamma^2} = \frac{4\mathrm{N}^2(\Delta\rho^0)^2\alpha^4}{\hbar^2\gamma^2}\left[1 + \left(\frac{\gamma}{2\delta}\right)^2\right]. \tag{8.157}$$

For the case where $\gamma \ll \delta$, Eq. (8.157) reduces to

$$\chi_3^2 = \frac{4N^2(\Delta\rho^0)^2\alpha^4}{\hbar^2\gamma^2}. \tag{8.158}$$

As expected, the system exhibits two resonant features at $\Omega = \pm\delta$, and the square of the CARS susceptibility is inversely proportional to the rate of rotational transfer collisions. For the case where $\gamma \gg \delta$, however, Eq. (8.157) reduces to

$$\chi_3^2 = \frac{4N^2(\Delta\rho^0)^2\alpha^4}{\hbar^2\delta^2}. \tag{8.159}$$

The CARS intensity now depends only on the detuning factor δ and is independent of the rate of rotational transfer collisions.

Now consider the resonance behavior for $\Omega = 0$. In this case Eq. (8.156) reduces to

$$\chi_3^2 = \frac{4N^2(\Delta\rho^0)^2\alpha^4}{\hbar^2\delta^2}\left(\frac{\gamma}{\delta}\right)^2. \tag{8.160}$$

The square of the CARS susceptibility is now proportional to the square of the rate of rotational transfer collisions. For the case where $\gamma \gg \delta$, the strength of the CARS signal at $\Omega = 0$ will be $(\gamma/\delta)^2$-times stronger than the signal at $\Omega = \pm\delta$.

Figure 8.8 shows the results of calculations for a system of two CARS resonances separated by 4 cm^{-1} corresponding to $\delta = 3.77 \times 10^{11}$ s^{-1}. The time dependence of the Raman polarization is shown for $\gamma = 5 \times 10^{10}$, 2×10^{11}, 10^{12}, and 4×10^{12} s^{-1} in Figure 8.8a–d, respectively, and the normalized CARS spectra as calculated from Eq. (8.156) are shown in Figure 8.9. The time dependence of the Raman polarization was calculated assuming that the initial frequency of the transition was ω_1 and that rotational transfer collisions occurred randomly at the rate γ. The transition frequencies are 2,322 and 2,326 cm^{-1}, and the Raman polarization is calculated with a carrier frequency of $\omega_0/2\pi c = 2,320$ cm^{-1} subtracted. For the time slice shown in Figure 8.8a, the Q-branch resonance is oscillating initially with frequency $(\omega_1 - \omega_0)/2\pi c = 2$ cm^{-1}. Just after $t = 5.00$ ns, a rotational transfer collision occurs, and the Q-branch resonance is now oscillating at $(\omega_2 - \omega_0)/2\pi c = 6$ cm^{-1}. For $\gamma = 5 \times 10^{10}$ s^{-1}, only two collisions occur within the 0.12 ns time shown in Figure 8.8a (the laser pulse is 10 ns in duration). Consequently, the two separate frequencies at 2 and 6 cm^{-1} are clearly evident, and the Fourier transform of the Raman polarization closely resembles the $\gamma = 5 \times 10^{10}$ s^{-1} spectrum in Figure 8.9, with two clearly distinguishable peaks. For $\gamma = 2 \times 10^{11}$ s^{-1}, the two peaks have broadened and started to merge. For $\gamma = 10^{12}$ s^{-1}, the spectrum now consists of a single peak at the center frequency $(\omega_1 + \omega_2)/2$, and the phenomenon of collisional narrowing is clearly evident. For $\gamma = 4 \times 10^{12}$ s^{-1}, the rotational transfer collisions are so frequent that the Raman polarization in Figure 8.8d appears to exhibit pure sinusoidal oscillation at the single frequency of $(\omega_1 + \omega_2)/2$, and the spectral peak at this frequency has narrowed considerably compared to the peak for $\gamma = 10^{12}$ s^{-1}.

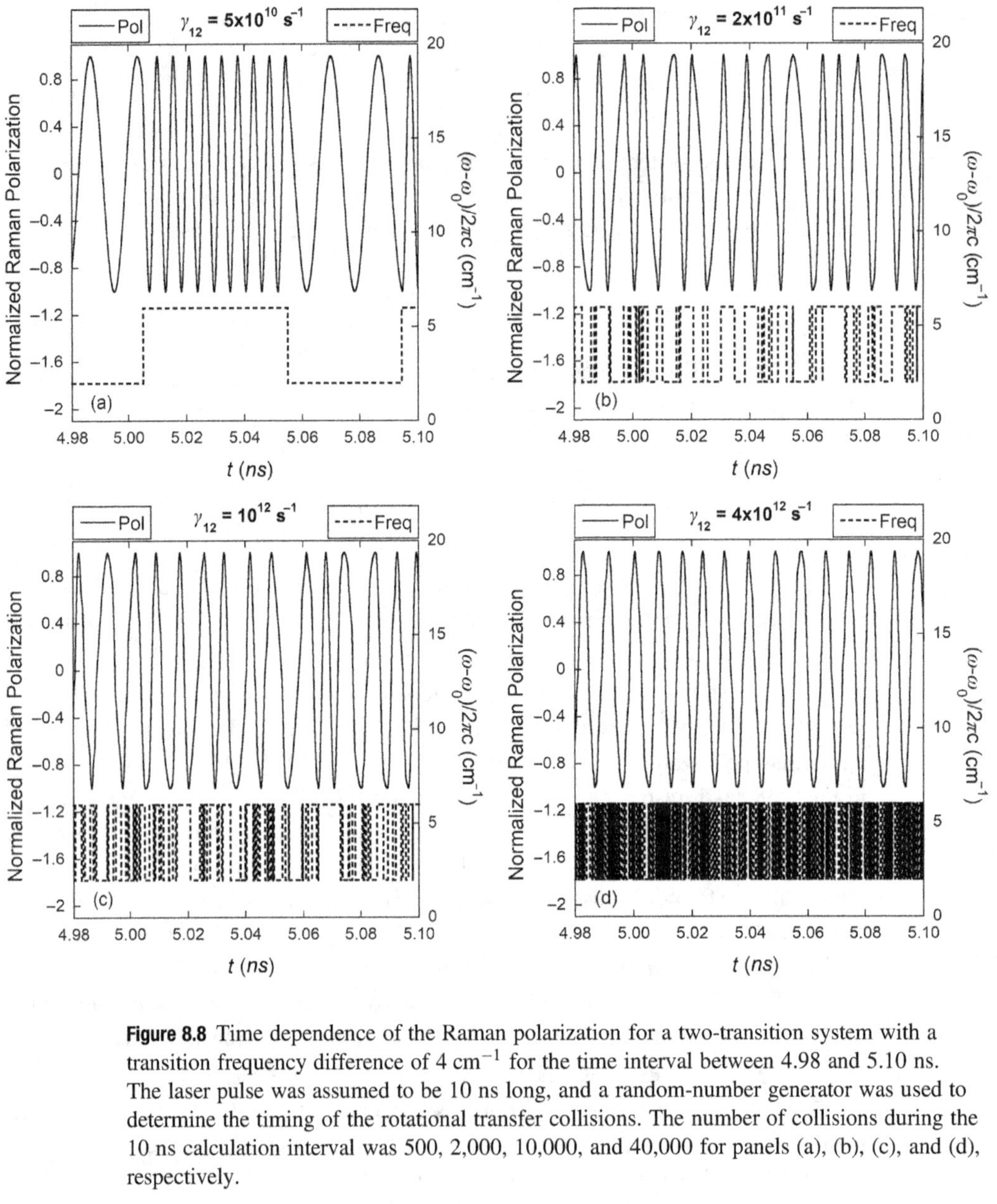

Figure 8.8 Time dependence of the Raman polarization for a two-transition system with a transition frequency difference of 4 cm^{-1} for the time interval between 4.98 and 5.10 ns. The laser pulse was assumed to be 10 ns long, and a random-number generator was used to determine the timing of the rotational transfer collisions. The number of collisions during the 10 ns calculation interval was 500, 2,000, 10,000, and 40,000 for panels (a), (b), (c), and (d), respectively.

For actual molecules like N_2, there are, of course, numerous Q-branch transitions that must be considered, and the inversion of the G matrix is a complex and time-consuming numerical process. A streamlined approach to the inversion of the G matrix is discussed by Koszykowski et al. (1985), and this approach was incorporated in the Sandia CARSFT code (Palmer, 1989) for N_2. The results of CARSFT calculations of the spectrum of N_2 at 300 K and pressures of 0.1, 0.5, 1.0, and 10.0 MPa are shown in Figure 8.10. The collapse of the spectrum due to collisional

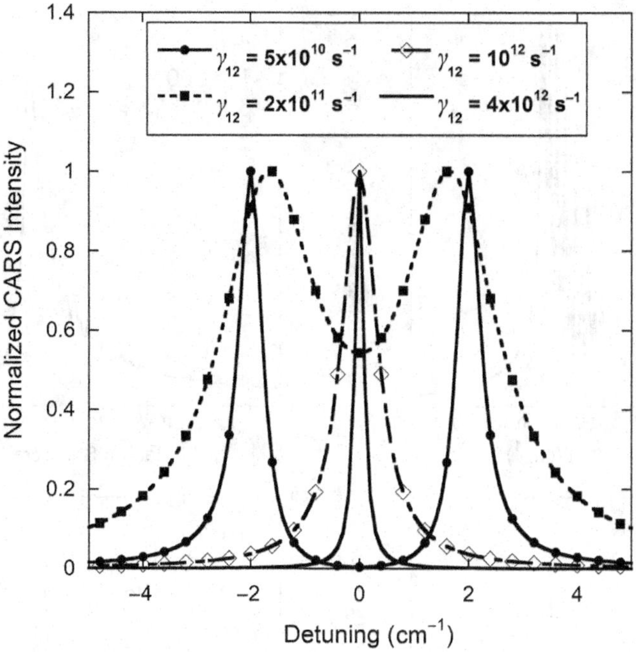

Figure 8.9 Calculated, normalized CARS spectra for a two-transition system with a transition frequency difference of 4 cm^{-1}. The spectra were calculated using Eq. (8.156) for γ_{12} values of 5×10^{10}, 2×10^{11}, 10^{12} and 4×10^{12} s^{-1}, respectively.

narrowing is clearly evident. Rahn et al. (1987) examined the use of two different rotational transfer models and concluded that the MEG model was superior to a polynomial energy-gap (PEG) model for calculations of high-pressure Raman spectra. Dreier et al. (1994) measured CARS lineshapes at temperatures from room temperature to 850 K and pressures up to 250 MPa. They compared their experimental results in detail with different collisional models.

8.9 Femtosecond and Picosecond CARS

During the last 20 years, the use of femtosecond and picosecond laser sources for performing CARS measurements in combustion has become much more frequent. The use of femtosecond pump and Stokes lasers for impulsive excitation of Raman coherences followed by generation of a CARS signal using a picosecond probe beam is the most common mode of implementing "femtosecond" CARS. Femtosecond CARS has a number of potentially significant advantages when compared to nanosecond CARS:

(1) The femtosecond pump and Stokes beams, despite their significant spectral widths, are very effective at exciting Raman coherences in gas-phase molecules,

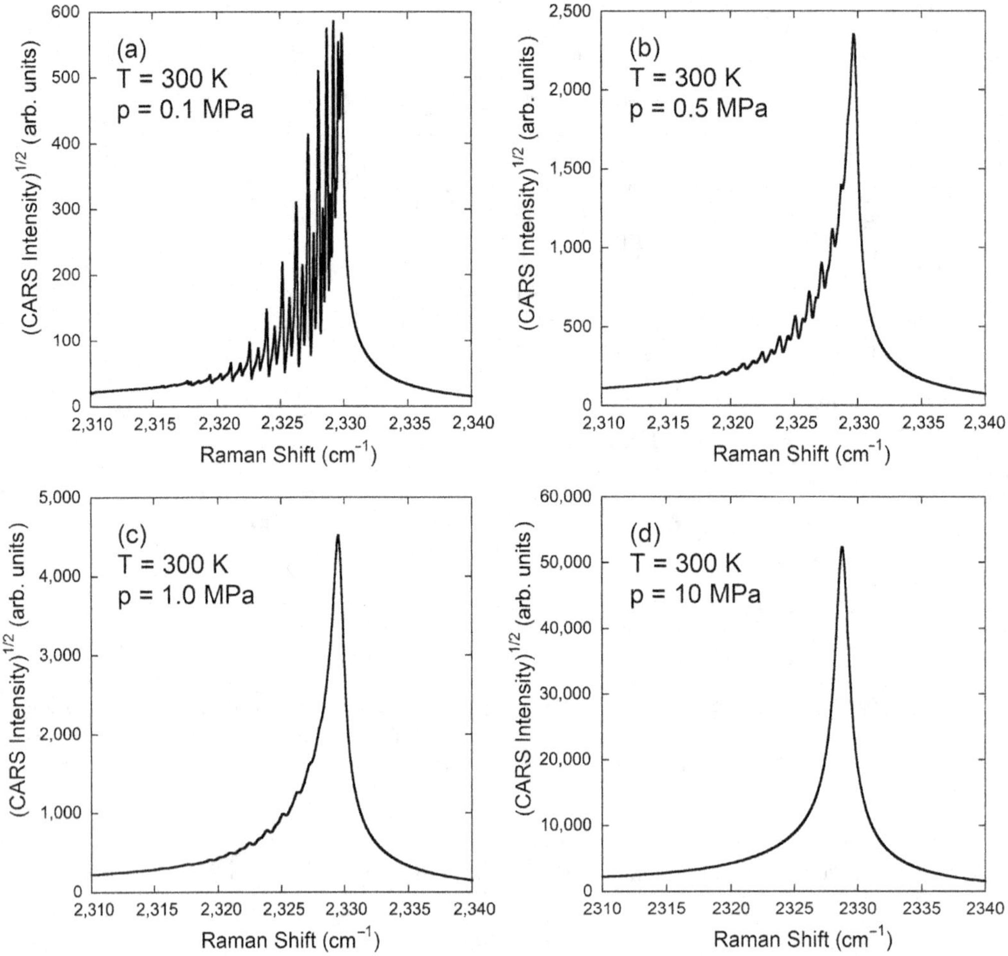

Figure 8.10 Calculated CARS spectra for nitrogen at $T = 300$ K and pressures of 0.1, 0.5, 1.0, and 10 MPa. The spectra were calculated using the Sandia CARFT code (Palmer, 1989).

even though the spectral width of the Raman resonances is on the order of three orders of magnitude less than the laser spectral widths. The pump and Stokes lasers, despite their large spectral widths, are single-mode lasers, and there are numerous spectral pairs under the pulse envelopes that contribute to excitation of the Raman resonances; this is depicted schematically in Figure 8.11. As long as the pump and Stokes pulses are Fourier-transform-limited, all of these frequency pairs contribute coherently to the excitation of the Raman coherence. The effect of chirp in the pump and Stokes beams is discussed in detail by Gu et al. (2020).

(2) If the spectral width of the pump and Stokes beams is large compared to the spectral extent of the Raman transitions, these different transitions will be excited with the same phase, giving rise to a giant Raman coherence at the conclusion of the impulsive pump–Stokes excitation. Once they are impulsively excited, the

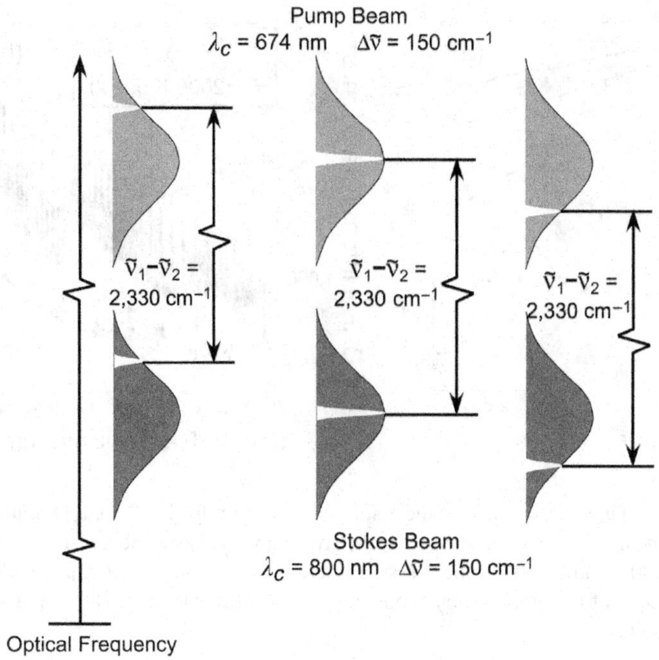

Figure 8.11 Energy-level diagram for the Raman excitation of N_2 Raman Q-branch resonances near $2,330\ cm^{-1}$. The pump and Stokes beams that are depicted have temporal widths of 100 fs and corresponding frequency bandwidths of $150\ cm^{-1}$.

different Raman transitions contributing to the initial coherence will oscillate with different natural frequencies. The time decay of this initial coherence is dependent on the spectral spread of Raman transitions with significant intensity, an effect referred to as "frequency-spread dephasing" (Lucht et al., 2006). The effect is illustrated in Figure 8.12a. As temperature increases, the spectral range over which Raman transitions have significant intensity increases, and the frequency-spread decay rate is thus dependent on temperature.

(3) If the Raman coherence can be probed within the first few picoseconds before collisions occur, the CARS signal will be completely independent of collisions, and no knowledge of Raman line widths is needed to analyze the spectra. Note that frequency-spread dephasing does not depend on the collision rate and is independent of pressure.

(4) The nonresonant background four-wave mixing signal that interferes with the CARS is generated only when the pump, Stokes, and probe beams overlap in time. If the probe beam is delayed by a few picoseconds with respect to the pump and Stokes beams, no nonresonant background will be generated. Even when it is generated, it is much less noisy than for nanosecond broadband CARS, and the resonant CARS signal can be heterodyned against the nonresonant background signal, increasing the signal-to-noise ratio of the measurements.

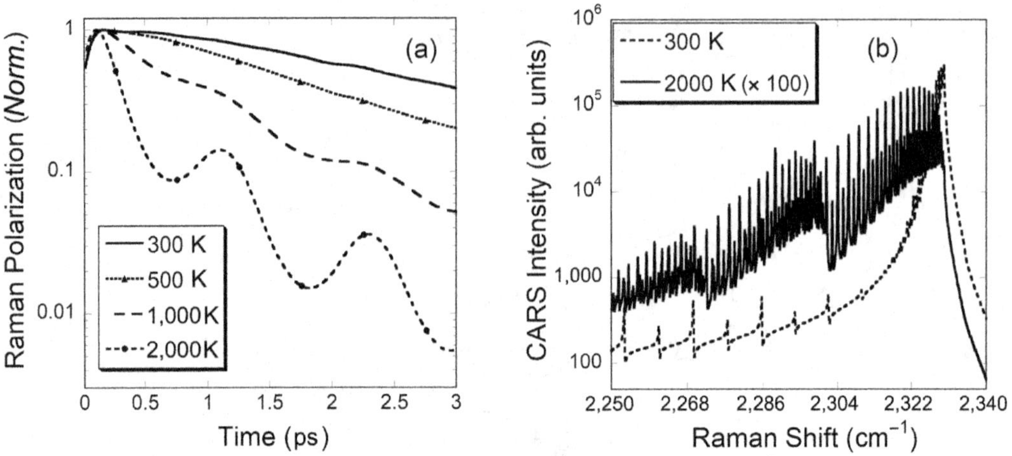

Figure 8.12 (a) Time dependence of the Raman polarization for N_2 Raman Q-branch resonances near $2,330\ cm^{-1}$ following impulsive pump-Stokes excitation for temperatures of 300, 500, 1,000, and 2,000 K. The pressure is assumed to be low enough that collisions do not affect the decay of the initial Raman coherence. (b) CARS spectra of N_2 at temperatures of 300 and 2,000 K.

Assuming that the probe intensity is very is weak compared to the pump and Stokes beam intensities, the CARS signal intensity is given by the product of the probe beam intensity and the resonant (Raman) and nonresonant polarizations in the medium. Assuming that the pump and Stokes beams have parallel polarizations,

$$E_{sig}(t) = E_{pr}(t)P_{res}(t) + E_{pr}(t)P_{nres}(t). \tag{8.161}$$

Assuming that the pump and Stokes beam bandwidths are very broad compared to the extent of the Raman shifts for the different Raman transitions, and that the temporal widths of the pump and Stokes beams are very small compared to $1/\Gamma_i$ for each transition, the Raman polarization in the medium is given by

$$P_{res}(t) = \beta\left[\int_{-\infty}^{t} E_{pump}(t')E_S(t')dt'\right] \times \left\{\sum_i \Delta n_i\left[\langle\alpha_{XX}\alpha_{YY}\rangle_i + \langle\alpha_{XY}\alpha_{XY}\rangle_i + \langle\alpha_{XY}\alpha_{YX}\rangle_i\right]\cos(\omega_i t + \phi)\exp(-\Gamma_i t)\right\}, \tag{8.162}$$

where β is a parameter used to scale the strengths of the resonant polarization term, ω_i is the Raman frequency of the transition, Γ_i (s^{-1}) is the Raman linewidth,

$$\Delta n_i = \frac{n_{Ai}^{(0)}}{(2J_i + 1)} - \frac{n_{Bi}^{(0)}}{(2J_i' + 1)}, \tag{8.163}$$

and

$$\langle \alpha_{XX}\alpha_{YY}\rangle_i + \langle \alpha_{XY}\alpha_{XY}\rangle_i + \langle \alpha_{XY}\alpha_{YX}\rangle_i = \frac{\hbar(v+1)}{2\mu_R\omega_{Ri}}\left[(a')^2(2J_i+1)\delta_{J'J} + \frac{4}{45}b^{(2)}_{J'J}(\gamma')^2(2J_i+1)\right]. \tag{8.164}$$

The nonresonant response of the medium is given by

$$P_{nres}(t) = \alpha\sigma E_{pump}(t)E_S(t)E_{pr}(t), \tag{8.165}$$

where $\sigma = 8\chi_{nr}/3$ and α is a parameter used to scale the strengths of the nonresonant polarization term. The α and β terms are needed due to the difficulty in quantitatively calculating the resonant response given the broad frequency bandwidth and short temporal duration of the pump and Stokes pulses (Lucht et al., 2007).

While Eqs. (8.162) and (8.165) are very useful for analysis of experimental fs CARS spectra, quantitative calculation of the interaction of the femtosecond pump and Stokes lasers requires solution of the time-dependent density matrix equations. The energy level system for the calculations is shown in Figure 8.13. Levels A and B are the lower and upper levels, respectively, for the Raman transition AB in the nitrogen molecule. Levels A and B are coupled by single-photon-allowed transitions in the fictitious $\kappa^1\Sigma_u^+$ and $\mu^1\Pi_u$ levels. The single-photon line strengths between the ground $X^1\Sigma_u^+$ level and these fictitious upper levels are adjusted to give the correct

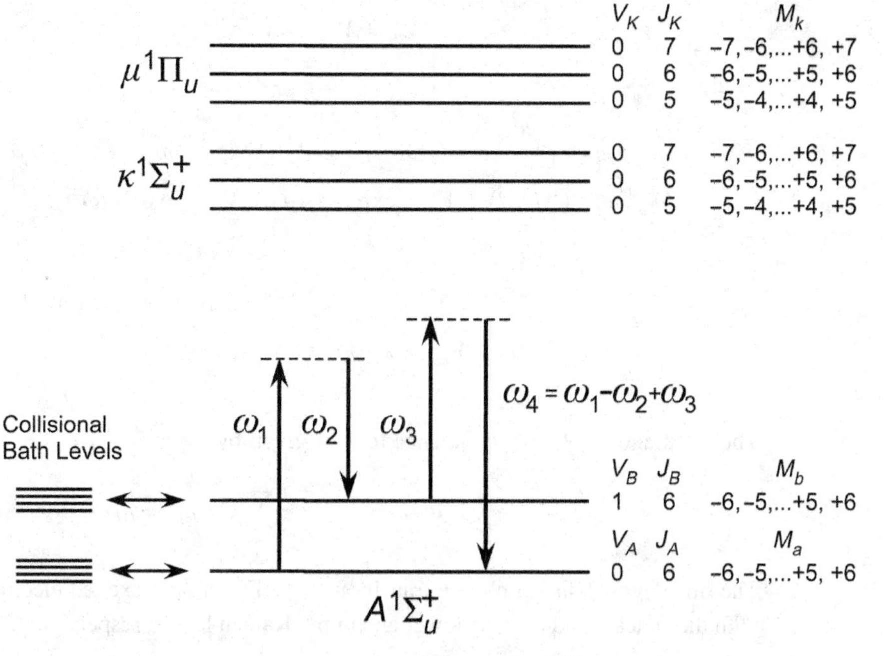

Figure 8.13 Energy level diagram for time-dependent density matrix analysis of the interaction of femtosecond pump and Stokes beams with the Raman resonances of N_2 near 2330 cm^{-1}. Reproduced from Lucht et al. (2007) with permission from AIP Publishing.

Raman isotropic and anisotropic tensor invariants. The density matrix equations for the lower and upper states of the Raman transition are given by

$$\frac{\partial \rho_{aa}}{\partial t} = -\Gamma_a \rho_{aa} + \sum_j \Gamma_{ja}\rho_{jj} - \frac{i}{\hbar}\sum_k (V_{ak}\rho_{ka} - \rho_{ak}V_{ka}), \tag{8.166}$$

$$\frac{\partial \rho_{bb}}{\partial t} = -\Gamma_b \rho_{bb} + \sum_j \Gamma_{jb}\rho_{jj} - \frac{i}{\hbar}\sum_k (V_{bk}\rho_{kb} - \rho_{bk}V_{kb}), \tag{8.167}$$

where the population transfer term $\Gamma_a = \sum_j \Gamma_{aj}$, where the symbol j in this case refers to all other levels in the nitrogen molecule. The electric dipole moment interaction terms such as V_{ak} are given by

$$V_{ak} = -\boldsymbol{\mu}_{ak} \bullet \boldsymbol{E}(t) = V_{ka}^*. \tag{8.168}$$

In the case of CARS, the input electric field has three frequency components:

$$\begin{aligned} \boldsymbol{E}(t) &= \frac{1}{2}\hat{\boldsymbol{e}}_1 A_1(t)\exp(-i\omega_1 t) + \frac{1}{2}\hat{\boldsymbol{e}}_2 A_2(t)\exp(-i\omega_2 t) \\ &\quad + \frac{1}{2}\hat{\boldsymbol{e}}_3 A_3(t)\exp(-i\omega_3 t) + \text{c.c.}, \end{aligned} \tag{8.169}$$

where it is assumed that the interaction occurs at $\boldsymbol{r} = 0$ and that the CARS phase-matching conditions are satisfied. The interaction term is written as the sum of terms associated with the plus and minus signs on the frequencies,

$$V_{ak} = V_{ak}^{(-)} + V_{ak}^{(+)}, \tag{8.170}$$

where

$$\begin{aligned} V_{ka}^{(-)} &= -\frac{1}{2}\vec{\mu}_{ka} \bullet [\hat{e}_1 A_1(t)\exp(-i\omega_1 t) + \hat{e}_2 A_2(t)\exp(-i\omega_2 t) + \hat{e}_3 A_3(t)\exp(-i\omega_3 t)] \\ &= V_{ka}^{(-1)}\exp(-i\omega_1 t) + V_{ka}^{(-2)}\exp(-i\omega_2 t) + V_{ka}^{(-3)}\exp(-i\omega_3 t). \end{aligned} \tag{8.171}$$

$$\begin{aligned} V_{ka}^{(+)} &= -\frac{1}{2}\vec{\mu}_{ka} \bullet \left[\hat{e}_1^* A_1^*(t)\exp(i\omega_1 t) + \hat{e}_2^* A_2^*(t)\exp(i\omega_2 t) + \hat{e}_3^* A_1^*(t)\exp(i\omega_3 t)\right] \\ &= V_{ka}^{(+1)}\exp(i\omega_1 t) + V_{ka}^{(+2)}\exp(i\omega_2 t) + V_{ka}^{(+3)}\exp(i\omega_3 t). \end{aligned} \tag{8.172}$$

The off-diagonal Raman coherence term is given by

$$\frac{\partial \rho_{ba}}{\partial t} = -\rho_{ba}(i\omega_{ba} + \Gamma_{ba}) - \frac{i}{\hbar}\sum_k (V_{bk}\rho_{ka} - \rho_{bk}V_{ka}). \tag{8.173}$$

The off-diagonal single-photon transition terms linking the excited electronic states k with the states a and b in the lower and upper Raman levels, respectively, are given by

$$\frac{\partial \rho_{ka}}{\partial t} = 0 = -\rho_{ka}(i\omega_{ka} + \gamma_{ka}) - \frac{i}{\hbar}\sum_{a'} V_{ka'}\rho_{a'a} - \frac{i}{\hbar}\sum_b V_{kb}\rho_{ba}. \tag{8.174}$$

$$\frac{\partial \rho_{kb}}{\partial t} = 0 = -\rho_{kb}(i\omega_{kb} + \gamma_{kb}) - \frac{i}{\hbar}\sum_a V_{ka}\rho_{ab} - \frac{i}{\hbar}\sum_{b'} V_{kb'}\rho_{b'b}. \tag{8.175}$$

The time derivative is set to zero in Eqs. (8.174) and (8.175) because it is assumed that the transitions *ka* and *kb* are far from resonance with any of the input laser beams. The next step in the analysis is to define slowly varying amplitudes for the frequency components in the density matrix elements that are important in the CARS process. This is a critical difference compared to the perturbative analysis in that these frequencies must be selected during the analysis rather than arising naturally as a result of the analysis. Referring to Figure 8.13, the important slowly varying amplitudes are given by

$$\rho_{ka} = \sigma_{ka}\exp(-i\omega_1 t) + \eta_{ka}\exp(-i\omega_4 t). \tag{8.176}$$

$$\rho_{kb} = \sigma_{kb}\exp(-i\omega_2 t) + \eta_{kb}\exp(-i\omega_3 t). \tag{8.177}$$

$$\rho_{ba} = \sigma_{ba}\exp[-i(\omega_1 - \omega_2)t]. \tag{8.178}$$

The next step in the analysis is to substitute Eqs. (8.176)–(8.178) and the expressions for the interaction terms V_{ij} into Eqs. (8.173)–(8.175) and to solve for the slowly varying amplitude terms σ_{ij} and n_{ij}. The amplitude and interactions terms are complex and are written as the sum of real and imaginary parts; for example, $V_{ka}^{(+1)} = V_{ka}^{(+1r)} + iV_{ka}^{(+1i)}$ and $\eta_{kb} = \eta_{kb}^r + i\eta_{kb}^i$. The equations are modified for solution of both the real and imaginary amplitude components. Terms in the equations that oscillate at optical frequencies are then neglected in the solution, the so-called rotating-wave approximation (although if the pulses become much shorter than 50 fs for visible wavelengths, the rotating-wave approximation will have to be re-examined). The result of this analysis for the Raman coherence term $\sigma_{ba} = \sigma_{ba}^r + i\sigma_{ba}^i$ is given by

$$\begin{aligned}\dot{\sigma}_{ba}^r &= \sigma_{ba}^i[\omega_{ba} - (\omega_1 - \omega_2)] - \sigma_{ba}^r\gamma_{ba} \\ &\quad + \frac{1}{\hbar}\sum_k \left[V_{bk}^{(+2r)}\sigma_{ka}^i + V_{bk}^{(+2i)}\sigma_{ka}^r + V_{bk}^{(+3r)}\eta_{ka}^i + V_{bk}^{(+3i)}\eta_{ka}^r - \sigma_{bk}^r V_{ka}^{(-1i)} - \sigma_{bk}^i V_{ka}^{(-1r)}\right].\end{aligned} \tag{8.179}$$

$$\begin{aligned}\dot{\sigma}_{ba}^i &= -\sigma_{ba}^r[\omega_{ba} - (\omega_1 - \omega_2)] - \sigma_{ba}^i\gamma_{ba} \\ &\quad + \frac{1}{\hbar}\sum_k \left[-V_{bk}^{(+2r)}\sigma_{ka}^r + V_{bk}^{(+2i)}\sigma_{ka}^i - V_{bk}^{(+3r)}\eta_{ka}^r + V_{bk}^{(+3i)}\eta_{ka}^i + \sigma_{bk}^r V_{ka}^{(-1r)} - \sigma_{bk}^i V_{ka}^{(-1i)}\right].\end{aligned} \tag{8.180}$$

The solution of the full set of equations is discussed in detail in Lucht et al. (2007). The results of some of the calculations for the nitrogen molecule at room temperature and pressure will now be discussed. Figure 8.14 shows the time dependence of the Raman coherence terms for the Q(2) and Q(20) Raman transitions in the fundamental band $v = 0 \rightarrow v' = 1$, as depicted in Figure 8.13. Immediately after the impulsive pump–Stokes Raman excitation, the Raman coherence amplitude term $\sigma_{BA}^i = \sum_b \sum_a \sigma_{ba}^i$ reaches a value of about $0.25\rho_{AA}^0$, where ρ_{AA}^0 is the population of level *A* before laser excitation, for both transitions. The peak intensity for the

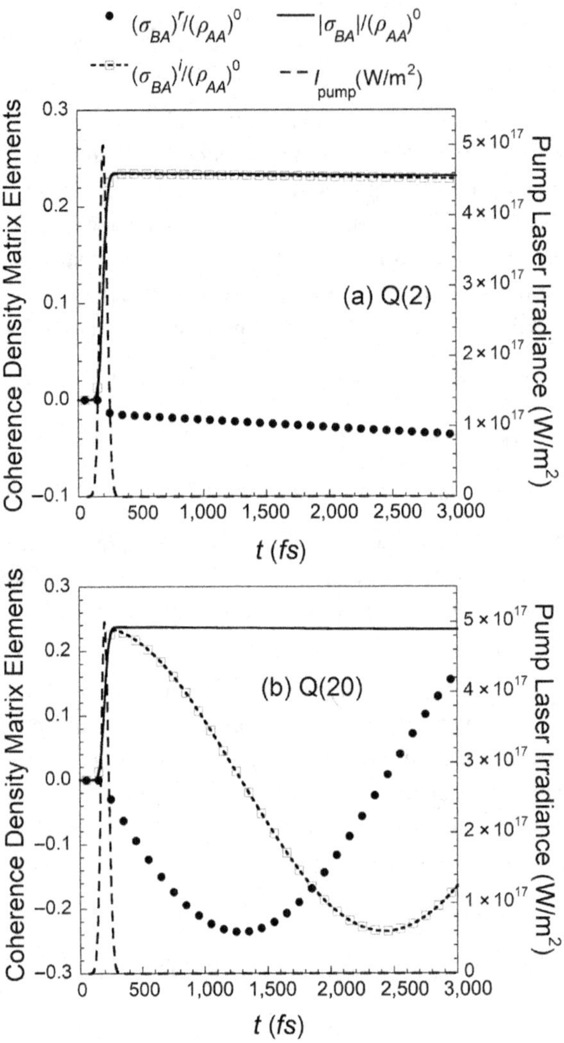

Figure 8.14 Temporal dependence of the real and imaginary components and the magnitude of the induced Raman coherence for the (a) Q(2) and (b) Q(20) transitions in the fundamental $v = 0 \rightarrow v' = 1$ band of N_2. The coherence matrix elements are normalized by dividing by the population of the ground level A prior to laser excitation. The difference $(\tilde{\nu}_1 - \tilde{\nu}_2)$ between the central frequencies of the pump and Stokes laser is 2330 cm^{-1}. The Raman frequencies for the Q(2) and Q(20) transitions are 2329.8 and 2322.7 cm^{-1}, respectively. The collisional dephasing 9 rate for each Raman transition is 5×10^9 s^{-1}, corresponding to a Raman line width of 0.053 cm^{-1}. The pump and Stokes laser pulses are both Gaussian with temporal widths of 70 fs (FWHM). The pump and Stokes pulses are overlapped exactly in time. The peak irradiance for both the pump and Stokes pulses is 5×10^{17} W/m^2. Reproduced from Lucht et al. (2007) with permission from AIP Publishing.

calculation of the pump and Stokes beams is 5×10^{17} W/m², which corresponds roughly to the peak intensity for 100 μJ, 70 fs beams focused to a diameter of 50 μm. The maximum possible value of $\sigma^i_{BA}/\rho^0_{AA}$ is 0.5, so the fs pump and Stokes beams are very effective at inducing the Raman coherence.

The frequency $(\tilde{\nu}_1 - \tilde{\nu}_2)$ between the central frequencies of the pump and Stokes laser beams in the calculation is 2,330 cm^{-1}, very close to the natural frequency of 2,329.8 cm^{-1} for the Q(2) transition. The term σ_{BA} is the slowly varying amplitude term at a frequency of 2,330 cm^{-1}, so for the Q(2) transition the phase of the oscillation is essentially unchanged for the 3 ps time period depicted in Figure 8.14a. This is not the case for the Q(20) transition with a Raman frequency of 2,322.7 cm^{-1}. The difference frequency of 7.3 cm^{-1} corresponds to a period of oscillation of 4.57 ps; in the figure the imaginary part of the Raman coherence reaches a minimum at approximately 2.3 ps after the impulsive pump–Stokes excitation. This is an illustration of frequency-spread dephasing. The Q(2) and Q(20) transitions initially contribute in-phase to the initial giant Raman coherence, but after approximately 2.5 ps they destructively interfere, and their contributions are canceled.

The degrees of excitation of several different Q-branch transitions for the nitrogen molecule are depicted in Figure 8.15. The frequency spread of the transitions from Q(2) to Q(20) shown in Figure 8.15 is only 7 cm^{-1}, which is much less than the 100 cm^{-1} frequency bandwidths of the pump and Stokes beams. In Eq. (8.162), it is assumed that the degree of excitation of each transition is proportional to the Raman polarizability for that transition, and this will be the case provided that the spread of the Raman transition frequencies is much less than the frequency bandwidths of the pump and Stokes beams. The case where the spread of the Raman frequencies is comparable to or greater than the

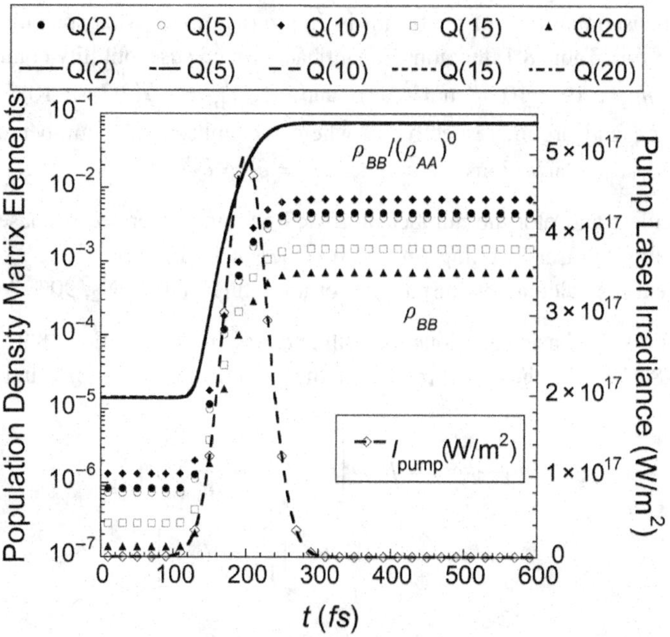

Figure 8.15 Temporal dependence of the excited state populations and the normalized excited populations for the Q(2), Q(5), Q(10), Q(15), and Q(20) transitions in the fundamental $v = 0 \rightarrow v' = 1$ Raman band of N_2. The collisional dephasing rate and the pump and Stokes pulse parameters are the same as given in the caption for **Fig. 8.14**.
Reproduced from Lucht et al. (2007) with permission from AIP Publishing.

frequency bandwidths of the pump and Stokes beams is discussed by Oron et al. (2003) and Gu et al. (2019). In this case the calculation of the resonant polarization would require a convolution over the pump and Stokes beam frequency bandwidths,

$$P_{res}(t) = \beta\left[\int_{-\infty}^{t} E_{pump}(t')E_S(t')dt'\right] \times \left\{\sum_i \Delta n_i \langle \alpha_{XX}\alpha_{YY}\rangle_i \cos(\omega_i t + \phi)\exp(-\Gamma_i t)\int_{-\infty}^{+\infty} E_{pump}(\omega)E_S^*(\omega - \omega_i)d\omega\right\}. \tag{8.181}$$

The effect of significant chirp in the pump and Stokes beams on the Raman excitation process is also discussed in detail by Gu et al. (2019).

8.10 CARS Spectroscopy Example Problems

Example Problem #1. A gas cell contains 80% nitrogen and 20% argon (by volume) at a total pressure of 150 kPa and a temperature of 600 K. Monochromatic pump and Stokes lasers are used to generate a CARS signal from the nitrogen O(12) line in the $0 \rightarrow 1$ vibrational band. The line width of the O(12) transition is $\Gamma_j/2\pi c = 0.15\ \text{cm}^{-1}$. The pulse energy of each of the two 532 nm pump beams and the Stokes beam is 10 mJ, the pulse lengths are 10 ns, the interaction length of the probe volume is 4 mm, and the focused diameter of the beams is 75 μm (assume top-hat spatial and temporal profiles). From Table 8.1, the nonresonant background susceptibility contributions are $(\chi_{nr})_{N_2} = n_{N_2}(2.19 \times 10^{-51}\ \text{m}^5\text{C}^2/\text{J}^2)$ and $(\chi_{nr})_{Ar} = n_{Ar}(2.74 \times 10^{-51}\ \text{m}^5\text{C}^2/\text{J}^2)$, for nitrogen and argon, respectively, where the units of the number densities are m^{-3}. The laser polarizations are parallel $(\hat{e}_1 = \hat{e}_2 = \hat{e}_3)$.

(a) Calculate and plot the number of CARS photons generated per laser pulse as a function of laser detuning $\Delta\omega_j$ from values of $-5\Gamma_j$ to $+5\Gamma_j$.
(b) Repeat the calculations of part (a) for a mixture of 10% N_2, 90% Ar.

Solution: (a) The basic equations for both parts (a) and (b) are Eqs. (8.92)–(8.96) and (8.116)–(8.119). For the O(12) transition, the parameters $a(\omega_1 - \omega_2)$ and $b(\omega_1 - \omega_2)$ are given by

$$a(\omega_1 - \omega_2) = \beta_{vJJ'}\left[(a')^2\delta_{JJ'} - \frac{2}{45}b_{J'J}(\gamma')^2\right], \tag{8.182}$$

$$b(\omega_1 - \omega_2) = \beta_{vJJ'}\left[\frac{2}{15}b_{J'J}(\gamma')^2\right], \tag{8.183}$$

where

$$\beta_{vJJ'} = \frac{\varepsilon_0(v+1)\left(n_{vJ} - \frac{2J+1}{2J'+1}n_{(v+1)J'}\right)}{\mu_R\,\omega_j}\frac{1}{2\Delta\omega_{j1} - i\Gamma_j}, \tag{8.184}$$

where

$$\Delta\omega_{j1} = \omega_j - (\omega_1 - \omega_2). \tag{8.185}$$

As a first step we will calculate the frequency of the Raman transition. For the O(12) line, $J = 12, J' = 10, \mathrm{v} = 0$, and $\mathrm{v}' = 1$. For the $^1\Sigma^+$ electronic level of N_2, we have the following spectroscopic parameters (Laurendeau, 2005):

$$\omega_e = 2{,}357.6\,\mathrm{cm}^{-1} \quad \omega_e x_e = 14.06\,\mathrm{cm}^{-1} \quad B_e = 1.998\,\mathrm{cm}^{-1} \quad \alpha_e = 0.0179\,\mathrm{cm}^{-1}$$
$$D_e = 5.76 \times 10^{-6}\,\mathrm{cm}^{-1}$$

$$G(\mathrm{v}) = \omega_e\left(\mathrm{v}+\frac{1}{2}\right) - \omega_e x_e\left(\mathrm{v}+\frac{1}{2}\right)^2 \Rightarrow G(0) = 1{,}175.3\,\mathrm{cm}^{-1},\ G(1) = 3{,}504.8\,\mathrm{cm}^{-1}$$

$$B_\mathrm{v} = B_e - \alpha_e\left(\mathrm{v}+\frac{1}{2}\right) \Rightarrow B_0 = 1.989\,\mathrm{cm}^{-1},\ B_1 = 1.971\,\mathrm{cm}^{-1}$$

$$F_\mathrm{v}(J) = B_\mathrm{v}J(J+1) - D_e J^2(J+1)^2 \Rightarrow F_0(12) = 310.1\,\mathrm{cm}^{-1},\ F_1(10) = 216.7\,\mathrm{cm}^{-1}.$$

For the O(12) line, the Raman frequency is given by

$$\tilde{\nu}_R = \frac{\varepsilon_{\mathrm{v}'J'}}{hc} - \frac{\varepsilon_{\mathrm{v}J}}{hc} = G(1) + F_1(10) - G(0) - F_0(12) = 2236.0\,\mathrm{cm}^{-1} \tag{8.186}$$

and

$$\omega_j = 2\pi c\tilde{\nu}_R = 2\pi\left(2.998 \times 10^8\ \mathrm{m/s}\right)\left(2.236 \times 10^5\ \mathrm{m}^{-1}\right) = 4.212 \times 10^{14}\ \mathrm{s}^{-1}. \tag{8.187}$$

The angular frequencies of the pump beam and CARS signals are given by

$$\omega_1 = \frac{2\pi c}{\lambda_1} = \frac{2\pi\left(2.998 \times 10^8\ \mathrm{m/s}\right)}{532 \times 10^{-9}\ \mathrm{m}} = 3.54 \times 10^{15}\ \mathrm{s}^{-1}. \tag{8.188}$$

$$\omega_4 = \omega_1 + \omega_R = 4.212 \times 10^{14} + 3.54 \times 10^{15}\ \mathrm{s}^{-1} = 3.96 \times 10^{15}\ \mathrm{s}^{-1}. \tag{8.189}$$

The number densities of N_2 molecules and argon atoms in the cell are given by

$$n_{N_2} = \frac{x_{N_2}p}{k_BT} = \frac{(0.8)\left(150 \times 10^3\ \mathrm{J/m^3}\right)}{\left(1.3806 \times 10^{-23}\ \mathrm{J/K}\right)(600\ \mathrm{K})} = 1.53 \times 10^{25}\ \mathrm{m}^{-3}. \tag{8.190}$$

$$n_{Ar} = \frac{x_{Ar}p}{k_BT} = \frac{(0.2)\left(150 \times 10^3\ \mathrm{J/m^3}\right)}{\left(1.3806 \times 10^{-23}\ \mathrm{J/K}\right)(600\ \mathrm{K})} = 3.62 \times 10^{24}\ \mathrm{m}^{-3}. \tag{8.191}$$

At 600 K, we can assume that all the N_2 molecules are in the $\mathrm{v} = 0$ level. The number of molecules in the $J = 12$ level is given by

$$\begin{aligned}
\frac{n_{J=12,\mathrm{v}=0}}{n_{\mathrm{v}=0}} = \frac{n_{J=12,\mathrm{v}=0}}{n_{N_2}} &= \frac{(NSSW)_{J=12}(2J+1)\exp\left[-hcB_eJ(J+1)/k_BT\right]}{Z_{nuc}Z_{rot}} \\
&= \frac{(NSSW)_{J=12}(2J+1)\exp\left[-(hc/k_B)B_eJ(J+1)/T\right]}{(2I_N+1)^2\{T/[2(hc/k_B)B_e]\}} \\
&= \frac{(6)(25)\exp\left[-(1.439\ \mathrm{K/cm^{-1}})(1.998\ \mathrm{cm^{-1}})(12)(13)/(600\ \mathrm{K})\right]}{(9)\{(600\ \mathrm{K})/[2(1.439\ \mathrm{K/cm^{-1}})(1.998\ \mathrm{cm^{-1}})]\}} \\
&= 0.0756 \\
n_{J=12,\mathrm{v}=0} = (0.0756)\left(1.53 \times 10^{25}\ \mathrm{m}^{-3}\right) &= 1.16 \times 10^{24}\ \mathrm{m}^{-3}.
\end{aligned} \tag{8.192}$$

For nitrogen, from Table 7.1 we find

$$(a')^2 = 3.79 \times 10^{-60}\ \frac{\mathrm{C^4\,m^2}}{\mathrm{J^2}} \qquad (\gamma')^2 = 5.15 \times 10^{-60}\ \frac{\mathrm{C^4\,m^2}}{\mathrm{J^2}} \qquad \rho = 0.022.$$

For the O (12) line, $\Gamma_j = 2\pi c(0.15\ \mathrm{cm^{-1}}) = 2.83 \times 10^{10}\ \mathrm{s^{-1}}$, and the Placzek coefficient is given by

$$b_{J'J} = b_{J-2,J} = \frac{3J(J-1)}{2(2J+1)(2J-1)} = \frac{3(12)(11)}{2(25)(23)} = 0.344 \qquad J \to J-2. \tag{8.193}$$

The parameter $\beta_{vJJ'} = \beta_r + i\beta_i$ can now be calculated for the O(12) Raman transitions,

$$\beta_{vJJ'} = \frac{(0+1)\left(n_{vJ} - \cancel{\frac{2J+1}{2J'+1} n_{(v+1)J'}}\right)}{\varepsilon_0 \mu_R \omega_j} \frac{1}{2\Delta\omega_{j1} - i\Gamma_j} = \frac{n_{vJ}}{\varepsilon_0 \mu_R \omega_j} \frac{2\Delta\omega_j - i\Gamma_j}{4(\Delta\omega_j)^2 + \Gamma_j^2}. \tag{8.194}$$

$$\beta_r = \frac{n_{vJ}}{\varepsilon_0 \mu_R \omega_j \Gamma_j} \frac{2(\Delta\omega_j/\Gamma_j)}{4(\Delta\omega_j/\Gamma_j)^2 + 1} = C_\beta \frac{2(\Delta\omega_j/\Gamma_j)}{4(\Delta\omega_j/\Gamma_j)^2 + 1}. \tag{8.195}$$

$$\beta_i = -\frac{n_{vJ}}{\varepsilon_0 \mu_R \omega_j \Gamma_j} \frac{1}{4(\Delta\omega_j/\Gamma_j)^2 + 1} = -C_\beta \frac{1}{4(\Delta\omega_j/\Gamma_j)^2 + 1}. \tag{8.196}$$

$$C_\beta = \frac{(1.16 \times 10^{24}\,\mathrm{m^{-3}})}{(8.854 \times 10^{-12}\,\mathrm{C^2/J{-}m})(1.169 \times 10^{-26}\,\mathrm{kg})(4.212 \times 10^{14}\,\mathrm{s^{-1}})(2.83 \times 10^{10}\,\mathrm{s^{-1}})}$$
$$= 9.40 \times 10^{35}\,\frac{\mathrm{s^2\,J}}{\mathrm{C^2\,m^2}}. \tag{8.197}$$

The parameters $a(\omega_1 - \omega_2), b(\omega_1 - \omega_2)$, and σ are given by

$$a(\omega_1 - \omega_2) = \beta_{vJJ'}\left[(a')^2 \cancel{\delta_{JJ'}} - \frac{2}{45} b_{J'J}(\gamma')^2\right] = -\beta_{vJJ'} \frac{2}{45} b_{J'J}(\gamma')^2. \tag{8.198}$$

$$b(\omega_1 - \omega_2) = \beta_{vJJ'} \frac{2}{15} b_{J'J}(\gamma')^2. \tag{8.199}$$

$$\sigma = \frac{8}{3}\chi_{nr} = \frac{8}{3}\left[(1.53 \times 10^{25}\ \mathrm{m^{-3}})(2.19 \times 10^{-51}\ \mathrm{m^5\,C^2/J^2})\right.$$
$$\left. + (3.62 \times 10^{24}\ \mathrm{m^{-3}})(2.74 \times 10^{-51}\ \mathrm{m^5\,C^2/J^2})\right] = 1.17 \times 10^{-25}\ \mathrm{m^2\,C^2/J^2}. \tag{8.200}$$

For parallel polarizations and for $\omega_3 = \omega_1$, Eq. (8.116) becomes

$$\begin{aligned}
\boldsymbol{E}(\omega_4) &= 3i\frac{\omega_4}{2c}\left[\chi^{(3)}_{1122}+\chi^{(3)}_{1212}+\chi^{(3)}_{1221}\right]E(\omega_1)E(\omega_1)E^*(\omega_2)L_{CARS}\hat{\boldsymbol{e}}_1 \\
&= 3i\frac{\omega_4}{48c}[3\sigma+4a(\omega_1-\omega_2)+4b(\omega_1-\omega_2)]E(\omega_1)E(\omega_1)E^*(\omega_2)L_{CARS} \\
&= i\frac{\omega_4}{16c}\left[3\sigma+\beta_{vJJ'}\frac{16}{45}b_{J'J}(\gamma')^2\right]E(\omega_1)E(\omega_1)E^*(\omega_2)L_{CARS} \\
&= i\frac{\omega_4}{16c}\left\{\left[3\sigma+\beta_r\frac{16}{45}b_{J'J}(\gamma')^2\right]+i\left[\beta_i\frac{16}{45}b_{J'J}(\gamma')^2\right]\right\}E(\omega_1)E(\omega_1)E^*(\omega_2)L_{CARS}.
\end{aligned} \tag{8.201}$$

The CARS intensity is given by

$$\begin{aligned}
I_4(\omega_4) &= \frac{1}{2}c\varepsilon_0\boldsymbol{E}(\omega_4)\bullet\boldsymbol{E}^*(\omega_4) = \frac{1}{2}c\varepsilon_0\frac{\omega_4^2}{256c^2}\left\{\left[3\sigma+\beta_r\frac{16}{45}b_{J'J}(\gamma')^2\right]^2\right. \\
&\quad\left.+\left[\beta_i\frac{16}{45}b_{J'J}(\gamma')^2\right]^2\right\}\left[\frac{2I_1(\omega_1)}{c\varepsilon_0}\right]^2\left[\frac{2I_2(\omega_2)}{c\varepsilon_0}\right]L^2_{CARS}.
\end{aligned} \tag{8.202}$$

Simplifying and using the assumption of monochromatic lasers, we obtain

$$I_4 = \frac{I_1^2 I_2 L^2_{CARS}\omega_4^2}{64\varepsilon_0^2 c^4}\left\{\left[3\sigma+\beta_r\frac{16}{45}b_{J'J}(\gamma')^2\right]^2+\left[\beta_i\frac{16}{45}b_{J'J}(\gamma')^2\right]^2\right\}. \tag{8.203}$$

The laser irradiances I_1 and $I_2$are equal and given by

$$I_1 = I_2 = \frac{E_L}{A_L\Delta t_L} = \frac{(10\times10^{-3}\ \text{J})}{\left(\frac{\pi}{4}\right)(75\times10^{-6}\ \text{m})^2(10^{-8}\ \text{s})} = 2.26\times10^{14}\ \frac{\text{W}}{\text{m}^2}. \tag{8.204}$$

Substituting numerical values into Eq. (8.203), we obtain

$$\begin{aligned}
I_4 &= \frac{\left(2.26\times10^{14}\ \frac{\text{J}}{\text{m}^2\text{s}}\right)^3(4\times10^{-3}\ \text{m})^2(3.96\times10^{15}\ \text{s}^{-1})^2}{64\left(8.854\times10^{-12}\ \frac{\text{C}^2}{\text{Jm}}\right)^2(2.998\times10^8\ \frac{\text{m}}{\text{s}})^4}\left\{\left[3\left(1.17\times10^{-25}\ \frac{\text{m}^2\text{C}^2}{\text{J}^2}\right)\right.\right. \\
&\quad\left.+\left(9.40\times10^{35}\ \frac{\text{s}^2\text{J}}{\text{C}^2\text{m}^2}\right)\frac{16}{45}(0.344)\left(5.15\times10^{-60}\ \frac{\text{C}^4\text{m}^2}{\text{J}^2}\right)\frac{2(\Delta\omega_j/\Gamma_j)}{4(\Delta\omega_j/\Gamma_j)^2+1}\right]^2 \\
&\quad\left.+\left[-\left(9.40\times10^{35}\ \frac{\text{s}^2\text{J}}{\text{C}^2\text{m}^2}\right)\frac{16}{45}(0.344)\left(5.15\times10^{-60}\ \frac{\text{C}^4\text{m}^2}{\text{J}^2}\right)\frac{1}{4(\Delta\omega_j/\Gamma_j)^2+1}\right]^2\right\}.
\end{aligned} \tag{8.205}$$

$$I_4 = \left(7.15 \times 10^{55}\,\frac{\mathrm{J}^5}{\mathrm{C}^4\,\mathrm{m}^6\,\mathrm{s}}\right)\left\{\left[3.51 \times 10^{-25} + 5.93 \times 10^{-25}\frac{2(\Delta\omega_j/\Gamma_j)}{4(\Delta\omega_j/\Gamma_j)^2+1}\,\frac{\mathrm{m}^2\,\mathrm{C}^2}{\mathrm{J}^2}\right]^2 + \left[-5.93 \times 10^{-25}\left(\frac{1}{4(\Delta\omega_j/\Gamma_j)^2+1}\right)\frac{\mathrm{m}^2\,\mathrm{C}^2}{\mathrm{J}^2}\right]^2\right\}. \tag{8.206}$$

The number of CARS photons per laser pulse is given by

$$N_{p4} = \frac{I_4 A_4 \Delta t_L}{\hbar\omega_4} = \frac{I_4\pi(75 \times 10^{-6}\ \mathrm{m})^2(10^{-8}\ \mathrm{s})}{4(1.055 \times 10^{-34}\ \mathrm{J-s})(3.96 \times 10^{15}\ \mathrm{s}^{-1})} = \left(105.8\ \frac{\mathrm{m}^2\,\mathrm{s}}{\mathrm{J}}\right)I_4$$
$$= (7.57 \times 10^{57})\left\{\left[3.51 \times 10^{-25} + 5.93 \times 10^{-25}\frac{2(\Delta\omega_j/\Gamma_j)}{4(\Delta\omega_j/\Gamma_j)^2+1}\right]^2 + \left[-5.93 \times 10^{-25}\left(\frac{1}{4(\Delta\omega_j/\Gamma_j)^2+1}\right)\right]^2\right\}. \tag{8.207}$$

(b) The cell composition is then changed to 10% N_2, 90% Ar. The laser parameters and CARS line width are unchanged. The number densities of N_2 molecules and Ar atoms in the cell are now given by

$$n_{N_2} = \frac{x_{N_2}p}{k_B T} = \frac{(0.1)(150 \times 10^3\ \mathrm{J/m^3})}{(1.3806 \times 10^{-23}\ \mathrm{J/K})(600\ \mathrm{K})} = 1.81 \times 10^{24}\ \mathrm{m}^{-3}. \tag{8.208}$$

$$n_{Ar} = \frac{x_{Ar}p}{k_B T} = \frac{(0.9)(150 \times 10^3\ \mathrm{J/m^3})}{(1.3806 \times 10^{-23}\ \mathrm{J/K})(600\ \mathrm{K})} = 1.63 \times 10^{25}\ \mathrm{m}^{-3}. \tag{8.209}$$

The nonresonant susceptibility parameter is now given by

$$\sigma = \frac{8}{3}\chi_{nr} = \frac{8}{3}[(1.81 \times 10^{24}\ \mathrm{m}^{-3})(2.19 \times 10^{-51}\ \mathrm{m}^5\,\mathrm{C}^2/\mathrm{J}^2) + (1.63 \times 10^{25}\ \mathrm{m}^{-3})(2.74 \times 10^{-51}\ \mathrm{m}^5\,\mathrm{C}^2/\mathrm{J}^2)] = 1.30 \times 10^{-25}\ \mathrm{m}^2\,\mathrm{C}^2/\mathrm{J}^2. \tag{8.210}$$

The numerical value of $\beta_{vJJ'}$ and consequently the values of $a(\omega_1 - \omega_2)$ and $b(\omega_1 - \omega_2)$ will be a factor of 8 smaller than for part (a) because the number density of N_2 is a factor of 8 smaller. The expression for the number of photons per laser pulse thus becomes

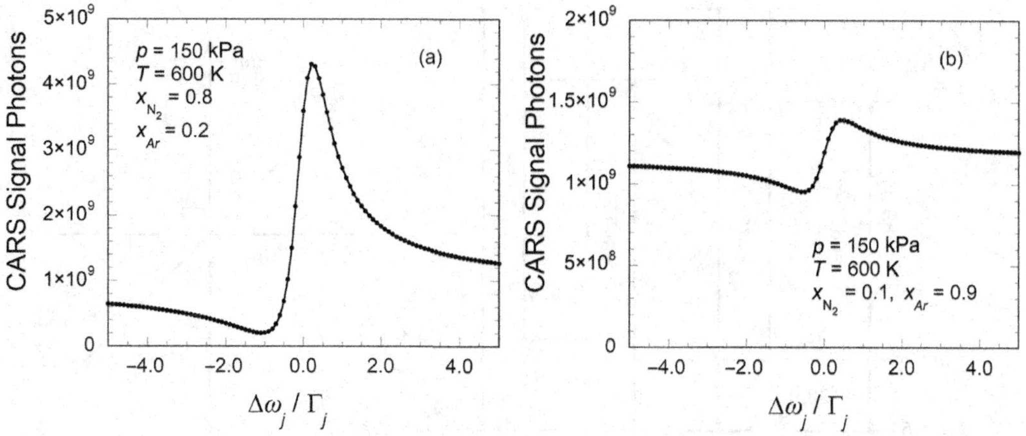

Figure 8.16 Calculated CARS lineshapes for the O(12) Raman transition in the $^1\Sigma^+$ v = 0 → v′ = 1 fundamental band of N_2. The pressure and temperature are 150 kPa and 600 K for both panels (a) and (b). For panel (a), the mole fractions of N_2 and Ar are 0.8 and 0.2, respectively. For panel (b), the mole fractions of N_2 and Ar are 0.1 and 0.9, respectively.

$$N_{p4} = \left(7.57 \times 10^{57}\right)\left\{\left[3.90 \times 10^{-25} + 7.41 \times 10^{-26}\frac{2(\Delta\omega_j/\Gamma_j)}{4(\Delta\omega_j/\Gamma_j)^2 + 1}\right]^2 + \left[-7.41 \times 10^{-26}\left(\frac{1}{4(\Delta\omega_j/\Gamma_j)^2 + 1}\right)\right]^2\right\}. \tag{8.211}$$

The lineshapes for the O (12) line for parts (a) and (b) are plotted in Figure 8.16.

Example Problem #2. A cell is filled with 50% nitrogen (N_2) and 50% hydrogen (H_2) by volume. The total cell pressure is 200 kPa and the temperature is 300 K. Three monochromatic lasers with *approximate* wavelengths of $\lambda_1 = 532$ nm, $\lambda_2 = 683$ nm, and $\lambda_3 = 589$ nm are focused to a common probe volume. The pulse energy of each of the two pump beams and the Stokes beam is 10 mJ, the pulse lengths are 10 ns, the interaction length of the probe volume is 2 mm, and the focused diameter of the beams is 100 μm. The linewidths of the two transitions are the same, $\Gamma_j/2\pi c = 0.05$ cm^{-1}. The frequency difference $\omega_1 - \omega_2$ is tuned to line center of the hydrogen Q(1) transition, and the frequency difference $\omega_3 - \omega_2$ is tuned to line center of the nitrogen Q(5) transition. All three laser beams are linearly polarized. The polarizations of the two pump lasers are perpendicular, and the polarization of the Stokes beam is at 45° with respect to the pump beams, as shown in Figure 8.17.

(a) Assume that the nitrogen Q(5) and hydrogen Q(1) transitions are isolated from surrounding transitions and that the nonresonant background susceptibility is negligible. Calculate the polarization angles of the CARS signal beams $\boldsymbol{E}_{H_2}(\omega_4)$ and $\boldsymbol{E}_{N_2}(\omega_4)$ with respect to the polarization direction of $\boldsymbol{E}(\omega_1)$.

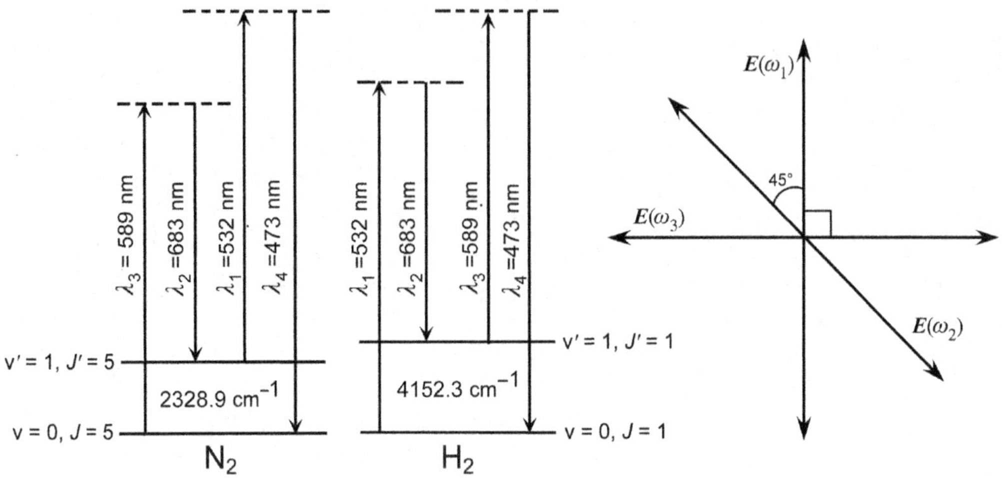

Figure 8.17 Energy-level diagram and polarization arrangement for the input CARS pump/probe and Stokes beams.

(b) A polarizer is placed in the CARS signal channel so that light polarized parallel to the polarization direction of $\boldsymbol{E}(\omega_1)$ is transmitted. Calculate the number of CARS signal photons due to the nitrogen Q(5) and hydrogen Q(1) transitions.

Solution: The basic equations for both parts (a) and (b) are again Eqs. (8.92)–(8.96) and (8.116)–(8.119). Referring to Figure 8.17, Eq. (8.116) reduces to

$$\boldsymbol{E}(\omega_4) = i\frac{\omega_4}{2c}\left[\chi^{(3)}_{1122}\hat{\boldsymbol{e}}_3(\hat{\boldsymbol{e}}_2 \bullet \hat{\boldsymbol{e}}_1) + \chi^{(3)}_{1212}\hat{\boldsymbol{e}}_2\cancel{(\hat{\boldsymbol{e}}_3 \bullet \hat{\boldsymbol{e}}_1)} + \chi^{(3)}_{1221}\hat{\boldsymbol{e}}_1(\hat{\boldsymbol{e}}_3 \bullet \hat{\boldsymbol{e}}_2)\right]$$
$$\times E(\omega_1)E^*(\omega_2)E(\omega_3)L_{CARS}$$
$$= i\frac{\omega_4}{2c}\left[\chi^{(3)}_{1122}\hat{\boldsymbol{e}}_3 \cos 45^\circ + \chi^{(3)}_{1221}\hat{\boldsymbol{e}}_1 \cos 45^\circ\right]E(\omega_1)E^*(\omega_2)E(\omega_3)L_{CARS}. \tag{8.212}$$

The susceptibility elements $\chi^{(3)}_{1122}$ and $\chi^{(3)}_{1221}$ are given by

$$\chi^{(3)}_{1122}(-\omega_4; \omega_3, -\omega_2, \omega_1) = \frac{1}{24}[\cancel{X} + 2a(\omega_1 - \omega_2) + b(\omega_3 - \omega_2)]. \tag{8.213}$$

$$\chi^{(3)}_{1221}(-\omega_4; \omega_3, -\omega_2, \omega_1) = \frac{1}{24}[\cancel{X} + b(\omega_1 - \omega_2) + 2a(\omega_3 - \omega_2)]. \tag{8.214}$$

For the Q-branch transitions, the susceptibility parameters are given by

$$a(\omega_1 - \omega_2) = \beta_{vJJ}\left[(a')^2 - \frac{2}{45}b_{JJ}(\gamma')^2\right]_{H_2}. \tag{8.215}$$

$$a(\omega_3 - \omega_2) = \beta_{vJJ}\left[(a')^2 - \frac{2}{45}b_{JJ}(\gamma')^2\right]_{N_2}. \tag{8.216}$$

$$b(\omega_1-\omega_2)=\beta_{vJJ}\left[\frac{2}{15}b_{JJ}(\gamma')^2\right]_{H_2}. \tag{8.217}$$

$$b(\omega_3-\omega_2)=\beta_{vJJ}\left[\frac{2}{15}b_{JJ}(\gamma')^2\right]_{N_2}. \tag{8.218}$$

The Placzek–Teller coefficient for Q-branch transitions is given by

$$b_{JJ}=\frac{J(J+1)}{(2J-1)(2J+3)}. \tag{8.219}$$

Now let's find the angles for $\boldsymbol{E}_{H_2}(\omega_4)$ and $\boldsymbol{E}_{N_2}(\omega_4)$ with respect to $\boldsymbol{E}(\omega_1)$. For the H_2 signal,

$$\boldsymbol{E}_{H_2}(\omega_3)=i\frac{\omega_3}{16c}\boldsymbol{E}(\omega_1)\boldsymbol{E}^*(\omega_2)\boldsymbol{E}(\omega_3)\left[2a(\omega_1-\omega_2)\cos 45^\circ\,\hat{\boldsymbol{e}}_3+b(\omega_1-\omega_2)\cos 45^\circ\,\hat{\boldsymbol{e}}_1\right]. \tag{8.220}$$

Now let θ_{H_2} be the angle of the H_2 CARS signal with respect to $\hat{e}_1$. Then

$$\tan\theta_{H_2}=\frac{2a(\omega_0-\omega_2)}{b(\omega_0-\omega_2)}=\frac{2\left[(a')^2-\frac{2}{45}b_{JJ}(\gamma')^2\right]}{\frac{2}{15}b_{JJ}(\gamma')^2}=\frac{15-2b_{JJ}(\gamma'/a')^2/3}{b_{JJ}(\gamma'/a')^2}. \tag{8.221}$$

From Eq. (8.219), for the Q(1) line of H_2, $b_{JJ}=0.4$, and from Table 7.1, $(\gamma'/a')^2=0.730$. Therefore

$$\tan\theta_{H_2}=\frac{15-2(0.4)(0.730)/3}{(0.4)(0.730)}=50.70\quad\Rightarrow\quad\theta_{H_2}=1.551\,\text{rad}=88.87^\circ. \tag{8.222}$$

Following similar procedures for the N_2 Q(5) Raman transition, we obtain

$$\boldsymbol{E}_{N_2}(\omega_4)=\frac{i\omega_4}{16c}E(\omega_1)E^*(\omega_2)E(\omega_3)\left[b(\omega_3-\omega_2)\cos 45^\circ\,\hat{\boldsymbol{e}}_3+2a(\omega_3-\omega_2)\cos 45^\circ\,\hat{\boldsymbol{e}}_1\right]. \tag{8.223}$$

The tangent of the angle between the N_2 CARS signal and $\hat{\boldsymbol{e}}_1$is given by

$$\tan\theta_{N_2}=\frac{b(\omega_3-\omega_2)}{2a(\omega_3-\omega_2)}=\frac{b_{JJ}(\gamma'/a')^2}{15-2b_{JJ}(\gamma'/a')^2/3}. \tag{8.224}$$

From Eq. (8.219), for the Q (5) line of N_2, $b_{JJ}=0.256$, and from Table 7.1, $(\gamma'/a')^2=1.36$. Therefore,

$$\tan\theta_{N_2}=\frac{(0.256)(1.36)}{15-2(0.256)(1.36)/3}=0.0236\quad\Rightarrow\quad\theta_{N_2}=0.0236\,\text{rad}=1.35^\circ. \tag{8.225}$$

The polarizations of the CARS signals from H_2 and N_2 with respect to the input beam polarizations are shown in Figure 8.18. Note that for these Q-branch transitions, the

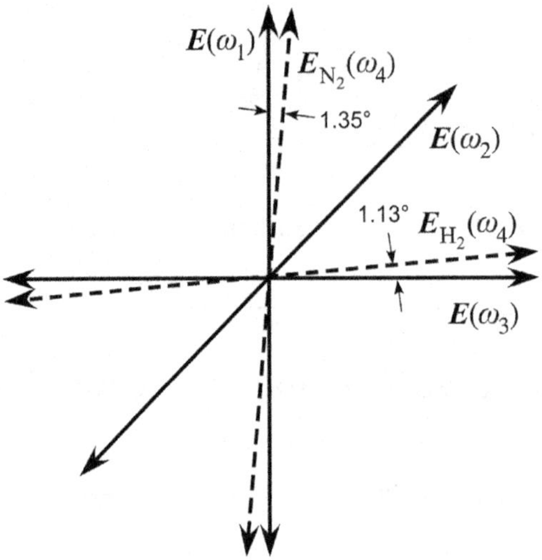

Figure 8.18 CARS signal beam polarizations with respect to the input CARS beam polarizations.

polarization of the signal beam is nearly parallel to the polarization of the probe beam for each transition. The departure from parallel with respect to the probe beam is a result of the (slight) depolarization of the Q-branch transitions.

Now we need to calculate the number of photons in the CARS signal channel for a polarizer aligned with $\hat{e}_1$. The calculations are very similar to the calculations outlined in the Section 7.6 and in the first example problem in this section. The results of the calculations of the number densities in the ground levels of the H_2 Q(1) transition and the N_2 Q(5) transition are

$$\frac{n_{J=1}}{n_{H_2}} = \frac{NSSW(2J+1)\exp\left[-hcF(J)/k_BT\right]}{Z_{rot}Z_{nuc}} = \frac{3(3)\exp\left[-(1.439\ \text{K cm})(118.5\ \text{cm}^{-1})/(300\ \text{K})\right]}{(1.76)(4)} = 0.724. \tag{8.226}$$

$$\frac{n_{J=5}}{n_{N_2}} = \frac{(3)(11)\exp[-(1.439)(59.7)/300]}{(9)(52.4)} = 0.0526. \tag{8.227}$$

$$n_{N_2} = n_{H_2} = \frac{\left(100 \times 10^3\ \text{J m}^{-3}\right)}{\left(1.3807 \times 10^{-23}\ \text{J K}^{-1}\right)(300\ \text{K})} = 2.41 \times 10^{25}\ \text{m}^{-3}. \tag{8.228}$$

Therefore

$$n_{J=1,H_2} = (0.724)\left(2.41 \times 10^{25}\ \text{m}^{-3}\right) = 1.75 \times 10^{25}\ \text{m}^{-3}. \tag{8.229}$$

$$n_{J=1,N_2} = (0.0526)\left(2.41 \times 10^{25}\ \text{m}^{-3}\right) = 1.27 \times 10^{24}\ \text{m}^{-3}. \tag{8.230}$$

The Raman frequency for the Q(1) line of H_2 is $\tilde{\nu}_R = 4,152.3\ \text{cm}^{-1}$ $\left(\omega_j = 7.8217 \times 10^{14}\ \text{s}^{-1}\right)$, as calculated in Section 7.6. The Raman frequency for the Q(5) line of N_2 is given by

$$\begin{aligned}\tilde{\nu}_R &= F_1(5) + G(1) - F_0(5) + G(0) \\ &= 59.1 + 3,504.8 - 59.7 - 1,175.3\ \text{cm}^{-1} = 2,328.9\ \text{cm}^{-1} \\ \omega_j &= 2\pi\tilde{\nu}_R = 4.3869 \times 10^{14}\ \text{s}^{-1}\end{aligned} \tag{8.231}$$

We can now calculate the parameters β_{vJJ,N_2} and β_{vJJ,H_2} at line center for the Raman transition,

$$\begin{aligned}\beta_{vJJ,N_2} &= \frac{n_{vJ}}{\varepsilon_0 \mu_R \omega_j \Gamma_j} i \\ &= \frac{i\left(1.27 \times 10^{24}\,\text{m}^{-3}\right)}{\left(8.854 \times 10^{-12}\,\frac{\text{C}^2}{\text{Jm}}\right)\left(1.169 \times 10^{-26}\,\text{kg}\right)\left(4.3869 \times 10^{14}\,\text{s}^{-1}\right)\left(9.42 \times 10^{9}\,\text{s}^{-1}\right)} \\ &= i\left(2.96 \times 10^{36}\,\text{C}^{-2}\right)\end{aligned} \tag{8.232}$$

and

$$\begin{aligned}\beta_{vJJ,H_2} &= \frac{i\left(1.75 \times 10^{25}\,\text{m}^{-3}\right)}{\left(8.854 \times 10^{-12}\,\frac{\text{C}^2}{\text{Jm}}\right)\left(0.835 \times 10^{-27}\,\text{kg}\right)\left(7.8217 \times 10^{14}\,\text{s}^{-1}\right)\left(9.42 \times 10^{9}\,\text{s}^{-1}\right)} \\ &= i\left(3.21 \times 10^{38}\,\text{C}^{-2}\right).\end{aligned} \tag{8.233}$$

When the CARS polarizer is aligned with the $\boldsymbol{E}(\omega_1)$ beam, only the χ_{1122} term contributes to the CARS signal. For H_2 the transmitted CARS amplitude is given by

$$E_{trans,H_2}(\omega_4) = \frac{i\omega_4}{16c} E(\omega_1)E^*(\omega_2)E(\omega_3)b(\omega_1 - \omega_2)\cos 45^\circ L_{CARS}. \tag{8.234}$$

$$\begin{aligned}b(\omega_1 - \omega_2)_{H_2} &= \beta_{vJJ,H_2}\left[\frac{2}{15} b_{JJ}(\gamma')^2\right] \\ &= i\left(3.21 \times 10^{39}\ \text{C}^{-2}\right)\left[\frac{2}{15}(0.4)\left(1.38 \times 10^{-60}\ \frac{\text{C}^4\,\text{m}^2}{\text{J}^2}\right)\right] \\ &= i\left(2.40 \times 10^{-23}\ \frac{\text{C}^2\,\text{m}^2}{\text{J}^2}\right).\end{aligned} \tag{8.235}$$

$$\begin{aligned}I_{4trans,H_2} &= \frac{1}{2}c\varepsilon_0\left|E_{trans,H_2}(\omega_4)\right|^2 = \frac{1}{2}c\varepsilon_0 \frac{\omega_4^2}{256c^2}\left|b(\omega_1 - \omega_2)\right|^2 \cos^2 45^\circ \\ &\times \left(\frac{2I_1}{c\varepsilon_0}\right)\left(\frac{2I_2}{c\varepsilon_0}\right)\left(\frac{2I_3}{c\varepsilon_0}\right)L_{CARS}^2 = \frac{\omega_4^2 L_{CARS}^2 I_1 I_2 I_3}{128\,c^4\varepsilon_0^2}\left|b(\omega_1 - \omega_2)\right|^2.\end{aligned} \tag{8.236}$$

The laser intensities are given by

$$I_0 = I_1 = I_2 = \frac{10^{-2}\ \text{J}}{\frac{\pi}{4}\left(10^{-4}\ \text{m}\right)^2\left(10^{-8}\ \text{s}\right)} = 1.27 \times 10^{14}\ \frac{\text{W}}{\text{m}^2}. \tag{8.237}$$

The Raman frequencies are given by

$$\omega_{4,\mathrm{H}_2} = (2\pi c)(\tilde{\nu}_3 + \tilde{\nu}_{R,\mathrm{H}_2}) = \omega_{4,\mathrm{N}_2} = (2\pi c)(\tilde{\nu}_1 + \tilde{\nu}_{R,\mathrm{N}_2})$$
$$= 2\pi(2.998 \times 10^{10}\ \mathrm{cm}^{-1})(18,797 + 2,328.9\ \mathrm{cm}^{-1}) = 3.98 \times 10^{15}\ \mathrm{s}^{-1}. \tag{8.238}$$

The transmitted CARS signal intensity for H_2 is given by

$$I_{4,\mathrm{H}_2} = \frac{(3.98\times10^{15}\ \mathrm{s}^{-1})^2(2\times10^{-3}\ \mathrm{m})^2\left(1.27\times10^{14}\ \frac{\mathrm{J}}{\mathrm{m}^2\,\mathrm{s}}\right)^3}{128\left(2.998\times10^{8}\ \frac{\mathrm{m}}{\mathrm{s}}\right)^4\left(8.854\times10^{-12}\ \frac{\mathrm{C}^2}{\mathrm{J\,m}}\right)^2}\left(2.40\times10^{-23}\ \frac{\mathrm{C}^2\,\mathrm{m}^2}{\mathrm{J}^2}\right)^2$$
$$= 9.22\times10^{8}\ \frac{\mathrm{W}}{\mathrm{m}^2}. \tag{8.239}$$

The number of photons in the H_2 CARS signal pulse, assuming top-hat spatial and temporal profiles, is given by

$$N_{p4trans,\mathrm{H}_2} = \frac{I_{4,trans}A_4\,\Delta t_L}{\hbar\omega_4} = \frac{\left(9.22\times10^{8}\ \frac{\mathrm{J}}{\mathrm{s\,m}^2}\right)\frac{\pi}{4}(10^{-4}\ \mathrm{m})^2(10^{-8}\ \mathrm{s})}{(1.055\times10^{-34}\ \mathrm{J\,s})(3.98\times10^{15}\ \mathrm{s}^{-1})} \tag{8.240}$$
$$= 1.73\times10^{11}\ \frac{photons}{laser\ pulse}.$$

For the N_2 CARS signal, the transmitted amplitude is given by

$$E_{trans,\mathrm{N}_2}(\omega_4) = \frac{i\omega_4}{16c}E(\omega_1)E^*(\omega_2)E(\omega_3)2a(\omega_1-\omega_2)\cos 45^\circ L_{CARS}, \tag{8.241}$$

and the transmitted intensity is given by

$$I_{4trans,\mathrm{N}_2} = \frac{1}{2}c\,\varepsilon_0|E_{trans,\mathrm{N}_2}(\omega_4)|^2 = \frac{\omega_4^2L_{CARS}^2I_1I_2I_3}{32\,c^4\varepsilon_0^2}|a(\omega_1-\omega_2)|^2. \tag{8.242}$$

The parameter $a(\omega_1-\omega_2)$ is given by

$$a(\omega_1-\omega_2) = \beta_{vJJ,\mathrm{N}_2}\left[(a')^2 - \frac{2}{45}b_{JJ}(\gamma')^2\right]_{N_2} = i(2.96\times10^{36}\ \mathrm{C}^{-2})\left[\left(3.79\times10^{-60}\ \frac{\mathrm{C}^4\mathrm{m}^2}{\mathrm{J}^2}\right)\right.$$
$$\left.-\frac{2}{45}(0.256)\left(5.15\times10^{-60}\ \frac{\mathrm{C}^4\mathrm{m}^2}{\mathrm{J}^2}\right)\right] = i1.10\times10^{-23}\ \frac{\mathrm{C}^2\mathrm{m}^2}{\mathrm{J}^2}. \tag{8.243}$$

The transmitted CARS signal intensity for N_2 is given by

$$I_{4,\mathrm{N}_2} = \frac{(3.98\times10^{15}\ \mathrm{s}^{-1})^2(2\times10^{-3}\ \mathrm{m})^2\left(1.27\times10^{14}\ \frac{\mathrm{J}}{\mathrm{m}^2\,\mathrm{s}}\right)^3}{32\left(2.998\times10^{8}\ \frac{\mathrm{m}}{\mathrm{s}}\right)^4\left(8.854\times10^{-12}\ \frac{\mathrm{C}^2}{\mathrm{J\,m}}\right)^2}\left(1.10\times10^{-23}\ \frac{\mathrm{C}^2\,\mathrm{m}^2}{\mathrm{J}^2}\right)^2$$
$$= 7.75\times10^{8}\ \frac{\mathrm{W}}{\mathrm{m}^2}. \tag{8.244}$$

The number of photons in the N_2 CARS signal pulse, assuming top-hat spatial and temporal profiles, is given by

$$N_{p4trans,N_2} = \frac{I_{4,trans} A_4 \Delta t_L}{\hbar \omega_4} = \frac{\left(7.75 \times 10^8 \ \frac{\mathrm{J}}{\mathrm{s\,m^2}}\right) \frac{\pi}{4} \left(10^{-4}\ \mathrm{m}\right)^2 \left(10^{-8}\ \mathrm{s}\right)}{\left(1.055 \times 10^{-34}\ \mathrm{J\,s}\right) \left(3.98 \times 10^{15}\ \mathrm{s^{-1}}\right)} \qquad (8.245)$$

$$= 1.45 \times 10^{11} \ \frac{photons}{laser\ pulse}.$$

Appendix 1 Spherical Harmonics and Radial Wavefunctions for One-Electron Atoms

Table A1.1 Spherical harmonic functions (Weissbluth, 1978)

l	m	$Y_{lm}(\theta, \varphi)$	$r^l Y_{lm}(x, y, z)$
0	0	$\sqrt{\frac{1}{4\pi}}$	$\sqrt{\frac{1}{4\pi}}$
1	0	$\sqrt{\frac{3}{4\pi}}\cos\theta$	$\sqrt{\frac{3}{4\pi}}z$
1	±1	$\mp\sqrt{\frac{3}{8\pi}}\sin\theta\exp(\pm i\varphi)$	$\mp\sqrt{\frac{3}{8\pi}}(x \pm iy)$
2	0	$\sqrt{\frac{5}{4\pi}}\sqrt{\frac{1}{4}}(3\cos^2\theta - 1)$	$\sqrt{\frac{5}{4\pi}}\sqrt{\frac{1}{4}}(3z^2 - r^2)$
2	±1	$\mp\sqrt{\frac{5}{4\pi}}\sqrt{\frac{3}{2}}\cos\theta\sin\theta\exp(\pm i\varphi)$	$\mp\sqrt{\frac{5}{4\pi}}\sqrt{\frac{3}{2}}z(x \pm iy)$
2	±2	$\sqrt{\frac{5}{4\pi}}\sqrt{\frac{3}{8}}\sin^2\theta\exp(\pm 2i\varphi)$	$\sqrt{\frac{5}{4\pi}}\sqrt{\frac{3}{8}}(x + iy)^2$
3	0	$\sqrt{\frac{7}{4\pi}}\sqrt{\frac{1}{4}}\left(2\cos^2\theta - 3\cos\theta\sin^2\theta\right)$	$\sqrt{\frac{7}{4\pi}}\sqrt{\frac{1}{4}}z(5z^2 - 3r^2)$
3	±1	$\mp\sqrt{\frac{7}{4\pi}}\sqrt{\frac{3}{16}}\left(4\cos^2\theta\sin\theta - \sin^3\theta\right)\exp(\pm i\phi)$	$\mp\sqrt{\frac{7}{4\pi}}\sqrt{\frac{3}{16}}(x + iy)(5z^2 - 3r^2)$
3	±2	$\sqrt{\frac{7}{4\pi}}\sqrt{\frac{15}{8}}\cos\theta\sin^2\theta\exp(\pm 2i\phi)$	$\sqrt{\frac{7}{4\pi}}\sqrt{\frac{15}{8}}z(x + iy)^2$
3	±3	$\mp\sqrt{\frac{7}{4\pi}}\sqrt{\frac{5}{16}}\sin^3\theta\exp(\pm 3i\phi)$	$\mp\sqrt{\frac{7}{4\pi}}\sqrt{\frac{5}{6}}(x + iy)^3$

Table A1.2 Radial components for the Schrödinger form of the one-electron atomic wavefunctions. For the hydrogen atom, $Z = 1$

n	l	$R_{nl}(r)$
1	0	$2\left(\frac{Z}{a_0}\right)^{3/2} \exp\left(-\frac{Zr}{a_0}\right)$
2	0	$\frac{1}{2\sqrt{2}}\left(\frac{Z}{a_0}\right)^{3/2}\left(2-\frac{Zr}{a_0}\right)\exp\left(-\frac{Zr}{2a_0}\right)$
2	1	$\frac{1}{2\sqrt{6}}\left(\frac{Z}{a_0}\right)^{3/2}\left(\frac{Zr}{a_0}\right)\exp\left(-\frac{Zr}{2a_0}\right)$
3	0	$\frac{2}{81\sqrt{3}}\left(\frac{Z}{a_0}\right)^{3/2}\left[27-18\frac{Zr}{a_0}+2\left(\frac{Zr}{a_0}\right)^2\right]\exp\left(-\frac{Zr}{3a_0}\right)$
3	1	$\frac{2\sqrt{2}}{81\sqrt{3}}\left(\frac{Z}{a_0}\right)^{3/2}\left[6\frac{Zr}{a_0}-\left(\frac{Zr}{a_0}\right)^2\right]\exp\left(-\frac{Zr}{3a_0}\right)$
3	2	$\frac{2\sqrt{2}}{81\sqrt{15}}\left(\frac{Z}{a_0}\right)^{3/2}\left(\frac{Zr}{a_0}\right)^2\exp\left(-\frac{Zr}{3a_0}\right)$

Appendix 2 Clebsch–Gordan Coefficients, Dipole Moments, and Spontaneous Emission Coefficients for the 2*p*–1*s* Transition in Atomic Hydrogen

Table A2.1 Clebsch–Gordan coefficients for the 2*p* level in atomic hydrogen

	$J=\frac{1}{2}$ $m_J=-\frac{1}{2}$	$J=\frac{1}{2}$ $m_J=\frac{1}{2}$	$J=\frac{3}{2}$ $m_J=-\frac{3}{2}$	$J=\frac{3}{2}$ $m_J=-\frac{1}{2}$	$J=\frac{3}{2}$ $m_J=\frac{1}{2}$	$J=\frac{3}{2}$ $m_J=\frac{3}{2}$
$m_l=1$ $m_s=-\frac{1}{2}$	0	$\sqrt{\frac{2}{3}}$	0	0	$\frac{1}{\sqrt{3}}$	0
$m_l=1$ $m_s=\frac{1}{2}$	0	0	0	0	0	1
$m_l=0$ $m_s=-\frac{1}{2}$	$\frac{1}{\sqrt{3}}$	0	0	$\sqrt{\frac{2}{3}}$	0	0
$m_l=0$ $m_s=\frac{1}{2}$	0	$-\frac{1}{\sqrt{3}}$	0	0	$\sqrt{\frac{2}{3}}$	0
$m_l=-1$ $m_s=-\frac{1}{2}$	0	0	1	0	0	0
$m_l=-1$ $m_s=\frac{1}{2}$	$-\sqrt{\frac{2}{3}}$	0	0	$\frac{1}{\sqrt{3}}$	0	0

Table A2.2 Electric dipole moment matrix elements for the *2p–1s* transition in the coupled representation: $-e\langle 21\tfrac{1}{2}J'm'_J|\boldsymbol{r}|10\tfrac{1}{2}Jm_J\rangle = \langle 21\tfrac{1}{2}J'm'_J|\boldsymbol{\mu}|10\tfrac{1}{2}Jm_J\rangle$

2p \ 1s	$J=\frac{1}{2}$ $m_J=-\frac{1}{2}$	$J=\frac{1}{2}$ $m_J=\frac{1}{2}$
$J'=\frac{1}{2}$ $m'_J=-\frac{1}{2}$	$-\frac{1.290\,ea_0}{3}\hat{z}$	$\frac{1.290\sqrt{2}ea_0}{3}\left(\frac{\hat{x}+i\hat{y}}{\sqrt{2}}\right)$

Table A2.2 *(cont.)*

2p \ 1s	$J = \frac{1}{2}$, $m_J = -\frac{1}{2}$	$J = \frac{1}{2}$, $m_J = \frac{1}{2}$
$J' = \frac{1}{2}$, $m'_J = \frac{1}{2}$	$\frac{1.290\sqrt{2}ea_0}{3}\left(\frac{\hat{x}-i\hat{y}}{\sqrt{2}}\right)$	$\frac{1.290\,ea_0}{3}\hat{z}$
$J' = \frac{3}{2}$, $m'_J = -\frac{3}{2}$	$-\frac{1.290\,ea_0}{\sqrt{3}}\left(\frac{\hat{x}+i\hat{y}}{\sqrt{2}}\right)$	0
$J' = \frac{3}{2}$, $m'_J = -\frac{1}{2}$	$-\frac{1.290\sqrt{2}ea_0}{3}\hat{z}$	$-\frac{1.290\,ea_0}{3}\left(\frac{\hat{x}+i\hat{y}}{\sqrt{2}}\right)$
$J' = \frac{3}{2}$, $m'_J = \frac{1}{2}$	$\frac{1.290ea_0}{3}\left(\frac{\hat{x}-i\hat{y}}{\sqrt{2}}\right)$	$-\frac{1.290\sqrt{2}ea_0}{3}\hat{z}$
$J' = \frac{3}{2}$, $m'_J = \frac{3}{2}$	0	$\frac{1.290\,ea_0}{\sqrt{3}}\left(\frac{\hat{x}-i\hat{y}}{\sqrt{2}}\right)$

Table A2.3 Spontaneous emission coefficients for the 2*p*–1*s* transition in the coupled representation

	A_{ba} $\left(\mathrm{s}^{-1}\right)$	A_{ba} $\left(\mathrm{s}^{-1}\right)$	A_{bA} $\left(\mathrm{s}^{-1}\right)$
2p \ 1s	$J = \frac{1}{2}$, $m_J = -\frac{1}{2}$	$J = \frac{1}{2}$, $m_J = \frac{1}{2}$	
$J' = \frac{1}{2}$, $m'_J = -\frac{1}{2}$	2.085×10^8	4.170×10^8	6.255×10^8
$J' = \frac{1}{2}$, $m'_J = \frac{1}{2}$	4.170×10^8	2.085×10^8	6.255×10^8
$J' = \frac{3}{2}$, $m'_J = -\frac{3}{2}$	6.255×10^8	0	6.255×10^8
$J' = \frac{3}{2}$, $m'_J = -\frac{1}{2}$	4.170×10^8	2.085×10^8	6.255×10^8

Table A2.3 (*cont.*)

	A_{ba} (s^{-1})	A_{ba} (s^{-1})	A_{bA} (s^{-1})
1s \ **2p**	$J = \frac{1}{2}$ $m_J = -\frac{1}{2}$	$J = \frac{1}{2}$ $m_J = \frac{1}{2}$	
$J' = \frac{3}{2}$ $m'_J = \frac{1}{2}$	2.085×10^8	4.170×10^8	6.255×10^8
$J' = \frac{3}{2}$ $m'_J = \frac{3}{2}$	0	6.255×10^8	6.255×10^8

Appendix 3 Properties and Values for Selected *3j* Symbols

Permutation properties

$$\begin{pmatrix} j_1 & j_2 & j_3 \\ m_1 & m_2 & m_3 \end{pmatrix} = \begin{pmatrix} j_3 & j_1 & j_2 \\ m_3 & m_1 & m_2 \end{pmatrix} = \begin{pmatrix} j_2 & j_3 & j_1 \\ m_2 & m_3 & m_1 \end{pmatrix} \qquad \text{(A3.1)}$$

$$(-1)^{j_1+j_2+j_3} \begin{pmatrix} j_1 & j_2 & j_3 \\ m_1 & m_2 & m_3 \end{pmatrix} = \begin{pmatrix} j_2 & j_1 & j_3 \\ m_2 & m_1 & m_3 \end{pmatrix} = \begin{pmatrix} j_1 & j_3 & j_2 \\ m_1 & m_3 & m_2 \end{pmatrix}$$
$$= \begin{pmatrix} j_3 & j_2 & j_1 \\ m_3 & m_2 & m_1 \end{pmatrix} = \begin{pmatrix} j_1 & j_2 & j_3 \\ -m_1 & -m_2 & -m_3 \end{pmatrix} \qquad \text{(A3.2)}$$

Summation properties

$$\sum_{m_1, m_2} \begin{pmatrix} j_1 & j_2 & k_3 \\ m_1 & m_2 & n_3 \end{pmatrix} \begin{pmatrix} j_1 & j_2 & j_3 \\ m_1 & m_2 & m_3 \end{pmatrix} = \frac{\delta_{j_3 k_3} \delta_{m_3 n_3}}{2j_3 + 1} \qquad \text{(A3.3)}$$

$$\sum_{j_3, m_3} (2j_3 + 1) \begin{pmatrix} j_1 & j_2 & j_3 \\ m_1 & m_2 & m_3 \end{pmatrix} \begin{pmatrix} j_1 & j_2 & j_3 \\ n_1 & n_2 & m_3 \end{pmatrix} = \delta_{m_1 n_1} \delta_{m_2 n_2} \qquad \text{(A3.4)}$$

Selected *3j* symbols

$$\begin{pmatrix} j_1 & j_2 & j_3 \\ 0 & 0 & 0 \end{pmatrix} \neq 0 \quad \text{for even } j_1 + j_2 + j_3, \ \beta = \frac{j_1 + j_2 + j_3}{2}$$
$$\begin{pmatrix} j_1 & j_2 & j_3 \\ 0 & 0 & 0 \end{pmatrix} = (-1)^{\beta} \sqrt{\frac{(j_1 + j_2 - j_3)!(j_3 + j_1 - j_2)!(j_2 + j_3 - j_1)!}{(2\beta + 1)!}}$$
$$\times \frac{\beta!}{(\beta - j_3)!(\beta - j_2)!(\beta - j_1)!} \qquad \text{(A3.5)}$$

$$\begin{pmatrix} j_1 & j_2 & j_3 \\ 0 & 0 & 0 \end{pmatrix} = 0 \quad \text{for odd } j_1 + j_2 + j_3 \qquad \text{(A3.6)}$$

$$\begin{pmatrix} j & j & 0 \\ m & -m & 0 \end{pmatrix} = \frac{(-1)^{j-m}}{\sqrt{(2j+1)}} \qquad \text{(A3.7)}$$

$$\begin{pmatrix} j & j & 1 \\ m & -m & 0 \end{pmatrix} = (-1)^{j-m} \frac{m}{\sqrt{(2j+1)j(j+1)}} \tag{A3.8}$$

$$\begin{pmatrix} j+1 & j & 1 \\ m & -m & 0 \end{pmatrix} = (-1)^{j-m-1} \sqrt{\frac{2(j+m+1)(j-m+1)}{(2j+3)(2j+2)(2j+1)}} \tag{A3.9}$$

$$\begin{pmatrix} j & j+1 & 1 \\ m & -m & 0 \end{pmatrix} = (-1)^{3j+m-1} \sqrt{\frac{2(j-m+1)(j+m+1)}{(2j+3)(2j+2)(2j+1)}} \tag{A3.10}$$

$$\begin{pmatrix} j & j-1 & 1 \\ m & -m & 0 \end{pmatrix} = (-1)^{j-m} \sqrt{\frac{2(j+m)(j-m)}{(2j+1)2j(2j-1)}} \tag{A3.11}$$

$$\begin{pmatrix} j+\frac{1}{2} & j & \frac{1}{2} \\ m & -\left(m+\frac{1}{2}\right) & \frac{1}{2} \end{pmatrix} = (-1)^{j-m-\frac{1}{2}} \sqrt{\frac{j-m+\frac{1}{2}}{(2j+2)(2j+1)}} \tag{A3.12}$$

$$\begin{pmatrix} j & j+\frac{1}{2} & \frac{1}{2} \\ m & -\left(m+\frac{1}{2}\right) & \frac{1}{2} \end{pmatrix} = (-1)^{3j+m+1} \sqrt{\frac{j+m+1}{(2j+2)(2j+1)}} \tag{A3.13}$$

$$\begin{pmatrix} j & j-\frac{1}{2} & \frac{1}{2} \\ m & -\left(m+\frac{1}{2}\right) & \frac{1}{2} \end{pmatrix} = (-1)^{j-m-1} \sqrt{\frac{j-m}{(2j+1)2j}} \tag{A3.14}$$

$$\begin{pmatrix} j-\frac{1}{2} & j & \frac{1}{2} \\ m & -\left(m+\frac{1}{2}\right) & \frac{1}{2} \end{pmatrix} = (-1)^{3j+m-\frac{1}{2}} \sqrt{\frac{j+m+\frac{1}{2}}{(2j+1)2j}} \tag{A3.15}$$

$$\begin{pmatrix} j & j & 1 \\ m & -(m+1) & 1 \end{pmatrix} = (-1)^{j-m} \sqrt{\frac{2(j-m)(j+m+1)}{(2j+2)(2j+1)(2j)}} \tag{A3.16}$$

$$\begin{pmatrix} j+1 & j & 1 \\ m & -(m+1) & 1 \end{pmatrix} = (-1)^{j-m-1} \sqrt{\frac{(j-m)(j-m+1)}{(2j+3)(2j+2)(2j+1)}} \tag{A3.17}$$

$$\begin{pmatrix} j & j+1 & 1 \\ -(m+1) & m & 1 \end{pmatrix} = (-1)^{3j-m+1} \sqrt{\frac{(j-m)(j-m+1)}{(2j+3)(2j+2)(2j+1)}} \tag{A3.18}$$

$$\begin{pmatrix} j & j+1 & 1 \\ m & -(m+1) & 1 \end{pmatrix} = (-1)^{3j+m} \sqrt{\frac{(j+m+1)(j+m+2)}{(2j+3)(2j+2)(2j+1)}} \tag{A3.19}$$

$\begin{pmatrix} j & j-1 & 1 \\ m & -(m+1) & 1 \end{pmatrix} = (-1)^{j-m}\sqrt{\frac{(j-m-1)(j-m)}{(2j+1)2j(2j-1)}}$	(A3.20)
$\begin{pmatrix} j & j & 2 \\ m & -m & 0 \end{pmatrix} = (-1)^{j-m}\frac{2[3m^2-j(j+1)]}{\sqrt{(2j+3)(2j+2)(2j+1)2j(2j-1)}}$	(A3.21)
$\begin{pmatrix} j+1 & j & 2 \\ m & -m & 0 \end{pmatrix} = (-1)^{j-m+1}2m\sqrt{\frac{6(j+m+1)(j-m+1)}{(2j+4)(2j+3)(2j+2)(2j+1)2j}}$	(A3.22)
$\begin{pmatrix} j & j+1 & 2 \\ m & -m & 0 \end{pmatrix} = (-1)^{3j+m}2m\sqrt{\frac{6(j-m+1)(j+m+1)}{(2j+4)(2j+3)(2j+2)(2j+1)2j}}$	(A3.23)
$\begin{pmatrix} j & j-1 & 2 \\ m & -m & 0 \end{pmatrix} = (-1)^{j-m}2m\sqrt{\frac{6(j+m)(j-m)}{(2j+2)(2j+1)2j(2j-1)(2j-2)}}$	(A3.24)
$\begin{pmatrix} j+2 & j & 2 \\ m & -m & 0 \end{pmatrix} = (-1)^{j-m}\sqrt{\frac{6(j+m+2)(j+m+1)(j-m+2)(j-m+1)}{(2j+5)(2j+4)(2j+3)(2j+2)(2j+1)}}$	(A3.25)
$\begin{pmatrix} j & j+2 & 2 \\ m & -m & 0 \end{pmatrix} = (-1)^{3j+m}\sqrt{\frac{6(j-m+2)(j-m+1)(j+m+2)(j+m+1)}{(2j+5)(2j+4)(2j+3)(2j+2)(2j+1)}}$	(A3.26)
$\begin{pmatrix} j & j-2 & 2 \\ m & -m & 0 \end{pmatrix} = (-1)^{j-m}\sqrt{\frac{6(j-m)(j-m-1)(j+m)(j+m-1)}{(2j+1)2j(2j-1)(2j-2)(2j-3)}}$	(A3.27)
$\begin{pmatrix} j & j & 2 \\ m & -(m+1) & 1 \end{pmatrix} = (-1)^{j-m}(2m+1)\sqrt{\frac{6(j+m+1)(j-m)}{(2j+3)(2j+2)(2j+1)2j(2j-1)}}$	(A3.28)
$\begin{pmatrix} j+1 & j & 2 \\ m & -(m+1) & 1 \end{pmatrix} = (-1)^{j-m+1}2(j+2m+2)\sqrt{\frac{(j-m+1)(j-m)}{(2j+4)(2j+3)(2j+2)(2j+1)2j}}$	(A3.29)
$\begin{pmatrix} j & j+1 & 2 \\ m & -(m+1) & 1 \end{pmatrix} = (-1)^{3j+m+1}2(j-2m)\sqrt{\frac{(j+m)(j+m+1)}{(2j+4)(2j+3)(2j+2)(2j+1)2j}}$	(A3.30)
$\begin{pmatrix} j & j-1 & 2 \\ m & -(m+1) & 1 \end{pmatrix} = (-1)^{j-m}2(j+2m+1)\sqrt{\frac{(j-m)(j-m-1)}{(2j+2)(2j+1)2j(2j-1)(2j-2)}}$	(A3.31)
$\begin{pmatrix} j+2 & j & 2 \\ m & -(m+1) & 1 \end{pmatrix} = (-1)^{j-m}\sqrt{\frac{4(j+m+2)(j-m+2)(j-m+1)(j-m)}{(2j+5)(2j+4)(2j+3)(2j+2)(2j+1)}}$	(A3.32)

$$\begin{pmatrix} j & j+2 & 2 \\ m & -(m+1) & 1 \end{pmatrix} = (-1)^{3j+m+1}\sqrt{\frac{4(j-m+1)(j+m+3)(j+m+2)(j+m+1)}{(2j+5)(2j+4)(2j+3)(2j+2)(2j+1)}} \quad \text{(A3.33)}$$

$$\begin{pmatrix} j & j-2 & 2 \\ m & -(m+1) & 1 \end{pmatrix} = (-1)^{j-m}\sqrt{\frac{4(j+m)(j-m)(j-m-1)(j-m-2)}{(2j+1)2j(2j-1)(2j-2)(2j-3)}} \quad \text{(A3.34)}$$

$$\begin{pmatrix} j & j & 2 \\ m & -(m+2) & 2 \end{pmatrix} = (-1)^{j-m}\sqrt{\frac{6(j-m-1)(j-m)(j+m+1)(j+m+2)}{(2j+3)(2j+2)(2j+1)2j(2j-1)}} \quad \text{(A3.35)}$$

$$\begin{pmatrix} j+1 & j & 2 \\ m & -(m+2) & 2 \end{pmatrix} = (-1)^{j-m+1}\sqrt{\frac{4(j-m-1)(j-m)(j-m+1)(j+m+2)}{(2j+4)(2j+3)(2j+2)(2j+1)2j}} \quad \text{(A3.36)}$$

$$\begin{pmatrix} j & j+1 & 2 \\ m & -(m+2) & 2 \end{pmatrix} = (-1)^{3j+m}\sqrt{\frac{4(j+m+1)(j+m+2)(j+m+3)(j-m)}{(2j+4)(2j+3)(2j+2)(2j+1)2j}} \quad \text{(A3.37)}$$

$$\begin{pmatrix} j & j-1 & 2 \\ m & -(m+2) & 2 \end{pmatrix} = (-1)^{j-m}\sqrt{\frac{4(j-m-2)(j-m-1)(j-m)(j+m+1)}{(2j+2)(2j+1)2j(2j-1)(2j-2)}} \quad \text{(A3.38)}$$

$$\begin{pmatrix} j+2 & j & 2 \\ m & -(m+2) & 2 \end{pmatrix} = (-1)^{j-m}\sqrt{\frac{(j-m-1)(j-m)(j-m+1)(j-m+2)}{(2j+5)(2j+4)(2j+3)(2j+2)(2j+1)}} \quad \text{(A3.39)}$$

$$\begin{pmatrix} j & j+2 & 2 \\ m & -(m+2) & 2 \end{pmatrix} = (-1)^{3j+m}\sqrt{\frac{(j+m+1)(j+m+2)(j+m+3)(j+m+4)}{(2j+5)(2j+4)(2j+3)(2j+2)(2j+1)}} \quad \text{(A3.40)}$$

$$\begin{pmatrix} j & j-2 & 2 \\ m & -(m+2) & 2 \end{pmatrix} = (-1)^{j-m}\sqrt{\frac{(j-m-3)(j-m-2)(j-m-1)(j-m)}{(2j+1)2j(2j-1)(2j-2)(2j-3)}} \quad \text{(A3.41)}$$

Appendix 4 Properties and Values for Selected 6*j* Symbols (Weissbluth, 1978)

Permutation properties	
$\begin{Bmatrix} j_1 & j_2 & j_3 \\ k_1 & k_2 & k_3 \end{Bmatrix} = \begin{Bmatrix} j_1 & j_3 & j_2 \\ k_1 & k_3 & k_2 \end{Bmatrix} = \begin{Bmatrix} j_2 & j_3 & j_1 \\ k_2 & k_3 & k_1 \end{Bmatrix} = \begin{Bmatrix} j_2 & j_1 & j_3 \\ k_2 & k_1 & k_3 \end{Bmatrix} = \begin{Bmatrix} j_3 & j_1 & j_2 \\ k_3 & k_1 & k_2 \end{Bmatrix} = \begin{Bmatrix} j_3 & j_2 & j_1 \\ k_3 & k_2 & k_1 \end{Bmatrix}$	(A4.1)
$\begin{Bmatrix} j_1 & j_2 & j_3 \\ k_1 & k_2 & k_3 \end{Bmatrix} = \begin{Bmatrix} k_1 & k_2 & j_3 \\ j_1 & j_2 & k_3 \end{Bmatrix} = \begin{Bmatrix} k_1 & j_2 & k_3 \\ j_1 & k_2 & j_3 \end{Bmatrix} = \begin{Bmatrix} j_1 & k_2 & k_3 \\ k_1 & j_2 & j_3 \end{Bmatrix}$	(A4.2)
Triangle condition for $\begin{Bmatrix} j_1 & j_2 & j_3 \\ k_1 & k_2 & k_3 \end{Bmatrix}$	
$\begin{matrix} \delta(j_1j_2j_3) & \delta(j_1k_2k_3) & \delta(k_1j_2k_3) & \delta(k_1k_2j_3) \\ j_1+j_2 \ge j_3 & j_1+k_2 \ge k_3 & k_1+j_2 \ge k_3 & k_1+k_2 \ge j_3 \\ j_1+j_3 \ge j_2 & j_1+k_3 \ge k_2 & k_1+k_3 \ge j_2 & k_1+j_3 \ge k_2 \\ j_2+j_3 \ge j_1 & k_2+k_3 \ge j_1 & j_2+k_3 \ge k_1 & k_2+j_3 \ge k_1 \end{matrix}$	(A4.3)
Selected 6*j* symbols $s = a + b + c$	
$\begin{Bmatrix} a & b & c \\ \frac{1}{2} & c-\frac{1}{2} & b+\frac{1}{2} \end{Bmatrix} = \begin{Bmatrix} a & c & b \\ \frac{1}{2} & b+\frac{1}{2} & c-\frac{1}{2} \end{Bmatrix} = (-1)^s \left[\frac{(s-2b)(s-2c+1)}{(2b+1)(2b+2)2c(2c+1)}\right]^{½}$	(A4.4)
$\begin{Bmatrix} a & b & c \\ \frac{1}{2} & c-\frac{1}{2} & b-\frac{1}{2} \end{Bmatrix} = \begin{Bmatrix} a & c-\frac{1}{2} & b-\frac{1}{2} \\ \frac{1}{2} & b & c \end{Bmatrix} = (-1)^s \left[\frac{(s+1)(s-2a)}{2b(2b+1)2c(2c+1)}\right]^{½}$	(A4.5)
$\begin{Bmatrix} a & b & c \\ 1 & c-1 & b-1 \end{Bmatrix} = (-1)^s \left[\frac{s(s+1)(s-2a-1)(s-2a)}{(2b-1)2b(2b+1)(2c-1)2c(2c+1)}\right]^{½}$	(A4.6)
$\begin{Bmatrix} a & b & c \\ 1 & c-1 & b \end{Bmatrix} = (-1)^s \left[\frac{2(s+1)(s-2a)(s-2b)(s-2c+1)}{2b(2b+1)(2b+2)(2c-1)2c(2c+1)}\right]^{½}$	(A4.7)

$$\begin{Bmatrix} a & b & c \\ 1 & c-1 & b+1 \end{Bmatrix} = (-1)^s \left[\frac{(s-2b-1)(s-2b)(s-2c+1)(s-2c+2)}{(2b+1)(2b+2)(2b+3)(2c-1)2c(2c+1)}\right]^{1/2} \quad \text{(A4.8)}$$

$$\begin{Bmatrix} a & b & c \\ 1 & c & b \end{Bmatrix} = (-1)^s \frac{2[a(a+1)-b(b+1)-c(c+1)]}{[2b(2b+1)(2b+2)2c(2c+1)(2c+2)]^{1/2}} \quad \text{(A4.9)}$$

Appendix 5 Allowed *LS* Coupling Terms for Equivalent d^2 Electrons

The following table shows all possible values of m_{l1}, m_{s1}, m_{l2}, and m_{s2} that are allowed given the restrictions imposed by the Pauli exclusion principle for equivalent d^2 electrons.

m_{l1}	m_{s1}	m_{l2}	m_{s2}	$M_L = m_{l1} + m_{l2}$	$M_S = m_{s1} + m_{s2}$	$M = M_L + M_S$
−2	−1/2	−2	+1/2	−4	0	−4
		−1	+1/2	−3	0	−3
		−1	−1/2	−3	−1	−4
		0	+1/2	−2	0	−2
		0	−1/2	−2	−1	−3
		+1	+1/2	−1	0	−1
		+1	−1/2	−1	−1	−2
		+2	+1/2	0	0	0
		+2	−1/2	0	−1	−1
−2	+1/2	−1	+1/2	−3	+1	−2
		−1	−1/2	−3	0	−3
		0	+1/2	−2	+1	−1
		0	−1/2	−2	0	−2
		+1	+1/2	−1	+1	0
		+1	−1/2	−1	0	−1
		+2	+1/2	0	+1	+1
		+2	−1/2	0	0	0

(*cont.*)

m_{l1}	m_{s1}	m_{l2}	m_{s2}	$M_L = m_{l1} + m_{l2}$	$M_S = m_{s1} + m_{s2}$	$M = M_L + M_S$
−1	−1/2	−1	+1/2	−2	0	−2
		0	+1/2	−1	0	−1
		0	−1/2	−1	−1	−2
		+1	+1/2	0	0	0
		+1	−1/2	0	−1	−1
		+2	+1/2	+1	0	+1
		+2	−1/2	+1	−1	0
−1	+1/2	0	+1/2	−1	+1	−0
		0	−1/2	−1	0	−1
		+1	+1/2	0	+1	+1
		+1	−1/2	0	0	0
		+2	+1/2	+1	+1	+2
		+2	−1/2	+1	0	+1
0	−1/2	0	+1/2	0	0	0
		+1	+1/2	+1	0	+1
		+1	−1/2	+1	−1	0
		+2	+1/2	+2	0	+2
		+2	−1/2	+2	−1	+1
0	+1/2	+1	+1/2	+1	+1	+2
		+1	−1/2	+1	0	+1
		+2	+1/2	+2	+1	+3
		+2	−1/2	+2	0	+2
+1	−1/2	+1	+1/2	+2	0	+2
		+2	+1/2	+3	0	+3
		+2	−1/2	+3	−1	+2
+1	+1/2	+2	+1/2	+3	+1	+4
		+2	−1/2	+3	0	+3
+2	−1/2	+2	+1/2	+4	0	+4

Counting the number of cases with particular values of M_L and M_S, we obtain the following matrix:

M_L \ M_S	**−1**	**0**	**+1**
−4		1	
−3	1	2	1
−2	1	3	1
−1	2	4	2
0	2	5	2
+1	2	4	2
+2	1	3	2
+3	1	2	1
+4		1	

This matrix in turn can be represented as the sum of the following submatrices:

M_L \ M_S	**−1**	**0**	**+1**
−4		1	
−3		1	
−2		1	
−1		1	
0		1	
+1		1	
+2		1	
+3		1	
+4		1	

M_L \ M_S	**−1**	**0**	**+1**
−4			
−3	1	1	1
−2	1	1	1
−1	1	1	1
0	1	1	1
+1	1	1	1
+2	1	1	1
+3	1	1	
+4			

M_L \ M_S	**−1**	**0**	**+1**
−4			
−3			
−2		1	
−1		1	
0		1	
+1		1	
+2		1	
+3			
+4			

M_L \ M_S	**−1**	**0**	**+1**
−4			
−3			
−2			
−1	1	1	1
0	1	1	1
+1	1	1	1
+2			
+3			
+4			

M_L \ M_S	**−1**	**0**	**+1**
−4			
−3			
−2			
−1			
0		1	
+1			
+2			
+3			
+4			

These submatrices in order correspond to ^1G, ^3F, ^1D, ^3P, and ^1S terms.

Appendix 6 Derivation of the Higher-Order Density Matrix Elements for Doublet and Triplet Electronic Levels

A6.1 Doublet Electronic Levels

Rotational Energy: Second-Order Correction for Centrifugal Distortion

$$\frac{H_{11R}}{B_v}=F_{11R}=z+\Lambda \quad F_{22R}=z-\Lambda \quad F_{12R}=F_{21R}=-\left(z-\Lambda^2\right)^{1/2}$$
$$\frac{H_{11CD}}{-D_v}=F_{11CD}=z^2+(1+2\Lambda)z,\ F_{22CD}=z^2+(1-2\Lambda)z,$$
$$F_{12CD}=F_{21CD}=-2z\left(z-\Lambda^2\right)^{1/2}$$

$$\begin{aligned}H_{11CH}&=\frac{1}{2}H_v\left[(F_{CD}F_R)_{11}+(F_RF_{CD})_{11}\right]=H_v(F_{SR}F_R)_{11}=H_v(F_{11CD}F_{11R}+F_{12CD}F_{21R})\\&=H_v\left\{\left[z^2+(1+2\Lambda)z\right](z+\Lambda)+2z\left(z-\Lambda^2\right)\right\}=H_v\left[z^3+(3+3\Lambda)z^2+\Lambda z\right]\end{aligned}$$

$$\begin{aligned}H_{22CH}&=\frac{1}{2}H_v\left[(F_{CD}F_R)_{22}+(F_RF_{CD})_{22}\right]=H_v(F_{CD}F_R)_{22}=H_v(F_{22CD}F_{22R}+F_{21CD}F_{12R})\\&=H_v\left\{\left[z^2+(1-2\Lambda)z\right](z-\Lambda)+2z\left(z-\Lambda^2\right)\right\}=H_v\left[z^3+(3-3\Lambda)z^2-\Lambda z\right]\end{aligned}$$

$$\begin{aligned}H_{12CH}&=\frac{1}{2}H_v(F_{12CD}F_{22R}+F_{11CD}F_{12R}+F_{12R}F_{22CD}+F_{11R}F_{12CD})\\&=\frac{1}{2}H_v\left(z-\Lambda^2\right)^{1/2}\left\{-2z(z-\Lambda)-\left[z^2+(1+2\Lambda)z\right]-\left[z^2+(1-2\Lambda)z\right]-2z(z+\Lambda)\right\}\\&=-H_v\left(z-\Lambda^2\right)^{1/2}\left[3z^2+z\right].\end{aligned}\tag{A6.1}$$

Rotational Energy: Third -Order Correction for Centrifugal Distortion

$$\frac{H_{11R}}{B_v}=F_{11R}=z+\Lambda \quad F_{22R}=z-\Lambda \quad F_{12R}=F_{21R}=-\left(z-\Lambda^2\right)^{1/2}$$
$$\frac{H_{11CH}}{-D_v}=F_{11CD}=z^3+(3+3\Lambda)z^2+\Lambda z,\ F_{22CH}=z^3+(3-3\Lambda)z^2-\Lambda z,$$
$$F_{12CH}=F_{21CH}=-\left(z-\Lambda^2\right)^{1/2}\left(3z^2+z\right)$$

$$
\begin{aligned}
H_{11CL} &= -\frac{1}{2}L_v\left[(F_{CH}F_R)_{11} + (F_RF_{CH})_{11}\right] = -L_v(F_{CH}F_R)_{11} \\
&= -L_v(F_{11CH}F_{11R} + F_{12CH}F_{21R}) \\
&= -L_v\left\{\left[z^3 + (3+3\Lambda)z^2 + \Lambda z\right](z+\Lambda) + (3z^2+z)(z-\Lambda^2)\right\} \\
&= -L_v\left[z^4 + (6+3\Lambda)z^3 + (1+4\Lambda)z^2\right]
\end{aligned}
$$

$$
\begin{aligned}
H_{22CL} &= -\frac{1}{2}L_v\left[(F_{CH}F_R)_{22} + (F_RF_{CH})_{22}\right] = -L_v(F_{CH}F_R)_{22} \\
&= -L_v(F_{22CH}F_{22R} + F_{21CH}F_{12R}) \\
&= -L_v\left\{\left[z^3 + (3+3\Lambda)z^2 - \Lambda z\right](z-\Lambda) + (3z^2+z)(z-\Lambda^2)\right\} \\
&= -L_v\left[z^4 + (6+3\Lambda)z^3 + (1+3\Lambda-7\Lambda^2)z^2\right]
\end{aligned}
$$

$$
\begin{aligned}
H_{12CL} &= -\frac{1}{2}L_v\left[(F_{CH}F_R)_{12} + (F_RF_{CH})_{12}\right] \\
&= -\frac{1}{2}L_v(F_{11CH}F_{12R} + F_{12CH}F_{22R} + F_{11R}F_{12CH} + F_{12R}F_{22CH}) \\
&= -\frac{1}{2}L_v(z-\Lambda^2)^{1/2}\left\{-\left[z^3 + (3+3\Lambda)z^2 + \Lambda z\right] - (3z^2+z)(z-\Lambda)\right\} \\
&\quad -\frac{1}{2}L_v(z-\Lambda^2)^{1/2}\left\{-(z+\Lambda)(3z^2+z) - \left[z^3 + (3-3\Lambda)z^2 - \Lambda z\right]\right\} \\
&= L_v(z-\Lambda^2)^{1/2}\left[4z^3 + 4z^2\right].
\end{aligned}
\tag{A6.2}
$$

Spin–Rotation Interaction: First-Order Correction for Centrifugal Distortion

$$
\frac{H_{11R}}{B_v} = F_{11R} = z+\Lambda \quad F_{22R} = z-\Lambda \quad F_{12R} = F_{21R} = -(z-\Lambda^2)^{1/2}
$$

$$
\frac{H_{11SR}}{\gamma_v} = F_{11SR} = -\frac{1}{2}(\Lambda+1) \quad F_{22SR} = \frac{1}{2}(\Lambda-1) \quad F_{12SR} = F_{21SR} = \frac{1}{2}(z-\Lambda^2)^{1/2}
$$

$$
\begin{aligned}
H_{11SRD} &= \frac{1}{2}\gamma_{Dv}\left[(F_{SR}F_R)_{11} + (F_RF_{SR})_{11}\right] = \gamma_{Dv}(F_{11SR}F_{11R} + F_{12SR}F_{21R}) \\
&= \gamma_{Dv}\left[-\frac{1}{2}(\Lambda+1)(z+\Lambda) - \frac{1}{2}(z-\Lambda^2)\right] = -\frac{1}{2}\gamma_{Dv}(2z + \Lambda z + \Lambda)
\end{aligned}
$$

$$
\begin{aligned}
H_{22SRD} &= \frac{1}{2}\gamma_{Dv}\left[(F_{SR}F_R)_{22} + (F_RF_{SR})_{22}\right] = \gamma_{Dv}(F_{22SR}F_{22R} + F_{21SR}F_{12R}) \\
&= \gamma_{Dv}\left[\frac{1}{2}(\Lambda-1)(z-\Lambda) - \frac{1}{2}(z-\Lambda^2)\right] = \frac{1}{2}\gamma_{Dv}(-2z + z\Lambda + \Lambda)
\end{aligned}
$$

$$
\begin{aligned}
H_{12SRD} &= \frac{1}{2}\gamma_{Dv}(F_{12SR}F_{22R} + F_{11SR}F_{12R} + F_{12R}F_{22SR} + F_{11R}F_{12SR}) \\
&= \frac{1}{2}\gamma_{Dv}(z-\Lambda^2)^{1/2}\left[\frac{1}{2}(z-\Lambda) + \frac{1}{2}(\Lambda+1) - \frac{1}{2}(\Lambda-1) + \frac{1}{2}(z+\Lambda)\right] \\
&= \frac{1}{2}\gamma_{Dv}(z-\Lambda^2)^{1/2}(z+1)
\end{aligned}
\tag{A6.3}
$$

Spin–Rotation Interaction: Second-Order Correction for Centrifugal Distortion

$$\frac{H_{11SRD}}{\gamma_v} = F_{11SRD} = -\frac{1}{2}(2z + \Lambda z + \Lambda) \quad F_{22SRD} = \frac{1}{2}(-2z + z\Lambda + \Lambda),$$
$$F_{12SR} = F_{21SR} = \frac{1}{2}\left(z - \Lambda^2\right)^{1/2}(z+1)$$

$$\begin{aligned} H_{11SRH} &= \frac{1}{2}\gamma_{Hv}\left[(F_{SRD}F_R)_{11} + (F_RF_{SRD})_{11}\right] = \gamma_{Hv}(F_{11SRD}F_{11R} + F_{12SRD}F_{21R}) \\ &= \gamma_{Hv}\left[-\frac{1}{2}(2z + \Lambda z + \Lambda)(z+\Lambda) - \frac{1}{2}(z - \Lambda^2)(z+1)\right] \\ &= -\frac{1}{2}\gamma_{Hv}\left[z^2(3+\Lambda) + z(1+3\Lambda)\right] \end{aligned}$$

$$\begin{aligned} H_{22SRH} &= \frac{1}{2}\gamma_{Hv}\left[(F_{SRD}F_R)_{22} + (F_RF_{SRD})_{22}\right] = \gamma_{Hv}(F_{22SRD}F_{22R} + F_{21SRD}F_{12R}) \\ &= \gamma_{Hv}\left[-\frac{1}{2}(-2z + \Lambda z + \Lambda)(z-\Lambda) - \frac{1}{2}(z - \Lambda^2)(z+1)\right] \\ &= -\frac{1}{2}\gamma_{Hv}\left[z^2(-1+\Lambda) + z\left(1+3\Lambda-2\Lambda^2\right) - 2\Lambda^2\right] \end{aligned}$$

$$\begin{aligned} H_{12SRH} &= \frac{1}{2}\gamma_{Hv}(F_{12SRD}F_{22R} + F_{11SRD}F_{12R} + F_{12R}F_{22SRD} + F_{11R}F_{12SRD}) \\ &= \frac{1}{4}\gamma_{Hv}\left(z - \Lambda^2\right)^{1/2}\left[(z+1)(z-\Lambda) - (2z + \Lambda z + \Lambda) - (-2z + z\Lambda + \Lambda)\right. \\ &\quad \left. + (z+\Lambda)(z+1)\right] = \frac{1}{2}\gamma_{Hv}\left(z - \Lambda^2\right)^{1/2}\left[z^2 + z(1-\Lambda) - \Lambda\right]. \end{aligned} \tag{A6.4}$$

Spin–Orbit Interaction: First-Order Correction for Centrifugal Distortion

$$\frac{H_{11R}}{B_v} = F_{11R} = z + \Lambda \quad F_{22R} = z - \Lambda \quad F_{12R} = F_{21R} = -\left(z - \Lambda^2\right)^{1/2}$$

$$\frac{H_{11SO}}{A_v} = F_{11SO} = -\frac{1}{2}\Lambda \quad F_{22SO} = \frac{1}{2}\Lambda \quad F_{12SO} = F_{21S}) = 0$$

$$H_{11SOD} = \frac{1}{2}A_{Dv}\left[(F_{SO}F_R)_{11} + (F_RF_{SO})_{11}\right] = A_{Dv}F_{11SO}F_{11R} = -\frac{1}{2}A_{Dv}\Lambda(z+\Lambda)$$

$$H_{22SOD} = \frac{1}{2}A_{Dv}\left[(F_{SO}F_R)_{22} + (F_RF_{SO})_{22}\right] = A_{Dv}\left[\frac{1}{2}\Lambda(z-\Lambda)\right] = \frac{1}{2}A_{Dv}\Lambda(z-\Lambda)$$

$$\begin{aligned} H_{12SOD} &= \frac{1}{2}A_{Dv}(F_{12SO}F_{22R} + F_{11SO}F_{12R} + F_{12R}F_{22SO} + F_{11R}F_{12SO}) \\ &= -\frac{1}{2}A_{Dv}\left(z - \Lambda^2\right)^{1/2}\left(-\frac{1}{2}\Lambda + \frac{1}{2}\Lambda\right) = 0. \end{aligned} \tag{A6.5}$$

Spin–Orbit Interaction: Second-Order Correction for Centrifugal Distortion

$$\frac{H_{11R}}{B_v}=F_{11R}=z+\Lambda \quad F_{22R}=z-\Lambda \quad F_{12R}=F_{21R}=-\left(z-\Lambda^2\right)^{1/2}$$
$$\frac{H_{11SOD}}{A_{Dv}}=F_{11SOD}=-\frac{1}{2}\Lambda(z+\Lambda) \quad F_{22SOD}=\frac{1}{2}\Lambda(z-\Lambda) \quad F_{12SOD}=F_{21SOD}=0$$
$$H_{11SOH}=\frac{1}{2}A_{Hv}\left[(F_{SOD}F_R)_{11}+(F_RF_{SOD})_{11}\right]=A_{Hv}F_{11SOD}F_{11R}=-\frac{1}{2}A_{Hv}\Lambda(z+\Lambda)^2$$
$$H_{22SOH}=\frac{1}{2}A_{Hv}\left[(F_{SOD}F_R)_{22}+(F_RF_{SOD})_{22}\right]=A_{Hv}F_{22SO}F_{22R}=\frac{1}{2}A_{Hv}\Lambda(z-\Lambda)^2$$
$$\begin{aligned}H_{12SOH}&=\frac{1}{2}A_{Hv}(F_{12SOD}F_{22R}+F_{11SOD}F_{12R}+F_{12R}F_{22SOD}+F_{11R}F_{12SOD})\\&=\frac{1}{4}A_{Hv}\left(z-\Lambda^2\right)^{1/2}[\Lambda(z+\Lambda)-\Lambda(z-\Lambda)]=\frac{1}{2}A_{Hv}\left(z-\Lambda^2\right)^{1/2}\Lambda^2.\end{aligned}$$

(A6.6)

Lambda-Doubling: First-Order Correction for Centrifugal Distortion

$$F_{11q\Lambda e/f}=\mp\sqrt{z} \quad F_{12q\Lambda e/f}=\pm\frac{1}{2}\sqrt{z(z-1)} \quad F_{11p\Lambda e/f}=\mp\frac{1}{2}\sqrt{z}$$
$$\frac{H_{11R}}{B_v}=F_{11R}=z+\Lambda \quad F_{22R}=z-\Lambda \quad F_{12R}=F_{21R}=-\left(z-\Lambda^2\right)^{1/2}$$
$$\begin{aligned}H_{11qD\Lambda e/f}&=\frac{1}{2}q_{Dv}\left[\left(F_{q\Lambda e/f}F_R\right)_{11}+\left(F_RF_{q\Lambda e/f}\right)_{11}\right]\\&=q_{Dv}\left(F_{11q\Lambda e/f}F_{11R}+F_{12q\Lambda e/f}F_{21R}\right)\\&=q_{Dv}\left[\mp\frac{1}{2}\sqrt{z}(z+1)\mp\frac{1}{2}\sqrt{z(z-1)}(z-1)^{1/2}\right]\\&=\mp\frac{1}{2}q_{Dv}\sqrt{z}[(z+1)+(z-1)]\end{aligned}$$
$$\begin{aligned}H_{11pD\Lambda e/f}&=\frac{1}{2}p_{Dv}\left[\left(F_{p\Lambda e/f}F_R\right)_{11}+\left(F_RF_{p\Lambda e/f}\right)_{11}\right]=p_{Dv}F_{11q\Lambda e/f}F_{11R}\\&=p_{Dv}\left[\mp\frac{1}{2}\sqrt{z}(z+1)\right]=\mp\frac{1}{2}p_{Dv}\sqrt{z}(z+1)\end{aligned}$$
$$\begin{aligned}H_{12qD\Lambda e/f}&=\frac{1}{2}q_{Dv}\left[\left(F_{q\Lambda e/f}F_R\right)_{12}+\left(F_RF_{q\Lambda e/f}\right)_{12}\right]\\&=\frac{1}{2}q_{Dv}\left(F_{11q\Lambda e/f}F_{12R}+F_{12q\Lambda e/f}F_{22R}+F_{11R}F_{12q\Lambda e/f}\right)\\&=\frac{1}{2}q_{Dv}\left[\pm\sqrt{z(z-1)}\pm\frac{1}{2}\sqrt{z(z-1)}(z-1)\pm\frac{1}{2}\sqrt{z(z-1)}(z+1)\right]\\&=\pm\frac{1}{2}q_{Dv}\sqrt{z(z-1)}(1+z)\end{aligned}$$
$$\begin{aligned}H_{12pD\Lambda e/f}&=\frac{1}{2}p_{Dv}\left[\left(F_{p\Lambda e/f}F_R\right)_{12}+\left(F_RF_{p\Lambda e/f}\right)_{12}\right]\\&=p_{Dv}F_{11p\Lambda e/f}F_{12R}=\pm\frac{1}{4}p_{Dv}\sqrt{z(z-1)}\end{aligned}$$
$$\begin{aligned}H_{22qD\Lambda e/f}&=\frac{1}{2}q_{Dv}\left[\left(F_{p\Lambda e/f}F_R\right)_{22}+\left(F_RF_{p\Lambda e/f}\right)_{22}\right]=q_{Dv}F_{21q\Lambda e/f}F_{12R}\\&=q_{Dv}\left[\mp\frac{1}{2}\sqrt{z}(z-1)\right]=\mp\frac{1}{2}q_{Dv}\sqrt{z}(z-1).\end{aligned}$$

(A6.7)

A6.2 Triplet Electronic Levels

Rotational Energy: First-Order Correction for Centrifugal Distortion

$$\frac{H_{11R}}{B_v} = F_{11R} = x + 2\Lambda,\ F_{22R} = x + 2,\ F_{33R} = x - 2\Lambda$$
$$F_{12R} = -[2x - 2\Lambda(\Lambda - 1)]^{1/2},\ F_{23R} = -[2x - 2\Lambda(\Lambda + 1)]^{1/2},\ F_{13R} = 0$$

$$\begin{aligned} H_{11CD} &= -D_v[F_{11R}F_{11R} + F_{12R}F_{21R}] = -D_v\left[(x + 2\Lambda)^2 + 2x - 2\Lambda(\Lambda - 1)\right] \\ &= -D_v\left[x^2 + 4\Lambda x + 4\Lambda^2 + 2x - 2\Lambda(\Lambda - 1)\right] \\ &= -D_v\left[x^2 + (4\Lambda + 2)x + 2\Lambda^2 + 2\Lambda\right] \end{aligned}$$

$$\begin{aligned} H_{22CD} &= -D_v(F_{22R}F_{22R} + F_{21R}F_{12R} + F_{32R}F_{32R}) \\ &= -D_v\left[(x + 2)^2 + 2x - 2\Lambda(\Lambda - 1)\right. \\ &\quad \left. + 2x - 2\Lambda(\Lambda + 1)\right] = -D_v\left[x^2 + 8x + 4 - 4\Lambda^2\right] \end{aligned}$$

$$\begin{aligned} H_{33CD} &= -D_v(F_{33R}F_{33R} + F_{32R}F_{23R}) = -D_v\left[(x - 2\Lambda)^2 + 2x - 2\Lambda(\Lambda + 1)\right] \\ &= -D_v\left[x^2 + (-4\Lambda + 2)x + 2\Lambda^2 - 2\Lambda\right] \end{aligned}$$

$$\begin{aligned} H_{12CD} &= -D_v(F_{11R}F_{22R} + F_{12R}F_{22R}) = D_v\sqrt{2x - 2\Lambda(\Lambda - 1)}[x + 2\Lambda + x + 2] \\ &= 2D_v\sqrt{2x - 2\Lambda(\Lambda - 1)}[x + (\Lambda + 1)] \end{aligned}$$

$$\begin{aligned} H_{23CD} &= -D_v(F_{22R}F_{23R} + F_{23R}F_{33R}) = D_v\sqrt{2x - 2\Lambda(\Lambda + 1)}[x + 2 + x - 2\Lambda] \\ &= 2D_v\sqrt{2x - 2\Lambda(\Lambda + 1)}[x - (\Lambda - 1)] \end{aligned}$$

$$\begin{aligned} H_{13CD} &= -D_vF_{12R}F_{23R} = -D_v\sqrt{2x - 2\Lambda(\Lambda + 1)}\sqrt{2x - 2\Lambda(\Lambda - 1)} \\ &= -2D_v\sqrt{x^2 - 2x\Lambda^2 + \Lambda^2(\Lambda^2 - 1)}. \end{aligned} \tag{A6.8}$$

Rotational Energy: Second-Order Correction for Centrifugal Distortion

$$\frac{H_{11R}}{B_v} = F_{11R} = x + 2\Lambda,\ F_{22R} = x + 2,\ F_{33R} = x - 2\Lambda$$
$$F_{12R} = -\sqrt{2x - 2\Lambda(\Lambda - 1)},\ F_{23R} = -\sqrt{2x - 2\Lambda(\Lambda + 1)},\ F_{13R} = 0$$
$$\frac{H_{11CD}}{-D_v} = F_{11CD} = x^2 + (4\Lambda + 2)x + 2\Lambda(\Lambda + 1),\ F_{22CD} = x^2 + 8x + 4(1 - \Lambda^2),$$

$$F_{33CD} = x^2 + (-4\Lambda + 2)x + 2\Lambda(\Lambda - 1), F_{12CD} = -2\sqrt{2x - 2\Lambda(\Lambda - 1)}[x + \Lambda + 1],$$
$$F_{23R} = -2\sqrt{2x - 2\Lambda(\Lambda + 1)}[x - \Lambda + 1],\ F_{13R} = 2\sqrt{x^2 - 2x\Lambda^2 + \Lambda^2(\Lambda^2 - 1)}$$

$$
\begin{aligned}
H_{11CH} &= H_{\rm v}[F_{11R}F_{11CD} + F_{12R}F_{21CD} + F_{11CD}F_{11R} + F_{12CD}F_{21R}] \\
&= H_{\rm v}\{(x+2\Lambda)\left[x^2 + (4\Lambda+2)x + 2\Lambda(\Lambda+1)\right] \\
&\quad + 2[2x - 2\Lambda(\Lambda-1)][x + (\Lambda+1)]\} \\
&= H_{\rm v}\{\left[x^3 + (6\Lambda+2)x^2 + \left(10\Lambda^2 + 6\Lambda\right)x + 4\Lambda^2(\Lambda+1)\right] \\
&\quad + \left[4x^2 + x\left(4 + 8\Lambda - 4\Lambda^2\right) - 4\Lambda^3 + 4\Lambda\right]\} \\
&= H_{\rm v}\left[x^3 + (6\Lambda+6)x^2 + \left(6\Lambda^2 + 14\Lambda + 4\right)x + 4\Lambda^2 + 4\Lambda\right]
\end{aligned}
$$

$$
\begin{aligned}
H_{22CH} &= H_{\rm v}(F_{22CD}F_{22R} + F_{21CD}F_{12R} + F_{32CD}F_{32R}) \\
&= H_{\rm v}\{(x+2)\left[x^2 + 8x + 4\left(1-\Lambda^2\right)\right] + 2[2x - 2\Lambda(\Lambda-1)][x + \Lambda + 1] \\
&\quad + 2[2x - 2\Lambda(\Lambda+1)][x - \Lambda + 1] \\
&= H_{\rm v}\{\left[x^3 + 10x^2 + \left(-4\Lambda^2 + 20\right)x - 8\Lambda^2 + 8\right] \\
&\quad + \left[4x^2 + x\left(4 + 8\Lambda - 4\Lambda^2\right) - 4\Lambda^3 + 4\Lambda\right] \\
&\quad + \left[4x^2 + x\left(4 - 8\Lambda - 4\Lambda^2\right) + 4\Lambda^3 - 4\Lambda\right]\} \\
&= H_{\rm v}\left[x^3 + 18x^2 + \left(-12\Lambda^2 + 28\right)x - 8\Lambda^2 + 8\right]
\end{aligned}
$$

$$
\begin{aligned}
H_{33CH} &= H_{\rm v}[F_{33CD}F_{33R} + F_{32R}F_{23CD}] \\
&= H_{\rm v}\{(x-2\Lambda)\left[x^2 + (-4\Lambda+2)x + 2\Lambda(\Lambda-1)\right] \\
&\quad + 2[2x - 2\Lambda(\Lambda+1)][x - (\Lambda-1)]\} \\
&= H_{\rm v}\{\left[x^3 + (-6\Lambda+2)x^2 + \left(10\Lambda^2 - 6\Lambda\right)x - 4\Lambda^2(\Lambda-1)\right] \\
&\quad + \left[4x^2 + \left(-4\Lambda^2 - 8\Lambda + 4\right)x + 4\Lambda^3 - 4\Lambda\right]\} \\
&= H_{\rm v}\left[x^3 + (-6\Lambda+6)x^2 + \left(6\Lambda^2 - 14\Lambda + 4\right)x\right]
\end{aligned}
$$

$$
\begin{aligned}
H_{12CH} &= \frac{1}{2}H_{\rm v}(F_{12CD}F_{22R} + F_{11R}F_{12CD} + F_{12R}F_{22CD} + F_{11CD}F_{12R}) \\
&= H_{\rm v}\{-4\sqrt{2x - 2\Lambda(\Lambda-1)}[x + (\Lambda+1)]^2 \\
&\quad - \sqrt{2x - 2\Lambda(\Lambda-1)}\left[2x^2 + (4\Lambda+10)x - 2\Lambda^2 + 2\Lambda + 4\right] \\
&= -\frac{1}{2}H_{\rm v}\sqrt{2x - 2\Lambda(\Lambda-1)}\left[6x^2 + (12\Lambda+18)x + 2\Lambda^2 + 10\Lambda + 8\right] \\
&= -H_{\rm v}\sqrt{2x - 2\Lambda(\Lambda-1)}\left[3x^2 + (6\Lambda+9)x + \Lambda^2 + 5\Lambda + 4\right]
\end{aligned}
$$

$$
\begin{aligned}
H_{23CH} &= \frac{1}{2}H_{\rm v}(F_{23CD}F_{33R} + F_{22R}F_{23CD} + F_{23R}F_{33CD} + F_{22CD}F_{23R}) \\
&= \frac{1}{2}H_{\rm v}\{-4\sqrt{2x - 2\Lambda(\Lambda+1)}[x - (\Lambda-1)]^2 \\
&\quad - \sqrt{2x - 2\Lambda(\Lambda+1)}\left[2x^2 + (-4\Lambda+10)x - 2\Lambda^2 - 2\Lambda + 4\right] \\
&= -\frac{1}{2}H_{\rm v}\sqrt{2x - 2\Lambda(\Lambda+1)}\left[6x^2 + (-12\Lambda+18)x + 2\Lambda^2 - 10\Lambda + 8\right] \\
&= -H_{\rm v}\sqrt{2x - 2\Lambda(\Lambda+1)}\left[3x^2 + (-6\Lambda+9)x + \Lambda^2 - 5\Lambda + 4\right]
\end{aligned}
$$

$$
\begin{aligned}
H_{13CH} &= \frac{1}{2}H_{\rm v}(F_{13CD}F_{33R} + F_{11R}F_{13CD} + F_{12R}F_{23CD} + F_{12CD}F_{23R}) \\
&= \frac{1}{2}H_{\rm v}\left\{4x\sqrt{x^2 - 2x\Lambda^2 + \Lambda^2\left(\Lambda^2 - 1\right)} + 2\sqrt{2x - 2\Lambda(\Lambda-1)}\sqrt{2x - 2\Lambda(\Lambda+1)}[2x+2]\right\} \\
&= H_{\rm v}\sqrt{x^2 - 2x\Lambda^2 + \Lambda^2\left(\Lambda^2 - 1\right)}(6x+4).
\end{aligned}
\tag{A6.9}
$$

Spin–Rotation Interaction: First-Order Correction for Centrifugal Distortion

$$\frac{H_{11R}}{B_v} = F_{11R} = x + 2\Lambda,\ F_{22R} = x + 2,\ F_{33R} = x - 2\Lambda$$
$$F_{12R} = -\sqrt{2x - 2\Lambda(\Lambda - 1)},\ F_{23R} = -\sqrt{2x - 2\Lambda(\Lambda + 1)},\ F_{13R} = 0$$
$$\frac{H_{11SR}}{\gamma_v} = F_{11SR} = -(\Lambda + 1),\ F_{22SR} = -2,\ F_{33SR} = (\Lambda - 1),$$
$$F_{12SR} = \frac{1}{2}\sqrt{2x - 2\Lambda(\Lambda - 1)},\ F_{23SR} = \frac{1}{2}\sqrt{2x - 2\Lambda(\Lambda + 1)},\ F_{13SR} = 0$$

$$H_{11SRD} = \gamma_{Dv}(F_{11SR}F_{11R} + F_{12SR}F_{21R}) = \gamma_{Dv}[-(\Lambda + 1)(x + 2\Lambda) \\ -x + \Lambda(\Lambda - 1)] = -[(\Lambda + 2)x + \Lambda^2 + 3\Lambda]$$

$$H_{22SRD} = \gamma_{Dv}(F_{22SR}F_{22R} + F_{21SR}F_{12R} + F_{23SR}F_{32R}) = \gamma_{Dv}[-2x - 4 \\ -x + \Lambda(\Lambda - 1) - x + \Lambda(\Lambda + 1)] = -[4x - 2\Lambda^2 + 4]$$

$$H_{33SRD} = \gamma_{Dv}(F_{33SR}F_{33R} + F_{32SR}F_{23R}) = \gamma_{Dv}[(\Lambda - 1)(x - 2\Lambda) - x + \Lambda(\Lambda + 1)] \\ = [(\Lambda - 2)x - \Lambda^2 + 3\Lambda]$$

$$H_{12SRD} = \frac{1}{2}\gamma_{Dv}(F_{11SR}F_{12R} + F_{12R}F_{22SR} + F_{11R}F_{12SR} + F_{12SR}F_{22R}) \\ = \frac{1}{2}\gamma_{Dv}\sqrt{2x - 2\Lambda(\Lambda - 1)}[(\Lambda + 1) + 2 + x + (\Lambda + 1)] \\ = \frac{1}{2}\gamma_{Dv}\sqrt{2x - 2\Lambda(\Lambda - 1)}[x + 2\Lambda + 4]$$

$$H_{23SRD} = \frac{1}{2}\gamma_{Dv}(F_{22SR}F_{23R} + F_{23R}F_{33SR} + F_{22R}F_{23SR} + F_{23SR}F_{33R}) \\ = \frac{1}{2}\gamma_{Dv}\sqrt{2x - 2\Lambda(\Lambda + 1)}[-\Lambda + 3 + x - (\Lambda - 1)] \\ = \frac{1}{2}\gamma_{Dv}\sqrt{2x - 2\Lambda(\Lambda + 1)}[x + 4 - 2\Lambda]$$

$$H_{13SRD} = \frac{1}{2}\gamma_{Dv}(F_{12SR}F_{23R} + F_{12R}F_{23SR}) = -\gamma_{Dv}\sqrt{[x - \Lambda(\Lambda + 1)][x - \Lambda(\Lambda - 1)]}. \tag{A6.10}$$

Spin–Spin Interaction: First-Order Correction for Centrifugal Distortion

$$\frac{H_{11R}}{B_v} = F_{11R} = x + 2\Lambda,\ F_{22R} = x + 2,\ F_{33R} = x - 2\Lambda$$
$$F_{12R} = -\sqrt{2x - 2\Lambda(\Lambda - 1)},\ F_{23R} = -\sqrt{2x - 2\Lambda(\Lambda + 1)},\ F_{13R} = 0$$
$$\frac{H_{11SS}}{\lambda_v} = F_{11SS} = 2/3,\ F_{22SS} = -4/3\ F_{33SS} = 2/3, F_{12SS} = 0,\ F_{23SS} = 0,\ F_{13SS} = 0$$
$$H_{11SSD} = \gamma_{Dv}(F_{11SS}F_{11R}) = \lambda_{Dv}[2(x + 2\Lambda)/3]$$
$$H_{22SSD} = \lambda_{Dv}(F_{22SS}F_{22R}) = \lambda_{Dv}[-4(x + 2)/3]$$
$$H_{33SSD} = \lambda_{Dv}(F_{33SS}F_{33R}) = \lambda_{Dv}[2(x - 2\Lambda)/3]$$

$$H_{12SSD} = \frac{1}{2}\lambda_{Dv}(F_{11SS}F_{12R} + F_{12R}F_{22SS} + F_{11R}F_{12SS} + F_{12SS}F_{22R})$$
$$= -\frac{1}{2}\lambda_{Dv}\sqrt{2x - 2\Lambda(\Lambda - 1)}[2/3 - 4/3] = \lambda_{Dv}\sqrt{2x - 2\Lambda(\Lambda - 1)}/3$$

$$H_{23SSD} = \frac{1}{2}\lambda_{Dv}(F_{22SS}F_{23R} + F_{23R}F_{33SS} + F_{22R}F_{23SS} + F_{23SS}F_{33R})$$
$$= \frac{1}{2}\lambda_{Dv}\sqrt{2x - 2\Lambda(\Lambda + 1)}[-4/3 + 2/3] = \lambda_{Dv}\sqrt{2x - 2\Lambda(\Lambda + 1)}/3$$

$$H_{13SSD} = \frac{1}{2}\lambda_{Dv}(F_{13SS}F_{33R} + F_{13R}F_{33SS} + F_{12SR}F_{23R} + F_{12R}F_{23SR}) = 0. \tag{A6.11}$$

Appendix 7 Einstein Coefficients for Spontaneous Emission for the $X^2\Pi$–$A^2\Sigma^+$ (0,0) Bands of OH and NO and the $X^3\Sigma^-$–$A^3\Pi$ (0,0) Band of NH

OH $X^2\Pi - A^2\Sigma^+$ $(0,0)$ Band: $F_1 \rightarrow F_1'$ and $F_1 \rightarrow F_2'$ Transitions

NX	PP11	QQ11	RR11	QP21	RQ21	SR21
1.0	0.86480E+06	0.41369E+06	0.81134E+05	0.57544E+06	0.18188E+06	0.23045E+05
2.0	0.58034E+06	0.51913E+06	0.14075E+06	0.26484E+06	0.15768E+06	0.27407E+05
3.0	0.49710E+06	0.57453E+06	0.18343E+06	0.16115E+06	0.12500E+06	0.25363E+05
4.0	0.45755E+06	0.60728E+06	0.21382E+06	0.10911E+06	0.98052E+05	0.21743E+05
5.0	0.43311E+06	0.62640E+06	0.23533E+06	0.78434E+05	0.77503E+05	0.18173E+05
6.0	0.41516E+06	0.63641E+06	0.25036E+06	0.58719E+05	0.62056E+05	0.15103E+05
7.0	0.40038E+06	0.64002E+06	0.26056E+06	0.45312E+05	0.50385E+05	0.12585E+05
8.0	0.38726E+06	0.63890E+06	0.26710E+06	0.35807E+05	0.41458E+05	0.10548E+05
9.0	0.37504E+06	0.63415E+06	0.27077E+06	0.28844E+05	0.34530E+05	0.89036E+04
10.0	0.36329E+06	0.62649E+06	0.27213E+06	0.23604E+05	0.29071E+05	0.75697E+04
11.0	0.35177E+06	0.61646E+06	0.27161E+06	0.19570E+05	0.24709E+05	0.64796E+04
12.0	0.34032E+06	0.60442E+06	0.26950E+06	0.16404E+05	0.21175E+05	0.55811E+04
13.0	0.32885E+06	0.59065E+06	0.26604E+06	0.13879E+05	0.18277E+05	0.48340E+04
14.0	0.31730E+06	0.57538E+06	0.26141E+06	0.11834E+05	0.15873E+05	0.42074E+04
15.0	0.30564E+06	0.55879E+06	0.25576E+06	0.10157E+05	0.13858E+05	0.36776E+04
16.0	0.29386E+06	0.54103E+06	0.24921E+06	0.87657E+04	0.12154E+05	0.32261E+04
17.0	0.28195E+06	0.52225E+06	0.24187E+06	0.76007E+04	0.10700E+05	0.28385E+04
18.0	0.26992E+06	0.50255E+06	0.23384E+06	0.66161E+04	0.94499E+04	0.25037E+04
19.0	0.25778E+06	0.48207E+06	0.22519E+06	0.57774E+04	0.83672E+04	0.22127E+04
20.0	0.24555E+06	0.46090E+06	0.21601E+06	0.50578E+04	0.74235E+04	0.19582E+04
21.0	0.23326E+06	0.43917E+06	0.20637E+06	0.44365E+04	0.65963E+04	0.17347E+04
22.0	0.22093E+06	0.41698E+06	0.19635E+06	0.38969E+04	0.58671E+04	0.15374E+04
23.0	0.20859E+06	0.39442E+06	0.18600E+06	0.34261E+04	0.52213E+04	0.13625E+04
24.0	0.19626E+06	0.37161E+06	0.17541E+06	0.30133E+04	0.46468E+04	0.12069E+04
25.0	0.18400E+06	0.34865E+06	0.16462E+06	0.26499E+04	0.41338E+04	0.10679E+04
26.0	0.17181E+06	0.32563E+06	0.15372E+06	0.23289E+04	0.36740E+04	0.94354E+03
27.0	0.15976E+06	0.30266E+06	0.14275E+06	0.20446E+04	0.32607E+04	0.83189E+03
28.0	0.14786E+06	0.27984E+06	0.13178E+06	0.17919E+04	0.28881E+04	0.73145E+03
29.0	0.13616E+06	0.25726E+06	0.12087E+06	0.15670E+04	0.25514E+04	0.64097E+03
30.0	0.12469E+06	0.23502E+06	0.11009E+06	0.13663E+04	0.22465E+04	0.55935E+03
31.0	0.11349E+06	0.21323E+06	0.99492E+05	0.11871E+04	0.19702E+04	0.48570E+03
32.0	0.10259E+06	0.19197E+06	0.89136E+05	0.10270E+04	0.17194E+04	0.41921E+03

OH $X^2\Pi - A^2\Sigma^+$ (0, 0) Band: $F_2 \rightarrow F_1'$ and $F_2 \rightarrow F_2'$ Transitions

NX	OP12	PQ12	QR12	PP22	QQ22	RR22
1.0	0.00000E+00	0.48394E+06	0.12105E+06	0.00000E+00	0.48546E+06	0.12181E+06
2.0	0.11257E+06	0.26756E+06	0.11324E+06	0.39994E+06	0.49961E+06	0.17214E+06
3.0	0.74833E+05	0.18735E+06	0.93176E+05	0.38841E+06	0.54857E+06	0.20728E+06
4.0	0.52960E+05	0.13683E+06	0.74225E+05	0.38696E+06	0.58579E+06	0.23297E+06
5.0	0.38835E+05	0.10243E+06	0.58870E+05	0.38527E+06	0.61080E+06	0.25152E+06
6.0	0.29296E+05	0.78347E+05	0.46999E+05	0.38191E+06	0.62612E+06	0.26462E+06
7.0	0.22645E+05	0.61122E+05	0.37924E+05	0.37695E+06	0.63412E+06	0.27352E+06
8.0	0.17877E+05	0.48540E+05	0.30963E+05	0.37075E+06	0.63660E+06	0.27917E+06
9.0	0.14373E+05	0.39158E+05	0.25570E+05	0.36358E+06	0.63481E+06	0.28222E+06
10.0	0.11739E+05	0.32027E+05	0.21340E+05	0.35563E+06	0.62966E+06	0.28319E+06
11.0	0.97183E+04	0.26508E+05	0.17978E+05	0.34706E+06	0.62175E+06	0.28240E+06
12.0	0.81396E+04	0.22168E+05	0.15273E+05	0.33795E+06	0.61157E+06	0.28015E+06
13.0	0.68860E+04	0.18704E+05	0.13069E+05	0.32838E+06	0.59944E+06	0.27662E+06
14.0	0.58762E+04	0.15902E+05	0.11254E+05	0.31840E+06	0.58563E+06	0.27198E+06
15.0	0.50521E+04	0.13608E+05	0.97441E+04	0.30805E+06	0.57033E+06	0.26636E+06
16.0	0.43716E+04	0.11711E+05	0.84759E+04	0.29738E+06	0.55374E+06	0.25987E+06
17.0	0.38037E+04	0.10127E+05	0.74016E+04	0.28642E+06	0.53600E+06	0.25260E+06
18.0	0.33254E+04	0.87919E+04	0.64846E+04	0.27520E+06	0.51724E+06	0.24465E+06
19.0	0.29190E+04	0.76585E+04	0.56962E+04	0.26376E+06	0.49759E+06	0.23608E+06
20.0	0.25710E+04	0.66893E+04	0.50143E+04	0.25214E+06	0.47716E+06	0.22697E+06
21.0	0.22708E+04	0.58553E+04	0.44208E+04	0.24036E+06	0.45607E+06	0.21738E+06
22.0	0.20103E+04	0.51335E+04	0.39018E+04	0.22847E+06	0.43442E+06	0.20740E+06
23.0	0.17828E+04	0.45056E+04	0.34457E+04	0.21649E+06	0.41232E+06	0.19707E+06
24.0	0.15830E+04	0.39569E+04	0.30432E+04	0.20447E+06	0.38988E+06	0.18646E+06
25.0	0.14068E+04	0.34754E+04	0.26866E+04	0.19244E+06	0.36719E+06	0.17564E+06
26.0	0.12506E+04	0.30512E+04	0.23697E+04	0.18045E+06	0.34436E+06	0.16467E+06
27.0	0.11115E+04	0.26764E+04	0.20872E+04	0.16852E+06	0.32149E+06	0.15360E+06
28.0	0.98726E+03	0.23442E+04	0.18348E+04	0.15671E+06	0.29868E+06	0.14250E+06
29.0	0.87586E+03	0.20492E+04	0.16087E+04	0.14504E+06	0.27603E+06	0.13143E+06
30.0	0.77570E+03	0.17865E+04	0.14059E+04	0.13355E+06	0.25363E+06	0.12044E+06
31.0	0.68537E+03	0.15524E+04	0.12238E+04	0.12229E+06	0.23159E+06	0.10959E+06
32.0	0.60373E+03	0.13434E+04	0.10601E+04	0.11129E+06	0.21001E+06	0.98954E+05

NO $X^2\Pi - A^2\Sigma^+$ (0, 0) Band: $F_1 \rightarrow F_1'$ and $F_1 \rightarrow F_2'$ Transitions

NX	PP11	QQ11	RR11	QP21	RQ21	SR21
1.0	0.17992E+06	0.27646E+06	0.10236E+06	0.15188E+06	0.25412E+06	0.96815E+05
2.0	0.16605E+06	0.27494E+06	0.11268E+06	0.13252E+06	0.23668E+06	0.10075E+06
3.0	0.16203E+06	0.27914E+06	0.11995E+06	0.12227E+06	0.22614E+06	0.10140E+06
4.0	0.16126E+06	0.28481E+06	0.12581E+06	0.11508E+06	0.21759E+06	0.10059E+06
5.0	0.16194E+06	0.29096E+06	0.13090E+06	0.10932E+06	0.20988E+06	0.98995E+05
6.0	0.16337E+06	0.29728E+06	0.13551E+06	0.10435E+06	0.20263E+06	0.96960E+05

(*cont.*)

NX	PP11	QQ11	RR11	QP21	RQ21	SR21
7.0	0.16522E+06	0.30361E+06	0.13978E+06	0.99880E+05	0.19568E+06	0.94660E+05
8.0	0.16731E+06	0.30989E+06	0.14379E+06	0.95765E+05	0.18896E+06	0.92200E+05
9.0	0.16954E+06	0.31608E+06	0.14761E+06	0.91917E+05	0.18246E+06	0.89646E+05
10.0	0.17184E+06	0.32215E+06	0.15126E+06	0.88283E+05	0.17614E+06	0.87045E+05
11.0	0.17417E+06	0.32809E+06	0.15476E+06	0.84831E+05	0.17000E+06	0.84427E+05
12.0	0.17650E+06	0.33389E+06	0.15813E+06	0.81539E+05	0.16405E+06	0.81817E+05
13.0	0.17881E+06	0.33953E+06	0.16137E+06	0.78391E+05	0.15827E+06	0.79231E+05
14.0	0.18109E+06	0.34500E+06	0.16449E+06	0.75377E+05	0.15268E+06	0.76681E+05
15.0	0.18331E+06	0.35031E+06	0.16750E+06	0.72489E+05	0.14726E+06	0.74178E+05
16.0	0.18548E+06	0.35545E+06	0.17040E+06	0.69720E+05	0.14202E+06	0.71730E+05
17.0	0.18758E+06	0.36041E+06	0.17320E+06	0.67064E+05	0.13696E+06	0.69341E+05
18.0	0.18962E+06	0.36520E+06	0.17590E+06	0.64518E+05	0.13208E+06	0.67017E+05
19.0	0.19158E+06	0.36982E+06	0.17849E+06	0.62078E+05	0.12737E+06	0.64759E+05
20.0	0.19348E+06	0.37426E+06	0.18100E+06	0.59738E+05	0.12283E+06	0.62570E+05
21.0	0.19530E+06	0.37854E+06	0.18340E+06	0.57497E+05	0.11847E+06	0.60451E+05
22.0	0.19705E+06	0.38265E+06	0.18572E+06	0.55349E+05	0.11427E+06	0.58403E+05
23.0	0.19872E+06	0.38659E+06	0.18795E+06	0.53293E+05	0.11023E+06	0.56426E+05
24.0	0.20032E+06	0.39038E+06	0.19010E+06	0.51324E+05	0.10635E+06	0.54519E+05
25.0	0.20185E+06	0.39401E+06	0.19216E+06	0.49440E+05	0.10263E+06	0.52681E+05
26.0	0.20331E+06	0.39748E+06	0.19414E+06	0.47637E+05	0.99058E+05	0.50911E+05
27.0	0.20471E+06	0.40081E+06	0.19604E+06	0.45912E+05	0.95630E+05	0.49208E+05
28.0	0.20603E+06	0.40400E+06	0.19787E+06	0.44261E+05	0.92343E+05	0.47571E+05
29.0	0.20730E+06	0.40705E+06	0.19963E+06	0.42683E+05	0.89191E+05	0.45997E+05
30.0	0.20850E+06	0.40996E+06	0.20132E+06	0.41173E+05	0.86170E+05	0.44484E+05
31.0	0.20964E+06	0.41275E+06	0.20294E+06	0.39729E+05	0.83275E+05	0.43032E+05
32.0	0.21072E+06	0.41541E+06	0.20450E+06	0.38349E+05	0.80500E+05	0.41637E+05
33.0	0.21175E+06	0.41795E+06	0.20599E+06	0.37028E+05	0.77842E+05	0.40298E+05
34.0	0.21272E+06	0.42038E+06	0.20743E+06	0.35766E+05	0.75295E+05	0.39013E+05
35.0	0.21364E+06	0.42270E+06	0.20881E+06	0.34558E+05	0.72854E+05	0.37779E+05
36.0	0.21452E+06	0.42492E+06	0.21014E+06	0.33402E+05	0.70516E+05	0.36596E+05
37.0	0.21534E+06	0.42703E+06	0.21141E+06	0.32297E+05	0.68275E+05	0.35460E+05
38.0	0.21612E+06	0.42905E+06	0.21264E+06	0.31240E+05	0.66128E+05	0.34370E+05
39.0	0.21686E+06	0.43098E+06	0.21381E+06	0.30227E+05	0.64070E+05	0.33324E+05
40.0	0.21756E+06	0.43281E+06	0.21494E+06	0.29259E+05	0.62098E+05	0.32320E+05
41.0	0.21821E+06	0.43457E+06	0.21603E+06	0.28332E+05	0.60207E+05	0.31356E+05
42.0	0.21883E+06	0.43624E+06	0.21708E+06	0.27444E+05	0.58394E+05	0.30431E+05
43.0	0.21941E+06	0.43783E+06	0.21808E+06	0.26593E+05	0.56655E+05	0.29543E+05
44.0	0.21996E+06	0.43935E+06	0.21904E+06	0.25779E+05	0.54987E+05	0.28691E+05
45.0	0.22048E+06	0.44080E+06	0.21997E+06	0.24998E+05	0.53387E+05	0.27872E+05
46.0	0.22096E+06	0.44218E+06	0.22087E+06	0.24250E+05	0.51851E+05	0.27085E+05
47.0	0.22141E+06	0.44350E+06	0.22173E+06	0.23532E+05	0.50377E+05	0.26329E+05
48.0	0.22184E+06	0.44476E+06	0.22255E+06	0.22844E+05	0.48962E+05	0.25603E+05
49.0	0.22223E+06	0.44595E+06	0.22335E+06	0.22184E+05	0.47603E+05	0.24905E+05
50.0	0.22260E+06	0.44709E+06	0.22412E+06	0.21551E+05	0.46297E+05	0.24234E+05
51.0	0.22295E+06	0.44817E+06	0.22485E+06	0.20943E+05	0.45042E+05	0.23589E+05
52.0	0.22327E+06	0.44921E+06	0.22557E+06	0.20359E+05	0.43836E+05	0.22968E+05
53.0	0.22356E+06	0.45019E+06	0.22625E+06	0.19798E+05	0.42676E+05	0.22371E+05
54.0	0.22384E+06	0.45112E+06	0.22691E+06	0.19259E+05	0.41561E+05	0.21796E+05

(*cont.*)

NX	PP11	QQ11	RR11	QP21	RQ21	SR21
55.0	0.22409E+06	0.45201E+06	0.22754E+06	0.18741E+05	0.40488E+05	0.21242E+05
56.0	0.22433E+06	0.45285E+06	0.22815E+06	0.18244E+05	0.39455E+05	0.20709E+05
57.0	0.22454E+06	0.45365E+06	0.22874E+06	0.17765E+05	0.38461E+05	0.20196E+05
58.0	0.22474E+06	0.45442E+06	0.22931E+06	0.17304E+05	0.37504E+05	0.19702E+05
59.0	0.22491E+06	0.45514E+06	0.22986E+06	0.16861E+05	0.36582E+05	0.19225E+05
60.0	0.22508E+06	0.45582E+06	0.23038E+06	0.16434E+05	0.35694E+05	0.18765E+05
61.0	0.22522E+06	0.45647E+06	0.23089E+06	0.16023E+05	0.34838E+05	0.18322E+05
62.0	0.22535E+06	0.45708E+06	0.23138E+06	0.15627E+05	0.34012E+05	0.17895E+05
63.0	0.22546E+06	0.45766E+06	0.23185E+06	0.15246E+05	0.33216E+05	0.17482E+05
64.0	0.22556E+06	0.45821E+06	0.23230E+06	0.14878E+05	0.32449E+05	0.17084E+05
65.0	0.22564E+06	0.45873E+06	0.23274E+06	0.14524E+05	0.31708E+05	0.16700E+05
66.0	0.22571E+06	0.45921E+06	0.23316E+06	0.14182E+05	0.30992E+05	0.16329E+05
67.0	0.22577E+06	0.45967E+06	0.23356E+06	0.13852E+05	0.30302E+05	0.15970E+05
68.0	0.22581E+06	0.46010E+06	0.23395E+06	0.13533E+05	0.29635E+05	0.15624E+05
69.0	0.22585E+06	0.46050E+06	0.23433E+06	0.13226E+05	0.28990E+05	0.15289E+05
70.0	0.22587E+06	0.46087E+06	0.23469E+06	0.12929E+05	0.28368E+05	0.14965E+05

NO $X^2\Pi - A^2\Sigma^+$ (0, 0) Band: $F_2 \rightarrow F_1'$ and $F_2 \rightarrow F_2'$ Transitions

NX	OP12	PQ12	QR12	PP22	QQ22	RR22
1.0	0.00000E+00	0.00000E+00	0.00000E+00	0.00000E+00	0.00000E+00	0.00000E+00
2.0	0.47886E+06	0.18626E+06	0.30191E+05	0.50684E+06	0.20867E+06	0.35802E+05
3.0	0.27868E+06	0.20649E+06	0.47075E+05	0.31217E+06	0.24483E+06	0.59079E+05
4.0	0.21437E+06	0.20850E+06	0.56796E+05	0.25411E+06	0.26159E+06	0.75425E+05
5.0	0.18168E+06	0.20564E+06	0.62531E+05	0.22785E+06	0.27297E+06	0.87855E+05
6.0	0.16115E+06	0.20099E+06	0.65867E+05	0.21378E+06	0.28219E+06	0.97884E+05
7.0	0.14657E+06	0.19556E+06	0.67667E+05	0.20560E+06	0.29035E+06	0.10633E+06
8.0	0.13535E+06	0.18978E+06	0.68437E+05	0.20071E+06	0.29787E+06	0.11368E+06
9.0	0.12622E+06	0.18387E+06	0.68492E+05	0.19780E+06	0.30497E+06	0.12023E+06
10.0	0.11851E+06	0.17792E+06	0.68039E+05	0.19617E+06	0.31175E+06	0.12615E+06
11.0	0.11181E+06	0.17202E+06	0.67221E+05	0.19542E+06	0.31825E+06	0.13160E+06
12.0	0.10585E+06	0.16619E+06	0.66137E+05	0.19526E+06	0.32452E+06	0.13664E+06
13.0	0.10048E+06	0.16048E+06	0.64861E+05	0.19552E+06	0.33058E+06	0.14135E+06
14.0	0.95584E+05	0.15489E+06	0.63448E+05	0.19609E+06	0.33642E+06	0.14577E+06
15.0	0.91072E+05	0.14944E+06	0.61938E+05	0.19688E+06	0.34206E+06	0.14995E+06
16.0	0.86888E+05	0.14414E+06	0.60364E+05	0.19781E+06	0.34751E+06	0.15390E+06
17.0	0.82987E+05	0.13900E+06	0.58751E+05	0.19886E+06	0.35277E+06	0.15765E+06
18.0	0.79335E+05	0.13403E+06	0.57117E+05	0.19998E+06	0.35783E+06	0.16121E+06
19.0	0.75905E+05	0.12921E+06	0.55478E+05	0.20114E+06	0.36271E+06	0.16460E+06
20.0	0.72676E+05	0.12456E+06	0.53846E+05	0.20233E+06	0.36740E+06	0.16783E+06
21.0	0.69628E+05	0.12008E+06	0.52231E+05	0.20352E+06	0.37192E+06	0.17092E+06
22.0	0.66748E+05	0.11576E+06	0.50640E+05	0.20471E+06	0.37626E+06	0.17386E+06
23.0	0.64022E+05	0.11160E+06	0.49078E+05	0.20589E+06	0.38043E+06	0.17668E+06
24.0	0.61439E+05	0.10759E+06	0.47551E+05	0.20705E+06	0.38442E+06	0.17937E+06

(*cont.*)

NX	OP12	PQ12	QR12	PP22	QQ22	RR22
25.0	0.58989E+05	0.10375E+06	0.46061E+05	0.20818E+06	0.38826E+06	0.18194E+06
26.0	0.56665E+05	0.10006E+06	0.44611E+05	0.20928E+06	0.39194E+06	0.18440E+06
27.0	0.54457E+05	0.96509E+05	0.43203E+05	0.21035E+06	0.39546E+06	0.18675E+06
28.0	0.52359E+05	0.93107E+05	0.41838E+05	0.21138E+06	0.39883E+06	0.18900E+06
29.0	0.50365E+05	0.89843E+05	0.40516E+05	0.21237E+06	0.40206E+06	0.19116E+06
30.0	0.48467E+05	0.86714E+05	0.39238E+05	0.21332E+06	0.40515E+06	0.19322E+06
31.0	0.46662E+05	0.83715E+05	0.38003E+05	0.21424E+06	0.40811E+06	0.19520E+06
32.0	0.44944E+05	0.80840E+05	0.36810E+05	0.21511E+06	0.41093E+06	0.19709E+06
33.0	0.43308E+05	0.78086E+05	0.35661E+05	0.21595E+06	0.41363E+06	0.19890E+06
34.0	0.41749E+05	0.75447E+05	0.34552E+05	0.21675E+06	0.41622E+06	0.20064E+06
35.0	0.40263E+05	0.72920E+05	0.33485E+05	0.21751E+06	0.41868E+06	0.20230E+06
36.0	0.38846E+05	0.70498E+05	0.32456E+05	0.21823E+06	0.42104E+06	0.20390E+06
37.0	0.37496E+05	0.68178E+05	0.31467E+05	0.21892E+06	0.42329E+06	0.20543E+06
38.0	0.36207E+05	0.65956E+05	0.30514E+05	0.21957E+06	0.42545E+06	0.20690E+06
39.0	0.34977E+05	0.63827E+05	0.29598E+05	0.22019E+06	0.42750E+06	0.20830E+06
40.0	0.33803E+05	0.61787E+05	0.28717E+05	0.22077E+06	0.42946E+06	0.20965E+06
41.0	0.32682E+05	0.59832E+05	0.27869E+05	0.22133E+06	0.43134E+06	0.21095E+06
42.0	0.31611E+05	0.57959E+05	0.27053E+05	0.22185E+06	0.43312E+06	0.21219E+06
43.0	0.30587E+05	0.56163E+05	0.26269E+05	0.22234E+06	0.43483E+06	0.21339E+06
44.0	0.29608E+05	0.54441E+05	0.25515E+05	0.22280E+06	0.43646E+06	0.21454E+06
45.0	0.28672E+05	0.52790E+05	0.24789E+05	0.22324E+06	0.43802E+06	0.21564E+06
46.0	0.27777E+05	0.51207E+05	0.24091E+05	0.22365E+06	0.43950E+06	0.21670E+06
47.0	0.26920E+05	0.49688E+05	0.23420E+05	0.22403E+06	0.44092E+06	0.21772E+06
48.0	0.26099E+05	0.48230E+05	0.22774E+05	0.22439E+06	0.44227E+06	0.21870E+06
49.0	0.25313E+05	0.46831E+05	0.22153E+05	0.22472E+06	0.44356E+06	0.21964E+06
50.0	0.24560E+05	0.45487E+05	0.21555E+05	0.22503E+06	0.44479E+06	0.22054E+06
51.0	0.23838E+05	0.44197E+05	0.20980E+05	0.22532E+06	0.44596E+06	0.22142E+06
52.0	0.23146E+05	0.42958E+05	0.20426E+05	0.22559E+06	0.44708E+06	0.22225E+06
53.0	0.22482E+05	0.41768E+05	0.19892E+05	0.22584E+06	0.44814E+06	0.22306E+06
54.0	0.21845E+05	0.40623E+05	0.19378E+05	0.22607E+06	0.44916E+06	0.22384E+06
55.0	0.21233E+05	0.39523E+05	0.18884E+05	0.22628E+06	0.45012E+06	0.22459E+06
56.0	0.20646E+05	0.38465E+05	0.18407E+05	0.22648E+06	0.45104E+06	0.22531E+06
57.0	0.20082E+05	0.37447E+05	0.17947E+05	0.22665E+06	0.45192E+06	0.22600E+06
58.0	0.19540E+05	0.36468E+05	0.17505E+05	0.22681E+06	0.45275E+06	0.22667E+06
59.0	0.19019E+05	0.35525E+05	0.17078E+05	0.22696E+06	0.45355E+06	0.22731E+06
60.0	0.18518E+05	0.34617E+05	0.16666E+05	0.22708E+06	0.45430E+06	0.22793E+06
61.0	0.18036E+05	0.33743E+05	0.16269E+05	0.22720E+06	0.45501E+06	0.22853E+06
62.0	0.17572E+05	0.32901E+05	0.15885E+05	0.22730E+06	0.45569E+06	0.22911E+06
63.0	0.17125E+05	0.32089E+05	0.15515E+05	0.22738E+06	0.45634E+06	0.22966E+06
64.0	0.16695E+05	0.31306E+05	0.15158E+05	0.22746E+06	0.45695E+06	0.23020E+06
65.0	0.16281E+05	0.30551E+05	0.14813E+05	0.22752E+06	0.45752E+06	0.23071E+06
66.0	0.15882E+05	0.29823E+05	0.14480E+05	0.22756E+06	0.45807E+06	0.23121E+06
67.0	0.15497E+05	0.29121E+05	0.14158E+05	0.22760E+06	0.45859E+06	0.23169E+06
68.0	0.15126E+05	0.28443E+05	0.13847E+05	0.22762E+06	0.45907E+06	0.23215E+06
69.0	0.14768E+05	0.27788E+05	0.13547E+05	0.22764E+06	0.45953E+06	0.23259E+06
70.0	0.14422E+05	0.27156E+05	0.13256E+05	0.22764E+06	0.45996E+06	0.23302E+06

NH $X^3\Sigma^- - A^3\Pi\,(0,0)$ Band: $F_1 \rightarrow F_1'$, and $F_1 \rightarrow F_2'$ Transitions

JX	PP11	QQ11	RR11	QP21	RQ21	SR21
0.0	0.000000E+00	0.000000E+00	0.000000E+00	0.000000E+00	0.000000E+00	0.000000E+00
1.0	0.000000E+00	0.000000E+00	0.166744E+07	0.000000E+00	0.171493E+07	0.113378E+06
2.0	−0.000000E+00	0.146958E+07	0.153316E+07	0.960432E+06	0.916719E+06	0.732531E+05
3.0	0.460819E+06	0.200150E+07	0.148358E+07	0.657816E+06	0.529834E+06	0.470450E+05
4.0	0.749199E+06	0.224597E+07	0.145553E+07	0.419323E+06	0.332492E+06	0.319788E+05
5.0	0.911519E+06	0.236899E+07	0.143519E+07	0.278113E+06	0.224239E+06	0.229564E+05
6.0	0.100454E+07	0.243400E+07	0.141879E+07	0.194386E+06	0.160185E+06	0.172481E+05
7.0	0.105940E+07	0.246778E+07	0.140436E+07	0.142265E+06	0.119681E+06	0.134472E+05
8.0	0.109180E+07	0.248252E+07	0.139056E+07	0.108081E+06	0.926187E+05	0.108024E+05
9.0	0.111040E+07	0.248539E+07	0.137713E+07	0.846384E+05	0.737273E+05	0.889352E+04
10.0	0.111984E+07	0.247994E+07	0.136364E+07	0.679412E+05	0.600725E+05	0.748108E+04
11.0	0.112274E+07	0.246795E+07	0.134963E+07	0.556462E+05	0.498769E+05	0.640223E+04
12.0	0.112110E+07	0.245141E+07	0.133526E+07	0.463596E+05	0.420859E+05	0.556336E+04
13.0	0.111570E+07	0.243044E+07	0.132000E+07	0.391742E+05	0.359989E+05	0.489979E+04
14.0	0.110744E+07	0.240594E+07	0.130393E+07	0.335103E+05	0.311620E+05	0.436811E+04
15.0	0.109698E+07	0.237863E+07	0.128716E+07	0.289724E+05	0.272569E+05	0.393544E+04
16.0	0.108437E+07	0.234811E+07	0.126922E+07	0.252778E+05	0.240591E+05	0.358050E+04
17.0	0.107003E+07	0.231490E+07	0.125026E+07	0.222333E+05	0.214092E+05	0.328602E+04
18.0	0.105418E+07	0.227915E+07	0.123024E+07	0.196961E+05	0.191900E+05	0.303985E+04
19.0	0.103695E+07	0.224097E+07	0.120911E+07	0.175619E+05	0.173187E+05	0.283458E+04
20.0	0.101847E+07	0.220043E+07	0.118684E+07	0.157510E+05	0.157277E+05	0.266289E+04
21.0	0.998826E+06	0.215762E+07	0.116342E+07	0.142011E+05	0.143623E+05	0.251768E+04
22.0	0.978245E+06	0.211288E+07	0.113897E+07	0.128677E+05	0.131866E+05	0.239599E+04
23.0	0.956367E+06	0.206536E+07	0.111301E+07	0.117089E+05	0.121655E+05	0.229409E+04
24.0	0.933660E+06	0.201599E+07	0.108599E+07	0.107006E+05	0.112780E+05	0.220977E+04
25.0	0.910028E+06	0.196451E+07	0.105775E+07	0.981797E+04	0.105039E+05	0.214126E+04
26.0	0.885632E+06	0.191122E+07	0.102842E+07	0.904302E+04	0.982716E+04	0.208672E+04
27.0	0.860249E+06	0.185559E+07	0.997707E+06	0.835855E+04	0.923406E+04	0.204533E+04
28.0	0.834034E+06	0.179793E+07	0.965753E+06	0.775234E+04	0.871231E+04	0.201510E+04
29.0	0.807008E+06	0.173824E+07	0.932554E+06	0.721529E+04	0.825718E+04	0.199723E+04

NH $X^3\Sigma^- - A^3\Pi\,(0,0)$ Band: $F_1 \rightarrow F_3'$ and $F_2 \rightarrow F_1'$ Transitions

JX	RP31	SQ31	TR31	OP12	PQ12	QR12
0.0	0.000000E+00	0.000000E+00	0.000000E+00	0.000000E+00	0.000000E+00	0.000000E+00
1.0	0.177433E+07	0.898523E+05	0.317609E+03	0.000000E+00	−0.000000E+00	0.112449E+07
2.0	0.198837E+06	0.232870E+05	0.110935E+03	0.000000E+00	0.449804E+06	0.601893E+06
3.0	0.460934E+05	0.749466E+04	0.390592E+02	0.500569E+05	0.342343E+06	0.364991E+06
4.0	0.152979E+05	0.298686E+04	0.139963E+02	0.430410E+05	0.238444E+06	0.241324E+06
5.0	0.630492E+04	0.136687E+04	0.490962E+01	0.313567E+05	0.169360E+06	0.169900E+06
6.0	0.300250E+04	0.686321E+03	0.156257E+01	0.226655E+05	0.124452E+06	0.125353E+06
7.0	0.158281E+04	0.365843E+03	0.363080E+00	0.167550E+05	0.944366E+05	0.958630E+05

(*cont.*)

JX	RP31	SQ31	TR31	OP12	PQ12	QR12
8.0	0.898291E+03	0.201630E+03	0.205800E−01	0.127209E+05	0.736526E+05	0.753985E+05
9.0	0.542529E+03	0.114900E+03	0.119239E−01	0.989982E+04	0.587717E+05	0.606483E+05
10.0	0.337511E+03	0.622728E+02	0.189561E+00	0.787170E+04	0.477986E+05	0.496811E+05
11.0	0.218815E+03	0.335483E+02	0.361862E+00	0.637501E+04	0.394983E+05	0.413128E+05
12.0	0.146609E+03	0.173944E+02	0.460457E+00	0.524421E+04	0.330798E+05	0.347865E+05
13.0	0.997466E+02	0.783382E+01	0.577039E+00	0.437191E+04	0.280215E+05	0.296007E+05
14.0	0.679105E+02	0.244475E+01	0.772764E+00	0.368668E+04	0.239680E+05	0.254129E+05
15.0	0.472982E+02	0.343652E+00	0.833518E+00	0.313962E+04	0.206725E+05	0.219833E+05
16.0	0.324192E+02	0.123861E+00	0.995023E+00	0.269665E+04	0.179585E+05	0.191394E+05
17.0	0.224355E+02	0.127998E+01	0.105464E+01	0.233339E+04	0.156976E+05	0.167549E+05
18.0	0.158673E+02	0.304414E+01	0.966658E+00	0.203209E+04	0.137949E+05	0.147359E+05
19.0	0.106583E+02	0.581688E+01	0.103465E+01	0.177957E+04	0.121787E+05	0.130111E+05
20.0	0.661138E+01	0.976464E+01	0.124476E+01	0.156594E+04	0.107942E+05	0.115256E+05
21.0	0.419801E+01	0.133530E+02	0.122095E+01	0.138363E+04	0.959904E+04	0.102368E+05
22.0	0.260674E+01	0.168880E+02	0.112416E+01	0.122679E+04	0.855996E+04	0.911099E+04
23.0	0.141802E+01	0.210870E+02	0.110893E+01	0.109083E+04	0.765049E+04	0.812143E+04
24.0	0.719718E+00	0.249288E+02	0.999958E+00	0.972099E+03	0.684955E+04	0.724650E+04
25.0	0.262624E+00	0.293169E+02	0.948145E+00	0.867700E+03	0.613996E+04	0.646863E+04
26.0	0.646081E−01	0.330345E+02	0.797382E+00	0.775276E+03	0.550773E+04	0.577354E+04
27.0	0.473752E−02	0.383935E+02	0.820383E+00	0.692885E+03	0.494140E+04	0.514938E+04
28.0	0.597198E−01	0.415622E+02	0.610604E+00	0.618971E+03	0.443142E+04	0.458625E+04
29.0	0.304923E+00	0.476878E+02	0.665684E+00	0.552221E+03	0.396975E+04	0.407595E+04

NH $X^3\Sigma^- - A^3\Pi\,(0,0)$ Band: $F_2 \to F_2'$ and $F_2 \to F_3'$ Transitions

JX	PP22	QQ22	RR22	QP32	RQ32	SR32
0.0	0.000000E+00	0.000000E+00	0.000000E+00	0.000000E+00	0.000000E+00	0.000000E+00
1.0	−0.000000E+00	0.368230E+06	0.707963E+06	−0.000000E+00	0.979010E+06	0.396873E+05
2.0	0.365432E+06	0.124599E+07	0.102336E+07	0.971707E+06	0.531597E+06	0.332328E+05
3.0	0.660154E+06	0.177346E+07	0.117335E+07	0.528237E+06	0.324856E+06	0.252389E+05
4.0	0.854192E+06	0.206385E+07	0.124931E+07	0.325235E+06	0.216665E+06	0.193024E+05
5.0	0.968341E+06	0.222756E+07	0.128957E+07	0.218987E+06	0.153991E+06	0.151149E+05
6.0	0.103638E+07	0.232262E+07	0.131100E+07	0.156868E+06	0.114765E+06	0.121320E+05
7.0	0.107747E+07	0.237808E+07	0.132158E+07	0.117587E+06	0.887101E+05	0.995941E+04
8.0	0.110195E+07	0.240908E+07	0.132535E+07	0.912490E+05	0.705754E+05	0.833899E+04
9.0	0.111560E+07	0.242392E+07	0.132450E+07	0.727643E+05	0.574707E+05	0.710353E+04
10.0	0.112179E+07	0.242746E+07	0.132029E+07	0.593105E+05	0.477056E+05	0.614303E+04
11.0	0.112257E+07	0.242263E+07	0.131345E+07	0.492229E+05	0.402416E+05	0.538350E+04
12.0	0.111927E+07	0.241131E+07	0.130443E+07	0.414702E+05	0.344128E+05	0.477391E+04
13.0	0.111277E+07	0.239473E+07	0.129352E+07	0.353869E+05	0.297771E+05	0.427848E+04
14.0	0.110366E+07	0.237371E+07	0.128092E+07	0.305279E+05	0.260320E+05	0.387141E+04
15.0	0.109238E+07	0.234884E+07	0.126673E+07	0.265871E+05	0.229651E+05	0.353390E+04
16.0	0.107923E+07	0.232054E+07	0.125105E+07	0.233481E+05	0.204236E+05	0.325191E+04
17.0	0.106444E+07	0.228912E+07	0.123392E+07	0.206548E+05	0.182956E+05	0.301492E+04

(*cont.*)

JX	PP22	QQ22	RR22	QP32	RQ32	SR32
18.0	0.104819E+07	0.225479E+07	0.121538E+07	0.183920E+05	0.164976E+05	0.281485E+04
19.0	0.103062E+07	0.221773E+07	0.119547E+07	0.164737E+05	0.149663E+05	0.264544E+04
20.0	0.101183E+07	0.217808E+07	0.117419E+07	0.148342E+05	0.136527E+05	0.250186E+04
21.0	0.991896E+06	0.213595E+07	0.115155E+07	0.134229E+05	0.125193E+05	0.238023E+04
22.0	0.970901E+06	0.209140E+07	0.112758E+07	0.122002E+05	0.115360E+05	0.227757E+04
23.0	0.948896E+06	0.204453E+07	0.110227E+07	0.111349E+05	0.106792E+05	0.219149E+04
24.0	0.925929E+06	0.199538E+07	0.107563E+07	0.102020E+05	0.992986E+04	0.212008E+04
25.0	0.902040E+06	0.194400E+07	0.104766E+07	0.938137E+04	0.927255E+04	0.206188E+04
26.0	0.877264E+06	0.189044E+07	0.101838E+07	0.865679E+04	0.869493E+04	0.201570E+04
27.0	0.851630E+06	0.183473E+07	0.987788E+06	0.801491E+04	0.818664E+04	0.198067E+04
28.0	0.825163E+06	0.177693E+07	0.955891E+06	0.744467E+04	0.773933E+04	0.195618E+04
29.0	0.797888E+06	0.171705E+07	0.922701E+06	0.693698E+04	0.734623E+04	0.194182E+04

NH $X^3\Sigma^- - A^3\Pi\,(0,0)$ Band: $F_3 \rightarrow F_1'$, *and* $F_3 \rightarrow F_2'$ Transitions

JX	NP13	OQ13	PR13	OP23	PQ23	QR23
0.0	0.000000E+00	−0.000000E+00	−0.000000E+00	0.000000E+00	0.000000E+00	0.129639E+07
1.0	−0.000000E+00	−0.000000E+00	0.820426E+05	0.000000E+00	0.598788E+06	0.618174E+06
2.0	−0.000000E+00	0.169149E+05	0.295000E+05	0.539317E+05	0.388006E+06	0.371707E+06
3.0	0.933315E+03	0.798857E+04	0.120875E+05	0.370556E+05	0.256624E+06	0.245029E+06
4.0	0.519784E+03	0.372675E+04	0.567857E+04	0.258318E+05	0.179207E+06	0.172144E+06
5.0	0.266210E+03	0.189006E+04	0.296659E+04	0.186183E+05	0.130822E+06	0.126768E+06
6.0	0.142426E+03	0.104212E+04	0.167997E+04	0.138746E+05	0.990349E+05	0.968127E+05
7.0	0.802979E+02	0.617057E+03	0.101214E+04	0.106470E+05	0.772018E+05	0.760742E+05
8.0	0.472102E+02	0.387798E+03	0.639996E+03	0.837344E+04	0.616192E+05	0.611426E+05
9.0	0.277441E+02	0.252633E+03	0.417034E+03	0.672749E+04	0.501632E+05	0.500649E+05
10.0	0.173365E+02	0.175845E+03	0.284046E+03	0.549564E+04	0.414931E+05	0.416202E+05
11.0	0.103720E+02	0.124477E+03	0.195909E+03	0.455840E+04	0.347942E+05	0.350402E+05
12.0	0.581606E+01	0.895105E+02	0.136399E+03	0.383052E+04	0.295214E+05	0.298227E+05
13.0	0.309311E+01	0.664193E+02	0.964990E+02	0.325267E+04	0.252894E+05	0.256083E+05
14.0	0.156451E+01	0.513736E+02	0.696850E+02	0.278688E+04	0.218443E+05	0.221581E+05
15.0	0.575566E+00	0.396409E+02	0.498726E+02	0.240829E+04	0.190109E+05	0.193024E+05
16.0	0.141821E+00	0.319778E+02	0.363680E+02	0.209467E+04	0.166443E+05	0.169057E+05
17.0	0.520456E−05	0.257847E+02	0.261675E+02	0.183344E+04	0.146526E+05	0.148776E+05
18.0	0.150792E+00	0.204282E+02	0.182345E+02	0.161386E+04	0.129611E+05	0.131461E+05
19.0	0.458647E+00	0.171849E+02	0.130036E+02	0.142636E+04	0.115090E+05	0.116543E+05
20.0	0.778461E+00	0.153861E+02	0.955603E+01	0.126504E+04	0.102530E+05	0.103597E+05
21.0	0.132141E+01	0.131835E+02	0.654183E+01	0.112635E+04	0.916228E+04	0.922981E+04
22.0	0.201574E+01	0.112500E+02	0.425218E+01	0.100589E+04	0.820862E+04	0.823813E+04
23.0	0.268244E+01	0.100492E+02	0.274545E+01	0.899785E+03	0.736522E+04	0.735901E+04
24.0	0.350738E+01	0.884699E+01	0.159442E+01	0.806451E+03	0.661923E+04	0.657890E+04
25.0	0.427123E+01	0.812729E+01	0.880191E+00	0.723389E+03	0.595379E+04	0.588172E+04
26.0	0.522671E+01	0.729151E+01	0.369074E+00	0.649423E+03	0.535874E+04	0.525691E+04
27.0	0.581990E+01	0.726124E+01	0.156433E+00	0.582382E+03	0.482080E+04	0.469220E+04
28.0	0.692141E+01	0.655709E+01	0.988265E−02	0.522164E+03	0.433525E+04	0.418141E+04
29.0	0.731605E+01	0.699898E+01	0.733480E−03	0.466751E+03	0.389172E+04	0.371624E+04

NH $X^3\Sigma^- - A^3\Pi\,(0,0)$ Band: $F_3 \rightarrow F_3'$ Transitions

JX	PP33	QQ33	RR33
0.0	0.000000E+00	0.000000E+00	0.494821E+06
1.0	0.000000E+00	0.162106E+07	0.100536E+07
2.0	0.100951E+07	0.211738E+07	0.118977E+07
3.0	0.109231E+07	0.232342E+07	0.127192E+07
4.0	0.113764E+07	0.242577E+07	0.131291E+07
5.0	0.116251E+07	0.247889E+07	0.133370E+07
6.0	0.117580E+07	0.250607E+07	0.134380E+07
7.0	0.118173E+07	0.251765E+07	0.134744E+07
8.0	0.118238E+07	0.251844E+07	0.134649E+07
9.0	0.117939E+07	0.251203E+07	0.134250E+07
10.0	0.117359E+07	0.250010E+07	0.133610E+07
11.0	0.116535E+07	0.248337E+07	0.132752E+07
12.0	0.115536E+07	0.246321E+07	0.131737E+07
13.0	0.114354E+07	0.243938E+07	0.130542E+07
14.0	0.113025E+07	0.241254E+07	0.129196E+07
15.0	0.111578E+07	0.238324E+07	0.127725E+07
16.0	0.109992E+07	0.235096E+07	0.126093E+07
17.0	0.108292E+07	0.231617E+07	0.124324E+07
18.0	0.106483E+07	0.227895E+07	0.122417E+07
19.0	0.104571E+07	0.223936E+07	0.120374E+07
20.0	0.102560E+07	0.219745E+07	0.118195E+07
21.0	0.100455E+07	0.215326E+07	0.115881E+07
22.0	0.982705E+06	0.210714E+07	0.113448E+07
23.0	0.959687E+06	0.205822E+07	0.110847E+07
24.0	0.935933E+06	0.200741E+07	0.108128E+07
25.0	0.911324E+06	0.195443E+07	0.105273E+07
26.0	0.886000E+06	0.189960E+07	0.102299E+07
27.0	0.859724E+06	0.184236E+07	0.991761E+06
28.0	0.832639E+06	0.178305E+07	0.959202E+06
29.0	0.804760E+06	0.172167E+07	0.925325E+06

Appendix 8 Effect of Hyperfine Splitting on Radiative Transition Rates

The effects of the hyperfine interaction are commonly ignored in gas-phase spectroscopy because the magnitude of the energy-level splitting is typically much less than either the homogeneous or Doppler frequency widths of transitions. In this example we will show that the spontaneous emission coefficient from the upper level of a transition and the absorption coefficient from the lower level of a transition are unaffected by the hyperfine interaction. The reduced matrix elements for transitions between the hyperfine energy levels F and F' and energy levels J and J' are related by the following equation:

$$\langle \alpha JIF \| \boldsymbol{\mu}_1 \| \alpha' J' IF' \rangle = (-1)^{J+I+F'+1} \langle \alpha J \| \boldsymbol{\mu}_1 \| \alpha' J' \rangle \sqrt{(2F+1)(2F'+1)} \begin{Bmatrix} J & F & I \\ F' & J' & 1 \end{Bmatrix}. \tag{A8.1}$$

The reduced matrix element for a radiative transition coupling levels J and J' is given by

$$\langle \alpha' J' \| \boldsymbol{\mu}_1 \| \alpha J' - 1 \rangle = \pm \sqrt{\frac{A_{J'J}(2J'+1) 3\pi \varepsilon_0 \hbar c^3}{\omega_{J'J}^3}}. \tag{A8.2}$$

By analogy the reduced matrix element for a transition coupling hyperfine energy levels F and F' is given by

$$\langle \alpha JIF \| \boldsymbol{\mu}_1 \| \alpha' J' IF' \rangle = \pm \sqrt{\frac{A_{F'F}(2F'+1) 3\pi \varepsilon_0 \hbar c^3}{\omega_{J'J}^3}}. \tag{A8.3}$$

Consequently, the spontaneous emission coefficient $A_{F'F}$ for a transition coupling hyperfine energy levels F and F' is related to the spontaneous emission coefficient $A_{J'J}$ by

$$A_{F'F} = A_{J'J} \frac{(2J'+1) \langle \alpha JIF \| \boldsymbol{\mu}_1 \| \alpha' J' IF' \rangle^2}{(2F'+1) \langle \alpha' J' \| \boldsymbol{\mu}_1 \| \alpha J' - 1 \rangle^2}. \tag{A8.4}$$

The absorption coefficient for the transition from energy level J to J' is given by

$$\alpha_{BA}(\omega_L - \omega_{BA}) = \frac{\lambda_{BA}^2 A_{J'J} g_H(\omega_L - \omega_{BA})}{4} \left(n_A \frac{2J'+1}{2J+1} - n_B \right). \tag{A8.5}$$

Assuming that the population in level B is negligible compared to the population in level A, Eqn. (6.31) reduces to

$$\alpha_{BA}(\omega_L - \omega_{BA}) = \frac{\lambda_{BA}^2\, n_A\, g_H(\omega_L - \omega_{BA})}{4}\left(A_{J'J}\frac{2J'+1}{2J+1}\right). \tag{A8.6}$$

The effect of the hyperfine interaction will now be analyzed for the $^{14}N^{16}O$ $R_1(10)$ transition. For this molecule, $I = 1$ due to the spin of the ^{14}N nucleus; the ^{16}O nucleus has zero spin. The hyperfine structure of the transition is depicted schematically in Figure A8.1.

$\underline{J = 10.5 \quad F = 11.5}$

$\underline{J' = 11.5 \quad F' = 12.5}$

$$\begin{Bmatrix} J & F & I \\ F' & J' & 1 \end{Bmatrix} = \begin{Bmatrix} 10.5 & 11.5 & 1 \\ 12.5 & 11.5 & 1 \end{Bmatrix} = \begin{Bmatrix} 1 & 10.5 & 11.5 \\ 1 & 12.5 & 11.5 \end{Bmatrix} = \begin{Bmatrix} 1 & 12.5 & 11.5 \\ 1 & 10.5 & 11.5 \end{Bmatrix}$$

$s = 25 \quad a = 1 \quad b = 12.5 \quad c = 11.5$

$$\begin{Bmatrix} 1 & 12.5 & 11.5 \\ 1 & 10.5 & 11.5 \end{Bmatrix} = (-1)^s\left[\frac{s(s+1)(s-2a-1)(s-2a)}{(2b-1)2b(2b+1)(2c-1)2c(2c+1)}\right]^{1/2}$$

$$= (-1)^{25}\left[\frac{25(26)(22)(23)}{(24)(25)(26)(22)(23)(24)}\right]^{1/2} = -\frac{1}{24}$$

$$\frac{\langle \alpha JIF \| \boldsymbol{\mu}_1 \| \alpha' J' I F' \rangle^2}{\langle \alpha' J' \| \boldsymbol{\mu}_1 \| \alpha J' - 1 \rangle^2} = (2F+1)(2F'+1)\begin{Bmatrix} 1 & 12.5 & 11.5 \\ 1 & 10.5 & 11.5 \end{Bmatrix}^2 = \frac{(24)(26)}{(-24)^2} = 1.08333$$

$$A_{F'F} = A_{J'J}\frac{(2J'+1)\langle \alpha JIF \| \boldsymbol{\mu}_1 \| \alpha' J' I F' \rangle^2}{(2F'+1)\langle \alpha' J' \| \boldsymbol{\mu}_1 \| \alpha J' - 1 \rangle^2} = A_{J'J}\left(\frac{24}{26}\right)(1.08333) = 1.0000 A_{J'J}. \tag{A8.7}$$

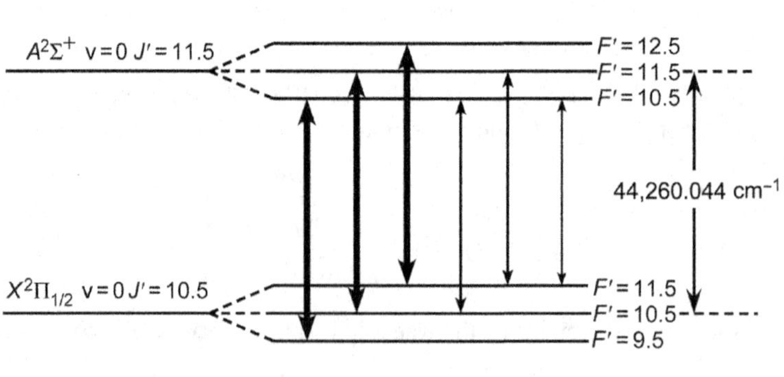

Figure A8.1 Hyperfine transitions in the $R_{11}(10)(J = 10.5 \leftrightarrow J' = 11.5)$ transition of $^{14}N^{16}O$. The width of the vertical lines indicates approximately the strength of the transition.

$\underline{J' = 11.5 \quad F' = 11.5}$

$$\begin{Bmatrix} J & F & I \\ F' & J' & 1 \end{Bmatrix} = \begin{Bmatrix} 10.5 & 11.5 & 1 \\ 11.5 & 11.5 & 1 \end{Bmatrix} = \begin{Bmatrix} 1 & 10.5 & 11.5 \\ 1 & 11.5 & 11.5 \end{Bmatrix} = \begin{Bmatrix} 1 & 11.5 & 11.5 \\ 1 & 10.5 & 11.5 \end{Bmatrix}$$

$s = 24 \quad a = 1 \quad b = 11.5 \quad c = 11.5$

$$\begin{Bmatrix} 1 & 11.5 & 11.5 \\ 1 & 10.5 & 11.5 \end{Bmatrix} = (-1)^{24}\left[\frac{2(s+1)(s-2a)(s-2b)(s-2c+1)}{2b(2b+1)(2b+2)(2c-1)2c(2c+1)}\right]^{1/2}$$

$$= (-1)^{24}\left[\frac{2(25)(22)(1)(2)}{(23)(24)(25)(22)(23)(24)}\right]^{1/2} = +3.6232 \times 10^{-3}$$

$$\frac{\langle \alpha JIF \| \boldsymbol{\mu}_1 \| \alpha' J' I F' \rangle^2}{\langle \alpha' J' \| \boldsymbol{\mu}_1 \| \alpha J' - 1 \rangle^2} = (2F+1)(2F'+1)\begin{Bmatrix} 1 & 11.5 & 11.5 \\ 1 & 10.5 & 11.5 \end{Bmatrix}^2$$

$$= (24)(24)\left(+3.6232 \times 10^{-3}\right)^2 = 7.5614 \times 10^{-3}$$

$$A_{F'F} = A_{J'J}\frac{(2J'+1)\langle \alpha JIF \| \boldsymbol{\mu}_1 \| \alpha' J' I F' \rangle^2}{(2F'+1)\langle \alpha' J' \| \boldsymbol{\mu}_1 \| \alpha J' - 1 \rangle^2}$$

$$= A_{J'J}\left(\frac{24}{24}\right)\left(7.5614 \times 10^{-3}\right) = 7.5614 \times 10^{-3} A_{J'J}. \tag{A8.8}$$

$\underline{J' = 11.5 \quad F' = 10.5}$

$$\begin{Bmatrix} J & F & I \\ F' & J' & 1 \end{Bmatrix} = \begin{Bmatrix} 10.5 & 11.5 & 1 \\ 10.5 & 11.5 & 1 \end{Bmatrix} = \begin{Bmatrix} 1 & 10.5 & 11.5 \\ 1 & 10.5 & 11.5 \end{Bmatrix}$$

$s = 23 \quad a = 1 \quad b = 10.5 \quad c = 11.5$

$$\begin{Bmatrix} 1 & 10.5 & 11.5 \\ 1 & 10.5 & 11.5 \end{Bmatrix} = (-1)^{s}\left[\frac{(s-2b-1)(s-2b)(s-2c+1)(s-2c+2)}{(2b+1)(2b+2)(2b+3)(2c-1)2c(2c+1)}\right]^{1/2}$$

$$= (-1)^{23}\left[\frac{(1)(2)(1)(2)}{(22)(23)(24)(22)(23)(24)}\right]^{1/2} = -1.6469 \times 10^{-4}$$

$$\frac{\langle \alpha JIF \| \boldsymbol{\mu}_1 \| \alpha' J' I F' \rangle^2}{\langle \alpha' J' \| \boldsymbol{\mu}_1 \| \alpha J' - 1 \rangle^2}(2F+1)(2F'+1)\begin{Bmatrix} 1 & 10.5 & 11.5 \\ 1 & 10.5 & 11.5 \end{Bmatrix}^2$$

$$= (11.5)(10.5)\left(-1.6469 \times 10^{-4}\right)^2 = 1.4321 \times 10^{-5}$$

$$A_{F'F} = A_{J'J}\frac{(2J'+1)\langle \alpha JIF \| \boldsymbol{\mu}_1 \| \alpha' J' I F' \rangle^2}{(2F'+1)\langle \alpha' J' \| \boldsymbol{\mu}_1 \| \alpha J' - 1 \rangle^2}$$

$$= A_{J'J}\left(\frac{24}{22}\right)\left(1.4321 \times 10^{-5}\right) = 1.5623 \times 10^{-5} A_{J'J}. \tag{A8.9}$$

$\underline{J = 10.5 \quad F = 10.5}$

$\underline{J' = 11.5 \quad F' = 12.5}$

$$\begin{Bmatrix} J & F & I \\ F' & J' & 1 \end{Bmatrix} = \begin{Bmatrix} 10.5 & 10.5 & 1 \\ 12.5 & 11.5 & 1 \end{Bmatrix} = \begin{Bmatrix} 1 & 10.5 & 10.5 \\ 1 & 12.5 & 11.5 \end{Bmatrix} = \begin{Bmatrix} 1 & 12.5 & 11.5 \\ 1 & 10.5 & 10.5 \end{Bmatrix} = 0.$$

from triangle rule

(A8.10)

$\underline{J' = 11.5 \quad F' = 11.5}$

$$\begin{Bmatrix} J & F & I \\ F' & J' & 1 \end{Bmatrix} = \begin{Bmatrix} 10.5 & 10.5 & 1 \\ 11.5 & 11.5 & 1 \end{Bmatrix} = \begin{Bmatrix} 1 & 10.5 & 10.5 \\ 1 & 11.5 & 11.5 \end{Bmatrix} = \begin{Bmatrix} 1 & 11.5 & 11.5 \\ 1 & 10.5 & 10.5 \end{Bmatrix}$$

$s = 24 \quad a = 1 \quad b = 11.5 \quad c = 11.5$

$$\begin{Bmatrix} 1 & 11.5 & 11.5 \\ 1 & 10.5 & 10.5 \end{Bmatrix} = (-1)^s \left[\frac{s(s+1)(s-2a-1)(s-2a)}{(2b-1)2b(2b+1)(2c-1)2c(2c+1)} \right]^{1/2}$$

$$= (-1)^{24} \left[\frac{(24)(25)(21)(22)}{(22)(23)(24)(22)(23)(24)} \right]^{1/2} = +4.3355 \times 10^{-2}$$

$$\frac{\langle \alpha JIF \| \boldsymbol{\mu}_1 \| \alpha' J' IF' \rangle^2}{\langle \alpha' J' \| \boldsymbol{\mu}_1 \| \alpha J' - 1 \rangle^2} = (2F+1)(2F'+1) \begin{Bmatrix} 1 & 11.5 & 11.5 \\ 1 & 10.5 & 10.5 \end{Bmatrix}^2$$

$$= (24)(24)\left(+4.3355 \times 10^{-2}\right)^2 = 1.08266$$

$$A_{F'F} = A_{J'J} \frac{(2J'+1)\langle \alpha JIF \| \boldsymbol{\mu}_1 \| \alpha' J' IF' \rangle^2}{(2F'+1)\langle \alpha' J' \| \boldsymbol{\mu}_1 \| \alpha J' - 1 \rangle^2} = A_{J'J}\left(\frac{22}{24}\right)(1.08266) = 0.99244 A_{J'J}.$$

(A8.11)

$\underline{J' = 11.5 \quad F' = 10.5}$

$$\begin{Bmatrix} J & F & I \\ F' & J' & 1 \end{Bmatrix} = \begin{Bmatrix} 10.5 & 10.5 & 1 \\ 10.5 & 11.5 & 1 \end{Bmatrix} = \begin{Bmatrix} 1 & 10.5 & 10.5 \\ 1 & 10.5 & 11.5 \end{Bmatrix} = \begin{Bmatrix} 1 & 10.5 & 11.5 \\ 1 & 10.5 & 10.5 \end{Bmatrix}$$

$s = 23 \quad a = 1 \quad b = 10.5 \quad c = 11.5$

$$\begin{Bmatrix} 1 & 10.5 & 11.5 \\ 1 & 10.5 & 10.5 \end{Bmatrix} = (-1)^s \left[\frac{2(s+1)(s-2a)(s-2b)(s-2c+1)}{2b(2b+1)(2b+2)(2c-1)2c(2c+1)} \right]^{1/2}$$

$$= (-1)^{23} \left[\frac{(2)(24)(21)(2)(1)}{(21)(22)(23)(22)(23)(24)} \right]^{1/2} = -3.95257 \times 10^{-3}$$

$$\frac{\langle \alpha JIF \| \boldsymbol{\mu}_1 \| \alpha' J' IF' \rangle^2}{\langle \alpha' J' \| \boldsymbol{\mu}_1 \| \alpha J' - 1 \rangle^2} (2F+1)(2F'+1) \begin{Bmatrix} 1 & 10.5 & 11.5 \\ 1 & 10.5 & 11.5 \end{Bmatrix}^2$$

$$= (22)(22)\left(-3.95257 \times 10^{-3}\right)^2 = 7.56144 \times 10^{-3}$$

$$A_{F'F} = A_{J'J} \frac{(2J'+1)\langle \alpha JIF \| \boldsymbol{\mu}_1 \| \alpha' J' IF' \rangle^2}{(2F'+1)\langle \alpha' J' \| \boldsymbol{\mu}_1 \| \alpha J' - 1 \rangle^2}$$

$$= A_{J'J}\left(\frac{24}{22}\right)\left(7.56144 \times 10^{-3}\right) = 8.2488 \times 10^{-3} A_{J'J}.$$

(A8.12)

$\underline{J = 10.5 \ \ F = 9.5}$

$\underline{J' = 11.5 \ \ F' = 12.5}$

$$\begin{Bmatrix} J & F & I \\ F' & J' & 1 \end{Bmatrix} = \begin{Bmatrix} 10.5 & 9.5 & 1 \\ 12.5 & 11.5 & 1 \end{Bmatrix} = \begin{Bmatrix} 1 & 10.5 & 9.5 \\ 1 & 12.5 & 11.5 \end{Bmatrix} = \begin{Bmatrix} 1 & 12.5 & 11.5 \\ 1 & 10.5 & 9.5 \end{Bmatrix} = 0.$$

from triangle rule

(A8.13)

$\underline{J' = 11.5 \ \ F' = 11.5}$

$$\begin{Bmatrix} J & F & I \\ F' & J' & 1 \end{Bmatrix} = \begin{Bmatrix} 10.5 & 9.5 & 1 \\ 11.5 & 11.5 & 1 \end{Bmatrix} = \begin{Bmatrix} 1 & 10.5 & 9.5 \\ 1 & 11.5 & 11.5 \end{Bmatrix} = \begin{Bmatrix} 1 & 11.5 & 11.5 \\ 1 & 10.5 & 9.5 \end{Bmatrix} = 0.$$

from the triangle rule

(A8.14)

$\underline{J' = 11.5 \ \ F' = 10.5}$

$$\begin{Bmatrix} J & F & I \\ F' & J' & 1 \end{Bmatrix} = \begin{Bmatrix} 10.5 & 9.5 & 1 \\ 10.5 & 11.5 & 1 \end{Bmatrix} = \begin{Bmatrix} 1 & 10.5 & 9.5 \\ 1 & 10.5 & 11.5 \end{Bmatrix} = \begin{Bmatrix} 1 & 10.5 & 11.5 \\ 1 & 10.5 & 9.5 \end{Bmatrix}$$

$s = 23 \quad a = 1 \quad b = 10.5 \quad c = 11.5$

$$\begin{Bmatrix} 1 & 10.5 & 11.5 \\ 1 & 10.5 & 9.5 \end{Bmatrix} = (-1)^s \left[\frac{s(s+1)(s-2a-1)(s-2a)}{(2b-1)2b(2b+1)(2c-1)2c(2c+1)} \right]^{1/2}$$

$$= (-1)^{23} \left[\frac{(23)(24)(20)(21)}{(20)(21)(22)(22)(23)(24)} \right]^{1/2} = -4.54545 \times 10^{-2}$$

$$\frac{\langle \alpha JIF \| \boldsymbol{\mu}_1 \| \alpha' J' I F' \rangle^2}{\langle \alpha' J' \| \boldsymbol{\mu}_1 \| \alpha J' - 1 \rangle^2} (2F+1)(2F'+1) \begin{Bmatrix} 1 & 10.5 & 11.5 \\ 1 & 10.5 & 9.5 \end{Bmatrix}^2$$

$$= (20)(22)\left(-4.54545 \times 10^{-2}\right)^2 = 0.909091$$

$$A_{F'F} = A_{J'J} \frac{(2J'+1)\langle \alpha JIF \| \boldsymbol{\mu}_1 \| \alpha' J' I F' \rangle^2}{(2F'+1)\langle \alpha' J' \| \boldsymbol{\mu}_1 \| \alpha J' - 1 \rangle^2} = A_{J'J} \left(\frac{24}{22} \right) (0.909091) = 0.99174 A_{J'J}$$

(A8.15)

Note from Table A8.1 that the total spontaneous emission coefficient $\sum_F A_{F'F}$ from each upper level F' is equal to the $A_{J'J}$. In addition, the factor $\sum_{F'} A_{F'F} \frac{(2F'+1)}{(2F+1)}$, proportional to the total absorption coefficient for level F, is equal to $A_{J'J} \frac{(2J'+1)}{(2J+1)}$. Consequently, the spontaneous emission rate coefficient and the absorption coefficients are not affected by the hyperfine interaction for measurements where the hyperfine splitting is not resolved.

Table A8.1 Comparison of spontaneous emission rate coefficients and absorption coefficients for the hyperfine transitions in the $R_1(10)$ transition in the $A^2\Sigma^+ - X^2$ (0, 0) band of $^{14}N^{16}O$.

$A_{F'F}/A_{J'J}$	$F = 9.5$	$F = 10.5$	$F = 11.5$	$\sum_F A_{F'F}/A_{J'J}$
$F' = 10.5$	0.99174	8.2488×10^{-3}	1.5263×10^{-5}	1.0000
$F' = 11.5$	0	0.99244	7.5614×10^{-3}	1.0000
$F' = 12.5$	0	0	1.00000	1.0000
$\sum_{F'} \frac{A_{F'F}(2F'+1)}{A_{J'J}(2F+1)}$	1.0909	1.0909	1.0909	

Appendix 9 Voigt Function Values

x	a = 0	a = 0.1	a = 0.2	a = 0.3	a = 0.4	a = 0.5	a = 0.6	a = 0.7
0.00	1.000000	0.896457	0.809020	0.734599	0.671788	0.615690	0.567805	0.525930
0.10	0.990050	0.888479	0.802567	0.729337	0.666463	0.612109	0.564818	0.523423
0.20	0.960789	0.864983	0.783538	0.713801	0.653680	0.601513	0.555974	0.515991
0.30	0.913931	0.827246	0.752895	0.688720	0.632996	0.584333	0.541605	0.503896
0.40	0.852144	0.777267	0.712146	0.655244	0.605295	0.561252	0.522246	0.487556
0.50	0.778801	0.717588	0.663223	0.614852	0.571717	0.533157	0.498591	0.467521
0.60	0.697676	0.651076	0.608322	0.569238	0.533581	0.501079	0.471453	0.444434
0.70	0.612626	0.580698	0.549739	0.520192	0.492289	0.466127	0.441712	0.418998
0.80	0.527292	0.509299	0.489710	0.469480	0.449244	0.429418	0.410264	0.391936
0.90	0.444858	0.439421	0.430271	0.418736	0.405763	0.392021	0.377977	0.363957
1.00	0.367879	0.373170	0.373153	0.369386	0.363020	0.354900	0.345649	0.335721
1.10	0.298197	0.312136	0.319717	0.322586	0.321993	0.318884	0.313978	0.307816
1.20	0.236928	0.257374	0.270928	0.279199	0.283443	0.284638	0.283540	0.280740
1.30	0.184520	0.209431	0.227362	0.239793	0.247908	0.252654	0.254784	0.254895
1.40	0.140858	0.168407	0.189247	0.204662	0.215711	0.223262	0.228026	0.230578
1.50	0.105399	0.134049	0.156521	0.173865	0.186984	0.196636	0.203461	0.207990
1.60	0.077305	0.105843	0.128895	0.147272	0.161702	0.172820	0.181177	0.187245
1.70	0.055576	0.083112	0.105929	0.124612	0.139717	0.151751	0.161171	0.168379
1.80	0.039164	0.065099	0.087090	0.105522	0.120793	0.133288	0.143369	0.151366
1.90	0.027052	0.051038	0.071811	0.089592	0.104641	0.117233	0.127644	0.136134
2.00	0.018316	0.040201	0.059531	0.076396	0.090944	0.103359	0.113836	0.122574
2.10	0.012155	0.031936	0.049726	0.065521	0.079385	0.091422	0.101765	0.110558
2.20	0.007907	0.025678	0.041927	0.056586	0.069655	0.081182	0.091245	0.099943
2.30	0.005042	0.020958	0.035728	0.049248	0.061473	0.072408	0.082092	0.090585
2.40	0.003151	0.017397	0.030792	0.043211	0.054585	0.064890	0.074132	0.082345
2.50	0.001930	0.014698	0.026841	0.038226	0.048773	0.058437	0.067205	0.075088
2.60	0.001159	0.012635	0.023653	0.034087	0.043849	0.052885	0.061167	0.068691
2.70	0.000682	0.011037	0.021057	0.030626	0.039656	0.048090	0.055890	0.063043
2.80	0.000394	0.009778	0.018918	0.027707	0.036064	0.043930	0.051264	0.058046
2.90	0.000223	0.008769	0.017134	0.025225	0.032967	0.040304	0.047194	0.053611
3.00	0.000123	0.007943	0.015627	0.023095	0.030279	0.037126	0.043598	0.049665
3.10	0.000067	0.007254	0.014338	0.021250	0.027929	0.034328	0.040407	0.046141
3.20	0.000036	0.006670	0.013225	0.019639	0.025862	0.031849	0.037565	0.042983
3.30	0.000019	0.006167	0.012252	0.018222	0.024032	0.029643	0.035022	0.040144
3.40	0.000010	0.005728	0.011394	0.016966	0.022403	0.027670	0.032738	0.037582
3.50	0.000005	0.005340	0.010633	0.015846	0.020944	0.025897	0.030677	0.035263
3.60	0.000002	0.004995	0.009952	0.014841	0.019632	0.024297	0.028812	0.033158
3.70	0.000001	0.004685	0.009339	0.013935	0.018446	0.022847	0.027118	0.031239
3.80	0.000000	0.004406	0.008786	0.013115	0.017370	0.021529	0.025574	0.029486

(*cont.*)

x	$a = 0$	$a = 0.1$	$a = 0.2$	$a = 0.3$	$a = 0.4$	$a = 0.5$	$a = 0.6$	$a = 0.7$
3.90	0.000000	0.004153	0.008282	0.012368	0.016389	0.020326	0.024162	0.027880
4.00	0.000000	0.003922	0.007824	0.011687	0.015493	0.019225	0.022867	0.026405
5.00	0.000000	0.002407	0.004807	0.007194	0.009560	0.011900	0.014208	0.016477
6.00	0.000000	0.001637	0.003271	0.004899	0.006518	0.008125	0.009717	0.011292
7.00	0.000000	0.001188	0.002375	0.003559	0.004738	0.005910	0.007075	0.008231
8.00	0.000000	0.000903	0.001805	0.002705	0.003603	0.004497	0.005386	0.006271
9.00	0.000000	0.000710	0.001419	0.002127	0.002834	0.003538	0.004239	0.004938
10.00	0.000000	0.000573	0.001145	0.001717	0.002288	0.002857	0.003424	0.003990

x	$a = 0.8$	$a = 0.9$	$a = 1.0$	$a = 1.1$	$a = 1.2$	$a = 1.3$	$a = 1.4$	$a = 1.5$
0.00	0.489101	0.456532	0.427584	0.401730	0.378537	0.357643	0.338744	0.321585
0.10	0.486982	0.454731	0.426044	0.400406	0.377393	0.356649	0.337876	0.320825
0.20	0.480697	0.449383	0.421468	0.396470	0.373989	0.353691	0.335294	0.318561
0.30	0.470452	0.440655	0.413989	0.390028	0.368412	0.348839	0.331054	0.314839
0.40	0.456579	0.428808	0.403818	0.381250	0.360800	0.342206	0.325248	0.309736
0.50	0.439512	0.414191	0.391234	0.370363	0.351335	0.333942	0.318001	0.303355
0.60	0.419766	0.397216	0.376571	0.357637	0.340241	0.324229	0.309463	0.295820
0.70	0.397906	0.378341	0.360200	0.343375	0.327766	0.313273	0.299804	0.287274
0.80	0.374518	0.358043	0.342511	0.327900	0.314176	0.301294	0.289208	0.277869
0.90	0.350182	0.336799	0.323899	0.311537	0.299741	0.288519	0.277865	0.267766
1.00	0.325446	0.315064	0.304744	0.294606	0.284731	0.275174	0.265967	0.257128
1.10	0.300807	0.293259	0.285402	0.277407	0.269401	0.261476	0.253967	0.246112
1.20	0.276693	0.271752	0.266189	0.260213	0.253985	0.247628	0.241233	0.234870
1.30	0.253461	0.250858	0.247381	0.243266	0.238695	0.233813	0.228733	0.223542
1.40	0.231385	0.230826	0.229205	0.226767	0.223710	0.220192	0.216340	0.212253
1.50	0.210664	0.211846	0.211837	0.210881	0.209182	0.206902	0.204177	0.201115
1.60	0.191423	0.194049	0.195407	0.195734	0.195228	0.194053	0.192347	0.190222
1.70	0.173725	0.177513	0.180002	0.181414	0.181938	0.181733	0.180933	0.179651
1.80	0.157578	0.162268	0.165667	0.167977	0.169373	0.170003	0.169997	0.169465
1.90	0.142949	0.148310	0.152418	0.155452	0.157569	0.158906	0.159585	0.159709
2.00	0.129768	0.135600	0.140240	0.143840	0.146541	0.148466	0.149725	0.150415
2.10	0.117948	0.124081	0.129097	0.133125	0.136286	0.138689	0.140432	0.141604
2.20	0.107383	0.113679	0.118941	0.123277	0.126788	0.129570	0.131709	0.133284
2.30	0.097963	0.104309	0.109709	0.114251	0.118019	0.121092	0.123548	0.125454
2.40	0.089576	0.095884	0.101336	0.105999	0.109942	0.113232	0.115935	0.118109
2.50	0.082112	0.088317	0.093751	0.098466	0.102518	0.105960	0.108848	0.111233
2.60	0.075467	0.081521	0.086885	0.091598	0.095702	0.099243	0.102264	0.104811
2.70	0.069548	0.075416	0.080670	0.085338	0.089451	0.093044	0.096155	0.098820
2.80	0.064266	0.069927	0.075043	0.079632	0.083718	0.087328	0.090492	0.093239
2.90	0.059543	0.064986	0.069944	0.074431	0.078462	0.082059	0.085245	0.088044
3.00	0.055311	0.060529	0.065318	0.069685	0.073641	0.077202	0.080385	0.083210
3.10	0.051509	0.056501	0.061114	0.065350	0.069216	0.072722	0.075883	0.078712
3.20	0.048083	0.052854	0.057289	0.061387	0.065151	0.068589	0.071711	0.074529
3.30	0.044989	0.049544	0.053801	0.057757	0.061413	0.064773	0.067844	0.070636

(*cont.*)

x	$a = 0.8$	$a = 0.9$	$a = 1.0$	$a = 1.1$	$a = 1.2$	$a = 1.3$	$a = 1.4$	$a = 1.5$
3.40	0.042185	0.046532	0.050615	0.054428	0.057970	0.061246	0.064258	0.067012
3.50	0.039637	0.043785	0.047698	0.051370	0.054798	0.057984	0.060928	0.063637
3.60	0.037316	0.041274	0.045023	0.048556	0.051869	0.054962	0.057835	0.060491
3.70	0.035195	0.038974	0.042565	0.045962	0.049161	0.052159	0.054958	0.057557
3.80	0.033253	0.036861	0.040301	0.043567	0.046653	0.049558	0.052279	0.054819
3.90	0.031469	0.034916	0.038212	0.041352	0.044328	0.047319	0.049783	0.052260
4.00	0.029827	0.033122	0.036281	0.039299	0.042169	0.044888	0.047455	0.049868
5.00	0.018703	0.020880	0.023003	0.025069	0.027075	0.029016	0.030890	0.032696
6.00	0.012846	0.014378	0.015885	0.017365	0.018815	0.020234	0.021620	0.022972
7.00	0.009377	0.010510	0.011630	0.012735	0.013825	0.014898	0.015953	0.016989
8.00	0.007149	0.008020	0.008884	0.009739	0.010585	0.011421	0.012246	0.013060
9.00	0.005632	0.006322	0.007008	0.007688	0.008363	0.009031	0.009693	0.010347
10.00	0.004553	0.005113	0.005670	0.006224	0.006773	0.007319	0.007861	0.008397

x	$a = 1.6$	$a = 1.7$	$a = 1.8$	$a = 1.9$	$a = 2.0$	$a = 2.1$	$a = 2.2$	$a = 2.3$
0.00	0.305953	0.291663	0.278560	0.266509	0.255396	0.245119	0.235593	0.226742
0.10	0.305284	0.291072	0.278035	0.266042	0.254978	0.244745	0.235256	0.226438
0.20	0.303290	0.289309	0.276470	0.264648	0.253732	0.243628	0.234251	0.225531
0.30	0.300009	0.286406	0.273892	0.262350	0.251677	0.241783	0.232592	0.224033
0.40	0.295506	0.282417	0.270346	0.259186	0.248844	0.239239	0.230300	0.221963
0.50	0.289866	0.277412	0.265890	0.255205	0.245176	0.236031	0.227407	0.219347
0.60	0.283192	0.271479	0.260598	0.250469	0.241025	0.232204	0.223952	0.216219
0.70	0.275602	0.264718	0.254554	0.245050	0.236152	0.227810	0.219978	0.212616
0.80	0.267228	0.257237	0.247851	0.239027	0.230724	0.222905	0.215535	0.208581
0.90	0.258203	0.249151	0.240586	0.232482	0.224813	0.217552	0.210676	0.204160
1.00	0.248665	0.240578	0.232861	0.225503	0.218493	0.211816	0.205457	0.199402
1.10	0.238752	0.231635	0.224775	0.218176	0.211839	0.205760	0.199935	0.194356
1.20	0.228592	0.222436	0.216428	0.210587	0.204926	0.199452	0.194166	0.189072
1.30	0.218309	0.213086	0.207912	0.202818	0.197827	0.192953	0.188208	0.183599
1.40	0.208014	0.203684	0.199315	0.191947	0.190608	0.186324	0.182112	0.177985
1.50	0.197806	0.194320	0.190717	0.187043	0.183335	0.179623	0.175930	0.172276
1.60	0.187772	0.185073	0.182189	0.179172	0.176064	0.172901	0.169710	0.166513
1.70	0.177983	0.176008	0.173792	0.171390	0.168849	0.166206	0.163493	0.160735
1.80	0.168500	0.167183	0.165579	0.163746	0.161733	0.159580	0.157320	0.154982
1.90	0.159369	0.158641	0.157593	0.156282	0.154757	0.153059	0.151224	0.149281
2.00	0.150622	0.150418	0.149870	0.149032	0.147953	0.146675	0.145234	0.143660
2.10	0.142283	0.142540	0.142434	0.142021	0.141347	0.140453	0.139375	0.138145
2.20	0.134367	0.135021	0.135305	0.135269	0.134959	0.134414	0.133669	0.132755
2.30	0.126877	0.127873	0.128495	0.128792	0.128805	0.128574	0.128130	0.127506
2.40	0.119812	0.121096	0.122010	0.122597	0.122897	0.122945	0.122773	0.122411
2.50	0.113165	0.114690	0.115851	0.116689	0.117239	0.117534	0.117606	0.117481
2.60	0.106925	0.108647	0.110016	0.111067	0.111834	0.112347	0.112635	0.112723
2.70	0.101076	0.102957	0.101198	0.105730	0.106683	0.107386	0.107864	0.108140
2.80	0.095601	0.097608	0.099288	0.100671	0.101783	0.102649	0.103293	0.103737
2.90	0.090482	0.092584	0.094376	0.095882	0.097127	0.098133	0.098922	0.099513

(*cont.*)

x	$a = 1.6$	$a = 1.7$	$a = 1.8$	$a = 1.9$	$a = 2.0$	$a = 2.1$	$a = 2.2$	$a = 2.3$
3.00	0.085697	0.087870	0.089749	0.091355	0.092711	0.093835	0.094748	0.095467
3.10	0.081229	0.083450	0.085394	0.087080	0.088525	0.089749	0.090767	0.091597
3.20	0.077055	0.079306	0.081297	0.083044	0.084562	0.085867	0.086974	0.087900
3.30	0.073758	0.075423	0.077445	0.079236	0.080811	0.082182	0.083364	0.084370
3.40	0.069518	0.071785	0.073823	0.075646	0.077263	0.078687	0.079930	0.081004
3.50	0.066116	0.068374	0.070419	0.072260	0.073908	0.075373	0.076666	0.077796
3.60	0.062936	0.065176	0.067217	0.069068	0.070736	0.072232	0.073563	0.074739
3.70	0.059962	0.062177	0.064206	0.066058	0.067738	0.069254	0.070615	0.071829
3.80	0.057179	0.059362	0.061374	0.063219	0.064903	0.066433	0.067815	0.069058
3.90	0.054572	0.056720	0.058708	0.060540	0.062222	0.063759	0.065156	0.066420
4.00	0.052128	0.054237	0.056198	0.058013	0.059687	0.061224	0.062630	0.063909
5.00	0.034430	0.036093	0.037683	0.039200	0.040644	0.042014	0.043312	0.044538
6.00	0.024288	0.025566	0.026807	0.028008	0.029170	0.030292	0.031373	0.032413
7.00	0.018005	0.019000	0.019973	0.020925	0.021853	0.022759	0.023640	0.024497
8.00	0.013862	0.014652	0.015429	0.016193	0.016942	0.017677	0.018397	0.019103
9.00	0.010994	0.011950	0.012265	0.012887	0.013501	0.014105	0.014699	0.015284
10.00	0.008929	0.009456	0.009977	0.010492	0.011002	0.011505	0.012002	0.012492

x	$a = 2.4$	$a = 2.5$	$a = 2.6$	$a = 2.7$	$a = 2.8$	$a = 2.9$	$a = 3.0$	$a = 4.0$
0.00	0.218499	0.210806	0.203613	0.196874	0.190549	0.184602	0.179001	0.136999
0.10	0.218224	0.210557	0.203387	0.196668	0.190360	0.184429	0.178842	0.136925
0.20	0.217404	0.209813	0.202710	0.196050	0.189796	0.183912	0.178368	0.136701
0.30	0.216047	0.208582	0.201589	0.195028	0.188861	0.183056	0.177581	0.136331
0.40	0.214172	0.206879	0.200039	0.193613	0.187566	0.181868	0.176491	0.135814
0.50	0.211800	0.204723	0.198074	0.191818	0.185924	0.180361	0.175105	0.135156
0.60	0.208961	0.202139	0.195717	0.189664	0.183950	0.178549	0.173437	0.134359
0.70	0.205686	0.199155	0.192992	0.187170	0.181662	0.176447	0.171502	0.133428
0.80	0.202013	0.195804	0.189928	0.184362	0.179084	0.174074	0.169315	0.132369
0.90	0.197982	0.192120	0.186554	0.181265	0.176237	0.171452	0.166895	0.131187
1.00	0.193634	0.188139	0.182903	0.177910	0.173147	0.168602	0.164261	0.129888
1.10	0.189014	0.183901	0.179008	0.174324	0.169840	0.165546	0.161434	0.128480
1.20	0.184165	0.179444	0.174903	0.170538	0.166342	0.162310	0.158435	0.126970
1.30	0.179131	0.174805	0.170623	0.166582	0.162681	0.158916	0.155285	0.125365
1.40	0.173954	0.170024	0.166201	0.162487	0.158883	0.155389	0.152005	0.123672
1.50	0.168674	0.165136	0.161669	0.158281	0.154975	0.151753	0.148618	0.121901
1.60	0.163330	0.160175	0.157060	0.153993	0.150981	0.148030	0.145144	0.120059
1.70	0.157958	0.155175	0.152402	0.149649	0.146927	0.144243	0.141602	0.118154
1.80	0.152591	0.150165	0.147722	0.145274	0.142834	0.140411	0.138012	0.116194
1.90	0.147256	0.145172	0.143045	0.140892	0.138725	0.136555	0.134391	0.114187
2.00	0.141982	0.140220	0.138395	0.136523	0.134619	0.132693	0.130757	0.112139
2.10	0.136789	0.135331	0.133791	0.132187	0.130533	0.128842	0.127125	0.110047
2.20	0.131699	0.130524	0.129252	0.127900	0.126483	0.125016	0.123510	0.107955
2.30	0.126726	0.125814	0.124792	0.123676	0.122484	0.121229	0.119922	0.105826
2.40	0.121884	0.121215	0.120424	0.119530	0.118548	0.117492	0.116375	0.103696
2.50	0.117184	0.116737	0.116160	0.115471	0.114685	0.113816	0.112878	0.101554

(*cont.*)

x	$a = 2.4$	$a = 2.5$	$a = 2.6$	$a = 2.7$	$a = 2.8$	$a = 2.9$	$a = 3.0$	$a = 4.0$
2.60	0.112633	0.112389	0.112008	0.111508	0.110904	0.110210	0.109439	0.099411
2.70	0.108238	0.108177	0.107975	0.107648	0.107213	0.106682	0.106067	0.097278
2.80	0.104002	0.104105	0.104066	0.103898	0.103617	0.103236	0.102767	0.095145
2.90	0.099925	0.100177	0.100284	0.100261	0.100122	0.099879	0.099544	0.093040
3.00	0.096010	0.096393	0.096632	0.096739	0.096729	0.096613	0.096402	0.090934
3.10	0.092255	0.092754	0.093110	0.093360	0.093442	0.093442	0.093345	0.088871
3.20	0.088657	0.089259	0.089719	0.090050	0.090863	0.090368	0.090375	0.086808
3.30	0.085213	0.085905	0.086458	0.086883	0.087190	0.087391	0.087493	0.084800
3.40	0.081921	0.082690	0.083324	0.083832	0.084225	0.084511	0.084700	0.082793
3.50	0.078774	0.079611	0.080316	0.080898	0.081366	0.081730	0.081996	0.080849
3.60	0.075770	0.076664	0.077430	0.078076	0.078612	0.079044	0.079381	0.078906
3.70	0.072902	0.073845	0.074663	0.075366	0.075961	0.076455	0.076855	0.077034
3.80	0.070166	0.071149	0.072013	0.072764	0.073411	0.073959	0.074415	0.075162
3.90	0.067556	0.068572	0.069474	0.070267	0.070959	0.071555	0.072061	0.073366
4.00	0.065067	0.066110	0.067043	0.067872	0.068603	0.069240	0.069791	0.071570
5.00	0.045693	0.046779	0.047798	0.048750	0.049637	0.050462	0.051226	0.055997
6.00	0.033412	0.034370	0.035287	0.036164	0.037000	0.037797	0.038555	0.044141
7.00	0.025329	0.026137	0.026919	0.027676	0.028408	0.029115	0.029796	0.035264
8.00	0.019792	0.020466	0.021124	0.021766	0.022392	0.023001	0.023593	0.028603
9.00	0.015859	0.016423	0.016977	0.017520	0.018053	0.018574	0.019084	0.023550
10.00	0.012975	0.013451	0.013920	0.014382	0.014836	0.015283	0.015722	0.019663

x	$a = 5.0$	$a = 6.0$	$a = 7.0$	$a = 8.0$	$a = 9.0$	$a = 10.0$
0.00	0.110705	0.092777	0.079800	0.069985	0.062308	0.056141
0.10	0.110664	0.092752	0.079785	0.069975	0.062300	0.056136
0.20	0.110543	0.092680	0.079738	0.069943	0.062278	0.056119
0.30	0.110342	0.092560	0.079661	0.069891	0.062241	0.056092
0.40	0.110062	0.092392	0.079553	0.069817	0.062189	0.056054
0.50	0.109703	0.092177	0.079414	0.069723	0.062122	0.056004
0.60	0.109268	0.091915	0.079245	0.069608	0.062040	0.055944
0.70	0.108757	0.091608	0.079047	0.069473	0.061944	0.055874
0.80	0.108174	0.091256	0.078819	0.069317	0.061834	0.055793
0.90	0.107520	0.090859	0.078562	0.069142	0.061709	0.055701
1.00	0.106798	0.090421	0.078277	0.068947	0.061570	0.055598
1.10	0.106010	0.089940	0.077965	0.068733	0.061417	0.055486
1.20	0.105160	0.089420	0.077625	0.068500	0.061251	0.055363
1.30	0.104250	0.088861	0.077259	0.068248	0.061071	0.055230
1.40	0.103284	0.088264	0.076868	0.067979	0.060877	0.055087
1.50	0.102265	0.087632	0.076451	0.067691	0.060671	0.054934
1.60	0.101196	0.086966	0.076011	0.067387	0.060452	0.054772
1.70	0.100082	0.086267	0.075548	0.067065	0.060221	0.054600
1.80	0.098925	0.085537	0.075062	0.066728	0.059978	0.054419
1.90	0.097729	0.084779	0.074556	0.066374	0.059722	0.054229
2.00	0.096498	0.083994	0.074029	0.066006	0.059455	0.054030
2.10	0.095221	0.083171	0.073474	0.065616	0.059172	0.053819

(*cont.*)

x	$a = 5.0$	$a = 6.0$	$a = 7.0$	$a = 8.0$	$a = 9.0$	$a = 10.0$
2.20	0.093944	0.082348	0.072918	0.065226	0.058889	0.053607
2.30	0.092617	0.081482	0.072328	0.064809	0.058584	0.053379
2.40	0.091290	0.080616	0.071738	0.064391	0.058280	0.053150
2.50	0.089927	0.079714	0.071118	0.063950	0.057956	0.052907
2.60	0.088564	0.078811	0.070497	0.063508	0.057632	0.052663
2.70	0.087176	0.077880	0.069850	0.063044	0.057290	0.052404
2.80	0.085789	0.076948	0.069203	0.062580	0.056948	0.052146
2.90	0.084388	0.075993	0.068533	0.062096	0.056589	0.051874
3.00	0.082988	0.075039	0.067863	0.061613	0.056231	0.051602
3.10	0.081585	0.074068	0.067174	0.061111	0.055857	0.051317
3.20	0.080182	0.073097	0.066485	0.060610	0.055484	0.051033
3.30	0.078785	0.072115	0.065781	0.060094	0.055096	0.050736
3.40	0.077389	0.071134	0.065077	0.059578	0.054709	0.050440
3.50	0.076007	0.070147	0.064362	0.059048	0.054310	0.050133
3.60	0.074625	0.069161	0.063646	0.058519	0.053911	0.049826
3.70	0.073265	0.068174	0.062923	0.057980	0.053501	0.049509
3.80	0.071904	0.067187	0.062199	0.057440	0.053091	0.049193
3.90	0.070570	0.066205	0.061470	0.056337	0.052672	0.048867
4.00	0.069236	0.065223	0.060742	0.055235	0.052254	0.048542
5.00	0.059261	0.055771	0.053487	0.050744	0.047887	0.045098
6.00	0.046756	0.047335	0.046647	0.045230	0.043439	0.041492
7.00	0.038546	0.040128	0.040502	0.040070	0.039133	0.037904
8.00	0.032032	0.034115	0.035144	0.035398	0.035110	0.034459
9.00	0.026873	0.029152	0.030553	0.031259	0.031440	0.031238
10.00	0.022768	0.025069	0.026656	0.027642	0.028147	0.028280

References

Alekseyev, V., Grasiuk, A., Ragulsky, V., Sobel'man, I., and Faizulov, F. (1968). S-6-Stimulated Raman scattering in gases and gain pressure dependence. *IEEE Journal of Quantum Electronics*, **QE-4**, 654–656.

Amiot C., Bacis, A., and Guelachvili, G. (1978). Infrared study of the $X^2\Pi$ v = 0, 1, 2 levels of $^{14}N^{16}O$. Preliminary results on the v = 0, 1 levels of $^{14}N^{17}O$, $^{14}N^{18}O$, and $^{15}N^{16}O$. *Canadian Journal of Physics*, **56**, 251–265.

Bassani, F., Forney, J. J., and Quattropani, A. (1977). Choice of gauge in two-photon transitions: 1s-2s transitions in atomic hydrogen. *Physical Review Letters*, **39**, 1070–1073.

Beaud, P., Frey, H.-M., Lang, T., and Motzkus, M. (2001). Flame thermometry by femtosecond CARS. *Chemical Physics Letters*, **344**, 407–412.

Becker, R. (1964). *Electromagnetic Fields and Interactions*. New York: Dover Publications.

Becker, R. (2006). *Electromagnetic Fields and Interactions*. New York: Dover.

Bennett, R. J. M. (1970). Hönl–London factors for doublet transitions in diatomic molecules. *Monthly Notices of the Royal Astronomical Society*, **147**, 35–46.

Bérard,M., Lallemand, P., Cebe, J. P., and Giraud, M. (1983). Experimental and theoretical analysis of the temperature dependence of the rotational Raman linewidths of oxygen. *Journal of Chemical Physics*, **78**, 672–687.

Bernath, P. F. (2016). *Spectra of Atoms and Molecules*, 3rd edition. New York: Oxford University Press.

Bethe, H. A. and Saltpeter, E. E. (1957). *Quantum Mechanics of One- and Two-Electron Atoms*. Berlin: Springer-Verlag.

Beyer, W. H. (1978). *Handbook of Mathematical Sciences*, 5th edition. Boca Raton, FL: CRC Press.

Billoux, T., Cressault, Y., and Gleizes, A. (2014). Tables of radiative transition probabilities for the main diatomic molecular systems of OH, CH, CH^+, CO, and CO^+ occurring in CO-H_2 syngas-type plasma. *Journal of Quantitative Spectroscopy and Radiative Transfer*, **133**, 434–444. DOI: https://doi.org/10.1016/j.jqsrt.2013.09.005.

Bohlin, A., Bengtsson, P.-E., and Marrocco, M. (2011). On the sensitivity of rotational CARS N_2 thermometry to the Herman–Wallis factor. *Journal of Raman Spectroscopy*, **42**, 1843–1847.

Bohlin, A. and Kliewer, C. J. (2013). Communication: Two-dimensional gas-phase coherent anti-Stokes Raman spectroscopy (2D-CARS): Simultaneous planar imaging and multiplex spectroscopy in a single laser shot. *Journal of Chemical Physics*, **138**, Article No. 221101. DOI: https://doi.org/10.1063/1.4815927.

Bohlin, A. and Kliewer, C. J. (2014). Two-beam ultrabroadband coherent anti-Stokes Raman spectroscopy for high resolution gas-phase multiplex imaging. *Applied Physics Letters*, **104**, Article No. 031107. DOI: https://doi.org/10.1063/1.4862980.

Bohlin, A., Jainski, C., Patterson, B. D., Dreizler, A., and Kliewer, C. J. (2017). Multiparameter spatio-thermochemical probing of flame-wall interactions with coherent Raman imaging. *Proceedings of the Combustion Institute*, **36**, 4557–4564. DOI: https://doi.org/10.1016/j.proci.2016.07.062.

Boyd, R. W. (2008). *Nonlinear Optics*, 3rd edition. Burlington, MA: Academic Press.

Brazier, C. R., Ram, R. S., and Bernath, P. F. (1986). Fourier transform spectroscopy of the $A^3\Pi$-$X^3\Sigma^-$ transition of NH. *Journal of Molecular Spectroscopy,* **120**, 381–402.

Brooke, J. S. A., Bernath, P. F., Western, C. M., Sneden, C., Afşar, M., Li, G., and Gordon, I. E. (2016). Line strengths of rovibrational and rotational transitions in the $X^2\Pi$ ground state of OH. *Journal of Quantitative Spectroscopy and Radiative Transfer*, **168**, 142–157. DOI: https://doi.org/10.1016/j.jqsrt.2015. 07.021

Brooke, J. S. A., Ram, R. S., Western, C. M., Schwenke, D. W., Li, G., and Bernath, P. F. (2014). Einstein A coefficients and oscillator strengths for the $A^2\Pi - X^2\Sigma^+$ (red) and $B^2\Sigma^+ - X^2\Sigma^+$ (violet) systems and rovibrational transitions in the $X^2\Sigma^+$ state of CN. *The Astrophysical Journal Supplement Series,* **210**, 23 (15 pp.). DOI: https://doi.org/10.1088/0067-0049/210/2/23.

Brown, J. M., Hougen, J. T., Huber, K.-P., Johns, J. W. C., Kopp, I., Lefebvre-Brion, H., Merer, A. J., Ramsay, D. A., Rostas, J., and Zare, R. N. (1975). The labeling of parity doublet levels in linear molecules. *Journal of Molecular Spectroscopy*, **55**: 500–503.

Brown, J. M., Kaise, M., Kerr, C. M. L., and Milton, D. J. (1978). A determination of fundamental Zeeman parameters of the OH radical. *Molecular Physics*, **36**, 553–582.

Brown, J. M. and Merer, A. J. (1979). Lambda-type doubling parameters for molecules in Π electronic states of triplet and higher multiplicity. *Journal of Molecular Spectroscopy*, **74**, 488–494.

Brown, J. M., Cheung, A. S.-C., and Merer, A. J. (1987). Λ-type doubling parameters for molecules in Δ electronic states. *Journal of Molecular Spectroscopy*, **124**, 464–475.

Brown, J. M. and Carrington, A. (2003). *Rotational Spectroscopy of Diatomic Molecules*. Cambridge: Cambridge University Press.

Brouard, M., Chadwick, H. Chang, Y.-P., Howard, B. J., Marinakis, S., Screen, N., Seamons, S. A., and La Via, A. (2012). The hyperfine structure of NO $\left(A^2\Sigma^+\right)$. *Journal of Molecular Spectroscopy*, **282**, 42–49. DOI: https://doi.org/10.1016/j.jms.2012.11.003

Bouanich, J. P. and Brodbeck, C. (1976). Vibration-rotation matrix elements for diatomic molecules; vibration-rotation interaction functions $F_{\mathrm{v}}^{\mathrm{v}'}(m)$ for CO. *Journal of Quantitative Spectroscopy and Radiative Transfer,* **16**, 153–163.

Buldakov, M. A., Chrepanov, V. N., Korolev, B. V., and Matrosov, I. I. (2003). Role of intramolecular interactions in Raman spectra of N_2 and O_2 molecules. *Journal of Molecular Spectroscopy*, **217**, 1–8.

Butcher, P. N. and Cotter, D. (1990). *The Elements of Nonlinear Optics*. Cambridge: Cambridge University Press.

Chang, A. Y., DiRosa, M. D., and Hanson, R. K. (1992). Temperature dependence of collision broadening and shift in the NO $A \leftarrow X$ $(0,0)$ band in the presence of argon and nitrogen. *Journal of Quantitative Spectroscopy and Radiative Transfer*, **47**, 375–390.

Chou, S.-I., Baer, D. S., and Hanson, R. K. (1999a). Spectral intensity and lineshape measurements in the first overtone band of HF using tunable diode lasers. *Journal of Molecular Spectroscopy*, **195**, 123–131.

Chou, S.-I., Baer, D. S., and Hanson, R. K. (1999b). Diode-laser measurements of He-, Ar-, and N_2-broadened HF lineshapes in the first overtone band. *Journal of Molecular Spectroscopy*, **196**, 70–76.

Condon, E. U. and Shortley, G. H. (1951). *The Theory of Atomic Spectra*. New York: Cambridge University Press.

Cowan, R. D. (1981). *The Theory of Atomic Structure and Spectra*. Berkeley, CA: University of California Press.

Craig, D. P. and Thirunamachandran, T. (1984). *Molecular Quantum Electrodynamics: An Introduction to Radiation-Molecule Interactions.* London: Academic Press.

Dedic, C. E., Miller, J. D., and Meyer, T. R. (2014). Dual-pump vibrational/rotational femtosecond/picosecond coherent anti-Stokes Raman scattering temperature and species measurements. *Optics Letters*, **39**, 6608–6611. DOI: https://doi.org/10.1364/OL39.006608.

Dennis, C. N., Satija, A., and Lucht, R. P. (2016a). High dynamic range thermometry at 5 kHz in hydrogen–air diffusion flame using chirped-probe-pulse femtosecond coherent anti-Stokes Raman scattering. *Journal of Raman Spectroscopy,* **47**, 177–188. DOI: https://doi.org/10.1002/jrs.4773

Dennis, C. N., Slabaugh, C. D., Boxx, I. G., Meier, W., and Lucht, R. P. (2016b). 5 kHz thermometry in a swirl-stabilized gas turbine model combustor using chirped probe pulse femtosecond CARS. Part 1: Temporally resolved swirl-flame thermometry. *Combustion and Flame,* **173**, 441–453. DOI: https://doi.org/10.1016/j.combustflame.2016.02.033

Dion, P. and May, A. D. (1973). Motional narrowing and other effects in the Q branch of HD. *Canadian Journal of Physics*, **51**, 36–39.

Dirac, P. A. M. (1958). *The Principles of Quantum Mechanics*, 4th edition. Oxford: Oxford University Press.

Dreier, T., Schiff, G. and Suvernev, A. A. (1994). Collisional effects in *Q* branch coherent anti-Stokes Raman spectra of N_2 and O_2 at high pressure and high temperature. *Journal of Chemical Physics*, **100**, 6275–6289. DOI: https://doi.org/10.1063/1.467090

Duxbury, G., Kelly, J. F., Blake, T. A., and Langford, N. (2012). Sub-Doppler spectra of infrared hyperfine transitions of nitric oxide using a pulse modulated quantum cascade laser: Rapid passage, free induction decay, and the ac Stark effect. *Journal of Chemical Physics*, **136**, Art. No. 174319. DOI: https://doi.org/10.1063/1.4710542

Earls, L. T. (1935). Intensities in the $^2\Pi - ^2\Sigma$ transitions in diatomic molecules. *Physical Review*, **48**, 423–424

Eckbreth, A. C. (1996). *Laser Diagnostics for Combustion Temperature and Species,* 2nd edition. Amsterdam: Gordon and Breach Publishers.

Edmonds, A. R. (1960). *Angular Momentum in Quantum Mechanics*. Princeton, NJ: Princeton University Press.

Engleman Jr. R. and Rouse P. E. (1971). The β and γ bands of nitric oxide observed during the flash photolysis of nitrosyl chloride. *Journal of Molecular Spectroscopy,* **37**, 240–251.

Fernando A. M., Bernath P. F., Hodges J. N., and Masseron, T. (2018). A new line list for the $A^3\Pi - X^3\Sigma^-$ transition of the NH free radical. *Journal of Quantitative Spectroscopy and Radiative Transfer*, **217**, 29–34. DOI: https://doi.org/10.1016/j.jqsrt.2018.05.021

Forney, J. J., Quattropani, A., and Bassani, F. (1977). Choice of gauge in optical transitions. *Il Nuovo Cimento*, **37**, 78–88.

Galatry, L. (1961). Simultaneous effect of Doppler and foreign gas broadening on spectral lines. *Physical Review,* **122**, 1218–1223.

Gersten, J. I. and Foley, H. M. (1968). Combined Doppler and collision broadening. *Journal of the Optical Society of America*, **58**, 933–937.

Goldsmith, J. E. M. and Rahn, L. A. (1988). Doppler-free two-photon-excited fluoresecence spectroscopy of OH in flames. *Journal of the Optical Society of America B*, **5**, 749–755.

Gottfried, K. and Yan, T.-M. (2003). *Quantum Mechanics: Fundamentals*, 2nd edition. New York: Springer-Verlag.

Gu, M., Satija, A., and Lucht, R. P. (2019). Effects of self-phase modulation (SPM) on femtosecond coherent anti-Stokes Raman scattering spectroscopy. *Optics Express,* **27**, 33955–33967. DOI: https://doi.org/10.1364/OE.27.033954

Gu, M., Satija, A., and Lucht, R. P. (2020). Impact of moderate pump-Stokes chirp on femtosecond coherent anti-Stokes Raman scattering spectra. *Journal of Raman Spectroscopy,* **51**, 115–124. DOI: https://doi.org/10.1002/jrs.5754

Hall, R. J., Verdieck, J. F., and Eckbreth, A. C. (1980). Pressure-induced narrowing of the CARS spectrum of N_2. *Optics Communications*, **35**, 69–75.

Hall, R. J. (1983). Coherent anti-Stokes Raman spectroscopic modeling for combustion diagnostics. *Optical Engineering*, **22**, 322–329

Hanson, R. K., Spearrin, R. M., and Goldenstein, C. S. (2016). *Spectroscopy and Optical Diagnostics for Gases.* New York: Springer.

Herbert, F. (1974). Spectrum line profiles: A generalized Voigt function including collisional narrowing. *Journal of Quantitative Spectroscopy and Radiation Transfer*, **14**, 943–951.

Herman, R. and Wallis, R. F. (1955). Influence of vibration-rotation interaction on line intensities in vibration-rotation bands of diatomic molecules. *Journal of Chemical Physics*, **23**, 637–646.

Hilborn, R. C. (1982). Einstein coefficients, cross sections, *f* values, dipole moments, and all that. *American Journal of Physics*, **50**, 982–986.

Hougen, J. T. (1970). "The Calculation of Rotational Energy Levels and Rotational Intensities in Diatomic Molecules," N.B.S. Monograph 115, Washington, D. C.

Incropera, Frank P. (1974). *Introduction to Molecular Structure and Thermodynamics.* New York: John Wiley and Sons.

James, T. C. and Klemperer, W. (1959). Line intensities in the Raman effect of ${}^1\Sigma$ diatomic molecules. *Journal of Chemical Physics,* **31**, 130–134.

James, T. C. (1963). Intensity distribution in the forbidden ${}^1\Sigma{-}{}^3\Pi_r$transition of SiO. *Journal of Chemical Physics*, **38**, 1094–1097.

James, T. C. (1971). Transition moments, Franck–Condon factors, and lifetimes of forbidden transitions. Calculation of the intensity of the Cameron system of CO. *Journal of Chemical Physics*, **55**, 4118–4124.

Judd, B. R. (1975). *Angular Momentum Theory for Diatomic Molecules.* New York: Academic Press.

Kataoka, H., Maeda, S., and Hirose, C. (1982). Effects of laser linewidth on the coherent anti-Stokes Raman spectroscopy spectral profile. *Applied Spectroscopy*, **36**, 565–569.

Kearney, S. P. and D. J. Scoglietti, D. J. (2013). Hybrid femtosecond/picosecond rotational coherent anti-Stokes Raman scattering at flame temperatures using a second-harmonic bandwidth-compressed probe. *Optics Letters*, **38**, 833–835.

Ketter R. L. and Prawel, S. P. (1969). *Modern Methods of Engineering Computation.* New York: McGraw-Hill.

Kleiman, Valeria D., Park, Hongkun, Gordon, Robert J., and Zare, Richard N. (1998). *Companion to Angular Momentum.* New York: John Wiley and Sons.

Kobe, D. H., and Smirl, A. L. (1978). Gauge invariant formulation of the interaction of electromagnetic radiation and matter. *American Journal of Physics*, **46**, 624–633.

Koszykowski, M. L., Farrow, R. L., and Palmer, R. E. (1985). Calculation of collisionally narrowed coherent anti-Stokes Raman spectroscopy spectra. *Optics Letters*, **10**, 478–480.

Kovacs, I. (1960). Intensities in $\Sigma-\Delta$ and $\Pi-\Delta$ transitions in diatomic molecules. *Canadian Journal of Physics*, **38**, 955–963.

Kovacs, I. (1969). *Rotational Structure in the Spectra of Diatomic Molecules*. London: Hilger.

Kramida, A., Ralchenko, Yu., Reader, J., and NIST ASD Team (2022). *NIST Atomic Spectra Database* (version 5.10), [Online]. Available: https://physics.nist.gov/asd [Fri Apr 14 2023]. National Institute of Standards and Technology, Gaithersburg, MD. DOI: https://doi.org/10.18434/T4W30F

Kulatilaka, W. D., Hsu, P. S., Stauffer, H. U., Gord, J. R., and Roy, S. (2010). Direct measurement of rotationally resolved H_2 Q-branch Raman coherence lifetimes using time-resolved picosecond coherent anti-Stokes Raman scattering. *Applied Physics Letters*, **97**, Art. No. 081112.

Kulatilaka W. D. and Lucht, R. P. (2017). Two-photon-absorption line strengths for nitric oxide: Comparison of theory and sub-Doppler, laser-induced fluorescence measurements. *Journal of Chemical Physics*, **146**, Article No. 124311. DOI: https://doi.org/10.1063/1.4978921

Lamb, W. E. Jr. (1952). Fine structure of the hydrogen atom. III. *Physical Review*, 85, 259–276.

Lamb, W. E. Jr., Schlicher, R. R., and Scully, M. O. (1987). Matter–field interaction in atomic physics and quantum optics. *Physical Review A*, **36**, 2763–2772.

Lang, T., Motzkus, M., Frey, H. M., and Beaud, P. (2001). High resolution coherent anti-Stokes Raman scattering: Determination of rotational constants, molecular anharmonicity, collisional line shifts, and temperature. *Journal of Chemical Physics*, **115**, 5418–5426.

Lang, T. and Motzkus, M. (2002). Single-shot femtosecond coherent anti-Stokes Raman-scattering thermometry. *Journal of the Optical Society of America, B*, **19**, 340–344.

Lapp, M. (1980). Lecture notes, Raman short course. Sandia National Laboratories, Livermore, CA.

Laurendeau, Normand M. (2005). *Statistical Thermodynamics: Fundamentals and Applications*. Cambridge: Cambridge University Press.

Lavorel, B., Millot, G., Saint-Loup, R., Wenger, C., Berger, H., Sala, J. P., Bonamy, J., and Robert, D. (1986). Rotational collisional line broadening at high temperatures in the N_2 fundamental Q-branch studied with stimulated Raman spectroscopy. *Journal of Physics France*, **47**, 417–425.

Lempert, W., Rosasco, G. J., and Hurst, W. S. (1984). Rotational collisional narrowing in the NO fundamental Q branch, studied with cw stimulated Raman spectroscopy. *Journal of Chemical Physics*, **81**, 4241–4245.

Le Roy, R. J. (2017a). RKR1: A computer program implementing the first-order RKR method for determining diatomic molecule potential energy functions. *Journal of Quantitative Spectroscopy and Radiative Transfer*, **186**, 158–167. DOI: https://doi.org/10.1016/j.jqsrt.2016.03.030.

Le Roy, R. J. (2017b). LEVEL: A computer program for solving the radial Schrödinger equation for bound and quasibound levels. *Journal of Quantitative Spectroscopy and Radiative Transfer*, **186**, 167–178. DOI: https://doi.org/10.1016/j.jqsrt.2016.05.028.

Lepard, D. W. (1970). Theoretical calculations of electronic Raman effects in the NO and O_2 molecules. *Canadian Journal of Physics*, **48**, 1664–1674.

Li, H., Farooq, A., Jeffries, J. B., and Hanson, R. K. (2008). Diode laser measurements of temperature-dependent collisional–narrowing and broadening parameters of Ar-perturbed H_2O transitions at 1391.7 and 1397.8 nm. *Journal of Quantitative Spectroscopy and Radiative Transfer*, **109**, 132–143.

Long, Derek A. (2002). *The Raman Effect: A Unified Treatment of the Theory of Raman Scattering by Molecules*. Chichester: John Wiley and Sons.

Lofthus, A. and Krupenie, P. H. (1977). The spectrum of molecular nitrogen. *Journal of Physical and Chemical Reference Data*, **6**, 113–307 DOI: https://doi.org/10.1063/1.555546

Lorentz, H. A. (2011). *The Theory of Electrons and Its Applications to the Phenomena of Light and Radiant Heat*, 2nd edition. New York: Dover Publications.

Lucht, R. P. (1987). Three-laser coherent anti-Stokes Raman scattering measurements of two species. *Optics Letters,* **12**, 78–80.

Lucht, R. P., Roy, S., Meyer, T. R., and Gord, J. R. (2006). Femtosecond coherent anti-Stokes Raman scattering measurement of gas temperatures from frequency-spread dephasing of the Raman coherence. *Applied Physics Letters,* **89**, Art. No. 251112.

Lucht, R. P., Kinnius, P. J., Roy, S., and Gord, J. R. (2007). Theory of femtosecond coherent anti-Stokes Raman scattering for gas-phase transitions. *Journal of Chemical Physics,* **127**, 044316. DOI: https://doi.org/10.1063/1.2751184

Luque, J. and Crosley, D. R. (1996a). Electronic transition moment and rotational transition in CH. I. $A^2\Delta$-$X^2\Pi$ system. *Journal of Chemical Physics,* **104**, 2146–2155.

Luque, J. and Crosley, D. R. (1996b). Electronic transition moment and rotational transition in CH. II. $B^2\Sigma^-$-$X^2\Pi$ system. *Journal of Chemical Physics*, **104**, 3907–3913.

Luque, J. and Crosley, D. R. (1998). Transition probabilities in the $A^2\Sigma^+$-$X^2\Pi_i$ electronic system of OH. *Journal of Chemical Physics*, **109**, 439–448.

Luque, J. and Crosley, D. R. (1999a). LIFBASE: database and spectral simulation program (v. 2.0). SRI International Report MP, 99-009. https://www.sri.com/engage/products-solution/lifbase.

Luque, J. and Crosley, D. R. (1999b). Transition probabilities and electronic transition moments of the $A^2\Sigma^+$-$X^2\Pi$ and $D^2\Sigma^+$-$X^2\Pi$ Systems of NO. *Journal of Chemical Physics*, **111**, 7405–7415.

Luthe, J. C., Beiting, E. J., and Yueh, F. Y. (1986). Algorithms for calculating coherent anti-Stokes Raman spectra: Application to several small molecules. *Computer Physics Communications*, **42**, 73–92.

Marcuse, D. (1980). *Principles of Quantum Electronics*. New York: Academic Press.

Marion, Jerry B. and Heald, Mark A. (1980). *Classical Electromagnetic Radiation,* 2nd edition. New York: Academic Press.

Maroulis, G. (2003). Accurate electric multipole moment, static polarizability and hyperpolarizability derivatives for N_2. *Journal of Chemical Physics*, **118**, 2673–2687.

Martin, W. C. and Wiese, W. L. (2002), *Atomic, Molecular, and Optical Physics Handbook* (version 2.2). [Online]. Available:https://www.nist.gov/pml/atomic-spectroscopy-compendium-basic-ideas-notation-data-and-formulas [*2023, April 12*]. National Institute of Standards and Technology, Gaithersburg, MD.

Marrocco, M. (2007). A quantitative approach to evaluate the problem of coherence of spectral components of the third-order susceptibility generating coherent anti-Stokes Raman signals. *Journal of Raman Spectroscopy*, **38**, 452–459. DOI: https://doi.org/10.1002/jrs.2201

Marrocco, M. (2009). Comparative analysis of Herman–Wallis factor for uses in coherent anti-Stokes Raman spectra of light molecules. *Journal of Raman Spectroscopy*, **40**, 741–747.

Marrocco, M. (2010). CARS thermometry revisited in light of the intramolecular perturbation. *Journal of Raman Spectroscopy*, **41**, 870–874.

Marrocco, M., Magnotti, G., and Cutler, A. D. (2012). Herman–Wallis corrections in dual-pump CARS intensities for combustion temperature and species. *Journal of Raman Spectroscopy*, **43**, 595–598.

Marrocco, M. (2012). Vibration-rotation interaction in time-resolved coherent anti-Stokes Raman scattering for gas-phase thermometry. *Journal of Raman Spectroscopy*, **43**, 621–626.

Mavrodineanu, Radu, and Boiteux, Henri (1965). *Flame Spectroscopy*. New York: John Wiley and Sons.

Miller, J. D., Slipchenko, M. N., and Meyer, T. R. (2011a). Probe-pulse optimization for nonresonant suppression in hybrid fs/ps coherent anti-Stokes Raman scattering at high temperature. *Optics Express*, **19**, 13326–13333.

Miller, J. D., Roy, S., Slipchenko, M. N., Gord, J. R., and Meyer, T. R. (2011b). Single-shot gas phase thermometry using pure rotational hybrid femtosecond/picosecond coherent anti-Stokes Raman scattering. *Optics Express*, **19**, 15627–15640.

Morse, P. M. (1929). Diatomic molecules according to the wave mechanics. II. Vibrational levels. *Physical Review,* **34**, 57–64.

Nicholls, R. W. and Stewart, A. L. (1962). Allowed transitions. In: *Atomic and Molecular Processes* (edited by Bates, D. R.). New York: Academic Press.

Noda, C. and Zare, R. N. (1982). Relation between classical and quantum formulations of the Franck–Condon principle: The generalized r-centroid approximation. *Journal of Molecular Spectroscopy*, **95**, 254–270.

Oron, D., Dudovich, N., and Silberberg, Y. (2003). Femotsecond phase-and-polarization control for background-free coherent anti-Stokes Raman spectroscopy. *Physical Review Letters*, **90**, Art. No. 213902. DOI: 10.1103/PhysRevLett.90.213902

Ouyang, X. and Varghese, P. L. (1989). Reliable and efficient program for fitting Galatry and Voigt profiles to spectral data on multiple lines. *Applied Optics*, **28**, 1538–1545.

Owono, L. C. O., Abdallah, D. B., Jaidane, N. and Lakhdar, Z. B. (2008). Theoretical radiative properties between states of the triplet manifold of NH radical. *Journal of Chemical Physics*, **128**, 084309. DOI: https://doi.org/10.1063/1.2884923

Owyoung, A. (1978). High-resolution cw stimulated Raman spectroscopy in molecular hydrogen. *Optics Letters*, **2**, 91–93.

Owyoung, A., Patterson, C. W. and McDowell, R. S. (1978). Cw stimulated Raman gain spectroscopy of the ν_1 fundamental of methane. *Chemical Physics Letters*, **59**, 156–162.

Palmer, R. E. (1989). The CARSFT computer code for calculating coherent anti-Stokes Raman spectra: user and programmer information. Sandia Report SAND89–8206, Sandia National Laboratories.

Parigger, Christian G. and Hornkohl, J. O. (2010). Diatomic molecular spectroscopy with standard and anomalous commutators. *International Review of Atomic and Molecular Physics*, **1**, 25–43.

Paul, P. H. (1997). Calculation of transition frequencies and rotational line strengths in the γ-bands of nitric oxide. *Journal of Quantitative Spectroscopy and Radiative Transfer*, **57**, 581–589. DOI: https://doi.org/10.1016/S0022-4073(96)00158-6.

Pekeris, C. L. (1934). The rotation-vibration coupling in diatomic molecules. *Physical Review*, **45**, 98–103.

Peticolas, W. L., Norris, R., and Rieckhoff, K. E. (1965). Polarization effects in the two-photon excitation of anthracene. *Journal of Chemical Physics*, **42**, 4164–4169.

Pine, A. S. (1980). Collisional narrowing of HF fundamental band spectral lines by neon and argon. *Journal of Molecular Spectroscopy*, **82**, 435–448.

Prince, B. D., Chakraborty, A., Prince, B. M., and Stauffer, H. U. (2006). Development of simultaneous frequency- and time-resolved coherent anti-Stokes Raman scattering for ultrafast detection of molecular Raman spectra. *Journal of Chemical Physics,* **125**, 044502. DOI: https://doi.org/10.1063/1.2219439.

Qin, Z., Zhao, J. M., and Liu, L. H. (2017). Radiative transition probabilities for the main diatomic electronic systems of $N_2, N_2^+, NO, O_2, CO, CO^+, CN, C_2$ and H_2 produced in the plasma of atmospheric entry. *Journal of Quantitative Spectroscopy and Radiative Transfer*, **202**, 286–301. DOI: https://doi.org/10.1016/j.jqsrt.2016.08.010.

Rahn, L. A. and Palmer, R. E. (1986). Studies of nitrogen self-broadening at high temperature with inverse Raman spectroscopy. *Journal of the Optical Society of America B*, **3**, 1164–1169.

Rahn, L. A., Palmer, R. E., Koszykowski, M. L., and Greenhalgh, D. A. (1987). Comparison of rotationally inelastic collision models for Q-branch Raman spectra of N_2. *Chemical Physics Letters*, **133**, 513–516.

Rahn, L. A., Farrow, R. L., and Rosasco, G. J. (1991). Measurement of the self-broadening of the H_2 Q(0–5) Raman transitions from 295 to 1000 K. *Phyiscal Review A*, **43**, 6075–6088.

Rakestraw, D. J., Lucht, R. P., and Dreier, T. (1989). Use of a charge-coupled device camera for broadband coherent anti-Stokes Raman scattering measurements. *Applied Optics*, **28**, 4116–4120.

Ralchenko, Yu., Kramida, A. E., Reader, J., and NIST ASD Team (2010). *NIST Atomic Spectra Database* (version 4.0), [Online]. Available: http://physics.nist.gov/asd

Ram, R. S. and Bernath, P. F. (2010). Revised molecular constants and term values for the $X^3\Sigma^-$ and $A^3\Pi$ states of NH. *Journal of Molecular Spectroscopy*, **260**, 115–119.

Rautian, S. G. and Sobel'man, I. I. (1967). The effect of collisions on the Doppler broadening of spectral lines. *Soviet Physics Uspekhi*, **9**, 701–716.

Reisel, J. R., Carter, C. D., and Laurendeau, N. M. (1992). Einstein coefficients for rotational lines of the (0,0) band of the NO $A^2\Sigma^+ - X^2\Pi$ system. *Journal of Quantitative Spectroscopy and Radiative Transfer*, **47**, 43–54.

Reitz, J. R. and Milford, F. J. (1967). *Foundations of Electromagnetic Theory*, 2nd edition. Reading, MA: Addison-Wesley.

Renschler, D. L., Hunt, J. L., McCubbin Jr., T. K., and Polo, S. R. (1969). Triplet structure of the rotational Raman spectrum of oxygen. *Journal of Molecular Spectroscopy*, **31**, 173–176.

Richardson, D. R., Lucht, R. P., Roy, S., Kulatilaka, W. D., and Gord, J. R. (2013). Chirped-probe-pulse femtosecond coherent anti-Stokes Raman scattering concentration measurements. *Journal of the Optical Society of America B*, **30**, 188–196.

Rosasco, G. J., Rahn, L. A., Hurst, W. S., Palmer, R. E., and Dohne, S. M. (1989). Measurement and prediction of Raman Q-branch line self-broadening coefficients for CO from 400 to 1500 K. *Journal of Chemical Physics*, **90**, 4059–4068.

Rose, M. E. (1957). *Elementary Theory of Angular Momentum*. New York: Dover.

Roy, S., Richardson, D. R., Kulatilaka, W. D., Lucht, R. P., and Gord, J. R. (2009). Gas-phase thermometry at 1 kHz using femtosecond coherent anti-Stokes Raman scattering (fs-CARS) spectroscopy. *Optics Letters*, **34**, 3857–3859.

Rychlewski, J. (1980). An accurate calculation of the polarizability of the hydrogen molecule and its dependence on rotation, vibration and isotopic substitution. *Molecular Physics*, **41**, 833–842. DOI: https://doi.org/10.1080/00268978000103191

Satija A., Chai, N., Arendt, M. T., and Lucht, R. P. (2020). Pure rotational coherent anti-Stokes Raman scattering spectroscopy of nitric oxide: Determination of Raman tensor invariants. *Journal of Raman Spectroscopy*, **51**, 807–828. DOI: https://doi.org/10.1002/jrs.5836

Schiff, L. I. (1968). *Quantum Mechanics*, 2nd edition. New York: McGraw-Hill.

Schrötter, H. W. and Klöckner, H. W. (1979). Raman scattering cross sections in gases and liquids. In: *Raman Spectroscopy of Gases and Liquids* (Edited by Weber, A.). Berlin: Springer-Verlag.

Schneider, H. and Barker, G. P. (1973). *Matrices and Linear Algebra.* 2nd edition. New York: Holt, Rinehart, and Winston.

Scully, M. O. and Zubairy, M. S. (1997). *Quantum Optics.* Cambridge: Cambridge University Press.

Settersten, T. B., Patterson, B. D., and Humphries IV, W. H. (2009). Radiative lifetimes of NO $A^2\Sigma^+(v' = 0, 1, 2)$ and the electronic transition moment of the $A^2\Sigma^+ - X^2\Pi$ system. *Journal of Chemical Physics*, **131**, Article No. 104309.

Siegman, A. E. (1986). *Lasers.* Mill Valley, CA: University Science Books.

Shore, B. W. and Menzel, D. H. (1968). *Principles of Atomic Spectra.* New York: John Wiley and Sons.

Slater, J. C. (1960a). *Quantum Theory of Atomic Structure: Volume I.* New York: McGraw-Hill.

Slater, J. C. (1960b). *Quantum Theory of Atomic Structure: Volume II.* New York: McGraw-Hill.

Sobelman, I. I. (1992). *Atomic Spectra and Radiative Transitions,* 2nd edition. Berlin: Springer-Verlag.

Stark, G., Brault, J. W., and Abrams, M. C. (1994). Fourier-transform spectra of the $A^2\Sigma^+ - X^2\Pi\ \Delta v = 0$ bands of OH and OD. *Journal of the Optical Society of America B*, **11**, 3–32.

Stauffer, H. U., Roy, S., Schmidt, J. B., Wrzesinski, P. J., and Gord, J. R. (2016). Two-color vibrational, femtosecond fully resonant electronically enhanced CARS (FREE-CARS) of gas-phase nitric oxide. *Journal of Chemical Physics,* **145**, Article No. 124308. DOI: https://doi.org/10.1063/1.4962834.

Struve, W. S. (1988). *Fundamentals of Molecular Spectroscopy.* New York: John Wiley and Sons.

Teets, R. E. (1984). Accurate convolutions of coherent anti-Stokes Raman spectra. *Optics Letters*, **9**, 226–228.

Thomas, L. M., Lowe, A., Satija, A., Masri, A. R., and Lucht, R. P. (2019). 5 kHz thermometry in turbulent spray flames using chirped-probe-pulse femtosecond coherent anti-Stokes Raman scattering, Part I: Processing and interference analysis. *Combustion and Flame,* **200**, 405–416. DOI: https://doi.org/10.1016/j.combustflame.2018.11.004

Tipping, R. H. and Bouanic, J.-P. (2001). On the use of Herman–Wallis factors for diatomic molecules. *Journal of Quantitative Spectroscopy and Radiative Transfer*, **71**, 99–103.

Tipping, R. H. and Ogilvie, J. F. (1984). Herman–Wallis factors for Raman transitions of $^1\Sigma$-state diatomic molecules. *Journal of Raman Spectroscopy*, **15**, 38–40.

Utsav, K. and Varghese, P. L. (2013). Accurate temperature measurements in flames with high spatial resolution using Stokes Raman scattering from nitrogen in a multiple-pass cell. *Applied Optics*, **52**, 5007–5021.

Varberg, T. D., Stroh, F., Evenson, K. M. (1999). Far-infrared rotational and fine structure transition frequencies and molecular constants of ^{14}NO and ^{15}NO in the $X^2\Pi$ (v=0) state. *Journal of Molecular Spectroscopy*, **196**, 5–13.

Varghese, P. L. and Hanson, R. K. (1984). Collisional narrowing effects on spectral line shapes measured at high resolution. *Applied Optics*, **23**, 2376–2385.

Walecka, J. D. (2103). *Topics in Modern Physics: Theoretical Foundations.* Hackensack, NJ: World Scientific Publishing.

Weissbluth, M. (1978). *Atoms and Molecules.* New York: Academic Press.

Weissbluth, M. (1989). *Photon-Atom Interactions.* London: Academic Press.

Werner, H. J., Knowles, P. J., Knizia, G., Manby, F. R., and Schütz, M. (2012). MOLPRO: a general-purpose quantum chemistry program package. *Wiley Interdisciplinary Reviews – Computational Molecular Science*, **2**, 242–253. DOI: https://doi.org/10.1002/wcms.82.

Whiting, E. E. and Nicholls, R. M. (1974). Reinvestigation of rotational line intensity factors. *The Astrophysical Journal Supplement Series No. 235*, **27**, 1–19.

Wolniewicz, L. (1993). Relativistic energies of the ground state of the hydrogen molecule. *Journal of Chemical Physics*, **99**, 1851–1868. DOI: https://doi.org/10.1063/1.465303.

Yang, K.-H. 1976. Gauge transformations and quantum mechanics. I. Gauge invariant interpretation of quantum mechanics. *Annals of Physics*, **101**, 62–96.

Yousefi, M., Bernath, P. F., Hodges, J., and Masseron, T. (2018). A new line list for the $A^2\Sigma^+ - X^2\Pi$ electronic transition of OH. *Journal of Quantitative Spectroscopy and Radiative Transfer*, **217**, 416–424. DOI: https://doi.org/10.1016/j.jqsrt.2018.06.016.

Yuratich, M. A. (1979). Effects of laser linewidth on coherent anti-Stokes Raman spectroscopy. *Molecular Physics*, **38**, 625–655.

Zare, R. N., Schmeltekopf, A. L., Harrop, W. J., and Albritton, D. L. (1973). A direct approach for the reduction of diatomic spectra to molecular constants for the construction of RKR potentials. *Journal of Molecular Spectroscopy*, **46**, 37–66.

Zare, Richard N. (1988). *Angular Momentum: Understanding Spatial Aspects in Chemistry and Physics.* New York: John Wiley and Sons.

Index

3j symbol, 17, 365–368
6j symbol, 154, 156–158, 369–370, 394–397

Absorption
  Circularly polarized radiation, 242–244
  Coefficient, 239–241
  Cross section, 243
  Induced absorption rate, 238–239
  Linearly polarized radiation, 241–242
  Lineshape models, 247–253
  Rate coefficient, 239
Angular momentum operator, generalized, 30–32
Atomic hydrogen
  Spin–orbit splitting, 29
Atomic hydrogen
  Electric dipole moment, coupled representation, 143–146
  Electric dipole moment, uncoupled representation, 139–143
  Radial integrals, 140
  Solution, Schrödinger wave equation, 22–26
  Spontaneous emission coefficients, 147–148
Atomic nitrogen, 38–39

Band strength, diatomic molecules, 165
Born–Oppenheimer approximation, 56, 58–60

CARS
  Broadband, 330–332
  Collisional narrowing, 334–339
  Electric field modeling, 326–327
  Example problem, dual-pump CARS, 353–359
  Example problem, parallel polarization, 348–353
  Femtosecond, time-dependent equations, 342–343
  Nonresonant susceptibility, 322
  Polarizability, orientation averaging, 317–318
  Polarizability, pure rotational, 319–320
  Polarizability, vibrational, 320–321
  Scanning, monochromatic laser fields, 327–328
  Scanning, narrowband lasers, 329–330
  Scanning, narrowband Stokes, 328–329
  Signal amplitude formula, 325
  Signal irradiance formula, 325
  Susceptibility, general expression, 321–322
  Transition linewidths, 332–334
Center of mass coordinates, 57
Centrifugal stretching, 62
Central-field approximation, 33
Central potential, 22
Classical electron oscillator model, 3–11
Clebsch–Gordan coefficients, 17, 28, 45, 138
  Relation to *3j* symbols, 28, 47
Closure relations, 264
Collisional broadening, 239

Density matrix, 134–136
  Coherence decay terms, 136
  Energy interaction terms, 134–136
  Multistate system, 236–237
  Two-state system, 237–238
Diatomic molecules
  Electronic transitions, 164–166
Dielectric permittivity, 6
Dirac
  Bra and ket notation, 17
  Inner product, 18
Dirac wave equation, 16, 22, 27
Doppler broadening, 248–249
Doppler linewidth, 257
Doublet electronic levels
  $^2\Lambda$, $\Lambda>0$ levels, 82–96
  $^2\Pi$ levels, 82–96
  $^2\Sigma$, 90–100
  $\Lambda>0$, matrix elements, including higher order, 91–92

Effective Hamiltonian, 68
Electric dipole interaction
  Long wavelength approximation, 125–127
Electric dipole moment, 1–4
  Atomic hydrogen, 136–146
  Reduced density matrix element, 148
Electric dipole oscillator
  Radiative decay, 12–15
Electric quadrupole moment, 1, 3

Electromagnetic field
Plane wave approximation, 11
Propagation in dielectric medium, 11
Electronic transition moment
MOLPRO code, 203
NH $X^3\Sigma^-$-$A^3\Pi$, 230
NO $X^2\Pi$-$A^2\Sigma^+$, 203
OH $X^2\Pi$-$A^2\Sigma^+$, 203
Electronic transitions
Diatomic molecules, 164–166
Doublet transitions, diatomic molecules, $^2\Pi$-$^2\Sigma$, 198–199
Doublet transitions, diatomic molecules, $^2\Sigma$-$^2\Pi$, 191–194
Doublet transitions, diatomic molecules, $^2\Sigma$-$^2\Sigma$, 184–186
Doublet transitions, diatomic molecules, F_1-F_1, 178–182
Doublet transitions, diatomic molecules, $\Delta\Lambda$ = -1, 196–198
Doublet transitions, diatomic molecules, $\Delta\Lambda$=+1, Λ>0, 188–191, 193–195
Doublet transitions, diatomic molecules, $\Delta\Lambda$=0, Λ,Λ'>0, 186–189
Forbidden transitions, diatomic molecules, 232–233
Singlet transitions, diatomic molecules, $^1\Pi$-$^{1\Sigma,}$ 170–174
Singlet transitions, diatomic molecules, $^1\Sigma$-$^{1\Sigma,}$ 166–170
Singlet transitions, diatomic molecules, $\Lambda > 0$, $\Lambda' > 0$, 174–177
Triplet transitions, diatomic molecules, $^3\Sigma$-$^{3\Pi,}$ 216–222
Triplet transitions, diatomic molecules, $^3\Sigma$-$^{3\Sigma,}$ 211–216
Triplet transitions, diatomic molecules, $\Lambda > 0$, $\Lambda' > 0$, 222–229
Euler angles, 42–43

Franck–Condon factors, 166

Galatry profile, 252–253
Gauge invariance, 121–125

Hard collision model, 253
Harmonic oscillator potential, 57
Harmonic oscillator wavefunction, 56–61
Homogeneous broadening, 239
Hönl–London factors, 165–166
Singlet transitions, diatomic molecule, $^1\Sigma$-$^{1\Sigma,}$ 169–170
Singlet transitions, diatomic molecules, $^1\Pi$-$^{1\Sigma,}$ 173–174
Singlet transitions, diatomic molecules, $\Lambda > 0$, $\Lambda' > 0$, 175–177
Hund's case (a), 56–118
Basis wavefunctions, 65–66
Eigenvalues, 66–67
Hamiltonian, 79
Matrix elements, 66–67
Matrix elements, centrifugal distortion, rotation, 72–75
Matrix elements, centrifugal distortion, spin–orbit interaction, 76
Matrix elements, lambda-doubling, 76–77
Matrix elements, rotation, 68–71
Matrix elements, spin–orbit interaction, 71
Matrix elements, spin–rotation interaction, 71–72
Matrix elements, spin–spin interaction, 72
Parity of wavefunctions, 65–66
Hund's case (b), 56–118
Hyperfine interaction
Effect on absorption coefficient, 393–397
Effect on spontaneous emission rate coefficient, 393–397
Hyperfine splitting
Atomic hydrogen, 29

Inhomogeneous broadening, 248–249
Irreducible spherical tensor operators, 49–50
Scalar product, 52
Irreducible spherical tensors, 49–50, 148–150

Jacobi polynomial, 44
jj coupling, 16, 39

Lamb shift, 29
Lorentz classical electron oscillator model, 3
Lorentzian lineshape, 239
Lowering operator, 30
LS coupling, 16, 38
Radiative transitions, atomic oxygen, 156–159
Single-photon transition selection rules, 155

Macroscopic polarization, 5
Magnetic permeability, 11
Morse potential, 61–62
Multielectron atoms, 16
Electron configuration, 33–34
Orbital angular momentum, 35–36
Quantum numbers, 33
Spin angular momentum, 35–36
Wavefunction, 34–35

NH
$A^3\Pi$ (v=0) Hund's case (a) coefficients, 56–113
$A^3\Pi$ (v=0) level energies, 112
$A^3\Pi$ (v=0) spectroscopic parameters, 111
$X^3\Sigma^-$-$A^3\Pi$ vibrational band strengths, 229–232
$X^3\Sigma^-$-$A^3\Pi$ (0,0) Einstein coefficients, 384–392
$X^3\Sigma^-$-$A^3\Pi$ Einstein coefficients, 232
NIST Atomic Spectra Database, 38
NO
$X^2\Pi$ (v=0) Hund's case(a) coefficients, 93–94

$X^2\Pi$ (v=0) level energies, 93–94
$X^2\Pi$ (v=0) spectroscopic parameters, 92
$X^2\Pi$-$A^2\Sigma^+$ (0,0) Einstein coefficients, 206–210
$X^2\Pi$-$A^2\Sigma^+$ electronic transition moment, 203
$X^2\Pi$-$A^2\Sigma^+$ vibrational band strengths, 201–206
Nuclear rotation angular momentum, 64–119

OH
$X^2\Pi$ (v=0) Hund's case (a) coefficients, 93–96
$X^2\Pi$ (v=0) level energies, 93–96
$X^2\Pi$ (v=0) spectroscopic parameters, 95
$X^2\Pi$-$A^2\Sigma^+$ (0,0) Einstein coeffcients, 206–210
$X^2\Pi$-$A^2\Sigma^+$ electronic transition moment, 203
$X^2\Pi$-$A^2\Sigma^+$ vibrational band strengths, 201–206
Operators, quantum mechanical, 17–18
Hermitian, 18
Orbital angular momentum, 23
Oscillator strength, 244

Pair coupling, 16, 39–40
Parity, 39
Pauli exclusion principle, 32–33
Pauli spinor, 27
Perturbation analysis
Multiphoton resonances, 131–134
Raman resonances, 133–134
Single-photon resonances, 128–131
Two-photon absorption, 133
Poynting vector, 14
Principal moments of inertia, 62
Pure dephasing collisions, 5–6

Quantum Liouville equation, 134–136

Rabi frequency, 238
Radial wavefunctions
Atomic hydrogen, 26
Radiative transitions
Doublet transitions, diatomic molecules, $^2\Pi$-$^{2\Sigma,}$ 198–199
Doublet transitions, diatomic molecules, $^2\Sigma$-$^{2\Pi,}$ 191–194
Doublet transitions, diatomic molecules, $^2\Sigma$-$^{2\Sigma,}$ 184–186
Doublet transitions, diatomic molecules, F_1-$F_{1,}$ 178–182
Doublet transitions, diatomic molecules, $\Delta\Lambda$ = -1, 196–198
Doublet transitions, diatomic molecules, $\Delta\Lambda$=+1, Λ>0, 188–191, 193–195
Doublet transitions, diatomic molecules, $\Delta\Lambda$=0, Λ,Λ'>0, 186–189
Doublet transitions, diatomic molecules, $\Delta\Lambda$=-1, Λ>1, 193–200
Effect of hyperfine splitting, 153–154
Forbidden transitions, diatomic molecules, 232–233
Hund's case (a) basis states, 163–164
LS coupling multiplet terms, 156–159
LS coupling selection rules, 155
Oscillator strength, 244
Pure rotational, 161–162
Singlet transitions, diatomic molecules, $^1\Pi$-$^{1\Sigma,}$ 170–174
Singlet transitions, diatomic molecules, $^1\Sigma$-$^{1\Sigma,}$ 166–170
Singlet transitions, diatomic molecules, $\Lambda > 0$, $\Lambda' > 0$, 174–177
Spontaneous emission rate coefficient, 152–153
Triplet transitions, diatomic molecules, $^3\Sigma$-$^{3\Pi,}$ 216–222
Triplet transitions, diatomic molecules, $^3\Sigma$-$^{3\Sigma,}$ 211–216
Triplet transitions, diatomic molecules, $\Lambda > 0$, $\Lambda' > 0$, 222–229
Vibrational, 162–163
Raising operator, 30
Raman polarizability
Irreducible spherical tensor representation, 264–266
Operator, 264
Placzek theory, 262–276
Relation between space-fixed and molecule-fixed tensor components, 266–268
Rotation operator, 266–267
Taylor series expansion, 270
Raman scattering
Cross section, 290–291
Cross section, Stokes Q-branch, 291–305
Depolarization ratio, 275–276
Example problem, 298–305
$H_2Q(1)$ line, 300–302
Intensity formulae, 273–275
$N_2O(6)$ line, 302–305
N_2Q-branch, 298–300
Polarizability, definition, 259–262
Scattering geometry, 259–262
Raman transitions
$O_2\,^3\Sigma^-$ transitions, 276–290
Placzek–Teller coefficients, 272–273
Pure rotational Raman transitions, 273
Pure rotational tensor components, 271–273
Vibrational tensor components, 273
Vibrational transitions, 273
Reduced matrix element, 50–51
Coupled angular momenta, 51
RKR calculations
LEVEL code, 201
OH and NO $X^2\Pi$-$A^2\Sigma^+$, 201–203
Rotating wave approximation, 238
Rotation operator, 41–45
Integral relations, 46–48
Relation with spherical harmonics, 45–48
Russell–Saunders coupling, 16

Schrödinger wave equation, 16
  Cartesian coordinates, 19
  External electromagnetic fields, 120–125
  Separation of variables, 19–20
  Solution atomic hydrogen, 22–26
  Two-particle, 60
Secular matrix
  Doublet levels, $\Lambda>0$, 85
  Triplet levels, $\Lambda>0$, 101
Singlet electronic levels
  $^1\Pi$ levels, 80–82
  $^1\Sigma$ levels, 79–80
Soft collision model, 252–253
Speed of light, 11
Spherical harmonics, 25–26
  Relation with rotation operators, 45
Spin orbital function, 137
Spin wavefunction, 27, 137
Spin–orbit interaction
  Atomic hydrogen, 29
  Calcium atom, 52–55
  Multielectron atoms, 33
  Spherical tensor analysis, 52–55
Spontaneous emission
  Angular distribution, 253–255
  Polarization, 253–255
  Zeeman-state dependence, 253–255
Spontaneous emission rate coefficient, 147, 152–153, 165, 182
Stimulated emission rate, 238–239
Susceptibility, 6
  Linear, 244–247, 306–309
  Second-order, 309–311
  Third-order, 311–317
Symmetric top molecule wavefunction, 62–64
Symmetry operation σ_v, 191

Term symbol, 36
Triplet electronic levels
  $\Lambda>0$, 99–109
  $\Lambda>0$, diagonal matrix elements, 109
  $\Lambda>0$, off-diagonal matrix elements, 110

Velocity-changing collisions, 250–253
Voigt function
  Table of values, 399–404
Voigt lineshape, 249–250, 257

Wigner–Eckart theorem, 17, 52, 120, 150
  Electric dipole transitions, 148–153
  Reduced density matrix elements, 148–153

Printed in the United States
by Baker & Taylor Publisher Services

Printed in the United States
by Baker & Taylor Publisher Services